Ocean Mixing
Drivers, Mechanisms and Impacts

Ocean Mixing
Drivers, Mechanisms and Impacts

Edited by
Michael Meredith
Alberto Naveira Garabato

ELSEVIER

Elsevier
Radarweg 29, PO Box 211, 1000 AE Amsterdam, Netherlands
The Boulevard, Langford Lane, Kidlington, Oxford OX5 1GB, United Kingdom
50 Hampshire Street, 5th Floor, Cambridge, MA 02139, United States

Copyright © 2022 Elsevier Inc. All rights reserved.

No part of this publication may be reproduced or transmitted in any form or by any means, electronic or mechanical, including photocopying, recording, or any information storage and retrieval system, without permission in writing from the publisher. Details on how to seek permission, further information about the Publisher's permissions policies and our arrangements with organizations such as the Copyright Clearance Center and the Copyright Licensing Agency, can be found at our website: www.elsevier.com/permissions.

This book and the individual contributions contained in it are protected under copyright by the Publisher (other than as may be noted herein).

Notices

Knowledge and best practice in this field are constantly changing. As new research and experience broaden our understanding, changes in research methods, professional practices, or medical treatment may become necessary.

Practitioners and researchers must always rely on their own experience and knowledge in evaluating and using any information, methods, compounds, or experiments described herein. In using such information or methods they should be mindful of their own safety and the safety of others, including parties for whom they have a professional responsibility.

To the fullest extent of the law, neither the Publisher nor the authors, contributors, or editors, assume any liability for any injury and/or damage to persons or property as a matter of products liability, negligence or otherwise, or from any use or operation of any methods, products, instructions, or ideas contained in the material herein.

Library of Congress Cataloging-in-Publication Data
A catalog record for this book is available from the Library of Congress

British Library Cataloguing-in-Publication Data
A catalogue record for this book is available from the British Library

ISBN: 978-0-12-821512-8

For information on all Elsevier publications
visit our website at https://www.elsevier.com/books-and-journals

Publisher: Candice Janco
Acquisitions Editor: Louisa Munro
Editorial Project Manager: Leticia M. Lima
Production Project Manager: Joy Christel Neumarin Honest Thangiah
Designer: Greg Harris

Typeset by VTeX

Contents

List of contributors ix
Editors' biographies xi
Acknowledgements xiii

1. Ocean mixing: oceanography at a watershed
Alberto Naveira Garabato and Michael Meredith

References 3

2. The role of ocean mixing in the climate system
Angélique V. Melet, Robert Hallberg and David P. Marshall

2.1 Introduction 5
2.2 The role of ocean mixing in shaping the contemporary climate mean state 11
    2.2.1 Meridional overturning circulation and heat transport 11
    2.2.2 Southern Ocean 14
    2.2.3 Mixing in exchanges between marginal seas and the open ocean 16
    2.2.4 Mixing and marine ecosystems 17
2.3 Ocean mixing and transient climate change 18
    2.3.1 Ocean anthropogenic heat and carbon uptake 18
    2.3.2 Contemporary and future sea level rise 19
    2.3.3 Changes in nutrient fluxes 20
    2.3.4 Changes in ocean mixing sources 21
2.4 Ocean mixing in past climate states 21
    2.4.1 The Early Pliocene 23
    2.4.2 The Last Glacial Maximum (LGM) 23
2.5 Summary and conclusion 23
References 25

3. The role of mixing in the large-scale ocean circulation
Casimir de Lavergne, Sjoerd Groeskamp, Jan Zika and Helen L. Johnson

3.1 Introduction 35
3.2 Flavours of mixing 36
3.3 Non-dissipative theories of ocean circulation 40
    3.3.1 Ekman pumping 40
    3.3.2 Momentum redistribution by geostrophic turbulence 41
3.4 How can mixing shape circulation? 42
    3.4.1 By altering surface wind and buoyancy forcing 42
    3.4.2 By altering density gradients 42
    3.4.3 By producing and consuming water masses 43
3.5 Where is mixing most effective at shaping circulation? 43
    3.5.1 Isotropic mixing, from top to bottom 44
    3.5.2 Mesoscale stirring, from top to bottom 47
3.6 Some impacts on basin-scale overturning circulation 49
    3.6.1 Abyssal overturning cell 49
    3.6.2 North Atlantic Deep Water circulation 51
    3.6.3 Southern Ocean upwelling: adiabatic or diabatic? 52
    3.6.4 The return flow to the North Atlantic 52
    3.6.5 Shallow hemispheric cells 53
3.7 Some impacts on basin-scale horizontal circulation 53
    3.7.1 Upper-ocean gyres 53
    3.7.2 The Stommel and Arons circulation 55
    3.7.3 The Antarctic Circumpolar Current 55
3.8 Conclusions 56
References 57

4. Ocean near-surface layers
Baylor Fox-Kemper, Leah Johnson and Fangli Qiao

4.1 Introduction 65
4.2 Mixing layers and mixed layers in theory 66
    4.2.1 Mixing and surface layers: Monin–Obukhov scaling 68
    4.2.2 Near-surface distinctions from M–O theory and each other 70
    4.2.3 Mixed layers: boundary layer memory 76

		4.2.4 A home for submesoscales	78
4.3	Observing the surface layers and their processes		79
	4.3.1	Observing mixing	79
	4.3.2	Wave-driven turbulence	81
	4.3.3	Laboratory experiments	81
4.4	Modelling surface layers and their processes		82
	4.4.1	Large eddy simulations	82
	4.4.2	1D boundary layer models	82
	4.4.3	Ocean and climate models	84
4.5	Global perspective		85
	4.5.1	Energy and forcing	85
	4.5.2	Surface layers, weather, and climate	85
4.6	Outlook		85
Acknowledgements			86
References			86

5. The lifecycle of surface-generated near-inertial waves

Leif N. Thomas and Xiaoming Zhai

5.1	Introduction		95
5.2	Generation of near-inertial waves at the surface		95
5.3	Propagation of near-inertial waves out of the mixed layer		98
	5.3.1	Refraction	99
	5.3.2	Straining	100
	5.3.3	Interaction with frontal vertical circulations	101
5.4	Interactions of near-inertial waves with variable stratification, other internal waves, and mean flows in the interior		102
	5.4.1	Variable stratification	102
	5.4.2	Interactions with other internal waves	103
	5.4.3	Interactions with mean flows	104
5.5	Dissipation of near-inertial waves		107
	5.5.1	Near-surface dissipation	107
	5.5.2	Interior dissipation	108
	5.5.3	Near-bottom dissipation	109
5.6	Discussion		110
	5.6.1	Vertical mixing	110
	5.6.2	Lateral mixing	111
5.7	Conclusions and outstanding questions		111
Acknowledgements			112
References			112

6. The lifecycle of topographically-generated internal waves

Ruth Musgrave, Friederike Pollmann, Samuel Kelly and Maxim Nikurashin

6.1	Introduction	117
6.2	Generation	118

	6.2.1	Internal tides	118
	6.2.2	Quasi-steady lee waves	123
6.3	Internal tide propagation and an integral estimate of decay		125
6.4	Wave-wave interactions		126
	6.4.1	Theoretical background	126
	6.4.2	Parametric subharmonic instability of the internal tide	128
	6.4.3	Wave-wave interactions in finestructure methods, mixing parameterisations, and numerical simulations	129
	6.4.4	Global perspective	130
6.5	Wave-mean flow interactions		130
	6.5.1	Theoretical background	130
	6.5.2	Mean-flow effects on wave propagation	131
	6.5.3	Mean-flow effects on wave energy	132
	6.5.4	Global perspective	133
6.6	Wave-topography interaction		133
	6.6.1	Theoretical background & observational estimates	134
	6.6.2	Global distribution	135
6.7	Conclusions and outstanding questions		136
Acknowledgements			138
References			138

7. Mixing at the ocean's bottom boundary

Kurt L. Polzin and Trevor J. McDougall

7.1	Introduction		145
7.2	Common ground		147
	7.2.1	Equations	147
	7.2.2	Boundary conditions	148
	7.2.3	Coordinate transformations and the one-dimensional model	149
	7.2.4	Integration	151
	7.2.5	Energetics and mixing	152
7.3	Implications of the bottom intensification of ocean mixing for upwelling: buoyancy budgets for bottom-intensified mixing		155
	7.3.1	Abyssal ocean circulation models are sensitive to bottom topography	156
	7.3.2	One-dimensional solutions for flow near a sloping bottom boundary	156
	7.3.3	Expressions for the upwelling in the BBL and downwelling in the SML	157
	7.3.4	How much larger is the upwelling in the BBL than the net upwelling?	159
	7.3.5	Net upwelling in the abyss depends mainly on the shape of the ocean floor	161
	7.3.6	What can be learned from purposefully released tracers?	161
	7.3.7	Implications for the circulation of the abyssal ocean	162
	7.3.8	Summary remarks	163

7.4 Production mechanisms for boundary mixing — 164
   7.4.1 Internal wave reflection / internal tide generation — 164
   7.4.2 Sub-inertial flow and topography — 166
   7.4.3 Friction and sub-inertial flows — 168
7.5 Discussion — 171
Acknowledgements — 176
References — 176

8. Submesoscale processes and mixing
Jonathan Gula, John Taylor, Andrey Shcherbina and Amala Mahadevan

8.1 Introduction — 181
8.2 Life-cycle of submesoscale fronts — 184
   8.2.1 Frontogenesis — 184
   8.2.2 Instability of surface boundary layer fronts — 185
   8.2.3 Submesoscale processes at the bottom of the ocean — 193
   8.2.4 The influence of vertical mixing on the evolution of a submesoscale front — 198
   8.2.5 Frontal arrest and routes to dissipation — 200
8.3 Redistribution of density and restratification at the submesoscale — 201
   8.3.1 Restratification induced by submesoscale processes — 201
   8.3.2 Competition between destratification and restratification of a front — 202
   8.3.3 Bottom boundary layer mixing and restratification — 203
8.4 Redistribution of passive tracers and particles — 204
   8.4.1 Conservative tracers — 204
   8.4.2 Mixing and transport of reactive tracers — 205
   8.4.3 Impacts on the dispersion of buoyant material — 205
   8.4.4 Dispersion by the deep submesoscale currents — 205
8.5 Conclusion and future directions — 206
Acknowledgements — 207
References — 207

9. Isopycnal mixing
Ryan Abernathey, Anand Gnanadesikan, Marie-Aude Pradal and Miles A. Sundermeyer

9.1 Introduction — 215
9.2 Background concepts — 216
   9.2.1 What is mixing? — 216
   9.2.2 What is isopycnal? — 218
   9.2.3 What then is isopycnal mixing? — 220
   9.2.4 Lateral mixing near boundaries — 221
9.3 Mechanisms of isopycnal stirring and dissipation — 221
   9.3.1 Mesoscale turbulence — 221
   9.3.2 Transport by coherent structures — 222
   9.3.3 Chaotic advection — 223
   9.3.4 Shear-driven mixing — 224
   9.3.5 Additional submesoscale isopycnal mixing processes — 225
   9.3.6 Diapycnal dissipation of isopycnal tracer variance — 226
   9.3.7 Frontogenesis and loss of balance — 227
9.4 Frameworks for thinking about isopycnal mixing — 227
   9.4.1 Reynolds-averaged tracer equations — 227
   9.4.2 Mixing-length theory — 228
   9.4.3 Spectral-space view of turbulence and mixing — 229
   9.4.4 Isopycnal mixing in numerical models — 231
9.5 Observational estimates of isopycnal mixing — 233
   9.5.1 Tracer-based methods — 234
   9.5.2 Drifter and float-based methods — 238
9.6 Simulation-based estimates — 240
   9.6.1 Inverse methods — 240
   9.6.2 Direct simulation — 241
9.7 Impacts of isopycnal mixing — 242
   9.7.1 Physical circulation — 243
   9.7.2 Passive tracers — 244
   9.7.3 Isopycnal mixing and ocean biogeochemical cycles — 244
9.8 Summary and future directions — 245
Acknowledgements — 247
References — 247

10. Mixing in equatorial oceans
James N. Moum, Andrei Natarov, Kelvin J. Richards, Emily L. Shroyer and William D. Smyth

10.1 Introduction — 257
10.2 Ocean turbulence peaks at the equator, or does it? — 259
10.3 Mixing in the cold tongues: diurnal forcing of turbulence below the mixed layer — 259
10.4 The concepts of marginal instability and self-organised criticality and how they apply to mixing in the cold tongues — 260
10.5 The importance of inertia-gravity waves and flow instabilities — 262
10.6 Westerly wind bursts in the Indian Ocean and western Pacific — 264
10.7 Variations on subseasonal, seasonal and interannual timescales — 264
10.8 Equatorial mixing in large-scale models — 265
10.9 Shortcomings, surprises and targets for future investigation — 267
References — 269

11. Mixing in the Arctic Ocean
Yueng-Djern Lenn, Ilker Fer, Mary-Louise Timmermans and Jennifer A. MacKinnon

11.1 Introduction	275
11.2 Foundations	276
11.3 Key findings	279
11.3.1 Ice-ocean interactions	280
11.3.2 Tidal mixing	281
11.3.3 Near-inertial motions and the internal wave continuum	283
11.3.4 Eddies	284
11.3.5 Double diffusion	287
11.3.6 Thermohaline intrusions	288
11.4 Grand challenges	290
11.5 Conclusions	292
Acknowledgements	292
References	292

12. Mixing in the Southern Ocean
Sarah T. Gille, Katy L. Sheen, Sebastiaan Swart and Andrew F. Thompson

12.1 Introduction	301
12.2 Large-scale context: foundations	301
12.3 Upper cell: mixed-layer transformations	303
12.3.1 Foundations and setting	303
12.3.2 Mixing in the surface boundary layer and connection to subsurface adiabatic stirring	304
12.4 Interior mixing: regional and mesoscale processes	307
12.4.1 Foundations: Southern Ocean eddy pathways	307
12.4.2 Mixing and coherent structures: adiabatic recipes for Southern Ocean mixing	308
12.5 Interior mixing: closing the budgets through turbulence at the smallest scales	312
12.5.1 Foundations	312
12.5.2 Recent findings: sub-surface diapycnal mixing pathways in the Southern Ocean	313
12.6 Grand challenges	315
Acknowledgements	318
References	318

13. The crucial contribution of mixing to present and future ocean oxygen distribution
Marina Lévy, Laure Resplandy, Jaime B. Palter, Damien Couespel and Zouhair Lachkar

13.1 Introduction	329
13.2 Role of mixing in oxygen minimum zones	330
13.3 Role of mixing on global deoxygenation	334
13.4 Response of OMZ to global warming	339
13.5 Conclusions and grand challenges	340
Acknowledgements	342
References	342

14. New technological frontiers in ocean mixing
Eleanor Frajka-Williams, J. Alexander Brearley, Jonathan D. Nash and Caitlin B. Whalen

14.1 Introduction	345
14.2 Current and historical measurements of mixing	345
14.3 Recent technological developments: novel methods	348
14.3.1 Shear microstructure on AUVs	348
14.3.2 Temperature microstructure on new platforms	351
14.3.3 Finescale parameterisations from autonomous platforms	353
14.3.4 Large-eddy method using autonomous platforms and moorings	354
14.4 Future outlook	358
14.5 Conclusions	358
References	359

Index   363

List of contributors

Ryan Abernathey, Lamont Doherty Earth Observatory of Columbia University, New York, NY, United States

J. Alexander Brearley, British Antarctic Survey, Cambridge, United Kingdom

Damien Couespel, Sorbonne Université, LOCEAN-IPSL (CNRS/IRD/MNHN), Paris, France

Casimir de Lavergne, LOCEAN Laboratory, Sorbonne University-CNRS-IRD-MNHN, Paris, France

Ilker Fer, Geophysical Institute, University of Bergen, Bergen, Norway

Baylor Fox-Kemper, Brown University, Department of Earth, Environmental, and Planetary Sciences, Providence, RI, United States

Eleanor Frajka-Williams, National Oceanography Centre, Southampton, United Kingdom

Sarah T. Gille, Scripps Institution of Oceanography, University of California San Diego, La Jolla, CA, United States

Anand Gnanadesikan, Johns Hopkins University, Baltimore, MD, United States

Sjoerd Groeskamp, NIOZ Royal Netherlands Institute for Sea Research, Den Burg, Texel, the Netherlands

Jonathan Gula, Univ Brest, CNRS, IRD, Ifremer, Laboratoire d'Océanographie Physique et Spatiale (LOPS), IUEM, Brest, France
Institut Universitaire de France (IUF), Paris, France

Robert Hallberg, NOAA/Geophysical Fluid Dynamics Laboratory, Princeton, NJ, United States

Helen L. Johnson, Department of Earth Sciences, University of Oxford, Oxford, United Kingdom

Leah Johnson, Brown University, Department of Earth, Environmental, and Planetary Sciences, Providence, RI, United States

Samuel Kelly, Large Lakes Observatory, Physics & Astronomy Department, University of Minnesota Duluth, Duluth, MN, United States

Zouhair Lachkar, Center for Prototype Climate Modelling, New-York University, Abu Dhabi, United Arab Emirates

Yueng-Djern Lenn, School of Ocean Sciences, Bangor University, Wales, United Kingdom

Marina Lévy, Sorbonne Université, LOCEAN-IPSL (CNRS/IRD/MNHN), Paris, France

Jennifer A. MacKinnon, Scripps Institution of Oceanography, University of California San Diego, La Jolla, CA, United States

Amala Mahadevan, Woods Hole Oceanographic Institution, Woods Hole, MA, United States

David P. Marshall, University of Oxford, Oxford, United Kingdom

Trevor J. McDougall, School of Mathematics and Statistics, University of New South Wales, Sydney, NSW, Australia

Angélique V. Melet, Mercator Ocean International, Ramonville St Agne, France

Michael Meredith, British Antarctic Survey, Cambridge, United Kingdom

James N. Moum, College of Earth, Ocean, and Atmospheric Sciences, Oregon State University, Corvallis, OR, United States

Ruth Musgrave, Department of Oceanography, Dalhousie University, Halifax, NS, Canada

Jonathan D. Nash, College of Earth, Ocean and Atmospheric Science, Oregon State University, Corvallis, OR, United States

Andrei Natarov, IPRC, University of Hawaii at Manoa, Honolulu, HI, United States

Alberto Naveira Garabato, Ocean and Earth Science, University of Southampton, National Oceanography Centre, Southampton, United Kingdom

Maxim Nikurashin, Institute of Marine and Antarctic Studies, University of Tasmania, Hobart, TAS, Australia

Jaime B. Palter, Graduate School of Oceanography, University of Rhode Island, Narragansett, RI, United States

Friederike Pollmann, Universität Hamburg, Center for Earth System Research and Sustainability, Hamburg, Germany

Kurt L. Polzin, Department of Physical Oceanography, Woods Hole Oceanographic Institution, Woods Hole, MA, United States

Marie-Aude Pradal, Johns Hopkins University, Baltimore, MD, United States

Fangli Qiao, First Institute of Oceanography, Ministry of Natural Resources, Qingdao, China

Laure Resplandy, Department of Geosciences and Princeton Environmental Institute, Princeton University, Princeton, NJ, United States

Kelvin J. Richards, IPRC, University of Hawaii at Manoa, Honolulu, HI, United States

Andrey Shcherbina, Applied Physics Laboratory, University of Washington, Seattle, WA, United States

Katy L. Sheen, Penryn Campus, University of Exeter, Cornwall, England, United Kingdom

Emily L. Shroyer, College of Earth, Ocean, and Atmospheric Sciences, Oregon State University, Corvallis, OR, United States

William D. Smyth, College of Earth, Ocean, and Atmospheric Sciences, Oregon State University, Corvallis, OR, United States

Miles A. Sundermeyer, University of Massachusetts Dartmouth, North Dartmouth, MA, United States

Sebastiaan Swart, Department of Marine Sciences, University of Gothenburg, Gothenburg, Sweden
Department of Oceanography, University of Cape Town, Rondebosch, South Africa

John Taylor, Department of Applied Mathematics and Theoretical Physics, University of Cambridge, Cambridge, United Kingdom

Leif N. Thomas, Department of Earth System Science, Stanford University, Stanford, CA, United States

Andrew F. Thompson, Environmental Science and Engineering, California Institute of Technology, Pasadena, CA, United States

Mary-Louise Timmermans, Department of Earth and Planetary Sciences, Yale University, New Haven, CT, United States

Caitlin B. Whalen, Applied Physics Laboratory, University of Washington, Seattle, WA, United States

Xiaoming Zhai, Centre for Ocean and Atmospheric Sciences, School of Environmental Sciences, University of East Anglia, Norwich, United Kingdom

Jan Zika, School of Mathematics and Statistics, University of New South Wales, Sydney, NSW, Australia

Editors' biographies

Professor Mike Meredith

Professor Mike Meredith is an oceanographer and Science Leader at the British Antarctic Survey (BAS) in Cambridge, UK. He is head of the Polar Oceans team at BAS, which has research foci on determining the role of the polar oceans on global climate, the ice sheets, and the interdisciplinary ocean system. He is an Honorary Professor at the University of Bristol, a Fellow of the Royal Geographical Society, and a NERC Individual Merit Promotion (Band 2) scientist. He has published more than 200 papers in international journals, and was the inaugural Chair of the Southern Ocean Observing System. He led the design and delivery of the multi-institute ORCHESTRA programme, which is unravelling the role of the Southern Ocean in controlling global climate. He was recently Coordinating Lead Author for the IPCC Special Report on Oceans and Cryosphere in a Changing Climate. In 2018, Mike was awarded the Tinker-Muse Prize for Science and Policy in Antarctica, in recognition of his contributions to the study of the Southern Ocean and its global impacts, and the Challenger Medal, for his contributions to marine science.

Professor Alberto Naveira Garabato

Professor Alberto Naveira Garabato is an oceanographer interested in the processes governing ocean circulation and its role in climate. His group's research focuses on unravelling the dynamics connecting the breadth of scales of oceanic flow—from small-scale turbulence to the basin-scale circulation—through the development and application of new approaches to measure the ocean. He holds a Chair of Physical Oceanography at the University of Southampton, and is an Honorary Fellow of the British Antarctic Survey. His work has been recognised with the Outstanding Early Career Scientist Award of the European Geosciences Union (2008), an Honorary Fellowship of the Challenger Society (2010), a Philip Leverhulme Prize (2010) and a Royal Society Wolfson Research Merit award (2014). He was the lead proponent of the RoSES programme, which is assessing the role of the Southern Ocean in the global carbon cycle. He is the founding director of the NEXUSS Centre of Doctoral Training, which is training 45 PhD students at 10 UK institutions in the use of cutting-edge sensor and autonomous system technologies for environmental science.

Acknowledgements

We thank the very many people who have contributed to the production of this book. These include Leticia Lima, Louisa Munro, Sathya Narayanan and Joy C. Neumarin, who worked tirelessly on its production. We are grateful to all who provided scientific guidance and advice throughout the drafting process, and to the expert reviewers who donated their time and effort to ensure the veracity of the text: we thus sincerely thank Ryan Abernathey, Peter Brandt, Matthew Alford, Jörn Callies, Eric D'Asaro, Marcus Dengler, Ilker Fer, David Ferreira, Sarah Gille, Stephen Griffies, Andy Hogg, Phil Hosegood, Casimir de Lavergne, Jim Ledwell, Marina Lévy, Angélique Melet, Igor Polyakov, Kurt Polzin, Eric Skyllingstad, Kevin Speer, Seb Swart, John Taylor, Leif Thomas, Andrew Thompson, Stephanie Waterman, Ric Williams, Carl Wunsch, Xiaoming Zhai and Jan Zika.

Chapter 1

Ocean mixing: oceanography at a watershed

Alberto Naveira Garabato[a] and Michael Meredith[b]

[a]*Ocean and Earth Science, University of Southampton, National Oceanography Centre, Southampton, United Kingdom,* [b]*British Antarctic Survey, Cambridge, United Kingdom*

Ocean mixing – the three-dimensional turbulent interleaving and blending of oceanic waters with different properties (Eckart, 1948; Welander, 1955) – plays a critical part in defining almost every aspect of the ocean's structure and role within the Earth system. This is, perhaps, unsurprising given the intrinsically turbulent nature of oceanic flows, as characterised by very large Reynolds numbers (i.e., an overwhelming dominance of inertial forces over viscous forces). Yet despite its widely recognised prevalence and pre-eminence today, ocean mixing was a relative latecomer in ocean science, only beginning to garner wider attention in the 1960s as marine physics was moving from adolescence into maturity.[1] Prior to then, most oceanographers had conceptualised the ocean as a fundamentally laminar system, with a behaviour and governing dynamics that could be adequately expressed by linear approximations to the equations of motion. This view was, of course, founded on the technological limitations of the early decades of observational oceanography; since the pioneering expeditions of the HMS *Challenger* and the FS *Meteor* in the late 19th and early 20th centuries, sampling of the ocean had consisted mainly of spatially coarse snapshots that missed the fine spatio-temporal scales on which ocean mixing operates.

Much has occurred since oceanography encountered this first critical juncture in the 1960s, from which ocean mixing emerged as a key ingredient of the ocean circulation recipe. A detailed history of this major transition is beyond the scope of this book, so we refer the interested reader to Gregg's (1991) excellent account. Suffice it to say that the explosive entrance of ocean mixing into the oceanographic arena was sparked initially by inferences about the underlying causes of the ocean's large-scale behaviour – the outstanding example of which is Munk's (1966) seminal work highlighting the role of diapycnal mixing (mixing across density surfaces) in sustaining the deep-ocean stratification and overturning circulation. This was subsequently amplified and extended by a range of ingenious experiments and technological advances to measure small-scale turbulence and mixing, both in oceanic and laboratory settings. A singular outcome of this body of work was a suite of comprehensive theoretical descriptions of the signatures of small-scale turbulence and its parent flows (notably, internal waves) on oceanic properties and currents (e.g., Batchelor, 1959; Kraichnan, 1968; Nasmyth, 1970; Garrett and Munk, 1972) and of the effects of turbulence on mixing (e.g., Osborn and Cox, 1972; Thorpe, 1977; Osborn, 1980). With modest adjustments, we still rely on these fundamental descriptions to this day.

While this small-scale world was coming into view, a largely distinct community of oceanographers was initiating their own transformative incursion into the field of ocean mixing. This foray came from a very different angle, and one that provided the remainder of the pieces for the ocean mixing jigsaw that we have assembled to date: the discovery of the ocean's mesoscale eddy field. Such discovery had been spurred by theoretical predictions of the existence of mesoscale eddies from the 1960s (see e.g., Stommel, 1963), and was ultimately accomplished through a series of pioneering Russian and Anglo-American experiments during the following decade (see Robinson (1983) for an overview of results from those programmes, and Munk and Day (2008) for an insightful personal chronicle of how the experiments came to be). The realisation soon emerged that mesoscale eddies – accounting for over three quarters of all oceanic kinetic energy (Ferrari and Wunsch, 2009) – are efficient lateral (i.e., isopycnal, or along density surfaces) stirrers of ocean properties on large horizontal scales (up to many tens of kilometres), and can thereby promote mixing by generating the finer-scale gradients on which small-scale turbulence acts.

What followed this dichotomous early rise of ocean mixing was half a century of rapid – and occasionally vertiginous – progress, which has both unravelled and integrated the small-scale and mesoscale perspectives of mixing through major

---

1. Note, though, that with the benefit of hindsight we can find inferential evidence on the significance of ocean mixing as far back as at least 1754, when Immanuel Kant published his work on the role of tidal dissipation within the Earth–Moon–Sun system (Kant, 1754).

2 Ocean Mixing

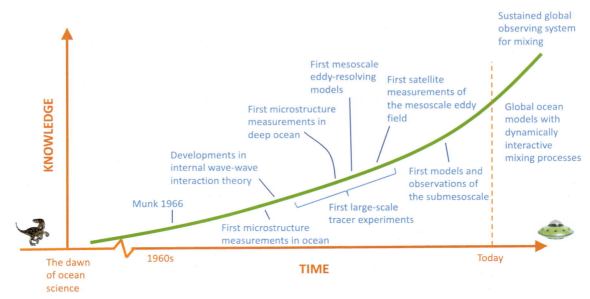

FIGURE 1.1 Some of the key milestones in the progression of understanding of ocean mixing, from the landmark studies of the 1960s through to the present and future challenges.

advances in process understanding. This has been achieved by the development and uptake of new theories, observations and models, in similar measure; an overview of some of the milestones in this journey is given in Fig. 1.1. Such milestones include (but are certainly not limited to):

- developments in theoretical understanding of the mechanisms by which interactions between internal waves generate turbulent mixing (e.g., Müller et al., 1986), and their observational assessment and advancement (e.g., Gregg, 1989; Polzin et al., 1995), which underpin some of our most widely-applied approaches to assess diapycnal mixing and its drivers today;
- the advent of mesoscale eddy-resolving models of the large-scale ocean circulation (Holland, 1985; see also Bryan (2008)), eddy parameterisations (Redi, 1982; Gent and McWilliams, 1990; Griffies, 1998) and satellite altimetric measurements of the eddy field (Chelton and Schlax, 1996; Stammer, 1997; Wunsch and Stammer, 1998) in the 1980s and 1990s, which revealed a swathe of mechanisms via which mesoscale eddies shape the way in which the ocean mixes;
- landmark ocean mixing experiments that provided some of the first direct estimates of diapycnal mixing rates in the pycnocline (the part of the water column characterised by an enhanced vertical density gradient) (Gregg, 1989; Ledwell et al., 1993) and the abyssal ocean (Polzin et al., 1997; Ledwell et al., 2000). These used pioneering tracer releases and full-depth microstructure measurements, and generated our present consensus view of weak cross-pycnocline mixing and intensified mixing over rough seafloor topography;
- the recent unveiling of submesoscale turbulence, which is characterised by horizontal scales of a few hundred metres to a few kilometres, and thus dwells at scales intermediate between the small-scale and mesoscale perspectives on ocean mixing. This has been achieved through progress in dynamical theory (see overviews by Thomas et al. (2008) and McWilliams (2016)), advances in model resolution (e.g., Capet et al., 2008; Su et al., 2018) and breakthroughs in observational technology (e.g., D'Asaro et al., 2011; Poje et al., 2014; Thompson et al., 2016), and has reset our understanding of how mixing works near the ocean's boundaries.

Yet, as stimulating and enlightening as the ocean mixing voyage has been, there is one basic premise that remains untested from its inception, and that still frames our vision of ocean mixing today: the assumption that ocean mixing processes are, to a large extent, dynamically passive – an assumption that has its roots in the outdated descriptions of the ocean as a laminar system from before the 1960s and 1970s. This premise implies that mixing does not significantly affect how the ocean adjusts to (or feeds back on) changes in climatic forcing – except at the largest and longest scales of the 'background' ocean state, which motivated the emergence of ocean mixing science in the first place (e.g., Munk, 1966). In this lingering and pervasive view, the action of ocean mixing is essentially conceptualised as a steady and gradual modification of the ocean's scalar properties (e.g., temperature, density, dissolved oxygen, etc.) that must be incorporated conceptually to understand and explain the distribution of such properties, and the circulation across their gradients, on time scales of centuries and longer. This view is embedded deeply in many of the numerical models that we use to investigate the operation and wider impacts of ocean circulation, which regularly represent the effects of ocean mixing with a set of unchanging – or, at most, minimally interactive – diffusive coefficients.

Evidence is mounting, however, that such a picture of ocean mixing as a largely dynamically passive phenomenon does not provide an adequate explanation of many climatically important aspects of the ocean's behaviour. These range from the variability in the transport of major ocean currents (Marshall et al., 2017; Sasaki et al., 2018) to changes in deep-ocean heat content (Naveira Garabato et al., 2019; Spingys et al., 2021), and from the ventilation of the oceanic pycnocline (Su et al., 2018; Yu et al., 2019; Bachman and Klocker, 2020) to the ocean's pacing of major climatic modes (Goswami et al., 2016; Warner and Moum, 2019). All of these notable elements of the circulation – and many more – are governed by agile dynamical interactions between ocean mixing processes and the large-scale ocean state, and are thus not well captured by the current generation of state-of-the-art Earth system-class ocean models. The conclusion seems inevitable that, to raise our understanding and modelling capability to the next level, we must challenge the lingering foundational premise of ocean mixing, and recognise that its circulation-shaping role may be much richer than anticipated by the field's pioneers.

This is, in our opinion, the next critical juncture facing ocean mixing, the passing of which is likely to enable the field to approach a resolution of many of the headline questions that spurred its growth half a century ago (such as Munk's (1966) iconic deep-ocean density budget problem), and evolve rapidly into a significantly different, more sophisticated and advanced discipline with a new set of frontier questions. It thus seems timely to take stock of our contemporary understanding of ocean mixing and its cutting-edge issues, debate the pathways to closure of some of its key foundational problems, and reflect on the challenges to which the field must rise if it is to expand its scope and transformative impact on ocean and Earth system science – as the discipline's state of maturity demands. To spur this, we have invited contributions from many of the leading practitioners in the field of ocean mixing research, with the intention of assembling a set of interlinked chapters that collectively define the state of the art of the field, and the next-level grand challenges to be tackled. The chapters span the totality of the global ocean volume, from the surface, through the interior, to the seabed, and from the equator to the poles. They assess each of the key relevant processes by which ocean mixing is driven, modulated and controlled, as well as its impacts across disciplines and on scales up to planetary. They also highlight ongoing methodological and technological advances, and draw a roadmap for the field's future direction of travel.

References

Bachman, S.D., Klocker, A., 2020. Interaction of jets and submesoscale dynamics leads to rapid ocean ventilation. J. Phys. Oceanogr. 50, 2873–2883.

Batchelor, G.K., 1959. Small-scale variation of convected quantities like temperature in turbulent fluid. Part 1: general discussion and the case of small conductivity. J. Fluid Mech. 5, 113–133.

Bryan, F.O., 2008. Introduction: ocean modelling – eddy or not. In: Hecht, M.W., Hasumi, H. (Eds.), Ocean Modeling in an Eddying Regime. In: AGU Geophysical Monograph Series, vol. 177.

Capet, X., McWilliams, J.C., Molemaker, M.J., Shchepetkin, A.F., 2008. Mesoscale to submesoscale transition in the California current system. Part I: flow structure, eddy flux, and observational tests. J. Phys. Oceanogr. 38, 29–43.

Chelton, D.B., Schlax, M.G., 1996. Global observations of oceanic Rossby waves. Science 272, 234–238.

D'Asaro, E., Lee, C., Rainville, L., Harcourt, R., Thomas, L., 2011. Enhanced turbulence and energy dissipation at ocean fronts. Science 332, 318–322.

Eckart, C., 1948. An analysis of the stirring and mixing processes in incompressible fluids. J. Mar. Res. 7, 265–275.

Ferrari, R., Wunsch, C., 2009. Ocean circulation kinetic energy: reservoirs, sources and sinks. Annu. Rev. Fluid Mech. 41, 253–282.

Garrett, C., Munk, W., 1972. Space-time scales of internal waves. Geophys. Fluid Dyn. 2, 255–264.

Gent, P.R., McWilliams, J.C., 1990. Isopycnal mixing in ocean circulation models. J. Phys. Oceanogr. 20, 150–155.

Goswami, B.N., Rao Suryachandra, A., Sengupta, D., Chakravorty, S., 2016. Monsoons to mixing in the Bay of Bengal: multiscale air-sea interactions and monsoon predictability. Oceanography 29, 18–27.

Gregg, M.C., 1989. Scaling turbulent dissipation in the thermocline. J. Geophys. Res. 94, 9686–9698.

Gregg, M.C., 1991. The study of mixing in the ocean: a brief history. Oceanography 4, 39–45.

Griffies, S.M., 1998. The Gent – McWilliams skew flux. J. Phys. Oceanogr. 28, 831–841.

Holland, W.R., 1985. Simulation of mesoscale variability in midlatitude gyres. In: Manabe, S. (Ed.), Issues in Atmospheric and Oceanic Modeling. Part A: Climate Dynamics, pp. 479–523.

Kant, I., 1754. Whether the Earth Has Undergone an Alteration of Its Axial Rotation. Wöchentliche Frag- und Anzeigungs-Nachrichten, Königsberg, pp. 23–24; English translation in W. Hastie, Kant's Cosmogony. James Maclehose, Glasgow, 1900.

Kraichnan, R.H., 1968. Small-scale structure of a scalar field convected by turbulence. Phys. Fluids 11, 945.

Ledwell, J.R., Montgomery, E.T., Polzin, K.L., St. Laurent, L.C., Schmitt, R.W., Toole, J.M., 2000. Evidence for enhanced mixing over rough topography in the abyssal ocean. Nature 403, 179–182.

Ledwell, J.R., Watson, A.J., Law, C.S., 1993. Evidence for slow mixing across the pycnocline from an open-ocean tracer-release experiment. Nature 364, 701–703.

Marshall, D.P., Ambaum, M.H.P., Maddison, J.R., Munday, D.R., Novak, L., 2017. Eddy saturation and frictional control of the Antarctic Circumpolar Current. Geophys. Res. Lett. 44, 286–292.

McWilliams, J.C., 2016. Submesoscale currents in the ocean. Proc. R. Soc. A 472. https://doi.org/10.1098/rspa.2016.0117.
Müller, P., Holloway, G., Henyey, F., Pomphrey, N., 1986. Nonlinear interactions among internal gravity waves. Rev. Geophys. 24, 493–536.
Munk, W.H., 1966. Abyssal recipes. Deep-Sea Res. 13, 707–730.
Munk, W., Day, D., 2008. Glimpses of oceanography in the postwar period. Oceanography 21, 14–21.
Nasmyth, P.W., 1970. Oceanic turbulence. Ph.D. thesis. Institute of Oceanography, University of British Columbia.
Naveira Garabato, A.C., Frajka-Williams, E.E., Spingys, C.P., Legg, S., Polzin, K.L., Forryan, A., Abrahamsen, E.P., Buckingham, C.E., Griffies, S.M., McPhail, S.D., Nicholls, K.W., Thomas, L.N., Meredith, M.P., 2019. Rapid mixing and exchange of deep-ocean waters in an abyssal boundary currents. Proc. Natl. Acad. Sci. 116, 13233–13238.
Osborn, T.R., 1980. Estimates of the local rate of vertical diffusion from dissipation measurements. J. Phys. Oceanogr. 10, 83–89.
Osborn, T.R., Cox, C.S., 1972. Oceanic fine structure. Geophys. Fluid Dyn. 3, 321–345.
Poje, A.C., Özgökmen, T.M., Lipphardt, B.L., Haus, B.K., Ryan, E.H., Haza, A.C., Jacobs, G.A., Reniers, A.J.H.M., Olascoaga, M.J., Novelli, G., Griffa, A., Beron-Vera, F.J., Chen, S.S., Coelho, E., Hogan, P.J., Kirwan, A.D., Huntley, H.S., Mariano, A.J., 2014. Submesoscale dispersion in the vicinity of the Deepwater Horizon spill. Proc. Natl. Acad. Sci. 111, 12693–12698.
Polzin, K.L., Toole, J.M., Ledwell, J.R., Schmitt, R.W., 1997. Spatial variability of turbulent mixing in the abyssal ocean. Science 276, 93–96.
Polzin, K.L., Toole, J.M., Schmitt, R.W., 1995. Finescale parameterizations of turbulent dissipation. J. Phys. Oceanogr. 25, 306–328.
Redi, M.H., 1982. Oceanic isopycnal mixing by coordinate rotation. J. Phys. Oceanogr. 12, 1154–1158.
Robinson, A.R., 1983. Overview and summary of eddy science. In: Robinson, A.R. (Ed.), Eddies in Marine Science. Springer-Verlag.
Sasaki, H., Kida, S., Furue, R., Nonaka, M., Matsumoto, Y., 2018. An increase of the Indonesian Throughflow by internal tidal mixing in a high-resolution quasi-global ocean simulation. Geophys. Res. Lett. 45, 8416–8424.
Spingys, C.P., Naveira Garabato, A.C., Legg, S., Polzin, K.L., Abrahamsen, E.P., Buckingham, C.E., Forryan, A., Frajka-Williams, E.E., 2021. Mixing and transformation in a deep western boundary current: a case study. J. Phys. Oceanogr. 51 (4), 1205–1222.
Stammer, D., 1997. Global characteristics of ocean variability estimated from regional TOPEX / Poseidon altimeter measurements. J. Phys. Oceanogr. 27, 1743–1769.
Stommel, H., 1963. Varieties of oceanographic experience. Science 139, 572–576.
Su, Z., Wang, J., Klein, P., Thompson, A.F., Menemenlis, D., 2018. Ocean submesoscales as a key component of the global heat budget. Nat. Commun. 9, 775.
Thomas, L.N., Tandon, A., Mahadevan, A., 2008. Submesoscale processes and dynamics. In: Hecht, M.W., Hasumi, H. (Eds.), Ocean Modeling in an Eddying Regime. In: AGU Geophysical Monograph Series, vol. 177. American Geophysical Union.
Thompson, A.F., Lazar, A., Buckingham, C., Naveira Garabato, A.C., Damerell, G., Heywood, K.J., 2016. Open-ocean submesoscale motions: a full seasonal cycle of mixed layer instabilities from gliders. J. Phys. Oceanogr. 46, 1285–1307.
Thorpe, S.A., 1977. Turbulence and mixing in a Scottish loch. Philos. Trans. R. Soc. A 286. https://doi.org/10.1098/rsta.1977.0112.
Warner, S.J., Moum, J.N., 2019. Feedback of mixing to ENSO phase change. Geophys. Res. Lett. 46, 13920–13927.
Welander, P., 1955. Studies on the general development of motion in a 2-dimensional, ideal fluid. Tellus 7, 141–156.
Wunsch, C., Stammer, D., 1998. Satellite altimetry, the marine geoid, and the oceanic general circulation. Annu. Rev. Earth Planet. Sci. 26, 219–253.
Yu, X., Naveira Garabato, A.C., Martin, A.P., Buckingham, C.E., Brannigan, L., Su, Z., 2019. An annual cycle of submesoscale vertical flow and restratification in the upper ocean. J. Phys. Oceanogr. 49, 1439–1461.

Chapter 2

The role of ocean mixing in the climate system

Angélique V. Melet[a], Robert Hallberg[b,d] and David P. Marshall[c]
[a]Mercator Ocean International, Ramonville St Agne, France, [b]NOAA/Geophysical Fluid Dynamics Laboratory, Princeton, NJ, United States,
[c]University of Oxford, Oxford, United Kingdom

2.1 Introduction

Many different physical processes contribute to mixing in the ocean. Mixing plays a significant role in shaping the mean state of the ocean and its response to a changing climate. This chapter provides a review of some recent work on the processes driving mixing in the ocean, on techniques for parameterizing the various mixing processes in climate models, and on the role of ocean mixing in the climate system. For the latter, this chapter illustrates how ocean mixing shapes the contemporary mean climate state by focusing on key ocean features influencing the climate (such as the meridional overturning circulation and heat transport, ocean heat and carbon uptake, ocean ventilation, and overflows from marginal seas), how ocean mixing participates in shaping the transient climate change (including anthropogenic ocean heat and carbon uptake, sea level rise and changes in nutrient fluxes that impact marine ecosystems), and how ocean mixing is projected to change under future climate change and how tides and related mixing differed for paleoclimates. Improving our collective understanding of the dynamics of mixing processes and their interactions with the large-scale state of the ocean will lead to greater confidence in projections of how the climate system will evolve under climate change and to a better understanding of the feedbacks that will act to regulate this evolution.

The ocean is mixed by a variety of turbulent processes, and stirred by a rich field of geostrophic eddies. In the stratified ocean, the kinetic energy associated with turbulence is in part converted irreversibly to potential energy via the mixing of waters of different density, and in part converted into heat through viscous dissipation. Turbulent mixing occurs on small spatial and temporal scales, yet mixing events influence a wide range of oceanic motions, including the global thermohaline circulation, and have broader implications for the Earth's climate. For instance, the ocean would be very different without mixing across the different neutral density layers of the ocean, with an overly simplified picture of a warm surface layer sharply overtopping a vast and cool ocean. Instead, ocean turbulence mixes together denser and lighter water parcels, blending their properties, contributing to a net downward heat flux and to the upwelling of dense water from the deep ocean to shallower depths. These mixing processes have the effects of resupplying the ocean interior with potential energy, contributing to a stably stratified deep ocean and to the large-scale global overturning circulation.

At the air-sea interface, there is an exchange of heat and quantities, including carbon and oxygen, which play a key role in the climate system. Turbulent ocean mixing contributes to the redistribution of these tracers and of other quantities, such as freshwater and nutrients, that are of importance for marine ecosystems and for shaping the climate. Mixing also links waters that have been in recent contact with the atmosphere with the deeper ocean, thereby participating in the ventilation of the ocean and in the uptake and sequestration of heat and carbon within the ocean over long timescales, offering a large storage capacity and a large inertia in the climate system. As such, ocean mixing is also influential for shaping the equilibrium climate state and the climate transient response to forcing changes (such as the current one induced by anthropogenic emissions of greenhouse gases).

The energetics of ocean mixing, and the global distribution and amplitude of ocean mixing, are not static but rather evolve with the state of the climate, influencing it in return. In addition, ocean turbulence is patchy and intermittent, leading to a rich global distribution of mixing (e.g., MacKinnon, 2013; Waterhouse et al., 2014). Ocean turbulence can be decomposed into two main regimes: geostrophic turbulence and small-scale, three-dimensional isotropic turbulence (Chapter 3). Geostrophic turbulence is mediated by mesoscale eddies that stir ocean tracers along isopycnals in the ocean interior. Three-dimensional turbulent mixing generally occurs with events of spatial scales of order 0.1–100 m and lasting

---

d. Dr. Hallberg's contribution to the chapter is in the Public Domain as he is an US Government employee.

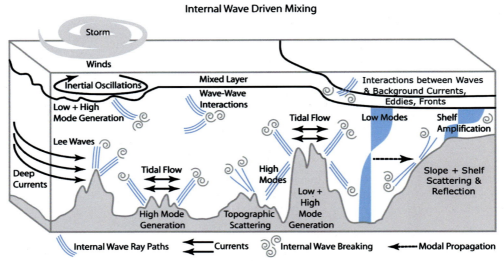

FIGURE 2.1 Schematic of the main internal-wave-mixing processes in the open ocean. Tides interact with topographic features to generate internal waves. Deep currents flowing over topography can generate lee waves. Storms cause inertial oscillations in the mixed layer, which can generate near-inertial waves. In the open ocean, these internal waves can scatter off of rough topography and potentially interact with the background currents, mesoscale fronts, and eddies until they ultimately dissipate through wave-wave interactions. Internal waves that reach the continental slope can be reflected, scattered at the slope, or transmitted to the shelf. Adapted from MacKinnon et al. (2017).

minutes to hours. The contribution of three-dimensional, small-scale turbulence along isopycnal surfaces is negligible compared to the contribution of geostrophic turbulence. For that reason, even though small-scale mixing is isotropic, it is often referred to as diapycnal (across density surfaces) mixing. In this chapter, an emphasis will be placed on three-dimensional, small-scale, diapycnal mixing. Some comments are also provided on the role of geostrophic turbulence on properties along density surfaces and the adiabatic redistribution of water masses by mesoscale eddies (the reader is referred to Chapter 9 for a more extensive discussion of along-isopycnal mixing processes and their impacts).

Diapycnal mixing in the stratified ocean interior is remarkably weak, so that clearly identifiable water masses with distinctive properties span thousands of kilometres along density surfaces, but just a few hundred metres to a kilometre or so in the vertical, and can take decades to many centuries to be replenished. The paucity of diapycnal mixing makes the dynamics of the mixing that does occur all the more significant to the climate system. In the stratified ocean interior, diapycnal mixing is mostly associated with breaking internal waves; the distribution of mixing is thus set by the detailed geography of the generation, propagation, and dissipation of internal waves (MacKinnon et al. (2017); Whalen et al. (2020), Chapters 5 and 6). Internal waves are generated by many different processes, including interactions between the large-scale flows and topography, direct forcing, or the adjustment of unbalanced large-scale flows (discussed in Chapter 6), as illustrated schematically in Fig. 2.1. Once generated, internal waves can propagate far in the vertical or horizontal, breaking and depositing their energy into three-dimensional turbulence, where the waves themselves become critically steep, through focusing by the ocean properties, interactions with topography or superposition with other internal waves or the mean flow. The flow of internal wave energy through the ocean, from generation to wave breaking and turbulent mixing (addressed in Chapters 5, 6), is therefore important for understanding the rich geographic variations in ocean mixing, the response of mixing to changing ocean conditions, and how mixing modifies those conditions in return (see Fig. 2.2).

Climate models are the primary tools used to study past and future changes in the climate system. Since ocean mixing and much of the spectrum of geostrophic turbulence occur on spatial scales that are too small to be resolved explicitly in ocean climate models and given that mixing has large-scale consequences, the different processes leading to ocean mixing have to be parameterised in such models. A description of parameterisations for various physics responsible for ocean mixing is provided in the stand-alone box.

This chapter illustrates the role of mixing in the climate system by focusing on the main aspects through which mixing influences the climate state. We start by illustrating how ocean mixing shapes the contemporary mean climate state (Section 2.2). After reviewing the role of mixing in the meridional overturning circulation and heat transport, we focus on the Southern Ocean, because of its critical role in shaping the contemporary mean climate state and buffering changes by connecting the near-surface ocean with the interior (further specifics relating to mixing in the Southern Ocean are provided in Chapter 12). We then turn to the role of mixing for the contemporary and future transient climate change due to anthropogenic emissions of greenhouse gases (Section 2.3). Mixing influences ocean heat and carbon uptake, thereby

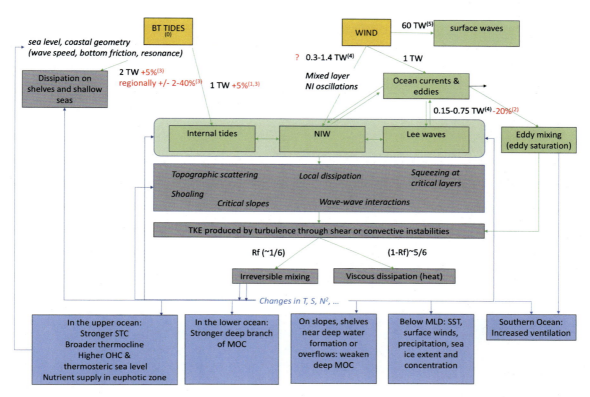

FIGURE 2.2 Schematic of energy pathways from sources (yellow boxes); ocean dynamic features (in green); processes leading to dissipation (in grey); corresponding changes and impacts on the ocean, sea-ice, and atmosphere (in blue), and feedback on energy pathways (blue arrows). Numbers in red indicate estimates of percentage changes in energy fluxes in 2100 under RCP8.5 (or a high-end scenario of 2 m sea level rise, Bamber et al. (2019)) compared to the historical baseline, based on the following studies indicated by numbers in parentheses: (0) M2 tidal amplitude changes: ∼2–20% per metre of sea level rise (Pickering et al., 2017; Schindelegger et al., 2018). (1) Scaling from Wilmes et al. (2017) for a 2 m sea level rise. (2) Melet et al. (2015). (3) Wilmes et al. (2017) (scaling from their Table 1 for a 2 m uniform sea level rise). (4) Watanabe and Hibiya (2002); Jiang et al. (2005); Rimac et al. (2013); Simmons and Alford (2012); Furuichi et al. (2008); Song et al. (2019b). (5) Wang and Huang (2004).

slowing the rate of atmospheric warming and buffering atmospheric heat and carbon over centuries, and rising sea levels. A further illustration of the role of mixing under transient climate change is given for changes in nutrient fluxes. As the climate changes, so do mixing sources and distribution, with feedback on the climate state. This is illustrated first for the contemporary transient climate change in Section 2.3.4, and then for paleo climates in Section 2.4.

Stand-alone box on parameterisations of ocean mixing in climate models

Ocean mixing occurs on spatial scales that are too small to be resolved explicitly in ocean climate models. Turbulence generated by internal waves, for instance, is patchy and intermittent, occurring with events of spatial scales of order 0.1–100 m and lasting minutes to hours. Yet, because these small-scale events have large-scale climatic implications, representing the effects of ocean mixing through physically based parameterisations is crucial for realistic ocean and climate simulations, and to let mixing evolve in a changing ocean. Various parameterisations are needed to represent the many subgrid scale processes with distinct governing physics that are responsible for ocean mixing. Interior ocean diapycnal mixing is often represented in ocean models through a diapycnal diffusivity, using a vertical Fickian diffusion framework that is analogous to molecular diffusion. However it should be noted that in the turbulent surface boundary layer, there can be non-local net fluxes that are independent of the average gradient (e.g., Large et al., 1994). The ocean buoyancy b is defined as $b = -g\rho/\rho_0$, where ρ is the ocean density, ρ_0 a reference ocean density, and g the gravitational acceleration. The temporal tendency of the ocean buoyancy b from diapycnal mixing is given by the convergence of the buoyancy fluxes. An upward net buoyancy flux represents an increase in the potential energy of the water column; a net upward buoyancy flux can be thought of as being achieved by turbulent motions raising dense water parcels and lowering light ones, and then stirring them together before molecular diffusion homogenises the parcels, but the movement of the parcels takes work. Turbulent mixing in stratified water is sustained by the irreversible conversion of some of the turbulent kinetic energy to potential energy. The vertical buoyancy flux $\overline{w'b'}$ can thus be expressed as the fraction R_f, of the total turbulent kinetic energy dissipation ϵ_T, that sustains this potential energy conversion, or $\overline{w'b'} = R_f \epsilon_T$. Focusing on vertical motions, the time evolution of buoyancy due to turbulent mixing under the approximation of a linear equation of state and using a Fickian diffusion framework can thus be expressed as

$$\left.\frac{\partial b}{\partial t}\right|_{mix} = \left.\frac{-g}{\rho_0}\frac{\partial \rho}{\partial t}\right|_{mix} = -\left.\frac{\partial(\overline{w'b'})}{\partial z}\right|_{mix} = \frac{\partial(R_f \epsilon_T)}{\partial z} = \frac{\partial}{\partial z}\left(K_d \frac{\partial b}{\partial z}\right) = \frac{\partial(K_d N^2)}{\partial z}, \tag{2.1}$$

where N is the Brunt Vaisala frequency ($N^2 = \partial b/\partial z$), and K_d is the diapycnal diffusivity. The overbar denotes a time average, whereas primes denote temporal anomalies.

A focus will be made here on parameterisations of internal wave-driven diapycnal mixing, since (i) away from ocean boundaries, breaking internal waves supply the majority of power for diapycnal mixing and set both the background mixing in the ocean interior and locally enhanced turbulence, and (ii) much progress has been made over the last decades on that topic of active research (e.g., MacKinnon et al., 2017; de Lavergne et al., 2020; Whalen et al., 2020). The representation of internal wave-driven mixing in models has largely evolved with our understanding of the various processes responsible for the generation, propagation and dissipation of internal waves (e.g., MacKinnon et al., 2017).

The ocean turbulent diapycnal mixing was first parameterised using an ad hoc homogeneous time-invariant diapycnal diffusivity (usually O(10^{-5} m^2 s^{-1}) in the main thermocline), adjusted so that the model's meridional transports of heat and mass match with observations, since these are important metrics of ocean models and are sensitive to the value of K_d (Bryan, 1987; Vallis, 2000; Park and Bryan, 2000). Direct measurements of mixing indeed reveal turbulent buoyancy fluxes that are consistent with turbulent diffusivities of the order of 1-2 × 10^{-5} m^2 s^{-1} in the ocean thermocline, and over the full water column over smooth abyssal plains, a value therefore representative of a background level of turbulence supported by the internal wave continuum Munk (1981). However, micro-structure observations show that diapycnal diffusivities can be orders of magnitude above that background level in the deep ocean above rough topography (e.g., Polzin et al., 1997; Ledwell et al., 2000). In these regions, where internal waves are generated, elevated diffusivities were observed hundreds of metres above the ocean floor. In an effort to take into account the bottom-enhanced turbulent mixing in regions of internal tides generation, parameterisations using fixed diffusivity vertical profiles were developed (e.g., Bryan and Lewis, 1979; Huang, 1999; Tsujino et al., 2000).

Near-field internal-tide dissipation. An important breakthrough was made when St. Laurent et al. (2002) formulated a new, semi-empirical parameterisation of mixing, that was subsequently used in several climate models (e.g. Simmons et al., 2004b; Saenko and Merryfield, 2005; Bessières et al., 2008; Jayne, 2009) of the CMIP5 generation (coupled model intercomparison project phase 5). An important feature of this parameterisation is that it relates mixing to energy dissipation with energetically constrained sources. Energy dissipation is translated to a diffusion acting on the model's local density stratification (Osborn, 1980), ensuring consistency between the power available for turbulence, the turbulent kinetic energy dissipation rate, and the potential energy increase due to ocean mixing. A fixed fraction of local dissipation of the internal-wave energy, q, and a vertical profile of dissipation $F(z)$, are empirically prescribed as

$$\epsilon = \frac{qE(x,y)}{\rho}F(z) = \frac{-\overline{w'b'}}{R_f}. \tag{2.2}$$

In this parameterisation, the energy flux into the baroclinic tide $E(x, y)$ is computed from the internal tide generation linear theory of Bell (1975). Yet, the subsequent dissipation of this energy remains ad hoc: the fraction of energy dissipated locally q, is set to 1/3, and the vertical profile of dissipation, $F(z)$, is set to be exponentially decaying above the topography with a uniform decay scale in St. Laurent et al. (2002), matching observations from a single transect in the Brazil basin. Polzin (2009) formulated a more dynamically based vertical profile for internal tide energy dissipation, based on analytical solutions to a radiation balance equation (Polzin, 2004), allowing the vertical profile of local energy dissipation to vary in time and space, and to evolve with a changing climate. This was implemented in a climate model in Melet et al. (2013), which showed that the meridional overturning circulation (MOC) is sensitive to the vertical distribution of the internal tide local energy dissipation. Lefauve et al. (2015) also derived profiles of energy dissipation of internal tides generated by abyssal hills, highlighting the existence of two main types of vertical profiles.

The fraction of local dissipation is strongly heterogeneous (Waterhouse et al., 2014; Lefauve et al., 2015; Vic et al., 2019), so additional parameterisations are required instead of a uniform value for a more realistic representation of local dissipation. Moreover, a substantial fraction of the dissipation, occurring away from the site of generation, is not represented through the parameterisation (Eq. (2.2)).

Far-field internal-tide dissipation. The fraction of internal tide energy that is dissipated away from the generation site was historically represented in ocean models via the background diffusivity, which also accounts for other missing physics. With the simplest approach, this background diffusivity is constant in space and time (e.g., Jayne, 2009). Direct observations of the rate of kinetic energy dissipation show a dependence on stratification (Gargett and Holloway, 1984) and latitude due to wave-wave interactions (Henyey et al., 1986; Gregg et al., 2003), inspiring more elaborate formulations of the background diapycnal diffusivities that have been used successfully in global models (Gregg et al., 2003; Harrison and Hallberg, 2008; Jochum, 2009; Danabasoglu et al., 2012; Dunne et al., 2012). A large community effort was subsequently undertaken to improve our understanding and parameterisations of internal-wave-driven mixing (see a review by MacKinnon et al. (2017)), with new parameterisations being implemented in several CMIP6 models (e.g., Adcroft et al., 2019; Voldoire et al., 2019; Boucher et al., 2020). Using idealised scenarios for the horizontal and vertical distribution of low-mode internal tide energy dissipation, Melet et al. (2016) highlighted the sensitivity of the ocean state to the heterogeneity of remote internal tide energy dissipation. Oka and Niwa (2013) parameterised remote internal tide dissipation assuming $q = 1/3$, using maps of vertically integrated internal tide dissipation provided by a 3D-tidal model, and assuming a vertically uniform profile of energy dissipation. A framework for accounting for the geography of internal tide energy propagation and dissipation with a ray-tracing approach was depicted in MacKinnon et al. (2017) and implemented in de Lavergne et al. (2019), building upon Eden and Olbers (2014). In de Lavergne et al. (2019) and de Lavergne et al. (2020), internal tide beams for the first five vertical modes (higher modes are supposed to dissipate locally) of the three main tidal constituents are tracked individually with a Lagrangian scheme, and four dissipative processes (wave-wave interactions, scattering by abyssal hills, interaction with topographic slopes and shoaling) act as energy sinks to produce climatological 2D maps of low-mode internal tide energy dissipation, then associated with different vertical profiles of dissipation. Assessing the climatic impacts of this parameterisation is ongoing work. Building on these ideas to allow for time-varying maps of the column-integrated energy dissipation in the context of an evolving ocean state and non-linear interactions between beams may introduce feedbacks that act to regulate the ocean water mass structure.

It should be noted that even ocean general circulation models with explicit tides typically do not resolve the generation of high-mode internal tides, scattering of low-mode energy into higher modes, and various processes leading to internal tide energy dissipation, so parameterisations are still required to get realistic internal tides and dissipation (e.g., Arbic et al., 2010; Ansong et al., 2017).

Lee waves. Lee waves are another class of internal waves for which our understanding has evolved over the last decade. Although the energy flux into lee waves is thought to be smaller than that into internal tides (0.16-0.75 TW, Nikurashin et al. (2014); Nikurashin and Ferrari (2011); Scott et al. (2011); Wright et al. (2014), Fig. 2.2), the spatial distributions of mixing induced by these two classes of internal waves differ and parameterizing lee-wave driven mixing makes a significant impact on the ocean state (Melet et al., 2014). In coarse resolution ocean models, a full coupling between wind power, currents from parameterised and resolved eddies as well as larger-scale geostrophic circulations, ocean stratification, lee-wave drag and induced mixing has yet to be achieved (Melet et al., 2014). Improving our understanding of lee-wave energy dissipation is needed for parameterisations (e.g., Legg, 2021). That includes wave-mean flow interactions and energy transfer (e.g., Kunze and Lien, 2019); the role of critical layers (when the vertical wavelength gets very small) for lee-wave energy dissipation or reabsorption by the mean flow (e.g., Booker and Bretherton, 1967).

Using a 2D non-hydrostatic numerical model configured to be representative of Drake passage flow-topography interactions, Zheng and Nikurashin (2019) found that 30–40% of lee-wave energy is dissipated locally, while the remainder is dissipated downstream of the generation site with an e-folding length scale of order 20–30 km, which is still close to the generation site from the perspective of basin-scale ocean or global climate models. Parameterisations of wave drag have been developed for lee waves. In addition to the generation of lee waves and their subsequent dissipation, wave drag includes topographic flow blocking and splitting (thought to occur where the steepness parameter exceeds 0.4, Nikurashin et al. (2014)) that lead to a momentum flux and direct local conversion of flow kinetic energy to dissipation (e.g., Trossman et al., 2015). Implementing such a wave drag parameterisation allows for feedback between the wave drag and the low-frequency flow that produce lee waves. Another interplay could be the redistribution of lee-wave energy with a reinjection of energy into the mean flow (Kunze and Lien, 2019).

Near-inertial waves. Variable surface wind stresses excite inertial oscillations in the surface mixed layer. Part of the inertial oscillation energy is dissipated within the mixed layer, supplying it with potential energy and modifying the mixed-layer depth. The reminder radiates downward from the mixed-layer depth and equatorward as near-inertial internal waves (NIW). Several parameterisations of mixing in and below the mixed-layer induced by NIW have been developed (e.g., Jochum et al., 2013; Eden and Olbers, 2014), building on existing schemes for internal tides or on radiation balance equations. As for internal tides, different aspects of the parameterisations remain uncertain, such as the power available for propagating NIW (e.g., Alford, 2020) (Fig. 2.2), the partitioning between locally dissipated and far-field propagating NIW and the vertical structure and decay scale of the energy dissipation. Another difficulty in parameterizing NIW induced mixing is related to the isolation of near-inertial currents in ocean models (e.g., Jochum et al., 2013; Jing et al., 2016; Song et al., 2019b). In addition, the redistribution of NIW energy is affected by interactions between the waves and the background currents (e.g., Balmforth et al., 1998; Danioux et al., 2015; Jeon et al., 2019), fronts (e.g., Grisouard and Thomas, 2016), and the eddy field (e.g. Kunze, 1985; Zhai et al., 2009; Whalen et al., 2018; Martínez-Marrero et al., 2019; Lelong et al., 2020). Efforts to account for these interactions in parameterisations have been made (e.g., Olbers and Eden, 2017; Eden and Olbers, 2017; Jing et al., 2017).

Shear-driven mixing. Sufficiently strong velocity shears are subject to various instabilities that can overcome the suppressing effects of stable density stratification. Kelvin–Helmholtz instabilities occur when the vertical shears of the horizontal velocities are large enough relative to the vertical density stratification. Small fluctuations in the vertical velocities will spontaneously grow, drawing energy from the spatial mean horizontal velocities and wrapping up dense and light fluid into prominent turbulent billows with order O(1) aspect ratio before devolving into vigorously turbulent layers. The necessary condition for these instabilities to grow is that the local shear Richardson number,

$$Ri = \frac{N^2}{\left|\frac{\partial \vec{u}}{\partial z}\right|^2}, \tag{2.3}$$

where $\vec{u}$ is the horizontal velocity and N^2, is the squared buoyancy frequency, has to drop below a critical value of 1/4. When the flow is unstable, Kelvin–Helmholtz instabilities grow very rapidly, with characteristic growth rates proportional to the velocity shear, or a growth timescale of order a few minutes in vigorous oceanic overflows or breaking internal waves. Symmetric and centrifugal instabilities are driven by horizontal shears, and drive an exchange along tilting surfaces, but can also lead to mixing via secondary instabilities (e.g., Yankovsky and Legg, 2019). Shear instabilities are one of the pathways by which sufficiently strong internal waves (especially at near-inertial frequencies) break in the interior ocean, leading to irreversible mixing (e.g., Thorpe, 2018).

Mean-shear driven turbulence, which is particularly important for the climatic state in overflows and the Pacific equatorial undercurrent, can be parameterised using different formulations (Pacanowski and Philander, 1981; Gaspar et al., 1990; Large et al., 1994; Legg et al., 2006; Jackson et al., 2008; Danabasoglu et al., 2010).

Other interior diapycnal processes. Apart from internal waves, other processes contribute to interior ocean mixing and need to be parameterised in ocean models. In hydrostatic models, convective processes, for instance, responsible for deep water formation, induced by static instabilities of the water column (heavy water parcel above lighter one) cannot be explicitly represented. Gravitational instabilities giving rise to vertical convection can be accounted for through, e.g., a large vertical diffusivity (Large et al., 1994; Klinger et al., 1996; Lazar et al., 1999) or a convective adjustment scheme (Rahmstorf, 1995; Madec et al., 1991; Giordani et al., 2020).

Isopycnal eddy mixing (see Chapter 9) in coarse, non-eddy-permitting models has been widely parameterised with expressions that combine along-isopycnal diffusion of tracers by geostrophic eddy motions (Solomon, 1971; Redi, 1982; Griffies et al., 1998) with an additional advection of tracers associated with the extraction of available potential energy and flattening of isopycnals (referred to as GM eddy parameterisation) (Gent and McWilliams, 1990; Gent et al., 1995). If the lateral diffusivities used by the parameterisations of along-isopycnal diffusion and advection of tracers are the same, these two processes can be combined in the same operator (Griffies, 1998), although more recent work suggests that the vertical structure of the two diffusivities differ (Smith and Marshall, 2009; Stanley et al., 2020). Although the GM eddy parameterisation is advective and does not directly dissipate tracer variance, it is associated largely with the irreversible mixing of the dynamical tracer, potential vorticity, along isopycnals (e.g., Greatbatch, 1998; Marshall et al., 1999). Hallberg (2013) suggests how to transition from this parameterisation of eddy effects in areas where eddies are unresolved to an explicit representation by the model in regions where mesoscale eddies are resolved. The transition of the GM parameterisation in the ocean interior to horizontal tracer mixing in the mixed layer is tackled by Danabasoglu et al. (2008) and Ferrari et al. (2010). Most recently, the GEOMETRIC eddy parameterisation (Marshall et al., 2012; Mak et al., 2018) extends GM to include an explicit eddy energy budget, with the dissipation of geostrophic eddy energy playing a vital role in setting the magnitude of the geostrophic eddy fluxes. Thus GEOMETRIC introduces an interplay between the parameterisations of isopycnal eddy mixing and lee-wave mixing (discussed above). Finally, eddy mixing along isopycnals can lead to changes in the density of fluid parcels due to processes such as cabbeling and thermobaricity arising from non-linearities in the equation of state (Nycander et al., 2015).

Surface boundary layer turbulence. Winds, waves, and buoyancy fluxes drive vigorous turbulence in a near-surface region that nearly homogenises properties in a "mixed layer". There is a long and successful history of parameterisations of ocean boundary layer mixing based on considerations of an explicit turbulent kinetic energy budget balancing energy inputs from forcing with the potential energy changes from mixing into stratified water (see the review by Niiler and Krauss (1977)), or other related considerations (e.g., Large et al., 1994). As is reflected in the recent review paper by Li et al. (2019), there are extensive ongoing efforts to refine and improve parameterisations of boundary layer mixing across a wide range of wind, wave, and surface buoyancy forcing conditions.

In the surface layers, parameterisations of Langmuir turbulence have been introduced to explicitly capture the interactions between the surface wave fields and the dynamics of the boundary layer (e.g., Li et al., 2019).

Submesoscale processes in the boundary layer (Chapter 8) mediate the cascade of energy and tracer variance from mesoscales to the smaller scales of three-dimensional turbulence, in some circumstances leading to mixing and exchange of mixed fluid between the boundary layer and the ocean interior by symmetric and centrifugal instabilities (Bachman et al., 2017). The re-stratification of the surface mixed-layer by submesoscale eddies has been parameterised in Fox-Kemper et al. (2008, 2011); Calvert et al. (2020). Submesoscale processes can also be generated at the bottom of the ocean along sloping topography and contribute to diapycnal mixing and upwelling of deep waters (e.g., Naveira Garabato et al., 2019; Yankovsky and Legg, 2019; Wenegrat and Thomas, 2020). For example, both along-isopycnal transport and the diapycnal mixing by secondary shear instabilities figure prominently in a recently developed parameterisation of submesoscale symmetric instability in dense flows along topography (Yankovsky et al., 2020).

Mixing efficiency. Parameterisations of ocean mixing have largely focused on the turbulent kinetic energy produced by turbulence, but the efficiency at which this energy is irreversibly converted to potential energy by raising dense water parcels and lowering lighter ones to mix them together, thereby fluxing buoyancy downwards, is of fundamental importance to estimates of diapycnal mixing (e.g., Osborn, 1980; Mashayek and Peltier, 2013; Gregg et al., 2018; Caulfield, 2021). Early estimates of the mixing efficiency range around 15%–20% (Osborn, 1980). Although viscous dissipation converts the remaining majority of the turbulent kinetic energy to heat, this source of heating is mostly negligible in the ocean as it is 3–4 orders of magnitude smaller than the leading order terms in the ocean's thermal energy budget (e.g., McDougall, 2003). The mixing efficiency is not a constant, but rather it depends on the flow properties. In particular, the mixing efficiency is lower in actively mixing and weakly stratified waters (e.g., Caulfield, 2021). The impact of a variable mixing efficiency on the abyssal overturning was explored in de Lavergne et al. (2016b) and Mashayek et al. (2017), suggesting that buoyancy-Reynolds number-based parameterisations might be needed. Co-variations between ocean-mixing efficiency and the fraction of local dissipation q are also non-trivial and should be considered to produce more accurate estimates of diapycnal upwelling or downwelling rates (Cimoli et al., 2019).

Time-evolving mixing. Processes underlying ocean mixing are non-steady, which calls for time-evolving mixing parameterisations that change along with the climate state. For instance, the energy flux into lee waves exhibits a clear annual cycle in the Southern ocean, and a decreasing trend in the global energy flux is projected under climate change (Melet et al., 2015). Annual cycles related to near-inertial waves are also identified (Whalen et al., 2012), as well as clear annual cycles in global upper ocean dissipation rates (Whalen et al., 2012; Alford and Whitmont, 2007; Silverthorne and Toole, 2009). The parameterisation of mixing due to internal tides and lee waves developed by de Lavergne et al. (2020) allows for a time-evolving mixing by using the modelled stratification in the different vertical profiles of energy dissipation; however, in this case the 3D energy-dissipation maps also rely on static 2D maps of vertically integrated energy dissipation per mode class. These maps of vertically integrated energy dissipation were derived from static energy fluxes into internal tides and lee waves, from a static map for the fraction of the internal-wave energy that is dissipated locally, and from a static, climatological ocean stratification for the propagation of low modes, and for the split of internal tide energy at topographic slopes between shoaling, scattering into high modes and reflection to the deep ocean.

Despite progress, advances are still needed for a more complete representation of the time-varying energy sources and dissipation mechanisms (see also the discussion of lee-wave-driven mixing above). Dynamical parameterisations are required to allow mixing to evolve in space and time, depending on the state of the ocean (e.g., hydrography, stratification, circulation, eddy fields, fronts, tides, and sea level) and of the climate (e.g., surface winds, sea-ice extent). Dynamically-based energetically-constrained parameterisations of ocean mixing in climate models should allow for more credible simulations of a changing climate (e.g., Huang, 1999; Eden and Olbers, 2014; Melet et al., 2016; MacKinnon et al., 2017; de Lavergne et al., 2020).

Spurious numerical mixing. Another difficulty in accurately representing mixing in ocean models relates to the presence of spurious numerical mixing due to numerics and truncation errors. For instance, the discretisation of equations can lead to tracer advection occurring in directions that are not perfectly aligned with isopycnals, which can induce substantial numerical mixing (Griffies et al., 2000; Ilicak et al., 2012). Reducing spurious mixing is needed to be able to explicitly add mixing induced by physical processes without ending up with an overly diffusive model.

2.2 The role of ocean mixing in shaping the contemporary climate mean state

2.2.1 Meridional overturning circulation and heat transport

Decadal to millennial changes in the ocean overturning circulation and meridional heat transports have widespread climate impacts. A few examples stemming from weakening Atlantic overturning include a widespread cooling throughout the North Atlantic and northern hemisphere; changes in precipitation and a strengthening of the North Atlantic storm track (e.g., Jackson et al., 2015), changes in sea levels along the northeast coasts of North America (e.g., Goddard et al., 2015), or abrupt climate changes (e.g., Dansgaard–Oeschger events, Kageyama et al., 2010; Lynch-Stieglitz, 2017). The meridional overturning circulation has been classically simplified and schematised with a mostly adiabatic upper cell, corresponding to North Atlantic deep water (NADW) formation and overturning, and a lower, abyssal cell, corresponding to the formation of Antarctic bottom water (AABW) and its largely diabatic overturning pathway (e.g., Gordon, 1986).

The actual picture of the global overturning circulation (the role of mixing in the global overturning circulation and its associated heat balance) is far more intricate (e.g., Talley, 2013) (Chapter 3). Although the relative importance of diabatic and adiabatic processes in controlling the return flow of surface ventilated deep water has been long debated, a community consensus view has emerged over the last decade (e.g., Marshall and Speer (2012); Talley (2013), Chapters 3, 7).

NADW is formed through surface densification. Although this chapter mainly focuses on diapycnal mixing, isopycnal mixing can influence the intensity of wintertime deep convection in the subpolar gyre by changing the transport of salt and altering static stability (Pradal and Gnanadesikan, 2014, see also Chapter 9). During its descent and deep pole-to-pole southward journey, mixing at overflows determines the density of NADW, which in turn is key in setting the depth of this water mass in the global ocean (Chapter 3, Galbraith and de Lavergne (2019); Sun et al. (2020)). NADW is also mixed with the underlying abyssal AABW through bottom intensified internal-wave mixing (Fig. 2.3). The NADW return journey

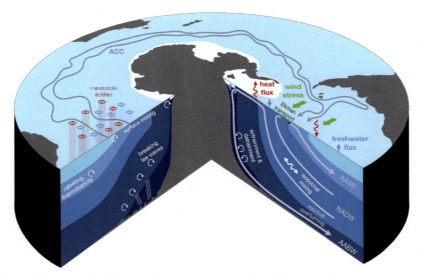

FIGURE 2.3 A schematic illustration of the circulation, forcing, water masses, and mixing processes in the Southern ocean, inspired by Meredith et al. (2019). The zonal winds (green arrows) provide forcing to the Antarctic circumpolar current (wiggly dark grey lines with arrows), which exhibits significant anticyclonic and cyclonic mesoscale eddy variability (red and blue whorls), especially where the ACC interacts with significant topographic obstacles, which in turn drives along-isopycnal mixing (wiggly white arrows). The two major Atlantic overturning cells described in the text (light grey lines with arrows) are connected to water mass modification by net heat and freshwater fluxes (red and blue arrows) at the surface and turbulent mixing (white circular arrows) near the surface and where it is driven by breaking internal waves (short white lines), or flow instabilities.

to the North Atlantic involves a (mostly) adiabatic, wind-driven upwelling along isopycnals in the Southern ocean (e.g., Toggweiler and Samuels, 1998; Marshall and Speer, 2012) (Fig. 2.4). There, NADW contributes to the formation of both the very cold and dense AABW and the fresher and lighter Antarctic intermediate water (AAIW) (see Fig. 2.3). Nonlinearities in the equation of state play a vital role in the formation and descent of AAIW in the open ocean (Nycander et al., 2015), notably via the process of cabbeling, whereby two fluids with the same neutral density but with different potential temperatures and salinities attain a higher density after mixing together (see also Chapter 9, Fig. 2.3).

AABW sinks to the abyssal ocean and spreads northward in all ocean basins. Mixing at abyssal sills and straits lightens the densest AABW, while internal-wave-driven mixing and geothermal heating also reduce the density of AABW along its northward journey (Chapters 3, 7). Secondary circulations driven by this spatially variable mixing help maintain a finite abyssal stratification (Chapter 7). In the Indian and Pacific oceans, diapycnal mixing of AABW with overlying waters leads to the diffusive formation of Indian deep water (IDW) and Pacific deep water (PDW). These deep water masses are further lightened through mixing in the low latitudes Indian and Pacific oceans, and are subsequently upwelled along the sloping neutral density surfaces north of the Antarctic circumpolar current (ACC) in the Southern ocean due to the Ekman divergence. There, together with air-sea flux induced water mass transformation, they form the bulk of the lighter subantarctic mode water (SAMW). Diapycnal mixing in the low latitude Pacific and Atlantic beneath and across the main thermocline contributes to further lightening of the SAMW (Chapter 10), enabling its return to the surface circulation and North Atlantic (e.g., Talley (2013), Fig. 2.4), although some upwelling of deep cold water to the ocean surface in coastal upwelling zones has also been hypothesised (Toggweiler et al., 2019).

In this description of the global MOC, diapycnal mixing is central for setting the heat content of water masses and for lightening the AABW (and NADW) through the formation of IDW and PDW, allowing a return pathway to their surface formation places. The relationships between the MOC, abyssal stratification, eddies, diapycnal mixing, and wind forcing have been studied since at least Munk (1966) using simple conceptual models (e.g., Stommel and Arons, 1960; Pedlosky, 1992; Munk and Wunsch, 1998; Gnanadesikan, 1999; Ito and Marshall, 2008; Nikurashin and Vallis, 2011, 2012). In the limit of overturning being diffusively-driven, the overturning strength would scale with the diapycnal diffusivity K_d to the power of 2/3 for the NADW overturning, and to the power of 1/2 for the AABW overturning. These scalings indicate an increase in overturning strengths with a uniform increase in diffusivities. As meridional heat transport (MHT) is connected to the MOC, it also increases with increases of uniform diffusivities (e.g., Vallis and Farneti, 2009).

The sensitivity of the MOC and MHT to diapycnal mixing has also been studied in various realistic modelling studies (e.g., Schmittner and Weaver, 2001; Simmons et al., 2004a; Saenko and Merryfield, 2005; Jayne, 2009; Melet et al., 2013, 2014, 2016; Hieronymus et al., 2019). Hieronymus et al. (2019) explores the dependence of the strength of the lower

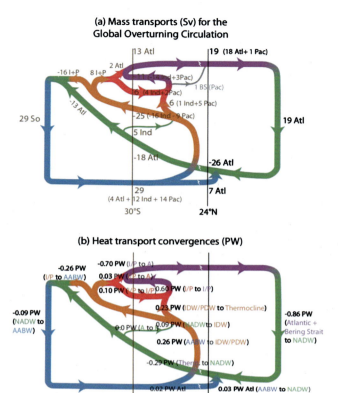

FIGURE 2.4 (a) Mass transports (in Sverdrups; 1 Sv = 10^6 m^3 s^{-1}) for each branch of the global overturning circulation, where the assumed conversions from one water mass to another are provided. (b) Heat transport convergence (in petawatts; 1 PW = 10^{15} W) for each mass-balanced conversion shown in (a). Each number shows the net air-sea heat flux within the ocean sector that is associated with the conversion. Negative is heat loss; positive is heat gain. For instance, in the Southern ocean, south of 30°S, the heat transport convergence of "−0.09 PW (NADW to AABW)" means that 0.09 PW is lost from the ocean to the atmosphere south of 30°S associated with converting North Atlantic deep water (NADW) to Antarctic bottom water (AABW). PDW stands for Pacific deep water and IDW for Indian deep water. Meridional heat transports associated with the upper ocean subtropical gyres are not included in (b). From Talley (2013) based on Talley (2008).

and upper cells of the MOC with the level of diapycnal mixing in fully coupled climate model simulations, including atmospheric feedbacks. In this study, the levels of background diapycnal diffusivities are uniformly increased by values ranging from zero, through modest values, up to implausibly large values (that are comparable to numerically induced diapycnal diffusivities in some climate models) in a series of fully spun-up simulations. The larger added diffusivities favour the diapycnal upwelling of deep water compared to the along-isopycnal pathway. The climate model encompassed otherwise classical parameterisations of diapycnal mixing (e.g., St. Laurent et al., 2002). The set of simulations show that increasing the level of background diapycnal mixing (K_d) indeed increases the strength of the upper and lower MOC cells (Hieronymus et al., 2019). Ocean MHT also increases, although not as a simple and direct consequence of stronger overturning, as would be the case in ocean-only simulations.

In addition to the overall magnitude of background diapycnal mixing, the spatial distribution of ocean mixing, both horizontally and vertically, influences the MOC (Scott and Marotzke, 2002; Jayne, 2009; Melet et al., 2013, 2014, 2016). Using the same energy input into internal tides but different vertical profiles of dissipation, Melet et al. (2013) show that not only the energy input to the internal tides matters, but also where in the vertical it is dissipated, because it is the convergence of the buoyancy flux that changes density (also see the stand alone box). A monotonic bottom-intensified dissipation leads to denser water except in the bottom layer, where a no-flux bottom boundary condition has to be satisfied, leading to lighter water there. Bottom-intensified dissipation thus drives upwelling along the sloping topography, downwelling above, with the net abyssal upwelling being a small residual of these transports (see also Chapter 7). The implications of bottom-intensified dissipation for the abyssal overturning circulation have been explored in several studies (e.g., Melet et al., 2013; Ferrari et al., 2016; de Lavergne et al., 2016a; McDougall and Ferrari, 2017; Cimoli et al., 2019). However, if the vertical profile of dissipation changes with variable stratification, a convergence of the buoyancy flux in the water column where

14 Ocean Mixing

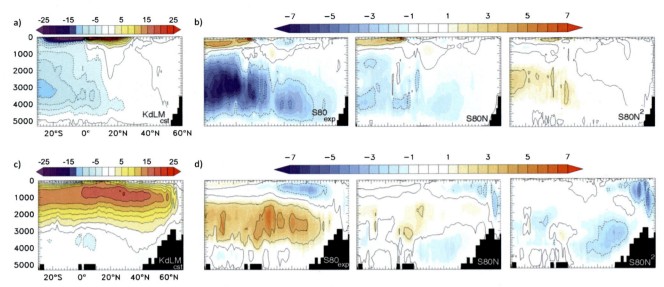

FIGURE 2.5 Climate model experiments (GFDL-ESM2G) showing changes in the Indo-Pacific MOC (upper panels) and Atlantic MOC (AMOC, lower panels) due to changes in spatial and vertical location the internal-tide remote dissipation (80% of internal tide energy in all cases). (a) The Indo-Pacific MOC and (c) the AMOC for the last century of 1000-yr simulations, in Sv, in a reference experiment, where the remote dissipation of internal tide is parameterised with a spatially uniform vertical diffusivity, whose value is set so that work against stratification done by low-mode internal-tide dissipation (KdLM) is consistent with the global energy input into low modes. (b) and (d) show anomalies from (a) and (c) respectively for the Indo-Pacific MOC and AMOC when the remote dissipation of internal tides is set to occur only on slopes, using either an exponentially decaying vertical profile (left panel), a vertical profile scaling with N (middle panel) or N^2 (right panel). All four simulations have the same total energy input and dissipation for internal tides. After Melet et al. (2016).

stratification strongly increases can occur, leading to water becoming lighter there. The differences in the location of layers becoming lighter or denser cause changes most prominently in the abyssal cell of the MOC, as increased mixing in the abyssal ocean tends to enhance the AABW overturning (Fig. 2.2).

The spatial distribution of energy fluxes into internal waves and their local bottom-intensified dissipation has been shown to impact the abyssal ocean and the abyssal cell of the MOC. In a climate model, adding lee-wave-driven mixing on top of tidal mixing leads to a stronger AABW overturning, a result mostly assigned to the different spatial distribution of the two energy sources, with lee waves being dominant in the Southern ocean (Melet et al., 2014). The importance of the spatial distribution of mixing is further illustrated in a set of climate experiments using the same total energy input into the internal tide field and the same magnitude and location of near-field dissipation, but different vertical profiles (exponentially decaying from the bottom or scaling with N or N^2) and horizontal locations (deep basins, continental slopes, coastal shelves) for the remote dissipation of low mode internal tides (Melet et al. (2016), Fig. 2.5, see also de Lavergne et al. (2016a)). In this set of experiments, increased dissipation in the upper thermocline (N^2 scaling) leads to stronger overturning of the shallow subtropical cells and to a broader thermocline, with consequences for heat uptake (Section 2.3.1, Fig. 2.2). Increased stratification in the abyssal ocean leads to a stronger AABW overturning. The horizontal location of remote mixing (deep oceans or continental slopes or coastal shelves) influences the circulation most prominently when it impacts the properties of dense waters in deep water formation regions (Nordic seas, Antarctic margins) leading to less dense deep waters and reduced deep overturning (Fig. 2.2, 2.5).

2.2.2 Southern Ocean

The strongest and largest ocean current systems in the world are found in the Southern ocean. As illustrated in Fig. 2.3, the Southern ocean plays a critical role in the climate system, due to the outcropping to the surface of dense waters that are found at great depths elsewhere, and due to its interactions with the Antarctic coastal waters that form cold bottom water that fills the deep basins in the Atlantic, Pacific, and Indian Oceans (Section 2.2.1). It plays a disproportionately large role in the uptake of anthropogenic heat and carbon from the atmosphere and in their sequestration into water masses that spread into the ocean interior globally, both at intermediate depths and in the abyss.

Mixing by geostrophic eddies and three-dimensional turbulence plays a uniquely important role in the Southern ocean. The absence of continental barriers over the circumpolar belt means that the eastward wind stress at the surface is balanced by a zonal pressure gradient across abyssal topographic barriers, resulting in an Ekman overturning cell that extends to great depth in the Southern ocean. This Ekman overturning cell steepens the neutral density surfaces, resulting in a baroclinic ACC through thermal wind balance (assuming relatively weak circumpolar flow at depth). The steepening of the neutral density surfaces and acceleration of the ACC is arrested by the generation of intense mesoscale eddies through baroclinic instability, the eddies transporting zonal momentum down to the depths of the topographic ridges (in a process known as eddy form stress), while also transporting relatively warm and light surface waters southwards in opposition to the northward mean wind-driven Ekman transport.

The baroclinic structure of the water masses in the Southern ocean tends to have meridional slopes of density surfaces that are just above the threshold for the growth of baroclinic eddies. Increasing the strength of the Southern ocean winds does not lead to substantial changes in the baroclinic volume transport of the ACC, but primarily to more intense baroclinic eddies, an idea that has been termed "eddy saturation", and which is supported by numerous numerical simulations and by oceanic observations (Meredith and Hogg, 2006; Hallberg and Gnanadesikan, 2006; Munday et al., 2013). Recent work has suggested that eddy saturation arises through a balance between the generation of eddy energy through baroclinic instability and the dissipation of eddy energy through processes such as bottom drag and lee wave generation Marshall et al. (2017). Thus increased generation of lee waves might lead not only to increased diapycnal mixing and an increase in the transport of the abyssal overturning cell (Broadbridge et al., 2016), but also to an increase in the baroclinic volume transport of the ACC and the slope of neutral density surfaces to maintain the balance between the sources and sinks of eddy energy (Marshall et al., 2017).

The total transport of the ACC can considerably vary on short timescales with changes in the wind-forcing. Observational estimates of the ACC total transport have differed greatly as new observations have been brought into consideration (Donohue et al., 2016). However, these are changes related to variations in the near-bottom velocity, and not the geostrophically balanced shear throughout the water column. On the one hand, the baroclinic water mass structure of the ACC is strongly connected with the mean stratification (and heat content) of all the major ocean basins to the north of the ACC, and hence is remarkably stable, varying on centennial and longer time scales Allison et al. (2011). On the other hand, the near-bottom flows that help set the total ACC transport are determined in part by the northward flow of modest volumes of abyssal waters that are formed around Antarctica. Differences in these abyssal properties are a primary source of the wide range of total ACC transports in climate models (Russell et al., 2018). The net transport of the largest ocean current in the world is thus sensitive to the details of the mixing that regulates the creation of dense water masses around Antarctica.

The open Southern ocean is also of critical importance for ventilating the Indian and Pacific deep waters (Fig. 2.4, also collectively referred to as common deep water) that have been sequestered for centuries in the deep ocean and fill much of the deep Pacific and Indian basins. The same westerly winds that drive the ACC lead to a northward Ekman transport in the Southern ocean that draws dense waters up to the surface (see Fig. 2.3), where turbulent mixing and net precipitation and ice melt transform this deep water into Antarctic intermediate waters and mode waters that return northward into the Pacific, Atlantic, and Indian Oceans at depths of about 1000 m (Sverdrup et al., 1942). Because the upwelling water is largely devoid of anthropogenic carbon and is rich in nutrients, the Southern ocean plays a disproportionately large role in the carbon cycle (Frölicher et al., 2015). Changes in the intensity of the Southern ocean winds, and their peak latitude relative to the Drake passage gap between South America and Antarctica, can both elicit a significant response in overturning and water mass creation (Russell et al., 2006). However, as with the strength of the ACC, eddy mixing also plays a strong role in regulating the response of the Southern ocean overturning circulation to forcing changes. A strengthening of the winds leads to an increased intensity of mesoscale eddy transports across the ACC that partially offsets the change in the Ekman transport (Hallberg and Gnanadesikan, 2006; Morrison and Hogg, 2013). However, because the Ekman transport is largely confined to the turbulent surface mixed layers that are substantially shallower than the mesoscale eddy overturning response, this "eddy compensation" is imperfect and changing winds drive significant changes in the shallow overturning cells in the top few hundred metres of the Southern ocean that drive an exchange between the turbulent mixed layer and the interior ocean (Bishop et al., 2016). Due to interactions with a uniquely deep-reaching upwelling and overturning circulation, Southern ocean mixing in the surface boundary layer and ocean interior can be particularly effective at modifying water mass properties worldwide.

The Southern ocean's intense eddy fields, and the dynamical requirement that they interact with the topography in order to balance the wind forcing, lead to significant sources of eddy-driven energy to the internal wave field and related three-dimensional mixing. This topic has been the subject of intensive observational scrutiny following the pioneering work of Naveira Garabato et al. (2004), as is discussed by Gille et al. in Chapter 12.

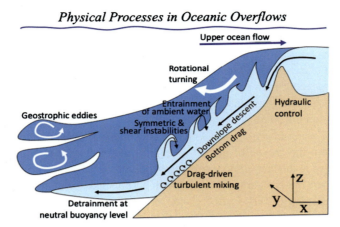

FIGURE 2.6 A schematic illustration of the physical processes acting in oceanic overflows, adapted from Legg et al. (2009).

2.2.3 Mixing in exchanges between marginal seas and the open ocean

Marginal seas and estuaries create water masses that have properties that are distinct from those found in the open ocean. Much of the interior ocean is filled with water whose properties are directly influenced by outflows from these marginal seas, and changes in the forcing of these marginal seas or their watersheds is one of the pathways by which climate change can be manifest in the interior ocean. Riverine inflows introduce fresh water, leading to near-surface fresh plumes (e.g., Hickey et al., 1998; Coles et al., 2013; Hopkins et al., 2013), with the properties of these plumes as they enter the open ocean strongly influenced by shear-, wind-, and tide-driven mixing in their estuaries (Geyer and MacCready, 2014; MacCready et al., 2018). Conversely, net evaporation in seas like the Mediterranean and Red sea, or seasonal or annual net ice formation in the Arctic and Antarctic amplifies the salinity in marginal seas. These marginal sea source water masses are usually heavily modified by a variety of mixing processes before forming distinct water masses that are readily identifiable in the interior open ocean. The exchange of water between these marginal seas and the interior ocean may be regulated by one or more points of internal hydraulic control (Pratt and Lundberg, 1991), as is observationally demonstrated for the Baltic outflow (Nielsen et al., 2017), or they can be strongly influenced by rectified tidal exchanges (e.g., Gibraltar, Sannino (2004)).

Marine-terminating tidewater glaciers in Greenland, Alaska, Patagonia, the Antarctic peninsula, and elsewhere are collectively providing significant contributions to global mean sea level rise (Oppenheimer et al., 2019). Mixing within the fjords of tidewater glaciers is of particular importance, because the warmth of the subsurface ocean waters supports subsurface melting and contributes to the calving at the glacier face, while also contributing to the rich dynamics of the circulation in the fjords (Straneo and Cenedese, 2015). For example, the buoyant plumes driven by subsurface melting and subglacial discharge exhibit strong entrainment, which in turn helps drive the replenishment across sills of the warm and salty oceanic waters in the bottom of fjords (Carroll et al., 2017).

Dense overflows from marginal seas exhibit significant diapycnal mixing from a variety of processes, as is illustrated schematically in Fig. 2.6. The most prominent of these mixing processes is stratified shear-driven mixing related to Kelvin–Helmholtz instabilities (see also the stand alone box), which broadens stratified shear zones and reduces the density contrast and increases the Richardson number. It is common to find shear Richardson numbers close to 1/4 in the stratified region separating dense gravity currents from the ambient ocean water (Peters et al., 2005; Barringer and Price, 1997; Voet and Quadfasel, 2010). Bottom drag and flow over a rough ocean bottom can also provide a source of energy for 3-dimensional turbulence that drives homogenisation of the bottom boundary layer and entrainment of water from above (Peters and Johns, 2005). This mixing substantially modifies the water masses that ultimately result from the overflows. Different magnitudes of mixing in and immediately downstream of overflows can be a primary factor in determining such properties as the density and depth of the lower branch of the AMOC in different coupled climate models (Wang et al., 2015). It should also be noted that in the commonly used depth-coordinate ocean models, spurious numerical mixing in overflows is often much larger even than the vigorous physical mixing that should be occurring (e.g., Legg et al., 2006; Colombo et al., 2020).

Dense outflows undergo adjustment to rotation over a few inertial periods, often creating currents that follow topography cyclonically. The isopycnal surfaces separating these dense currents are often observed to mirror the slope of the bottom, in which case the geostrophically balanced velocity differences between the plumes and the interior ocean is given by $\partial u = g' k \nabla D / |f|$ (Nof, 1983), where $g' = g(\rho_{plume} - \rho_{ambient})/\rho_0$ is the reduced gravity associated with the density difference between the dense plume and the ambient ocean water above the plume, D is the bathymetric depth, and f is the Coriolis parameter. As a result, enhanced geostrophic shears tend to occur where plumes cross regions with a local steepening of the bathymetric slopes, resulting in enhanced mixing. In addition, because the geostrophic velocity shear and squared buoyancy frequency are both proportional to g', the local shear Richardson number in geostrophic plumes tends to decrease inversely with g', making vigorous shear mixing more likely. Increasing the density contrast in an overflow plume tends to lead to greater entrainment by the plume, so the final density with which the resulting water flows into the ocean interior depends only weakly on the density of the source water, although the magnification in the volume of the final water mass is a stronger function of the source density (e.g., Legg et al., 2006).

Non-linearities of the equation of state also figure prominently in the dynamics of mixing and entrainment in overflows, because density becomes increasingly sensitive to temperature with increasing pressure (Legg et al., 2009). A relatively warm and salty plume (like the Gibraltar outflow or the Red sea outflow plume through Bab el-Mandeb) entrains vigorously because of its large density contrast and strong velocity shears near the surface, but then tends to equilibrate at an interior depth as the relative importance of temperature to density increases. By contrast, a cold and fresh overflow plume, such as those found around Antarctica, tends to have a relatively small density contrast with ambient water near the surface and hence descends with weaker shears and less vigorous shear-driven turbulent mixing, but gains in density contrast with the ambient water as the pressure increases and the thermal differences have an increasing contribution to density anomalies relative to salinity. These effects of the non-linear equation of state are among the main reasons why the deepest ocean is filled with cold slightly fresh waters, even though the dense waters flowing directly out of the Mediterranean and Red seas are much denser (by virtue of their high salinity) than anything found at comparable depths in the Arctic or around Antarctica (Legg et al., 2009).

Although tides are often thought of as a periodic and reversing flow, they can have a prominent imprint on the exchanges between marginal seas and the open ocean. The water carried by tides out past an exchange point can be swept downstream by a lateral mean current or descending gravity current before the next inflowing tide. This tidal flushing operates effectively when there are important dynamics that occur at spatial scales that are smaller than the distance the flow covers in a half tidal period (about 6 or 12 hours), and in places where the tidal velocity is large compared with the mean flow. Precisely these conditions are met at Gibraltar, for example, where the critical exchange over the Camarinal sill occurs over just a few tens of kilometres, and the peak tidal velocities are of order 1 m s^{-1}. Tidally enhanced flows increase the energy input to turbulence at the seafloor, where the turbulent kinetic energy flux scales as the cube of the near-bottom velocity.

2.2.4 Mixing and marine ecosystems

The biological productivity of the ocean is largely regulated by the supply of nutrients to the well-lit near-surface euphotic zone. To grow, phytoplankton require sufficiently intense sunlight as well as nutrients, including phosphate, nitrate, and micronutrients, such as iron. Because sunlight in the wavelengths that are useful for photosynthesis is absorbed by seawater with an extinction length of order tens of metres, phytoplankton grow and take up nutrients only within the oceanic depth range that receives sunlight. But these nutrients are lost from the euphotic zone in the fecal pellets of grazing zooplankton and higher trophic levels, and in other sinking detritus. The nutrients are returned to the water column as this detritus remineralises, but this is typically far below the surface euphotic zone. The supply of nutrients into the euphotic zone from the interior ocean is thus controlled by a spatially varying combination of turbulent mixing across the base of surface mixed layer, the seasonal cycle of the mixed layer depth, and net upwelling. The role of isopycnal mixing on nutrient and micronutrient transport and supply to phytoplankton is discussed in Chapter 9.

Processes that change mixing can change biological productivity. One of the robust oceanic changes that is anticipated from a warming climate is an increase in the density stratification of the upper few hundred metres of the ocean as the ocean's surface warms more than the interior ocean and net melting and changing hydrologic cycle intensify high-latitude haloclines (Bindoff et al., 2019). In fact, a careful analysis of historical observations reveals that this increasing average ocean stratification has been going on over the past 50 years (Yamaguchi and Suga, 2019; Li et al., 2020). As discussed above, the intensity of turbulent diffusion is inversely proportional to the local density stratification (see also the stand-alone box), so an increase in the stratification suppresses vertical nutrient fluxes from mixing, even as it also inhibits upwelling. These considerations of the nutrient supply are the foundation of the robust projection from an ensemble of comprehensive Earth system models that global ocean net primary productivity would very likely decrease by between 4% and 11% by 2081–2100 relative to 2006–2015 with unabated carbon emissions, with even larger changes in total animal biomass and the maximum fisheries catch potential (Bindoff et al., 2019). However, there are aspects of the changing ocean mixing that are not yet incorporated into all of the Earth system models that contributed to these estimates. For example, some of these models use temporally fixed diffusivities instead of diffusivities that respond to energetic considerations (see stand-alone box). These estimates of changes in ocean productivity could be larger if all the known processes that could induce changes in ocean mixing (especially energetic constraints on mixing) were fully taken into account.

Although the interior ocean distribution of oxygen is primarily set by the balance between the advective transport of oxygen-rich waters from the surface and the consumption of oxygen by biological respiration, along isopycnal eddy mixing and secondarily diapycnal mixing are important for setting the properties of regions of the low-oxygen regions of the ocean without direct advective pathways from the surface. The impacts of ocean mixing on the distribution of well-oxygenated, hypoxic, suboxic, and anoxic waters in the ocean is explored in detail in Chapter 13 (Lévy et al.).

2.3 Ocean mixing and transient climate change

In addition to shaping the contemporary climate mean state (Section 2.2), ocean mixing also contributes to shape transient climate change. The transient climate response (TCR) is defined as the global mean surface air temperature increase at time of atmospheric CO_2 concentration, doubling from pre-industrial values under an idealised scenario of 1%/yr increase of CO_2 in climate models. The equilibrium climate sensitivity (ECS) corresponds to the global mean surface air temperature increase when the Earth's climate has reached an equilibrium to CO_2 concentration doubling. Ocean mixing influences the TCR and ECS to Earth's increased radiative forcing due to anthropogenic greenhouse gases emissions (e.g., Flato et al., 2013; Bouttes et al., 2013; Zickfeld et al., 2017; Ehlert et al., 2017), notably because of the large inertia and long equilibrium timescales (centuries to millennia) of the deep ocean response to increased radiative forcing.

2.3.1 Ocean anthropogenic heat and carbon uptake

With increasing CO_2 emissions and the corresponding climate warming, air-sea fluxes of heat and carbon are evolving to partly compensate for these changes. Oceans have absorbed about 25% of the anthropogenic CO_2 (Sabine, 2004; Le Quéré et al., 2018) and 93% of the resulting heat excess of the climate system (Levitus et al., 2012; Oppenheimer et al., 2019), leading to ocean warming and acidification (e.g., Bindoff et al., 2019).

The efficiency of ocean heat uptake (OHU, the temporal rate of change of the ocean heat content) depends on how quickly heat gained at the ocean surface is transported to depth, away from any contact with the atmosphere. The faster heat is exported to the interior ocean, the less the surface air temperature warms as more excess heat is taken up by the ocean and the lower the TCR (Marshall and Zanna, 2014; Krasting et al., 2018). OHU is therefore a significant source of uncertainty in climate model projections, for both the TCR and ECS.

Ocean mixing influences the OHU efficiency and carbon uptake notably through its impact on the stratification, the distribution of heat and carbon within the ocean interior and in the surface mixed layer, and their corresponding influence on air-sea fluxes of heat and carbon (e.g., Tatebe et al., 2018). Enhanced diapycnal mixing erodes the stratification in the upper ocean, leading to a weaker and more diffuse thermocline, increases the stratification at depth (depending on the vertical profile of K_d), enables more downward heat transport (e.g., Kuhlbrodt and Gregory, 2012) by mixing warm upper waters with cooler deep water and enables more carbon uptake (Schmittner et al., 2009; Melet et al., 2016; Ehlert et al., 2017). Modelling studies suggest a correlation between ocean heat and carbon uptake and the amplitude of the diapycnal diffusivity (Ehlert et al. (2017), Fig. 2.7).

OHU under transient climate change is substantial in the weakly stratified Southern ocean where deep-waters are formed (Sections 2.2.1, 2.2.2, Marshall and Zanna (2014); Zanna et al. (2019)), providing a direct link between ocean surface layers warmed by the atmosphere and transport of these water masses to the ocean depths. The spread in OHU in climate models under transient climate change is partly related to the models' differing representation of both diapycnal and isopycnal mixing (e.g., Kuhlbrodt and Gregory, 2012; Griffies et al., 2015; Ehlert et al., 2017; Krasting et al., 2018; Saenko et al., 2021), although the representation of clouds, aerosols, and their feedbacks are also major sources of variation between models (e.g., Melet and Meyssignac, 2015). Using an ensemble of climate models, Saenko et al. (2021) show that OHU under transient climate change is dominated by advective processes and can be explained broadly by air-sea heat fluxes into outcropping density layers and near-adiabatic distribution of heat within those layers. In this study, the contribution of diapycnal mixing processes to OHU is found to be largest in the upper ocean and at latitudes poleward of 40°N/S. However, rather small values of diapycnal diffusivity (10^{-5} m^2 s^{-1}) were prescribed in the main pycnocline in the climate model ensemble, which favours an ocean interior circulation and OHU along isopycnals rather than across them.

Under transient climate change, ocean carbon uptake differs from heat uptake for several reasons. First, air-sea fluxes of tracers depend on the air-sea differences in the tracer values or concentrations. While atmospheric surface temperatures exhibit strong meridional gradients, CO_2 is well mixed in the atmosphere and has a nearly uniform distribution. Second, temperature changes in the ocean impact the dynamics through changes in ocean density, stratification and pressure, while carbon is dynamically inert in the ocean. In turn, changes in ocean circulations and stratification can alter ocean heat and carbon uptake (e.g., Banks and Gregory, 2006; Winton et al., 2013). Mixing then contributes to the redistribution of carbon

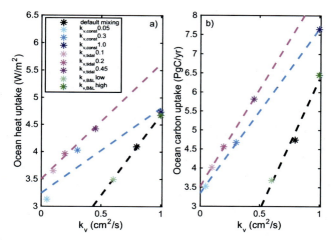

FIGURE 2.7 (a) OHU and (b) ocean carbon uptake at year 140 as a function of diapycnal mixing in a suite of climate model simulations under a standard idealised climate-warming scenario with an increase of CO_2 of 1% yr^{-1}. At year 140, CO_2 has quadrupled compared to pre-industrial levels. Different values of diapycnal mixing schemes are used (constant vertical mixing in blue, deep ocean tidal mixing in purple, Bryan-Lewis in green) together with different overall magnitudes of diffusivities. Extracted from Ehlert et al. (2017).

in the ocean, depending on the intensity of the diffusivity and on the gradients of carbon concentration. As is the case for OHU, climate models exhibit a substantial range in the ocean uptake of anthropogenic CO_2. Part of the uncertainty can be related to differences in mixing parameterisations in climate models. For example, accounting for near inertial wave driven diapycnal mixing in climate models has a substantial impact on air-sea CO_2 fluxes in the Southern ocean (Song et al., 2019a). In addition to diapycnal mixing, ocean isopycnal mixing due to eddies (Chapter 9), parameterised in climate models (see the stand-alone box), influences the uptake of anthropogenic CO_2. This is especially the case in the Southern ocean (Gnanadesikan et al., 2015), which also dominates ocean carbon uptake (as is the case for OHU), imparting to it a great importance in determining Earth's climatic mean state, TCR, and ECS (e.g., Frölicher et al., 2015).

Enhanced diapycnal mixing thus increases ocean heat and carbon uptake in a warming climate, both of which act to slow the rate of atmospheric warming due to anthropic emissions of CO_2 (e.g., Dalan et al. (2005); Schmittner et al. (2009); Ehlert et al. (2017), Fig. 2.7).

2.3.2 Contemporary and future sea level rise

As discussed in Section 2.3.1, ocean mixed-layer processes and interior ocean diapycnal and isopycnal mixing are important factors for the ocean's capacity to export heat at depth and for the rate of this export. As such, ocean mixing exerts a control over long (centennial) timescales on the ocean's uptake of heat (e.g., Dalan et al., 2005; Bouttes et al., 2013; Ehlert et al., 2017). The global mean thermosteric sea level (GMTSL), the component of sea level changes due to changes in ocean temperature and thermal expansion of sea water, can be directly related to OHU (Section 2.3.1) through the expansion efficiency of heat, ϵ (Russell et al., 2000): $GMTSL = \epsilon \times OHU$. For the observed and expected patterns of warming, the value of ϵ is close to 0.12 m YJ^{-1} in both observations and climate models (Melet and Meyssignac, 2015).

Several studies have analysed the sensitivity of climate models' projected sea level rise in response to increased greenhouse gas emissions as a function of diapycnal diffusivity (K_d) in the model. Dalan et al. (2005) show that increasing K_d from 10^{-5} to 10^{-4} $m^2 s^{-1}$ leads to an increase in GMTSL rise of about 4 cm after 70 years in a 1% yr^{-1} CO_2 concentration increase idealised scenario (at the time of CO_2 doubling). As enhanced diapycnal mixing increases the heat uptake in the ocean, the surface air temperature increase (corresponding to the TCR, Section 2.3) is lowered (by 0.4 K) correspondingly. Hallberg et al. (2013) illustrate the sensitivity of projected global mean thermosteric sea level rise during the 21st century to ocean model formulations by comparing climate projections from two CMIP5 Earth system models (ESM2G, ESM2M, Dunne et al. (2012)) that differ only in their ocean formulation. In particular, the broader main thermocline, deeper penetration of heat, and overall warmer state in ESM2M compared to ESM2G leads to a scenario-consistent 18% higher global mean thermosteric sea level rise in 2100 in ESM2M compared to ESM2G. Different levels of effective diapycnal mixing in the two models can explain much of the projected GMTSL rise differences.

Although the prescribed and effective diapycnal mixing can substantially differ in ocean models, the spread in modelled GMTSL rise is only partly explained by inter-model differences in the ocean heat uptake efficiency (Melet and Meyssignac,

20 Ocean Mixing

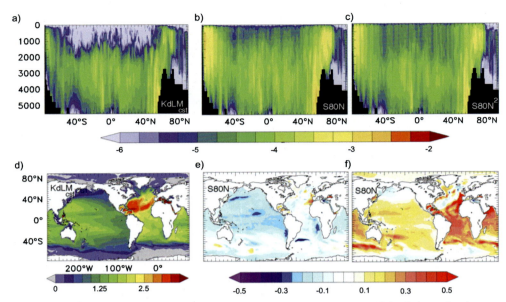

FIGURE 2.8 Diapycnal diffusivities (top row) and regional sea level (bottom row) in three climate model simulations using the same total magnitude of internal-tide dissipation but different spatial distribution. In all 3 simulations, 80% of internal-tide energy is dissipated remotely from their generation sites, but with a constant and spatially uniform diapycnal diffusivity of 1.4×10^{-5} m^2 s^{-1} in panels (a), and (d), on slopes with remote internal-tide dissipation scaling with the buoyancy frequency for panels (b) and (e) and with the squared buoyancy frequency for panels (c) and (f). (Top row) Global zonal mean of diapycnal diffusivities due to internal-tide mixing parameterisations ($\log_{10}$ m^2 s^{-1} so that a value of -6 means 10^{-6} m^2 s^{-1}) averaged over the last century of the 1000-yr simulations. In panel (a), the reference simulation, the constant and spatially uniform diapycnal diffusivity of 1.4×10^{-5} m^2 s^{-1} representative of unresolved remote internal-tide mixing has been excluded to help illustrate the structure from the model's other mixing parameterisations. (Bottom row) Thermosteric sea level referenced at the ocean bottom (in m). Panels (e) and (f) show differences in thermosteric sea level (again in m) from the constant diffusivity simulation (panel (d)). After Melet et al. (2016).

2015). A dominant source of uncertainties in climate models' projected GMTSL rise is related to inter-model spread in the climate feedback parameter and radiative forcing, as for OHU (Section 2.3.1) (Melet and Meyssignac, 2015). The response of clouds to climate change and cloud-climate interactions, as well as cloud-aerosol-radiation interactions, are especially uncertain (Boucher et al., 2013). Changes in the radiative budget of the Earth directly affect OHU. Indeed, the Earth's energy imbalance due to GHG emissions is almost entirely expressed as a heat excess and more than 90% of this is stored in the ocean (e.g., Levitus et al., 2012; Cheng et al., 2017), resulting in ocean heat uptake, a warmer ocean, and thermosteric sea level rise.

Ocean mixing also contributes to the heat and salt redistribution within the ocean. As such, it influences the heat and salinity patterns, and hence steric sea level changes induced by changes in seawater density. In addition to the overall magnitude of diapycnal mixing, its 3D spatial pattern matters, especially the vertical distribution of mixing, as it is the convergence of the buoyancy flux that can change water masses temperature or salinity. Changes in steric sea level in three climate model experiments that have the same total magnitude of internal-tide dissipation but different spatial distributions are illustrated in Fig. 2.8. When remote internal-tide dissipation is proportional to the buoyancy frequency (Fig. 2.8 b and e), so that diffusion is inversely proportional to buoyancy frequency, the thermocline is sharper, oceanic heat uptake is reduced, and thermosteric sea level is lower than when the energetically equivalent constant background diffusivity is used (Fig. 2.8 a and d). In contrast, using a remote internal-tide dissipation that is proportional to the squared buoyancy frequency (Fig. 2.8 c and f) leads to increased diffusivities in the upper ocean and reduced diffusivities at depth relative to the constant diffusivity control run (Fig. 2.8 a and d), consequently leading to relatively more heat uptake, a broader thermocline, warmer ocean, and higher steric sea level (Fig. 2.8, Melet et al. (2016)). Although these simulations were run using pre-industrial radiative forcing, they provide insights into the importance of the spatial distribution of ocean diapycnal mixing for simulating regional steric sea level changes.

2.3.3 Changes in nutrient fluxes

Stratification inhibits the vertical mixing of nutrients. One of the robust oceanic signals of climate change is an increase in upper ocean stratification (Bindoff et al., 2019), both from increased thermal stratification and increased salinity stratification in high latitudes. Turbulent mixing is important for supplying the nutrients to the biologically productive upper ocean, and it helps to regulate the net primary productivity of the open ocean (see also Section 2.2.4). Projections suggest that the net primary productivity of the ocean will very likely decrease by between 4% and 11% by 2100 under an unmitigated climate change scenario, mostly because of the effects of enhanced upper ocean stratification in reducing the turbulent mixing supply of nutrients (Bindoff et al., 2019).

2.3.4 Changes in ocean mixing sources

Ocean mixing is not static, but evolves in time. For instance, changes in winds induce changes in near inertial wave generation. Near inertial waves dissipation is also influenced by the dynamic eddy field (Whalen et al., 2018). Changes in geostrophic currents also impact the energy flux into lee waves. The dissipation of internal waves can be influenced by the eddy field and by mean currents (Fig. 2.1), as well as by changes in ocean stratification. In this subsection, we illustrate the time-evolving nature of ocean mixing by focusing on changes in two sources of ocean mixing: internal tides and lee waves.

2.3.4.1 Tides

Tides are often thought of as perfectly periodic and predictable. But tides are actually responding to changes in Earth's gravity field, rotation rate, and in changes in the geometry of ocean basins and shelves. Because tides propagate as shallow-water waves, their propagation speeds are strongly affected by sea level changes. Changes in ocean geometry and bathymetry also control tidal resonance in the open-ocean and in coastal regions (Garrett, 1972; Arbic et al., 2007, 2009; Arbic and Garrett, 2010). Striking examples of tidal changes over past climates are discussed in Section 2.4.

Contemporary and future sea level changes in response to climate change are altering the water depth and coastal topography. As tides are in near resonance in many locations, small changes in sea level and bay geometries can change the barotropic tides significantly (e.g., Harker et al., 2019; Pickering et al., 2017; Pelling et al., 2013). Changes in ocean stratification, to which ocean mixing contributes, could also alter barotropic tides (e.g., Müller, 2012). Internal tide generation and dissipation also respond to sea level rise and to ocean stratification, as is already found in the seasonal and interannual non-stationarity of internal tides (e.g., Zaron, 2017; Shriver et al., 2014).

In a high-end, multi-centennial sea level rise scenario, where the Greenland and West-Antarctic ice-sheets would have completely collapsed, global mean sea level would rise by 12 m (Mitrovica et al., 2011), but with a non-uniform pattern (e.g., Clark et al., 2016; Mitrovica et al., 2009). In this case, Wilmes et al. (2017) find that tides overall become larger, and the M2 dissipation in the deep ocean ($\geq$500 m depth) increases by 30%, but the changes are spatially heterogeneous (Fig. 2.9). Such tidal changes can, in turn, impact the extent of seasonal stratification, ocean dynamics, and ecosystems in shelf seas (Wilmes et al., 2017).

2.3.4.2 Lee waves

Lee waves are generated by the interaction of geostrophic flows with rough topography in the stratified ocean (see the stand-alone box). The energy flux that is transferred from geostrophic flows to lee waves (stand-alone box) is thus expected to depend on the ocean state. The sensitivity of the lee-wave energy flux to changes in both bottom stratification and geostrophic flows under different climate change scenarios is explored in Melet et al. (2015). They highlight a decrease of the global lee-wave energy flux over the next two centuries (Fig. 2.10). The lee-wave energy flux over the Southern ocean is also suggested to exhibit a clear annual cycle, mostly explained by the annual cycle of the Antarctic circumpolar current and related eddy-induced bottom velocities.

2.4 Ocean mixing in past climate states

Over geological timescales, plate tectonics have strongly altered continental and ocean basin configurations and subsequent bathymetric resonance effects on tides. In turn, variations in tidal dissipation have altered the orbital and precession motion of the Earth and the Earth-Moon system (e.g., Laskar et al., 2004; Munk and MacDonald, 1960). A review of changes in M2 due to the evolution of the ocean basins geometry is provided by Gotlib and Kagnan (1985) from the early Cambrian (~540–510 Ma) to present day and by Green et al. (2017) for the past 250 Ma. A review of past and future changes in tides is also provided by Haigh et al. (2020). A dramatic example of tidal changes in response to geometric changes is that of the Early Eocene (~56–50 million years ago). During the Early Eocene, ocean basin configurations were very different from present day ones, and the climate was much warmer. The large global mean tidal potential energy during that period led to much larger velocities in the deep ocean, with a strong influence on the global overturning circulation (Weber and Thomas, 2017a). The Last Interglacial (LIG, ~125 kyr ago) is often considered informative for the state of the climate in

22 Ocean Mixing

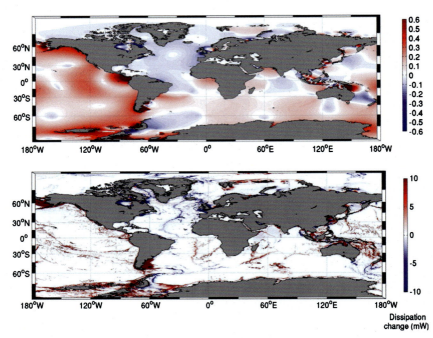

FIGURE 2.9 (a) M2 amplitude change from present day due to the fingerprint sea level change due to a complete Greenland ice sheet collapse (rising SL by 7 m) and to a collapse of the Western Antarctic ice sheet (rising SL by an additional 5 m). (b) Changes in the energy flux into M2 internal tides (in mW) due to the fingerprint sea level rise of (a). Adapted from Wilmes et al. (2017). (Also indicative of changes during the Last Interglacial (125 kyr BP), Section 2.4.)

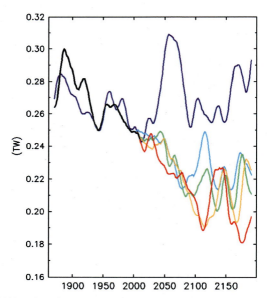

FIGURE 2.10 Times series over the 1860–2200 period of the globally integrated energy flux into lee waves (in TW) for the pi-control simulation (dark blue, using pre-industrial radiative forcing from year 1860), historical (black), RCP2.6 (blue), RCP4.5 (green) RCP6.0 (orange), and RCP8.5 (red) climate scenario. The energy flux was computed using linear theory and bottom stratification, resolved and parameterised bottom velocities from a climate model (GFDL CM2G). Time series were low-pass filtered using a Parzen filter with a 19-yr window. From Melet et al. (2015).

the next few millennia. Changes in tides described in Section 2.3.4.1 are therefore also representative of these of the LIG, during which sea level was 6–9 m higher than today due to the collapse of the West Antarctic ice sheet and of parts of the Greenland ice sheet (e.g., Dutton et al., 2015).

Major changes in ocean basin configurations and sea levels occurred during past climates of the Earth related to changes in Earth's orbit around the Sun, with the succession of glacial and interglacial cycles and associated ice-sheets growth and decay (e.g., Lambeck et al., 2002).

We will illustrate in this section corresponding changes in tides and induced effects on the climate during two contrasted paleo periods: the Early Pliocene (∼3.5 millions of years ago) and the Last glacial maximum (LGM, ∼23,000–19,000 years ago). Changes in tidal energy and mixing are dominated by changes in bathymetry due to different ocean configurations during the Early Pliocene, whereas they are dominated by sea level changes during the LGM.

2.4.1 The Early Pliocene

During the Early Pliocene, North and South America were not yet connected. This affected tidal resonance in the Atlantic and Pacific oceans, with – for instance – an increase of tidal amplitudes of 50 cm off the Brazilian coast compared to present conditions (Weber and Thomas, 2017b). Sea level was also approximately 20 m higher than the present-day value during the Pliocene (e.g., Dutton et al., 2015).

The zonal gradient of sea surface temperature along the equatorial Pacific was dramatically reduced during the Early Pliocene (e.g., Fedorov, 2006). In a sensitivity analysis, Brierley and Fedorov (2011) showed that the maintenance of this peculiar climate state could be explained by local and remote impacts of increased levels of internal-tide-driven mixing in the seas surrounding the Maritime continent, a key region for the climate.

Including tidal dynamics and shear-driven mixing, induced by tidal currents in paleo simulations of the Early Pliocene, results in an equatorward shift of the ACC in the Atlantic sector and poleward shift in the Indian and Pacific sector (Weber and Thomas, 2017b). Feedback loops between ocean circulation, sea surface temperature and sea ice concentrations were simulated. Tidal forcing locally reduces sea-ice concentration by up to 30% and local atmospheric near-surface temperatures by up to 4 °C (Weber and Thomas, 2017b). Although these results are to be taken only qualitatively due to large uncertainties in reconstructed paleo bathymetries and shoreline positions and the lack of representation of internal-tide-driven mixing, they highlight that tides can change under different climate conditions and, in turn, alter the climate state.

2.4.2 The Last Glacial Maximum (LGM)

During the LGM (∼23,000–19,000 years ago), large ice-sheets covered northern America and Eurasia. To form them, an immense amount of water was transferred from the ocean to the ice-sheets. As a result, sea level was 130 m below today's, and most of the shallow continental shelves were exposed. This sea-level induced change in the ocean basin configuration altered tidal distribution and dissipation patterns (e.g., Schmittner et al., 2015). Tides were larger in amplitude. Tidal dissipation on shelves through bottom friction, which represents two thirds of today's tidal energy dissipation, was greatly reduced during the LGM and instead shifted to the deep ocean. The energy flux into internal tides was dominated at depths and estimated to be up to three times greater than at present (Schmittner et al., 2015; Wilmes et al., 2019), with a particular increase in the North Atlantic due to resonance effects (Egbert et al., 2004; Green, 2010). The energy dissipation for M2 alone was globally ∼50% larger during the LGM (e.g., Wilmes and Green, 2014). Modelling studies suggest that these changes in tidal mixing, with larger vertical diffusivities, substantially accelerated and deepened the MOC (Fig. 2.11) (Schmittner et al., 2015; Wilmes et al., 2019), with consequences for the global heat, carbon, and nutrient fluxes (as explained in Section 2.2).

2.5 Summary and conclusion

In this chapter, we have illustrated how ocean mixing contributes to shape the climate state considering a range of important climate features from short (e.g., months) to long (millennia) timescales, and considering the location of mixing in the ocean (Fig. 2.2). Enhanced mixing in and below the surface mixed layer directly influences the mixed layer depth, upwelling, and surface winds (e.g., Friedrich et al., 2011; Jochum et al., 2013), the supply of nutrients in the euphotic zone supporting the ocean primary production, sea-surface temperature (e.g., Jochum, 2009; Hummels et al., 2020), thereby modifying precipitation, sea-ice extent, and concentration (e.g., Jochum et al., 2013) and air-sea fluxes of climate important tracers, such as heat and carbon. On longer timescales, stronger mixing in the upper ocean broadens the main thermocline, allowing for more heat uptake and leading to higher sea level, and strengthens the meridional overturning circulation in the subtropical cells. Mixing both along and across isopycnal surfaces and over the water column contributes to the oceanic uptake of heat and carbon and their redistribution in the ocean interior, to the ventilation of the ocean, to the meridional overturning circulation, and to the meridional heat transport (Fig. 2.2), which are all key for the climate state. Finally, mixing on shelves, slopes, and sills can alter deep-water formation and the water density in overflows, with impacts on long

24 Ocean Mixing

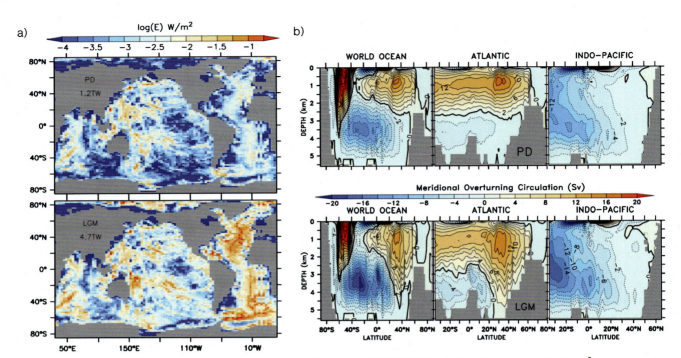

FIGURE 2.11 (a) Maps of tidal energy dissipation due to internal tide generation for M2, S2, K1, and O1 tidal constituents (in W m^{-2}, integral over the global ocean on Asia) for (top) present day (PD) and (bottom) the Last glacial maximum (LGM, around 20,000 years ago). (b) Meridional overturning circulation stream functions of the World ocean, Atlantic ocean and Indo-Pacific for (top) present day, and (bottom) the Last glacial maximum, accounting for the different power available for tidal mixing. Adapted from Schmittner et al. (2015).

timescales. In turn, changes in the meridional overturning circulation alter the ventilation of the ocean and the uptake and distribution of tracers, including key ones for the climate system, such as heat and carbon (e.g., Fontela et al., 2016; Winton et al., 2013).

In the interior ocean, diapycnal mixing is mostly supported by breaking internal waves, including internal tides, lee waves, and near-inertial waves (e.g., MacKinnon et al., 2017; Whalen et al., 2020). Diapycnal mixing is intermittent and highly heterogeneous in the horizontal and vertical, reflecting the variability of internal-wave generation and dissipation. The propagation and dissipation of internal waves is influenced by wave-mean flow, wave-eddy, and wave-front interactions. The eddy field supports isopycnal mixing, so that eddy induced mixing also evolves in space and time together with the eddy field. Mixing cannot therefore be considered static.

As the climate changes, mixing patterns are affected. Regarding diapycnal mixing, the responses of wind-forced near-inertial waves, lee waves, and internal tides to climate change are expected to be different because of differing climate change impacts on the energy source, propagation, and dissipation mechanisms of these different classes of internal waves (Fig. 2.2). This has been illustrated in this chapter both for contemporary climate change (Section 2.3) and for paleoclimates (Section 2.4). Feedbacks between ocean mixing and climate change can be exemplified by changes in sea level and internal tides. Mixing influences sea level notably through its role in ocean heat uptake and the redistribution of heat in the ocean. In turn, sea level changes influence mixing by changing barotropic tides, their dissipation on shelves and shallow seas, and the conversion of barotropic tides to internal tides (Fig. 2.2). For instance, a sea level rise of 2 metres compared to the pre-industrial period, which is considered a high-end scenario in 2100 under a high-emission, low-mitigation climate change scenario, could globally increase by 5% the energy flux into internal tides (Fig. 2.2). Another example was given for wind and ocean circulation changes in the Southern ocean in response to climate change, which could decrease the energy flux into lee waves by up to 20% in 2100. Finally, a dynamic representation of mixing, evolving with a changing climate, is needed for more realistic projections of changes in the net primary production and other aspects of marine ecosystems. Indeed, the upper ocean is stratifying because of climate change, due to surface intensified heat uptake and the addition of freshwater from the cryosphere (Yamaguchi and Suga, 2019). Such changes in stratification affect the nutrient supply to the euphotic zone through upwelling and mixing, with repercussions for the primary production, ocean carbon uptake, and marine ecosystems (Bindoff et al., 2019). Yet, many Earth system models still use incomplete or static representations of mixing, which contributes to uncertainties in ensemble-based projections of ocean primary productivity.

There remain a number of key challenges for fully understanding the role of ocean mixing in the climate system, and for projecting all of the consequences of the dynamic interactions of ocean mixing with the climate system. Mixing in the interior ocean is exceedingly anisotropic, with diapycnal mixing orders or magnitude smaller than mixing along density surfaces, even when the aspect ratio of the ocean is taken into account; this weak interior ocean diapycnal mixing is fundamental to shaping both the observed structure of the ocean and its response to a changing climate. Devising numerical ocean models that can capture this nearly adiabatic nature of much of the interior ocean remains a persistent challenge. As described above, the most vigorous oceanic mixing processes are brief, intermittent, and patchy, but have large-scale consequences for regulating the state of the ocean and the climate system as a whole. This mismatch of scales is at the heart of many of the greatest challenges with ocean mixing. Substantial improvements in the temporal and spatial coverage and mixing-relevant instrumentation of the ocean observing system would be invaluable for better characterising ocean mixing, especially in extreme environments, such as the deepest abyssal ocean, the near-surface ocean during storms or within ice-shelf cavities. Although our collective understanding of ocean mixing processes, what controls them, and their feedbacks on the climate system has dramatically improved in recent decades, there is still a compelling need for combined process-oriented observational campaigns and numerical simulations to continue to refine the understanding of how various mixing processes interact with their environment. Finally, there is a compelling societal need to fully incorporate our improved understanding of the interactions between the wide variety of ocean mixing processes and the evolving state of the ocean in our projections of climate change. Although there has been notable progress in developing physically-based and energetically-constrained parameterisations for climate models (e.g., Huang, 1999; Eden and Olbers, 2014; Melet et al., 2016; MacKinnon et al., 2017; de Lavergne et al., 2020), this effort to portray the full range of interactions across scales and ocean processes is far from complete. Modern parameterisations have made great strides in capturing the ways that ocean mixing changes in response to the state of the ocean (e.g., hydrography, stratification, circulation, eddy fields, fronts, surface waves, and sea level) and its external forcing (e.g., surface winds and sea-ice cover), and that should apply even as the oceans morph into parts of parameter space that people have never before observed. However, fully accounting for the interactive and dynamically-evolving roles of the many processes contributing to ocean mixing remains one of the great challenges in projecting how our planet's climate system will evolve in response to our collective choices in how we manage it.

References

Adcroft, A., Anderson, W., Balaji, V., Blanton, C., Bushuk, M., Dufour, C.O., Dunne, J.P., Griffies, S.M., Hallberg, R., Harrison, M.J., Held, I.M., Jansen, M.F., John, J.G., Krasting, J.P., Langenhorst, A.R., Legg, S., Liang, Z., McHugh, C., Radhakrishnan, A., Reichl, B.G., Rosati, T., Samuels, B.L., Shao, A., Stouffer, R., Winton, M., Wittenberg, A.T., Xiang, B., Zadeh, N., Zhang, R., 2019. The GFDL global ocean and sea ice model OM4.0: model description and simulation features. J. Adv. Model. Earth Syst. 11, 3167–3211. https://doi.org/10.1029/2019MS001726. https://onlinelibrary.wiley.com/doi/abs/10.1029/2019MS001726.

Alford, M.H., 2020. Revisiting near-inertial wind work: slab models, relative stress and mixed-layer deepening. J. Phys. Oceanogr. 1–54. https://doi.org/10.1175/JPO-D-20-0105.1. https://journals.ametsoc.org/jpo/article/354285/Revisiting-nearinertial-wind-work-slab-models.

Alford, M.H., Whitmont, M., 2007. Seasonal and spatial variability of near-inertial kinetic energy from historical moored velocity records. J. Phys. Oceanogr. 37, 2022–2037. https://doi.org/10.1175/JPO3106.1. https://journals.ametsoc.org/jpo/article/37/8/2022/10665/Seasonal-and-Spatial-Variability-of-NearInertial.

Allison, L.C., Johnson, H.L., Marshall, D.P., 2011. Spin-up and adjustment of the Antarctic Circumpolar Current and global pycnocline. J. Mar. Res. 69, 167–189. https://doi.org/10.1357/002224011798765330. http://www.ingentaconnect.com/content/10.1357/002224011798765330.

Ansong, J.K., Arbic, B.K., Alford, M.H., Buijsman, M.C., Shriver, J.F., Zhao, Z., Richman, J.G., Simmons, H.L., Timko, P.G., Wallcraft, A.J., Zamudio, L., 2017. Semidiurnal internal tide energy fluxes and their variability in a Global Ocean Model and moored observations: internal tide energy flux. J. Geophys. Res., Oceans 122, 1882–1900. https://doi.org/10.1002/2016JC012184. http://doi.wiley.com/10.1002/2016JC012184.

Arbic, B.K., Garrett, C., 2010. A coupled oscillator model of shelf and ocean tides. Cont. Shelf Res. 30, 564–574. https://doi.org/10.1016/j.csr.2009.07.008. https://linkinghub.elsevier.com/retrieve/pii/S0278434309002271.

Arbic, B.K., Karsten, R.H., Garrett, C., 2009. On tidal resonance in the global ocean and the back-effect of coastal tides upon open-ocean tides. Atmos.-Ocean 47, 239–266. https://doi.org/10.3137/OC311.2009. https://www.tandfonline.com/doi/full/10.3137/OC311.2009.

Arbic, B.K., St-Laurent, P., Sutherland, G., Garrett, C., 2007. On the resonance and influence of the tides in Ungava Bay and Hudson Strait. Geophys. Res. Lett. 34, L17606. https://doi.org/10.1029/2007GL030845. http://doi.wiley.com/10.1029/2007GL030845.

Arbic, B.K., Wallcraft, A.J., Metzger, E.J., 2010. Concurrent simulation of the eddying general circulation and tides in a global ocean model. Ocean Model. 32, 175–187.

Bachman, S., Fox-Kemper, B., Taylor, J., Thomas, L., 2017. Parameterization of frontal symmetric instabilities. I: theory for resolved fronts. Ocean Model. 109, 72–95. https://doi.org/10.1016/j.ocemod.2016.12.003. https://linkinghub.elsevier.com/retrieve/pii/S1463500316301482.

Balmforth, N.J., Llewellyn, S.G., Young, W.R., 1998. Enhanced dispersion of near-inertial waves in an idealized geostrophic flow. J. Mar. Res. 56, 1–40. https://doi.org/10.1357/002224098321836091. http://www.ingentaselect.com/rpsv/cgi-bin/cgi?ini=xref&body=linker&reqdoi=10.1357/002224098321836091.

Bamber, J.L., Oppenheimer, M., Kopp, R.E., Aspinall, W.P., Cooke, R.M., 2019. Ice sheet contributions to future sea-level rise from structured expert judgment. Proc. Natl. Acad. Sci. 116, 11195–11200. https://doi.org/10.1073/pnas.1817205116. http://www.pnas.org/lookup/doi/10.1073/pnas.1817205116.

Banks, H.T., Gregory, J.M., 2006. Mechanisms of ocean heat uptake in a coupled climate model and the implications for tracer based predictions of ocean heat uptake. Geophys. Res. Lett. 33, L07608. https://doi.org/10.1029/2005GL025352. http://doi.wiley.com/10.1029/2005GL025352.

Barringer, M., Price, J.F., 1997. Mixing and spreading of the Mediterranean Overflow. J. Phys. Oceanogr. 27, 1654–1677.

Bell, T.H., 1975. Topographically generated internal waves in the open ocean. J. Geophys. Res. 80, 320–327.

Bessières, L., Madec, G., Lyard, F., 2008. Global tidal residual mean circulation: does it affect a climate OGCM? Geophys. Res. Lett. 35, L03609.

Bindoff, N.L., Cheung, W.W.L., Kairo, J.C., Aristegui, J., Guinder, V.A., Hallberg, R.W., Hilmi, N., Jiao, N., Karim, M.S., Levin, L., O'Donoghue, S., Purca Cuicapusa, S.R., Rinkevich, B., Suga, T., Tagliabue, A., Williamson, P., 2019. Changing ocean, marine ecosystems, and dependent communities. In: Pörtner, H.O., Roberts, D.C., Masson-Delmotte, V., Zhai, P., Tignor, M., Poloczanska, E., Mintenbeck, K., Alegria, A., Nicolai, M., Okem, A., Petzold, J., Rama, B., Weyer, N.M. (Eds.), IPCC Special Report on the Ocean and Cryosphere in a Changing Climate. Cambridge University Press, Cambridge, United Kingdom and New York, NY, USA.

Bishop, S.P., Gent, P.R., Bryan, F.O., Thompson, A.F., Long, M.C., Abernathey, R., 2016. Southern Ocean overturning compensation in an eddy-resolving climate simulation. J. Phys. Oceanogr. 46, 1575–1592. https://doi.org/10.1175/JPO-D-15-0177.1. https://journals.ametsoc.org/jpo/article/46/5/1575/342931/Southern-Ocean-Overturning-Compensation-in-an.

Booker, J.R., Bretherton, F.P., 1967. The critical layer for internal gravity waves in a shear flow. J. Fluid Mech. 27, 513–539. https://doi.org/10.1017/S0022112067000515. https://www.cambridge.org/core/product/identifier/S0022112067000515/type/journal_article.

Boucher, O., Randall, D., Artaxo, P., Bretherton, C., Feingold, G., Forster, P.M., Kerminen, V.M., Kondo, Y., Liao, H., Lohmann, U., Rasch, P., Satheesh, S., Sherwood, S., Stevens, B., Zhang, X., 2013. Clouds and aerosols. In: Stocker, T., Qin, D., Plattner, G.K., Tignor, M., Allen, S.K., Boschung, J., Nauels, A., Xia, Y., Bex, V., Midgley, P.M. (Eds.), Climate Change 2013: the Physical Science Basis. Contribution of Working Group I to the Fifth Assessment Report of the Intergovernmental Panel on Climate Change. Cambridge University Press, Cambridge, United Kingdom and New York, NY, USA.

Boucher, O., Servonnat, J., Albright, A.L., Aumont, O., Balkanski, Y., Bastrikov, V., Bekki, S., Bonnet, R., Bony, S., Bopp, L., Braconnot, P., Brockmann, P., Cadule, P., Caubel, A., Cheruy, F., Codron, F., Cozic, A., Cugnet, D., D'Andrea, F., Davini, P., Lavergne, C., Denvil, S., Deshayes, J., Devilliers, M., Ducharne, A., Dufresne, J., Dupont, E., Éthé, C., Fairhead, L., Falletti, L., Flavoni, S., Foujols, M., Gardoll, S., Gastineau, G., Ghattas, J., Grandpeix, J., Guenet, B., Guez, E.L., Guilyardi, E., Guimberteau, M., Hauglustaine, D., Hourdin, F., Idelkadi, A., Joussaume, S., Kageyama, M., Khodri, M., Krinner, G., Lebas, N., Levavasseur, G., Lévy, C., Li, L., Lott, F., Lurton, T., Luyssaert, S., Madec, G., Madeleine, J., Maignan, F., Marchand, M., Marti, O., Mellul, L., Meurdesoif, Y., Mignot, J., Musat, I., Ottlé, C., Peylin, P., Planton, Y., Polcher, J., Rio, C., Rochetin, N., Rousset, C., Sepulchre, P., Sima, A., Swingedouw, D., Thiéblemont, R., Traore, A.K., Vancoppenolle, M., Vial, J., Vialard, J., Viovy, N., Vuichard, N., 2020. Presentation and evaluation of the IPSL-CM6A-LR climate model. J. Adv. Model. Earth Syst. 12. https://doi.org/10.1029/2019MS002010. https://onlinelibrary.wiley.com/doi/abs/10.1029/2019MS002010, 2020.

Bouttes, N., Gregory, J.M., Lowe, J.A., 2013. The reversibility of sea level rise. J. Climate 26, 2502–2513. https://doi.org/10.1175/JCLI-D-12-00285.1. https://journals.ametsoc.org/jcli/article/26/8/2502/33355/The-Reversibility-of-Sea-Level-Rise.

Brierley, C.M., Fedorov, A.V., 2011. Tidal mixing around Indonesia and the Maritime continent: Implications for paleoclimate simulations: paleotidal mixing. Geophys. Res. Lett. 38. https://doi.org/10.1029/2011GL050027. http://doi.wiley.com/10.1029/2011GL050027.

Broadbridge, M.B., Naveira Garabato, A.C., Nurser, A.J.G., 2016. Forcing of the overturning circulation across a circumpolar channel by internal wave breaking: forcing of the MOC by internal waves. J. Geophys. Res., Oceans 121, 5436–5451. https://doi.org/10.1002/2015JC011597. http://doi.wiley.com/10.1002/2015JC011597.

Bryan, F.O., 1987. Parameter sensitivity of primitive equation ocean general circulation models. J. Phys. Oceanogr. 17, 970–985.

Bryan, K., Lewis, L., 1979. A water mass model of the world ocean. J. Geophys. Res. 84, 2503–2517.

Calvert, D., Nurser, G., Bell, M.J., Fox-Kemper, B., 2020. The impact of a parameterisation of submesoscale mixed layer eddies on mixed layer depths in the NEMO ocean model. Ocean Model. 154, 101678. https://doi.org/10.1016/j.ocemod.2020.101678. https://linkinghub.elsevier.com/retrieve/pii/S1463500320301803.

Carroll, D., Sutherland, D.A., Shroyer, E.L., Nash, J.D., Catania, G.A., Stearns, L.A., 2017. Subglacial discharge-driven renewal of tidewater glacier fjords: subglacial discharge-driven renewal. J. Geophys. Res., Oceans 122, 6611–6629. https://doi.org/10.1002/2017JC012962. http://doi.wiley.com/10.1002/2017JC012962.

Caulfield, C., 2021. Layering, instabilities, and mixing in turbulent stratified flows. Annu. Rev. Fluid Mech. 53, 113–145. https://doi.org/10.1146/annurev-fluid-042320-100458. https://www.annualreviews.org/doi/10.1146/annurev-fluid-042320-100458.

Cheng, L., Trenberth, K.E., Fasullo, J., Boyer, T., Abraham, J., Zhu, J., 2017. Improved estimates of ocean heat content from 1960 to 2015. Sci. Adv. 3, e1601545. https://doi.org/10.1126/sciadv.1601545. https://advances.sciencemag.org/lookup/doi/10.1126/sciadv.1601545.

Cimoli, L., Caulfield, C.P., Johnson, H.L., Marshall, D.P., Mashayek, A., Naveira Garabato, A.C., Vic, C., 2019. Sensitivity of deep ocean mixing to local internal tide breaking and mixing efficiency. Geophys. Res. Lett. 46, 14622–14633. https://doi.org/10.1029/2019GL085056. https://onlinelibrary.wiley.com/doi/abs/10.1029/2019GL085056.

Clark, P.U., Shakun, J.D., Marcott, S.A., Mix, A.C., Eby, M., Kulp, S., Levermann, A., Milne, G.A., Pfister, P.L., Santer, B.D., Schrag, D.P., Solomon, S., Stocker, T.F., Strauss, B.H., Weaver, A.J., Winkelmann, R., Archer, D., Bard, E., Goldner, A., Lambeck, K., Pierrehumbert, R.T., Plattner, G.K., 2016. Consequences of twenty-first-century policy for multi-millennial climate and sea-level change. Nat. Clim. Change 6, 360–369. https://doi.org/10.1038/nclimate2923. http://www.nature.com/articles/nclimate2923.

Coles, V.J., Brooks, M.T., Hopkins, J., Stukel, M.R., Yager, P.L., Hood, R.R., 2013. The pathways and properties of the Amazon River Plume in the tropical North Atlantic Ocean: Amazon river plume. J. Geophys. Res., Oceans 118, 6894–6913. https://doi.org/10.1002/2013JC008981. http://doi.wiley.com/10.1002/2013JC008981.

Colombo, P., Barnier, B., Penduff, T., Chanut, J., Deshayes, J., Molines, J.M., Le Sommer, J., Verezemskaya, P., Gulev, S., Treguier, A.M., 2020. Representation of the Denmark Strait overflow in a <i>z</i>-coordinate eddying configuration of the NEMO (v3.6) ocean model: resolution and parameter impacts. Geosci. Model Dev. 13, 3347–3371. https://doi.org/10.5194/gmd-13-3347-2020. https://gmd.copernicus.org/articles/13/3347/2020/.

Dalan, F., Stone, P.H., Sokolov, A.P., 2005. Sensitivity of the ocean's climate to diapycnal diffusivity in an EMIC. Part II: global warming scenario. J. Climate 18, 2482–2496. https://doi.org/10.1175/JCLI3412.1. https://journals.ametsoc.org/jcli/article/18/13/2482/30782/Sensitivity-of-the-Oceans-Climate-to-Diapycnal.

Danabasoglu, G., Bates, S.C., Briegleb, B.P., Jayne, S.R., Jochum, M., Large, W.G., Peacock, S., Yeager, S.G., 2012. The CCSM4 ocean component. J. Climate 25, 1361–1389. https://doi.org/10.1175/JCLI-D-11-00091.1. http://journals.ametsoc.org/doi/10.1175/JCLI-D-11-00091.1.

Danabasoglu, G., Ferrari, R., McWilliams, J.C., 2008. Sensitivity of an ocean general circulation model to a parameterization of near-surface eddy fluxes. J. Climate 21, 1192–1208. https://doi.org/10.1175/2007JCLI1508.1. https://journals.ametsoc.org/jcli/article/21/6/1192/31795/Sensitivity-of-an-Ocean-General-Circulation-Model.

Danabasoglu, G., Large, W.G., Briegleb, B.P., 2010. Climate impacts of parameterized Nordic Sea overflows. J. Geophys. Res. 115, C11005. https://doi.org/10.1029/2010JC006243. http://doi.wiley.com/10.1029/2010JC006243.

Danioux, E., Vanneste, J., Bühler, O., 2015. On the concentration of near-inertial waves in anticyclones. J. Fluid Mech. 773, R2. https://doi.org/10.1017/jfm.2015.252. https://www.cambridge.org/core/product/identifier/S0022112015002529/type/journal_article.

Donohue, K.A., Tracey, K.L., Watts, D.R., Chidichimo, M.P., Chereskin, T.K., 2016. Mean Antarctic Circumpolar Current transport measured in Drake Passage. Geophys. Res. Lett. 43. https://doi.org/10.1002/2016GL070319. https://onlinelibrary.wiley.com/doi/abs/10.1002/2016GL070319.

Dunne, J.P., John, J., Adcroft, A., Griffies, S.M., Hallberg, R.W., Shevliakova, E., Stouffer, R.J., Cooke, W.F., Dunne, K.A., Harrison, M.J., Krasting, J.P., Malyshev, S., Milly, P.C.D., Phillipps, P., Sentman, L.T., Samuels, B.L., Spelman, M.J., Winton, M., Wittenberg, A.T., Zadeh, N., 2012. GFDL's ESM2 global coupled climate-carbon Earth System Models Part I: physical formulation and baseline simulation characteristics. J. Climate 25, 6646–6665.

Dutton, A., Carlson, A.E., Long, A.J., Milne, G.A., Clark, P.U., DeConto, R., Horton, B.P., Rahmstorf, S., Raymo, M.E., 2015. Sea-level rise due to polar ice-sheet mass loss during past warm periods. Science 349, aaa4019. https://doi.org/10.1126/science.aaa4019. https://www.sciencemag.org/lookup/doi/10.1126/science.aaa4019.

Eden, C., Olbers, D., 2014. An energy compartment model for propagation, nonlinear interaction, and dissipation of internal gravity waves. J. Phys. Oceanogr. 44, 2093–2106.

Eden, C., Olbers, D., 2017. A closure for internal wave–mean flow interaction. Part II: wave drag. J. Phys. Oceanogr. 47, 1403–1412. https://doi.org/10.1175/JPO-D-16-0056.1. https://journals.ametsoc.org/jpo/article/47/6/1403/44444/A-Closure-for-Internal-WaveMean-Flow-Interaction.

Egbert, G.D., Ray, R.D., Bills, B.G., 2004. Numerical modeling of the global semidiurnal tide in the present day and in the last glacial maximum: global tidal modeling. J. Geophys. Res., Oceans 109. https://doi.org/10.1029/2003JC001973. http://doi.wiley.com/10.1029/2003JC001973.

Ehlert, D., Zickfeld, K., Eby, M., Gillett, N., 2017. The sensitivity of the proportionality between temperature change and cumulative CO_2 emissions to ocean mixing. J. Climate 30, 2921–2935. https://doi.org/10.1175/JCLI-D-16-0247.1. http://journals.ametsoc.org/doi/10.1175/JCLI-D-16-0247.1.

Fedorov, A.V., 2006. The Pliocene paradox (mechanisms for a permanent El Nino). Science 312, 1485–1489. https://doi.org/10.1126/science.1122666. https://www.sciencemag.org/lookup/doi/10.1126/science.1122666.

Ferrari, R., Griffies, S.M., Nurser, A.G., Vallis, G.K., 2010. A boundary-value problem for the parameterized mesoscale eddy transport. Ocean Model. 32, 143–156. https://doi.org/10.1016/j.ocemod.2010.01.004. https://linkinghub.elsevier.com/retrieve/pii/S1463500310000065.

Ferrari, R., Mashayek, A., McDougall, T.J., Nikurashin, M., Campin, J.M., 2016. Turning ocean mixing upside down. J. Phys. Oceanogr. 46, 2239–2261. https://doi.org/10.1175/JPO-D-15-0244.1. https://journals.ametsoc.org/jpo/article/46/7/2239/44704/Turning-Ocean-Mixing-Upside-Down.

Flato, G., Marotzke, J., Abiodun, A., Braconnot, P., Chou, S.C., Collins, W., Cox, P., Driouech, F., Emori, S., Eyring, V., Forest, C.E., Gleckler, P.J., Guilyardi, E., Jakob, C., Kattsov, V., Reason, C., Rummukainen, M., 2013. Evaluation of climate models. In: Stocker, T., Qin, D., Plattner, G.K., Tignor, M., Allen, S.K., Boschung, J., Nauels, A., Xia, Y., Bex, V., Midgley, P.M. (Eds.), Climate Change 2013: the Physical Science Basis. Contribution of Working Group I to the Fifth Assessment Report of the Intergovernmental Panel on Climate Change. Cambridge University Press, Cambridge, United Kingdom and New York, NY, USA.

Fontela, M., García-Ibáñez, M.I., Hansell, D.A., Mercier, H., Pérez, F.F., 2016. Dissolved organic carbon in the North Atlantic meridional overturning circulation. Sci. Rep. 6, 26931. https://doi.org/10.1038/srep26931. http://www.nature.com/articles/srep26931.

Fox-Kemper, B., Danabasoglu, G., Ferrari, R., Griffies, S., Hallberg, R., Holland, M., Maltrud, M., Peacock, S., Samuels, B., 2011. Parameterization of mixed layer eddies. III: implementation and impact in global ocean climate simulations. Ocean Model. 39, 61–78. https://doi.org/10.1016/j.ocemod.2010.09.002. https://linkinghub.elsevier.com/retrieve/pii/S1463500310001290.

Fox-Kemper, B., Ferrari, R., Hallberg, R., 2008. Parameterization of mixed layer eddies. Part I: theory and diagnosis. J. Phys. Oceanogr. 38, 1145–1165. https://doi.org/10.1175/2007JPO3792.1. https://journals.ametsoc.org/jpo/article/38/6/1145/10836/Parameterization-of-Mixed-Layer-Eddies-Part-I.

Friedrich, T., Timmermann, A., Decloedt, T., Luther, D.S., Mouchet, A., 2011. The effect of topography-enhanced diapycnal mixing on ocean and atmospheric circulation and marine biogeochemistry. Ocean Model. 39, 262–274.

Frölicher, T.L., Sarmiento, J.L., Paynter, D.J., Dunne, J.P., Krasting, J.P., Winton, M., 2015. Dominance of the Southern Ocean in anthropogenic carbon and heat uptake in CMIP5 models. J. Climate 28, 862–886. https://doi.org/10.1175/JCLI-D-14-00117.1. https://journals.ametsoc.org/jcli/article/28/2/862/106296/Dominance-of-the-Southern-Ocean-in-Anthropogenic.

Furuichi, N., Hibiya, T., Niwa, Y., 2008. Model-predicted distribution of wind-induced internal wave energy in the world's oceans. J. Geophys. Res. 113, C09034. https://doi.org/10.1029/2008JC004768. http://doi.wiley.com/10.1029/2008JC004768.

Galbraith, E., de Lavergne, C., 2019. Response of a comprehensive climate model to a broad range of external forcings: relevance for deep ocean ventilation and the development of late Cenozoic ice ages. Clim. Dyn. 52, 653–679. https://doi.org/10.1007/s00382-018-4157-8. http://link.springer.com/10.1007/s00382-018-4157-8.

Gargett, A.E., Holloway, G., 1984. Dissipation and diffusion by internal wave breaking. J. Mar. Res. 42, 15–27.
Garrett, C., 1972. Tidal resonance in the Bay of Fundy and Gulf of Maine. Nature 238, 441–443. https://doi.org/10.1038/238441a0. http://www.nature.com/articles/238441a0.
Gaspar, P., Grégoris, Y., Lefevre, J.M., 1990. A simple eddy kinetic energy model for simulations of the oceanic vertical mixing: tests at station Papa and long-term upper ocean study site. J. Geophys. Res. 95, 16179. https://doi.org/10.1029/JC095iC09p16179. http://doi.wiley.com/10.1029/JC095iC09p16179.
Gent, P.R., McWilliams, J., 1990. Isopycnal mixing in ocean circulation models. J. Phys. Oceanogr. 20 (1), 150–155.
Gent, P.R., Willebrand, J., McDougall, T.J., McWilliams, J.C., 1995. Parameterizing eddy-induced tracer transports in ocean circulation models. J. Phys. Oceanogr. 25, 463–474. https://doi.org/10.1175/1520-0485(1995)025<0463:PEITTI>2.0.CO;2.
Geyer, W.R., MacCready, P., 2014. The estuarine circulation. Annu. Rev. Fluid Mech. 46, 175–197. https://doi.org/10.1146/annurev-fluid-010313-141302. http://www.annualreviews.org/doi/10.1146/annurev-fluid-010313-141302.
Giordani, H., Bourdallé-Badie, R., Madec, G., 2020. An eddy-diffusivity mass-flux parameterization for modeling oceanic convection. J. Adv. Model. Earth Syst. 12. https://doi.org/10.1029/2020MS002078. https://onlinelibrary.wiley.com/doi/10.1029/2020MS002078.
Gnanadesikan, A., 1999. A simple predictive model for the structure of the oceanic pycnocline. Science 283, 2077–2079.
Gnanadesikan, A., Pradal, M., Abernathey, R., 2015. Isopycnal mixing by mesoscale eddies significantly impacts oceanic anthropogenic carbon uptake. Geophys. Res. Lett. 42, 4249–4255. https://doi.org/10.1002/2015GL064100. https://onlinelibrary.wiley.com/doi/abs/10.1002/2015GL064100.
Goddard, P.B., Yin, J., Griffies, S.M., Zhang, S., 2015. An extreme event of sea-level rise along the Northeast coast of North America in 2009–2010. Nat. Commun. 6, 6346.
Gordon, A.L., 1986. Is there a global scale ocean circulation? Eos Trans. AGU 67, 109. https://doi.org/10.1029/EO067i009p00109. http://doi.wiley.com/10.1029/EO067i009p00109.
Gotlib, V.Y., Kagnan, B.A., 1985. A reconstruction of the tides in the paleocean: results of a numerical simulation. Dtsch. Hydrogr. Z. H2, 43–67.
Greatbatch, R.J., 1998. Exploring the relationship between eddy-induced transport velocity, vertical momentum transfer, and the isopycnal flux of potential vorticity. J. Phys. Oceanogr. 28, 422–432. https://doi.org/10.1175/1520-0485(1998)028<0422:ETRBEI>2.0.CO;2.
Green, J., Huber, M., Waltham, D., Buzan, J., Wells, M., 2017. Explicitly modelled deep-time tidal dissipation and its implication for Lunar history. Earth Planet. Sci. Lett. 461, 46–53. https://doi.org/10.1016/j.epsl.2016.12.038. https://linkinghub.elsevier.com/retrieve/pii/S0012821X16307518.
Green, J.A.M., 2010. Ocean tides and resonance. Ocean Dyn. 60, 1243–1253. https://doi.org/10.1007/s10236-010-0331-1. http://link.springer.com/10.1007/s10236-010-0331-1.
Gregg, M., D'Asaro, E., Riley, J., Kunze, E., 2018. Mixing efficiency in the ocean. Annu. Rev. Mar. Sci. 10, 443–473. https://doi.org/10.1146/annurev-marine-121916-063643. http://www.annualreviews.org/doi/10.1146/annurev-marine-121916-063643.
Gregg, M.C., Sanford, T.B., Winkel, D.P., 2003. Reduced mixing from the breaking of internal waves in equatorial waters. Nature 422, 477–478.
Griffies, S.M., 1998. The Gent–McWilliams skew flux. J. Phys. Oceanogr. 28, 831–841. https://doi.org/10.1175/1520-0485(1998)028,0831:TGMSF.2.0.CO;2.
Griffies, S.M., Gnanadesikan, A., Pacanowski, R.C., Larichev, V., Dukowicz, J.K., Smith, R.D., 1998. Isoneutral diffusion in a z-coordinate ocean model. J. Phys. Oceanogr. 28, 805–830. https://doi.org/10.1175/1520-0485(1998)028<0805:IDIAZC>2.0.CO;2.
Griffies, S.M., Pacanowski, R.C., Hallberg, R.W., 2000. Spurious diapycnal mixing associated with advection in a z-coordinate ocean model. Mon. Weather Rev. 128, 538–564.
Griffies, S.M., Winton, M., Anderson, W.G., Benson, R., Delworth, T.L., Dufour, C.O., Dunne, J.P., Goddard, P., Morrison, A.K., Rosati, A., Wittenberg, A.T., Yin, J., Zhang, R., 2015. Impacts on ocean heat from transient mesoscale eddies in a hierarchy of climate models. J. Climate 28, 952–977. https://doi.org/10.1175/JCLI-D-14-00353.1. https://journals.ametsoc.org/jcli/article/28/3/952/106719/Impacts-on-Ocean-Heat-from-Transient-Mesoscale.
Grisouard, N., Thomas, L.N., 2016. Energy exchanges between density fronts and near-inertial waves reflecting off the ocean surface. J. Phys. Oceanogr. 46, 501–516. https://doi.org/10.1175/JPO-D-15-0072.1. https://journals.ametsoc.org/jpo/article/46/2/501/12555/Energy-Exchanges-between-Density-Fronts-and.
Haigh, I.D., Pickering, M.D., Green, J.A.M., Arbic, B.K., Arns, A., Dangendorf, S., Hill, D.F., Horsburgh, K., Howard, T., Idier, D., Jay, D.A., Jänicke, L., Lee, S.B., Müller, M., Schindelegger, M., Talke, S.A., Wilmes, S., Woodworth, P.L., 2020. The tides they are a-changin': a comprehensive review of past and future nonastronomical changes in tides, their driving mechanisms, and future implications. Rev. Geophys. 58. https://doi.org/10.1029/2018RG000636. https://onlinelibrary.wiley.com/doi/abs/10.1029/2018RG000636.
Hallberg, R., Gnanadesikan, A., 2006. The role of eddies in determining the structure and response of the wind-driven Southern Hemisphere overturning: results from the modeling eddies in the Southern Ocean (MESO) project. J. Phys. Oceanogr. 36 (12), 2232–2252.
Hallberg, R.W., 2013. Using a resolution function to regulate parameterizations of oceanic mesoscale eddy effects. Ocean Model. 72, 92–103.
Hallberg, R.W., Adcroft, A., Dunne, J.P., Krasting, J.P., Stouffer, R.J., 2013. Sensitivity of twenty-first-century global-mean steric sea level rise to ocean model formulation. J. Climate 26 (9), 2947–2956.
Harker, A., Green, J.A.M., Schindelegger, M., Wilmes, S.B., 2019. The impact of sea-level rise on tidal characteristics around Australia. Ocean Sci. 15, 147–159. https://doi.org/10.5194/os-15-147-2019. https://os.copernicus.org/articles/15/147/2019/.
Harrison, M.J., Hallberg, R.W., 2008. Pacific subtropical cell response to reduced equatorial dissipation. J. Phys. Oceanogr. 38, 1894–1912.
Henyey, F.S., Wright, J., Flatté, S.M., 1986. Energy and action flow through the internal wave field: an eikonal approach. J. Geophys. Res. 91, 8487–8496.
Hickey, B.M., Pietrafesa, L.J., Jay, D.A., Boicourt, W.C., 1998. The Columbia River Plume Study: subtidal variability in the velocity and salinity fields. J. Geophys. Res., Oceans 103, 10339–10368. https://doi.org/10.1029/97JC03290. http://doi.wiley.com/10.1029/97JC03290.
Hieronymus, M., Nycander, J., Nilsson, J., Döös, K., Hallberg, R., 2019. Oceanic overturning and heat transport: the role of background diffusivity. J. Climate 32, 701–716. https://doi.org/10.1175/JCLI-D-18-0438.1. http://journals.ametsoc.org/doi/JCLI-D-18-0438.1.

Hopkins, J., Lucas, M., Dufau, C., Sutton, M., Stum, J., Lauret, O., Channelliere, C., 2013. Detection and variability of the Congo River plume from satellite derived sea surface temperature, salinity, ocean colour and sea level. Remote Sens. Environ. 139, 365–385. https://doi.org/10.1016/j.rse.2013.08.015. https://linkinghub.elsevier.com/retrieve/pii/S0034425713002733.

Huang, R.X., 1999. Mixing and energetics of the oceanic thermohaline circulation. J. Phys. Oceanogr. 29, 727–746.

Hummels, R., Dengler, M., Rath, W., Foltz, G.R., Schütte, F., Fischer, T., Brandt, P., 2020. Surface cooling caused by rare but intense near-inertial wave induced mixing in the tropical Atlantic. Nat. Commun. 11, 3829. https://doi.org/10.1038/s41467-020-17601-x. http://www.nature.com/articles/s41467-020-17601-x.

Ilicak, M., Adcroft, A., Griffies, S.M., Hallberg, R., 2012. Spurious dianeutral mixing and the role of momentum closure. Ocean Model. 45–46, 37–58.

Ito, T., Marshall, J., 2008. Control of lower-limb overturning circulation in the Southern Ocean by diapycnal mixing and mesoscale eddy transfer. J. Phys. Oceanogr. 38, 2832–2845.

Jackson, L., Hallberg, R., Legg, S., 2008. A parameterization of shear-driven turbulence for ocean climate models. J. Phys. Oceanogr. 38, 1033–1053.

Jackson, L.C., Kahana, R., Graham, T., Ringer, M.A., Woollings, T., Mecking, J.V., Wood, R.A., 2015. Global and European climate impacts of a slowdown of the AMOC in a high resolution GCM. Clim. Dyn. 45, 3299–3316. https://doi.org/10.1007/s00382-015-2540-2. http://link.springer.com/10.1007/s00382-015-2540-2.

Jayne, S.R., 2009. The impact of abyssal mixing parameterizations in an ocean general circulation model. J. Phys. Oceanogr. 39 (7), 1756–1775.

Jeon, C., Park, J.H., Nakamura, H., Nishina, A., Zhu, X.H., Kim, D.G., Min, H.S., Kang, S.K., Na, H., Hirose, N., 2019. Poleward-propagating near-inertial waves enabled by the western boundary current. Sci. Rep. 9, 9955. https://doi.org/10.1038/s41598-019-46364-9. http://www.nature.com/articles/s41598-019-46364-9.

Jiang, J., Lu, Y., Perrie, W., 2005. Estimating the energy flux from the wind to ocean inertial motions: the sensitivity to surface wind fields. Geophys. Res. Lett. 32, L15610.

Jing, Z., Wu, L., Ma, X., 2016. Sensitivity of near-inertial internal waves to spatial interpolations of wind stress in ocean generation circulation models. Ocean Model. 99, 15–21. https://doi.org/10.1016/j.ocemod.2015.12.006. https://linkinghub.elsevier.com/retrieve/pii/S1463500315002425.

Jing, Z., Wu, L., Ma, X., 2017. Energy exchange between the mesoscale oceanic eddies and wind-forced near-inertial oscillations. J. Phys. Oceanogr. 47, 721–733. https://doi.org/10.1175/JPO-D-16-0214.1. https://journals.ametsoc.org/jpo/article/47/3/721/44610/Energy-Exchange-between-the-Mesoscale-Oceanic.

Jochum, M., 2009. Impact of latitudinal variations in vertical diffusivity on climate simulations. J. Geophys. Res. 114, C01010. https://doi.org/10.1029/2008JC005030. http://doi.wiley.com/10.1029/2008JC005030.

Jochum, M., Briegleb, B.P., Danabasoglu, G., Large, W.G., Norton, N.J., Jayne, S.R., Alford, M.H., Bryan, F.O., 2013. The impact of oceanic near-inertial waves on climate. J. Climate 26, 2833–2844. https://doi.org/10.1175/JCLI-D-12-00181.1. https://journals.ametsoc.org/jcli/article/26/9/2833/33558/The-Impact-of-Oceanic-NearInertial-Waves-on.

Kageyama, M., Paul, A., Roche, D.M., Van Meerbeeck, C.J., 2010. Modelling glacial climatic millennial-scale variability related to changes in the Atlantic meridional overturning circulation: a review. Quat. Sci. Rev. 29, 2931–2956. https://doi.org/10.1016/j.quascirev.2010.05.029. https://linkinghub.elsevier.com/retrieve/pii/S0277379110001745.

Klinger, B.A., Marshall, J., Send, U., 1996. Representation of convective plumes by vertical adjustment. J. Geophys. Res., Oceans 101, 18175–18182. https://doi.org/10.1029/96JC00861. http://doi.wiley.com/10.1029/96JC00861.

Krasting, J.P., Stouffer, R.J., Griffies, S.M., Hallberg, R.W., Malyshev, S.L., Samuels, B.L., Sentman, L.T., 2018. Role of ocean model formulation in climate response uncertainty. J. Climate 31, 9313–9333. https://doi.org/10.1175/JCLI-D-18-0035.1. https://journals.ametsoc.org/jcli/article/31/22/9313/91919/Role-of-Ocean-Model-Formulation-in-Climate.

Kuhlbrodt, T., Gregory, J.M., 2012. Ocean heat uptake and its consequences for the magnitude of sea level rise and climate change. Geophys. Res. Lett. 39, L18608.

Kunze, E., 1985. Near-inertial wave propagation in geostrophic shear. J. Phys. Oceanogr. 15, 544–565. https://doi.org/10.1175/1520-0485(1985)015<0544:NIWPIG>2.0.CO;2.

Kunze, E., Lien, R.C., 2019. Energy sinks for lee waves in shear flow. J. Phys. Oceanogr. 49, 2851–2865. https://doi.org/10.1175/JPO-D-19-0052.1. http://journals.ametsoc.org/doi/10.1175/JPO-D-19-0052.1.

Lambeck, K., Esat, T.M., Potter, E.K., 2002. Links between climate and sea levels for the past three million years. Nature 419, 199–206. https://doi.org/10.1038/nature01089. http://www.nature.com/articles/nature01089.

Large, W.G., McWilliams, J.C., Doney, S.C., 1994. Oceanic vertical mixing: a review and a model with a nonlocal boundary layer parameterization. Rev. Geophys. 32, 363. https://doi.org/10.1029/94RG01872. http://doi.wiley.com/10.1029/94RG01872.

Laskar, J., Robutel, P., Joutel, F., Gastineau, M., Correia, A.C.M., Levrard, B., 2004. A long-term numerical solution for the insolation quantities of the Earth. Astron. Astrophys. 428, 261–285. https://doi.org/10.1051/0004-6361:20041335. http://www.aanda.org/10.1051/0004-6361:20041335.

de Lavergne, C., Falahat, S., Madec, G., Roquet, F., Nycander, J., Vic, C., 2019. Toward global maps of internal tide energy sinks. Ocean Model. 137, 52–75. https://doi.org/10.1016/j.ocemod.2019.03.010. https://linkinghub.elsevier.com/retrieve/pii/S1463500318302890.

de Lavergne, C., Gurvan, M., Le Sommer, J., Nurser, G.A.J., Naveira Garabato, A.C., 2016a. On the consumption of Antarctic Bottom Water in the abyssal ocean. J. Phys. Oceanogr. 46, 635–661.

de Lavergne, C., Madec, G., Le Sommer, J., Nurser, A.J.G., Naveira Garabato, A.C., 2016b. The impact of a variable mixing efficiency on the abyssal overturning. J. Phys. Oceanogr. 46, 663–681. https://doi.org/10.1175/JPO-D-14-0259.1. https://journals.ametsoc.org/view/journals/phoc/46/2/jpo-d-14-0259.1.xml.

de Lavergne, C., Vic, C., Madec, G., Roquet, F., Waterhouse, A.F., Whalen, C.B., Cuypers, Y., Bouruet-Aubertot, P., Ferron, B., Hibiya, T., 2020. A parameterization of local and remote tidal mixing. J. Adv. Model. Earth Syst. 12. https://doi.org/10.1029/2020MS002065. https://onlinelibrary.wiley.com/doi/abs/10.1029/2020MS002065.

Lazar, A., Madec, G., Delecluse, P., 1999. The deep interior downwelling, the veronis effect, and mesoscale tracer transport parameterizations in an OGCM. J. Phys. Oceanogr. 29, 2945–2961.

Le Quéré, C., Andrew, R.M., Friedlingstein, P., Sitch, S., Hauck, J., Pongratz, J., Pickers, P.A., Korsbakken, J.I., Peters, G.P., Canadell, J.G., Arneth, A., Arora, V.K., Barbero, L., Bastos, A., Bopp, L., Chevallier, F., Chini, L.P., Ciais, P., Doney, S.C., Gkritzalis, T., Goll, D.S., Harris, I., Haverd, V., Hoffman, F.M., Hoppema, M., Houghton, R.A., Hurtt, G., Ilyina, T., Jain, A.K., Johannessen, T., Jones, C.D., Kato, E., Keeling, R.F., Goldewijk, K.K., Landschützer, P., Lefèvre, N., Lienert, S., Liu, Z., Lombardozzi, D., Metzl, N., Munro, D.R., Nabel, J.E.M.S., Nakaoka, S.i., Neill, C., Olsen, A., Ono, T., Patra, P., Peregon, A., Peters, W., Peylin, P., Pfeil, B., Pierrot, D., Poulter, B., Rehder, G., Resplandy, L., Robertson, E., Rocher, M., Rödenbeck, C., Schuster, U., Schwinger, J., Séférian, R., Skjelvan, I., Steinhoff, T., Sutton, A., Tans, P.P., Tian, H., Tilbrook, B., Tubiello, F.N., van der Laan-Luijkx, I.T., van der Werf, G.R., Viovy, N., Walker, A.P., Wiltshire, A.J., Wright, R., Zaehle, S., Zheng, B., 2018. Global carbon budget 2018. Earth Syst. Sci. Data 10, 2141–2194. https://doi.org/10.5194/essd-10-2141-2018. https://essd.copernicus.org/articles/10/2141/2018/.

Ledwell, J.R., Montgomery, E.T., Polzin, K.L., Laurent, L.C.S., Schmitt, R.W., Toole, J.M., 2000. Mixing over rough topography in the Brazil Basin. Nature 403, 179–182.

Lefauve, A., Muller, C., Melet, A., 2015. A three-dimensional map of tidal dissipation over abyssal hills: global map of tidal dissipation. J. Geophys. Res., Oceans 120, 4760–4777. https://doi.org/10.1002/2014JC010598. http://doi.wiley.com/10.1002/2014JC010598.

Legg, S., 2021. Mixing by oceanic lee waves. Annu. Rev. Fluid Mech. 53, 173–201. https://doi.org/10.1146/annurev-fluid-051220-043904. https://www.annualreviews.org/doi/10.1146/annurev-fluid-051220-043904.

Legg, S., Briegleb, B., Chang, Y., Chassignet, E.P., Danabasoglu, G., Ezer, T., Gordon, A.L., Griffies, L., Hallberg, R., Jackson, L., Large, W., Ozgokmen, T.M., Peters, H., Price, J., Riemenschneider, U., Wu, W., Xu, X., Yang, J., 2009. Improving oceanic overflow representation in climate models: the gravity current entrainment climate process team. Bull. Am. Meteorol. Soc. 90, 657–670. https://doi.org/10.1175/2008BAMS2667.1.

Legg, S., Hallberg, R.W., Girton, J.B., 2006. Comparison of entrainment in overflows simulated by z-coordinate, isopycnal and nonhydrostatic models. Ocean Model. 11, 69–97.

Lelong, M.P., Cuypers, Y., Bouruet-Aubertot, P., 2020. Near-inertial energy propagation inside a Mediterranean anticyclonic eddy. J. Phys. Oceanogr. 50, 2271–2288. https://doi.org/10.1175/JPO-D-19-0211.1. https://journals.ametsoc.org/view/journals/phoc/50/8/jpoD190211.xml.

Levitus, S., Antonov, J.I., Boyer, T.P., Baranova, O.K., Garcia, H.E., Locarnini, R.A., Mishonov, A.V., Reagan, J.R., Seidov, D., Yarosh, E.S., Zweng, M.M., 2012. World ocean heat content and thermosteric sea level change (0–2000 m), 1955–2010. Geophys. Res. Lett. 39, L10603.

Li, G., Cheng, L., Zhu, J., Trenberth, K.E., Mann, M.E., Abraham, J.P., 2020. Increasing ocean stratification over the past half-century. Nat. Clim. Change 10, 1116–1123. https://doi.org/10.1038/s41558-020-00918-2. http://www.nature.com/articles/s41558-020-00918-2.

Li, Q., Reichl, B.G., Fox-Kemper, B., Adcroft, A.J., Belcher, S.E., Danabasoglu, G., Grant, A.L.M., Griffies, S.M., Hallberg, R., Hara, T., Harcourt, R.R., Kukulka, T., Large, W.G., McWilliams, J.C., Pearson, B., Sullivan, P.P., Van Roekel, L., Wang, P., Zheng, Z., 2019. Comparing ocean surface boundary vertical mixing schemes including Langmuir turbulence. J. Adv. Model. Earth Syst. 11, 3545–3592. https://doi.org/10.1029/2019MS001810. https://onlinelibrary.wiley.com/doi/abs/10.1029/2019MS001810.

Lynch-Stieglitz, J., 2017. The Atlantic meridional overturning circulation and abrupt climate change. Annu. Rev. Mar. Sci. 9, 83–104. https://doi.org/10.1146/annurev-marine-010816-060415. http://www.annualreviews.org/doi/10.1146/annurev-marine-010816-060415.

MacCready, P., Geyer, W.R., Burchard, H., 2018. Estuarine exchange flow is related to mixing through the salinity variance budget. J. Phys. Oceanogr. 48, 1375–1384. https://doi.org/10.1175/JPO-D-17-0266.1. https://journals.ametsoc.org/jpo/article/48/6/1375/42551/Estuarine-Exchange-Flow-Is-Related-to-Mixing.

MacKinnon, J.A., 2013. Mountain waves in the deep ocean. Nature 501, 321–322.

MacKinnon, J.A., Zhao, Z., Whalen, C.B., Waterhouse, A.F., Trossman, D.S., Sun, O.M., St. Laurent, L.C., Simmons, H.L., Polzin, K., Pinkel, R., Pickering, A., Norton, N.J., Nash, J.D., Musgrave, R., Merchant, L.M., Melet, A.V., Mater, B., Legg, S., Large, W.G., Kunze, E., Klymak, J.M., Jochum, M., Jayne, S.R., Hallberg, R.W., Griffies, S.M., Diggs, S., Danabasoglu, G., Chassignet, E.P., Buijsman, M.C., Bryan, F.O., Briegleb, B.P., Barna, A., Arbic, B.K., Ansong, J.K., Alford, M.H., 2017. Climate process team on internal wave–driven ocean mixing. Bull. Am. Meteorol. Soc. 98, 2429–2454. https://doi.org/10.1175/BAMS-D-16-0030.1. https://journals.ametsoc.org/bams/article/98/11/2429/216009/Climate-Process-Team-on-Internal-WaveDriven-Ocean.

Madec, G., Chartier, M., Delecluse, P., Crépon, M., 1991. A three-dimensional numerical study of deep water formation in the northwestern Mediterranean Sea. J. Phys. Oceanogr. 21, 1349–1371.

Mak, J., Maddison, J.R., Marshall, D.P., Munday, D.R., 2018. Implementation of a geometrically informed and energetically constrained mesoscale eddy parameterization in an ocean circulation model. J. Phys. Oceanogr. 48, 2363–2382. https://doi.org/10.1175/JPO-D-18-0017.1. https://journals.ametsoc.org/jpo/article/48/10/2363/44331/Implementation-of-a-Geometrically-Informed-and.

Marshall, D.P., Ambaum, M.H.P., Maddison, J.R., Munday, D.R., Novak, L., 2017. Eddy saturation and frictional control of the Antarctic circumpolar current: eddy saturation of the ACC. Geophys. Res. Lett. 44, 286–292. https://doi.org/10.1002/2016GL071702. http://doi.wiley.com/10.1002/2016GL071702.

Marshall, D.P., Maddison, J.R., Berloff, P.S., 2012. A framework for parameterizing eddy potential vorticity fluxes. J. Phys. Oceanogr. 42, 539–557. https://doi.org/10.1175/JPO-D-11-048.1. https://journals.ametsoc.org/jpo/article/42/4/539/11348/A-Framework-for-Parameterizing-Eddy-Potential.

Marshall, D.P., Williams, R.G., Lee, M., 1999. The relation between eddy-induced transport and isopycnic gradients of potential vorticity. J. Phys. Oceanogr. 29, 1571–1578. https://doi.org/10.1175/1520-0485(1999)029<1571:TRBEIT>2.0.CO;2.

Marshall, D.P., Zanna, L., 2014. A conceptual model of ocean heat uptake under climate change. J. Climate 27, 8444–8465. https://doi.org/10.1175/JCLI-D-13-00344.1. https://journals.ametsoc.org/jcli/article/27/22/8444/34820/A-Conceptual-Model-of-Ocean-Heat-Uptake-under.

Marshall, J., Speer, K.G., 2012. Closure of the meridional overturning circulation through Southern Ocean upwelling. Nat. Geosci. 5, 171–180.

Martínez-Marrero, A., Barceló-Llull, B., Pallàs-Sanz, E., Aguiar-González, B., Estrada-Allis, S.N., Gordo, C., Grisolía, D., Rodríguez-Santana, A., Arístegui, J., 2019. Near-inertial wave trapping near the base of an anticyclonic mesoscale eddy under normal atmospheric conditions. J. Geophys. Res., Oceans 124, 8455–8467. https://doi.org/10.1029/2019JC015168. https://onlinelibrary.wiley.com/doi/abs/10.1029/2019JC015168.

Mashayek, A., Peltier, W.R., 2013. Shear-induced mixing in geophysical flows: does the route to turbulence matter to its efficiency? J. Fluid Mech. 725, 216–261. https://doi.org/10.1017/jfm.2013.176. https://www.cambridge.org/core/product/identifier/S0022112013001766/type/journal_article.

Mashayek, A., Salehipour, H., Bouffard, D., Caulfield, C.P., Ferrari, R., Nikurashin, M., Peltier, W.R., Smyth, W.D., 2017. Efficiency of turbulent mixing in the abyssal ocean circulation: mixing efficiency in ocean circulation. Geophys. Res. Lett. 44, 6296–6306. https://doi.org/10.1002/2016GL072452. http://doi.wiley.com/10.1002/2016GL072452.

McDougall, T.J., 2003. Potential enthalpy: a conservative oceanic variable for evaluating heat content and heat fluxes. J. Phys. Oceanogr. 33, 945–963. https://doi.org/10.1175/1520-0485(2003)033<0945:PEACOV>2.0.CO;2.

McDougall, T.J., Ferrari, R., 2017. Abyssal upwelling and downwelling driven by near-boundary mixing. J. Phys. Oceanogr. 47, 261–283. https://doi.org/10.1175/JPO-D-16-0082.1. https://journals.ametsoc.org/view/journals/phoc/47/2/jpo-d-16-0082.1.xml.

Melet, A., Hallberg, R., Adcroft, A., Nikurashin, M., Legg, S., 2015. Energy flux into internal lee waves: sensitivity to future climate changes using linear theory and a climate model. J. Climate 28, 2365–2384. https://doi.org/10.1175/JCLI-D-14-00432.1. https://journals.ametsoc.org/jcli/article/28/6/2365/35364/Energy-Flux-into-Internal-Lee-Waves-Sensitivity-to.

Melet, A., Hallberg, R., Legg, S., Nikurashin, M., 2014. Sensitivity of the ocean state to lee wave–driven mixing. J. Phys. Oceanogr. 44, 900–921. https://doi.org/10.1175/JPO-D-13-072.1. https://journals.ametsoc.org/jpo/article/44/3/900/12157/Sensitivity-of-the-Ocean-State-to-Lee-WaveDriven.

Melet, A., Hallberg, R., Legg, S., Polzin, K., 2013. Sensitivity of the ocean state to the vertical distribution of internal-tide-driven mixing. J. Phys. Oceanogr. 43, 602–615. https://doi.org/10.1175/JPO-D-12-055.1. https://journals.ametsoc.org/jpo/article/43/3/602/11650/Sensitivity-of-the-Ocean-State-to-the-Vertical.

Melet, A., Legg, S., Hallberg, R., 2016. Climatic impacts of parameterized local and remote tidal mixing. J. Climate 29, 3473–3500. https://doi.org/10.1175/JCLI-D-15-0153.1. https://journals.ametsoc.org/jcli/article/29/10/3473/35577/Climatic-Impacts-of-Parameterized-Local-and-Remote.

Melet, A., Meyssignac, B., 2015. Explaining the spread in global mean thermosteric sea level rise in CMIP5 climate models*. J. Climate 28, 9918–9940. https://doi.org/10.1175/JCLI-D-15-0200.1. https://journals.ametsoc.org/jcli/article/28/24/9918/106774/Explaining-the-Spread-in-Global-Mean-Thermosteric.

Meredith, M., Sommerkorn, M., Cassotta, S., Derksen, C., Hollowed, A., Kofinas, G., Mackintosh, A., Melbourne-Thomas, J., Muelbert, M.M.C., Ottersen, G., Pritchard, H., Schuur, E.A.G., 2019. Polar regions. In: Pörtner, H.O., Roberts, D.C., Masson-Delmotte, V., Zhai, P., Tignor, M., Poloczanska, E., Mintenbeck, K., Alegría, A., Nicolai, M., Okem, A., Petzold, J., Rama, B., Weyer, N.M. (Eds.), IPCC Special Report on the Ocean and Cryosphere in a Changing Climate. Cambridge University Press, Cambridge, United Kingdom and New York, NY, USA.

Meredith, M.P., Hogg, A.M., 2006. Circumpolar response of Southern Ocean eddy activity to a change in the Southern Annular Mode. Geophys. Res. Lett. 33, L16608.

Mitrovica, J.X., Gomez, N., Clark, P.U., 2009. The sea-level fingerprint of West Antarctic collapse. Science 323, 753. https://doi.org/10.1126/science.1166510. https://www.sciencemag.org/lookup/doi/10.1126/science.1166510.

Mitrovica, J.X., Gomez, N., Morrow, E., Hay, C., Latychev, K., Tamisiea, M.E., 2011. On the robustness of predictions of sea level fingerprints: on predictions of sea-level fingerprints. Geophys. J. Int. 187, 729–742. https://doi.org/10.1111/j.1365-246X.2011.05090.x. https://academic.oup.com/gji/article-lookup/doi/10.1111/j.1365-246X.2011.05090.x.

Morrison, A.K., Hogg, A.M., 2013. On the relationship between Southern Ocean overturning and ACC transport. J. Phys. Oceanogr. 43, 140–148.

Müller, M., 2012. The influence of changing stratification conditions on barotropic tidal transport and its implications for seasonal and secular changes of tides. Cont. Shelf Res. 47, 107–118. https://doi.org/10.1016/j.csr.2012.07.003. https://linkinghub.elsevier.com/retrieve/pii/S027843431200177X.

Munday, D.R., Johnson, H.L., Marshall, D.P., 2013. Eddy saturation of equilibrated circumpolar currents. J. Phys. Oceanogr. 43, 507–532. https://doi.org/10.1175/JPO-D-12-095.1. https://journals.ametsoc.org/jpo/article/43/3/507/11702/Eddy-Saturation-of-Equilibrated-Circumpolar.

Munk, W., 1981. Evolution of Physical Oceanography - Scientific Surveys in Honor of Henry Stommel. MIT Press, pp. 264–291. Section: Internal waves and small-scale processes.

Munk, W., Wunsch, C., 1998. Abyssal recipes II: energetics of tidal and wind mixing. Deep-Sea Res. 45, 1977–2010.

Munk, W.H., 1966. Abyssal recipes. Deep-Sea Res. Oceanogr. Abstr. 13, 707–730. https://doi.org/10.1016/0011-7471(66)90602-4. https://linkinghub.elsevier.com/retrieve/pii/0011747166906024.

Munk, W.H., MacDonald, C.J.F., 1960. The Rotation of the Earth: A Geophysical Discussion. Cambridge University Press, New York.

Naveira Garabato, A.C., Frajka-Williams, E.E., Spingys, C.P., Legg, S., Polzin, K.L., Forryan, A., Abrahamsen, E.P., Buckingham, C.E., Griffies, S.M., McPhail, S.D., Nicholls, K.W., Thomas, L.N., Meredith, M.P., 2019. Rapid mixing and exchange of deep-ocean waters in an abyssal boundary current. Proc. Natl. Acad. Sci. 116, 13233–13238. https://doi.org/10.1073/pnas.1904087116. http://www.pnas.org/lookup/doi/10.1073/pnas.1904087116.

Naveira Garabato, A.C., Polzin, K.L., King, B.A., Heywood, K.J., Visbeck, M., 2004. Widespread intense turbulent mixing in the Southern Ocean. Science 303, 210–213.

Nielsen, M.H., Lund-Hansen, L.C., Vang, T., 2017. Internal hydraulic control in the Little Belt, Denmark. Observations of flow configurations and water mass formation. preprint. In: Situ Observations/Temperature, Salinity and Density Fields/Shelf-sea depth/Baltic Sea. https://os.copernicus.org/preprints/os-2017-22/os-2017-22.pdf.

Niiler, P., Krauss, E., 1977. One-dimensional models of the upper ocean. In: Krauss, E. (Ed.), Modelling and Prediction of the Upper Layers of the Ocean. Pergamon Press, pp. 143–172.

Nikurashin, M., Ferrari, R., 2011. Global energy conversion rate from geostrophic flows into internal lee waves in the deep ocean. Geophys. Res. Lett. 38, L08610.

Nikurashin, M., Ferrari, R., Grisouard, N., Polzin, K., 2014. The impact of finite amplitude bottom topography on internal wave generation in the Southern Ocean. J. Phys. Oceanogr. 44, 2938–2950.

Nikurashin, M., Vallis, G., 2011. A theory of deep stratification and overturning circulation in the ocean. J. Phys. Oceanogr. 41, 485–502. https://doi.org/10.1175/2010JPO4529.1. https://journals.ametsoc.org/jpo/article/41/3/485/11493/A-Theory-of-Deep-Stratification-and-Overturning.

Nikurashin, M., Vallis, G., 2012. A theory of the interhemispheric meridional overturning circulation and associated stratification. J. Phys. Oceanogr. 42, 1652–1667.

Nof, D., 1983. The translation of isolated cold eddies on a sloping bottom. Deep-Sea Res., A, Oceanogr. Res. Pap. 30, 171–182. https://doi.org/10.1016/0198-0149(83)90067-5. https://linkinghub.elsevier.com/retrieve/pii/0198014983900675.

Nycander, J., Hieronymus, M., Roquet, F., 2015. The nonlinear equation of state of sea water and the global water mass distribution. Geophys. Res. Lett. 42, 7714–7721. https://doi.org/10.1002/2015GL065525. https://onlinelibrary.wiley.com/doi/abs/10.1002/2015GL065525.

Oka, A., Niwa, Y., 2013. Pacific deep circulation and ventilation controlled by tidal mixing away from the sea bottom. Nat. Commun. 4, 2419.

Olbers, D., Eden, C., 2017. A closure for internal wave–mean flow interaction. Part I: energy conversion. J. Phys. Oceanogr. 47, 1389–1401. https://doi.org/10.1175/JPO-D-16-0054.1. https://journals.ametsoc.org/jpo/article/47/6/1389/44477/A-Closure-for-Internal-WaveMean-Flow-Interaction.

Oppenheimer, M., Glavovic, B., Hinkel, J., van de Wal, R.S.W., Magnan, A., Abd-Elgawad, A., Cai, R., Cifuentes-Jara, M., DeConto, R.M., Ghosh, T., Hay, J., Isla, F.I., Marzeion, B., Meyssignac, B., Sebesvari, Z., 2019. Sea level rise and implications for low lying islands, coasts and communities. In: Pörtner, H.-O., et al. (Eds.), IPCC Special Report on the Ocean and Cryosphere in a Changing Climate. Cambridge University Press, Cambridge, UK.

Osborn, T.R., 1980. Estimates of the local rate of vertical diffusion from dissipation measurements. J. Phys. Oceanogr. 10, 83–89.

Pacanowski, R.C., Philander, S.G.H., 1981. Parameterization of vertical mixing in numerical models of tropical oceans. J. Phys. Oceanogr. 11, 1443–1451.

Park, Y.G., Bryan, K., 2000. Comparison of thermally driven circulations from a depth-coordinate model and an isopycnal-layer model. Part I: scaling-law sensitivity to vertical diffusivity. J. Phys. Oceanogr. 30, 590–605.

Pedlosky, J., 1992. The baroclinic structure of the abyssal circulation. J. Phys. Oceanogr. 22, 652–659.

Pelling, H.E., Mattias Green, J., Ward, S.L., 2013. Modelling tides and sea-level rise: to flood or not to flood. Ocean Model. 63, 21–29. https://doi.org/10.1016/j.ocemod.2012.12.004. https://linkinghub.elsevier.com/retrieve/pii/S1463500312001813.

Peters, H., Johns, W.E., 2005. Mixing and entrainment in the Red Sea outflow plume. Part II: turbulence characteristics. J. Phys. Oceanogr. 35, 584–600. https://doi.org/10.1175/JPO2689.1. https://journals.ametsoc.org/jpo/article/35/5/584/10623/Mixing-and-Entrainment-in-the-Red-Sea-Outflow.

Peters, H., Johns, W.E., Bower, A.S., Fratantoni, D.M., 2005. Mixing and entrainment in the Red Sea outflow plume. Part I: plume structure. J. Phys. Oceanogr. 35, 569–583. https://doi.org/10.1175/JPO2679.1. https://journals.ametsoc.org/jpo/article/35/5/569/10621/Mixing-and-Entrainment-in-the-Red-Sea-Outflow.

Pickering, M., Horsburgh, K., Blundell, J., Hirschi, J.M., Nicholls, R., Verlaan, M., Wells, N., 2017. The impact of future sea-level rise on the global tides. Cont. Shelf Res. 142, 50–68. https://doi.org/10.1016/j.csr.2017.02.004. https://linkinghub.elsevier.com/retrieve/pii/S0278434316304824.

Polzin, K.L., 2004. Idealized solutions for the energy balance of the finescale internal wave field. J. Phys. Oceanogr. 34, 231–246.

Polzin, K.L., 2009. An abyssal recipe. Ocean Model. 30, 298–309. https://doi.org/10.1016/j.ocemod.2009.07.006. http://www.sciencedirect.com/science/article/pii/S1463500309001565.

Polzin, K.L., Toole, J.M., Ledwell, J.R., Schmitt, R.W., 1997. Spatial variability of turbulent mixing in the abyssal ocean. Science 276, 93–96.

Pradal, M.A., Gnanadesikan, A., 2014. How does the Redi parameter for mesoscale mixing impact global climate in an Earth System Model? J. Adv. Model. Earth Syst. 6, 586–601. https://doi.org/10.1002/2013MS000273. http://doi.wiley.com/10.1002/2013MS000273.

Pratt, L.J., Lundberg, P.A., 1991. Hydraulics of rotating strait and sill flow. Annu. Rev. Fluid Mech. 23, 81–106. https://doi.org/10.1146/annurev.fl.23.010191.000501. http://www.annualreviews.org/doi/10.1146/annurev.fl.23.010191.000501.

Rahmstorf, S., 1995. Multiple convection patterns and thermohaline flow in an idealized OGCM. J. Climate 8, 3028–3039.

Redi, M.H., 1982. Oceanic isopycnal mixing by coordinate rotation. J. Phys. Oceanogr. 12, 1154–1158. https://doi.org/10.1175/1520-0485(1982)012<1154:OIMBCR>2.0.CO;2.

Rimac, A., Storch, J., Eden, C., Haak, H., 2013. The influence of high-resolution wind stress field on the power input to near-inertial motions in the ocean. Geophys. Res. Lett. 40, 4882–4886. https://doi.org/10.1002/grl.50929. https://onlinelibrary.wiley.com/doi/abs/10.1002/grl.50929.

Russell, G.L., Gornitz, V., Miller, J.R., 2000. Regional sea-level changes projected by the NASA/GISS atmosphere-ocean model. Clim. Dyn. 16, 789–797.

Russell, J.L., Dixon, K.W., Gnanadesikan, A., Stouffer, R.J., Toggweiler, J.R., 2006. The southern hemisphere westerlies in a warming world: propping open the door to the deep ocean. J. Climate 19, 6382–6390. https://doi.org/10.1175/JCLI3984.1. https://journals.ametsoc.org/jcli/article/19/24/6382/30968/The-Southern-Hemisphere-Westerlies-in-a-Warming.

Russell, J.L., Kamenkovich, I., Bitz, C., Ferrari, R., Gille, S.T., Goodman, P.J., Hallberg, R., Johnson, K., Khazmutdinova, K., Marinov, I., Mazloff, M., Riser, S., Sarmiento, J.L., Speer, K., Talley, L.D., Wanninkhof, R., 2018. Metrics for the evaluation of the Southern Ocean in coupled climate models and Earth system models. J. Geophys. Res., Oceans 123, 3120–3143. https://doi.org/10.1002/2017JC013461. https://onlinelibrary.wiley.com/doi/abs/10.1002/2017JC013461.

Sabine, C.L., 2004. The oceanic sink for anthropogenic CO2. Science 305, 367–371. https://doi.org/10.1126/science.1097403. https://www.sciencemag.org/lookup/doi/10.1126/science.1097403.

Saenko, O.A., Gregory, J.M., Griffies, S.M., Couldrey, M.P., Dias, F.B., 2021. Contribution of ocean physics and dynamics at different scales to heat uptake in low-resolution AOGCMs. J. Climate 34, 2017–2035. https://doi.org/10.1175/JCLI-D-20-0652.1. https://journals.ametsoc.org/view/journals/clim/34/6/JCLI-D-20-0652.1.xml.

Saenko, O.A., Merryfield, W.J., 2005. On the effect of topographically enhanced mixing on the global ocean circulation. J. Phys. Oceanogr. 35, 826–834.

Sannino, G., 2004. Numerical modeling of the semidiurnal tidal exchange through the Strait of Gibraltar. J. Geophys. Res. 109, C05011. https://doi.org/10.1029/2003JC002057. http://doi.wiley.com/10.1029/2003JC002057.

Schindelegger, M., Green, J.A.M., Wilmes, S., Haigh, I.D., 2018. Can we model the effect of observed sea level rise on tides? J. Geophys. Res., Oceans 123, 4593–4609. https://doi.org/10.1029/2018JC013959. https://onlinelibrary.wiley.com/doi/abs/10.1029/2018JC013959.

Schmittner, A., Green, J.A.M., Wilmes, S., 2015. Glacial ocean overturning intensified by tidal mixing in a global circulation model. Geophys. Res. Lett. 42, 4014–4022. https://doi.org/10.1002/2015GL063561. https://onlinelibrary.wiley.com/doi/abs/10.1002/2015GL063561.

Schmittner, A., Urban, N.M., Keller, K., Matthews, D., 2009. Using tracer observations to reduce the uncertainty of ocean diapycnal mixing and climate-carbon cycle projections: reducing uncertainties in ocean mixing. Glob. Biogeochem. Cycles 23. https://doi.org/10.1029/2008GB003421. http://doi.wiley.com/10.1029/2008GB003421.

Schmittner, A., Weaver, A.J., 2001. Dependence of multiple climate states on ocean mixing parameters. Geophys. Res. Lett. 28, 1027–1030.

Scott, J.R., Marotzke, J., 2002. The location of diapycnal mixing and the meridional overturning circulation. J. Phys. Oceanogr. 32, 3578–3595.

Scott, R.B., Goff, J.A., Garabato, A.C.N., Nurser, A.J., 2011. Global rate and spectral characteristics of internal gravity wave generation by geostrophic flow over topography. J. Geophys. Res. 116, C09029.

Shriver, J.F., Richman, J.G., Arbic, B.K., 2014. How stationary are the internal tides in a high-resolution global ocean circulation model? J. Geophys. Res., Oceans 119, 2769–2787. https://doi.org/10.1002/2013JC009423. http://doi.wiley.com/10.1002/2013JC009423.

Silverthorne, K.E., Toole, J.M., 2009. Seasonal kinetic energy variability of near-inertial motions. J. Phys. Oceanogr. 39, 1035–1049. https://doi.org/10.1175/2008JPO3920.1. https://journals.ametsoc.org/jpo/article/39/4/1035/11012/Seasonal-Kinetic-Energy-Variability-of.

Simmons, H., Alford, M.H., 2012. Simulating the long-range swell of internal waves generated by ocean storms. Oceanography 25 (2), 30–41.

Simmons, H.L., Hallberg, R.W., Arbic, B.K., 2004a. Internal wave generation in a global baroclinic tide model. Deep-Sea Res. II 51, 3043–3068.

Simmons, H.L., Jayne, S.R., Laurent, L.C.S., Weaver, A.J., 2004b. Tidally driven mixing in a numerical model of the ocean general circulation. Ocean Model. 6, 245–263. https://doi.org/10.1016/S1463-5003(03)00011-8. http://www.sciencedirect.com/science/article/pii/S1463500303000118.

Smith, K.S., Marshall, J., 2009. Evidence for enhanced eddy mixing at middepth in the Southern Ocean. J. Phys. Oceanogr. 39, 50–69. https://doi.org/10.1175/2008JPO3880.1.

Solomon, H., 1971. On the representation of isentropic mixing in ocean circulation models. J. Phys. Oceanogr. 1, 233–234. https://doi.org/10.1175/1520-0485(1971)001<0233:OTROIM>2.0.CO;2.

Song, H., Marshall, J., Campin, J., McGillicuddy, D.J., 2019a. Impact of near-inertial waves on vertical mixing and air-sea CO_2 fluxes in the Southern Ocean. J. Geophys. Res., Oceans 124, 4605–4617. https://doi.org/10.1029/2018JC014928. https://onlinelibrary.wiley.com/doi/abs/10.1029/2018JC014928.

Song, Z., Jing, Z., Wu, L., 2019b. Online isolation of near-inertial internal waves in ocean general circulation models. Ocean Model. 134, 30–41. https://doi.org/10.1016/j.ocemod.2018.12.004. https://linkinghub.elsevier.com/retrieve/pii/S1463500318302671.

St. Laurent, L.C., Simmons, H.L., Jayne, S.R., 2002. Estimating tidally driven mixing in the deep ocean. Geophys. Res. Lett. 29 (23), 2106.

Stanley, Z., Bachman, S.D., Grooms, I., 2020. Vertical structure of ocean mesoscale eddies with implications for parameterizations of tracer transport. J. Adv. Model. Earth Syst. 12. https://doi.org/10.1029/2020MS002151. https://onlinelibrary.wiley.com/doi/abs/10.1029/2020MS002151.

Stommel, H., Arons, A.B., 1960. On the abyssal circulation of the World Ocean-I. Stationary planetary flow patterns on a sphere. Deep-Sea Res., Part 1, Oceanogr. Res. Pap. 6, 140–154.

Straneo, F., Cenedese, C., 2015. The dynamics of Greenland's Glacial Fjords and their role in climate. Annu. Rev. Mar. Sci. 7, 89–112. https://doi.org/10.1146/annurev-marine-010213-135133. http://www.annualreviews.org/doi/10.1146/annurev-marine-010213-135133.

Sun, S., Eisenman, I., Zanna, L., Stewart, A.L., 2020. Surface constraints on the depth of the Atlantic meridional overturning circulation: Southern Ocean versus North Atlantic. J. Climate 33, 3125–3149. https://doi.org/10.1175/JCLI-D-19-0546.1. https://journals.ametsoc.org/jcli/article/33/8/3125/345076/Surface-Constraints-on-the-Depth-of-the-Atlantic.

Sverdrup, H.U., Fleming, R.H., Johnson, M.W., 1942. The Oceans, Their Physics, Chemistry, and General Biology. Prentice-Hall, New York. http://ark.cdlib.org/ark:/13030/kt167nb66r/.

Talley, L.D., 2008. Freshwater transport estimates and the global overturning circulation: shallow, deep and throughflow components. Prog. Oceanogr. 78, 257–303.

Talley, L.D., 2013. Closure of the global overturning circulation through the Indian, Pacific, and Southern Oceans: schematics and transports. Oceanography 26 (1), 80–97.

Tatebe, H., Tanaka, Y., Komuro, Y., Hasumi, H., 2018. Impact of deep ocean mixing on the climatic mean state in the Southern Ocean. Sci. Rep. 8, 14479. https://doi.org/10.1038/s41598-018-32768-6. http://www.nature.com/articles/s41598-018-32768-6.

Thorpe, S.A., 2018. Models of energy loss from internal waves breaking in the ocean. J. Fluid Mech. 836, 72–116. https://doi.org/10.1017/jfm.2017.780. https://www.cambridge.org/core/product/identifier/S0022112017007807/type/journal_article.

Toggweiler, J.R., Druffel, E.R.M., Key, R.M., Galbraith, E.D., 2019. Upwelling in the ocean basins North of the ACC: 1. On the upwelling exposed by the surface distribution of \Delta^{14}C. J. Geophys. Res., Oceans 124, 2591–2608. https://doi.org/10.1029/2018JC014794. https://onlinelibrary.wiley.com/doi/abs/10.1029/2018JC014794.

Toggweiler, J.R., Samuels, B., 1998. On the ocean's large-scale circulation near the limit of no vertical mixing. J. Phys. Oceanogr. 28, 1832–1852.

Trossman, D.S., Waterman, S., Polzin, K.L., Arbic, B.K., Garner, S.T., Naveira-Garabato, A.C., Sheen, K.L., 2015. Internal lee wave closures: parameter sensitivity and comparison to observations. J. Geophys. Res., Oceans 120, 7997–8019.

Tsujino, H., Hasumi, H., Suginohara, N., 2000. Deep Pacific circulation controlled by vertical diffusivity at the lower thermocline depths. J. Phys. Oceanogr. 30, 2853–2865.

Vallis, G., 2000. Large-scale circulation and production of stratification: effects of wind, geometry, and diffusion. J. Phys. Oceanogr. 30, 933–954.

Vallis, G.K., Farneti, R., 2009. Meridional energy transport in the coupled atmosphere-ocean system: scaling and numerical experiments. Q. J. R. Meteorol. Soc. 135, 1643–1660. https://doi.org/10.1002/qj.498. http://doi.wiley.com/10.1002/qj.498.

Vic, C., Naveira Garabato, A.C., Green, J.A.M., Waterhouse, A.F., Zhao, Z., Melet, A., de Lavergne, C., Buijsman, M.C., Stephenson, G.R., 2019. Deep-ocean mixing driven by small-scale internal tides. Nat. Commun. 10, 2099. https://doi.org/10.1038/s41467-019-10149-5. http://www.nature.com/articles/s41467-019-10149-5.

Voet, G., Quadfasel, D., 2010. Entrainment in the Denmark Strait overflow plume by meso-scale eddies. Ocean Sci. 6, 301–310. https://doi.org/10.5194/os-6-301-2010. https://os.copernicus.org/articles/6/301/2010/.

Voldoire, A., Saint-Martin, D., Sénési, S., Decharme, B., Alias, A., Chevallier, M., Colin, J., Guérémy, J., Michou, M., Moine, M., Nabat, P., Roehrig, R., Salas y Mélia, D., Séférian, R., Valcke, S., Beau, I., Belamari, S., Berthet, S., Cassou, C., Cattiaux, J., Deshayes, J., Douville, H., Ethé, C., Franchistéguy, L., Geoffroy, O., Lévy, C., Madec, G., Meurdesoif, Y., Msadek, R., Ribes, A., Sanchez-Gomez, E., Terray, L., Waldman, R., 2019. Evaluation of CMIP6 DECK experiments with CNRM-CM6-1. J. Adv. Model. Earth Syst. 11, 2177–2213. https://doi.org/10.1029/2019MS001683. https://onlinelibrary.wiley.com/doi/abs/10.1029/2019MS001683.

Wang, H., Legg, S.A., Hallberg, R.W., 2015. Representations of the Nordic Seas overflows and their large scale climate impact in coupled models. Ocean Model. 86, 76–92. https://doi.org/10.1016/j.ocemod.2014.12.005. https://linkinghub.elsevier.com/retrieve/pii/S1463500314001802.

Wang, W., Huang, R.X., 2004. Wind energy input to the surface waves. J. Phys. Oceanogr. 34, 1276–1280. https://doi.org/10.1175/1520-0485(2004)034<1276:WEITTS>2.0.CO;2.

Watanabe, M., Hibiya, T., 2002. Global estimates of the wind-induced energy flux to inertial motions in the surface mixed layer: wind-induced global inertial energy. Geophys. Res. Lett. 29, 64-1–64-3. https://doi.org/10.1029/2001GL014422. http://doi.wiley.com/10.1029/2001GL014422.

Waterhouse, A.F., MacKinnon, J.A., Nash, J.D., Alford, M.H., Kunze, E., Simmons, H.L., Polzin, K.L., St. Laurent, L.C., Sun, O.M., Pinkel, R., Talley, L.D., Whalen, C.B., Huussen, T.N., Carter, G.S., Fer, I., Waterman, S., Naveira Garabato, A.C., Sanford, T.B., Lee, C.M., 2014. Global patterns of diapycnal mixing from measurements of the turbulent dissipation rate. J. Phys. Oceanogr. 44, 1854–1872. https://doi.org/10.1175/JPO-D-13-0104.1.

Weber, T., Thomas, M., 2017a. Influence of ocean tides on the general ocean circulation in the early Eocene. Paleoceanography 32, 553–570. https://doi.org/10.1002/2016PA002997. http://doi.wiley.com/10.1002/2016PA002997.

Weber, T., Thomas, M., 2017b. Tidal dynamics and their influence on the climate system from the Cretaceous to present day. Glob. Planet. Change 158, 173–183. https://doi.org/10.1016/j.gloplacha.2017.09.019. https://linkinghub.elsevier.com/retrieve/pii/S0921818116301552.

Wenegrat, J.O., Thomas, L.N., 2020. Centrifugal and symmetric instability during Ekman adjustment of the bottom boundary layer. J. Phys. Oceanogr. 50, 1793–1812. https://doi.org/10.1175/JPO-D-20-0027.1. https://journals.ametsoc.org/view/journals/phoc/50/6/JPO-D-20-0027.1.xml.

Whalen, C.B., de Lavergne, C., Klymak, J.M., Naveira Garabato, A.C., Sheen, K.L., MacKinnon, J.A., 2020. Internal wave driven mixing: governing processes and consequences for climate. Nat. Rev. Earth Environ. 1, 606–621. https://doi.org/10.1038/s43017-020-0097-z.

Whalen, C.B., MacKinnon, J.A., Talley, L.D., 2018. Large-scale impacts of the mesoscale environment on mixing from wind-driven internal waves. Nat. Geosci. 11, 842–847. https://doi.org/10.1038/s41561-018-0213-6. http://www.nature.com/articles/s41561-018-0213-6.

Whalen, C.B., Talley, L.D., MacKinnon, J.A., 2012. Spatial and temporal variability of global ocean mixing inferred from Argo profiles: global ocean mixing inferred from Argo. Geophys. Res. Lett. 39. https://doi.org/10.1029/2012GL053196. http://doi.wiley.com/10.1029/2012GL053196.

Wilmes, S., Green, J.A.M., Gomez, N., Rippeth, T.P., Lau, H., 2017. Global tidal impacts of large-scale ice sheet collapses. J. Geophys. Res., Oceans 122, 8354–8370. https://doi.org/10.1002/2017JC013109. https://onlinelibrary.wiley.com/doi/abs/10.1002/2017JC013109.

Wilmes, S., Schmittner, A., Green, J.A.M., 2019. Glacial ice sheet extent effects on modeled tidal mixing and the global overturning circulation. Paleoceanogr. Paleoclimatol. 34, 1437–1454. https://doi.org/10.1029/2019PA003644. https://onlinelibrary.wiley.com/doi/abs/10.1029/2019PA003644.

Wilmes, S.B., Green, J.A.M., 2014. The evolution of tides and tidal dissipation over the past 21,000 years. J. Geophys. Res., Oceans 119, 4083–4100. https://doi.org/10.1002/2013JC009605. http://doi.wiley.com/10.1002/2013JC009605.

Winton, M., Griffies, S.M., Samuels, B.L., Sarmiento, J.L., Frölicher, T.L., 2013. Connecting changing ocean circulation with changing climate. J. Climate 26, 2268–2278. https://doi.org/10.1175/JCLI-D-12-00296.1. https://journals.ametsoc.org/jcli/article/26/7/2268/33101/Connecting-Changing-Ocean-Circulation-with.

Wright, C.J., Scott, R.B., Ailliot, P., Furnival, D., 2014. Lee wave generation rates in the deep ocean. Geophys. Res. Lett. 41. https://doi.org/10.1002/2013GL059087.

Yamaguchi, R., Suga, T., 2019. Trend and variability in global upper-ocean stratification since the 1960s. J. Geophys. Res., Oceans 124, 8933–8948. https://doi.org/10.1029/2019JC015439. https://onlinelibrary.wiley.com/doi/abs/10.1029/2019JC015439.

Yankovsky, E., Legg, S., 2019. Symmetric and baroclinic instability in dense shelf overflows. J. Phys. Oceanogr. 49, 39–61. https://doi.org/10.1175/JPO-D-18-0072.1. https://journals.ametsoc.org/view/journals/phoc/49/1/jpo-d-18-0072.1.xml.

Yankovsky, E., Legg, S., Hallberg, R.W., 2020. Parameterization of submesoscale symmetric instability in dense flows along topography. preprint. Oceanography. https://doi.org/10.1002/essoar.10503774.1. http://www.essoar.org/doi/10.1002/essoar.10503774.1.

Zanna, L., Khatiwala, S., Gregory, J.M., Ison, J., Heimbach, P., 2019. Global reconstruction of historical ocean heat storage and transport. Proc. Natl. Acad. Sci. 116, 1126–1131. https://doi.org/10.1073/pnas.1808838115. http://www.pnas.org/lookup/doi/10.1073/pnas.1808838115.

Zaron, E.D., 2017. Mapping the nonstationary internal tide with satellite altimetry: mapping the nonstationary internal tide. J. Geophys. Res., Oceans 122, 539–554. https://doi.org/10.1002/2016JC012487. http://doi.wiley.com/10.1002/2016JC012487.

Zhai, X., Greatbatch, R.J., Eden, C., Hibiya, T., 2009. On the loss of wind-induced near-inertial energy to turbulent mixing in the upper ocean. J. Phys. Oceanogr. 39, 3040–3045. https://doi.org/10.1175/2009JPO4259.1. https://journals.ametsoc.org/jpo/article/39/11/3040/11214/On-the-Loss-of-WindInduced-NearInertial-Energy-to.

Zheng, K., Nikurashin, M., 2019. Downstream propagation and remote dissipation of internal waves in the Southern Ocean. J. Phys. Oceanogr. 49, 1873–1887. https://doi.org/10.1175/JPO-D-18-0134.1. http://journals.ametsoc.org/doi/10.1175/JPO-D-18-0134.1.

Zickfeld, K., Solomon, S., Gilford, D.M., 2017. Centuries of thermal sea-level rise due to anthropogenic emissions of short-lived greenhouse gases. Proc. Natl. Acad. Sci. 114, 657–662. https://doi.org/10.1073/pnas.1612066114. http://www.pnas.org/lookup/doi/10.1073/pnas.1612066114.

Chapter 3

The role of mixing in the large-scale ocean circulation

Casimir de Lavergne[a], Sjoerd Groeskamp[b], Jan Zika[c] and Helen L. Johnson[d]

[a]*LOCEAN Laboratory, Sorbonne University-CNRS-IRD-MNHN, Paris, France,* [b]*NIOZ Royal Netherlands Institute for Sea Research, Den Burg, Texel, the Netherlands,* [c]*School of Mathematics and Statistics, University of New South Wales, Sydney, NSW, Australia,* [d]*Department of Earth Sciences, University of Oxford, Oxford, United Kingdom*

3.1 Introduction

At low latitudes, the Earth receives more heat from the Sun than it radiates out to space over an annual orbit. The opposite occurs at high latitudes, where outgoing longwave radiation exceeds incoming shortwave insolation. This latitudinal contrast is the primary driving force of coupled atmosphere and ocean circulations. It is imprinted on the ocean surface, which gains heat at low latitudes and loses heat to the atmosphere at higher latitudes. This differential surface heating of the ocean generates and amplifies the contrast between warm tropical waters and colder waters that fill higher latitude and deep basins (Fig. 3.1). To counteract the tendency to warm warm waters and cool cold waters, the ocean must transfer heat from low to high latitudes.

Such transfer can be achieved by moving warm surface water into the region of surface heat loss, thus evacuating heat from the warm bowl and reducing exposure to cooling of the cold pool (Fig. 3.1a, b). This advective scenario implies a permanent or transient deformation or displacement of the warm bowl so that it stretches into the cooling latitudes. Compensating equatorward flow of colder waters can occur in the horizontal plane, establishing a horizontal circulation, such as a gyre or an eddy (Fig. 3.1a). Alternatively, the compensating flow may take place in the vertical plane, leading to an overturning circulation, such as an inter-hemispheric overturning (Fig. 3.1b). By shifting water bodies along the meridional axis, horizontal and overturning circulations can thus mitigate the imbalance implied by the latitudinal contrast in surface heat fluxes.

The scenarios described above invoke exclusively advection (i.e., mass transport) and surface heat fluxes, implicitly assuming that the ocean interior is devoid of non-advective heat fluxes or is *adiabatic*. The opacity of seawater does prevent solar heating from penetrating significantly deeper than 200 m. However, heat exchanges are not restricted to the sunlit surface layer. Through temperature diffusion or *mixing*, heat can be transferred between water bodies in the ocean interior. If diffusion is able to transfer heat from the warm bowl to the cold pool at a rate similar to the net low-latitude surface heat gain, the idealised two-layer ocean of Fig. 3.1 may achieve heat balance via mixing rather than circulation (Fig. 3.1c).

A recent calculation of mixing-induced heat transfer across temperature classes in a realistic ocean model finds that the peak transfer, estimated at 1.6 PW across the 22 °C isotherm, is of similar magnitude to the peak poleward oceanic heat transport (Holmes et al., 2018a, 2019). This result suggests that mixing plays a central role in maintaining heat balance against differential surface heating. Does it imply that the diffusive scenario of Fig. 3.1c dominates over advective scenarios of Fig. 3.1a, b? The reality is more subtle, because advection and diffusion are not independent processes, and indeed combine to shape oceanic transports of mass and heat. For example, the presence of a diffusive heat supply to the cold pool may allow hemispheric overturning circulations to develop (Fig. 3.1d). These circulations depend upon the existence of mixing, and contribute to poleward heat transport.

Coupling between advection and diffusion is ubiquitous. Any mass transport that crosses isosurfaces of a conservative tracer (e.g., surfaces of constant salinity) below the ocean surface requires mixing. Mixing of momentum also exerts a profound, direct influence on ocean currents. Conversely, mixing rates often depend on surrounding currents, and more specifically on the turbulent motions that stir the ocean. Indeed, over most of the ocean, turbulence elevates effective mixing rates well above the weak molecular diffusivities. Mass transports thus contribute to amplify mixing via turbulence. This implies that mixing and advection rely on shared sources of kinetic energy, ultimately derived from surface winds, surface buoyancy fluxes, tides, and geothermal heating (Fig. 3.2). Understanding the role of mixing in global ocean circulation

36 Ocean Mixing

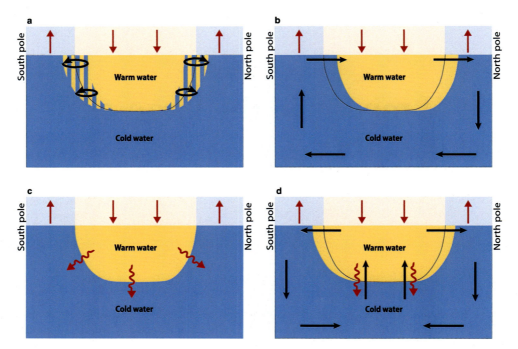

FIGURE 3.1 Idealised two-layer ocean composed of a warm bowl overlying a cold pool. Heating and cooling at the surface occurs within fixed latitudinal bands, represented by light blue (cooling) and light yellow (heating) colours. Salinity and freshwater fluxes are ignored. Straight and wiggly red arrows represent surface and interior heat fluxes, respectively. Black arrows represent mass transports. Each panel corresponds to a distinct scenario of poleward oceanic heat transport. Shown variations in the position and shape of the warm bowl are illustrative and partly arbitrary. **a**, Horizontal circulation moves warm water into the cooling latitudes and cold water into the warming latitudes. **b**, An inter-hemispheric overturning circulation shifts the warm bowl northward, reducing net heat gain (loss) of warm (cold) waters. **c**, Mixing transfers heat from the warm bowl to the cold pool. Note that we implicitly assume that mixing within each layer maintains temperature homogeneity, and thus connects surface and interior heat fluxes. **d**, Mixing converts cold waters into warm waters, allowing hemispheric overturning circulations to develop and transport heat poleward. Part of the diffusive heat gain of the lower layer may also offset surface cooling via intra-layer heat transports, as in **c**.

requires (i) quantifying mixing rates by tracking energy from external sources to turbulence to mixing, and (ii) identifying the upscale influence of turbulent mixing on large-scale current systems.

In this chapter, we will focus on current systems spanning thousands of kilometres: basin-scale ocean gyres, the Antarctic circumpolar current and subtropical and deep overturning circulations. We begin with an overview of mixing processes and their mathematical representation (Section 3.2). We next examine features of the large-scale ocean circulation that can be understood without consideration of mixing (Section 3.3). The discussion of the impacts of mixing focuses by turns on mechanisms (Section 3.4), their distribution (Section 3.5), and their consequences for observed overturning (Section 3.6) and horizontal (Section 3.7) circulations. The emerging view is that of a relatively adiabatic interior ocean circulation commanded by strongly mixing boundaries (Section 3.8).

3.2 Flavours of mixing

Ocean flows are governed by the Navier–Stokes equations on a rotating sphere. The equations determine the evolution of the three-dimensional velocity vector $\vec{u} = (u, v, w)$ as a function of inertial, pressure gradient, gravitational, Coriolis, and frictional forces:

$$\frac{\partial \vec{u}}{\partial t} + \vec{u} \cdot \nabla \vec{u} = -\frac{1}{\rho}\nabla p + \vec{g} - f\vec{z} \times \vec{u} + \nabla \cdot \nu \nabla \vec{u}, \qquad (3.1)$$

where we have made the traditional approximation; that is, we neglected the vertical component of the Coriolis force. In (3.1), p is pressure, ρ density, $\vec{g}$ the gravity vector, f the Coriolis parameter, $\vec{z}$ the vertical unit vector, ν the molecular kinematic viscosity of seawater, and ∇ the three-dimensional gradient operator. The last term represents the divergence of the down-gradient momentum transfer achieved by frictional interactions between water molecules. Assuming no sources and sinks of momentum at ocean boundaries, it can be shown that this term leaves the average momentum over the whole

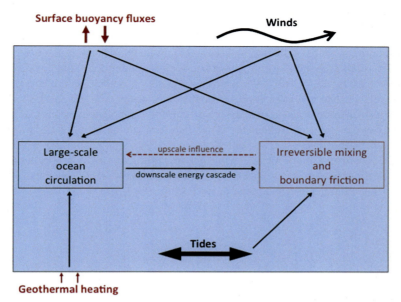

FIGURE 3.2 Simplified schematic of energy flows in the ocean, from forcing to dissipation. Forcing boils down to surface buoyancy fluxes, surface winds, tides (caused by gravitational interaction with the Moon and Sun), and geothermal heating along the seafloor. The large-scale ocean circulation is forced directly by large-scale wind and buoyancy forcing, and indirectly by irreversible mixing. Energy of the large-scale circulation is ultimately dissipated by boundary friction (drag) and irreversible mixing. Irreversible mixing includes momentum, temperature and salinity mixing at molecular scale. Mixing is energised directly by tides, winds, surface buoyancy fluxes and indirectly by the energy cascade from large-scale circulation to turbulence. Note that energy fuelling irreversible mixing and boundary friction is either lost as heat (momentum mixing and drag) or does work against gravity (mixing-driven buoyancy fluxes). The direct forcing of global ocean circulation by tides is thought to be secondary (Bessières et al., 2008) and is therefore not highlighted here, despite known contributions to regional circulation features (e.g., Thompson et al., 2018, Chapter 2).

ocean volume unchanged but decreases the global variance of momentum. These two integral properties are fundamental characteristics of diffusion.

Eq. (3.1) does not suffice to characterise the evolution of the fluid. It must be complemented by the continuity equation (expressing mass conservation), an equation of state relating density to pressure, temperature, and salinity, and evolution equations for conservative temperature Θ and absolute salinity S_A (McDougall, 2003; McDougall et al., 2012). The latter two equations are analogous to that governing the evolution of an arbitrary tracer C,

$$\frac{\partial C}{\partial t} + \vec{u} \cdot \nabla C = Q_C + \nabla \cdot \kappa_C \nabla C, \qquad (3.2)$$

where Q_C encapsulates interior sources and sinks, and κ_C is the molecular diffusivity of C. A conservative tracer has $Q_C \equiv 0$; this is the case for conservative temperature (except for subsurface solar heating) and absolute salinity. Boundary fluxes of C enter as boundary conditions on the last term in (3.2). In the absence of such fluxes, this diffusion term preserves the domain average C and decreases the domain-wide variance of C. The reduction of variance can be illustrated with the idealised heat balance of Fig. 3.1c: diffusion acts to reduce the temperature contrast between the two layers, hence to diminish global temperature variance, whereas surface forcing acts in the opposite sense. More generally, boundary fluxes are the only means to increase global variance of a conservative tracer, implying that mixing and boundary fluxes are intrinsically tied in this variance competition (Walin, 1977, 1982; Zika et al., 2015).

Can the diffusive terms $\nabla \cdot \nu \nabla \vec{u}$, $\nabla \cdot \kappa_\Theta \nabla \Theta$ and $\nabla \cdot \kappa_{S_A} \nabla S_A$ affect the large-scale distribution of momentum $\vec{u}$? Molecular diffusion coefficients ν, κ_Θ, and κ_{S_A} vary little about respective values of 1×10^{-6}, 1.4×10^{-7} and 1.4×10^{-9} m^2 s^{-1}. The characteristic time to diffuse a momentum anomaly from the surface to 4 km depth follows as $(4000 \text{ m})^2/\nu = 16 \times 10^{12}$ s or about 500,000 years. Equivalent timescales for heat and salt are respectively one and three order(s) of magnitude larger. It is immediately apparent that molecular diffusivities are far too weak to affect global circulation systems known to evolve on timescales of seasons to millennia. Scaling frictional and Coriolis terms in (3.1) further shows that, for a typical value of $f \sim 10^{-4}$ s^{-1}, the two terms are comparable at length scales of 10^{-1} m. Only at length scales smaller than 10 centimetres do molecular processes become considerable.

Using fast probes able to measure velocity and temperature variations at centimetre scale, it is possible to estimate the rate of molecular dissipation of kinetic energy $(-\nu(\nabla\vec{u})^2)$ and temperature variance $(-\kappa_\Theta(\nabla\Theta)^2)$ along a vertical cast in the ocean (Osborn and Cox, 1972; Osborn, 1978, Chapter 14). Such measurements show that the ocean is strewn with small patches of elevated shear and temperature microstructure (Gregg, 1987). The magnitude and distribution of the measured micro-scale gradients cannot be explained by large-scale momentum and temperature variations: they arise from intermittent turbulent motions. These motions stir large-scale gradients and produce small-scale variance that is ultimately dissipated by molecular interactions. As a result, molecular dissipation of momentum and temperature variance is typically several orders of magnitude larger than would be expected in a laminar ocean (Gargett and Osborn, 1981; Oakey, 1982).

Stirring by motions of intermediate scale—between major currents spanning thousands of kilometres and molecular processes acting over centimetres—thus accelerates the downscale cascade of variance and amplifies the gradients upon which molecular viscosity and diffusivities act. The ability of turbulence to amplify mixing becomes apparent in Eqs. (3.1) and (3.2) when a Reynolds decomposition of variables into mean and fluctuating components is performed: $\vec{u} = \overline{\vec{u}} + \vec{u}'$, $C = \overline{C} + C'$, and so on. The overbar indicates an average over spatio-temporal scales larger than those of turbulence. Evolution equations for the mean momentum and mean tracer become (Gill, 1982)

$$\frac{\partial \overline{\vec{u}}}{\partial t} + \overline{\vec{u}} \cdot \nabla \overline{\vec{u}} = -\frac{1}{\rho}\nabla \overline{p} + \vec{g} - f\vec{z} \times \overline{\vec{u}} + \nabla \cdot \nu \nabla \overline{\vec{u}} - \overline{\vec{u}' \cdot \nabla \vec{u}'}, \tag{3.3}$$

$$\frac{\partial \overline{C}}{\partial t} + \overline{\vec{u}} \cdot \nabla \overline{C} = \overline{Q_C} + \nabla \cdot \kappa_C \nabla \overline{C} - \overline{\vec{u}' \cdot \nabla C'}. \tag{3.4}$$

The last terms in (3.3) and (3.4) imply that correlations between fluctuations in velocity and velocity (or tracer) gradients can modify the mean velocity (or tracer) tendency. Using the continuity equation under Boussinesq approximation $\nabla \cdot \vec{u} = 0$, and assuming turbulent fluxes can be modelled by down-gradient Fickian diffusion, the system of equations is closed with

$$\overline{\vec{u}' \cdot \nabla C'} = \nabla \cdot \overline{\vec{u}' C'} = \nabla \cdot (-\vec{K}_C \cdot \nabla \overline{C}) \tag{3.5}$$

and analogous relations for each component of the mean velocity vector. In (3.5), we introduced the (a priori unknown) turbulent diffusivity vector $\vec{K}_C$. In most oceanic conditions, each component of the turbulent diffusivity vector will exceed its molecular counterpart, and molecular terms in (3.3) and (3.4) can be neglected. However, molecular processes are still necessary and operate to dissipate variance: it is via an increase of the gradients available to molecular diffusivities that stirring increases rates of irreversible mixing. The phrase *irreversible mixing* refers to mixing involving molecular interactions and diminishing the domain-wide variance, whereas *stirring* refers to a transfer of variance to smaller scales.

Eqs. (3.3)–(3.5) imply that the impact of mixing on mean ocean circulation depends on the magnitude of turbulent diffusivities. Key to estimating these diffusivities is an understanding of the rate-controlling processes that transfer variance to dissipation scales. Two pivotal regimes of oceanic turbulence have been identified: three-dimensional or small-scale turbulence, and geostrophic turbulence. Three-dimensional turbulence, triggered by gravitational and shear instabilities, is active on scales of 1–100 m. It induces isotropic turbulent mixing rates that range from molecular levels to 10–100 m^2 s^{-1}. Values above 10^{-2} m^2 s^{-1} occur mostly within surface and bottom boundary layers, where boundary conditions are immediately felt (Large et al., 1994; van Haren and Gostiaux, 2012). Away from boundaries, moderate turbulence levels are largely sustained by the breaking of internal waves generated by tides (Fig. 3.3a, b; Kunze et al., 2006; Waterhouse et al., 2014; de Lavergne et al., 2020). Three-dimensional turbulence mixes tracers and momentum alike, although the isotropic diffusivities of momentum and tracers can differ (Gaspar et al., 1990).

Geostrophic turbulence consists of mesoscale eddies, with a typical diameter of 10–100 km and a baroclinic velocity structure that spans the whole water column. Most of these eddies are generated by baroclinic instability of the large-scale flow environment (Charney, 1947; Eady, 1949; McWilliams and Chow, 1981). As opposed to three-dimensional turbulence, they stir background tracer gradients only along density surfaces in the ocean interior (Iselin, 1939; McDougall et al., 2014), and horizontally near the surface, at estimated rates varying from about 10 to 10^4 m^2 s^{-1} (Fig. 3.3c, d; Klocker and Abernathey, 2014; Cole et al., 2015; Groeskamp et al., 2020, Chapter 9). In the surface boundary layer and where tracer surfaces are not aligned with density surfaces, mesoscale eddies are thus able to produce finescale tracer variance (Klein et al., 1998). This variance is ultimately dissipated at molecular scale with the aid of background three-dimensional turbulence (Smith and Ferrari, 2009; Naveira Garabato et al., 2016). The cooperation of mesoscale, small-scale and molecular processes thus contributes to the homogenisation of all tracers along density surfaces. Density surfaces being called *isopycnals*, the phrase *isopycnal mixing* is often used as a shorthand for this suite of processes.

Rates of isopycnal mixing are thought to be set by mesoscale stirring, rather than by three-dimensional turbulence (Chapter 9). Stirring rates by mesoscale eddies are typically seven orders of magnitude larger than isotropic diffusivities. Consequently, mixing by three-dimensional turbulence is usually referred to as *diapycnal mixing*: its contribution to mixing

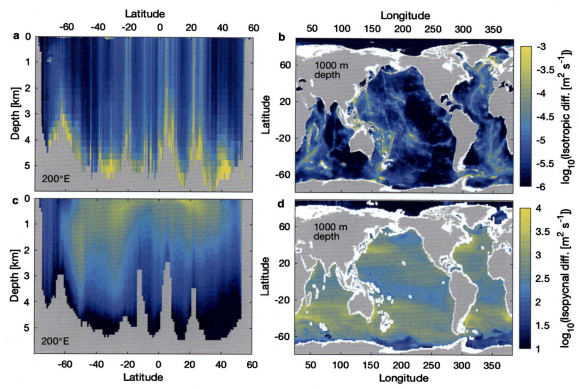

FIGURE 3.3 Estimated isotropic (**a, b**) and isopycnal (**c, d**) diffusivities, at 200°E/160°W (**a, c**) and at 1000 m depth (**b, d**). Both diffusivities are shown on a $\log_{10}$ scale that spans three orders of magnitude (see colourscales on the right). Isotropic diffusivity here only includes the contribution of internal waves energised by tides (de Lavergne et al., 2020). Isopycnal diffusivity quantifies rates of mesoscale stirring (Groeskamp et al., 2020).

along isopycnals being overwhelmed by that of geostrophic turbulence, only the contribution to mixing across isopycnals is considered. Because diapycnal gradients are well approximated by vertical gradients except at boundaries, diapycnal mixing is also frequently referred to as *vertical mixing*. We instead use the phrase *isotropic mixing*, for it is the most general and physical description of mixing by small-scale turbulence (McDougall et al., 2014).

In addition to tracer stirring, geostrophic turbulence causes efficient lateral and vertical redistribution of horizontal momentum (Rhines and Holland, 1979). Mathematically, these effects reside in the last term of Eq. (3.3). The lateral redistribution effect, termed Reynolds stress, is often modelled by down-gradient diffusion along geopotential surfaces (Munk, 1950; Smith and McWilliams, 2003). However, this diffusive representation of momentum fluxes is imperfect, because mesoscale eddies can transfer momentum up-gradient and accelerate or rectify the large-scale flow (Harrison, 1978; Rhines and Holland, 1979). The vertical redistributive effect, called eddy form stress, has a profound effect on large-scale currents (Johnson and Bryden, 1989; Olbers, 1998): even though mesoscale eddies are unable to flux tracers across isopycnals, they are adept at transferring momentum across isopycnals via pressure fluctuations. The induced vertical momentum transfers may be represented as Fickian vertical mixing, provided the diffusivity has an appropriate form (Rhines and Young, 1982; Gent et al., 1995). However, they are more commonly represented and interpreted as an eddy-induced mass transport due to correlations between mesoscale velocities and isopycnal layer thicknesses (Gent and McWilliams, 1990; McDougall and McIntosh, 2001).

The example of eddy form stress shows that the same physical process can have alternative diffusive and advective representations, illustrating the difficulty in defining mixing. Still, it is possible to distinguish processes that unambiguously contribute to irreversible mixing, and thereby dissipate whole-ocean kinetic energy and/or tracer variance. Reynolds and eddy form stresses exerted by geostrophic turbulence do not qualify: they communicate momentum via pressure gradient forces rather than molecular friction. By contrast, three-dimensional turbulence directly contributes to dissipate both kinetic energy and tracer variance. Similarly, tracer stirring by geostrophic eddies contributes to irreversible mixing by transferring variance down to dissipation scales. In the remainder of this chapter, we will reserve the word *mixing* for only those processes causing irreversible mixing, and discuss the role of such processes in basin-scale ocean circulation (Sections 3.4–3.7). In the next section, we briefly expose ocean circulation theories that do *not* appeal to mixing.

3.3 Non-dissipative theories of ocean circulation

In 1992, a purposeful tracer release experiment at 300 m depth in the northeastern Atlantic showed that the isotropic diffusivity in the region's pycnocline is close to 10^{-5} m^2 s^{-1} (Ledwell et al., 1993). Temperature microstructure in the stratified upper ocean indicates similar average rates of temperature mixing, despite integrating the additional contribution of geostrophic turbulence (Osborn and Cox, 1972; Gregg, 1987; Davis, 1994). At the rate of 10^{-5} m^2 s^{-1}, diffusive heat transfer over 1 km takes about 3000 years: too slow to compete with surface heat forcing and heat transport by upper-ocean currents. These observations motivated—or justified a posteriori—the neglect of molecular-scale dissipation in landmark models of ocean circulation.

3.3.1 Ekman pumping

The first such model is due to Sverdrup. Classical scaling of Eq. (3.1) shows that, away from the equator, ocean currents are close to geostrophic balance: pressure gradient and Coriolis forces set horizontal velocities. Cross-differentiation of zonal and meridional geostrophic equations and use of continuity yields

$$\beta v = f \frac{\partial w}{\partial z}, \qquad (3.6)$$

where $\beta = df/dy$. Eq. (3.6) states that poleward (equatorward) flow must be balanced by vertical stretching (squeezing) of fluid columns. Winds are a prominent force able to squeeze or stretch water columns: the curl of the surface wind stress $\vec{\tau}$ generates a pumping velocity w_{Ek} at the base of the thin surface Ekman layer. Vertical integration of (3.6) from a depth h_0 of assumed zero vertical motion up to the bottom of the Ekman layer (of thickness h_{Ek}) gives Sverdrup's prediction for the meridional circulation,

$$\beta \int_{z=-h_0}^{z=-h_{Ek}} v \, dz = f w_{Ek} = f \nabla \times (\vec{\tau}/f). \qquad (3.7)$$

Relation (3.7), called Sverdrup balance, proved very powerful to explain the broad equatorward (poleward) flow of subtropical (subpolar) gyres (Fig. 3.4). In this model, depth-integrated currents are shaped by direct wind forcing. This forcing relies on the existence of molecular friction, necessary for winds to transfer momentum to an initially still ocean surface. Yet wind stress effectively occurs via pressure forces onto surface waves (Plant, 1982), so that Ekman pumping and Sverdrup balance do not depend on viscosity.

Sverdrup balance is mute about the vertical structure of circulation and density in the ocean. The three-dimensional circulation problem is particularly complex, because surface buoyancy fluxes, interior density gradients and ocean currents all depend upon each other. A major advance owes to Luyten et al. (1983), who established an adiabatic theory for the large-scale currents and density structure of the ocean thermocline. They assumed that the thermocline consists of a few homogeneous isopycnal layers governed by geostrophic and hydrostatic balances (thus retaining the first three terms on the right-hand side of (3.1)). Specifying surface densities, where Ekman pumping is downward, they tracked the depth of isopycnals and the transport along isopycnals, progressing from the isopycnals' outcrop locations toward the equator. Their theory, called the ventilated thermocline theory, successfully predicts certain key features of the thermocline. It suggests that, underneath the direct influence of surface forcing, the density structure and ventilation of the upper ocean are essentially controlled by Ekman pumping—and largely independent of mixing.

An adiabatic model of circulation in the deep ocean was championed a decade later by Toggweiler and Samuels (1993, 1995, 1998). They proposed that a large proportion of dense waters that fill the ocean deeper than about 2 km are drawn to the surface by Ekman upwelling in the Southern ocean (Fig. 3.5). There, westerly winds drive a divergent northward Ekman transport, whereas a zonally unbounded channel at Drake passage latitudes (0–2 km; 56–60°S) results in the selection of deeper waters as the mass replacement for the surface divergence (Toggweiler and Samuels, 1995). Indeed, within this channel, the zonal mean longitudinal pressure gradient is zero, and so the net meridional geostrophic flow must also be zero. Numerical experiments with global ocean models further showed that the wind-driven upwelling of dense waters occurs along rising isopycnals (Fig. 3.5) and persists in the limit of zero mixing (Toggweiler and Samuels, 1998; Wolfe and Cessi, 2011). This implies that an inter-hemispheric overturning circulation, akin to that schematised in Fig. 3.1b, can exist without mixing: (i) gravitational sinking at northern high latitudes carries dense waters into the deep ocean; (ii) geostrophic southward flow brings these waters to the Ekman divergence south of 50°S, where they are lifted up to the surface; (iii) surface density transformations and northward upper-ocean currents close the circulation.

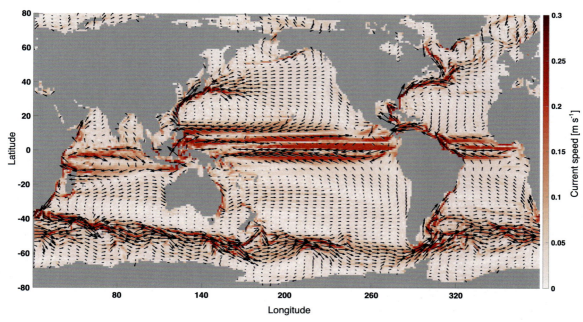

FIGURE 3.4 Annual mean surface geostrophic currents calculated using satellite observations. Surface geostrophic velocity obtained from the CMEMS (Copernicus marine environment monitoring service) operational delayed-time sea surface geostrophic velocity anomalies derived from satellite altimetry (Pujol et al., 2016; Taburet et al., 2019), using a β-plane approximation of the geostrophic equations in the equatorial band (Lagerloef et al., 1999). Daily, quarter degree resolution data since 1993 is averaged and smoothed into a mean for illustrative purposes. Colour is indicative of the speed, with darker colours being faster currents.

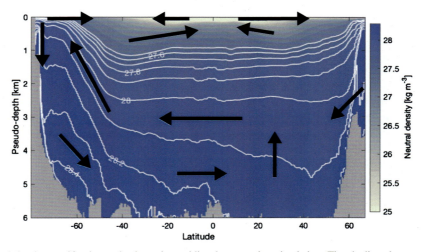

FIGURE 3.5 Global neutral density stratification and schematic meridional overturning circulation. The shading shows neutral density (Jackett and McDougall, 1997) mapped by Gouretski and Koltermann (2004) as a function of latitude and pseudo-depth. The pseudo-depth of density surfaces is found by filling each latitude band from the bottom up with ocean grid cells ordered from dense to light. The neutral density range 27.5–28.5 kg m^{-3} is contoured in white with a 0.1 kg m^{-3} interval. Black arrows give a simplified view of overturning flows.

3.3.2 Momentum redistribution by geostrophic turbulence

Sverdrup balance does not explain the closure of gyres, which occurs via a return flow focused along the western boundary of ocean basins (Fig. 3.4). This return flow was long thought to rely on lateral Reynolds stresses induced by the mesoscale eddy field (Munk, 1950; Pedlosky, 1996). Form stress exerted by sloping bottom topography has more recently been acknowledged as the principal force upsetting Sverdrup balance along western boundaries (Hughes and de Cuevas, 2001). Realistic ocean models indicate that both variable topography and geostrophic turbulence contribute to alter the balance (3.7) and shape the depth-integrated flow of major gyres (Le Corre et al., 2020). The same applies to the momentum balance of the Antarctic circumpolar current (ACC), the World ocean's largest current which flows eastward in the latitude range 40–70°S (Fig. 3.4). There, bottom form stress due to topographic obstacles along the ACC path provides the sink of zonal momentum necessary to balance that imparted at the surface by westerly winds (Munk and Palmén, 1951). In this balance, geostrophic turbulence plays an essential role by transferring momentum downward via eddy form stresses, connecting the surface source to the bottom sink of zonal momentum (Olbers, 1998; Ferreira et al., 2005).

Momentum redistribution by geostrophic eddies plays a similarly essential role in the ocean's overturning circulations. Surface wind and buoyancy forcing often produces relatively steep isopycnal slopes that are baroclinically unstable. Baroclinic instability then generates mesoscale eddies that act to flatten out the isopycnals (Gent et al., 1995). The slumping of isopycnals occurs via an eddy-induced baroclinic circulation, which generally opposes Ekman pumping velocities (Marshall, 1997; Marshall et al., 2002; Doddridge et al., 2016). This circulation alters the shallow overturning cells that span the subtropical thermocline (Doddridge et al., 2016) as well as the deep overturning circulation (Marshall and Speer, 2012). In particular, mesoscale eddies can induce southward isopycnal mass fluxes within the zonally continuous ACC (Marshall, 1997; Marshall and Radko, 2003), thus overcoming the constraint on upper ocean southward flow at Drake passage latitudes suggested by Toggweiler and Samuels (1995). Nevertheless, simulations of the Southern ocean, including realistic topography and a rich eddy field, suggest only weak southward crossing of the ACC via eddy-induced mass transport (Zika et al., 2012; Mazloff et al., 2013; Dufour et al., 2015).

Lateral and vertical stresses induced by geostrophic turbulence thus modulate the ocean's response to Ekman pumping and surface buoyancy fluxes, implying a role for the ocean's chaotic nature in setting its circulation and stratification. However, the theories outlined in this section include no explicit role for temperature and salinity modification by turbulent mixing. Instead, they rationalise many of the observed features of the ocean by invoking purely adiabatic dynamics, asserting a view of ocean circulation in step with the scenarios illustrated in Fig. 3.1a, b.

3.4 How can mixing shape circulation?

3.4.1 By altering surface wind and buoyancy forcing

The conceptual frameworks exposed in the previous section take the wind stress and surface buoyancy fluxes or surface densities as given. However, these surface boundary conditions, essential drivers of ocean circulation, depend on mixing and on the circulation itself.

First, the wind stress is a function of the difference between the wind velocity vector above the sea surface and the oceanic surface velocity vector (Pacanowski, 1987; Duhaut and Straub, 2006). Vertical momentum mixing near the surface acts to reduce the ocean surface velocity, generally augmenting the wind stress.

Second, surface heat and freshwater fluxes depend upon the sea surface temperature (SST), which is profoundly affected by mixing in the surface boundary layer. For example, SST cooling by surface heat loss in winter is generally damped by isotropic mixing, which redistributes the heat loss over the depth of the surface mixed layer. More generally, a given air-sea heat (or freshwater) flux induces a change in the temperature (or salinity) of the ocean surface that is inversely proportional to the mixed layer depth (MLD). Mixing processes controlling the MLD and its evolution thus play a major role in establishing surface densities and surface buoyancy fluxes. These processes include both momentum mixing (which affects the shear of Ekman and other currents, which in turn influences turbulence and MLD) and temperature and salinity mixing by mesoscale, submesoscale and three-dimensional turbulence (see Chapters 4 and 8).

3.4.2 By altering density gradients

On horizontal scales exceeding several kilometres, and away from frictional boundary layers, the ocean is in near geostrophic and hydrostatic balances. Combined, these balances give the thermal wind relationship

$$f\frac{\partial u}{\partial z} = g\frac{\partial \rho}{\partial y}, \quad f\frac{\partial v}{\partial z} = -g\frac{\partial \rho}{\partial x}, \tag{3.8}$$

showing that the vertical variation of horizontal velocities is controlled by horizontal density gradients. This implies that isotropic mixing can change horizontal circulation by altering horizontal variations in density. For example, localised deepening and densification of the surface mixed layer—via surface buoyancy loss and convective mixing—can stimulate horizontal motion around the convective chimney. On the other hand, isopycnal mixing catalysed by mesoscale eddies does not modify density except for effects related to the non-linearity of the equation of state (McDougall, 1984), and is therefore less able to modify horizontal circulation. In general, salinity and temperature modifications that have compensating effects on density have no impact on circulation, unless they influence surface buoyancy forcing. Mixing of passive tracers is equally neutral to circulation—unless it impacts phytoplankton concentrations and, via their modulation of albedo and light absorption, near-surface densities (Sweeney et al., 2005).

Mixing can also affect circulation by altering the vertical density distribution. The densest ocean waters are formed at the surface by heat and freshwater loss to the atmosphere and cryosphere. Their gravitational, ageostrophic sinking into the deep ocean relies on their higher density relative to underlying waters. This density difference (between newly formed dense waters and underlying waters) owes to the mixing of sinking waters with lighter waters en-route to the deep ocean. Mixing thus maintains relatively low densities in the deep ocean that sustain the downwelling of the densest waters. As incoming dense waters intrude below older waters, they drive a compensating upwelling of these older waters. This upwelling is often quantified by a diapycnal velocity ω, related to the divergence of the mixing-driven density flux F (Nurser et al., 1999),

$$\omega = \frac{\partial F}{\partial \rho}. \tag{3.9}$$

Eq. (3.9) states that local density loss due to a divergent density flux is balanced by upward advection of denser water—with the reverse true for a convergent density flux. The diapycnal velocity ω may be a true Eulerian velocity, as required at steady state, or merely a downward (upward) movement of the isopycnal due to local lightening (densification). Note that the term diapycnal upwelling (downwelling) is used when the velocity ω is directed toward lighter (denser) layers, even though the orientation of the velocity may vary.

Vertical velocities induced by mixing can then set up horizontal circulations (Stommel, 1958; Pedlosky, 1992). Indeed, Eq. (3.6) shows that, if the vertical velocity varies in the vertical, meridional geostrophic flow is expected to balance the local squeezing or stretching. For example, Stommel (1958) proposed that widespread upwelling across the 2 km depth interface generates broad interior poleward motion at depths greater than 2 km. Pedlosky (1992) further demonstrated that longitudinal variations in the upwelling rate can cause the deep meridional flow to have a sheared, baroclinic structure.

3.4.3 By producing and consuming water masses

In the above situation of gravitational sinking, enabled by mixing-driven reductions in density at depth, a circulation is established between the surface source and the bottom sink of dense waters. This perspective can be generalised as follows: Mixing both produces and consumes water masses; that is, it adds and removes mass from given density classes. If the ocean stratification is statistically steady, isopycnal mass transports (i.e., circulation) must connect the sources and sinks of mass within each isopycnal layer. Reciprocally, for an isopycnal circulation to be maintained, mass must be added at the starting point of the circulation and removed at its finish point. The mass gains and losses of isopycnal layers, referred to as *density transformations* or *water-mass transformations*, can occur at the surface via surface buoyancy fluxes, at the bottom via geothermal heating, or in the remainder of the ocean via mixing (Groeskamp et al., 2019).

Were mixing absent, circulation across density classes would be restricted to boundaries. This restriction is best illustrated by examining circulation in a density-depth coordinate system (Nycander et al., 2007; Fig. 3.6). In this space, the adiabatic circulations of Luyten et al. (1983) and Toggweiler and Samuels (1998) reduce to downward and upward motions at fixed density (Fig. 3.6a). They can be contrasted with the simplified overturning circulation of Munk and Wunsch (1998), whereby low-latitude upwelling from 4 to 1 km depth is enabled by mixing-driven buoyancy gain (Fig. 3.6b). By causing density transformations throughout the ocean volume, mixing thus confers an additional degree of freedom on the circulation. Whether observed overturning circulations are closer to the idealised scenarios of Figs. 3.1b and 3.6a, versus those of Figs. 3.1d and 3.6b, remains a matter of debate. We will argue that a more accurate depiction of the overturning involves a substantial decrease of density during the descent of dense waters, a modest decrease during their ascent along the seafloor, and upwelling at constant density from 2.5 km depth to the near surface (Fig. 3.6d).

3.5 Where is mixing most effective at shaping circulation?

The forces that set the ocean in motion (Fig. 3.2) and the mechanisms identified in Section 3.4 hint at the locations where mixing is most influential on circulation: near the surface, and near the bottom. Here we briefly survey observed distributions of isotropic mixing and mesoscale stirring to substantiate this proposal. We define the near-surface region as waters shallower than the local annual maximum MLD plus 100 m, and the near-bottom region as waters lying within 500 m of the seafloor. The intervening waters will be referred to as the ocean interior. Thus defined, the ocean interior makes up 83% of the global ocean volume.

44 Ocean Mixing

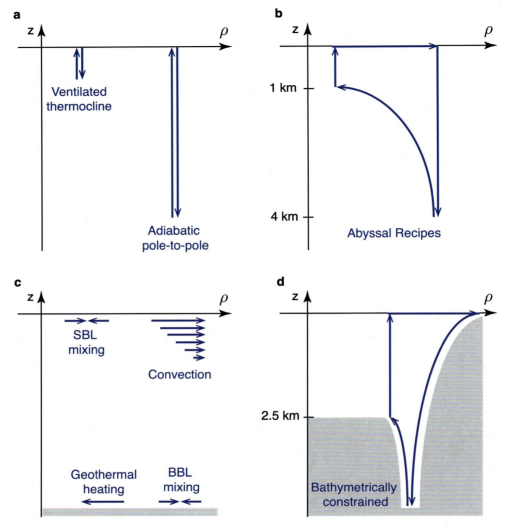

FIGURE 3.6 Idealised circulations viewed in density-depth coordinates. **a**, The ventilated thermocline (Luyten et al., 1983) and adiabatic pole-to-pole (Toggweiler and Samuels, 1998; Wolfe and Cessi, 2011) circulation frameworks involve flow along isopycnals only (with density transformations, i.e., movement along the x-axis, allowed only at the surface). **b**, The overturning circulation as modelled in the Abyssal recipes of Munk and Wunsch (1998): dense waters sink at high latitudes down to 4 km depth and return to 1 km depth via mixing-driven upwelling across the low-latitude stratification. **c**, Schematic view of water parcel movements associated with mixing in the surface boundary layer (SBL), convective mixing forced by surface buoyancy loss, mixing in the bottom boundary layer (BBL), and geothermal heating. **d**, Proposed view of the overturning circulation: dense waters sink along the ocean floor, losing a large fraction of their density excess as they descend to abyssal depths and mix with overlying waters; mixing near the bottom and geothermal heating allows them to return to lighter layers up to 2.5 km depth; adiabatic Southern ocean upwelling brings them from 2.5 km depth to the surface. The grey shading emphasises the role of bathymetric constraints, but does not imply that flow is disallowed within this phase space.

3.5.1 Isotropic mixing, from top to bottom

The canonical pycnocline isotropic diffusivity of 10^{-5} m^2 s^{-1} is usually deemed too small to be a leading-order control of circulation and tracer budgets (Ledwell et al., 1993; Davis, 1994; Munk and Wunsch, 1998). To evaluate this expectation, we apply a uniform mixing rate of 10^{-5} m^2 s^{-1} to the observed density distribution (Figs. 3.5 and 3.7; Gouretski and Koltermann, 2004), and deduce diapycnal velocities, according to Eq. (3.9). By summing these velocities along isopycnals within each ocean basin, we obtain diapycnal mass transports (Fig. 3.8a), measured in Sverdrups (Sv; 1 Sv $\equiv 10^6$ m^3 s^{-1}). The density measure employed from here on is the neutral density of Jackett and McDougall (1997), which does not depend on a reference pressure. The uniform 10^{-5} m^2 s^{-1} mixing rate generates a few Sv of diapycnal upwelling within the 25.5–28 kg m^{-3} density range, and a few Sv of downwelling at lower densities (Fig. 3.8a). Given that deep and subtropical overturning circulations are thought to carry a few tens of Sv (Ganachaud and Wunsch, 2000; Lumpkin and

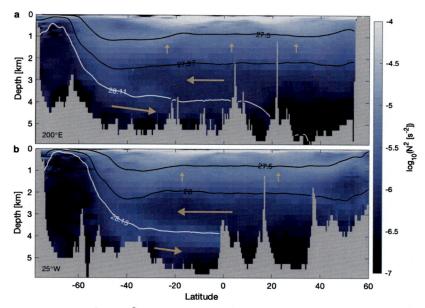

FIGURE 3.7 Squared buoyancy frequency ($N^2 = -\frac{g}{\rho}\frac{\partial \rho}{\partial z}$) along 200°E/160°W (a) and 25°W (b) transects cutting through the Pacific and Atlantic basins, respectively. Data from Gouretski and Koltermann (2004). Orange arrows illustrate the circulation implied by diapycnal transports diagnosed in Fig. 3.8. In the Atlantic, the southward mid-depth flow is stronger than the bottom northward flow due to North Atlantic deep water inflow (Talley, 2013; Lozier et al., 2019). White contours are the neutral density surfaces of meridional flow reversal, coinciding with a local stratification maximum. Black contours are density surfaces enclosing the Munk regime characterised by weak mixing-driven upwelling (see Section 3.6.1 and Fig. 3.8). This regime's density range overlies that of abundant seafloor (de Lavergne et al., 2017; Fig. 3.12) and underlies that of intermediate waters (Naveira Garabato et al., 2014).

Speer, 2007), we conclude that an average diffusivity one order of magnitude larger, i.e., 10^{-4} m^2 s^{-1}, would be required to make isotropic mixing a leading-order contributor to these circulations (Munk and Wunsch, 1998).

Caveats to the above conclusion include (i) the dependence of diapycnal transports on the three-dimensional distribution of density and diffusivity (a uniform diffusivity causes different transports than a varying diffusivity with the same average) and (ii) the role of diffusivity in shaping the ocean's density distribution, which was taken as given. Nevertheless, the calculation provides a useful rule of thumb: regions where isotropic mixing plays a direct and substantial role in circulation are likely to have average diffusivities close to or larger than 10^{-4} m^2 s^{-1}.

Winds and waves maintain high levels of mixing in the surface boundary layer (Chapter 4). Typically, this layer extends over a few tens to hundreds of metres, possesses quasi homogeneous properties and isotropic mixing rates in excess of 10^{-2} m^2 s^{-1} (Gaspar et al., 1990; Large et al., 1994; de Boyer Montégut et al., 2004). When the ocean surface loses buoyancy, mixing rates are further enhanced by convective instability. Convection leads to deepening and density gain of the mixed layer (Fig. 3.6c), and underpins much of the pronounced seasonal cycle of MLD. Together, these surface mixing processes shape global ocean circulation and stratification by modifying surface currents and wind stresses, SST and air-sea buoyancy exchanges, as well as water-mass transformation and subduction. In particular, winter convective mixing plays a primary role in setting the properties of water masses entering and establishing the ocean's permanent stratification (Iselin, 1939; Stommel, 1979).

Energetic three-dimensional turbulence often encroaches below the base of the mixed layer. Turbulence may be triggered by pronounced vertical shear in the mean currents, as occurs in the shallow thermocline (near 50 m depth) of the central and eastern equatorial Atlantic and Pacific (Gregg et al., 1985; Smyth and Moum, 2013; Hummels et al., 2013, Chapter 10). These mixing hotspots are outstanding: the shear of large-scale and mesoscale currents below the mixed layer is generally insufficient to set off such instabilities, particularly in the more sluggish ocean interior (Wunsch, 1997). Storm-induced inertial oscillations of the mixed layer produce shear and mixing at its base (Pollard et al., 1973; Price et al., 1986, Chapter 5). In addition, these oscillations often radiate internal waves able to catalyse mixing deeper down (Alford and Gregg, 2001; Jing and Wu, 2014). In mid-latitude oceans, a consequent wintertime increase of the average isotropic diffusivity by up to an order of magnitude is observed down to 2 km depth (Whalen et al., 2018). Nonetheless, the wintertime diffusivity averaged over a large region remains of order 10^{-5} m^2 s^{-1} (Whalen et al., 2018, Chapter 5), except in near-surface waters, where the bulk of the energy supply to mixing lies (Zhai et al., 2009; Alford, 2020).

46 Ocean Mixing

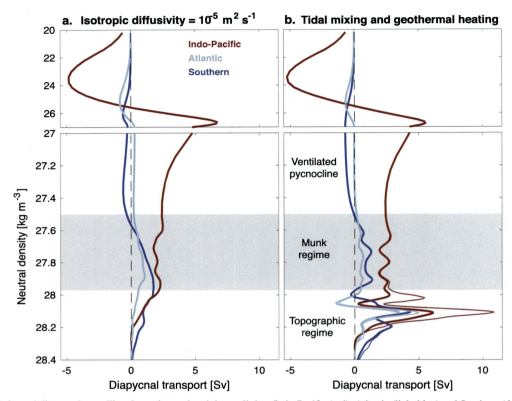

FIGURE 3.8 Estimated diapycnal upwelling due to isotropic mixing, split into Indo-Pacific (red), Atlantic (light blue) and Southern (dark blue) oceans. The Southern ocean is defined as south of 32°S. **a**, Constant isotropic diffusivity of 10^{-5} m^2 s^{-1}. **b**, Realistic tidal mixing (thick curves), with added contribution of geothermal heating (thin curves). Positive values correspond to transport toward smaller densities (diapycnal upwelling). Where the shown transports increase upward, mixing causes volume loss or consumption; where transports decrease upward, mixing causes volume gain or formation. Regimes are defined in Section 3.6 (see also Figs. 3.7 and 3.12). The employed climatological hydrography is that of Gouretski and Koltermann (2004). Tidal mixing rates are from de Lavergne et al. (2020) and geothermal heat fluxes from Lucazeau (2019). The methodology follows that of de Lavergne et al. (2016).

Tides constitute a leading source of three-dimensional turbulence outside mixed boundary layers (Munk, 1997; Waterhouse et al., 2014; de Lavergne et al., 2019; Vic et al., 2019). When tidal currents flow over uneven seafloor, they generate internal waves called *internal tides* that propagate and fuel three-dimensional turbulence throughout the global ocean (Garrett and Kunze, 2007, Chapter 6). Estimated mixing rates due to internal tides vary widely in the horizontal and vertical, from 10^{-6} m^2 s^{-1} up to 10^{-2} m^2 s^{-1} in localised hotspots (Polzin et al., 1997; Rudnick et al., 2003; Waterhouse et al., 2014). They were recently mapped over the global ocean using Lagrangian tracking of internal-tide energy from sources to sinks, accounting for local and remote pathways to mixing (Fig. 3.3a, b; de Lavergne et al., 2020). A zonal average of the thus estimated tidal diffusivity, weighted by $|\frac{\partial \rho}{\partial z}|$ so that mean values relate to density fluxes, shows a sharp increase near 2.5 km depth from order 10^{-5} m^2 s^{-1} above to order 10^{-4} m^2 s^{-1} below (Fig. 3.9a). The transition at 2.5 km corresponds to the typical depth to which topographic ridges rise (Fig. 3.7). This zonal mean distribution indicates that internal tides cannot account for average mixing rates nearing 10^{-4} m^2 s^{-1} above the depth range of major ridges. Given the broad agreement of the mapped tidal diffusivity with available observations of internal-wave-driven turbulence below 400 m (de Lavergne et al., 2020), we contend that breaking internal waves are unlikely to sustain large diapycnal flows between 400 m and 2.5 km depth.

High isotropic diffusivities deeper than 2.5 km do not necessarily imply large net diapycnal upwelling. This is because these diffusivities are sufficiently concentrated near the bottom (Fig. 3.9a, b) that they tend to homogenise abyssal waters, rather than lighten them by draining buoyancy from the upper ocean. Indeed, close to rough or steep topography, where elevated diffusivities are observed, turbulence is bottom-intensified (Toole et al., 1994; Polzin et al., 1997). As a result, the downward buoyancy flux increases toward the seafloor, except in a thin bottom layer, where it must dwindle to match the bottom boundary condition (Fig. 3.10b; St. Laurent et al., 2001; de Lavergne et al., 2016; Ferrari et al., 2016). Bottom-enhanced turbulence thus generates a dipole of density transformation: buoyancy gain along the seafloor, and buoyancy loss

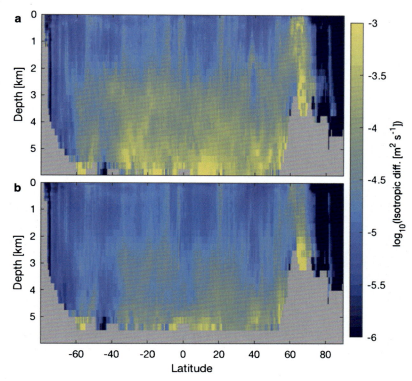

FIGURE 3.9 Zonal means of the isotropic diffusivity induced by internal tides, as mapped by de Lavergne et al. (2020). **a**, Global zonal mean diffusivity, where the average is weighted by $|\frac{\partial \rho}{\partial z}|$ so that mean values relate to density fluxes. **b**, Same as **a**, with the bottom 500 m of every water column excluded from the averaging.

immediately above (Fig. 3.10c; see also Chapter 7). The bottom lightening is associated with diapycnal upwelling, whereas the overlying densification implies diapycnal downwelling. Consequently, there is a substantial degree of cancellation between upwelling and downwelling that diminishes the ability of abyssal mixing to maintain a large-scale overturning circulation (de Lavergne et al., 2016; Ferrari et al., 2016; McDougall and Ferrari, 2017).

The dipole of density transformation also applies to the second major source of mixing at depth: downslope or constricted ocean currents carrying dense waters over sills or through straits (Fig. 3.10a; Polzin et al., 1996; Bryden and Nurser, 2003). In these locations, bottom-intensified turbulence draws energy from the flow itself, lightening waters hugging the seafloor while densifying and entraining overlying waters. Sills and straits host some of the largest deep-ocean turbulence levels and diffusivities (Ferron et al., 1998; MacKinnon et al., 2008; Voet et al., 2014). The induced diapycnal transports, although localised and of both signs, are responsible for step changes in bottom ocean properties (Fig. 3.10d) and abyssal stratification following dense water pathways (Mantyla and Reid, 1983; Bryden and Nurser, 2003).

3.5.2 Mesoscale stirring, from top to bottom

The near surface is where stirring by geostrophic turbulence is expected to be most efficient at shaping circulation, for four reasons. First, the competition between variance input by surface tracer fluxes and variance removal by mixing on the global scale (Section 3.2) implies that temperature, salinity and density contrasts tend to decrease with depth. Second, mesoscale stirring rates are surface-intensified (Fig. 3.3c; Ferreira et al., 2005; Cole et al., 2015; Groeskamp et al., 2017; Canuto et al., 2019). Third, isopycnal stirring near the surface can modify the mixed layer heat budget and air-sea buoyancy forcing (Guilyardi et al., 2001; Hieronymus and Nycander, 2013). Last, geometry demands that eddy stirring transitions from isopycnal to along-boundary directions within diabatic boundary layers (Treguier et al., 1997; Ferrari et al., 2008, Chapter 9). Over the depth of the surface mixed layer, mesoscale eddies thus directly affect the density field via horizontal stirring (Danabasoglu et al., 2008).

Horizontal stirring by geostrophic turbulence in the surface mixed layer has an important influence on the transformation and subduction of water masses in certain regions (Robbins et al., 2000; Price, 2001; Groeskamp et al., 2016), thus affecting circulation in the deeper ocean. At the very surface, mesoscale stirring exchanges water across gradients in air-sea fluxes,

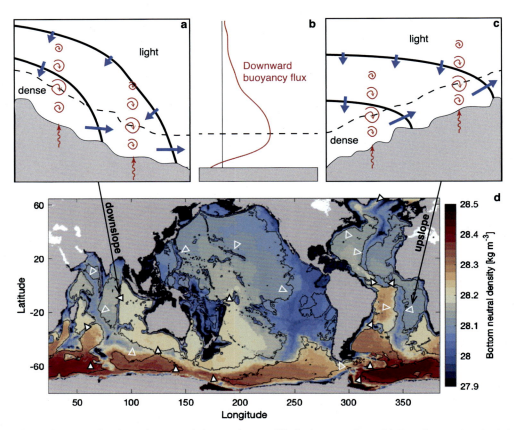

FIGURE 3.10 a–c, Downslope (a) and upslope (c) currents tied to near-bottom diffusive buoyancy fluxes (b). Boundary-catalysed turbulence (red spirals) and geothermal heat fluxes (red wiggly arrows) drive a convergent buoyancy flux within a thin bottom layer (dashed line) and a divergent buoyancy flux above it. The bottom buoyancy gain is balanced by along-slope flow, whereas the buoyancy loss above is balanced by sinking of interior waters (blue arrows). Thick black lines represent density surfaces. d, For illustration, some representative locations of intense cross-density flows (triangles; filled for downslope, empty for upslope) are shown on top of shaded bottom neutral density (Gouretski and Koltermann, 2004). The 4 km bathymetric contour is shown in black. Strong downslope flows filling the abyss are found downstream of major dense water formation sites and sills, whereas upslope cross-density transport is thought to be concentrated along the rough flanks of ridges.

modulating these fluxes and the ocean heat balance without reliance on irreversible mixing, much as sketched in Fig. 3.1a. In practice, however, active three-dimensional turbulence colludes with surface buoyancy forcing to dissipate the density variance produced by mesoscale currents in the surface mixed layer. Horizontal stirring and isotropic mixing thus interact to shape surface water mass transformations.

Below the surface mixed layer, isopycnal mixing can also accomplish density transformations, due to the nonlinear dependence of density on temperature and pressure. There are two separate effects, cabbeling and the thermobaric effect, which arise due to the dependence of the thermal expansion coefficient on temperature and pressure, respectively (McDougall, 1984). When mixing two water parcels with different temperatures, the mixed product is denser than the average of the two initial densities, a process known as cabbeling. Via this effect, isopycnal mixing can cause densification and attendant diapycnal downwelling. Thermobaricity, on the other hand, can cause both lightening and densification, and is active when water parcels move across a substantial pressure range. Hence, for isopycnal stirring to induce significant density transformations, it must mix across relatively large temperature and pressure contrasts. The ACC, whose steep isopycnals coincide with strong eddy activity and contrasting water masses, is one such region. Calculations suggest that isopycnal diffusivities of order 10^3 m^2 s^{-1} are sufficient to cause 5–10 Sv of downwelling in the ACC region (Figs. 3.11a and 3.13a, b; Iudicone et al., 2008; Klocker and McDougall, 2010; Nycander et al., 2015; Groeskamp et al., 2016).

At the bottom boundary, mesoscale stirring again must follow the direction of topography rather than that of isopycnals (Greatbatch and Li, 2000). Since isopycnals intersect the seafloor at right angles, due to the insulating boundary condition (Wunsch, 1970), mesoscale stirring must have a diapycnal component along the bottom topography. This effect, and its coupling with smaller scale turbulence in the bottom boundary layer, has received little attention to date. Whether near-

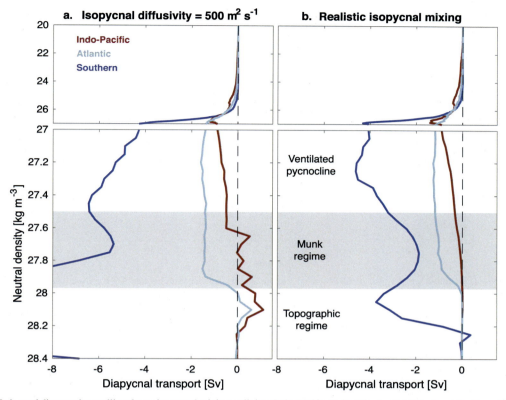

FIGURE 3.11 Estimated diapycnal upwelling due to isopycnal mixing, split into Indo-Pacific (red), Atlantic (light blue) and Southern (dark blue) oceans. The Southern ocean is defined as south of 32°S. **a**, Constant isopycnal diffusivity of 500 m² s⁻¹. **b**, Varying isopycnal diffusivity mapped by Groeskamp et al. (2020). The calculation follows the methodology of Groeskamp et al. (2016) and is based on monthly climatological hydrographic fields from World ocean atlas 2018 (Locarnini et al., 2018; Zweng et al., 2018). Downwelling outside the axis range in panel **a**, at densities greater than 27.8 kg m⁻³, occurs near the Antarctic continent and could be an artifact of poor observational coverage; more realistic diffusivities used in **b** eliminate these large transports.

bottom mesoscale diffusivities are sufficiently large to substantially alter density transformations in the deep ocean is presently unknown.

3.6 Some impacts on basin-scale overturning circulation

The role of mixing in basin-scale overturning circulations is discussed in this section, focusing on quantitative assessments of mixing-driven water mass transformations. We begin with the circulation of the densest ocean waters and move progressively toward lighter layers.

3.6.1 Abyssal overturning cell

Antarctic bottom water (AABW) is the densest global-scale water mass. It is produced around Antarctica, sinks along the Antarctic continental slope and spreads northward to fill most of the ocean deeper than 4 km (Orsi et al., 1999). The northward deepening and ultimate grounding of density surfaces at these depths indicate that AABW becomes progressively lighter as it flows northward (Fig. 3.5). Once lighter than about 28.1 kg m⁻³, it returns southward and ultimately upward in the Antarctic divergence (Toggweiler and Samuels, 1993; Ganachaud and Wunsch, 2000). This circulation loop is often referred to as the abyssal overturning.

What causes the lightening of AABW along its path? The first and principal cause is mixing at sills and straits (Bryden and Nurser, 2003). A map of neutral density at the ocean bottom shows that the density of bottom waters decreases from 28.4 kg m⁻³ or more near Antarctica to 28.1–28.15 kg m⁻³ at the northern end of the Indian, Atlantic and Pacific basins (Fig. 3.10d). This decrease occurs in steps that coincide with narrow passages connecting sub-basins, implicating intense mixing within constricted or overflowing currents (Polzin et al., 1996; Orsi et al., 1999). The second major cause is bottom-intensified mixing by breaking internal waves (de Lavergne et al., 2016; Ferrari et al., 2016), principally internal tides

50 Ocean Mixing

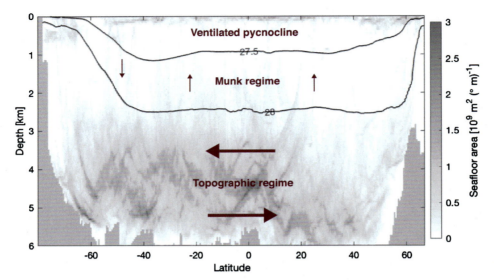

FIGURE 3.12 Proposed circulation regimes. The zonally summed seafloor area (in square metres per unit depth and per latitude degree) is shaded. Black curves represent the pseudo-depth of 27.5 and 28 kg m^{-3} neutral density surfaces, where the pseudo-depth of density surfaces is found by filling each latitude band from the bottom up with ocean grid cells ordered from dense to light. The Munk regime, between the two black contours, hosts moderate mixing-driven vertical circulation. The underlying topographic regime is characterised by northward abyssal flow and southward deep flow, both strongly influenced by topography and near-bottom mixing. The ventilated pycnocline hosts swift upper-ocean flows influenced by mixing near the surface.

(Ledwell et al., 2000; Vic et al., 2019; de Lavergne et al., 2020, Chapter 6). Using a realistic map of mixing fuelled by internal tides (de Lavergne et al., 2020), we estimate that tidal mixing converts about 15 Sv of 28.11–28.2 kg m^{-3} waters into lighter 28.05–28.11 kg m^{-3} waters (Fig. 3.8b). Of this conversion, a third occurs in the Southern ocean (south of 32°S), 6 Sv in the Indo-Pacific and 4 Sv in the Atlantic. A third primary cause of AABW lightening is geothermal heating (Adcroft et al., 2001; Emile-Geay and Madec, 2009). Incorporating the contribution of geothermal heat fluxes mapped by Lucazeau (2019) into diapycnal velocities along the seafloor, we calculate that geothermal heating augments the peak diapycnal upwelling by about 7 Sv globally (Fig. 3.8b). The bulk of this geothermal density transformation occurs in the wide Pacific basin and is focused around 28.11 and 28.03 kg m^{-3} densities, which cover a large fraction of the North Pacific and southeastern Pacific seafloor, respectively (Fig. 3.10d).

These three causes of AABW lightening are not equivalent. Mixing in constrictive passages accounts for most of the overall density contrast traversed by the abyssal overturning circulation. However, its contribution to AABW lightening is restricted to densities greater than 28.15 kg m^{-3}, and is reliant on the existence of the circulation itself. The peak diapycnal upwelling, which dictates the level of meridional flow reversal and the magnitude of the circulation, occurs at lighter densities and is largely driven by breaking internal waves and geothermal heating (de Lavergne et al., 2017). Hence, tidal mixing and geothermal heating are the actual engines of the circulation: they supply potential energy to the flow, whereas overflow mixing consumes the flow's potential energy (Huang, 1999; Bryden and Nurser, 2003). Still, overflow mixing profoundly affects the strength of the circulation by shaping the abyssal stratification, thus modifying internal wave-driven mixing and the upwelling rates induced by geothermal buoyancy gains (Emile-Geay and Madec, 2009; de Lavergne et al., 2016).

An additional difference between geothermal heating and abyssal mixing must be underlined. Geothermal heating provides a net buoyancy gain, moving water toward lighter layers only (Fig. 3.6c). By contrast, abyssal mixing is dominated by bottom-intensified turbulence, which causes near-compensating gains and losses of buoyancy near the bottom (Fig. 3.6c). This compensation limits the ability of bottom-enhanced turbulence to drive a net circulation toward lighter layers. In particular, bottom-intensified turbulence can cause net densification of a deep water mass (de Lavergne et al., 2016), even at the basin scale, as obtained here in the 28.02–28.09 kg m^{-3} density range within the Atlantic (Fig. 3.8b).

Geothermal heating and bottom-enhanced turbulence share one important characteristic: they cause buoyancy gain only along the seafloor. As a result, the net buoyancy gain of a density layer depends closely on its access to the ocean floor (de Lavergne et al., 2016, 2017; Holmes et al., 2018b). In turn, access to the seafloor is strongly constrained by the ocean's geometry (Fig. 3.12): 85% of the seafloor area lies deeper than 2.5 km; 8% lies between 1 and 2.5 km depth. This peculiar depth distribution of the seafloor largely restricts AABW upwelling to depths greater than 2.5 km and densities greater than 28 kg m^{-3} (Figs. 3.7 and 3.8b; de Lavergne et al., 2017). Diapycnal upwelling persists between 1 and 2.5 km depth (27.5–28 kg m^{-3}), but remains relatively constant at a magnitude of a few Sv (Fig. 3.8b). In this depth range, seafloor area is scarce (Fig. 3.12), density surfaces are relatively flat at low and middle latitudes (Figs. 3.5 and 3.7), and the isotropic diffusivity is relatively uniform in the vertical (Figs. 3.3a and 3.9). Circulation thus essentially abides by the one-dimensional recipe of Munk (1966),

$$\omega(z) = K_\rho \frac{\partial^2 \rho}{\partial z^2} \bigg/ \frac{\partial \rho}{\partial z} \approx \text{constant}, \tag{3.10}$$

noting that the (horizontally averaged) vertical velocity and diffusivity are an order of magnitude less than Munk originally proposed. The 1–2.5 km depth range, or equivalently the 27.5–28 kg m^{-3} neutral density range of the modern ocean, could therefore be named the "Munk regime". This regime hosts a relatively weak and weakly divergent diapycnal circulation, in contrast to the underlying "topographic regime", where circulation is shaped by basin geometry and near-bottom mixing (Fig. 3.12).

Would the abyssal overturning persist in the absence of mixing? The presence of geothermal heating implies that it should persist so long as dense AABW is produced around Antarctica. Everything else equal, the abyssal ocean would be expected to become very dense, almost homogeneous, and traversed by a relatively swift circulation necessary to balance the steady geothermal buoyancy gain (Emile-Geay and Madec, 2009). The abyssal circulation would have a structure broadly similar to that observed—constrained by topography and wind-driven upwelling—but would cross a much smaller density range. This thought experiment suggests that the primary impact of mixing along the AABW path is to reduce the density of bottom waters and to increase the abyssal stratification. These effects have global repercussions throughout the water column.

3.6.2 North Atlantic Deep Water circulation

The second major source of dense water in today's ocean is situated in the subpolar North Atlantic. Deep convective mixing in the Labrador sea forms small amounts of dense water that participate in regional circulation and ventilate the deep ocean, but contribute little to global overturning (Pickart and Spall, 2007; Lozier et al., 2019). Denser North Atlantic deep water (NADW) overflows at the submarine ridges that connect Scotland, Iceland and Greenland (Dickson and Brown, 1994). This relatively salty deep water mass then traverses the whole Atlantic, flows along and across the ACC and reaches the near surface at high southern latitudes (Talley, 2013; Tamsitt et al., 2017). A sinuous return flow in the upper ocean closes the circulation (Gordon, 1986; Talley, 2013), usually referred to as the Atlantic meridional overturning circulation (AMOC).

The pole-to-pole journey of NADW is generally conceptualised as a largely adiabatic circulation (Toggweiler and Samuels, 1998; Wolfe and Cessi, 2011; Nikurashin and Vallis, 2012). However, observation-based tracer and mass budgets suggest that NADW undergoes substantial mixing with overlying and underlying waters, both within and outside the Atlantic basin (Talley, 2013; Naveira Garabato et al., 2014). Nordic overflows are the first and primary mixing hotspot, and cause NADW to considerably increase in volume while decreasing in density (Dickson and Brown, 1994; Lumpkin and Speer, 2003). This reduction of NADW's density is critical, because the difference in density between AABW and NADW controls their respective volumes of influence and the depth of the AMOC (Galbraith and de Lavergne, 2019; Sun et al., 2020).

Mixing-driven density transformations further downstream in the Atlantic depend in large part on the depth of NADW and its proximity to rough topography. The bulk of the southward flow occurs at depths greater than 2 km between densities 28 and 28.15 kg m^{-3} (Fig. 3.7b; Cunningham et al., 2007; Talley, 2013), whereas the bulk of the Atlantic seafloor area lies at depths greater than 3 km or densities above 28.05 kg m^{-3} (de Lavergne et al., 2017). A sizeable portion of NADW is thus subject to strong near-bottom diapycnal transports. In the western Atlantic south of 45°N, bottom-enhanced turbulence is expected to make NADW colder and denser on average, because AABW covers most of the seafloor and tends to monopolise buoyancy gain. The strongest mixing of NADW with denser AABW may occur within the low-latitude fracture zones that channel flow from the western to the eastern Atlantic (Polzin et al., 1996; Mercier and Speer, 1998; Demidov et al., 2007). Outflow from these channels is dominated by NADW, so that NADW occupies the deep eastern Atlantic down to the bottom (Sarmiento et al., 2007; de Lavergne et al., 2017). Mixing and geothermal heating in the abyssal eastern Atlantic thus serve to convert some of the densest (>28.11 kg m^{-3}) NADW into lighter deep water. This conversion contributes to the temperature shift of NADW transport observed across the equator (Friedrichs et al., 1994).

What is the net effect of these density transformations on NADW? Over the whole Atlantic between 32°S and 60°N, we estimate that the impact of tidal mixing is to make NADW more homogeneous: about 5 Sv of 28.07–28.11 kg m^{-3} water are produced at the expense of denser and lighter categories (Fig. 3.8b). We find little upwelling of NADW across the mid-depth stratification and into the northward flowing upper branch of the circulation: only about 1 Sv of 28–28.03 kg m^{-3} water is converted into <27.6 kg m^{-3} water (Fig. 3.8b). The actual amount of NADW upwelling into the Atlantic pycnocline may be larger, notably due to non-tidal sources of mixing that may dominate near the basin's western boundary (Zhai et al., 2010; Clément et al., 2016). Nonetheless, presently estimated diapycnal transports back the notion that the vast majority of NADW is exported all the way to the Southern ocean (Toggweiler and Samuels, 1995; Gnanadesikan, 1999; Talley, 2013).

3.6.3 Southern Ocean upwelling: adiabatic or diabatic?

Deep waters flowing out of the Pacific, Indian and Atlantic basins are thought to upwell along the sloping isopycnals of the Southern ocean (Fig. 3.5; Toggweiler and Samuels, 1993; Marshall, 1997; Sloyan and Rintoul, 2001; Marshall and Speer, 2012). This upwelling is generally called adiabatic, in the sense that it is density-preserving (see Chapter 12). However, the isopycnal flow crosses isotherms and isohalines (Zika et al., 2009; Naveira Garabato et al., 2016; Tamsitt et al., 2018). This implicates mixing, enhanced by geostrophic and small-scale turbulence, in the temperature and salinity modifications of deep waters along their isopycnal upwelling path.

Diapycnal upwelling in the ACC has been suggested to play a role in returning deep waters to the surface. The main supporting evidence comes from the observed rapid spreading of passive tracers across isopycnals in the Atlantic sector of the ACC (Naveira Garabato et al., 2007; Watson et al., 2013). Within this sector, the measured spreading rate is consistent with a mid-depth isotropic diffusivity of order 10^{-4} m^2 s^{-1}. Applied to the climatological stratification of the Southern ocean, a uniform mixing rate of this magnitude would cause about 10 Sv of diapycnal upwelling south of 32°S (Fig. 3.8a). However, the actual diapycnal flow is most certainly only a fraction of this rate, because diapycnal tracer spreading is slower in less hilly sectors of the ACC (Ledwell et al., 2011; Watson et al., 2013), and because the effective mixing rate of a passive tracer can far exceed the effective mixing of density (Mashayek et al., 2017). Indeed, passive tracers tend to hover in regions of weak flow, elevated mixing and reduced stratification near topographic obstacles (Mashayek et al., 2017). Microstructure observations of turbulence in the Scotia sea suggest that the effective mixing rate of density is much weaker than the diffusivity inferred from passive tracer measurements (St. Laurent et al., 2012; Sheen et al., 2013).

Diapycnal downwelling of deep waters is equally plausible. Observations indicate that energetic turbulence in the ACC is often bottom-intensified. If dominated by bottom-enhanced turbulence, isotropic mixing could cause a few Sv of net diapycnal downwelling of circumpolar deep waters (Melet et al., 2014; de Lavergne et al., 2016). Estimated diapycnal transports due to tidal mixing south of 32°S indicate that upwelling dominates, but remains weak at densities less than 28.05 kg m^{-3} (Fig. 3.8b). Net diapycnal downwelling may also arise from isopycnal mixing and the non-linearity of the equation of state (Iudicone et al., 2008; Klocker and McDougall, 2010; Groeskamp et al., 2016). We estimate that cabbeling and thermobaricity combined cause diapycnal downwelling, varying between 2 and 4 Sv at densities larger than 27.5 kg m^{-3} in the Southern ocean (Fig. 3.11b). This downwelling may act to shift the upwelling toward larger densities or to increase the northward flow of AABW.

Hence, net diapycnal flow of deep waters in the ACC is most likely an order of magnitude weaker than the overall upwelling rate, thought to be between 20 and 30 Sv (Lumpkin and Speer, 2007; Naveira Garabato et al., 2014). Upwelling does become diapycnal near the surface, however: alongside surface buoyancy forcing, mixing by three-dimensional turbulence plays an essential role in the entrainment of deep waters into the surface mixed layer and the diabatic closure of the Southern ocean overturning (Gordon and Huber, 1990; Iudicone et al., 2008; Abernathey et al., 2016; Evans et al., 2018).

3.6.4 The return flow to the North Atlantic

NADW formation is the conversion of about 15 Sv of water from the ventilated pycnocline (densities <27.5 kg m^{-3}) into denser waters (Lumpkin and Speer, 2003). The compensating conversion can occur either through mixing-driven upwelling at low and middle latitudes (Munk and Wunsch, 1998), or through near-surface lightening in the Southern ocean (Toggweiler and Samuels, 1993). We estimate that isotropic mixing causes about 3 Sv of upwelling across the 27.5 kg m^{-3} isopycnal (Figs. 3.8b and 3.13c), whereas isopycnal mixing induces an opposite transport of similar magnitude (Figs. 3.11b and 3.13a, b). These estimates imply that the net mass gain of the ventilated pycnocline occurs almost exclusively near the surface at southern high latitudes (Gnanadesikan, 1999). A return flow of about 15 Sv must therefore exist from the Antarctic source of light water (<27.5 kg m^{-3}) to its sink in the northern North Atlantic.

The first stage of this return route is the formation of mode and intermediate waters on the northern side of the ACC. Mixing in the surface boundary layer and immediately below plays an essential role in the formation of these subantarctic waters (McCartney, 1977; Sloyan and Rintoul, 2001; Iudicone et al., 2008; Sloyan et al., 2010). Cabbeling has also been highlighted as an important contributor to the formation of intermediate waters (Urakawa and Hasumi, 2012; Nycander et al., 2015; Groeskamp et al., 2016). Along the northern flank of the ACC, cold and fresh southern waters come close to warmer and saltier subtropical waters within an active mesoscale eddy field (Abernathey et al., 2010). We estimate that the resultant mixing along isopycnals forms about 2 Sv of intermediate water in the 27.25–27.5 kg m^{-3} range (Figs. 3.11b and 3.13a, b). This number is somewhat lower than previous estimates. We attribute the difference primarily to the suppression of mesoscale stirring by mean currents (Ferrari and Nikurashin, 2010). This suppression effect, included in the employed map of mesoscale diffusivities (Groeskamp et al., 2020), limits the intensity of isopycnal mixing and associated density transformations in the upper kilometre of the ACC (Fig. 3.3c).

A large fraction of subducted subantarctic waters feeds the return branch of the AMOC (Schmitz, 1995). In the subtropical North Atlantic, this return branch is dominated by near-surface waters that are about 15 °C warmer than subantarctic waters (Schmitz and Richardson, 1991). This implies that substantial warming and lightening must occur along the journey (Fig. 3.5). Inverse box models suggest that the bulk of this transformation occurs at low latitudes in the eastern Pacific and Atlantic and relies largely on isotropic mixing (Lumpkin and Speer, 2003; Sloyan et al., 2003). Elevated mixing in the upper layers of the eastern equatorial Pacific could be an important contributor to diapycnal upwelling of subantarctic water (Gregg et al., 1985; Smyth and Moum, 2013; Holmes et al., 2018a). However, it remains unclear whether observed mixing is sufficient to accommodate the required cold to warm conversion. Toggweiler et al. (2019a,b) hypothesise that most of the conversion is actually achieved by direct atmospheric forcing near eastern margins. They propose that the AMOC indirectly draws warm water westward in the Pacific and Atlantic, exposing cool subantarctic water to surface heating at the eastern end of the basins. Their mechanism is well illustrated by the schematic of Fig. 3.1b (equating the cold water exposed to heating in the south with subantarctic water exposed at eastern margins) and alleviates the requirement for strong low-latitude mixing beneath and across the thermocline.

3.6.5 Shallow hemispheric cells

In addition to the return branch of the AMOC, the upper ocean hosts several closed overturning cells. The most prominent cells inhabit the top few hundred metres of subtropical oceans. These subtropical cells involve subduction poleward of about 20° and upwelling near the equator, connected by poleward surface Ekman flow and equatorward subsurface geostrophic flow (Roemmich, 1983; Luyten et al., 1983; McCreary and Lu, 1994; McPhaden and Zhang, 2002). Convective mixing due to surface buoyancy loss shapes the rates and patterns of subduction (Marshall and Nurser, 1991; McCreary and Lu, 1994; Qu et al., 2013), whereas shear-driven mixing above the equatorial undercurrent in the Pacific and Atlantic contributes to lighten and return to the surface the upwelling waters (Lu et al., 1998; Moum et al., 2009; Hummels et al., 2013). Overall, subsurface isotropic mixing plays a key role in regulating the net overturning transport, that is, the residual of opposing Ekman and eddy-driven circulations (Henning and Vallis, 2004; Doddridge et al., 2016).

Perhaps more importantly, climate model experiments have shown that SSTs, MLDs and the vertical structure of the thermocline are sensitive to the representation of isotropic mixing rates in the low-latitude upper ocean (Jochum et al., 2013; Melet et al., 2016; Zhu and Zhang, 2019; Hieronymus et al., 2019, Chapter 2). Mixing near the surface has a weighty influence on air-sea interactions (Moum et al., 2013; Jochum et al., 2013; Zhu and Zhang, 2019), whereas mixing in the interior affects the heat content of the thermocline and heat transport by the AMOC (Melet et al., 2016; Holmes et al., 2019; Hieronymus et al., 2019). Hence, the direct impacts of mixing on the ocean's temperature distribution have ripple-like effects on circulation within the thermocline and beyond.

3.7 Some impacts on basin-scale horizontal circulation

We define horizontal circulations as networks of zonal and meridional currents integrated over a chosen depth range. An overturning circulation generally has a signature in horizontal circulation: for example, the Gulf stream participates in both the AMOC and in the North Atlantic gyre circulation. The impacts outlined in the previous section thus have counterparts for circulation in the horizontal plane. We expand below on the consequences of mixing for the large-scale horizontal flow of the upper ocean, the abyss and the ACC.

3.7.1 Upper-ocean gyres

Sverdrup balance (3.7) suggests that depth-integrated flow is controlled by the wind stress curl. Munk (1950) first tested this prediction using observed winds. He introduced an additional Reynolds stress term (able to accommodate concentrated boundary currents) and showed that the main wind patterns create cyclonic subpolar gyres and anticyclonic subtropical gyres that resemble observed upper-ocean currents. This finding demonstrated that Ekman pumping is a major determinant of gyre circulations in upper layers of the global ocean. It was a sign that other factors, such as buoyancy forcing and mixing, are less important.

54 Ocean Mixing

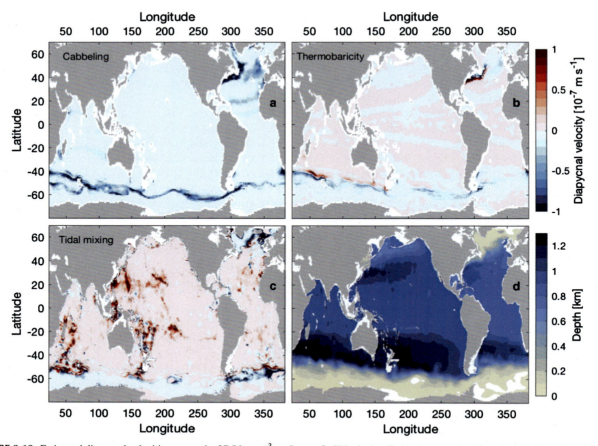

FIGURE 3.13 Estimated diapycnal velocities across the 27.5 kg m^{-3} surface. **a, b,** Velocity implied by isopycnal mixing via cabbeling (**a**) and thermobaricity effects (**b**), calculated according to Groeskamp et al. (2016), using World ocean atlas 2018 monthly hydrography (Locarnini et al., 2018; Zweng et al., 2018) and isopycnal diffusivities from Groeskamp et al. (2020). **c,** Velocity implied by tidal mixing calculated according to de Lavergne et al. (2016) using isotropic diffusivities from de Lavergne et al. (2020). **d,** Depth of the 27.5 kg m^{-3} neutral density surface in the climatology of Gouretski and Koltermann (2004). This density surface was chosen to separate the ventilated pycnocline from the Munk regime (Fig. 3.12).

When applying (3.7), Munk (1950) assumed that vertical motion vanishes near 1 km depth, so that squeezing and stretching of the ocean's top kilometre is set by Ekman pumping velocities at the base of the Ekman layer. Could deep vertical motion sustained by mixing and overturning alter the rates of squeezing and stretching that shape upper-ocean gyres? Ekman pumping velocities are of order 10^{-6} m s^{-1} or 30 m per year (Roquet et al., 2011). In Fig. 3.13, we map the velocity across the 27.5 kg m^{-3} isopycnal that is necessary to balance estimated mixing-driven density transformations. Diapycnal velocities reach values of order 10^{-6} m s^{-1} only in specific hotspots of water mass transformation or near the outcropping or incropping of the considered isopycnal. Over the vast majority of the upper ocean, Ekman pumping is therefore expected to dominate, suggesting that mesoscale stirring and tidal mixing play only a small role in driving gyres via squeezing and stretching of the ventilated pycnocline.

Regional hotspots of water mass transformation can nonetheless have major local and remote impacts on upper-ocean gyres, mediated by changes in regional density structure and mass balance. For example, mixing in the Nordic overflows has a first-order influence on the horizontal upper-ocean circulation of the whole North Atlantic (Lumpkin and Speer, 2003; Zhang et al., 2011). Large tidal mixing rates in the western Pacific and Indonesian archipelago impact the throughflow between—and the gyres within—Pacific and Indian basins (Koch-Larrouy et al., 2010; Sasaki et al., 2018). In general, impacts of mixing on the AMOC have repercussions for the basin-scale horizontal circulation of the upper ocean (Toggweiler et al., 2019b).

Mixing near the surface plays a direct and widespread role. Subtropical gyres are linked to subtropical overturning cells (McCreary and Lu, 1994; Samelson and Vallis, 1997) and are therefore influenced by near-surface mixing via the mechanisms described in Section 3.6.5. Subpolar gyres conform less to Sverdrup balance than their subtropical twins: they are more strongly influenced by non-linear dynamics and surface buoyancy forcing (Bryan et al., 1995; Su et al., 2014;

Le Corre et al., 2020). In the Southern Hemisphere, near-surface meridional density gradients are the leading control on the strength of subpolar gyres simulated by climate models (Wang and Meredith, 2008). These gradients, primarily set by freshwater and heat exchanges with the atmosphere and cryosphere, are modulated by mixing (Pellichero et al., 2017; Thompson et al., 2018).

3.7.2 The Stommel and Arons circulation

Horizontal circulation patterns in the deep ocean have received comparatively little attention. The reference theory for these patterns was proposed sixty years ago by Stommel (1958) and Stommel and Arons (1959a,b). This theory is based on Eq. (3.6), integrated from the seafloor ($z = -H$) to a chosen abyssal depth ($z = -h_{top}$):

$$\beta \int_{z=-H}^{z=-h_{top}} v\, dz = f[w(z=-h_{top}) - w(z=-H)]. \tag{3.11}$$

Stommel (1958) reasoned that the production and sinking of cold waters near Greenland and Antarctica should be balanced by downward diffusion of heat and upwelling across stratification at lower latitudes. Using (3.11) and assuming that the upwelling velocity at 2 km depth exceeds that at the bottom, Stommel mapped mass transports below 2 km as broadly poleward, except near western boundaries, where fast currents governed by different dynamics were assumed to close the mass balance.

The circulation patterns drawn by Stommel (1958) and Stommel and Arons (1959a,b) have been shown to hold in idealised flat-bottom model oceans, where a deep overturning is maintained by downward diffusion of buoyancy at a uniform rate (Samelson and Vallis, 1997). Observational inferences of deep circulation have invariably revealed a different and much more complex picture (Friedrichs and Hall, 1993; Hautala and Riser, 1993; Reid, 1997, 2003). Reasons are manifold. First, the theory assumes that the deep ocean hosts divergent vertical motion everywhere away from sinking regions. In reality, the area-integrated upwelling rate of deep waters at low and middle latitudes increases with height only up to about 4 km depth, then decreases with height until about 2.5 km depth (Fig. 3.8b; de Lavergne et al., 2017). Meanwhile, the ocean area increases markedly, so that the area-averaged upwelling tends to be less divergent than its area-integrated counterpart (McDougall, 1989; Rhines, 1993). Hence, convergent vertical motion favouring equatorward flow may be more common than the reverse. Moreover, upwelling is not horizontally uniform: it is a balance of upward and downward velocities that are largest near topography (Fig. 3.10). Bottom-intensified mixing entails complex patterns of squeezing and stretching that depend on local characteristics of topography and turbulence (St. Laurent et al., 2001; McDougall and Ferrari, 2017).

More importantly, horizontal flow in the abyss is strongly constrained by topography and is influenced by a range of dynamics not reflected in the combination of (3.9) and (3.11) (Holland, 1978; Garrett, 1991; Pedlosky, 1992; Callies, 2018; Naveira Garabato et al., 2019; Yang et al., 2020; see also Chapters 7 and 8). How mixing interacts with such dynamics to shape abyssal current systems is expected to vary between regions and remains little studied.

3.7.3 The Antarctic Circumpolar Current

The ACC transports over 130 Sv eastward as it circumnavigates Antarctica (Fig. 3.4; Meredith et al., 2011). The current has its largest speeds at the surface, but extends down to the seafloor (Peña-Molino et al., 2014). It is deemed to be driven in large part by Southern Hemisphere westerly winds (Munday et al., 2011; Howard et al., 2014). Indeed, if the global ocean were mixed to a single temperature and salinity, a full-depth ACC would likely persist as a conduit between momentum input at the surface and removal into the solid earth.

Strong contrasts in temperature, salinity and density are actually observed across the ACC. The zonal mass transport of the ACC can therefore be decomposed into a depth-independent component equal to the bottom velocity and a depth-dependent component obtained by vertical integration of the thermal current shear (3.8). This second component is directly linked to the meridional density difference across the ACC and accounts for about 85% of the total eastward transport (Peña-Molino et al., 2014). Any impact of mixing on meridional density gradients in the Southern ocean thus has implications for the strength and vertical structure of the ACC.

For instance, larger background rates of isotropic mixing tend to lower isopycnals north of the ACC, and thereby increase the thermal current shear and ACC transport (Munday et al., 2011). Likewise, deep convective mixing south of the ACC can enhance the north-south density gradient and accelerate the ACC (Behrens et al., 2016). Bottom-intensified mixing in the abyss has also been found to alter the ACC density structure and flow, in ways that depend on the latitudinal distribution of that mixing (Jayne, 2009; Melet et al., 2014).

Isopycnal mixing may also affect the ACC strength. Diapycnal downwelling due to cabbeling and thermobaricity, by acting as a sink of volume for the ventilated pycnocline, could contribute to raising isopycnals north of the ACC and slowing the circumpolar flow. Furthermore, isopycnal mixing can alter air-sea interactions at high latitudes and, via the induced changes in wind stress, convection and stratification, modify the ACC (Ragen et al., 2020).

Mixing can further influence the ACC by changing the strength of the meridional overturning circulation, without necessarily changing meridional density gradients within the current. Indeed, cross-stream flows affect the zonal momentum balance of the ACC (Gent et al., 2001; Howard et al., 2014; Stewart and Hogg, 2017). In particular, an increased abyssal overturning strength (such as may result from increased deep mixing) has the potential to accelerate the ACC as follows: the Coriolis force acts to deflect the northward AABW flow to the west, and the southward return flow to the east; the westward abyssal momentum is damped by topographic form stress and bottom friction; a net gain of depth-integrated eastward momentum ensues (Howard et al., 2014).

The vertical and zonal extent of the ACC makes it a crossroads of global ocean circulation (Rintoul and Naveira Garabato, 2013, Chapter 12). The ACC's structure and intensity are consequently tied to water mass transformations in all parts of the World ocean.

3.8 Conclusions

Munk (1966) first conjectured that mixing between water masses occurs mostly where these water masses meet the surface and the seafloor. Here we estimated that isotropic and isopycnal mixing in the ocean interior (defined as in Section 3.5) gives rise to diapycnal circulations of only a few Sverdrups. Rates of basin-scale overturning are thought to be an order of magnitude larger (Ganachaud and Wunsch, 2000; Lumpkin and Speer, 2007). This implies that Munk was right, and the diapycnal component of global ocean circulation is largely confined to near-surface and near-bottom regions. Although this notion is long established (Munk and Wunsch, 1998; Wunsch and Ferrari, 2004), its consequences for the structure of ocean circulation remain under-appreciated.

The first and foremost consequence of boundary-intensified mixing is the organisation of circulation by outcrop and incrop areas—that is, by water masses' access to boundary regions. Three main regimes can be identified (Fig. 3.12): (i) a ventilated pycnocline, where air-sea exchanges and mixing near surface outcrops govern the structure and rate of circulation; (ii) a topographic regime, where abundant seafloor deeper than 2.5 km leads to substantial near-bottom density transformations; and (iii) an intervening Munk regime, more isolated from boundaries, where interior mixing maintains a modest diapycnal circulation. In the topographic regime, circulation is both strong and strongly influenced by near-bottom mixing. In the Munk regime, basin-averaged circulation and isotropic mixing rates are relatively weak (Figs. 3.7–3.9). In the ventilated pycnocline, circulation is strong and strongly influenced by mixing near the surface, yet largely along-isopycnal in the interior.

A related consequence is the relationship between seafloor geometry and overturning circulation. Munk and Wunsch (1998), in their calculation of the effective diffusivity needed to upwell dense waters from 4 km to 1 km depth, did not account for the change with depth of the ocean's area. Actually, the small change of the ocean's area between 1 and 2.5 km depth impedes near-boundary diapycnal upwelling, whereas its rapid decrease at greater depths allows large diapycnal transports (de Lavergne et al., 2017). The depth distribution of the seafloor thus places a primary constraint on the structure of the overturning circulation. The compartmentalisation of the deep ocean into subbasins connected by sills and straits exerts an additional and essential constraint (Bryden and Nurser, 2003). A third crucial geometric ingredient is the interruption of north-south continental barriers in the Southern ocean, which favours deep southward flow across the ACC (Toggweiler and Samuels, 1995). Combined, these three ingredients lead to the simplified depiction in density-depth space of a bathymetrically constrained overturning (Fig. 3.6d).

Both NADW and AABW undergo larger density losses during their descent than their ascent, and both benefit from isopycnal Southern ocean upwelling to come back to the surface (Fig. 3.6d). Differences between the abyssal overturning cell and the AMOC do nevertheless exist. The abyssal overturning cell inhabits the topographic regime and can be considered essentially diabatic, in that its existence relies on the lightening of AABW at depth (Nikurashin and Vallis, 2012). NADW is partially embedded in the topographic regime and undergoes substantial transformation by near-bottom mixing (Section 3.6.2). However, density losses that are essential to the closure of the AMOC are believed to occur near the surface at southern high latitudes (Toggweiler and Samuels, 1998; Marshall and Speer, 2012) and at low latitudes (Toggweiler et al., 2019a,b). The AMOC may thus be considered as more adiabatic, insofar as its existence and structure depend less on mixing below the near-surface region.

We posited that mixing in the ocean interior has less of an influence on circulation than near-boundary mixing, because it causes comparatively weak diapycnal flows. However, interior mixing does impact circulation in several indirect and important ways. In particular, isopycnal mixing catalysed by mesoscale eddies modifies the temperature and salinity of water masses within the ocean interior. These modifications then affect air-sea interactions where the modified water masses outcrop (Guilyardi et al., 2001; Hieronymus and Nycander, 2013; Ragen et al., 2020). In addition, weak rates of isotropic mixing in the voluminous ocean interior exert an important influence on the stratification and heat balance of the upper ocean (Melet et al., 2016; Holmes et al., 2018a; Hieronymus et al., 2019). Altered density and temperature distributions then impact the structure and strength of horizontal and overturning circulations (Sasaki et al., 2018; Zhu and Zhang, 2019).

Mixing in the ocean interior is also essential for ventilation—that is, for the circulation of tracers (rather than the circulation of mass, which is the subject of this chapter). Tracer distributions are influenced by mixing via the impacts of mixing on ocean currents, but they are also directly impacted by mixing. These direct impacts are tracer-specific, since diffusive tracer fluxes depend on tracer gradients in addition to diffusivities. For example, small isotropic diffusivities can cause weak buoyancy fluxes, but large tracer fluxes. Weak mixing rates in the ocean interior can thus maintain important diapycnal tracer fluxes, even in the absence of a diapycnal circulation. Furthermore, isopycnal mixing in the interior plays a key role in shaping global tracer distributions (Ledwell et al., 1998; Jones and Abernathey, 2019) and is able to dominate over ventilation by the large-scale mean currents (Holzer and Primeau, 2006; Naveira Garabato et al., 2017). As such, interior mixing participates in setting the global state of climate and marine ecosystems in multiple and often underrated ways.

Understanding how circulation and mixing together establish the pathways and timescales of ocean ventilation constitutes a major and central challenge to this day. The recent advent of global three-dimensional maps of isotropic and isopycnal diffusivities (Fig. 3.3) opens up avenues for headway. However, these maps are incomplete and insufficiently constrained. New field measurements, coupled with research into the physics and energetics of turbulence across scales, are called for to incorporate all leading-order processes into comprehensive and realistic maps. Mechanistic understanding of the energy routes from forcing to circulation to mixing is essential to construct models that not only capture the observed mixing distributions, but also evolve these distributions consistently with changing boundary conditions (Eden et al., 2014). The path to faithful, conservative representation of the energy cycle (Fig. 3.2) in ocean models is long but vital to confidently probe and project ocean ventilation and its role in the climate system.

References

Abernathey, R., Marshall, J., Mazloff, M., Shuckburgh, E., 2010. Enhancement of mesoscale eddy stirring at steering levels in the Southern Ocean. J. Phys. Oceanogr. 40, 170–184.

Abernathey, R.P., Cerovecki, I., Holland, P.R., Newsom, E., Mazloff, M., Talley, L.D., 2016. Water-mass transformation by sea ice in the upper branch of the Southern Ocean overturning. Nat. Geosci. 9, 596–601.

Adcroft, A., Scott, J.R., Marotzke, J., 2001. Impact of geothermal heating on the global ocean circulation. Geophys. Res. Lett. 28, 1735–1738.

Alford, M.H., 2020. Revisiting near-inertial wind-work: slab models, relative stress, and mixed layer deepening. J. Phys. Oceanogr. 50, 3141–3156.

Alford, M.H., Gregg, M.C., 2001. Near-inertial mixing: modulation of shear, strain and microstructure at low latitude. J. Geophys. Res. 106, 16947–16968.

Behrens, E., Rickard, G., Morgenstern, O., Martin, T., Osprey, A., Joshi, M., 2016. Southern Ocean deep convection in global climate models: a driver for variability of subpolar gyres and Drake Passage transport on decadal timescales. J. Geophys. Res. 121, 3905–3925.

Bessières, L., Madec, G., Lyard, F., 2008. Global tidal residual mean circulation: does it affect a climate OGCM? Geophys. Res. Lett. 35, L03609.

Bryan, F.O., Böning, C.W., Holland, W.R., 1995. On the midlatitude circulation in a high-resolution model of the North Atlantic. J. Phys. Oceanogr. 25, 289–305.

Bryden, H.L., Nurser, A.J.G., 2003. Effects of strait mixing on ocean stratification. J. Phys. Oceanogr. 33, 1870–1872.

Callies, J., 2018. Restratification of abyssal mixing layers by submesoscale baroclinic eddies. J. Phys. Oceanogr. 48, 1995–2010.

Canuto, V.M., Cheng, Y., Howard, A.M., Dubovikov, M.S., 2019. Three-dimensional, space-dependent mesoscale diffusivity: derivation and implications. J. Phys. Oceanogr. 49, 1055–1074.

Charney, J.G., 1947. The dynamics of long waves in a baroclinic westerly current. J. Meteorol. 4, 136–162.

Clément, L., Frajka-Williams, E., Sheen, K.L., Brearley, J.A., Naveira Garabato, A.C., 2016. Generation of internal waves by eddies impinging on the western boundary of the North Atlantic. J. Phys. Oceanogr. 46, 1067–1079.

Cole, S.T., Wortham, C., Kunze, E., Owens, W.B., 2015. Eddy stirring and horizontal diffusivity from Argo float observations: geographic and depth variability. Geophys. Res. Lett. 42, 3989–3997.

Cunningham, S.A., Kanzow, T., Rayner, D., Baringer, M.O., Johns, W.E., Marotzke, J., Longworth, H.R., Grant, E.M., Hirschi, J.J.-M., Beal, L.M., Meinen, C.S., Bryden, H.L., 2007. Temporal variability of the Atlantic meridional overturning circulation at 26.5°N. Science 317, 935–938.

Danabasoglu, G., Ferrari, R., McWilliams, J.C., 2008. Sensitivity of an ocean general circulation model to a parameterization of near-surface eddy fluxes. J. Climate 21, 1192–1208.

Davis, R.E., 1994. Diapycnal mixing in the ocean: the Osborn–Cox Model. J. Phys. Oceanogr. 24, 2560–2576.

de Boyer Montégut, C., Madec, G., Fischer, A.S., Lazar, A., Iudicone, D., 2004. Mixed layer depth over the global ocean: an examination of profile data and a profile-based climatology. J. Geophys. Res. 109, C12003.

de Lavergne, C., Falahat, S., Madec, G., Roquet, F., Nycander, J., Vic, C., 2019. Toward global maps of internal tide energy sinks. Ocean Model. 137, 52–75.

de Lavergne, C., Madec, G., Le Sommer, J., Nurser, A.J.G., Naveira Garabato, A.C., 2016. On the consumption of Antarctic Bottom Water in the abyssal ocean. J. Phys. Oceanogr. 46, 635–661.

de Lavergne, C., Madec, G., Roquet, F., Holmes, R.M., McDougall, T.J., 2017. Abyssal ocean overturning shaped by seafloor distribution. Nature 551, 181–186.

de Lavergne, C., Vic, C., Madec, G., Roquet, F., Waterhouse, A.F., Whalen, C.B., Cuypers, Y., Bouruet-Aubertot, P., Ferron, B., Hibiya, T., 2020. A parameterization of local and remote tidal mixing. J. Adv. Model. Earth Syst. 12, e2020MS002065.

Demidov, A.N., Dobrolyubov, S.A., Morozov, E.G., Tarakanov, R.Y., 2007. Transport of bottom waters through the Vema Fracture Zone in the Mid-Atlantic ridge. Dokl. Earth Sci. 416, 1120–1124.

Dickson, R.R., Brown, J., 1994. The production of North Atlantic Deep Water: sources, rates, and pathways. J. Geophys. Res. 99, 12319–12341.

Doddridge, E.W., Marshall, D.P., Hogg, A.McC., 2016. Eddy cancellation of the Ekman cell in subtropical gyres. J. Phys. Oceanogr. 46, 2995–3010.

Dufour, C.O., Griffies, S.M., de Souza, G.F., Frenger, I., Morrison, A.K., Palter, J.B., Sarmiento, J.L., Galbraith, E.D., Dunne, J.P., Anderson, W.G., Slater, R.D., 2015. Role of mesoscale eddies in cross-frontal transport of heat and biogeochemical tracers in the Southern Ocean. J. Phys. Oceanogr. 45, 3057–3081.

Duhaut, T.H.A., Straub, D.N., 2006. Wind stress dependence on ocean surface velocity: implications for mechanical energy input to ocean circulation. J. Phys. Oceanogr. 36, 202–211.

Eady, E.T., 1949. Long waves and cyclone waves. Tellus 1, 33–52.

Eden, C., Czeschel, L., Olbers, D., 2014. Toward energetically consistent ocean models. J. Phys. Oceanogr. 44, 3160–3184.

Emile-Geay, J., Madec, G., 2009. Geothermal heating, diapycnal mixing and the abyssal circulation. Ocean Sci. 5, 203–217.

Evans, D.G., Zika, J.D., Garabato, A.C.N., Nurser, A.J.G., 2018. The cold transit of Southern Ocean upwelling. Geophys. Res. Lett. 45, 13386–13395.

Ferrari, R., Mashayek, A., McDougall, T.J., Nikurashin, M., Campin, J.-M., 2016. Turning ocean mixing upside down. J. Phys. Oceanogr. 46, 2239–2261.

Ferrari, R., McWilliams, J.C., Canuto, V.M., Dubovikov, M., 2008. Parameterization of eddy fluxes near oceanic boundaries. J. Climate 21, 2770–2789.

Ferrari, R., Nikurashin, M., 2010. Suppression of eddy diffusivity across jets in the Southern Ocean. J. Phys. Oceanogr. 40, 1501–1519.

Ferreira, D., Marshall, J., Heimbach, P., 2005. Estimating eddy stresses by fitting dynamics to observations using a residual-mean ocean circulation model and its adjoint. J. Phys. Oceanogr. 35, 1891–1910.

Ferron, B., Mercier, H., Speer, K., Gargett, A., Polzin, K., 1998. Mixing in the Romanche fracture zone. J. Phys. Oceanogr. 28, 1929–1945.

Friedrichs, M.A.M., Hall, M.M., 1993. Deep circulation in the tropical North Atlantic. J. Mar. Res. 51, 697–736.

Friedrichs, M.A.M., McCartney, M.S., Hall, M.M., 1994. Hemispheric asymmetry of deep water transport modes in the western Atlantic. J. Geophys. Res. 99, 25165–25179.

Galbraith, E., de Lavergne, C., 2019. Response of a comprehensive climate model to a broad range of external forcings: relevance for deep ocean ventilation and the development of late Cenozoic ice ages. Clim. Dyn. 52, 653–679.

Ganachaud, A., Wunsch, C., 2000. Improved estimates of global ocean circulation, heat transport and mixing from hydrographic data. Nature 408, 453–457.

Gargett, A.E., Osborn, T.R., 1981. Small-scale shear measurements during the fine and microstructure experiment (Fame). J. Geophys. Res. 86, 1929–1944.

Garrett, C., 1991. Marginal mixing theories. Atmos.-Ocean 29, 313–339.

Garrett, C., Kunze, E., 2007. Internal tide generation in the deep ocean. Annu. Rev. Fluid Mech. 39, 57–87.

Gaspar, P., Grégoris, Y., Lefevre, J.-M., 1990. A simple eddy kinetic energy model for simulations of the oceanic vertical mixing: tests at station Papa and Long-Term Upper Ocean Study site. J. Geophys. Res. 95, 16179–16193.

Gent, P.R., Large, W.G., Bryan, F.O., 2001. What sets the mean transport through Drake Passage? J. Geophys. Res. 106, 2693–2712.

Gent, P.R., McWilliams, J.C., 1990. Isopycnal mixing in ocean circulation models. J. Phys. Oceanogr. 20, 150–155.

Gent, P.R., Willebrand, J., McDougall, T.J., McWilliams, J.C., 1995. Parameterizing eddy-induced tracer transports in ocean circulation models. J. Phys. Oceanogr. 25, 463–474.

Gill, A.E., 1982. Atmosphere–Ocean Dynamics. Elsevier.

Gnanadesikan, A., 1999. A simple predictive model for the structure of the oceanic pycnocline. Science 283, 2077–2079.

Gordon, A.L., 1986. Interocean exchange of thermocline water. J. Geophys. Res. 91, 5037–5046.

Gordon, A.L., Huber, B.A., 1990. Southern Ocean winter mixed layer. J. Geophys. Res. 95, 11655.

Gouretski, V., Koltermann, K.P., 2004. WOCE global hydrographic climatology. Ber. BSH 35, 1–52.

Greatbatch, R.J., Li, G., 2000. Alongslope mean flow and an associated upslope bolus flux of tracer in a parameterization of mesoscale turbulence. Deep-Sea Res. 47, 709–735.

Gregg, M.C., 1987. Diapycnal mixing in the thermocline: a review. J. Geophys. Res. 92, 5249.

Gregg, M.C., Peters, H., Wesson, J.C., Oakey, N.S., Shay, T.J., 1985. Intensive measurements of turbulence and shear in the equatorial undercurrent. Nature 318, 140–144.

Groeskamp, S., Abernathey, R.P., Klocker, A., 2016. Water mass transformation by cabbeling and thermobaricity. Geophys. Res. Lett. 43, 2016GL070860.

Groeskamp, S., Griffies, S.M., Iudicone, D., Marsh, R., Nurser, A.J.G., Zika, J.D., 2019. The water mass transformation framework for ocean physics and biogeochemistry. Annu. Rev. Mar. Sci. 11, 271–305.

Groeskamp, S., LaCasce, J.H., McDougall, T.J., Rogé, M., 2020. Full-depth global estimates of ocean mesoscale eddy mixing from observations and theory. Geophys. Res. Lett. 47, e2020GL089425.

Groeskamp, S., Sloyan, B.M., Zika, J.D., McDougall, T.J., 2017. Mixing inferred from an ocean climatology and surface fluxes. J. Phys. Oceanogr. 47, 667–687.
Guilyardi, E., Madec, G., Terray, L., 2001. The role of lateral ocean physics in the upper ocean thermal balance of a coupled ocean-atmosphere GCM. Clim. Dyn. 17, 589–599.
Harrison, D.E., 1978. On the diffusion parameterization of mesoscale eddy effects from a numerical ocean experiment. J. Phys. Oceanogr. 8, 913–918.
Hautala, S.L., Riser, S.C., 1993. A nonconservative β-spiral determination of the deep circulation in the eastern South Pacific. J. Phys. Oceanogr. 23, 1975–2000.
Henning, C.C., Vallis, G.K., 2004. The effects of mesoscale eddies on the main subtropical thermocline. J. Phys. Oceanogr. 34, 2428–2443.
Hieronymus, M., Nycander, J., 2013. The budgets of heat and salinity in NEMO. Ocean Model. 67, 28–38.
Hieronymus, M., Nycander, J., Nilsson, J., Döös, K., Hallberg, R., 2019. Oceanic overturning and heat transport: the role of background diffusivity. J. Climate 32, 701–716.
Holland, W.R., 1978. The role of mesoscale eddies in the general circulation of the ocean–numerical experiments using a wind-driven quasi-geostrophic model. J. Phys. Oceanogr. 8, 363–392.
Holmes, R.M., de Lavergne, C., McDougall, T.J., 2018b. Ridges, seamounts, troughs, and bowls: topographic control of the dianeutral circulation in the abyssal ocean. J. Phys. Oceanogr. 48, 861–882.
Holmes, R.M., Zika, J.D., England, M.H., 2018a. Diathermal heat transport in a global ocean model. J. Phys. Oceanogr. 49, 141–161.
Holmes, R.M., Zika, J.D., Ferrari, R., Thompson, A.F., Newsom, E.R., England, M.H., 2019. Atlantic Ocean heat transport enabled by Indo-Pacific heat uptake and mixing. Geophys. Res. Lett. 46, 13939–13949.
Holzer, M., Primeau, F.W., 2006. The diffusive ocean conveyor. Geophys. Res. Lett. 33, L14618.
Howard, E., Hogg, A.McC., Waterman, S., Marshall, D.P., 2014. The injection of zonal momentum by buoyancy forcing in a Southern Ocean model. J. Phys. Oceanogr. 45, 259–271.
Huang, R.X., 1999. Mixing and energetics of the oceanic thermohaline circulation. J. Phys. Oceanogr. 29, 727–746.
Hughes, C.W., de Cuevas, B.A., 2001. Why western boundary currents in realistic oceans are inviscid: a link between form stress and bottom pressure torques. J. Phys. Oceanogr. 31, 2871–2885.
Hummels, R., Dengler, M., Bourlès, B., 2013. Seasonal and regional variability of upper ocean diapycnal heat flux in the Atlantic cold tongue. Prog. Oceanogr. 111, 52–74.
Iselin, C.O., 1939. The influence of vertical and lateral turbulence on the characteristics of the waters at mid-depths. Eos Trans. AGU 20, 414–417.
Iudicone, D., Madec, G., Blanke, B., Speich, S., 2008. The role of Southern Ocean surface forcings and mixing in the global conveyor. J. Phys. Oceanogr. 38, 1377–1400.
Jackett, D.R., McDougall, T.J., 1997. A neutral density variable for the World's Oceans. J. Phys. Oceanogr. 27, 237–263.
Jayne, S.R., 2009. The impact of abyssal mixing parameterizations in an ocean general circulation model. J. Phys. Oceanogr. 39, 1756–1775.
Jing, Z., Wu, L., 2014. Intensified diapycnal mixing in the midlatitude western boundary currents. Sci. Rep. 4.
Jochum, M., Briegleb, B.P., Danabasoglu, G., Large, W.G., Norton, N.J., Jayne, S.R., Alford, M.H., Bryan, F.O., 2013. The impact of oceanic near-inertial waves on climate. J. Climate 26, 2833–2844.
Johnson, G.C., Bryden, H.L., 1989. On the size of the Antarctic Circumpolar Current. Deep-Sea Res. 36, 39–53.
Jones, C.S., Abernathey, R.P., 2019. Isopycnal mixing controls deep ocean ventilation. Geophys. Res. Lett. 46, 13144–13151.
Klein, P., Treguier, A.-M., Hua, B.L., 1998. Three-dimensional stirring of thermohaline fronts. J. Mar. Res. 56, 589–612.
Klocker, A., Abernathey, R., 2014. Global patterns of mesoscale eddy properties and diffusivities. J. Phys. Oceanogr. 44, 1030–1046.
Klocker, A., McDougall, T.J., 2010. Influence of the nonlinear equation of state on global estimates of dianeutral advection and diffusion. J. Phys. Oceanogr. 40, 1690–1709.
Koch-Larrouy, A., Lengaigne, M., Terray, P., Madec, G., Masson, S., 2010. Tidal mixing in the Indonesian Seas and its effect on the tropical climate system. Clim. Dyn. 34, 891–904.
Kunze, E., Firing, E., Hummon, J.M., Chereskin, T.K., Thurnherr, A.M., 2006. Global abyssal mixing inferred from lowered ADCP shear and CTD strain profiles. J. Phys. Oceanogr. 36, 1553–1576.
Lagerloef, G.S.E., Mitchum, G.T., Lukas, R.B., Niiler, P.P., 1999. Tropical Pacific near-surface currents estimated from altimeter, wind, and drifter data. J. Geophys. Res. 104, 23313–23326.
Large, W.G., McWilliams, J.C., Doney, S.C., 1994. Oceanic vertical mixing: a review and a model with a nonlocal boundary layer parameterization. Rev. Geophys. 32, 363–403.
Le Corre, M., Gula, J., Tréguier, A.-M., 2020. Barotropic vorticity balance of the North Atlantic subpolar gyre in an eddy-resolving model. Ocean Sci. 16, 451–468.
Ledwell, J.R., Montgomery, E.T., Polzin, K.L., St Laurent, L.C., Schmitt, R.W., Toole, J.M., 2000. Evidence for enhanced mixing over rough topography in the abyssal ocean. Nature 403, 179–182.
Ledwell, J.R., St Laurent, L.C., Girton, J.B., Toole, J.M., 2011. Diapycnal mixing in the Antarctic Circumpolar Current. J. Phys. Oceanogr. 41, 241–246.
Ledwell, J.R., Watson, A.J., Law, C.S., 1993. Evidence for slow mixing across the pycnocline from an open-ocean tracer-release experiment. Nature 364, 701–703.
Ledwell, J.R., Watson, A.J., Law, C.S., 1998. Mixing of a tracer in the pycnocline. J. Geophys. Res. 103, 21499–21529.
Locarnini, R.A., Mishonov, A.V., Baranova, O.K., Boyer, T.P., Zweng, M.M., Garcia, H.E., Reagan, J.R., Seidov, D., Weathers, K., Paver, C.R., Smolyar, I., 2018. World Ocean Atlas 2018, Volume 1: Temperature. A. Mishonov, Technical Ed., NOAA Atlas NESDIS 81. 52 pp.

Lozier, M.S., Li, F., Bacon, S., Bahr, F., Bower, A.S., Cunningham, S.A., Jong, M.F. de, Steur, L. de, deYoung, B., Fischer, J., Gary, S.F., Greenan, B.J.W., Holliday, N.P., Houk, A., Houpert, L., Inall, M.E., Johns, W.E., Johnson, H.L., Johnson, C., Karstensen, J., Koman, G., Bras, I.A.L., Lin, X., Mackay, N., Marshall, D.P., Mercier, H., Oltmanns, M., Pickart, R.S., Ramsey, A.L., Rayner, D., Straneo, F., Thierry, V., Torres, D.J., Williams, R.G., Wilson, C., Yang, J., Yashayaev, I., Zhao, J., 2019. A sea change in our view of overturning in the subpolar North Atlantic. Science 363, 516–521.

Lu, P., McCreary, J.P., Klinger, B.A., 1998. Meridional circulation cells and the source waters of the Pacific equatorial undercurrent. J. Phys. Oceanogr. 28, 62–84.

Lucazeau, F., 2019. Analysis and mapping of an updated terrestrial heat flow data set. Geochem. Geophys. Geosyst. 20, 4001–4024.

Lumpkin, R., Speer, K., 2003. Large-scale vertical and horizontal circulation in the North Atlantic Ocean. J. Phys. Oceanogr. 33, 1902–1920.

Lumpkin, R., Speer, K., 2007. Global ocean meridional overturning. J. Phys. Oceanogr. 37, 2550–2562.

Luyten, J.R., Pedlosky, J., Stommel, H., 1983. The ventilated thermocline. J. Phys. Oceanogr. 13, 292–309.

MacKinnon, J.A., Johnston, T.M.S., Pinkel, R., 2008. Strong transport and mixing of deep water through the Southwest Indian Ridge. Nat. Geosci. 1, 755–758.

Mantyla, A.W., Reid, J.L., 1983. Abyssal characteristics of the World Ocean waters. Deep-Sea Res. 30, 805–833.

Marshall, D., 1997. Subduction of water masses in an eddying ocean. J. Mar. Res. 55, 201–222.

Marshall, J., Jones, H., Karsten, R., Wardle, R., 2002. Can eddies set ocean stratification? J. Phys. Oceanogr. 32, 26–38.

Marshall, J.C., Nurser, A.J.G., 1991. A continuously stratified thermocline model incorporating a mixed layer of variable thickness and density. J. Phys. Oceanogr. 21, 1780–1792.

Marshall, J., Radko, T., 2003. Residual-mean solutions for the Antarctic Circumpolar Current and its associated overturning circulation. J. Phys. Oceanogr. 33, 2341–2354.

Marshall, J., Speer, K., 2012. Closure of the meridional overturning circulation through Southern Ocean upwelling. Nat. Geosci. 5, 171–180.

Mashayek, A., Ferrari, R., Merrifield, S., Ledwell, J.R., Laurent, L.S., Naveira Garabato, A.C., 2017. Topographic enhancement of vertical turbulent mixing in the Southern Ocean. Nat. Commun. 8, 14197.

Mazloff, M.R., Ferrari, R., Schneider, T., 2013. The force balance of the Southern Ocean meridional overturning circulation. J. Phys. Oceanogr. 43, 1193–1208.

McCartney, M.S., 1977. Subantarctic mode water. In: A Voyage of Discovery. Deep-Sea Res. 24, 103–119.

McCreary, J.P., Lu, P., 1994. Interaction between the subtropical and equatorial ocean circulations: the subtropical cell. J. Phys. Oceanogr. 24, 466–497.

McDougall, T.J., 1984. The relative roles of diapycnal and isopycnal mixing on subsurface water mass conversion. J. Phys. Oceanogr. 14, 1577–1589.

McDougall, T.J., 1989. Dianeutral advection. In: Parameterization of Small-Scale Processes: Proc. 'Aha Huliko'a Hawaiian Winter Workshop. University of Hawaii at Manoa, Honolulu, HI, pp. 289–315.

McDougall, T.J., 2003. Potential enthalpy: a conservative oceanic variable for evaluating heat content and heat fluxes. J. Phys. Oceanogr. 33, 945–963.

McDougall, T.J., Ferrari, R., 2017. Abyssal upwelling and downwelling driven by near-boundary mixing. J. Phys. Oceanogr. 47, 261–283.

McDougall, T.J., Groeskamp, S., Griffies, S.M., 2014. On geometrical aspects of interior ocean mixing. J. Phys. Oceanogr. 44, 2164–2175.

McDougall, T.J., Jackett, D.R., Millero, F.J., Pawlowicz, R., Barker, P.M., 2012. A global algorithm for estimating Absolute Salinity. Ocean Sci. 8, 1123–1134.

McDougall, T.J., McIntosh, P.C., 2001. The temporal-residual-mean velocity. Part II: isopycnal interpretation and the tracer and momentum equations. J. Phys. Oceanogr. 31, 1222–1246.

McPhaden, M.J., Zhang, D., 2002. Slowdown of the meridional overturning circulation in the upper Pacific Ocean. Nature 415, 603–608.

McWilliams, J.C., Chow, J.H.S., 1981. Equilibrium geostrophic turbulence I: a reference solution in a β-plane channel. J. Phys. Oceanogr. 11, 921–949.

Melet, A., Hallberg, R., Legg, S., Nikurashin, M., 2014. Sensitivity of the ocean state to lee wave–driven mixing. J. Phys. Oceanogr. 44, 900–921.

Melet, A., Legg, S., Hallberg, R., 2016. Climatic impacts of parameterized local and remote tidal mixing. J. Climate 29, 3473–3500.

Mercier, H., Speer, K.G., 1998. Transport of bottom water in the Romanche fracture zone and the Chain fracture zone. J. Phys. Oceanogr. 28, 779–790.

Meredith, M.P., Woodworth, P.L., Chereskin, T.K., Marshall, D.P., Allison, L.C., Bigg, G.R., Donohue, K., Heywood, K.J., Hughes, C.W., Hibbert, A., Hogg, A.M., Johnson, H.L., Jullion, L., King, B.A., Leach, H., Lenn, Y.-D., Maqueda, M.A.M., Munday, D.R., Naveira Garabato, A.C., Provost, C., Sallée, J.-B., Sprintall, J., 2011. Sustained monitoring of the Southern Ocean at Drake Passage: past achievements and future priorities. Rev. Geophys. 49.

Moum, J.N., Lien, R.-C., Perlin, A., Nash, J.D., Gregg, M.C., Wiles, P.J., 2009. Sea surface cooling at the Equator by subsurface mixing in tropical instability waves. Nat. Geosci. 2, 761–765.

Moum, J.N., Perlin, A., Nash, J.D., McPhaden, M.J., 2013. Seasonal sea surface cooling in the equatorial Pacific cold tongue controlled by ocean mixing. Nature 500, 64–67.

Munday, D.R., Allison, L.C., Johnson, H.L., Marshall, D.P., 2011. Remote forcing of the Antarctic Circumpolar Current by diapycnal mixing. Geophys. Res. Lett. 38, L08609.

Munk, W., 1997. Once again: once again—tidal friction. Prog. Oceanogr. 40, 7–35.

Munk, W., Wunsch, C., 1998. Abyssal recipes II: energetics of tidal and wind mixing. Deep-Sea Res. 45, 1977–2010.

Munk, W.H., 1950. On the wind-driven ocean circulation. J. Meteorol. 7, 80–93.

Munk, W.H., 1966. Abyssal recipes. Deep-Sea Res. 13, 707–730.

Munk, W.H., Palmén, E., 1951. Note on the dynamics of the Antarctic Circumpolar Current. Tellus 3, 53–55.

Naveira Garabato, A.C., Frajka-Williams, E., Spingys, C.P., Legg, S., Polzin, K.L., Forryan, A., Abrahamsen, P., Buckingham, C., Griffies, S.M., McPhail, S., Nicholls, K., Thomas, L.N., Meredith, M., 2019. Rapid mixing and exchange of deep-ocean waters in an abyssal boundary current. Proc. Natl. Acad. Sci. 116, 13233–13238.

Naveira Garabato, A.C., MacGilchrist, G.A., Brown, P.J., Evans, D.G., Meijers, A.J.S., Zika, J.D., 2017. High-latitude ocean ventilation and its role in Earth's climate transitions. Philos. Trans. R. Soc. A 375, 20160324.

Naveira Garabato, A.C., Polzin, K.L., Ferrari, R., Zika, J.D., Forryan, A., 2016. A microscale view of mixing and overturning across the Antarctic Circumpolar Current. J. Phys. Oceanogr. 46, 233–254.

Naveira Garabato, A.C., Stevens, D.P., Watson, A.J., Roether, W., 2007. Short-circuiting of the overturning circulation in the Antarctic Circumpolar Current. Nature 447, 194–197.

Naveira Garabato, A.C., Williams, A.P., Bacon, S., 2014. The three-dimensional overturning circulation of the Southern Ocean during the WOCE era. Prog. Oceanogr. 120, 41–78.

Nikurashin, M., Vallis, G., 2012. A theory of the interhemispheric meridional overturning circulation and associated stratification. J. Phys. Oceanogr. 42, 1652–1667.

Nurser, A.J.G., Marsh, R., Williams, R.G., 1999. Diagnosing water mass formation from air-sea fluxes and surface mixing. J. Phys. Oceanogr. 29, 1468–1487.

Nycander, J., Hieronymus, M., Roquet, F., 2015. The nonlinear equation of state of sea water and the global water mass distribution. Geophys. Res. Lett. 42, 2015GL065525.

Nycander, J., Nilsson, J., Döös, K., Broström, G., 2007. Thermodynamic analysis of ocean circulation. J. Phys. Oceanogr. 37, 2038–2052.

Oakey, N.S., 1982. Determination of the rate of dissipation of turbulent energy from simultaneous temperature and velocity shear microstructure measurements. J. Phys. Oceanogr. 12, 256–271.

Olbers, D., 1998. Comments on "On the obscurantist physics of 'Form Drag' in theorizing about the Circumpolar Current". J. Phys. Oceanogr. 28, 1647–1654.

Orsi, A.H., Johnson, G.C., Bullister, J.L., 1999. Circulation, mixing, and production of Antarctic Bottom Water. Prog. Oceanogr. 43, 55–109.

Osborn, T.R., 1978. Measurements of energy dissipation adjacent to an island. J. Geophys. Res. 83, 2939.

Osborn, T.R., Cox, C.S., 1972. Oceanic fine structure. Geophys. Fluid Dyn. 3, 321–345.

Pacanowski, R.C., 1987. Effect of equatorial currents on surface stress. J. Phys. Oceanogr. 17, 833–838.

Pedlosky, J., 1992. The baroclinic structure of the abyssal circulation. J. Phys. Oceanogr. 22, 652–659.

Pedlosky, J., 1996. Ocean Circulation Theory. Springer, Berlin. 453 pp.

Pellichero, V., Sallée, J.-B., Schmidtko, S., Roquet, F., Charrassin, J.-B., 2017. The ocean mixed layer under Southern Ocean sea-ice: seasonal cycle and forcing. J. Geophys. Res. 122, 1608–1633.

Peña-Molino, B., Rintoul, S.R., Mazloff, M.R., 2014. Barotropic and baroclinic contributions to along-stream and across-stream transport in the Antarctic Circumpolar Current. J. Geophys. Res. 119, 8011–8028.

Pickart, R.S., Spall, M.A., 2007. Impact of Labrador Sea convection on the North Atlantic meridional overturning circulation. J. Phys. Oceanogr. 37, 2207–2227.

Plant, W.J., 1982. A relationship between wind stress and wave slope. J. Geophys. Res. 87, 1961–1967.

Pollard, R.T., Rhines, P.B., Thompson, R.O., 1973. The deepening of the wind-mixed layer. Geophys. Astrophys. Fluid Dyn. 4, 381–404.

Polzin, K.L., Speer, K.G., Toole, J.M., Schmitt, R.W., 1996. Intense mixing of Antarctic Bottom Water in the equatorial Atlantic Ocean. Nature 380, 54–57.

Polzin, K.L., Toole, J.M., Ledwell, J.R., Schmitt, R.W., 1997. Spatial variability of turbulent mixing in the abyssal ocean. Science 276, 93–96.

Price, J.F., 2001. Chapter 5.3 - Subduction. In: Siedler, G., Church, J., Gould, J. (Eds.), International Geophysics, Ocean Circulation and Climate. Academic Press, pp. 357–371.

Price, J.F., Weller, R.A., Pinkel, R., 1986. Diurnal cycling: observations and models of the upper ocean response to diurnal heating, cooling, and wind mixing. J. Geophys. Res. 91, 8411–8427.

Pujol, M.-I., Faugère, Y., Taburet, G., Dupuy, S., Pelloquin, C., Ablain, M., Picot, N., 2016. DUACS DT2014: the new multi-mission altimeter data set reprocessed over 20 years. Ocean Sci. 12, 1067–1090.

Qu, T., Gao, S., Fine, R.A., 2013. Subduction of South Pacific tropical water and its equatorward pathways as shown by a simulated passive tracer. J. Phys. Oceanogr. 43, 1551–1565.

Ragen, S., Pradal, M.-A., Gnanadesikan, A., 2020. The impact of parameterized lateral mixing on the Antarctic Circumpolar Current in a coupled climate model. J. Phys. Oceanogr. 50, 965–982.

Reid, J.L., 1997. On the total geostrophic circulation of the Pacific Ocean: flow patterns, tracers, and transports. Prog. Oceanogr. 39, 263–352.

Reid, J.L., 2003. On the total geostrophic circulation of the Indian Ocean: flow patterns, tracers, and transports. Prog. Oceanogr. 56, 137–186.

Rhines, P.B., 1993. Oceanic general circulation: wave and advection dynamics. In: Willebrand, J., Anderson, D.L.T. (Eds.), Modelling Oceanic Climate Interactions. In: NATO ASI Series. Springer, Berlin, Heidelberg, pp. 67–149.

Rhines, P.B., Holland, W.R., 1979. A theoretical discussion of eddy-driven mean flows. Dyn. Atmos. Ocean. 3, 289–325.

Rhines, P.B., Young, W.R., 1982. Homogenization of potential vorticity in planetary gyres. J. Fluid Mech. 122, 347–367.

Rintoul, S.R., Naveira Garabato, A.C., 2013. Chapter 18 - Dynamics of the Southern Ocean Circulation. In: Siedler, G., Griffies, S.M., Gould, J., Church, J.A. (Eds.), International Geophysics, Ocean Circulation and Climate. Academic Press, pp. 471–492.

Robbins, P.E., Price, J.F., Owens, W.B., Jenkins, W.J., 2000. The importance of lateral diffusion for the ventilation of the lower thermocline in the subtropical North Atlantic. J. Phys. Oceanogr. 30, 67–89.

Roemmich, D., 1983. The balance of geostrophic and Ekman transports in the tropical Atlantic Ocean. J. Phys. Oceanogr. 13, 1534–1539.

Roquet, F., Wunsch, C., Madec, G., 2011. On the patterns of wind-power input to the ocean circulation. J. Phys. Oceanogr. 41, 2328–2342.

Rudnick, D.L., Boyd, T.J., Brainard, R.E., Carter, G.S., Egbert, G.D., Gregg, M.C., Holloway, P.E., Klymak, J.M., Kunze, E., Lee, C.M., Levine, M.D., Luther, D.S., Martin, J.P., Merrifield, M.A., Moum, J.N., Nash, J.D., Pinkel, R., Rainville, L., Sanford, T.B., 2003. From tides to mixing along the Hawaiian ridge. Science 301, 355–357.

Samelson, R.M., Vallis, G.K., 1997. Large-scale circulation with small diapycnal diffusion: the two-thermocline limit. J. Mar. Res. 55, 223–275.

Sarmiento, J.L., Simeon, J., Gnanadesikan, A., Gruber, N., Key, R.M., Schlitzer, R., 2007. Deep ocean biogeochemistry of silicic acid and nitrate. Glob. Biogeochem. Cycles 21, GB1S90.

Sasaki, H., Kida, S., Furue, R., Nonaka, M., Masumoto, Y., 2018. An increase of the Indonesian Throughflow by internal tidal mixing in a high-resolution quasi-global ocean simulation. Geophys. Res. Lett. 45, 8416–8424.

Schmitz, W.J., 1995. On the interbasin-scale thermohaline circulation. Rev. Geophys. 33, 151–173.

Schmitz, W.J., Richardson, P.L., 1991. On the sources of the Florida Current. Deep-Sea Res. 38, S379–S409.

Sheen, K.L., Brearley, J.A., Naveira Garabato, A.C., Smeed, D.A., Waterman, S., Ledwell, J.R., Meredith, M.P., St. Laurent, L., Thurnherr, A.M., Toole, J.M., Watson, A.J., 2013. Rates and mechanisms of turbulent dissipation and mixing in the Southern Ocean: results from the Diapycnal and Isopycnal Mixing Experiment in the Southern Ocean (DIMES). J. Geophys. Res. 118, 2774–2792.

Sloyan, B.M., Johnson, G.C., Kessler, W.S., 2003. The Pacific cold tongue: a pathway for interhemispheric exchange. J. Phys. Oceanogr. 33, 1027–1043.

Sloyan, B.M., Rintoul, S.R., 2001. The Southern Ocean limb of the global deep overturning circulation. J. Phys. Oceanogr. 31, 143–173.

Sloyan, B.M., Talley, L.D., Chereskin, T.K., Fine, R., Holte, J., 2010. Antarctic Intermediate Water and Subantarctic Mode Water formation in the southeast Pacific: the role of turbulent mixing. J. Phys. Oceanogr. 40, 1558–1574.

Smith, K.S., Ferrari, R., 2009. The production and dissipation of compensated thermohaline variance by mesoscale stirring. J. Phys. Oceanogr. 39, 2477–2501.

Smith, R.D., McWilliams, J.C., 2003. Anisotropic horizontal viscosity for ocean models. Ocean Model. 5, 129–156.

Smyth, W.D., Moum, J.N., 2013. Marginal instability and deep cycle turbulence in the eastern equatorial Pacific Ocean. Geophys. Res. Lett. 40, 6181–6185.

St. Laurent, L., Naveira Garabato, A.C., Ledwell, J.R., Thurnherr, A.M., Toole, J.M., Watson, A.J., 2012. Turbulence and diapycnal mixing in Drake Passage. J. Phys. Oceanogr. 42, 2143–2152.

St. Laurent, L.C., Toole, J.M., Schmitt, R.W., 2001. Buoyancy forcing by turbulence above rough topography in the abyssal Brazil Basin. J. Phys. Oceanogr. 31, 3476–3495.

Stewart, A.L., Hogg, A.McC., 2017. Reshaping the Antarctic Circumpolar Current via Antarctic Bottom Water export. J. Phys. Oceanogr. 47, 2577–2601.

Stommel, H., 1958. The abyssal circulation. Deep-Sea Res. 5, 80–82.

Stommel, H., 1979. Determination of water mass properties of water pumped down from the Ekman layer to the geostrophic flow below. Proc. Natl. Acad. Sci. 76, 3051–3055.

Stommel, H., Arons, A.B., 1959a. On the abyssal circulation of the world ocean—II. An idealized model of the circulation pattern and amplitude in oceanic basins. Deep-Sea Res. 6, 217–233.

Stommel, H., Arons, A.B., 1959b. On the abyssal circulation of the world ocean—I. Stationary planetary flow patterns on a sphere. Deep-Sea Res. 6, 140–154.

Su, Z., Stewart, A.L., Thompson, A.F., 2014. An idealized model of Weddell Gyre export variability. J. Phys. Oceanogr. 44, 1671–1688.

Sun, S., Eisenman, I., Zanna, L., Stewart, A.L., 2020. Surface constraints on the depth of the Atlantic meridional overturning circulation: Southern Ocean versus North Atlantic. J. Climate 33, 3125–3149.

Sweeney, C., Gnanadesikan, A., Griffies, S.M., Harrison, M.J., Rosati, A.J., Samuels, B.L., 2005. Impacts of shortwave penetration depth on large-scale ocean circulation and heat transport. J. Phys. Oceanogr. 35, 1103–1119.

Taburet, G., Sanchez-Roman, A., Ballarotta, M., Pujol, M.-I., Legeais, J.-F., Fournier, F., Faugere, Y., Dibarboure, G., 2019. DUACS DT2018: 25 years of reprocessed sea level altimetry products. Ocean Sci. 15, 1207–1224.

Talley, L., 2013. Closure of the global overturning circulation through the Indian, Pacific, and Southern Oceans: schematics and transports. Oceanography 26, 80–97.

Tamsitt, V., Abernathey, R.P., Mazloff, M.R., Wang, J., Talley, L.D., 2018. Transformation of deep water masses along Lagrangian upwelling pathways in the Southern Ocean. J. Geophys. Res. 123, 1994–2017.

Tamsitt, V., Drake, H.F., Morrison, A.K., Talley, L.D., Dufour, C.O., Gray, A.R., Griffies, S.M., Mazloff, M.R., Sarmiento, J.L., Wang, J., Weijer, W., 2017. Spiraling pathways of global deep waters to the surface of the Southern Ocean. Nat. Commun. 8, 172.

Thompson, A.F., Stewart, A.L., Spence, P., Heywood, K.J., 2018. The Antarctic slope current in a changing climate. Rev. Geophys. 56, 741–770.

Toggweiler, J.R., Druffel, E.R.M., Key, R.M., Galbraith, E.D., 2019a. Upwelling in the ocean basins North of the ACC: 1. On the upwelling exposed by the surface distribution of $\Delta^{14}C$. J. Geophys. Res. 124, 2591–2608.

Toggweiler, J.R., Druffel, E.R.M., Key, R.M., Galbraith, E.D., 2019b. Upwelling in the ocean basins North of the ACC: 2. How cool subantarctic water reaches the surface in the tropics. J. Geophys. Res. 124, 2609–2625.

Toggweiler, J.R., Samuels, B., 1993. New radiocarbon constraints on the upwelling of abyssal water to the ocean's surface. In: Heimann, M. (Ed.), The Global Carbon Cycle. In: NATO ASI Series. Springer Berlin Heidelberg, pp. 333–366.

Toggweiler, J.R., Samuels, B., 1995. Effect of Drake Passage on the global thermohaline circulation. Deep-Sea Res. 42, 477–500.

Toggweiler, J.R., Samuels, B., 1998. On the ocean's large-scale circulation near the limit of no vertical mixing. J. Phys. Oceanogr. 28, 1832–1852.

Toole, J.M., Schmitt, R.W., Polzin, K.L., 1994. Estimates of diapycnal mixing in the abyssal ocean. Science 264, 1120–1123.

Treguier, A.M., Held, I.M., Larichev, V.D., 1997. Parameterization of quasigeostrophic eddies in primitive equation ocean models. J. Phys. Oceanogr. 27, 567–580.

Urakawa, L.S., Hasumi, H., 2012. Eddy-resolving model estimate of the cabbeling effect on the water mass transformation in the Southern Ocean. J. Phys. Oceanogr. 42, 1288–1302.

van Haren, H., Gostiaux, L., 2012. Detailed internal wave mixing above a deep-ocean slope. J. Mar. Res. 70, 173–197.

Vic, C., Naveira Garabato, A.C., Green, J.A.M., Waterhouse, A.F., Zhao, Z., Melet, A., Lavergne, C. de, Buijsman, M.C., Stephenson, G.R., 2019. Deep-ocean mixing driven by small-scale internal tides. Nat. Commun. 10, 2099.

Voet, G., Girton, J.B., Alford, M.H., Carter, G.S., Klymak, J.M., Mickett, J.B., 2014. Pathways, volume transport, and mixing of abyssal water in the Samoan Passage. J. Phys. Oceanogr. 45, 562–588.

Walin, G., 1977. A theoretical framework for the description of estuaries. Tellus 29, 128–136.

Walin, G., 1982. On the relation between sea-surface heat flow and thermal circulation in the ocean. Tellus 34, 187–195.

Wang, Z., Meredith, M.P., 2008. Density-driven southern hemisphere subpolar gyres in coupled climate models. Geophys. Res. Lett. 35, L14608.

Waterhouse, A.F., MacKinnon, J.A., Nash, J.D., Alford, M.H., Kunze, E., Simmons, H.L., Polzin, K.L., St. Laurent, L.C., Sun, O.M., Pinkel, R., Talley, L.D., Whalen, C.B., Huussen, T.N., Carter, G.S., Fer, I., Waterman, S., Naveira Garabato, A.C., Sanford, T.B., Lee, C.M., 2014. Global patterns of diapycnal mixing from measurements of the turbulent dissipation rate. J. Phys. Oceanogr. 44, 1854–1872.

Watson, A.J., Ledwell, J.R., Messias, M.-J., King, B.A., Mackay, N., Meredith, M.P., Mills, B., Naveira Garabato, A.C., 2013. Rapid cross-density ocean mixing at mid-depths in the Drake Passage measured by tracer release. Nature 501, 408–411.

Whalen, C.B., MacKinnon, J.A., Talley, L.D., 2018. Large-scale impacts of the mesoscale environment on mixing from wind-driven internal waves. Nat. Geosci. 11, 842.

Wolfe, C.L., Cessi, P., 2011. The adiabatic pole-to-pole overturning circulation. J. Phys. Oceanogr. 41, 1795–1810.

Wunsch, C., 1970. On oceanic boundary mixing. Deep-Sea Res. 17, 293–301.

Wunsch, C., 1997. The vertical partition of oceanic horizontal kinetic energy. J. Phys. Oceanogr. 27, 1770–1794.

Wunsch, C., Ferrari, R., 2004. Vertical mixing, energy, and the general circulation of the oceans. Annu. Rev. Fluid Mech. 36, 281–314.

Yang, X., Tziperman, E., Speer, K., 2020. Dynamics of deep ocean eastern boundary currents. Geophys. Res. Lett. 47, e2019GL085396.

Zhai, X., Greatbatch, R.J., Eden, C., Hibiya, T., 2009. On the loss of wind-induced near-inertial energy to turbulent mixing in the upper ocean. J. Phys. Oceanogr. 39, 3040–3045.

Zhai, X., Johnson, H.L., Marshall, D.P., 2010. Significant sink of ocean-eddy energy near western boundaries. Nat. Geosci. 3, 608–612.

Zhang, R., Delworth, T.L., Rosati, A., Anderson, W.G., Dixon, K.W., Lee, H.-C., Zeng, F., 2011. Sensitivity of the North Atlantic Ocean circulation to an abrupt change in the Nordic Sea overflow in a high resolution global coupled climate model. J. Geophys. Res. 116, C12024.

Zhu, Y., Zhang, R.-H., 2019. A modified vertical mixing parameterization for its improved ocean and coupled simulations in the tropical Pacific. J. Phys. Oceanogr. 49, 21–37.

Zika, J.D., Le Sommer, J., Dufour, C.O., Molines, J.-M., Barnier, B., Brasseur, P., Dussin, R., Penduff, T., Iudicone, D., Lenton, A., Madec, G., Mathiot, P., Orr, J., Shuckburgh, E., Vivier, F., 2012. Vertical eddy fluxes in the Southern Ocean. J. Phys. Oceanogr. 43, 941–955.

Zika, J.D., Skliris, N., Nurser, A.J.G., Josey, S.A., Mudryk, L., Laliberté, F., Marsh, R., 2015. Maintenance and broadening of the ocean's salinity distribution by the water cycle. J. Climate 28, 9550–9560.

Zika, J.D., Sloyan, B.M., McDougall, T.J., 2009. Diagnosing the Southern Ocean overturning from tracer fields. J. Phys. Oceanogr. 39, 2926–2940.

Zweng, M.M., Reagan, J.R., Seidov, D., Boyer, T.P., Locarnini, R.A., Garcia, H.E., Mishonov, A.V., Baranova, O.K., Weathers, K., Paver, C.R., Smolyar, I., 2018. World Ocean Atlas 2018, Volume 2: Salinity. A. Mishonov, Technical Ed., NOAA Atlas NESDIS 82. 50 pp.

Chapter 4

Ocean near-surface layers

Baylor Fox-Kemper[a], Leah Johnson[a] and Fangli Qiao[b]
[a]*Brown University, Department of Earth, Environmental, and Planetary Sciences, Providence, RI, United States,* [b]*First Institute of Oceanography, Ministry of Natural Resources, Qingdao, China*

4.1 Introduction

The most energy-dense turbulence in the ocean occurs in the upper boundary layer. This turbulence is driven by winds, waves, and buoyancy forcing, and its fate may be viscous dissipation, the creation of potential energy through mixing and entrainment, or energy loss through downward propagating internal waves. Typically all fates occur to some degree.

Fig. 4.1 shows an example budget of turbulent processes leading to dissipation of mesoscale kinetic energy in a global high-resolution (nominal 0.1°) model (Pearson et al., 2017). Fig. 4.1a shows vertical friction exceeds both horizontal dissipation and bottom drag, and vertical dissipation within the boundary layer is by far the most dissipative parameterisation of kinetic energy. Fig. 4.1b shows this dissipation is highly intermittent and fairly characterised by an approximately log-normal distribution, with the most likely value orders of magnitude higher than typical of turbulence values outside of the boundary layer. Global dissipation of turbulent kinetic energy is estimated to be 1.8 TW (Waterhouse et al., 2014), or 1.3×10^{-9} W kg^{-1} on average. Within the surface mixing layer, values of 1×10^{-5} W kg^{-1} are common.

The fundamental oceanographic notion of water mass formation typically relies upon surface ventilation, i.e., exposure to surface boundary-layer mixing, where waters are transformed by exchanges with the atmosphere, rapid vertical mixing that overcomes stratification, and exposure to radiative forcing. An example of this conceptualisation are the metrics of "ideal age" or transit time, framings that consider aging of water masses that start at zero when exposed to the surface (Hall and Haine, 2002). The only rival mixing processes, in terms of transforming water masses, are when neighbouring water masses sharing similar densities combine (e.g., at the Mediterranean outflow or Southern ocean fronts: Jones and Abernathey, 2019), or when intense bottom overflows entrain (e.g., leading to North Atlantic deep water and Antarctic bottom water: Snow et al., 2015; Yankovsky et al., 2021).

Ocean boundary layers, their atmospheric boundary layer partner, and turbulence within both play an important role in air-sea fluxes (Fig. 4.2). Mathematically, boundary layers are where the ocean boundary conditions are met: stress, heating, and freshwater fluxes are basic ingredients for establishing the surface layer setting, but so too are the pressure matching boundary condition and the surface cohesion of the dynamical boundary condition (or its violent restructuring in the case

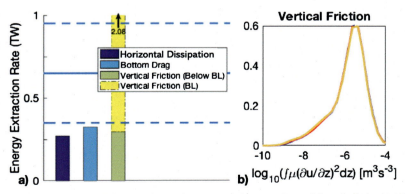

FIGURE 4.1 Global kinetic energy extraction rate by frictional processes in a global mesoscale-resolving simulation. (a) Global integrals of horizontal sub-grid dissipation, bottom drag, and vertical friction within and without the boundary layer. (b) Probability distribution (over horizontal location) of vertically-integrated energy dissipation by vertical friction (overlaid lines show results from three different simulations differing in horizontal dissipation schemes). Redrawn from Pearson et al. (2017).

of spray and bubbles) which are important for waves. On their distal side, they connect to deeper ocean waters and the free atmosphere. Regionally, they connect to the hydrosphere through rivers and the cryosphere (sea ice and ice shelves). There is also an important role for boundary layers in the biosphere: humans and other terrestrial species spend most of our lives in the atmospheric boundary layer, and the light-rich (euphotic) zone of the oceans is essentially collocated with the oceanic boundary layer. As these layers are attached to the atmosphere or cryosphere, it is difficult to disentangle these boundary layers from one another. Indeed, an approach sometimes taken is to denote a "wavy layer," which contains part atmosphere and part ocean (waves, spray, bubbles, foam, etc.), as effectively done in Janssen (2012). In summary, boundary layers mediate exchanges between the spheres.

These layers are active and evolve on a variety of timescales, and we are becoming increasingly aware of the importance of this variability. Particular focus had been on process interaction of vertical or isotropic turbulence with phenomena that occupy larger horizontal dimensions—chiefly at the submesoscale and mesoscale (see Chapters 8 and 9 of this book). The time it takes to mix or convect across the ocean boundary layer is measured in minutes to days, but many important impacts of these layers are manifested by their diurnal and seasonal variability. They are a critical player in the dominant modes of interannual and decadal variability.

Surface layers also play an important role in geochemistry, largely through exchanges of carbon and oxygen. Often, due to the rapid timescales of chemical reactions and stoichiometry in comparison to the relatively slow timescales (minutes to hours) of boundary layer mixing, and the relatively minimal vertical variations of properties within the mixed layer; the mixed layer can be assumed to be chemically homogeneous and relatively quick to arrive at a saturation level of chemical constituents, according to Henry's Law. However, it is possible for the finite timescale of reactions or piston velocity to be rivaled by ocean turbulence mixing timescales, in which case other effects can occur (Smith et al., 2018). Chapter 12 of this book discusses important aspects of the decreasing boundary layer saturation levels of oxygen under global warming. Bubbles and spray may also have interesting physical-chemical effects on gas transfer (e.g., Liang et al., 2013).

The surface layers are critical biologically, as they coexist with the euphotic zone and can govern how the essentials for life are distributed. The *critical depth hypothesis* (Sverdrup, 1953) posits that the spring phytoplankton bloom initiation in the North Atlantic can be understood by the increased light exposure of primary producers within a shoaled mixed layer. Taylor and Ferrari (2011) brought a more dynamical aspect to this theory, referred to the *critical turbulence hypothesis*, which considers the residence time of phytoplankton within the euphotic zone during mixing, and is in line with other elaborate theories linking zooplankton behaviour to mixing. Submesoscale restratification can accelerate the spring bloom, through its effect on suppressing turbulence and shoaling the mixed layer (Mahadevan et al., 2012), a beautiful example of submesoscale-boundary layer interaction with important consequences.

As a final example application of surface layers in real world problems, consider the location, detection, and capture of pollutants, such as oil and plastics. Often pollutants may be more buoyant than seawater, allowing them to rise through the boundary layer to the surface and accumulate where surface convergences lie (Taylor, 2018; D'Asaro et al., 2018). Alternatively, pollutants may gather, clump, flocculate, grow colonies, or be damaged to become negatively buoyant (Taylor et al., 2020). In both cases, the buoyancy of the pollutants can affect the dynamics as bubbles and spray do, or at the least (if pollutant concentrations are low) may be sorted into particular dynamical features, which influence their collection (e.g., van Sebille et al., 2020), detection (e.g., Brunner et al., 2015), and mitigation (e.g., Choy et al., 2019). Surface convergences and vertical velocities are intimately linked by the continuity equation, and surface convergences strongly affect the observation of boundary layer turbulence statistics of floats and drifters that accumulate much as pollution does (Chang et al., 2019; Pearson et al., 2019, 2020).

There are a number of recent review and broad perspective papers on the ocean surface layers (Umlauf and Burchard, 2005; Sullivan and McWilliams, 2010; Belcher et al., 2012; D'Asaro, 2014; Li et al., 2019) and waves (Cavaleri et al., 2012), as well as excellent treatment of key concepts in textbooks and monographs (Thorpe, 2007; Wyngaard, 2010; Csanady, 2000; Baumert et al., 2005). This chapter will therefore touch on basic theory (Section 4.2), but highlight recent developments in observations (Section 4.3), modelling (Section 4.4), and global perspectives (Section 4.5) emphasising processes, tools, and recent discoveries that are less well understood and do not appear commonly in textbooks.

4.2 Mixing layers and mixed layers in theory

Fig. 4.2 schematises the ocean surface layers. The atmospheric boundary layer (here idealised as a neutral (no stratification stable or unstable) boundary layer with a capping inversion), ocean mixing layer, ocean mixed layer ($0 > z > -h$, where temperature and salinity stratification is small), and the (assumed-stably) stratified thermocline and halocline that make up the upper pycnocline are illustrated. Solar radiation is noted to penetrate exponentially over roughly the mixed layer depth (depending on water clarity). Longwave (infrared, blackbody) radiation and turbulent fluxes of energy (latent, associated

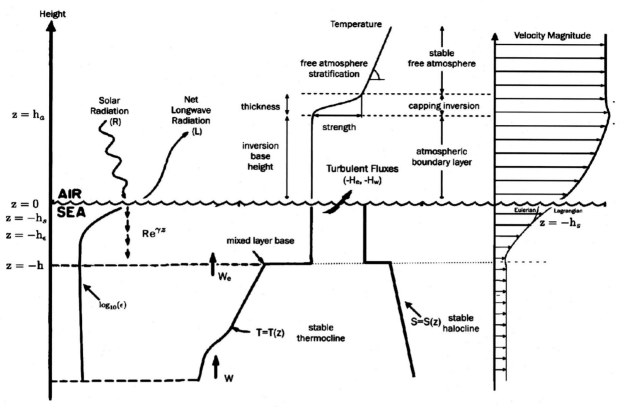

FIGURE 4.2 Schematic of the ocean surface layers, including entrainment velocity w_e, upwelling w, temperature T, salinity S, turbulent fluxes of thermal energy H_e, and freshwater H_w, dissipation ϵ, Stokes depth h_s, mixing-layer depth h_ϵ, and mixed-layer depth h.

with evaporation/precipitation, and sensible, associated with warm/cold air and water) can be totalled for air-sea energy exchanges that are realised within centimetres of the surface. Turbulent and radiative fluxes within the ocean and atmosphere tend to continue to the capping inversion and mixed layer base and beyond (during strong convective or shear forcing).

Waves present on the air-sea interface extend into the ocean ($0 > z > -h_s$), and are part of the exchanges and links between atmospheric (e.g., Suzuki et al., 2011) and oceanic (e.g. Teixeira and Belcher, 2002) turbulence. The ocean velocity profile on the right of Fig. 4.2 notes the difference between the Eulerian velocity (i.e., the velocity profile averaged over a wave period at a point in space, thus not including wave orbital velocities) and the Lagrangian velocity (the velocity averaged over a wave period following the path of the wave and current motion). As wave orbits carry parcels closer to the surface where the wave motion is faster; and then away from the surface, where it is slower, the Eulerian and Lagrangian velocities are not equal. This difference is the Stokes drift velocity, which decays rapidly away from the surface (depending on the wavelength of the waves, Webb and Fox-Kemper, 2011, as simplified in Fig. 4.2 by a single depth h_s). The waves shown here are not sinusoidal, which schematises the non-linear sharpening and breaking that occur from wave dynamics or whitecapping that would be expected under windy conditions.

The ocean mixing layer is the layer of active turbulence ($0 > z > -h_\epsilon$), where energy flow into turbulence and turbulence dissipation is elevated. When measurements or simulations of the dissipation rate of turbulence (ϵ) are available, this layer can be defined as the surface layer over which the value exceeds a typical background. The exceedance is typically many orders of magnitude above the background level in the mixing layer, which the schematic notes by showing a fast decay in $\log_{10}(\epsilon)$. Alternatively, the mixing layer can be defined based on the turbulent energy production or fluxes.

The mixed layer ($0 > z > -h$) may be deeper than the mixing layer during periods of relative calm after a deep mixing event. Alternatively, the mixing layer might be deeper than the mixed layer during active mixing and entrainment of stratified water into the mixed layer. Thus the mixed layer evolves semi-independently of the turbulence, lagging behind the active forcing to record a history of past forcing and restratification (the set of processes by which stratification, especially density stratification, is recovered after mixing). There are a variety of restratification processes, but the most important are solar (stronger heating near the surface) and dynamical (collectively processes—e.g., eddies, fronts, plumes, instabilities, filaments, geostrophic adjustment—that extract potential energy from the low to unstable stratification of the mixed layer by sorting light anomalies toward the surface and dense anomalies to depth).

The mixed layer base during convection or mixed layer deepening is punctured intermittently by plumes and jets of mixed layer water. The transition zone or entrainment layer, where this occurs, is partly diabatic, while the thermocline, halocline, and pycnocline below are subject to much lower levels of diabatic mixing. Some care is needed in defining this transition layer in practice, in a way that depends on scale: on small scales turbulence and diabatic entrainment determine the layer (Johnston and Rudnick, 2009; Li and Fox-Kemper, 2017, 2020); at submesoscales intrusions, ramps, and intra-thermocline eddies; and at mesoscales by isopycnal heaving and outcrops from eddies and fronts. The surface layer turbulence importantly responds and couples to the structures of eddies and fronts displacing the mixed layer base (Smith et al., 2018; Taylor, 2018; Taylor et al., 2020), and on large-scale average, such as used in climate models, diabatic and adiabatic processes are blended together in a general "transition zone" (Danabasoglu et al., 2008).

The atmospheric literature uses slightly different terminology (Holtslag and Nieuwstadt, 1986; Wyngaard, 2010). There is a surface layer, a mixed or upper layer, and then an entrainment or interfacial layer, where there might be a capping inversion. The mixed layer has similar meaning in both fluids, but the surface layer corresponds more directly to what is called the log-layer in the ocean, rather than the mixing layer. There does not appear to be a common use of a conceptual distinction between the mixing layer based on turbulent dissipation and the mixed layer in the atmosphere, even when dissipation is directly measured (e.g., Muñoz-Esparza et al., 2018). This may have to do with the rapid nightly restratification that keeps these two atmospheric layers tracking closely together.

In addition to vertical gradients, the surface layers also vary horizontally. Sometimes there is a strong dynamical coupling of these lateral gradients and the vertical structure of the surface layers, such as during submesoscale restratification or frontal slumping. These lateral gradients affect lateral transport, restratification, and air-sea fluxes. They are not the emphasis of this chapter, but are the focus of Chapters 8 and 9 of this book. Nonetheless, significant recent work has explored the relationships between lateral gradients and the surface layers, and these relationships will be discussed here.

4.2.1 Mixing and surface layers: Monin–Obukhov scaling

Monin and Obukhov (1954) proposed a scaling theory and similarity solutions for a dry atmospheric boundary layer over an infinite, rough surface, neglecting horizontal heterogeneities, but including heat fluxes through the surface, which distinguishes this theory from the earlier logarithmic "law of the wall" turbulent boundary layer (Von Kármán, 1930). Given the dynamical similarity of the dry atmosphere situation to an oceanic boundary layer, this same approach is often converted to an infinite, oceanic boundary layer under a rigid surface with cooling or heating applied at this surface. Here we will consider some aspects of coupling an M–O ocean underneath an M–O atmosphere.

4.2.1.1 Monin–Obukhov scaling

For simplicity, we will consider both the atmospheric and oceanic boundary layers to be Boussinesq fluids, with a background atmospheric density $\rho_a \approx 1.225$ kg m^{-3} and oceanic density $\rho_o \approx 1025$ kg m^{-3}, taking typical sea level values. This assumption oversimplifies the energy cycle slightly (see Section 4.2.2.3) and has implications for the relationships between salinity and freshwater fluxes (Nurser and Griffies, 2019).

The atmosphere and ocean are connected by an exchange of momentum (stress, τ) and thermal energy. In addition to the air-sea exchange of thermal energy ($Q_{as} = Q_L + Q_S$, positive for ocean warming and composed of latent and sensible components), there is also short-wave and long-wave radiation (R+L, positive for surface warming), which for the purposes of the M–O theory, can be considered to penetrate air without absorption in the boundary layer and penetrate seawater not at all. Reflections, albedo, and a balance between downwelling and upwelling longwave radiation is assumed, so that R+L represent only the energy transferred not the raw insolation or blackbody formulas. The total heating of the ocean applied at the surface in the M–O theory is thus $Q_{net} = R + L + Q_L + Q_S$, whereas the total heating of the atmosphere from the surface is $-Q_L - Q_S$ (assuming negligible net direct radiative heating or cooling of the atmosphere within the boundary layer).

M–O theory is normally framed in terms of a surface friction velocity scale, formed from the stress and used to scale the energy forcing at the surface and turbulent velocities within the boundary layer. To address both fluids, two velocities are needed, which we'll denote with subscript o for ocean and a for atmosphere as for the densities. Within the surface layers, these velocity scales are proportional to the turbulent velocities, and from this knowledge the profiles of mean velocity can be found.

The friction velocities are found from the surface stress,

$$\rho_o u_{*o}^2 = |\tau| = \rho_a u_{*a}^2. \tag{4.1}$$

In the absence of other vector quantities, Galilean invariance suggests that parameterisations of the vector stress should be oriented along the air-sea velocity difference. Taking the traditional sign convention of the stress exerted by the winds on the surface,

$$-\rho_o \overline{v'_o w'_o}^{0^-} = \rho_o u_{*o}^2 \frac{\overline{v}_a - \overline{v}_o}{|\overline{v}_a - \overline{v}_o|} = \tau = \rho_a u_{*a}^2 \frac{\overline{v}_a - \overline{v}_o}{|\overline{v}_a - \overline{v}_o|} = -\rho_a \overline{v'_a w'_a}^{0^+}, \quad (4.2)$$

where **v** indicates the *horizontal* velocity and w the vertical velocity component (later **u** will indicate 3D velocity, i.e., $\mathbf{u} = \mathbf{v} + w\hat{\mathbf{z}}$). The primes denote deviations from the horizontal mean, and the overbar denotes a horizontal mean. The superscript 0^+ and 0^- indicate that this condition holds in the limit as the surface is approached from above and below, respectively, but does not apply in a thin layer near the surface, which is taken to be a frictional sublayer with specified roughness in the standard M–O theory. The consequences of relative velocities, rather than neglecting the ocean velocity, can be large on mesoscales and submesoscales (Renault et al., 2016, 2018; Chassignet et al., 2020). Both of these assumptions are not as clear of a choice in the presence of waves as will be discussed below.

The M–O depth is a key parameter of the theory. In its most basic form, it is

$$L = \frac{1}{\kappa}\frac{-u_*^3}{\overline{w'b'}^0} = \frac{1}{\kappa}\frac{u_*^2}{b_*}, \quad (4.3)$$

where $\kappa = 0.4$ is the von Karman constant and b is the buoyancy of the fluid in question, which in the M–O theory depends only on the potential virtual temperature: atmosphere ($b = g(T_a - T_r)/T_r$) or ocean ($b = g\beta_T(T_o - T_r)$), with subscript r indicating reference values, such as the mixed layer average temperature. The flux buoyancy b_* is just a convenient scale formed from the buoyancy flux divided by $-u_*$; below we will relate it to surface heat and freshwater fluxes. Following Monin and Obukhov (1954), $L_a < 0$ is taken to indicate convective conditions in the atmosphere and $L_a > 0$ to indicate shear turbulence in a stable stratification. With L defined as above, $L_o > 0$ is convective in the ocean, whereas $L_o < 0$ occurs during restratification (following Reichl and Hallberg, 2018). In both the atmosphere and ocean, $z/L < 0$ indicates convective boundary forcing, and $z/L > 0$ indicates stabilising boundary forcing.

The basic interpretation of the M–O theory is that close to the surface, when $|z/L| \ll 1$, the buoyant forces and buoyant production of energy (scaled by $\overline{w'b'}^0$) are very small in comparison to the shear forces and production ($\overline{v'w'}\partial\overline{v}/\partial z$), which lead to the a near-surface that behaves as though all tracers are passive, just as in a neutral boundary layer, where $z/L = 0$. This assumption leads to scaling laws (Section 4.2.2.4) and similarity solutions for this surface layer that depend only on the surface fluxes and distance from the boundary in units of z/L. Recent work has begun to examine how well these ideas extend to more general circumstances and attach to other aspects of the air-sea near-surface layers.

4.2.1.2 Consistent drag laws

M–O theory is often combined with simple scalings for the fluxes between air and sea (Liu et al., 1979; Large and Pond, 1981, 1982; Fairall et al., 1996), although in principle these are independent parameterisations aside from the scaling theory (Monin and Obukhov, 1954). The basic form of these parameterisations is useful to consider here, although more advanced forms are widely considered (e.g., Fairall et al., 1996; Cronin et al., 2019).

$$\tau = -\rho_o \overline{v'_o w'_o}^{0^-} = -\rho_a \overline{v'_a w'_a}^{0^+} = \rho_a c_d |\overline{v}_a - \overline{v}_o|(\overline{v}_a - \overline{v}_o) \quad (4.4)$$

$$|\tau| = \rho_o u_{*o}^2 = \rho_a u_{*a}^2$$

$$Q_S = -\rho_o c_{po}\overline{w'_o T'_o}^{0^-} = -\rho_a c_{pa}\overline{w'_a T'_a}^{0^+} = \rho_a c_{pa}\sqrt{c_d}\sqrt{c_T}|\overline{v}_a - \overline{v}_o|\left(\overline{T_a}^{0^+} - \overline{T_o}^{0^-}\right) \quad (4.5)$$

$$= \rho_o c_{po} u_{*o} T_{*o} = \rho_a c_{pa} u_{*a} T_{*a}$$

$$Q_L = \rho_o L_e \overline{w'_o S'_o}^{0^-}/1000 = -\rho_a L_e \overline{w'_a q'_a}^{0^+} = \rho_a L_e \sqrt{c_d}\sqrt{c_q}|\overline{v}_a - \overline{v}_o|\left(\overline{q_a}^{0^+} - q_s\right) \quad (4.6)$$

$$= -\rho_o L_e u_{*o} S_{*o}/1000 = \rho_a L_e u_{*a} q_{*a}$$

The water vapour mass mixing ratio q underlies the turbulent flux of freshwater away from the surface. The interfacial value of the water vapour mixing ratio q_s is computed from the saturation mixing ratio corresponding to the sea surface temperature and a small correction due to salinity.

The factor of 1000 in the latent heat flux formula results from the practice of reporting salinity in parts per thousand (‰), rather than in mass fraction. The important balance in (4.6) is that in a coordinate system relative to the surface (i.e., sea level pressure), representing freshwater removed through the surface as a turbulent freshwater flux requires a downward salt flux of the same mass (Nurser and Griffies, 2019).

The new M–O flux parameters—T_{*a}, T_{*o}, q_{*a}, S_{*o}—like b_* are simply the surface limit of the turbulent fluxes divided by $-u_{*a}$ or $-u_{*o}$. The specific heat capacities at constant pressure are $c_{po} \approx 3986$ J kg^{-1} K^{-1} and $c_{pa} \approx 1000$ J kg^{-1} K^{-1}, whereas the latent heat of vaporisation is $L_e \approx 2.5 \times 10^6$ J kg^{-1}. The transfer coefficients c_d, c_T, c_q are in fact functions of the reference height, from which the atmospheric velocities are taken, scaled by L_a (Fairall et al., 1996), so these vary depending on the local value of the fluxes through L_a. In data-driven ocean model comparisons (Tsujino et al., 2020; Chassignet et al., 2020), calculating these bulk fluxes from observed atmospheric states is an important part of forcing models (Large and Yeager, 2004, 2009).

4.2.2 Near-surface distinctions from M–O theory and each other

While the M–O similarity theory is a fundamental grounding for basic relationships in the atmospheric and oceanic near-surface layers, and especially the relationships between turbulent velocity scales and boundary forcing, it is insufficient to describe many aspects of interest in these layers. In this section, the deviations from M–O theory are highlighted.

4.2.2.1 Salinity and humidity

Latent heating (i.e., evaporation) affects atmospheric moisture and oceanic salinity through freshwater removal. The M–O theory considers buoyancy driven by air-sea heat fluxes. However, the atmospheric boundary layer has the additional effects of humidity, which affect both the buoyancy of air and the thermodynamics. Dry, hot air leads to faster evaporation, whereas moister air is more buoyant, thus air-sea fluxes account for these effects of water vapour through the implementation of *virtual temperature*.

$$\overline{w'T'_v}^{0+} \approx \overline{w'T'}^{0+} + 0.61 T_r \overline{w'q'}^{0+} \tag{4.7}$$

The surface buoyancy flux needs to be adapted to include virtual temperature forcing, as does L_a:

$$L_a = \frac{1}{\kappa} \frac{-u_*^3}{\overline{w'b'}^{0+}} = \frac{T_r}{\kappa g} \frac{-u_*^3}{\overline{w'T'_v}^{0+}} = \frac{T_r}{\kappa g} \frac{-u_*^3}{\overline{w'T'}^{0+} + 0.61 T_r \overline{w'q'}^{0+}} = \frac{T_r}{\kappa g} \frac{u_*^2}{T_{*a} + 0.61 T_r q_{*a}}, \tag{4.8}$$

$$b_{*a} = g(T_{*a}/T_r + 0.61 q_{*o}). \tag{4.9}$$

Similarly, in the ocean, the buoyancy flux includes the salinity flux, so

$$L_o = \frac{1}{\kappa} \frac{-u_*^3}{\overline{w'b'}^{0-}} = \frac{1}{\kappa} \frac{-u_*^3}{\beta_T \overline{w'T'}^{0-} - \beta_S \overline{w'S'}^{0-}} = \frac{1}{\kappa} \frac{u_*^2}{\beta_T T_{*o} - \beta_S S_{*o}}, \tag{4.10}$$

$$b_{*o} = g(\beta_T T_{*o} - \beta_S S_{*o}). \tag{4.11}$$

The salinity flux is related to the evaporation minus precipitation at the sea surface. Here a linear equation of state for seawater is assumed, with approximate expansion/contraction coefficients $\beta_T \approx 1.67 \cdot 10^{-4} K^{-1}$, $\beta_S \approx 0.78 \cdot 10^{-4} ‰^{-1}$. Ordinarily, since the M–O theory is intended to be used under turbulent, neutral, convective (evaporating), and stabilising (condensing or precipitating) forcing conditions, we will assume that the freshwater flux (and thus latent heat) can be connected to the turbulent salt flux as in (4.6), rather than stabilising so strongly that the theory doesn't apply (e.g., the perfect fluid thought experiment of Nurser and Griffies, 2019).

Salinity is different dynamically from humidity in many ways. Salinity decreases buoyancy and humidity increases it, but salty water and humid air near the surface both tend to accelerate convection. Both are produced by evaporation, which implies a transfer of latent heat and freshwater from the ocean to the atmosphere (typically as the ocean absorbs excess radiative or sensible heating). For typical oceanic salinities, the evaporation rate is minimally affected by the salinity of the evaporating water. Likewise, the location where precipitation falls on the ocean surface does not depend on the salinity. Thus unlike temperature and humidity, salinity anomalies do not have a major feedback through air-sea processes, thus lateral freshwater sources can bring distant effects (MacKinnon et al., 2016). Salinity fronts often drive adjustments if they are not compensated by temperature gradients, in this form they can persist over long time scales, because they are more stable to air-sea interaction (Rudnick and Ferrari, 1999; Ferrari and Paparella, 2003). However, in the atmosphere, variability in sea surface temperature will drive variations in sensible and latent heat fluxes that may drive interactions within the coupled system, as may humidity gradients drive variations through their effect on latent heat fluxes (4.6).

4.2.2.2 Free convection layers and velocities

An extension of the M–O theory that has long been in use is the concept of free convection velocities. Higher in the atmosphere approaching the capping layer ($z/L_a \ll 1$), and deeper in the ocean layers, approaching the mixed layer base ($z/L_o \ll 1$), destabilising surface conditions lead to convective plumes that extend beyond the surface log-layers into a free convection region. This free convection is normally halted by the mixed layer base or a capping inversion, inducing entrainment. How big is the typical convective velocity?

The free convection velocities (w_*) are useful in describing motion where convection dominates over shear forcing. These velocities are not part of the M–O theory, as they involve a new parameter—the mixed layer depth—that is not prescribed by the surface forcing.

$$w_{*a}^3 = \overline{w_a'b_a'}^{0+} h_a, \qquad w_{*o}^3 = \overline{w_o'b_o'}^{0-} h. \tag{4.12}$$

These buoyancy fluxes should include the humidity and salinity contributions in addition to the buoyancy from anomalous temperatures of air and seawater. Many 1D parameterisations (Section 4.4.2) depend on both the friction and free convection velocities.

One of the smallest submesoscale instabilities, symmetric instability, links to convection within a well-mixed layer, along with more negative z/L transitions into along-isopycnal mixing of tracers and momentum. The depth at which this second effect ends is where the layer over which potential vorticity is well-mixed and opposite in sign to f ends (Thomas et al., 2013; Hamlington et al., 2014; Bachman et al., 2017; Dong et al., 2021). This symmetric instability or PV mixed layer is deeper than the convective (density) mixed layer when symmetric instability is active (Hamlington et al., 2014; Dong et al., 2021).

4.2.2.3 Penetrating radiation

A key difference between the atmosphere and the ocean is the impact of penetrating solar radiation in the ocean. In clear open ocean conditions, there is sunlight appreciably tens to hundreds of metres below the surface. This sunlight warms the ocean *below the surface*, and under certain conditions can trigger convection, because the surface is simultaneously cooled by longwave, sensible, and latent heat fluxes even when the net heating of the ocean is zero. Solar heating (R) occurs at depths (pressures) greater than the cooling (at the surface by $Q_L + Q_S + L$). Were this forcing steady, the situation would resemble the flux Rayleigh–Bénard problem studied by Otero et al. (2002). Under winds, penetrative surface fluxes can also enhance stabilising near-surface ocean boundary layers and result in strong diurnal warm layers (Hughes et al., 2020). The penetrative fluxes on their own tend to be stabilising, because they heat the near-surface more than at depth, but when combined with surface cooling the effect can trigger convection without net surface fluxes. This aspect of the oceanic boundary layer is often neglected in parameterisations and simulations of boundary layer mixing.

4.2.2.4 Budget equations

Consider a free atmosphere velocity $\mathbf{V}_a = (U_a, V_a, W_a)$ and an ocean pycnocline velocity $\mathbf{V}_o = (U_o, V_o, W_o)$. The near-surface layers link these two reservoirs of momentum together, but the turbulence within each boundary layer also allows for deviations in the mean velocity profile from these outer velocities. The horizontally homogeneous momentum equations for the velocities within the boundary layers ($\mathbf{v}_o, \mathbf{v}_a$) can be written as

$$\frac{\partial \bar{u}_a}{\partial t} + \frac{\partial \overline{u_a'w_a'}}{\partial z} = f(\bar{v}_a - V_a), \qquad \frac{\partial \bar{v}_a}{\partial t} + \frac{\partial \overline{v_a'w_a'}}{\partial z} = f(U_a - \bar{u}_a); \tag{4.13}$$

$$\frac{\partial \bar{u}_o}{\partial t} + \frac{\partial \overline{u_o'w_o'}}{\partial z} = f(\bar{v}_o - V_o), \qquad \frac{\partial \bar{v}_o}{\partial t} + \frac{\partial \overline{v_o'w_o'}}{\partial z} = f(U_o - \bar{u}_o). \tag{4.14}$$

If we integrate each of these equations from the surface to the extent of their boundary layers, where the turbulence vanishes (or slightly farther so as to have integral bounds fixed in time), multiply each equation by the corresponding background density, then

$$\frac{\partial}{\partial t}\int_0^{h_a} \rho_a \bar{u}_a \mathrm{d}z + \tau_u = f\int_0^{h_a} \rho_a(\bar{v}_a - V_a)\mathrm{d}z, \quad \frac{\partial}{\partial t}\int_0^{h_a} \rho_a \bar{v}_a \mathrm{d}z + \tau_v = f\int_0^{h_a} \rho_a(U_a - \bar{u}_a)\mathrm{d}z; \tag{4.15}$$

$$\frac{\partial}{\partial t}\int_{-h}^{0} \rho_o \bar{u}_o \mathrm{d}z - \tau_u = f\int_{-h}^{0} \rho_o(\bar{v}_o - V_o)\mathrm{d}z, \quad \frac{\partial}{\partial t}\int_{-h}^{0} \rho_o \bar{u}_o \mathrm{d}z - \tau_v = f\int_{-h}^{0} \rho_o(U_o - \bar{u}_o)\mathrm{d}z. \tag{4.16}$$

Note that τ transfers momentum from the atmosphere to the ocean. Relatedly, in steady state the (turbulent) Ekman transport is given by the right-hand integral in each equation; it is an anomaly from the free atmosphere and free ocean velocity, and the Ekman mass transport in the atmosphere is equal and opposite to the Ekman mass transport in the ocean. Considering the effect of the time dependence, it is clear that an inertial oscillation can be superimposed on the atmosphere or ocean without necessarily affecting the air-sea fluxes, unless τ at the surface changes to oscillate as well.

Note that although we have neglected horizontal variations in properties, often the free atmosphere velocity $\mathbf{V}_a$ and the ocean pycnocline velocity $\mathbf{V}_o$ are assumed to be geostrophic, which then implies that the pressure does vary in the horizontal direction. Taking this thinking farther, if these velocities are also hydrostatic, the vertical derivative of (4.13)–(4.14) results in a steady-state balance called the turbulent thermal wind balance (TTW; McWilliams (2016)). This balance is fundamental to submesoscale-boundary layer turbulence interactions at baroclinic fronts, and relates lateral gradients in density to baroclinic shear in the boundary layer—affected by turbulent vertical momentum fluxes as well. It is possible that the free atmosphere and ocean pycnocline velocities are in geostrophic balance with barotropic pressure anomalies, which would leave shear in the boundary layers as turbulent Ekman anomalies as in (4.15)–(4.16).

If the location in consideration is not in the centre of a submesoscale front, then lateral density gradients and baroclinicity are expected to be small. Monin and Obukhov (1954) go on to argue that the Coriolis parameter and time dependence may also be neglected near the surface. In this case, dimensional analysis proposes that the gradients in each profile property depend only on the fluxes, the distance from the boundary, and the M–O depth, or

$$\frac{\kappa z}{u_{*a}}\frac{\partial \overline{v}_a}{\partial z} = \phi_m\left(\frac{z}{L_a}\right), \qquad \frac{\kappa z}{T_{*a}}\frac{\partial \overline{T}_a}{\partial z} = \phi_h\left(\frac{z}{L_a}\right), \qquad \frac{\kappa z}{q_{*a}}\frac{\partial \overline{q}_a}{\partial z} = \phi_c\left(\frac{z}{L_a}\right); \qquad (4.17)$$

$$\frac{\kappa z}{u_{*o}}\frac{\partial \overline{v}_o}{\partial z} = \phi_m\left(\frac{z}{L_o}\right), \qquad \frac{\kappa z}{T_{*o}}\frac{\partial \overline{T}_o}{\partial z} = \phi_h\left(\frac{z}{L_o}\right), \qquad \frac{\kappa z}{S_{*o}}\frac{\partial \overline{S}_o}{\partial z} = \phi_c\left(\frac{z}{L_o}\right). \qquad (4.18)$$

The log layer behaviour is found by recognising all these similarity functions should go to one as z approaches zero. These similarity functions are universal, so long as the assumptions of M–O theory hold, so the same functions should apply in the atmosphere and ocean. The functions do differ depending on the sign of z/L, so there is different behaviour under convective ($z/L < 0$), neutral ($z/L = 0$), and stabilising ($z/L > 0$) forcing. Högström (1996) and Wyngaard (2010) provide many examples of empirical fits based on observations.

The fluxes of buoyancy, salinity, and temperature consistent with the similarity functions above can be found from the conservation equations, as can equivalent eddy diffusivity and viscosity functions. This approach provides a partial basis for many of the 1D parameterisations (Section 4.4.2).

Another primary basis for parameterisations and observation are the equations for conservation of energy, thermal, potential, and mechanical. In Boussinesq fluids, the thermal energy equation is decoupled from the others (and amounts to an equivalent of the conservation of temperature). The kinetic energy, temperature variance, and buoyancy transport equations are (written here for the ocean, but equivalent in the atmosphere as well), again assuming no lateral variations of any statistics:

$$\underbrace{\frac{\partial}{\partial t}\frac{\overline{\mathbf{u}'_o \cdot \mathbf{u}'_o}}{2}}_{\text{TKE tend.}} + \underbrace{\overline{\mathbf{u}'_o w'_o}\frac{\rho_o \partial \overline{\mathbf{u}}_o}{\partial z}}_{\text{shear prod.}} + \underbrace{\frac{\partial}{\partial z}\frac{\overline{\mathbf{u}'_o \cdot \mathbf{u}'_o w'_o}}{2}}_{\text{TKE transport}} = \underbrace{-\frac{1}{\rho_o}\frac{\partial}{\partial z}\overline{p'w'}}_{\text{pressure work}} + \underbrace{\overline{w'b'}}_{\text{buoy. prod.}} - \underbrace{\epsilon}_{\text{diss.}}, \qquad (4.19)$$

$$\underbrace{\frac{\partial}{\partial t}\frac{\overline{T'^2_o}}{2}}_{\text{Var. tend.}} + \underbrace{\overline{w'_o T'_o}\frac{\partial \overline{T}_o}{\partial z}}_{\text{prod.}} = \underbrace{\chi}_{\text{diss.}}, \qquad (4.20)$$

$$\underbrace{\frac{\partial}{\partial t}\frac{\overline{w'_o b'_o}}{2}}_{\text{Flux tend.}} + \underbrace{\overline{w'^2_o}\frac{\partial \overline{b}_o}{\partial z}}_{\text{prod.}} + \underbrace{\overline{b'_o w'_o}\frac{\partial \overline{w}_o}{\partial z}}_{\text{prod.}} = -\overline{b'^2_o} + \underbrace{\upsilon}_{\text{diss.}}. \qquad (4.21)$$

These are examples of second-moment equations, of the type underlying a particular class of parameterisations (Section 4.4.2.3). For examples of how these equations may be scaled by extending the M–O theory, see Grant and Belcher (2009) and Belcher et al. (2012).

Potential energy, e.g., from unstable stratification, including humidity (leading to convective available potential energy, or CAPE) and salinity effects can be converted into kinetic energy through the buoyancy production term. Alternatively, stratification can be destroyed (i.e., mixed) at the cost of buoyancy production. Note that penetrating solar radiation can lead to positive buoyancy production, thus making the ocean act as a heat engine or Carnot cycle with solar radiation and surface cooling as the thermal reservoirs. Precipitation also removes moisture from the air and adds it to the ocean, although the latent energy released during condensation before precipitation tends to affect the cloud layer rather than the surface. The sensible heating or cooling of the ocean surface due to precipitation at a different temperature is typically neglected. Latent heat during snow falling on the ocean can be significant (e.g., Ledley, 1991).

The temperature variance equation is also a useful budget to consider when making measurements of dissipation (Section 4.3). The buoyancy flux tendency equation is a bit unfamiliar perhaps, but it is formed in a similar manner to the other two equations and it serves to illustrate that fluxes also have a budget, although they are not materially conserved in perfect fluids.

4.2.2.5 Time-varying forcing

While the M–O theory and many modelling studies use constant-in-time forcing and statistically steady budgets, in most ocean regions diurnal, synoptic, and seasonal variability is critical to understanding the upper ocean layers. Only for brief moments in the morning and afternoon are the conditions within the whole atmospheric boundary layer approximately neutral, instead convective or stabilising boundary conditions are typical. Similarly, the ocean varies back and forth through stabilising and convective conditions, but even when the surface flux is zero the near-surface ocean can be driven toward convection by the penetrating solar radiation. Interestingly, the ocean is destabilised at night and stabilised during the day, whereas the atmosphere is opposite. Note that just using the daily-averaged solar radiation is not a good approximation, as mixing and restratification are distinct processes with significant irreversibility. The atmospheric boundary layer is also significantly different under convection (including rainfall) and subsidence due to the precipitation on updrafts and the related thermodynamic and buoyancy implications.

Additionally, over the past decade significant resources have been exerted into understanding the forcing and implications of near-inertial oscillations. These phenomena are readily excited by high-frequency winds and can drive significant shear near the mixed layer base and entrainment (Jochum et al., 2013; Alford et al., 2016). Cyclones drive a rapid mixed layer deepening to one side of the eye of the storm due to wind circulations that resonate with inertial excitations (Sanford et al., 2007).

While the M–O scaling parameters are still useful quantifications of the magnitude of forcing, when time-dependence is involved other aspects of the problem enter and a similarity solution is not readily found. Both the initial conditions and present conditions affect the boundary layer in this case (Pearson et al., 2015).

4.2.2.6 Lateral variation

As in the case of time-varying forcing, when fluxes vary laterally the M–O theory fails and other parameters enter. The case above of baroclinicity arising from a submesoscale front is an excellent example. The response of the atmospheric boundary to this case is studied by Sullivan et al. (2020), but the oceanic front case is also interesting (e.g., Hamlington et al., 2014; Taylor, 2018; Bodner et al., 2020), especially the turbulent thermal wind balance and its consequences (McWilliams, 2016). Note that while the turbulent Ekman case described above is almost trivial, considering Ekman pumping requires lateral variations, vertical vorticity of free atmosphere winds or pycnocline velocity, and surface divergence within the layer. For this reason, Ekman pumping goes beyond the M–O theory (although it is illustrated in Fig. 4.2).

The turbulent thermal wind (TTW) is another approach as to how lateral variations and turbulent mixing can be combined. The basic principle is that a three-way balance that adds turbulent mixing into the thermal wind relation captures important aspects of turbulence acting on fronts. This theory has been applied to analysis of simulations (McWilliams et al., 2015; Sullivan and McWilliams, 2019) and in analytic work (Crowe and Taylor, 2018; Bodner et al., 2020) recently.

4.2.2.7 Waves

Perhaps most of the effort in pushing oceanic boundary layer theory outside of the reach of M–O theory in the past decade has involved surface waves. This section will discuss some ways that waves do not fit into the M–O theory.

A variety of papers have reworked the M–O approach to include aspects of waves. Large et al. (2019) is a recent example, where a separable similarity function assumption is made, with the standard similarity function multiplied by a function depending on the turbulent Langmuir number,

$$La_t^2 = \frac{u_{*o}}{u_s^{0^-}}, \qquad (4.22)$$

where $u_s^{0^-}$ is the surface Stokes drift speed of the surface waves. However, this is only one of many possible wave parameters to include from the wave field: in CESM2, the wave model has nearly 10 times as many degrees of freedom as the whole ocean model profile to depth on the same horizontal grid (Li et al., 2017).

In the past, because most wave energy is carried in wind waves that eventually equilibrate to τ, it was assumed that there was no need to include wave properties outside of τ which is already part of the M–O theory (e.g., McWilliams and Restrepo, 1999). However, when operational wave models were used to test this hypothesis, it was found that waves are rarely in equilibrium in direction (Hanley et al., 2010), or magnitude, frequency, and penetration depth (Webb and Fox-Kemper, 2011), or in uniformity of direction (Webb and Fox-Kemper, 2015). Instead, surface waves tend to spread out in direction and may propagate long distances from their origin as swell. What does this amount to in wave impacts on boundary layers? It is important to note that careful observations do detect systematic differences between M–O theory predictions and turbulence statistics that are consistent with some theories of wave-enhanced turbulence (D'Asaro, 2014; D'Asaro et al., 2014; Sutherland et al., 2013, 2014; Zheng et al., 2021). The Charnock (1955) scaling for wind stress given wind speed does include waves, but only in a modest dimensional analysis manner. Non-equilibrium waves (which would depend on initial wave state) and remotely-generated waves (which depends on conditions elsewhere, e.g., crossing swell) are not part of Charnock's scaling. So, a key question is to what extent disequilibrium waves are important.

A first series of efforts were carried out to examine the effects of breaking waves, albedo from whitecaps, spray, etc. (see Cavaleri et al., 2012, for a review). These effects are often small on average, but some are predictable and have been parameterised in their impact on air-sea fluxes primarily (Section 4.2.2.10) and secondarily on surface transport properties (see Lenain and Pizzo, 2020, for a recent example). These are not disequilibrium events; indeed they are part of what allows for fully developed wave states to occur (Sullivan et al., 2004). Zheng et al. (2021) show a variety of ways that observations, especially profiles of temperature, are inconsistent with M–O theory but are consistent with wave breaking and Langmuir turbulence theory.

The depth to which wave effects penetrate is a key aspect of the wave problem, and is a disequilibrium situation as it takes some time for the mean wave period to arrive at a fully developed value (Webb and Fox-Kemper, 2011). Like the penetrating solar problem, it violates the assumption of the M–O theory that all of the effects take place near the surface. One approach is to consider the vertical decay scale of the orbital velocity, which is proportional to the wavelength of the waves (Qiao et al., 2004, 2010). Other approaches use the Stokes drift decay scale, or more precisely, an average of Stokes drift effects over a fraction of the boundary layer (Harcourt and D'Asaro, 2008; Van Roekel et al., 2012; Li et al., 2016). Because the Stokes drift decay scale tracks closely over seasons and forcing with the orbital velocity decay scale, both approaches yield similar sensitivity to wave states.

Entrainment is an important result of deeply penetrating Langmuir turbulence or non-breaking wave turbulence. Langmuir jets penetrate much more deeply than the shallow Stokes drift that enhances them (Polton and Belcher, 2007). Langmuir entrainment has important consequences for passive tracers (Smith et al., 2016) and mixed layer deepening (Li et al., 2017). Entrainment is elevated even when Langmuir cells (as recognised by their form, anisotropy, and turbulence statistics) do not reach the mixed layer base (Li and Fox-Kemper, 2020). Thus observations of Langmuir cells remaining trapped in the upper mixed layer do not imply that wave-enhanced mixing does not accelerate entrainment (as suggested by, e.g., Weller and Price, 1988; Thorpe et al., 2003); it does so by enhancing the shear near the mixed layer base via pressure work.

Non-breaking surface waves had not been considered to be able generate turbulence, because the governing equation of non-breaking surface waves is simplified as irrotational mathematically (Phillips, 1961). However, the groundbreaking work of Teixeira and Belcher (2002) showed that if there is vorticity already in turbulence in the ocean, then passing irrotational waves can intensify it. This conception is consistent with both the theory of Langmuir turbulence, effectively defined to be the class of turbulence that arises in simulations of the Craik–Leibovich or wave-averaged equations that include a Stokes vortex or Stokes shear force (McWilliams et al., 1997; Holm, 1996; Suzuki et al., 2016; Holm and Hu, 2021), often by comparison to large eddy simulations of these equations (Section 4.4.1). An important aspect to understand about the wave-averaged equations is that the only wave information that appears in them is the Stokes drift profile, hence the approach of Large et al. (2019) and the focus on Stokes drift in Webb and Fox-Kemper (2011, 2015). The non-breaking surface waves theory seeks to similarly explain the enhanced turbulence due to the presence of surface waves. The non-breaking surface-wave-induced vertical mixing scheme (Bv), however, does not phase-average the effect and instead relies on high-frequency observations of laboratory and field sites to study the enhancement effect (e.g., Qiao et al., 2016). Qiao et al. (2016) provided solid experimental evidence of wave-turbulence interactions in the ocean directly for the first time, by employing the newly developed Holo–Hilbert spectral analysis (Huang et al., 2016). Pragmatically, the distinction between the two theories lies in whether mixing varies more directly in response to the orbital velocity characteristics (non-breaking

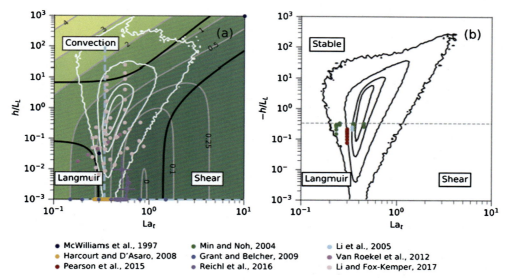

FIGURE 4.3 Regime diagram of Langmuir forcing of the world oceans based on JRA55-do (Tsujino et al., 2020) forcing reproduced with permission from (Li et al., 2019) following Belcher et al. (2012). The location where large eddy simulations of the wave-averaged equations have been performed are indicated by dots (McWilliams et al., 1997; Harcourt and D'Asaro, 2008; Pearson et al., 2015; Min and Noh, 2004; Grant and Belcher, 2009; Reichl et al., 2016; Li et al., 2005; Van Roekel et al., 2012; Li and Fox-Kemper, 2017). (a) Regions and seasons where convective forcing is present, indicating in solid lines, where 90% of the dissipated energy comes from convective, Langmuir (i.e., Stokes shear production), or shear production. (b) Regions and seasons where convective forcing is stabilising (e.g., surface warming). Above the dotted line, turbulence is expected to be only transient, whereas below the Langmuir or shear forcing exceeds the stabilising buoyancy forcing.

theory) or the Stokes drift statistics (Langmuir turbulence), although it can be challenging to distinguish them as they co-vary.

Another wave parameter that is important for the effects of Langmuir turbulence is the degree of misalignment of waves and winds. Hanley et al. (2010) showed that this was a common occurrence, and Van Roekel et al. (2012) showed that a simple approach of projecting waves and winds into the direction of the forming Langmuir cells collapses many misalignment cases onto the same profile. Van Roekel et al. (2012) go on to suggest that a redefined Langmuir number can capture both the misalignment and the depth effects (following Harcourt and D'Asaro, 2008) in one parameter, simplifying the dimensional analysis. Other studies of wave-wind misalignment have found that this simple approach is fairly robust, but could be improved (Sullivan et al., 2012; Wang et al., 2019a).

In Belcher et al. (2012), an important outcome was the establishment of a regime diagram based on the sources of energy, similar to scaling (4.19), but including Stokes forces. This same diagram was expanded upon and more carefully evaluated in Li et al. (2019) and is reproduced in Fig. 4.3. The vertical axis shows a dimensionless parameter that is similar to the boundary layer depth over the M–O length, but including wave effects. When this parameter is negative (right panel), the surface forcing is warming and stabilising. When it is positive, it is tending to convect—more strongly, the larger its magnitude. Along the horizontal axis is the turbulent Langmuir number, which indicates the relative amount of winds (to the right) versus waves (to the left). The regime diagram thus indicates the regions where convection, Langmuir forcing (i.e., wave energy), and wind forcing are most important. The white triangular contours map where realistic forcing of the world ocean places the highest likelihood. Coloured dots indicate where large eddy simulations have been carried out.

Li et al. (2019) also show that even among Langmuir schemes that purport to include the same effects and converge to the same LES there are considerable differences in the outcomes of applying the different schemes based on physical and numerical formulation. The same is likely true of the non-breaking wave approaches. Johnson et al. (2021) explore a dynamical systems approach to quantifying the divergent behaviours of different schemes under a variety of forcings.

4.2.2.8 Waves and momentum

Along with the wind-wave misalignment problem comes the general nature of the Stokes vortex or Stokes shear force (Suzuki and Fox-Kemper, 2016). This effect results in a downward force, acting like a density anomaly in the hydrostatic or nonhydrostatic equations, that is directional: only when the Lagrangian current (the combination of Eulerian mean current and Stokes drift) is oriented along the Stokes drift direction does this force maximise. When the current and waves are opposed, it is an upward force, and when they are perpendicular, it is zero.

This conceptualisation simplifies the mechanism for Langmuir turbulence. Langmuir cells converge when surface currents randomly flow in the Stokes drift direction. This causes convergences which strengthens the jets, forms turbulence and can accelerate frontogenesis (Suzuki et al., 2016) and filamentogenesis (Sullivan and McWilliams, 2019).

The air-sea transfer coefficient c_d is also likely directional. In the presence of crossing swell, there should be favoured directions for efficient momentum transfer (winds into swell) and others that are less so (winds behind swell or crossing across swell). The anisotropy of the drag coefficient (and likely also the heat and moisture coefficients, due to the changing near-surface airflow) implies that the relative velocity to stress relationship in (4.4) is probably an oversimplification, and the drag coefficient should instead be a second-rank tensor that depends on swell propagation direction.

4.2.2.9 Sea ice

As the sea ice interface sets the surface boundary conditions, stress at the ice-ocean interface depends on undersurface roughness, thereby altering u_*, whereas melting/freezing ice induce a negative/positive buoyancy flux, respectively. In the neutral limit, McPhee and Martinson (1994) show the importance of kinematic boundary stress and Ekman dynamics on turbulent mixing lengths, often incorporating rotation not included in the M–O theory. In the presence of stable (McPhee, 1994) or unstable (McPhee and Stanton, 1996) freshwater fluxes, similarity theory can be modified to account for distance from the ice, freshwater buoyancy flux, surface stress, and boundary layer stratification.

Sea ice leads to lateral intermittency in the ocean, as waves, fluxes, and albedo are strongly inhibited by the presence of sea ice. Under Arctic climate change, it is expected that a newly revealed ocean boundary layer will be in regular contact with the atmosphere during the Arctic summertime later this century. The opening up of the Arctic has consequences for planetary albedo (e.g., radiation arriving at the ocean), air-sea heat fluxes, and ocean heat uptake and surface waves in the Arctic. Recent studies have covered some aspects of these dynamics and their impact on expected changes (e.g., Horvat et al., 2016; Mensa and Timmermans, 2017; Manucharyan and Thompson, 2017; Roach et al., 2018).

4.2.2.10 Bubbles, spray, whitecaps, & foam

Like sea ice, bubbles, spray, and foam are multiphase phenomena that exist within the near-boundary layers. Bubbles and spray strongly affect the surface area, and thus can be important in air-sea fluxes, as well as in buoyancy effect (e.g., Liang et al., 2013, 2017). Foam, and other persistent surface anomalies, such as biological surfactants, can strongly affect surface roughness and wave dynamics (Cavaleri et al., 2012). Intriguingly, even the intermittent, stochastic forcing of rain droplets can interfere with the formation of coherent small-scale waves, thus affecting roughness and stress (Cavaleri et al., 2015; Zhao et al., 2017). At high wind speeds, bubbles and spray form some of the most challenging aspects of simulating and observing conditions (e.g., Soloviev et al., 2014). The fundamental scaling aspects of these "mushy layer" as opposed to surface problems remain highly speculative, as direct modelling of multiscale, multiphase fluids is not sufficiently advanced. Zhao et al. (2017) and Bao et al. (2020) demonstrate that parameterisations indicate a key role of sea spray in the simulation and prediction of typhoons and the climate system respectively.

4.2.2.11 Shallow water

In very shallow water, the ocean surface and bottom boundary layers merge, so that the whole depth is diabatic and turbulent. Studies on Langmuir super-cells (Gargett et al., 2004) follow the consequences into bottom boundary impacts of these surface-forced phenomena. Many shallow estuaries have essentially no vertical stratification during wintertime, which makes the lateral salinity and temperature gradients key to understanding the flow (e.g., Codiga and Ullman, 2010; O'Donnell et al., 2014; Jing et al., 2016).

4.2.3 Mixed layers: boundary layer memory

The mixed layer of the ocean in schematic form is straightforward (Fig. 4.2): how deep has the turbulent mixing homogenised the vertical temperature and salinity fields? But in practice the temporal and spatial variability and poor subsurface sampling lead to many issues in determining the mixed layer depth conclusively. The surface forcing in a particular place and time has little to do with the previous forcing and potentially also restratification processes that have brought the mixed layer to its present properties. There are thus a variety of definitions of mixed layer, which are very briefly summarised here. Some aspects of the dynamical significance of the different methods are also given.

The mixed layer is typical found based on one of two approaches: (1) a threshold for when deep properties have gone "far enough" away from their surface value to be different, and (2) when tracer values exceed a relative level of variability that is determined in the same region or cast. A critical breakthrough for either method was found by de Boyer Montégut

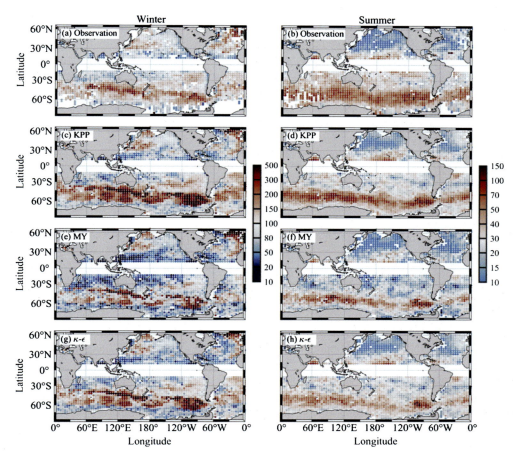

FIGURE 4.4 Mixed-layer depth m as determined using the (Huang et al., 2018) method from 2010–2019 Argo profiles in February and September (by summer and winter in both hemispheres) and 1D simulations of the generalised ocean turbulence model in the same seasons (Umlauf and Burchard, 2005) initialised from profiles and using surface forcing from a global MITgcm simulation (Su et al., 2018). Reproduced with permission from Dong et al. (2020).

et al. (2004). They noted that calculation of the mixed layer depth by most methods and lateral averaging of the properties (salinity, temperature, density) from observational profiles do not commute. That is, if you calculate the mixed layer depth on the mean properties (e.g., Monterey and Levitus, 1997) you do not get the same answer as when you calculate the mixed layer depth on each observational profile, and then laterally average. Similarly, when a simulation is initialised from observations, it should be initialised from a single profile, not an average (e.g., D'Asaro et al., 2014). The latter is clearly the more important approach for the dynamics discussed in this chapter. Other authors have continued to improve algorithms using the profile approach (Holte et al., 2017; Huang et al., 2018); quite a few others are reviewed in Dong et al. (2021) and Li et al. (2019).

The boundary layer can be observed in terms of turbulence dissipation with specialised observations, but the mixed layer relies on definitions of what scale of change indicates leaving the boundary. As noted by de Boyer Montégut et al. (2004) and Holte et al. (2017), there is no simple threshold on temperature or density change from the surface value that is reliable worldwide. Even what variable should be used is subject to interpretation: temperature, density, potential vorticity? The relative variance methods have the advantage that presumably the variability that is in the same profile cast gives some sense of the range of surface forcing variability typical of the location (indeed, that is the purpose of L). However, these algorithms tend to have a pattern recognition rather than a dynamical determination of relevant thresholds and limits, thus they too disagree among themselves.

Fortunately, (or unfortunately depending on your perspective) parameterisations of mixed layer processes tend to disagree more than definitions of mixed layers (Li et al., 2019), so for model-observation comparisons one is free to choose among reasonable definitions. Fig. 4.4 shows observed mixed layer depths versus 1D mixing models initialised from a high-resolution model, and then allowed to simulate ocean near-surface layer turbulent mixing (Dong et al., 2021).

4.2.3.1 Entrainment

This chapter has emphasised the air-sea exchange of properties, but perhaps even more societally significant is the exchange of properties across the mixed layer base to the deep ocean. Below the climatic implications of this statement will be described, but here it is important to lay out the dynamics involved in exchanges of properties between the mixed layer and the interior.

As described above, entrainment is a process whereby turbulence escapes the mixed layer to penetrate into the stratified fluids below. By homogenising the properties, the mixed layer deepens and becomes more dense because of this entrained fluid.

Entrainment is a highly complex process, however, and difficult to model and observe due to its intermittent character. Though wind-driven entrainment is a principle concept for OSBL evolution (e.g., Pollard et al., 1973; Price, 1979) and a foundation for upper-ocean mixing model validation (Mellor and Yamada, 1982; Large and Crawford, 1995), intermodel comparison reveals turbulence parameterisations fail to agree on how to deepen the mixed layer in practice (e.g., Li et al., 2019; Johnson et al., 2021). However, recent models (Li and Fox-Kemper, 2017; Van Roekel et al., 2018; Li and Fox-Kemper, 2020; Souza et al., 2020) and some observational analyses (D'Asaro, 2003) have managed to capture entrainment in a statistically and dynamically meaningful way. Given the important role of entrainment in the earth system, it has not received enough study.

4.2.3.2 Restratification

A restratifying mixed layer influences the climate reservoirs on both sides of its boundaries. With the atmospheric boundary layer through its transfer of heat and momentum–i.e., (4.15)–(4.16)–and with the ocean interior as the deeper mixed layer restratifies and joins the pycnocline. The implications of this latter process are featured in the Stommel (1979) demon idea: only the deepest wintertime mixed layer properties are permitted to stratify and join the stratified interior. In contrast to entrainment, which incorporates tracers and momentum from the underlying water, restratification decreases the volume of the mixed layer, but not its intrinsic properties.

Restratification in the ocean can be caused by the following: penetrating solar radiation (which decays exponentially, and thus heats the surface more quickly than deeper waters); surface heating by longwave or warm, moist air; winds that blow light water over dense (i.e., upfront winds); precipitation and condensation of fresh water; or dynamical restratification that extracts potential energy by sorting lighter waters to move above denser. Restratification is a determinative aspect of whether the mixed layer and mixing layer differ, as rapid restratification will tend to push the mixed layer toward the mixing layer, whereas leisurely restratification will allow patterns of mixing to become fossilised in mixed layers far deeper than presently active mixing.

Mixed layer instabilities, and baroclinic instabilities in general, extract energy from the potential energy inherent in stratification anomalies (Boccaletti et al., 2007; Fox-Kemper et al., 2008). The stratification anomalies do not have to be in contact with the surface for the mixed layer instabilities to arise. Submesoscale eddies formed from such instabilities are an important mechanism for restratification at the base of the mixed layer, where often solar restratification and Ekman buoyancy fluxes do not penetrate (Haney et al., 2012).

In addition to baroclinic instabilities, frontal restratification by other mechanisms can influence the upper ocean buoyancy budget. Processes involved include when horizontal buoyancy gradients are differentially advected by near inertial shear (Hosegood et al., 2006), time dependent Ekman dynamics (Dauhajre and McWilliams, 2018; Johnson et al., 2020), or gravity bores (Dale et al., 2008; Pham and Sarkar, 2018).

4.2.4 A home for submesoscales

The small scale of the submesoscale can be considered to be a consequence of the deformation radius ($L_d = Nh/f$) of the mixed layer (Thomas et al., 2008; Dong et al., 2020). The deformation radius of the mixed layer is much smaller than the deformation radius of the whole water column, as the mixed layer is shallower and less stratified.

Submesoscales are also often energised by the extraction of potential energy. This extraction implies a restratification, just as mixing implies a source of potential energy. Whereas submesoscale phenomena may not be the fastest agent of restratification, they are important in that they can act below the surface, remote from where winds and solar restratification are strong. Fig. 4.5 shows the buoyancy frequency, (i.e., $N = \sqrt{\partial b/\partial z}$) within the mixed layer for the same models. As argued in (Dong et al., 2021), the disagreement in stratification within the mixed layer has a larger impact on many aspects of submesoscale dynamics than disagreement in mixed layer depth.

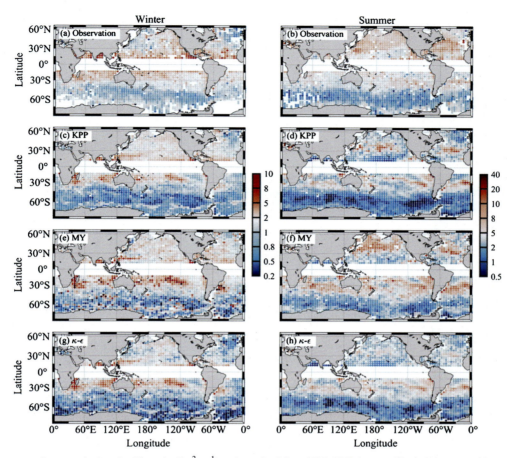

FIGURE 4.5 Buoyancy frequency in the mixed layer ($\times 10^{-3}$ s^{-1}) as determined from 2010–2019 Argo profiles in February and September (by summer and winter in both hemispheres) and 1D simulations of the generalised ocean turbulence model in the same seasons Umlauf et al. (2016), initialised from profiles and forced with surface forcing from a global MITgcm simulation (Su et al., 2018). Reproduced with permission from Dong et al. (2020).

4.3 Observing the surface layers and their processes

4.3.1 Observing mixing

Mixing in the ocean surface boundary layer (OSBL) can be inferred over a range of complexities depending on the topics of interest and/or measurement capabilities available. A description of technological advancements in ocean mixing observations are detailed in Chapter 14, but a quick review of observations specific to the OSBL layer are included here. Measurements of turbulent quantities within the OSBL are complicated, particularly by the span of turbulent motions from centimetres to the entire length of the mixed layer, and therefore require different approaches than methods used to measure turbulence in the ocean interior. As such, turbulence estimates can be drawn from a range of parameters, including vertical profiles of mean temperature, salinity, and velocity and their vertical gradients, turbulent fluxes, the reduction of temperature variance by turbulence χ, and dissipation of turbulent kinetic energy, ϵ. Additionally, turbulent quantities in the boundary layer are orders of magnitude larger than in the interior and coexist with a complicated wave field, both of which contribute to challenges in making meaningful measurements and require different platforms than the relatively quiescent interior.

4.3.1.1 Defining mixing from mean profiles

Routine and accessible metrics for mixing in the upper ocean can be obtained from profiles of mean tracers T, S, and velocities u, v. Most common is to assume quantities are homogeneous over a mixed layer depth h, using some of the methods in Section 4.2.3. In the atmosphere, profiles of mean shear are often used to constrain vertical fluxes, but this approach has not been successfully carried out in the ocean (except in large eddy simulations, Large et al., 2019). Yet, more active definitions isolate the processes of mixing. For example, shear driven turbulence can be identified through the ratio between shear and stratification as the Richardson number $Ri = N^2/|\partial_z u|^2$, where turbulence is suppressed at a critical value $Ri > 1/4$. Or in a mixed layer undergoing deep convection, the upper ocean may be unstably stratified ($N^2 < 0$), and ocean mixing can be inferred through the Thorpe scale (Thorpe, 2007).

All of these methods have been widely adopted to pertain to observations of the boundary layer with the common measurements collected by CTD instruments (conductivity, temperature, and depth) and ADCPs or ADVs (Acoustic doppler current profilers or Acoustic doppler velocimeters). Vertical profiles may be obtained through shipboard surveys with profiling instruments, Lagrangian floats (Harcourt et al., 2002), profiling drifters (e.g., Argo: Roemmich et al., 2019), gliders (e.g., Sea Glider, Slocum Javaid et al., 2014), or autonomous profilers (e.g., WireWalker: Pinkel et al., 2011). Whereas CTDs are standard on most observational platforms, obtaining accurate velocities from ADCPs are challenging in the near-surface layers, yet are becoming smaller and more common on autonomous platforms. A large contribution to boundary-layer mixing has come from very high resolution measurements of mean properties (such as thermistor chains and the Surf Otter), and platforms that can obtain near-surface meteorological data and subsurface measurements (Gentemann et al., 2020; Pinkel et al., 2011; Edson et al., 2007).

4.3.1.2 Estimating turbulent fluxes

Obtaining a more dynamical view of upper ocean mixing requires estimates of turbulent quantities v', w', C' and their turbulent covariances and fluxes $\langle v'^2 \rangle$, $\langle w'^2 \rangle$ and $\langle C'w' \rangle$. Yet turbulent velocity fluctuations are often obscured by the ~ 100 times larger surface wave induced velocities. To address this, turbulent vertical velocity can be estimated from a neutrally buoyant Lagrangian float (D'Asaro, 2003). In the OSBL, the float traverses the mixed layer along with the largest eddies, through which vertical velocity statistics can be calculated from the rate of pressure change. Based on this approach, D'Asaro et al. (2014) were able to show that over a wide range of wind and wave conditions, the vertical velocity statistics are *inconsistent* with the standard M–O scaling for oceanic boundary layers, but are consistent with the scaling of statistics expected based on one of the wave-driven turbulence theories (Harcourt and D'Asaro, 2008). Similarly, based on observations from two fixed offshore platforms, Chen et al. (2018, 2019) investigated the atmospheric boundary layer, particularly how surface waves modulate air-sea momentum flux, and explained the deviation of the wind stress from the wind direction under low to moderate wind conditions, indicating upward momentum transfer from the ocean to the atmosphere through surface waves.

Momentum flux (or turbulent Reynolds stress) in the OSBL is generally assumed to be equal to, at least no more than, the wind stress over the sea surface in ocean and climate models based on traditional boundary-layer theory. However, simultaneous in-situ observations conducted by Huang and Qiao (2021) show, for the first time, that the turbulent Reynolds stress in the OSBL is modulated greatly by surface waves and can be larger than the wind stress in certain circumstances. This observational result may have a potential impact on the development of ocean and climate models.

4.3.1.3 Dissipation and loss of shear and temperature variance

Values of kinetic energy dissipation ϵ and temperature variance dissipation χ are inferred through measurements of high-frequency fluctuations in velocity and temperature, respectively. A detailed discussion can be found in Chapter 14 and is summarised here. Dissipation of turbulent kinetic energy in the ocean is achieved by assuming isotropy and smooth v profiles, such that

$$\epsilon = \frac{15}{2} v \left\langle \left(\frac{\partial u}{\partial z} \right) \right\rangle^2. \tag{4.23}$$

This allows ϵ to be estimated from very fine resolution observations of velocity shear by microstructure profilers (e.g., Gregg, 1998; Ward et al., 2004), Lagrangian floats (Lien and D'Asaro, 2006), or through sophisticated analysis of acoustic Doppler measurements (Thomson et al., 2012; McCaffrey et al., 2015). Microstructure measurements in the mixed layer are generally collected by ship-based free falling profilers.

The decay of temperature variance in the ocean surface boundary layer can also be assumed under isotropy to provide an estimate of turbulence dissipation (Osborn, 1980; Oakey, 1982):

$$\epsilon = 6\kappa_T \left\langle \left(\frac{\partial T}{\partial z} \right) \right\rangle^2. \tag{4.24}$$

High-frequency and precise measurements of fine-scale temperature gradients in the ocean are easier to obtain than velocity shear, but may be aliased by internal waves that have strong T variance but are not turbulent. An early example of this approach in the mixed layer is Oakey and Elliott (1982), and a detailed comparison versus the dissipation method is Kocsis

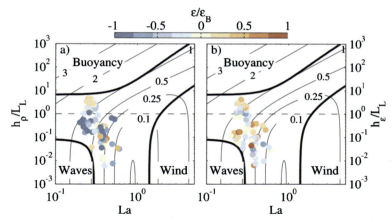

FIGURE 4.6 Values of ϵ from the North Atlantic obtained using microstructure measurements from the air-sea interaction profiler, non-dimensionalised by a dissipation scale ϵ_B (see Belcher et al. (2012) using a) mixed-layer depth h and (b) mixing-layer depth h_ϵ for the turbulent length scale, respectively. The black lines show the contours for $\log_{10}\epsilon_B$ with the heavy black lines denoting where the forcing accounts for 90% of the total dissipation. To be compared with Fig. 4.3. Reproduced with permission from Sutherland et al. (2014).

et al. (1999). Moulin et al. (2018) combined measurements of stratification of the upper 10 m with surface fluxes and turbulence measurements to show the importance of penetrating solar fluxes on turbulence suppression in diurnal warm layers, noting that penetrative solar fluxes not predicted by M–O theory play a role in upper ocean mixing (Section 4.2.2.3).

4.3.2 Wave-driven turbulence

Alongside measurements of turbulence, to assess wave-driven turbulence metrics of sea state conditions need to be collected alongside the turbulence measurements. The surface wave instrument float with tracking (SWIFT, Thomson, 2012) is an example of an instrument capturing both waves and dissipation. Surface buoys, stereo video capture (e.g., Vieira et al., 2020), X-band radar (Lund et al., 2016), radar combined with video (Benetazzo et al., 2018), or ultrasonic altimeters (Christensen et al., 2013) can all be used for wave observations.

Examples of the combined wave and turbulence observations show that the ocean surface mixed layer depends on wave state. Huang et al. (2012) observed the turbulence dissipation rate of the upper layer in the deep water of the South China sea, and compared the measurements with the non-breaking surface wave scheme and the law of wall, whereas the former fits the data well the latter is inconsistent. D'Asaro et al. (2014) collected a wide range of observations under varying sea state and wind conditions, and showed that a Langmuir turbulence scaling (Harcourt and D'Asaro, 2008) was consistent with the turbulence statistics, whereas a shear turbulence scaling was not. There are many difficulties in such measurements, since the energy density of surface waves is 3–5 orders of magnitude larger than that of boundary layer turbulence, and winds and waves often covary, because wind waves often dominate. Harcourt and D'Asaro (2008) performed experiments in the lee of a floating bridge to find calm seas in strong winds. Qiao et al. (2016) directly observed ocean velocity at a 64 Hz sampling rate under typhoon conditions and strong turbulence. Sutherland et al. (2013) compared the observed profiles of turbulent kinetic energy dissipation with different scalings and found that the non-breaking wave scheme proposed by Huang and Qiao (2010) is consistent. Sutherland et al. (2014) compared observed dissipation profiles versus scalings for Langmuir turbulence differing in whether the mixing layer or mixed layer depth is used, but again they found that wave enhancement is required for consistency with Fig. 4.3 as in Fig. 4.6 (Belcher et al., 2012). In all cases, experimental results indicate that the wave-turbulence interaction induces higher levels of background turbulence beyond the M–O theory for shear turbulence under a rigid boundary.

4.3.3 Laboratory experiments

Dai et al. (2010) designed laboratory experiments, where they cool down the bottom temperature to make a temperature stratification. It would take 20 hours to destroy the temperature stratification by molecular mixing. The stratification is destroyed within 30 minutes if non-breaking surface waves are generated in the tank. This experiment shows, as do other tank experiments (e.g., Babanin and Haus, 2009), that turbulent mixing is elevated in the presence of waves.

Similarly, there have been tank experiments claiming to set up mean flows and Stokes drift effects (e.g., Monismith et al., 2007). However, a recent important study (Calvert et al., 2019) makes it clear how difficult it is to control the production of spurious mean flows, which when studying turbulence that is much less energetic than surface waves, such as Langmuir turbulence and non-breaking wave turbulence, is of critical importance.

4.4 Modelling surface layers and their processes

There are a variety of models of upper-ocean and lower atmosphere boundary layers. This section provides a guide to the approaches taken and highlights some recent example studies.

4.4.1 Large eddy simulations

Large eddy simulations (LES) are carried out to study both the atmospheric and oceanic boundary layers, although rarely in coupled settings of atmosphere, ocean, and waves. Instead aspects of the wavy (Sullivan et al., 2014) and heterogeneous (Sullivan et al., 2020) surface conditions have been studied.

LES are defined by their ability to resolve partly into the 3D isotropic turbulence range, where turbulence parameterisations appropriate to that regime can be used (e.g., Sullivan et al., 1994; Fox-Kemper and Menemenlis, 2008). These simulations do not attempt to resolve the scales of molecular viscosity or diffusion, which would limit the scale of problem being considered to be many orders of magnitude smaller, but some sensitivity remains to which LES closures are chosen. Due to the relatively shallow spectral slopes of isotropic turbulence, this means that LES are not suitable for predicting vorticity or potential vorticity statistics (Bodner and Fox-Kemper, 2020). Resolving isotropic turbulence requires nonhydrostatic effects, which further constrain the choice of fluid dynamics codes away from most common ocean models.

This chapter has emphasised the effects of surface waves on boundary layer turbulence, and LES are one key tool in that study. Early LES (Skyllingstad and Denbo, 1995; McWilliams et al., 1997) simulated the wave-averaged equations (Lane et al., 2007) or generalised Lagrangian mean theory following the theoretical framing of Craik and Leibovich (1976). Ardhuin et al. (2017) offers a wide comparison of the theoretical approaches and their degree of consistency. Theoretical work focusing on understanding and solutions of these equations (Chini and Leibovich, 2005; Chini et al., 2009; Suzuki and Fox-Kemper, 2016; Holm and Hu, 2021) have emphasised how the Stokes vortex force, or equivalently the Stokes shear force, on driving turbulent motions in the boundary layer. These forces also affect larger lateral scale balances, such as Ekman (McWilliams et al., 2012), geostrophic balance and adjustment (McWilliams and Fox-Kemper, 2013), frontogenesis and filamentogenesis (Suzuki and Fox-Kemper, 2016; Sullivan and McWilliams, 2019), and submesoscale instabilities (Haney et al., 2015). Most of these studies were triggered by examination of LES in large enough domains to have submesoscale features present (Hamlington et al., 2014).

However, non-breaking wave theory suggests that it is not the average linkage between waves and turbulence that is important, but interaction of waves and turbulence enhances the pre-existing turbulence (Teixeira and Belcher, 2002; Babanin, 2006; Huang and Qiao, 2010). This effect is not captured fully in the wave-averaged equations, despite the high resolution of LES. LES with free surface, wave-phase-resolving physics are beginning to reveal aspects of these dynamics and whether they are importantly different from the wave-averaged theories of Langmuir turbulence (Wang and Ozgokmen, 2018; Fujiwara et al., 2018; Xuan et al., 2019). Computational limitations prevent large wave steepness and breaking in these simulations, which is one condition where the wave-averaged theory is expected to fail (Wang and Ozgokmen, 2018). Similarly, the large parameter space needing to be explored to compare different theories (Fujiwara et al., 2018) has only begun, and has not yet considered time-dependent wave trains or strongly sheared flows, where the wave-averaged theories are expected to break down. However, all of these studies find strong support for the interpretation of the wave-averaged theories as consistent with the phase-resolved vortex tilting and stretching mechanism (Teixeira and Belcher, 2002), but Xuan et al. (2019) find evidence for phase-resolved turbulent vorticity production potentially enhancing mixing *beyond what is predicted by the wave-averaged equations*.

4.4.2 1D boundary layer models

In most weather and climate applications, the lateral grid spacing of the ocean model is much wider than the boundary layer depth, so even the largest isotropic turbulence of the boundary layer cannot be resolved. Relatedly, these models are typically hydrostatic, which offers computational efficiency and little loss of accuracy when the discretisation is focused on wide, shallow features. In such simulations, all aspects of boundary layer turbulence must be parameterised, typically through providing a set of vertical fluxes based on the surface forcing and resolved profiles at each gridpoint. This 1D or "pencil" modelling approach is typically required for all realistic models.

There is a diversity of theoretical approaches to developing 1D models, and so recent software has offered the capability to switch among the different approaches. CVmix (Griffies et al., 2015) and GOTM (Umlauf and Burchard, 2005; Umlauf et al., 2016) offer community software to make switching between schemes easy. Recent comparison papers (Mukherjee et al., 2016; Van Roekel et al., 2018; Li et al., 2019; Dong et al., 2020) have exploited and expanded these codes to gain perspective on the ensemble span of these 1D models. Other comparison papers have also examined impacts of parameterisations without these tools (e.g., Reffray et al., 2015; Wu et al., 2015; Buckingham et al., 2019; Ali et al., 2019) or optimising parameters of the parameterisations (Souza et al., 2020).

4.4.2.1 Slab mixed layers

The simplest model of the oceanic mixed layer is the slab ocean model, which has a fixed-in-time mixed layer depth and fixed oceanic fluxes, but predicts sea surface temperature and salinity variability. This type of model is mainly used as a lower boundary condition for atmospheric models, but occasionally is used to show sensitivity of atmospheric or forecast systems to varying mixed layer depths (e.g., Samanta et al., 2018).

4.4.2.2 K-profile parameterisations

The K-profile parameterisation approach is used in both atmospheric (Troehn and Mahrt, 1986) and oceanic applications (Large et al., 1994). These 1D models specify profiles of diffusivity and viscosity following the predictions of M–O theory and other extensions (such as the free convective velocity from Section 4.2.2.2). In the ocean case, traditionally a critical Richardson number is used to find the boundary-layer depth and scale the vertical coordinate. Recent improvements on the algorithm underlying this boundary-layer depth calculation (Shchepetkin and McWilliams, 2009; Van Roekel et al., 2018) lead to surprisingly large impacts (Li et al., 2019). For this reason, numerical implementation remains important. Improvements to the physics of these schemes are many, because the underlying equations are so simply related to the M–O theory. Li et al. (2016) show that including extra wave-induced mixing based on Langmuir turbulence improves climate models. Li and Fox-Kemper (2017) shows that Langmuir-induced entrainment also improves the coupled system. Li et al. (2019) show that whereas Langmuir effects are modest on average mixed-layer properties, models without Langmuir mixing fail to reproduce LES of the wave-averaged equations.

4.4.2.3 Second-moment closures

Second-moment closure schemes build upon a closed set of prediction equations for all second moments, such as kinetic energy and fluxes, based on a simplifying assumptions, e.g., neglecting the contribution of third-moments (Umlauf and Burchard, 2005). Commonly, these schemes come in two-equation form, where two turbulent quantities (e.g., energy and length, or energy and frequency, or energy and dissipation) are solved for with PDEs in depth and time, or one-equation form, where only one such quantity is solved for (e.g., Blanke and Delecluse, 1993). These schemes have been a common choice for boundary-layer schemes in coastal modelling (Umlauf and Burchard, 2005), but in climate modelling, the coarse vertical grid scale and long-time steps tend to make them less robust. A recent development in the theory of second-moment closures is a major theoretical expansion on their treatment of Langmuir turbulence (Harcourt, 2013, 2015), including explicit treatment of the Stokes vortex force and the pressure-strain correlation, expanding the impact of wave-driven mixing beyond earlier approaches considering only the Stokes shear production (Kantha and Clayson, 2004). The non-breaking surface wave induced vertical mixing can improve the second-moment closure, and then simulation of the mixed layer in coastal ocean models (Lin et al., 2006), global ocean models (Qiao et al., 2004, 2010) and climate models (Fan and Griffies, 2014).

4.4.2.4 Bulk mixed layers

Bulk mixed layers integrate over the full mixed layer, and then close for momentum, energy (mechanical and thermal), and freshwater and salt budgets (Kraus and Turner, 1967; Niiler and Kraus, 1977; Price et al., 1986). These models have the great advantage that they are built on a solid theoretical basis of conserved properties and their relatively low level of detail provides them with robust numerical implementations over a wide range of time steps and vertical discretisations. However, because they do not provide properties in vertical profile, they are mainly used to analyse observations, and until recently they were not considered accurate enough for climate and weather modelling, However, the recent adaptation of the concept of energy constraints, but in a system using approximate vertical structures drawn from profiles second-moment schemes and LES after rescaling based on similarity theory and adapting a shear turbulence parameterisation to operate under longer-time steps (Jackson et al., 2008), has developed into the energetics-based planetary boundary layer (ePBL) scheme (Reichl and Hallberg, 2018). It shares energy conservation with bulk schemes, vertical profiles with second-moment schemes, and uses implicit numerics to allow for robust behaviour even under large time steps and grid spacings. Li et al. (2019) adapt the Langmuir mixing enhancements applied in the K-profile parameterisation (Li et al., 2016; Li and Fox-Kemper, 2017) to ePBL.

4.4.2.5 Prognostic vs. diagnostic boundary layer depth

Each of the schemes above has a method for determining the boundary-layer depth. The K-profile parameterisation tends to have a diagnostic boundary-layer depth that is found from profiles at each time step without memory. This approach makes the scheme capable of long-time steps. The second-moment schemes tend to predict evolving boundary-layer depths, which limit their time steps and can lead to numerical instabilities and oscillations. The ePBL scheme features an implicit diagnostic boundary-layer depth calculation. The OSMOSIS scheme (prototype presented in Li et al., 2019) features a prognostic boundary-layer depth, which gives it very different behaviour from other schemes especially in reproducing variability in deepening observed at the OSMOSIS mooring array (Buckingham et al., 2019).

4.4.2.6 Non-breaking wave mixing

Many models underestimate the ocean mixed-layer depth in summer and overestimate it in winter (Fox-Kemper et al., 2011; Sallée et al., 2013; Heuzé, 2017). This may indicate that the vertical mixing in the upper ocean is too weak in summer, whereas too strong in winter. Wave breaking can generate turbulence, however this process can only affect the depth in the order of wave amplitude, i.e., within several metres, so it contributes little to mixing the whole ocean mixed layer. Qiao et al. (2004, 2010) note that surface waves may affect the ocean on the order of their wave length (≈ 100 m), which coincides with the ocean mixed layer—deeper than the summertime depth and shallower than the wintertime—and were inspired to introduce a non-breaking wave-induced mixing theory (Bv theory). This theory relates enhanced turbulence to the scales of the waves orbital motion, and therefore provides more energy for mixing when waves are strong within the upper ocean, where orbital velocities are large. The Bv has been tested by different groups with different ocean circulation models; most models showed improvement in the simulation of the ocean mixed layer, such as in a lake model (Torma and Krámer, 2017), 2016, 1D models (Ghantous and Babanin, 2014; Wu et al., 2015), the coastal Princeton ocean model (Lin et al., 2006), a global instance of the AWI ocean model (Wang et al., 2019b), and other climate models (Qiao et al., 2013; Bao et al., 2020), which suggest that the physics represented by Bv are missing in many models. Fan and Griffies (2014) found that the Bv scheme mixed too much, as did some other schemes (McWilliams and Sullivan, 2000), but after limiting its effects below 100 m, they found that it deepened summertime mixed layers more effectively than the other schemes tested, which is a common challenge for nearly all ocean circulation models and climate models (Sallée et al., 2013). Bv can shoal the ocean mixed-layer simulation in winter (Chen et al., 2019) by transferring more heat to subsurface, and so enhance the ocean stratification in winter, resulting in a shallower mixed layer in winter.

The Bv scheme produces additional vertical viscosity and diffusivity where waves are strong and more deeply when waves are long. This additional viscosity is added into the other sources in the model's parameterisation (Qiao et al., 2010). The Langmuir schemes described above and reviewed and compared in (Li et al., 2019) also mix more where waves are strong and more deeply when wavelengths are long. These schemes thus differ only in their theoretical framing and their quantitative aspects, which makes it very difficult to use bias reduction as a discriminant. Indeed, as mentioned above, both theories rely on the heuristic of tilting and stretching of turbulent vorticity by passing waves (Teixeira and Belcher, 2002), but differ in whether or not these effects can be captured in wave-averaged equations, or if phase-locked phenomena go beyond the wave-averaged contributions. To conclude, a comparison of Langmuir 1D schemes and non-breaking schemes, such as the Bv scheme need to be compared against wave-phase-resolving simulations under a wide range of realistic conditions to better understand the flaws and oversimplifications in each approach.

4.4.3 Ocean and climate models

The previous section discusses a variety of new 1D parameterisations, but of course these parameterisations are developed to improve 3D ocean and climate modelling systems. Many of the new breakthroughs require wave information, which is often modelled as a new coupled system component: MASNUM in Qiao et al. (2010, 2013); Bao et al. (2020), WaveWatch-III in (Li et al., 2016, 2017), and WAM in ECMWF atmosphere-wave coupled system, which only alters air-sea fluxes not ocean mixing (Bidlot et al., 2007). The additional cost of adding the wave models can be considerable, and so statistical approaches have been attempted. For some applications, such as predicting Langmuir mixing, they are adequately accurate (Li et al., 2017). For more advanced purposes, such as interactions of waves and sea ice, future developments toward efficiency are needed (Roach et al., 2018).

A new earth system model (FIO-ESMv2.0) was developed by including four processes, including the non-breaking surface wave-induced mixing (Bv), sea spray from wave breaking, and the effects of Stokes drift in air-sea fluxes and the diurnal cycle of sea surface temperature. The mixed-layer depth in the upper ocean and historical sea surface temperature during 1850–2014 are well reproduced (Bao et al., 2020). For the CMIP5 generation of coupled models, only the FIO model included wave effects, and this FIO model is also part of the CMIP6 ensemble along with a few other models, including wave effects (e.g., CESM, Li et al., 2016).

Other models in the CMIP6 ensemble have had improvements in their boundary-layer schemes as well. The CESM now has a K-profile parameterisation, including Langmuir mixing, which improves summer and winter mixed-layer depths in the extratropical oceans and subsurface ventilation and temperatures (Li et al., 2016; Danabasoglu et al., 2020). The GFDL CM4.0 has ePBL (Reichl and Hallberg, 2018) as its boundary-layer scheme, brings strong improvement to Southern ocean summer and winter mixed-layer depths and temperatures (Dunne et al., 2020).

4.5 Global perspective

4.5.1 Energy and forcing

The ocean surface boundary layer is forced by winds, waves, and convection, which are partly represented in Fig. 4.3. This forcing, when combined with ocean currents and other feedbacks, leads to the patterns of mixed-layer depth and stratification observed and modelled, Figs. 4.4–4.5. Elimination of large-scale vertical shear in the surface boundary layer is the largest sink of kinetic energy in models (Fig. 4.1). Through the energy budget (4.19), geographical variation of ocean boundary-layer properties, forcing and turbulence is how potential energy enters the ocean (Fig. 4.4).

These layers respond to energetic forcing and also have the ability to shape it. As already noted, the phenomenon of surface cooling, with deeper warming by radiation, is the only solar heat engine in the ocean (Fig. 4.2).

4.5.2 Surface layers, weather, and climate

Since Hasselmann (1976), the ocean surface boundary layer has been considered an important agent of calming the weather and thereby creating a climate. The air-sea fluxes discussed here, and especially the entrainment and connections to the deeper ocean, are part of how this long-term climate variability occurs.

Goswami et al. (2016) and Orenstein (2018) are two good examples of how air-sea coupling affects not only the ocean but also the atmosphere through coupled modes of variability—in this case, intraseasonal variability of the Indian monsoon. These studies show observationally and through perturbed physics models that features in the upper ocean of the Bay of Bengal are associated with monsoon variability (and thus with societally important rainfall predictions).

The ocean surface boundary layers are also important for climate. This result is often shown by climate model sensitivity to perturbed physics (e.g., Danabasoglu and Gent, 2009), but a deeper connection to how our planet works has emerged as well. Many studies have begun to simplify the ocean's role in the planetary energy balance as a reservoir of energy, specifically akin to a simple two-layer model (Gregory, 2000; Gregory et al., 2004; Gregory and Forster, 2008; Geoffroy et al., 2013b,a; Held et al., 2010; Winton et al., 2010; Hall and Fox-Kemper, 2021). The upper layer of this model connects to both the atmosphere and the deep ocean, and this deep ocean connection has been shown to link to Atlantic meridional overturning (Kostov et al., 2014). However, a recent student project reveals intriguing connections between the dynamics of the mixed layer (dynamically defined as in this chapter) and the upper layer and upper to lower-layer connectivity in these simplified climate models (Hall, 2020). His results indicate that the mixed layer is an observable emergent phenomenon that constrains many properties of the climate sensitivity to forcing—natural and anthropogenic—on decadal to centennial timescales (Hall and Fox-Kemper, 2021). Continued study of the mixed layer and its dynamics are a crucial part of modelling the earth's weather and climate.

4.6 Outlook

As our understanding of the ocean surface layers (and the closely related atmospheric surface layers) has evolved, the M–O approach has served as a useful guide. However, as this chapter has noted, a variety of topics of present interest require approaches that go beyond the M–O theory: waves (Bv, Langmuir, breaking, spray, etc.), submesoscales, penetrating radiation, sea ice, and spatiotemporal variability in forcing. A variety of observational approaches and modelling practice have developed to establish "sea truth," and the treatment and impacts of these phenomena beyond M–O in weather and climate models. However, it remains challenging to perform comparative analysis and synthesis of these different approaches, although recent examples utilising community software (Umlauf and Burchard, 2005; Li et al., 2019), regimes of behaviour (Belcher et al., 2012), and evaluation of scalings (Sutherland et al., 2013, 2014) exemplify potential paths forward. Numerical sensitivity and model development experiments show great potential for model bias reduction through continued surface layer improvements.

Acknowledgements

BFK and LJ were supported by ONR N00014-17-1-2393. FLQ was supported by NFSC No. 41821004.

References

Alford, M.H., MacKinnon, J.A., Simmons, H.L., Nash, J.D., 2016. Near-inertial internal gravity waves in the ocean. Annu. Rev. Mar. Sci. 8, 95–123. https://doi.org/10.1146/annurev-marine-010814-015746.

Ali, A., Christensen, K.H., Breivik, Ø., Malila, M., Raj, R.P., Bertino, L., et al., 2019. A comparison of Langmuir turbulence parameterizations and key wave effects in a numerical model of the North Atlantic and Arctic oceans. Ocean Model. 137, 76–97. https://doi.org/10.1016/j.ocemod.2019.02.005.

Ardhuin, F., Suzuki, N., McWilliams, J.C., Aiki, H., 2017. Comments on "a combined derivation of the integrated and vertically resolved, coupled wave-current equations". J. Phys. Oceanogr. 47 (9), 2377–2385. https://doi.org/10.1175/JPO-D-17-0065.1.

Babanin, A., 2006. On a wave-induced turbulence and a wave-mixed upper ocean layer. Geophys. Res. Lett. 33, L20605. https://doi.org/10.1029/2006GL027308.

Babanin, A.V., Haus, B.K., 2009. On the existence of water turbulence induced by nonbreaking surface waves. J. Phys. Oceanogr. 39 (10), 2675–2679. https://doi.org/10.1175/2009JPO4202.1.

Bachman, S.D., Fox-Kemper, B., Taylor, J.R., Thomas, L.N., 2017. Parameterization of frontal symmetric instabilities. I: theory for resolved fronts. Ocean Model. 109, 72–95. https://doi.org/10.1016/j.ocemod.2016.12.003.

Bao, Y., Song, Z., Qiao, F., 2020. FIO-ESM version 2.0: model description and evaluation. J. Geophys. Res., Oceans 125 (6), e2019JC016036. https://doi.org/10.1029/2019JC016036.

Baumert, H.Z., Simpson, J., Simpson, J.H., Sundermann, J., Sündermann, J., 2005. Marine Turbulence: Theories, Observations, and Models. Cambridge University Press.

Belcher, S.E., Grant, A.A.L.M., Hanley, K.E., Fox-Kemper, B., Van Roekel, L., Sullivan, P.P., et al., 2012. A global perspective on Langmuir turbulence in the ocean surface boundary layer. Geophys. Res. Lett. 39 (18), L18605. https://doi.org/10.1029/2012GL052932.

Benetazzo, A., Serafino, F., Bergamasco, F., Ludeno, G., Ardhuin, F., Sutherland, P., et al., 2018. Stereo imaging and X-band radar wave data fusion: an assessment. Ocean Eng. 152, 346–352. https://doi.org/10.1016/j.oceaneng.2018.01.077.

Bidlot, J., Janssen, P., Abdalla, S., Hersbach, H., 2007. A Revised Formulation of Ocean Wave Dissipation and Its Model Impact. ECMWF, Reading, UK.

Blanke, B., Delecluse, P., 1993. Variability of the tropical Atlantic Ocean simulated by a general circulation model with two different mixed-layer physics. J. Phys. Oceanogr. 23 (7), 1363–1388.

Boccaletti, G., Ferrari, R., Fox-Kemper, B., 2007. Mixed layer instabilities and restratification. J. Phys. Oceanogr. 37 (9), 2228–2250. https://doi.org/10.1175/JPO3101.1.

Bodner, A., Fox-Kemper, B., 2020. A breakdown in potential vorticity estimation delineates the submesoscale-to-turbulence boundary in large eddy simulations. J. Adv. Model. Earth Syst. 12 (10), e2020MS002049. https://doi.org/10.1029/2020MS002049.

Bodner, A., Fox-Kemper, B., Van Roekel, L., McWilliams, J., Sullivan, P., 2020. A perturbation approach to understanding the effects of turbulence on frontogenesis. J. Fluid Mech. 883, A25. https://doi.org/10.1017/jfm.2019.804.

Brunner, K., Kukulka, T., Proskurowski, G., Law, K.L., 2015. Passive buoyant tracers in the ocean surface boundary layer: 2. observations and simulations of microplastic marine debris. J. Geophys. Res., Oceans 120 (11), 7559–7573. https://doi.org/10.1002/2015JC010840.

Buckingham, C.E., Lucas, N.S., Belcher, S.E., Rippeth, T.P., Grant, A.L.M., Sommer, J.L., et al., 2019. The contribution of surface and submesoscale processes to turbulence in the open ocean surface boundary layer. J. Adv. Model. Earth Syst. 11 (12), 4066–4094. https://doi.org/10.1029/2019MS001801.

Calvert, R., Whittaker, C., Raby, A., Taylor, P., Borthwick, A., Van Den Bremer, T., 2019. Laboratory study of the wave-induced mean flow and set-down in unidirectional surface gravity wave packets on finite water depth. Phys. Rev. Fluids 4 (11), 114801. https://doi.org/10.1103/PhysRevFluids.4.114801.

Cavaleri, L., Bertotti, L., Bidlot, J.-R., 2015. Waving in the rain. J. Geophys. Res., Oceans 120 (5), 3248–3260. https://doi.org/10.1002/2014JC010348.

Cavaleri, L., Fox-Kemper, B., Hemer, M., 2012. Wind waves in the coupled climate system. Bull. Am. Meteorol. Soc. 93 (11), 1651–1661. https://doi.org/10.1175/BAMS-D-11-00170.1.

Chang, H., Huntley, H.S., Kirwan Jr., A.D., Carlson, D.F., Mensa, J.A., Mehta, S., et al., 2019. Small-scale dispersion in the presence of Langmuir circulation. J. Phys. Oceanogr. 49 (12), 3069–3085. https://doi.org/10.1175/JPO-D-19-0107.1.

Charnock, H., 1955. Wind stress on a water surface. Q. J. R. Meteorol. Soc. 81, 639. https://doi.org/10.1002/qj.49708135027.

Chassignet, E.P., Yeager, S., Fox-Kemper, B., Bozec, A., Castruccio, F., Danabasoglu, G., et al., 2020. Impact of horizontal resolution on global ocean-sea-ice model simulations based on the experimental protocols of the Ocean Model Intercomparison Project phase 2 (OMIP-2). Geosci. Model Dev. 13. https://doi.org/10.5194/gmd-13-4595-2020.

Chen, S., Qiao, F., Huang, C.J., Zhao, B., 2018. Deviation of wind stress from wind direction under low wind conditions. J. Geophys. Res., Oceans 123 https://doi.org/10.1029/2018JC014137.

Chen, S., Qiao, F., Jiang, W., Guo, J., Dai, D., 2019. Impact of surface waves on wind stress under low to moderate wind conditions. J. Phys. Oceanogr. 49 (8), 2017–2028. https://doi.org/10.1175/JPO-D-18-0266.1.

Chini, G., Leibovich, S., 2005. Resonant Langmuir-circulation-internal-wave interaction. Part 2. Langmuir circulation instability. J. Fluid Mech. 524, 99–120. https://doi.org/10.1017/S002211200400182X.

Chini, G.P., Julien, K., Knobloch, E., 2009. An asymptotically reduced model of turbulent Langmuir circulation. Geophys. Astrophys. Fluid Dyn. 103 (2), 179–197. https://doi.org/10.1080/03091920802622236.

Choy, C.A., Robison, B.H., Gagne, T.O., Erwin, B., Firl, E., Halden, R.U., et al., 2019. The vertical distribution and biological transport of marine microplastics across the epipelagic and mesopelagic water column. Sci. Rep. 9 (1), 1–9. https://doi.org/10.1038/s41598-019-44117-2.

Christensen, K.H., Röhrs, J., Ward, B., Fer, I., Broström, G., Saetra, Ø., Breivik, Ø., 2013. Surface wave measurements using a ship-mounted ultrasonic altimeter. Methods Oceanogr. 6, 1–15. https://doi.org/10.1016/j.mio.2013.07.002.

Codiga, D.L., Ullman, D.S., 2010. Characterizing the physical oceanography of coastal waters off Rhode Island, Part 1: literature review, available observations, and a representative model simulation. In: Appendix to Rhode Island Ocean Special Area Management Plan, pp. 1–170.

Craik, A.D.D., Leibovich, S., 1976. Rational model for Langmuir circulations. J. Fluid Mech. 73, 401–426. https://doi.org/10.1017/S0022112076001420.

Cronin, M.F., Gentemann, C.L., Edson, J.B., Ueki, I., Bourassa, M., Brown, S., et al., 2019. Air-sea fluxes with a focus on heat and momentum. Front. Mar. Sci. 6, 430. https://doi.org/10.3389/fmars.2019.00430.

Crowe, M.N., Taylor, J.R., 2018. The evolution of a front in turbulent thermal wind balance. Part 1. Theory. J. Fluid Mech. 850, 179–211. https://doi.org/10.1017/jfm.2018.448.

Csanady, G.T., 2000. Air-Sea Interaction. Cambridge University Press, New York.

Dai, D., Qiao, F., Sulisz, W., Han, L., Babanin, A., 2010. An experiment on the nonbreaking surface-wave-induced vertical mixing. J. Phys. Oceanogr. 40 (9), 2180–2188. https://doi.org/10.1175/2010JPO4378.1.

Dale, A.C., Barth, J.A., Levine, M.D., Austin, J.A., 2008. Observations of mixed layer restratification by onshore surface transport following wind reversal in a coastal upwelling region. J. Geophys. Res. 113 (C1), C01010. https://doi.org/10.1029/2007JC004128. http://doi.wiley.com/10.1029/2007JC004128.

Danabasoglu, G., Ferrari, R., McWilliams, J.C., 2008. Sensitivity of an ocean general circulation model to a parameterization of near-surface eddy fluxes. J. Climate 21 (6), 1192–1208. https://doi.org/10.1175/2007JCLI1508.1.

Danabasoglu, G., Gent, P.R., 2009. Equilibrium climate sensitivity: is it accurate to use a slab ocean model? J. Climate 22 (9), 2494–2499. https://doi.org/10.1175/2008JCLI2596.1.

Danabasoglu, G., Lamarque, J.-F., Bacmeister, J., Bailey, D.A., DuVivier, A.K., Edwards, J., et al., 2020. The Community Earth System Model version 2 (CESM2). J. Adv. Model. Earth Syst. 12 (2), e2019MS001916. https://doi.org/10.1029/2019MS001916.

D'Asaro, E.A., 2003. The ocean boundary layer below Hurricane Dennis. J. Phys. Oceanogr. 33 (3), 561–579.

D'Asaro, E.A., 2014. Turbulence in the upper-ocean mixed layer. Annu. Rev. Mar. Sci. 6, 101–115. https://doi.org/10.1146/annurev-marine-010213-135138.

D'Asaro, E.A., Shcherbina, A.Y., Klymak, J.M., Molemaker, J., Novelli, G., Guigand, C.M., et al., 2018. Ocean convergence and the dispersion of flotsam. Proc. Natl. Acad. Sci. 115 (6), 1162–1167. https://doi.org/10.1073/pnas.1718453115.

D'Asaro, E.A., Thomson, J., Shcherbina, A.Y., Harcourt, R.R., Cronin, M.F., Hemer, M.A., Fox-Kemper, B., 2014. Quantifying upper ocean turbulence driven by surface waves. Geophys. Res. Lett. 41 (1), 102–107. https://doi.org/10.1002/2013GL058193.

Dauhajre, D.P., McWilliams, J.C., 2018. Diurnal evolution of submesoscale front and filament circulations. J. Phys. Oceanogr. 48 (10), 2343–2361. https://doi.org/10.1175/JPO-D-18-0143.1.

de Boyer Montégut, C., Madec, G., Fischer, A.S., Lazar, A., Iudicone, D., 2004. Mixed layer depth over the global ocean: an examination of profile data and a profile-based climatology. J. Geophys. Res., Oceans 109, C12. https://doi.org/10.1029/2004JC002378.

Dong, J., Fox-Kemper, B., Zhang, H., Dong, C., 2020. The scale of submesoscale baroclinic instability globally. J. Phys. Oceanogr. 50 (9), 2649–2667. https://doi.org/10.1175/JPO-D-20-0043.1.

Dong, J., Fox-Kemper, B., Zhang, H., Dong, C., 2021. The scale and activity of symmetric instability globally. J. Phys. Oceanogr. 51 (5), 1655–1670. https://doi.org/10.1175/JPO-D-20-0159.1.

Dunne, J., Horowitz, L., Adcroft, A., Ginoux, P., Held, I., John, J., et al., 2020. The GFDL Earth System Model Version 4.1 (GFDL-ESM 4.1): Overall coupled model description and simulation characteristics. J. Adv. Model. Earth Syst., e2019MS002015. https://doi.org/10.1029/2019MS002015.

Edson, J., Crawford, T., Crescenti, J., Farrar, T., Frew, N., Gerbi, G., et al., 2007. The coupled boundary layers and air-sea transfer experiment in low winds. Bull. Am. Meteorol. Soc. 88 (3), 341. https://doi.org/10.1175/BAMS-88-3-341.

Fairall, C.W., Bradley, E.F., Rogers, D.P., Edson, J.B., Young, G.S., 1996. Bulk parameterization of air-sea fluxes for Tropical Ocean-Global Atmosphere Coupled-Ocean Atmosphere Response Experiment. J. Geophys. Res., Oceans 101, 3747–3764. https://doi.org/10.1029/95JC03205.

Fan, Y., Griffies, S.M., 2014. Impacts of parameterized Langmuir turbulence and nonbreaking wave mixing in global climate simulations. J. Climate 27 (12), 4752–4775. https://doi.org/10.1175/JCLI-D-13-00583.1.

Ferrari, R., Paparella, F., 2003. Compensation and alignment of thermohaline gradients in the ocean mixed layer. J. Phys. Oceanogr. 33 (11), 2214–2223.

Fox-Kemper, B., Danabasoglu, G., Ferrari, R., Griffies, S.M., Hallberg, R.W., Holland, M.M., et al., 2011. Parameterization of mixed layer eddies. III: implementation and impact in global ocean climate simulations. Ocean Model. 39, 61–78. https://doi.org/10.1016/j.ocemod.2010.09.002.

Fox-Kemper, B., Ferrari, R., Hallberg, R.W., 2008. Parameterization of mixed layer eddies. Part I: theory and diagnosis. J. Phys. Oceanogr. 38 (6), 1145–1165. https://doi.org/10.1175/2007JPO3792.1.

Fox-Kemper, B., Menemenlis, D., 2008. Can large eddy simulation techniques improve mesoscale-rich ocean models? In: Hecht, M., Hasumi, H. (Eds.), Ocean Modeling in an Eddying Regime, Vol. 177. In: AGU Geophysical Monograph Series, pp. 319–338.

Fujiwara, Y., Yoshikawa, Y., Matsumura, Y., 2018. A wave-resolving simulation of Langmuir circulations with a nonhydrostatic free-surface model: comparison with Craik–Leibovich theory and an alternative Eulerian view of the driving mechanism. J. Phys. Oceanogr. 48 (8), 1691–1708. https://doi.org/10.1175/JPO-D-17-0199.1.

Gargett, A., Wells, J., Tejada-Martinez, A., Grosch, C., 2004. Langmuir supercells: a mechanism for sediment resuspension and transport in shallow seas. Science 306 (5703), 1925–1928. https://doi.org/10.1126/science.1100849.

Gentemann, C.L., Scott, J.P., Mazzini, P.L.F., Pianca, C., Akella, S., Minnett, P.J., et al., 2020. Saildrone: adaptively sampling the marine environment. Bull. Am. Meteorol. Soc. 101 (6), E744–E762. https://doi.org/10.1175/BAMS-D-19-0015.1.

Geoffroy, O., Saint-Martin, D., Bellon, G., Voldoire, A., Olivié, D., Tytéca, S., 2013a. Transient climate response in a two-layer energy-balance model. Part II: Representation of the efficacy of deep-ocean heat uptake and validation for CMIP5 AOGCMs. J. Climate 26 (6), 1859–1876. https://doi.org/10.1175/JCLI-D-12-00196.1.

Geoffroy, O., Saint-Martin, D., Olivié, D.J., Voldoire, A., Bellon, G., Tytéca, S., 2013b. Transient climate response in a two-layer energy-balance model. Part I: Analytical solution and parameter calibration using CMIP5 AOGCM experiments. J. Climate 26 (6), 1841–1857. https://doi.org/10.1175/JCLI-D-12-00195.1.

Ghantous, M., Babanin, A., 2014. One-dimensional modelling of upper ocean mixing by turbulence due to wave orbital motion. Nonlinear Process. Geophys. 21 (1), 325–338. https://doi.org/10.5194/npg-21-325-2014.

Goswami, B., Rao, S.A., Sengupta, D., Chakravorty, S., 2016. Monsoons to mixing in the Bay of Bengal: multiscale air-sea interactions and monsoon predictability. Oceanography 29 (2), 18–27. https://www.jstor.org/stable/24862666.

Grant, A.L.M., Belcher, S.E., 2009. Characteristics of Langmuir turbulence in the ocean mixed layer. J. Phys. Oceanogr. 39 (8), 1871–1887. https://doi.org/10.1175/2009JPO4119.1.

Gregg, M.C., 1998. Estimation and geography of diapycnal mixing in the stratified ocean. Coast. Estuar. Stud. 54, 305–338. https://doi.org/10.1029/CE054p0305.

Gregory, J., Forster, P., 2008. Transient climate response estimated from radiative forcing and observed temperature change. J. Geophys. Res., Atmos. 113 (D23). https://doi.org/10.1029/2008JD010405.

Gregory, J.M., 2000. Vertical heat transports in the ocean and their effect on time-dependent climate change. Clim. Dyn. 16 (7), 501–515. https://doi.org/10.1007/s003820000059.

Gregory, J.M., Ingram, W., Palmer, M., Jones, G., Stott, P., Thorpe, R., et al., 2004. A new method for diagnosing radiative forcing and climate sensitivity. Geophys. Res. Lett. 31 (3). https://doi.org/10.1029/2003GL018747.

Griffies, S.M., Levy, M., Adcroft, A.J., Danabasoglu, G., Hallberg, R.W., Jacobsen, D., et al., 2015. Theory and numerics of the Community Ocean Vertical Mixing (CVMix) Project. Tech. Rep.. Geophysical Fluid Dynamics Laboratory, National Center for Atmospheric Research, and Los Alamos National Laboratory. https://github.com/CVMix/CVMix-description.

Hall, G., 2020. Investigating the impacts of mixed layer depth on climate sensitivity. ScB thesis. Physics, Brown University. http://www.geo.brown.edu/research/Fox-Kemper/pubs/pdfs/HallThesis.pdf.

Hall, G., Fox-Kemper, B., 2021. Regional mixed layer depth as a climate diagnostic and emergent constraint. In preparation.

Hall, T.M., Haine, T.W.N., 2002. On ocean transport diagnostics: The idealized age tracer and the age spectrum. J. Phys. Oceanogr. 32 (6), 1987–1991. https://doi.org/10.1175/1520-0485(2002)032<1987:OOTDTI>2.0.CO;2.

Hamlington, P.E., Van Roekel, L.P., Fox-Kemper, B., Julien, K., Chini, G.P., 2014. Langmuir-submesoscale interactions: descriptive analysis of multiscale frontal spin-down simulations. J. Phys. Oceanogr. 44 (9), 2249–2272. https://doi.org/10.1175/JPO-D-13-0139.1.

Haney, S., Bachman, S., Cooper, B., Kupper, S., McCaffrey, K., Van Roekel, L., et al., 2012. Hurricane wake restratification rates of 1, 2 and 3-dimensional processes. J. Mar. Res. 70 (6), 824–850. https://doi.org/10.1357/002224012806770937.

Haney, S., Fox-Kemper, B., Julien, K., Webb, A., 2015. Symmetric and geostrophic instabilities in the wave-forced ocean mixed layer. J. Phys. Oceanogr. 45, 3033–3056. https://doi.org/10.1175/JPO-D-15-0044.1.

Hanley, K.E., Belcher, S.E., Sullivan, P.P., 2010. A global climatology of wind-wave interaction. J. Phys. Oceanogr. 40 (6), 1263–1282. https://doi.org/10.1175/2010JPO4377.1.

Harcourt, R., Steffen, E., Garwood, R., D'Asaro, E., 2002. Fully Lagrangian floats in Labrador Sea deep convection: Comparison of numerical and experimental results. J. Phys. Oceanogr. 32 (2), 493–510. https://doi.org/10.1175/1520-0485(2002)032<0493:FLFILS>2.0.CO;2.

Harcourt, R.R., 2013. A second-moment closure model of Langmuir turbulence. J. Phys. Oceanogr. 43 (4), 673–697. https://doi.org/10.1175/JPO-D-12-0105.1.

Harcourt, R.R., 2015. An improved second-moment closure model of Langmuir turbulence. J. Phys. Oceanogr. 45 (1), 84–103. https://doi.org/10.1175/JPO-D-14-0046.1.

Harcourt, R.R., D'Asaro, E.A., 2008. Large-eddy simulation of Langmuir turbulence in pure wind seas. J. Phys. Oceanogr. 38 (7), 1542–1562. https://doi.org/10.1175/2007JPO3842.1.

Hasselmann, K., 1976. Stochastic climate models. Part I. Theory. Tellus 28, 473–485. https://doi.org/10.3402/tellusa.v28i6.11316.

Held, I.M., Winton, M., Takahashi, K., Delworth, T., Zeng, F., Vallis, G.K., 2010. Probing the fast and slow components of global warming by returning abruptly to preindustrial forcing. J. Climate 23 (9), 2418–2427. https://doi.org/10.1175/2009JCLI3466.1.

Heuzé, C., 2017. North Atlantic Deep Water formation and AMOC in CMIP5 models. Ocean. Sci. 13 (4), 609–622. https://doi.org/10.5194/os-13-609-2017.

Högström, U., 1996. Review of some basic characteristics of the atmospheric surface layer. Bound.-Layer Meteorol. 78 (3), 215–246. https://doi.org/10.1007/BF00120937.

Holm, D., 1996. The ideal Craik-Leibovich equations. Physica D 98 (2–4), 415–441. https://doi.org/10.1016/0167-2789(96)00105-4.

Holm, D.D., Hu, R., 2021. Stochastic effects of waves on currents in the ocean mixed layer. J. Math. Phys. 62, 073102. https://doi.org/10.1063/5.0045010.

Holte, J., Talley, L.D., Gilson, J., Roemmich, D., 2017. An Argo mixed layer climatology and database. Geophys. Res. Lett. 44 (11), 5618–5626. https://doi.org/10.1002/2017GL073426.

Holtslag, A.A., Nieuwstadt, F.T., 1986. Scaling the atmospheric boundary layer. Bound.-Layer Meteorol. 36 (1–2), 201–209. https://doi.org/10.1007/BF00117468.

Horvat, C., Tziperman, E., Campin, J.-M., 2016. Interaction of sea ice floe size, ocean eddies, and sea ice melting. Geophys. Res. Lett. 43 (15), 8083–8090. https://doi.org/10.1002/2016GL069742.

Hosegood, P., Gregg, M.C., Alford, M.H., 2006. Sub-mesoscale lateral density structure in the oceanic surface mixed layer. Geophys. Res. Lett. 33 (22), L22604. https://doi.org/10.1029/2006GL026797.

Huang, C.J., Qiao, F., 2010. Wave-turbulence interaction and its induced mixing in the upper ocean. J. Geophys. Res., Oceans 115 (C4). https://doi.org/10.1029/2009JC005853.

Huang, C.J., Qiao, F., 2021. Simultaneous observations of turbulent Reynolds stress in the ocean surface boundary layer and wind stress over the sea surface. J. Geophys. Res., Oceans, e2020JC016839. https://doi.org/10.1029/2020JC016839.

Huang, C.J., Qiao, F., Dai, D., Ma, H., Guo, J., 2012. Field measurement of upper ocean turbulence dissipation associated with wave-turbulence interaction in the South China Sea. J. Geophys. Res., Oceans 117 (C11). https://doi.org/10.1029/2011JC007806.

Huang, N.E., Hu, K., Yang, A.C., Chang, H.-C., Jia, D., Liang, W.-K., et al., 2016. On Holo-Hilbert spectral analysis: A full informational spectral representation for nonlinear and non-stationary data. Philos. Trans. R. Soc. A, Math. Phys. Eng. Sci. 374 (2065), 20150206. https://doi.org/10.1098/rsta.2015.0206.

Huang, P.-Q., Cen, X.-R., Lu, Y.-Z., Guo, S.-X., Zhou, S.-Q., 2018. An integrated method for determining the oceanic bottom mixed layer thickness based on woce potential temperature profiles. J. Atmos. Ocean. Technol. 35 (11), 2289–2301. https://doi.org/10.1175/JTECH-D-18-0016.1.

Hughes, K.G., Moum, J.N., Shroyer, E.L., 2020. Heat transport through diurnal warm layers. J. Phys. Oceanogr. 50 (1), 2885–2905. https://doi.org/10.1175/JPO-D-20-0079.1.

Jackson, L., Hallberg, R., Legg, S., 2008. A parameterization of shear-driven turbulence for ocean climate models. J. Phys. Oceanogr. 38 (5), 1033–1053. https://doi.org/10.1175/2007JPO3779.1.

Janssen, P.A., 2012. Ocean wave effects on the daily cycle in SST. J. Geophys. Res., Oceans 117 (C11). https://doi.org/10.1029/2012JC007943.

Javaid, M.Y., Ovinis, M., Nagarajan, T., Hashim, F.B., 2014. Underwater gliders: a review. In: Matec Web of Conferences, Vol.13, p. 02020.

Jing, Z., Qi, Y., Fox-Kemper, B., Du, Y., Lian, S., 2016. Seasonal thermal fronts on the northern South China Sea shelf: satellite measurements and three repeated field surveys. J. Geophys. Res., Oceans 121 (3), 1914–1930. https://doi.org/10.1002/2015JC011222.

Jochum, M., Briegleb, B.P., Danabasoglu, G., Large, W.G., Norton, N.J., Jayne, S.R., et al., 2013. The impact of oceanic near-inertial waves on climate. J. Climate 26 (9), 2833–2844. https://doi.org/10.1175/JCLI-D-12-00181.1.

Johnson, L., Fox-Kemper, B., Li, Q., Pham, H., Sarkar, S., 2021. A dynamical systems approach to mixed layer model comparison. In preparation.

Johnson, L., Lee, C.M., D'Asaro, E.A., Wenegrat, J.O., Thomas, L.N., 2020. Restratification at a California current upwelling front. Part II: Dynamics. J. Phys. Oceanogr. 50 (5), 1473–1487. https://doi.org/10.1175/JPO-D-19-0204.1.

Johnston, T.S., Rudnick, D.L., 2009. Observations of the transition layer. J. Phys. Oceanogr. 39 (3), 780–797. https://doi.org/10.1175/2008JPO3824.1.

Jones, C., Abernathey, R.P., 2019. Isopycnal mixing controls deep ocean ventilation. Geophys. Res. Lett. 46 (22), 13144–13151. https://doi.org/10.1029/2019GL085208.

Kantha, L., Clayson, C., 2004. On the effect of surface gravity waves on mixing in the oceanic mixed layer. Ocean Model. 6 (2), 101–124. https://doi.org/10.1016/S1463-5003(02)00062-8.

Kocsis, O., Prandke, H., Stips, A., Simon, A., Wüest, A., 1999. Comparison of dissipation of turbulent kinetic energy determined from shear and temperature microstructure. J. Mar. Syst. 21, 67–84. https://doi.org/10.1016/S0924-7963(99)00006-8.

Kostov, Y., Armour, K.C., Marshall, J., 2014. Impact of the Atlantic Meridional Overturning Circulation on ocean heat storage and transient climate change. Geophys. Res. Lett. 41 (6), 2108–2116. https://doi.org/10.1002/2013GL058998.

Kraus, E., Turner, J., 1967. A one-dimensional model of the seasonal thermocline. II: The general theory and its consequences. Tellus 19, 98–106. https://doi.org/10.3402/tellusa.v19i1.9753.

Lane, E.M., Restrepo, J.M., McWilliams, J.C., 2007. Wave-current interaction: a comparison of radiation-stress and vortex-force representations. J. Phys. Oceanogr. 37 (5), 1122–1141. https://doi.org/10.1175/JPO3043.1.

Large, W.G., Crawford, G.B., 1995. Observations and simulations of upper-ocean response to wind events during the Ocean Storms Experiment. J. Phys. Oceanogr. 25. https://doi.org/10.1175/1520-0485(1995)025<2831:OASOUO>2.0.CO;2.

Large, W.G., McWilliams, J.C., Doney, S.C., 1994. Oceanic vertical mixing - a review and a model with a nonlocal boundary-layer parameterization. Rev. Geophys. 32 (4), 363–403. https://doi.org/10.1029/94RG01872.

Large, W.G., Patton, E.G., DuVivier, A.K., Sullivan, P.P., Romero, L., 2019. Similarity theory in the surface layer of large-eddy simulations of the wind-, wave-, and buoyancy-forced Southern Ocean. J. Phys. Oceanogr. 49 (8), 2165–2187. https://doi.org/10.1175/JPO-D-18-0066.1.

Large, W.G., Pond, S., 1981. Open ocean momentum flux measurements in moderate to strong winds. J. Phys. Oceanogr. 11 (3), 324–336. https://doi.org/10.1175/1520-0485(1981)011<0324:OOMFMI>2.0.CO;2.

Large, W.G., Pond, S., 1982. Sensible and latent heat flux measurements over the ocean. J. Phys. Oceanogr. 12 (5), 464–482. https://doi.org/10.1175/1520-0485(1982)012<0464:SALHFM>2.0.CO;2.

Large, W.G., Yeager, S.G., 2004. Diurnal to Decadal Global Forcing for Ocean and Sea-Ice Models: The Data Sets and Flux Climatologies. National Center for Atmospheric Research Boulder.

Large, W.G., Yeager, S.G., 2009. The global climatology of an interannually varying air–sea flux data set. Clim. Dyn 33 (2–3), 341–364. https://doi.org/10.1007/s00382-008-0441-3.

Ledley, T.S., 1991. Snow on sea ice: competing effects in shaping climate. J. Geophys. Res., Atmos. 96 (D9), 17195–17208. https://doi.org/10.1029/91JD01439.

Lenain, L., Pizzo, N., 2020. The contribution of high-frequency wind-generated surface waves to the Stokes drift. J. Phys. Oceanogr. 50 (12), 3455–3465. https://doi.org/10.1175/JPO-D-20-0116.1.

Li, M., Garrett, C., Skyllingstad, E., 2005. A regime diagram for classifying turbulent large eddies in the upper ocean. Deep-Sea Res., Part 1, Oceanogr. Res. Pap. 52, 259–278. https://doi.org/10.1016/j.dsr.2004.09.004.

Li, Q., Fox-Kemper, B., 2017. Assessing the effects of Langmuir turbulence on the entrainment buoyancy flux in the ocean surface boundary layer. J. Phys. Oceanogr. 47, 2863–2886. https://doi.org/10.1175/JPO-D-17-0085.1.

Li, Q., Fox-Kemper, B., 2020. Anisotropy of Langmuir turbulence and the Langmuir-enhanced mixed layer entrainment. Phys. Rev. Fluids 5, 013803. https://doi.org/10.1103/PhysRevFluids.5.013803.

Li, Q., Fox-Kemper, B., Breivik, O., Webb, A., 2017. Statistical models of global Langmuir mixing. Ocean Model. 113, 95–114. https://doi.org/10.1016/j.ocemod.2017.03.016.

Li, Q., Reichl, B.G., Fox-Kemper, B., Adcroft, A.J., Belcher, S., Danabasoglu, G., et al., 2019. Comparing ocean boundary vertical mixing schemes including Langmuir turbulence. J. Adv. Model. Earth Syst. 11 (11), 3545–3592. https://doi.org/10.1029/2019MS001810.

Li, Q., Webb, A., Fox-Kemper, B., Craig, A., Danabasoglu, G., Large, W.G., Vertenstein, M., 2016. Langmuir mixing effects on global climate: WAVEWATCH III in CESM. Ocean Model. 103, 145–160. https://doi.org/10.1016/j.ocemod.2015.07.020.

Liang, J.-H., Deutsch, C., McWilliams, J.C., Baschek, B., Sullivan, P.P., Chiba, D., 2013. Parameterizing bubble-mediated air-sea gas exchange and its effect on ocean ventilation. Glob. Biogeochem. Cycles 27 (3), 894–905. https://doi.org/10.1002/gbc.20080.

Liang, J.-H., Emerson, S.R., D'Asaro, E.A., McNeil, C.L., Harcourt, R.R., Sullivan, P.P., et al., 2017. On the role of sea-state in bubble-mediated air-sea gas flux during a winter storm. J. Geophys. Res., Oceans 122 (4), 2671–2685. https://doi.org/10.1002/2016JC012408.

Lien, R.-C., D'Asaro, E.A., 2006. Measurement of turbulent kinetic energy dissipation rate with a Lagrangian float. J. Atmos. Ocean. Technol. 23 (7), 964–976. https://doi.org/10.1175/JTECH1890.1.

Lin, X., Xie, S.-P., Chen, X., Xu, L., 2006. A well-mixed warm water column in the central Bohai Sea in summer: effects of tidal and surface wave mixing. J. Geophys. Res., Oceans 111 (C11). https://doi.org/10.1029/2006JC003504.

Liu, W.T., Katsaros, K.B., Businger, J.A., 1979. Bulk parameterization of air-sea exchanges of heat and water vapor including the molecular constraints at the interface. J. Atmos. Sci. 36 (9), 1722–1735. https://doi.org/10.1175/1520-0469(1979)036<1722:BPOASE>2.0.CO;2.

Lund, B., Collins III, C.O., Tamura, H., Graber, H.C., 2016. Multi-directional wave spectra from marine X-band radar. Ocean Dyn. 66, 973–988. https://doi.org/10.1007/s10236-016-0961-z.

MacKinnon, J.A., Nash, J.D., Alford, M.H., Lucas, A.J., Mickett, J.B., Shroyer, E., et al., 2016. A tale of two spicy seas. Oceanography. https://www.jstor.org/stable/24862669.

Mahadevan, A., D'Asaro, E., Lee, C., Perry, M.J., 2012. Eddy-driven stratification initiates North Atlantic spring phytoplankton blooms. Science 337 (6090), 54–58. https://doi.org/10.1126/science.1218740.

Manucharyan, G.E., Thompson, A.F., 2017. Submesoscale sea ice-ocean interactions in marginal ice zones. J. Geophys. Res., Oceans 122 (12), 9455–9475. https://doi.org/10.1002/2017JC012895.

McCaffrey, K., Fox-Kemper, B., Hamlington, P., Thomson, J., 2015. Characterization of turbulence anisotropy, coherence, and intermittency at a prospective tidal energy site: observational data analysis. Renew. Energy 76, 441–453. https://doi.org/10.1016/j.renene.2014.11.063.

McPhee, M.G., 1994. On the turbulent mixing length in the oceanic boundary layer. J. Phys. Oceanogr. 24 (9), 2014–2031. https://doi.org/10.1175/1520-0485(1994)024<2014:OTTMLI>2.0.CO;2.

McPhee, M.G., Martinson, D.G., 1994. Turbulent mixing under drifting pack ice in the Weddell Sea. Science 263 (5144), 218–221. https://doi.org/10.1126/science.263.5144.218.

McPhee, M.G., Stanton, T.P., 1996. Turbulence in the statically unstable oceanic boundary layer under Arctic leads. J. Geophys. Res., Oceans 101 (C3), 6409–6428. https://doi.org/10.1029/95JC03842.

McWilliams, J., Restrepo, J., 1999. The wave-driven ocean circulation. J. Phys. Oceanogr. 29 (10), 2523–2540. https://doi.org/10.1175/1520-0485(1999)029<2523:TWDOC>2.0.CO;2.

McWilliams, J.C., 2016. Submesoscale currents in the ocean. Proc. R. Soc. A 472 (2189), 20160117. https://doi.org/10.1098/rspa.2016.0117.

McWilliams, J.C., Fox-Kemper, B., 2013. Oceanic wave-balanced surface fronts and filaments. J. Fluid Mech. 730, 464–490. https://doi.org/10.1017/jfm.2013.348.

McWilliams, J.C., Gula, J., Molemaker, M.J., Renault, L., Shchepetkin, A.F., 2015. Filament frontogenesis by boundary layer turbulence. J. Phys. Oceanogr. 45 (8), 1988–2005. https://doi.org/10.1175/JPO-D-14-0211.1.

McWilliams, J.C., Huckle, E., Liang, J.-H., Sullivan, P.P., 2012. The wavy Ekman layer: Langmuir circulations, breaking waves, and Reynolds stress. J. Phys. Oceanogr. 42 (11), 1793–1816.

McWilliams, J.C., Sullivan, P.P., 2000. Vertical mixing by Langmuir circulations. Spill Sci. Technol. Bull. 6, 225–237. https://doi.org/10.1016/S1353-2561(01)00041-X.

McWilliams, J.C., Sullivan, P.P., Moeng, C.-H., 1997. Langmuir turbulence in the ocean. J. Fluid Mech. 334, 1–30. https://doi.org/10.1017/S0022112096004375.

Mellor, G.L., Yamada, T., 1982. Development of a turbulence closure model for geophysical fluid problems. Rev. Geophys. 20 (4), 851–875. https://doi.org/10.1029/RG020i004p00851.

Mensa, J.A., Timmermans, M.-L., 2017. Characterizing the seasonal cycle of upper-ocean flows under multi-year sea ice. Ocean Model. 113, 115–130. https://doi.org/10.1016/j.ocemod.2017.03.009.

Min, H.S., Noh, Y., 2004. Influence of the surface heating on Langmuir circulation. J. Phys. Oceanogr. 34 (12), 2630–2641. https://doi.org/10.1175/JPOJPO-2654.1.

Monin, A., Obukhov, A., 1954. Basic laws of turbulent mixing in the surface layer of the atmosphere. Contrib. Geophys. Inst. Acad. Sci. USSR 151 (163), e187.

Monismith, S.G., Cowen, E.A., Nepf, H.M., Magnadaudet, J., Thais, L., 2007. Laboratory observations of mean flows under surface gravity waves. J. Fluid Mech. 573, 131–147. https://doi.org/10.1017/S0022112006003594.

Monterey, G.I., Levitus, S., 1997. Seasonal Variability of Mixed Layer Depth for the World Ocean. US Department of Commerce, National Oceanic and Atmospheric Administration . . . .

Moulin, A.J., Moum, J.N., Shroyer, E.L., 2018. Evolution of turbulence in the Diurnal warm layer. J. Phys. Oceanogr. 48 (2), 383–396. https://doi.org/10.1175/JPO-D-17-0170.1.

Mukherjee, S., Ramachandran, S., Tandon, A., Mahadevan, A., 2016. Production and destruction of eddy kinetic energy in forced submesoscale eddy-resolving simulations. Ocean Model. 105, 44–59. https://doi.org/10.1016/j.ocemod.2016.07.002.

Muñoz-Esparza, D., Sharman, R.D., Lundquist, J.K., 2018. Turbulence dissipation rate in the atmospheric boundary layer: Observations and WRF mesoscale modeling during the XPIA field campaign. Mon. Weather Rev. 146 (1), 351–371. https://doi.org/10.1175/MWR-D-17-0186.1.

Niiler, P., Kraus, E., 1977. One-dimensional models of the upper ocean. In: Kraus, E. (Ed.), Modelling and Prediction of the Upper Layers of the Ocean. Pergamon Press, pp. 143–172.

Nurser, A.G., Griffies, S.M., 2019. Relating the diffusive salt flux just below the ocean surface to boundary freshwater and salt fluxes. J. Phys. Oceanogr. 49 (9), 2365–2376. https://doi.org/10.1175/JPO-D-19-0037.1.

Oakey, N.S., 1982. Determination of the rate of dissipation of turbulent energy from simultaneous temperature and velocity shear microstructure measurements. J. Phys. Oceanogr. 12, 256–271. https://doi.org/10.1175/1520-0485(1982)012<0256:DOTROD>2.0.CO;2.

Oakey, N.S., Elliott, J.A., 1982. Dissipation within the surface mixed layer. J. Phys. Oceanogr. 12, 171–185. https://doi.org/10.1175/1520-0485(1982)012<0171:DWTSML>2.0.CO;2.

O'Donnell, J., Wilson, R.E., Lwiza, K., Whitney, M., Bohlen, W.F., Codiga, D., et al., 2014. The physical oceanography of Long Island Sound. In: Long Island Sound. Springer, pp. 79–158.

Orenstein, P., 2018. Can we predict short term highs and lows in the Indian monsoon? Identifying and evaluating monsoon intraseasonal oscillations in climate data. ScB thesis. Geology-Physics/Mathematics, Brown University. http://www.geo.brown.edu/research/Fox-Kemper/pubs/pdfs/Orenstein18.pdf.

Osborn, T.R., 1980. Estimates of the local rate of vertical diffusion from dissipation measurements. J. Phys. Oceanogr. 10, 83–89. https://doi.org/10.1175/1520-0485(1980)010<0083:EOTLRO>2.0.CO;2.

Otero, J., Wittenberg, R.W., Worthing, R.A., Doering, C.R., 2002. Bounds on Rayleigh–Bénard convection with an imposed heat flux. J. Fluid Mech. 473, 191–199. https://doi.org/10.1017/S0022112002002410.

Pearson, B., Fox-Kemper, B., Bachman, S.D., Bryan, F.O., 2017. Evaluation of scale-aware subgrid mesoscale eddy models in a global eddy-rich model. Ocean Model. 115, 42–58. https://doi.org/10.1016/j.ocemod.2017.05.007.

Pearson, B., Grant, A.L., Polton, J.A., Belcher, S.E., 2015. Langmuir turbulence and surface heating in the ocean surface boundary layer. J. Phys. Oceanogr. 45 (12), 2897–2911. https://doi.org/10.1175/JPO-D-15-0018.1.

Pearson, J., Fox-Kemper, B., Barkan, R., Choi, J., Bracco, A., McWilliams, J.C., 2019. Impacts of Convergence on Structure Functions from Surface Drifters in the Gulf of Mexico. J. Phys. Oceanogr. 49 (3), 675–690. https://doi.org/10.1175/JPO-D-18-0029.1.

Pearson, J., Fox-Kemper, B., Pearson, B., Chang, H., Haus, B.K., Horstmann, J., et al., 2020. Biases in structure functions from observations of submesoscale flows. J. Geophys. Res., Oceans. https://doi.org/10.1029/2019JC015769.

Pham, H.T., Sarkar, S., 2018. Ageostrophic secondary circulation at a submesoscale front and the formation of gravity currents. J. Phys. Oceanogr. 48 (10), 2507–2529. https://doi.org/10.1175/JPO-D-17-0271.1.

Phillips, O., 1961. A note on the turbulence generated by gravity waves. J. Geophys. Res. 66 (9), 2889–2893. https://doi.org/10.1029/JZ066i009p02889.

Pinkel, R., Goldin, M.A., Smith, J.A., Sun, O.M., Aja, A.A., Bui, M.N., Hughen, T., 2011. The wirewalker: A vertically profiling instrument carrier powered by ocean waves. J. Atmos. Ocean. Technol. 28 (3), 426–435. https://doi.org/10.1175/2010JTECHO805.1.

Pollard, R.T., Rhines, P.B., Thompson, R.O.R.Y., 1973. The deepening of the wind-mixed layer. Geophys. Fluid Dyn. 4 (4), 381–404. https://doi.org/10.1080/03091927208236105.

Polton, J.A., Belcher, S.E., 2007. Langmuir turbulence and deeply penetrating jets in an unstratified mixed layer. J. Geophys. Res., Oceans 112 (C9). https://doi.org/10.1029/2007JC004205.

Price, J.F., 1979. On the scaling of stress-driven entrainment experiments. J. Fluid Mech. 90 (3), 509–529. https://doi.org/10.1017/S0022112079002366.

Price, J.F., Weller, R.A., Pinkel, R., 1986. Diurnal cycling: observations and models of the upper ocean response to diurnal heating, cooling, and wind mixing. J. Geophys. Res., Oceans 91, 8411–C7 8427. https://doi.org/10.1029/JC091iC07p08411.

Qiao, F., Song, Z., Bao, Y., Song, Y., Shu, Q., Huang, C., Zhao, W., 2013. Development and evaluation of an Earth System Model with surface gravity waves. J. Geophys. Res., Oceans 118 (9), 4514–4524. https://doi.org/10.1002/jgrc.20327.

Qiao, F., Yuan, Y., Deng, J., Dai, D., Song, Z., 2016. Wave–turbulence interaction-induced vertical mixing and its effects in ocean and climate models. Philos. Trans. R. Soc. A, Math. Phys. Eng. Sci. 374 (2065), 20150201. https://doi.org/10.1098/rsta.2015.0201.

Qiao, F., Yuan, Y., Ezer, T., Xia, C., Yang, Y., Lü, X., Song, Z., 2010. A three-dimensional surface wave–ocean circulation coupled model and its initial testing. Ocean Dyn. 60 (5), 1339–1355. https://doi.org/10.1007/s10236-010-0326-y.

Qiao, F., Yuan, Y., Yang, Y., Zheng, Q., Xia, C., Ma, J., 2004. Wave-induced mixing in the upper ocean: distribution and application to a global ocean circulation model. Geophys. Res. Lett. 31 (11). https://doi.org/10.1029/2004GL019824.

Reffray, G., Bourdalle-Badie, R., Calone, C., 2015. Modelling turbulent vertical mixing sensitivity using a 1-D version of NEMO. Geosci. Model Dev. 8 (1), 69–86. https://doi.org/10.5194/gmd-8-69-2015.

Reichl, B.G., Hallberg, R., 2018. A simplified energetics based Planetary Boundary Layer (ePBL) approach for ocean climate simulations. Ocean Model. 132, 112–129. https://doi.org/10.1016/j.ocemod.2018.10.004.

Reichl, B.G., Wang, D., Hara, T., Ginis, I., Kukulka, T., 2016. Langmuir turbulence parameterization in tropical cyclone conditions. J. Phys. Oceanogr. 46 (3), 863–886. https://doi.org/10.1175/JPO-D-15-0106.1.

Renault, L., McWilliams, J.C., Gula, J., 2018. Dampening of submesoscale currents by air-sea stress coupling in the Californian upwelling system. Sci. Rep. 8 (1), 1–8. https://doi.org/10.1038/s41598-018-31602-3.

Renault, L., Molemaker, M.J., McWilliams, J.C., Shchepetkin, A.F., Lemarié, F., Chelton, D., et al., 2016. Modulation of wind work by oceanic current interaction with the atmosphere. J. Phys. Oceanogr. 46 (6), 1685–1704. https://doi.org/10.1175/JPO-D-15-0232.1.

Roach, L.A., Horvat, C., Dean, S.M., Bitz, C.M., 2018. An emergent sea ice floe size distribution in a global coupled ocean-sea ice model. J. Geophys. Res., Oceans 123 (6), 4322–4337. https://doi.org/10.1029/2017JC013692.

Roemmich, D., Alford, M.H., Claustre, H., Johnson, K.S., King, B., Moum, J., et al., 2019. On the future of Argo: a global, full-depth, multi-disciplinary array. Front. Mar. Sci. 6, 439. https://doi.org/10.3389/fmars.2019.00439.

Rudnick, D.L., Ferrari, R., 1999. Compensation of horizontal temperature and salinity gradients in the ocean mixed layer. Science 283 (5401), 526–529. https://doi.org/10.1126/science.283.5401.526.

Sallée, J.-B., Shuckburgh, E., Bruneau, N., Meijers, A.J.S., Bracegirdle, T.J., Wang, Z., 2013. Assessment of Southern Ocean mixed-layer depths in CMIP5 models: historical bias and forcing response. J. Geophys. Res., Oceans 118 (4), 1845–1862. https://doi.org/10.1002/jgrc.20157.

Samanta, D., Hameed, S.N., Jin, D., Thilakan, V., Ganai, M., Rao, S.A., Deshpande, M., 2018. Impact of a narrow coastal Bay of Bengal sea surface temperature front on an Indian summer monsoon simulation. Sci. Rep. 8 (1), 17694. https://doi.org/10.1038/s41598-018-35735-3.

Sanford, T.B., Price, J.F., Girton, J.B., Webb, D.C., 2007. Highly resolved observations and simulations of the ocean response to a hurricane. Geophys. Res. Lett. 34, 13. https://doi.org/10.1029/2007GL029679.

Shchepetkin, A.F., McWilliams, J.C., 2009. Computational kernel algorithms for fine-scale, multiprocess, longtime oceanic simulations. In: Temam, R.M., Tribbia, J.J. (Eds.), Handbook of Numerical Analysis, Vol. 14. Elsevier, pp. 121–183.

Skyllingstad, E.D., Denbo, D.W., 1995. An ocean large-eddy simulation of Langmuir circulations and convection in the surface mixed layer. J. Geophys. Res., Oceans 100, 8501–8522. https://doi.org/10.1029/94JC03202.

Smith, K.M., Hamlington, P.E., Fox-Kemper, B., 2016. Effects of submesoscale turbulence on ocean tracers. J. Geophys. Res., Oceans 121 (1), 908–933. https://doi.org/10.1002/2015JC011089.

Smith, K.M., Hamlington, P.E., Niemeyer, K.E., Fox-Kemper, B., Lovenduski, N.S., 2018. Effects of Langmuir turbulence on upper ocean carbonate chemistry. J. Adv. Model. Earth Syst. 10. https://doi.org/10.1029/2018MS001486.

Snow, K., Hogg, A.M., Downes, S.M., Sloyan, B.M., Bates, M.L., Griffies, S.M., 2015. Sensitivity of abyssal water masses to overflow parameterisations. Ocean Model. 89, 84–103. https://doi.org/10.1016/j.ocemod.2015.03.004.

Soloviev, A.V., Lukas, R., Donelan, M.A., Haus, B.K., Ginis, I., 2014. The air-sea interface and surface stress under tropical cyclones. Sci. Rep. 4, 5306. https://doi.org/10.1038/srep05306.

Souza, A.N., Wagner, G., Ramadhan, A., Allen, B., Churavy, V., Schloss, J., et al., 2020. Uncertainty quantification of ocean parameterizations: application to the k-profile-parameterization for penetrative convection. J. Adv. Model. Earth Syst., e2020MS002108. https://doi.org/10.1029/2020MS002108.

Stommel, H., 1979. Determination of water mass properties of water pumped down from the Ekman layer to the geostrophic flow below. Proc. Natl. Acad. Sci. 76 (7), 3051–3055. https://doi.org/10.1073/pnas.76.7.3051.

Su, Z., Wang, J., Klein, P., Thompson, A.F., Menemenlis, D., 2018. Ocean submesoscales as a key component of the global heat budget. Nat. Commun. 9 (1), 1–8. https://doi.org/10.1038/s41467-018-02983-w.

Sullivan, P., McWilliams, J., Melville, W., 2004. The oceanic boundary layer driven by wave breaking with stochastic variability. Part 1. Direct numerical simulations. J. Fluid Mech. 507, 143–174. https://doi.org/10.1017/S0022112004008882.

Sullivan, P.P., McWilliams, J.C., 2010. Dynamics of winds and currents coupled to surface waves. Annu. Rev. Fluid Mech. 42, 19–42. https://doi.org/10.1146/annurev-fluid-121108-145541.

Sullivan, P.P., McWilliams, J.C., 2019. Langmuir turbulence and filament frontogenesis in the oceanic surface boundary layer. J. Fluid Mech. 879, 512–553. https://doi.org/10.1017/jfm.2019.655.

Sullivan, P.P., McWilliams, J.C., Moeng, C.H., 1994. A subgrid-scale model for Large-Eddy Simulation of planetary boundary-layer flows. Bound.-Layer Meteorol. 71 (3), 247–276. https://doi.org/10.1007/BF00713741.

Sullivan, P.P., McWilliams, J.C., Patton, E.G., 2014. Large-eddy simulation of marine atmospheric boundary layers above a spectrum of moving waves. J. Atmos. Sci. 71 (11), 4001–4027. https://doi.org/10.1175/JAS-D-14-0095.1.

Sullivan, P.P., McWilliams, J.C., Weil, J.C., Patton, E.G., Fernando, H.J.S., 2020. Marine boundary layers above heterogeneous SST: Across-front winds. J. Atmos. Sci. 77 (12), 4251–4275. https://doi.org/10.1175/JAS-D-20-0062.1.

Sullivan, P.P., Romero, L., McWilliams, J.C., Melville, W.K., 2012. Transient evolution of Langmuir turbulence in ocean boundary layers driven by hurricane winds and waves. J. Phys. Oceanogr. 42 (11), 1959–1980. https://doi.org/10.1175/JPO-D-12-025.1.

Sutherland, G., Christensen, K., Ward, B., 2014. Evaluating Langmuir turbulence parameterizations in the ocean surface boundary layer. J. Geophys. Res., Oceans 119 (3), 1899–1910. https://doi.org/10.1002/2013JC009537.

Sutherland, G., Ward, B., Christensen, K., 2013. Wave-turbulence scaling in the ocean mixed layer. Ocean. Sci. 9 (4), 597–608. https://doi.org/10.5194/os-9-597-2013.

Suzuki, N., Fox-Kemper, B., 2016. Understanding Stokes forces in the wave-averaged equations. J. Geophys. Res., Oceans 121, 1–18. https://doi.org/10.1002/2015JC011566.

Suzuki, N., Fox-Kemper, B., Hamlington, P.E., Van Roekel, L.P., 2016. Surface waves affect frontogenesis. J. Geophys. Res., Oceans 121, 1–28. https://doi.org/10.1002/2015JC011563.

Suzuki, N., Hara, T., Sullivan, P.P., 2011. Turbulent airflow at young sea states with frequent wave breaking events: large-eddy simulation. J. Atmos. Sci. 68 (6), 1290–1305. https://doi.org/10.1175/2011JAS3619.1.

Sverdrup, H., 1953. On conditions for the vernal blooming of phytoplankton. ICES J. Mar. Sci. 18 (3), 287–295.

Taylor, J.R., 2018. Accumulation and subduction of buoyant material at submesoscale fronts. J. Phys. Oceanogr. 48 (6), 1233–1241. https://doi.org/10.1175/JPO-D-17-0269.1.

Taylor, J.R., Ferrari, R., 2011. Shutdown of turbulent convection as a new criterion for the onset of spring phytoplankton blooms. Limnol. Oceanogr. 56 (6), 2293–2307. https://doi.org/10.4319/lo.2011.56.6.2293.

Taylor, J.R., Smith, K.M., Vreugdenhil, C.A., 2020. The influence of submesoscales and vertical mixing on the export of sinking tracers in large-eddy simulations. J. Phys. Oceanogr. 50 (5), 1319–1339. https://doi.org/10.1175/JPO-D-19-0267.1.

Teixeira, M., Belcher, S., 2002. On the distortion of turbulence by a progressive surface wave. J. Fluid Mech. 458, 229–267. https://doi.org/10.1017/S0022112002007838.

Thomas, L.N., Tandon, A., Mahadevan, A., 2008. Submesoscale processes and dynamics. In: Ocean Modeling in an Eddying Regime, Vol. 177, pp. 17–38.

Thomas, L.N., Taylor, J.R., Ferrari, R., Joyce, T.M., 2013. Symmetric instability in the Gulf Stream. Deep-Sea Res., Part 2, Top. Stud. Oceanogr. 91, 96–110. https://doi.org/10.1016/j.dsr2.2013.02.025.

Thomson, J., 2012. Wave breaking dissipation observed with "SWIFT" drifters. J. Atmos. Ocean. Technol. 29 (12), 1866–1882. https://doi.org/10.1175/JTECH-D-12-00018.1.

Thomson, J., Polagye, B., Durgesh, V., Richmond, M.C., 2012. Measurements of turbulence at two tidal energy sites in Puget Sound, WA. IEEE J. Ocean. Eng. 37 (3), 363–374. https://doi.org/10.1109/JOE.2012.2191656.

Thorpe, S., Osborn, T., Jackson, J., Hall, A., Lueck, R., 2003. Measurements of turbulence in the upper-ocean mixing layer using autosub. J. Phys. Oceanogr. 33 (1), 122–145. https://doi.org/10.1175/1520-0485(2003)033<0122:MOTITU>2.0.CO;2.

Thorpe, S.A., 2007. An Introduction to Ocean Turbulence. Cambridge University Press, Cambridge.

Torma, P., Krámer, T., 2017. Modeling the effect of waves on the diurnal temperature stratification of a shallow lake. Period. Polytech., Civil Eng. 61 (2), 165–175. https://doi.org/10.3311/PPci.8883.

Troehn, I., Mahrt, L., 1986. A simple model of the atmospheric boundary layer; sensitivity to surface evaporation. Bound.-Layer Meteorol. 37, 129–148. https://doi.org/10.1007/BF00122760.

Tsujino, H., Urakawa, L.S., Griffies, S.M., Danabasoglu, G., Adcroft, A.J., Amaral, A.E., et al., 2020. Evaluation of global ocean–sea-ice model simulations based on the experimental protocols of the Ocean Model Intercomparison Project phase 2 (OMIP-2). Geosci. Model Dev. 13, 3643–3708. https://doi.org/10.5194/gmd-13-3643-2020.

Umlauf, L., Burchard, H., 2005. Second-order turbulence closure models for geophysical boundary layers. A review of recent work. Cont. Shelf Res. 25, 795–827. https://doi.org/10.1016/j.csr.2004.08.004.

Umlauf, L., Burchard, H., Bolding, K., 2016. GOTM sourcecode and test case documentation. Tech. Rep.. Leibniz Institute for Baltic Sea Research. http://gotm.net.

Van Roekel, L., Adcroft, A.J., Danabasoglu, G., Griffies, S.M., Kauffman, B., Large, W., et al., 2018. The KPP boundary layer scheme for the ocean: revisiting its formulation and benchmarking one-dimensional simulations relative to LES. J. Adv. Model. Earth Syst. 10 (11), 2647–2685. https://doi.org/10.1029/2018MS001336.

Van Roekel, L.P., Fox-Kemper, B., Sullivan, P.P., Hamlington, P.E., Haney, S.R., 2012. The form and orientation of Langmuir cells for misaligned winds and waves. J. Geophys. Res., Oceans 117, C05001. https://doi.org/10.1029/2011JC007516.

van Sebille, E., Aliani, S., Law, K.L., Maximenko, N., Alsina, J.M., Bagaev, A., et al., 2020. The physical oceanography of the transport of floating marine debris. Environ. Res. Lett. 15. https://doi.org/10.1088/1748-9326/ab6d7d.

Vieira, M., Guimaraes, P.V., Violante-Carvalho, N., Benetazzo, A., Bergamasco, F., Pereira, H., 2020. A low-cost stereo video system for measuring directional wind waves. J. Mar. Sci. Eng. 8 (11), 831. https://doi.org/10.3390/jmse8110831.

Von Kármán, T., 1930. Mechanische änlichkeit und turbulenz. Nachr. Ges. Wiss. Gött., Math.-Phys. Kl. 1930, 58–76.

Wang, D., Kukulka, T., Reichl, B.G., Hara, T., Ginis, I., 2019a. Wind–wave misalignment effects on Langmuir turbulence in tropical cyclone conditions. J. Phys. Oceanogr. 49 (12), 3109–3126. https://doi.org/10.1175/JPO-D-19-0093.1.

Wang, P., Ozgokmen, T.M., 2018. Langmuir circulation with explicit surface waves from moving-mesh modeling. Geophys. Res. Lett. 45 (1), 216–226. https://doi.org/10.1002/2017GL076009.

Wang, S., Wang, Q., Shu, Q., Scholz, P., Lohmann, G., Qiao, F., 2019b. Improving the upper-ocean temperature in an ocean climate model (FESOM 1.4): Shortwave penetration versus mixing induced by nonbreaking surface waves. J. Adv. Model. Earth Syst. 11 (2), 545–557. https://doi.org/10.1029/2018MS001494.

Ward, B., Wanninkhof, R., Minnett, P.J., Head, M.J., 2004. SkinDeEP: a profiling instrument for upper-decameter sea surface measurements. J. Atmos. Ocean. Technol. 21 (2), 207–222. https://doi.org/10.1175/1520-0426(2004)021<0207:SAPIFU>2.0.CO;2.

Waterhouse, A.F., MacKinnon, J.A., Nash, J.D., Alford, M.H., Kunze, E., Simmons, H.L., et al., 2014. Global patterns of diapycnal mixing from measurements of the turbulent dissipation rate. J. Phys. Oceanogr. 44 (7), 1854–1872. https://doi.org/10.1175/JPO-D-13-0104.1.

Webb, A., Fox-Kemper, B., 2011. Wave spectral moments and Stokes drift estimation. Ocean Model. 40 (3–4), 273–288. https://doi.org/10.1016/j.ocemod.2011.08.007.

Webb, A., Fox-Kemper, B., 2015. Impacts of wave spreading and multidirectional waves on estimating Stokes drift. Ocean Model. 96 (1), 49–64. https://doi.org/10.1016/j.ocemod.2014.12.007.

Weller, R.A., Price, J., 1988. Langmuir circulation within the oceanic mixed layer. Deep-Sea Res., A, Oceanogr. Res. Pap. 35, 711–747. https://doi.org/10.1016/0198-0149(88)90027-1.

Winton, M., Takahashi, K., Held, I.M., 2010. Importance of ocean heat uptake efficacy to transient climate change. J. Climate 23 (9), 2333–2344. https://doi.org/10.1175/2009JCLI3139.1.

Wu, L., Rutgersson, A., Sahlée, E., 2015. Upper-ocean mixing due to surface gravity waves. J. Geophys. Res., Oceans 120 (12), 8210–8228. https://doi.org/10.1002/2015JC011329.

Wyngaard, J.C., 2010. Turbulence in the Atmosphere. Cambridge University Press.

Xuan, A., Deng, B.-Q., Shen, L., 2019. Study of wave effect on vorticity in Langmuir turbulence using wave-phase-resolved large-eddy simulation. J. Fluid Mech. 875, 175–224. https://doi.org/10.1017/jfm.2019.481.

Yankovsky, E., Legg, S., Hallberg, R.W., 2021. Parameterization of submesoscale symmetric instability in dense flows along topography. J. Adv. Model. Earth Syst. 13 (6), e2020MS002049. https://doi.org/10.1029/2020MS002049.

Zhao, B., Qiao, F., Cavaleri, L., Wang, G., Bertotti, L., Liu, L., 2017. Sensitivity of typhoon modeling to surface waves and rainfall. J. Geophys. Res., Oceans 122, 2017–2028. https://doi.org/10.1002/2016JC012262.

Zheng, Z., Harcourt, R.R., D'Asaro, E.A., 2021. Evaluating Monin-Obukhov scaling in the unstable oceanic surface layer. J. Phys. Oceanogr. 51 (3), 911–930. https://doi.org/10.1175/JPO-D-20-0201.1.

Chapter 5

The lifecycle of surface-generated near-inertial waves

Leif N. Thomas[a] and Xiaoming Zhai[b]

[a]Department of Earth System Science, Stanford University, Stanford, CA, United States, [b]Centre for Ocean and Atmospheric Sciences, School of Environmental Sciences, University of East Anglia, Norwich, United Kingdom

5.1 Introduction

The unstratified surface mixed layer does not support internal gravity waves, but it does permit inertial motions, currents that oscillate and rotate anticyclonically in time at the inertial frequency, $f = 2\Omega \sin\phi$ (where Ω is the angular velocity of the Earth and ϕ the latitude). Inertial motions are readily accelerated by the winds, especially in regions where the winds are variable, such as in the mid-latitude storm tracks. The rate of energy input into inertial motions by the winds was thought to be $\mathcal{O}(10^{12}$ W), and thus could represent a significant fraction of the power required to drive the mixing necessary for maintaining the abyssal stratification and meridional overturning circulation (Wunsch and Ferrari, 2004). Part of the energy in inertial motions can be transmitted to downward-propagating near-inertial waves (NIWs). In the upper ocean, around 50% of the energy contained in internal wave spectra is attributable to NIWs (Ferrari and Wunsch, 2009). In addition, since NIWs are the internal waves with the lowest frequencies, they have the strongest vertical shears and are observed to have shear wavenumber spectra that are bluer than other internal waves (e.g., Pinkel, 1985; Silverthorne and Toole, 2009; Alford, 2010; Alford et al., 2017), and are thus thought to be a major contributor to ocean mixing by generating turbulence via shear instabilities. For these reasons, the focus of this chapter will be exclusively on wind-driven NIWs, even though other surface-generated internal waves do exist, for example, those triggered by boundary-layer turbulence (e.g. Ansong and Sutherland, 2010; Czeschel and Eden, 2019).

The lifecycle of NIWs will be detailed—from their generation by the winds in the mixed layer, to their radiation and propagation into the stratified interior, and finally to their ultimate demise, which can come in a range of forms, from turbulent dissipation, absorption by mean-flows, or energy loss to other internal waves. The topics emphasised are no doubt biased by our expertise and interests in wind-driven generation of NIWs and wave-mean flow interactions, but we have tried to be as comprehensive as possible. The emphasis will be on NIWs with larger vertical wavenumbers, rather than on the low modes. We refer the interested reader to the articles by Fu (1981) and Alford et al. (2016) for more extensive reviews on NIWs driven by winds and other processes and on the low modes. In terms of structure, the chapter is organised in sections following the three stages of a NIW's lifecycle. Each section is self-contained, providing the necessary background material and motivating questions for readers unfamiliar with the subject, while emphasising recent key findings. The chapter is finalised by a discussion on the implications for ocean mixing (both in the vertical and horizontal directions) and on the outstanding questions related to NIWs and mixing.

5.2 Generation of near-inertial waves at the surface

Wind fluctuations, particularly those associated with midlatitude travelling winter storms, are long known to be important for exciting near-inertial motions in the ocean. This association with the winds and near-inertial motions is corroborated by the observed seasonal cycle of near-inertial energy, the amplitude of which is several times higher in wintertime both in the surface mixed layer (Chaigneau et al., 2008) and in the deep ocean (Alford and Whitmont, 2007). The work done by the wind on surface inertial motions has been estimated primarily based on numerical models, ranging from the simple slab model to three-dimensional ocean general circulation models (Pollard and Millard, 1970; D'Asaro, 1985; Alford, 2003b; Furuichi et al., 2008; Rimac et al., 2013), and recently from surface drifters (Liu et al., 2019). The global distribution of wind work on surface inertial motions can also be estimated from a combination of satellite wind measurements and near-surface velocities derived from surface drifters (Liu et al., 2019). Large values of wind work are generally found in

96 Ocean Mixing

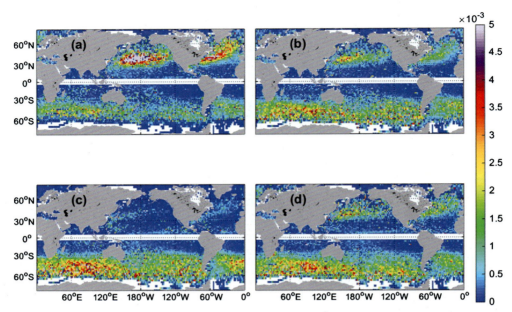

FIGURE 5.1 The global distribution of wind work on surface near-inertial motions (W m^{-2}) estimated from surface drifters and satellite wind measurements in (a) December–February, (b) March–May, (c) June–August, and (d) September–November. From Liu et al. (2019).

the midlatitude storm-track regions in the winter season (see Fig. 5.1), a common characteristic shared by all these studies. The global total of near-inertial wind work is estimated to be in the range of 0.3–1.5 TW (1 TW = 10^{12} W), with the spread largely explained by the different wind stress products used in different estimates (Jiang et al., 2005; Rimac et al., 2013). The amount of work done by the wind on near-inertial motions is therefore comparable to that of wind work on surface geostrophic currents (∼0.8 TW; Wunsch, 1998) and energy conversion rate from the barotropic tides to baroclinic tides (∼0.9 TW; Egbert and Ray, 2000).

The wind forcing of surface inertial currents is known to be highly intermittent in time; strong resonant wind events associated with a handful of winter storms typically account for the bulk of annual wind energy input to surface inertial motions (D'Asaro, 1985). In this context, storms often refer to a number of transient atmospheric phenomena, including synoptic weather systems, cold and warm fronts, and travelling lows, which cover a wide range of spatial scales. What scales of atmospheric motions are responsible for the majority of the near-inertial wind work? In the pioneering work of D'Asaro (1985), he examined wind forcing of surface inertial currents using time series of wind stress computed from meteorological buoy data off North America and found that near-inertial motions at these buoy locations are primarily excited by atmospheric features with scales of 100 km or less, such as translating lows and cold fronts. The larger-scale atmospheric phenomena, although they are more energetic, do not excite strong inertial oscillations. The dominance of smaller-scale winds in generating near-inertial motions was attributed to advection of midlatitude mesoscale atmospheric features by the background wind field, which has a typical speed of 10 m s^{-1}. As a result, features with a scale of about 100 km will be advected past a given location with a time scale close to $1/f$, enabling efficient resonant excitation of inertial oscillations in the surface layer. The dependence of near-inertial wind work on the scale of atmospheric motions was recently re-investigated by Zhai (2017) using a high-resolution reanalysis wind product. It was found that atmospheric systems with horizontal scales less than 1000 km are responsible for almost all the near-inertial wind work in the midlatitude North Atlantic and North Pacific. Transient atmospheric features with scales in the order of 100 km contribute significantly (about 25-30%) to this wind work, but they are not as dominant as traditionally thought. The implication is that wind forcing of surface inertial currents is a result of atmospheric features with scales less than 1000 km propagating at a range of speeds over given locations, rather than being dominated by features with scales of ∼100 km propagating at a speed of 10 m s^{-1}. Whether wind work on near-inertial motions is done by atmospheric features with scales of ∼100 km, or by those with larger scales, may have consequences for near-inertial mixing in the ocean interior, since the rate at which the wind-induced near-inertial energy radiates from the surface mixed layer into the ocean interior is inversely proportional to the length scale of atmospheric forcing squared (e.g. Gill, 1984; D'Asaro, 1989).

Midlatitude storm activities are known to vary significantly from one year to another, not only in storm intensity, but also in storm track. For example, the storm track in the North Atlantic is tilted northeastward, stretching from Cape Hatteras

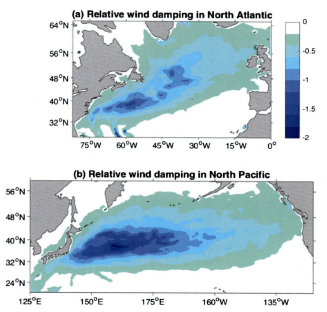

FIGURE 5.2 The relative wind damping of wind energy input to near-inertial motions (mW m^{-2}) averaged over the period of 2001–2010 in the midlatitude (a) North Atlantic and (b) North Pacific. From Zhai (2017).

into the Nordic seas, during the positive phase of the North Atlantic oscillation (NAO), whereas it shifts equatorward and becomes more zonally oriented during the negative phase (e.g., Hurrell, 1995; Rogers, 1997). Storm tracks are also predicted to intensify and shift poleward in the Southern Hemisphere under global warming (e.g., Chang et al., 2012). These observed and predicted storm-track shifts raise the question of latitudinal dependence of wind-generated near-inertial energy in the ocean. Dippe et al. (2015) recently compared wind work on near-inertial motions in the North Atlantic in a strongly positive NAO year with that in a strongly negative NAO year and found greater basin-wide near-inertial wind work in the negative NAO year, despite reduced storm intensity. They attributed this result to greater efficiency of wind forcing in exciting surface inertial currents at low latitudes. Motivated by the study of Dippe et al. (2015) and using a damped slab model, Zhai (2015) showed that, in the case of resonant inertial wind forcing, which dominates the overall wind work on near-inertial motions (e.g., Crawford and Large, 1996; Plueddemann and Farrar, 2006), both the near-inertial wind work and wind-induced near-inertial energy are independent of latitude. On the other hand, the observed surface wind stress spectra are red, with spectral slopes typically varying between -1 and -2 (e.g., Gille, 2005), which means longer inertial periods at lower latitudes are capable of sampling greater inertial wind stress forcing (Rath et al., 2014; Zhai, 2015). Therefore an equatorward shift of the storm track is likely to result in an increase in wind-induced near-inertial energy, whereas the opposite is true for a poleward shift.

Atmospheric winds are known to systematically damp ocean currents, such as ocean eddies, if the relative motion between the atmosphere and underlying surface ocean is taken into account in the wind stress calculation, i.e., the so-called relative wind damping effect (e.g., Duhaut and Straub, 2006; Zhai et al., 2012; Xu et al., 2016). However, the relative wind damping effect has received far less attention in the literature on wind generation of near-inertial motions and is not accounted for in most global estimates of wind work on surface inertial currents (e.g., Alford, 2003b; Furuichi et al., 2008; Rimac et al., 2013). The relative wind damping effect arises, because accounting for relative motions in the stress calculation systematically reduces the positive wind work when the wind and surface inertial current are aligned and enhances the negative wind work when they oppose each other. A recent numerical study by Rath et al. (2013) found that including ocean surface velocity in the wind stress calculation led to a noticeable reduction (about 22%) of wind work on near-inertial motions in the Southern Ocean. A similar percentage reduction in wind work on near-inertial motions was also found in the midlatitude North Atlantic and North Pacific (Zhai, 2017). Most recently, Alford (2020b) estimated that the global reduction of the wind work by this effect is 13%. Furthermore, this reduction in near-inertial wind work is most significant in the storm-track regions, where the winds are particularly strong (Fig. 5.2), since the relative wind damping effect is proportional to the magnitude of surface wind speed (Duhaut and Straub, 2006; Zhai et al., 2012; Rath et al., 2013). An even larger relative wind damping effect was reported by Liu et al. (2019), who estimated wind work on near-inertial oscillations using surface drifter data and found that neglecting relative motions in the stress calculation led to an overestimate of global near-inertial wind work by 60%.

Although progress has been made in recent decades, there are still a number of open questions regarding wind generation of near-inertial currents. First, regions of large wind work on surface inertial motions are also regions of strong mesoscale variability in the ocean (Zhai et al., 2005). The juxtaposition of atmospheric and oceanic storm tracks raises the question of how the presence of ocean mesoscale features modifies the resonance conditions for the wind generation of near-inertial motions, and how the ocean mesoscale may impact the net near-inertial wind work. An earlier modelling study by Klein et al. (2004a) showed that mesoscale vorticity can shift the resonant frequency from the local inertial frequency f to the effective inertial frequency $f_{eff} = f + \bar{\zeta}/2$, where $\bar{\zeta} = \partial \bar{v}/\partial x - \partial \bar{u}/\partial y$ is the relative vorticity of the mesoscale field (Kunze, 1985), and that mesoscale eddies can set the lateral scales of near-inertial motions in the surface mixed layer of an eddying ocean, which renders the rate at which near-inertial energy radiates downward into the ocean interior much less sensitive to the scales of atmospheric forcing. On the other hand, the presence of a turbulent eddy field was found to have little effect on the net wind-generated near-inertial energy in their model. With the increase in computing power and advances in remote and in situ observations, the question of how mesoscale and submesoscale features in the ocean affect wind generation of near-inertial motions deserves further investigation. Second, an important motivation for estimating global wind work on near-inertial motions is to quantify mechanical energy sources for maintaining deep ocean mixing. However, not all the wind work on surface near-inertial motions is available for deep ocean mixing. For example, Plueddemann and Farrar (2006) showed that during resonant wind events, the near-inertial wind energy input is primarily balanced by dissipation at the base of the surface mixed layer due to shear instability. Modelling studies by Furuichi et al. (2008) and Zhai et al. (2009) found 70–85% of wind-induced near-inertial energy at the surface was lost to turbulent mixing in the upper 200 m or so, and hence is not available to generate diapycnal mixing at greater depth. More recently, Alford (2020b) estimated that over 50% of the wind work goes into energising turbulence in the mixed layer and is unavailable for downward propagating NIWs. Other mechanisms for generating NIWs besides the winds, such as geostrophic adjustment (Ou, 1984) and spontaneous generation, i.e., internal waves emitted from balanced flows in the absence of external forcing (Alford et al., 2013; Vanneste, 2013; Shakespeare and Taylor, 2014; Nagai et al., 2015), should be considered.

5.3 Propagation of near-inertial waves out of the mixed layer

After NIWs are generated in the upper ocean by winds, their energy can either be dissipated locally (see Section 5.5) or lost through downward radiation of energy. The seminal work of Gill (1984) laid a theoretical framework that has been used to study the latter. In Gill's model, winds generate a slab-like inertial wave over the mixed layer with a horizontal wavenumber $\mathbf{k}_h$ that is set by the structure of the winds. Convergence/divergence of the inertial motions[1] generate "inertial pumping" at the base of the mixed layer, i.e., vertical velocities that displace isopycnals in the thermocline and lead to pressure anomalies that allow for vertical propagation and downward radiation of energy. The rate of energy loss in the mixed layer by this mechanism is proportional to the square of the horizontal wavenumber $\mathbf{k}_h$ and increases with mixed-layer depth (Gill, 1984). The dependence on $\mathbf{k}_h$ follows from the expression for the vertical component of the group velocity of inertia-gravity waves

$$c_{g,z} \approx -\frac{N^2 |\mathbf{k}_h|^2}{f m^3} \qquad (5.1)$$

(calculated here using the dispersion relation for NIWs in the absence of background flows). The group velocity, and hence energy flux, increases as the square of $\mathbf{k}_h$ and decreases with the vertical wavenumber as m^{-3}. The latter implies that lower modes flux energy more rapidly out of the mixed layer. Since deeper mixed layers have an initial slab-like inertial motion with a larger projection onto lower modes, they result in a more rapid transfer of energy to the thermocline.

Whereas Gill's theory provides a useful conceptual model to understand the loss of near-inertial energy from the mixed layer via radiation, it greatly overpredicts the time that it takes for the mixed-layer inertial energy to decay for NIWs generated by storms in the midlatitudes (i.e., months instead of days), since these storms typically span hundreds of kilometres. For winds with smaller-scales, however, such as fast-moving atmospheric fronts or tropical cyclones, the decay time is more in line with observations (Price, 1983), and can result in rapid radiation of NIW energy in a few days (see Fig. 5.3 for a particularly striking example of NIW radiation in the wake of a hurricane).

In the absence of small-scales in the winds, however, wind-driven near-inertial motions must experience an increase in the horizontal wavenumber to efficiently radiate energy from the surface ocean into the interior. A useful framework for

---

1. Note that we are using the informal definition of inertial motions here, i.e., currents in the mixed layer that oscillate at frequencies close to f. Technically, inertial motions should have a horizontal wavenumber of zero, no buoyancy or pressure anomalies, and hence do not propagate, unlike NIWs.

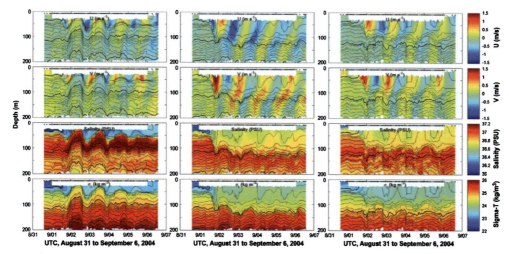

FIGURE 5.3 Observations from three EM-APEX floats made in the wake of Hurricane Frances (2004) illustrating the radiation of NIWs from the storm. The floats were deployed 0 km (left panels), 55 km (middle panels), and 110 km (right panels) to the right of the hurricane. The zonal (U) and meridional (V) components of the velocity are plotted in the upper panels, whereas the salinity and potential density are plotted beneath, all as a function of depth and time. Isotherms are contoured in each panel with a spacing of 0.5 °C. From Sanford et al. (2011).

understanding the processes that can lead to an increase in the horizontal wavevector is the ray-tracing equations, which assume that waves can be decomposed into a series of plane waves with variable wavevectors, i.e.,

$$\frac{D_g \mathbf{k}_h}{Dt} = \underbrace{-\sum_i \frac{\partial \hat{\Omega}}{\partial \lambda_i} \nabla_h \lambda_i}_{\text{refraction}} \underbrace{-\nabla_h \bar{\mathbf{u}}_h \cdot \mathbf{k}_h}_{\text{straining}} \underbrace{-m \nabla_h \bar{w}}_{\text{diff. vertical advection}}, \tag{5.2}$$

which establish how the wavevector changes following a wave group (as indicated by the subscript "g" on the material derivative) via (i) refraction caused by modulations in the parameters λ_i that describe the medium in which the wave propagates; for example, λ_i could represent the stratification, the Coriolis parameter, or properties of the mean flow, such as the vertical vorticity, where $\hat{\Omega}(\mathbf{k}, \lambda_i)$ is the dispersion relation of the wave; (ii) straining caused by horizontal advection of wave phase by a background flow $\bar{\mathbf{u}}_h$, and (iii) differential vertical advection of wave phase by a background vertical circulation $\bar{w}$ (Gill, 1982). How these three mechanisms can lead to the growth in horizontal wavevector of a NIW is described below.

5.3.1 Refraction

Variations in the medium in which a wave propagates lead to refraction. For NIWs, the characteristic of the medium that is critical to their dynamics is the net spin of the fluid, which is dependent on both the Coriolis parameter and the vertical vorticity $\bar{\zeta}$ of a background flow. Or in the framework of (5.2), the key parameters describing the medium for NIWs are $\lambda_1 = f$ and $\lambda_2 = \bar{\zeta}$. Therefore variations in either f (specifically, $df/dy \equiv \beta$) or $\bar{\zeta}$ can lead to NIW refraction, known as β and ζ-refraction, respectively, both of which can lead to the reduction in lateral scale of NIOs. The general idea is that gradients in f or $\bar{\zeta}$ set up lateral differences in wave phase, since NIOs separated a short distance from one another oscillate at slightly different frequencies. As a result, the NIOs develop a horizontal wavenumber, whose magnitude increases with time.

The theory for β-refraction was developed by D'Asaro (1989) and was tested during the groundbreaking Ocean Storms experiment (D'Asaro et al., 1995). The theory predicts that the meridional wavenumber l of a NIW decreases with time as $l = -\beta t$, where t is the time, since the NIW was generated by the winds. This prediction follows from taking the gradient of the phase θ of an inertial oscillation on a β-plane, which has a velocity of the form $e^{i\theta} = e^{-if(y)t}$, where the Coriolis parameter is $f(y) = f_o + \beta(y - y_o)$ (f_o and y_o are constants), and hence $l \equiv \partial \theta / \partial y = -\beta t$. In the Ocean Storms experiment, a combination of drifters and moorings distributed over a few degrees of latitude were used to track phase differences in near-surface inertial currents and the subsequent propagation of near-inertial (NI) energy into the interior after the passage of a storm. By fitting a plane wave to the inertial currents, an estimate of the meridional wavenumber was made. The resultant wavenumber was negative and decreased linearly with time at a rate consistent with β-refraction. By 20 days after the storm, the wavelength of the NIWs had shrunk to $\sim$50 km, and a beam of NIW energy was observed at 100 m that was accompanied by a decrease in NI energy in the mixed layer, in line with a flux of energy away from the surface. The observed rate of NI energy decay in the mixed layer was 20–40% faster than that predicted by linear theory, however, indicating that the behaviour of the waves was not entirely explained by β-refraction.

β-refraction leads to the equatorward propagation of NIWs. This phenomena has been observed in mooring records (Alford, 2003a) and simulated with global circulation models (Simmons and Alford, 2012). The energy in the far-reaching, equatorward-propagating NIWs is primarily contained in low modes. The high modes do not propagate as rapidly and are thought to contribute to mixing more localised to where the NIWs are generated by the winds (Alford, 2020a).

ζ-refraction of NIWs by mesoscale eddies can significantly enhance the radiative damping of NIW energy in the mixed layer. In fact, Balmforth and Young (1999) demonstrated that it can explain the discrepancy in the predicted and observed damping times in the Ocean Storms experiment. The process results in a horizontal wavevector that grows in time as $\mathbf{k}_{h,\zeta} = -\nabla_h \overline{\zeta} t/2$, which is orders of magnitude faster than β-refraction since gradients in vertical vorticity, even on the mesoscale, are much larger than β (van Meurs, 1998). Given the larger wavenumbers, and hence group velocities (5.1) that ζ-refraction can generate, the radiation it drives is thus much enhanced over β-refraction. Additionally, radiation associated with ζ-refraction is heterogeneous and leads to the preferential propagation of NIW energy into regions of anticyclonic vorticity, since $\mathbf{k}_{h,\zeta}$ is directed down the gradient in vorticity (Kunze, 1985; Young and Ben-Jelloul, 1997; Klein et al., 2004b). Once in an anticyclone, NIW energy is rapidly fluxed downwards through a process which was initially termed the *inertial chimney* effect and has recently been renamed as the *inertial drainpipe* effect, more evocative of the phenomenon (Lee and Niiler, 1998; Asselin and Young, 2020). The observational evidence for this effect has been sparse (see discussion below and Thomas et al., 2020), although the inference of an enhanced seasonal cycle in mixing in anticyclones relative to cyclones by Whalen et al. (2018) and the trapping of NIWs in anticyclones Kunze (1986); Kunze et al. (1995); Joyce et al. (2013) are consistent with this mechanism.

While the modulation and decoherence of NIOs by vertical vorticity have been observed and described in the literature (Weller, 1982; Elipot et al., 2010), the progression from NIO generation by winds, shrinking of NIO lateral scale by ζ-refraction, to propagation of NIWs into regions of anticyclonic vorticity has only recently been measured in the ocean on a field campaign to the Iceland Basin made as part of the Near-Inertial Shear and Kinetic Energy in the North Atlantic experiment (NISKINe) (Thomas et al., 2020). The observations were made from ship-based surveys conducted near the axis of a dipole vortex, where the gradient in vorticity was large, and hence ζ-refraction was prominent. Differences in NIW phase across the dipole were observed to grow in time, generating a lateral wavelength that shrank at a rate consistent with ζ-refraction, reaching $\sim$40 km in 1.5 days. Two days later, a NIW beam with a $\sim$13 km horizontal and $\sim$200 m vertical wavelength was detected at depth radiating energy downwards and towards the dipole's anticyclone (see Fig. 5.4). The rapidity of the NIW evolution and radiation can be contrasted with the waves observed in the Ocean Storms experiment described above. ζ-refraction is clearly more effective at radiating NIW energy away from the surface than β-refraction, and thus plays an important role in the lifecycle of wind-forced NIWs, given that mesoscale eddies are omnipresent in the ocean.

5.3.2 Straining

In addition to vorticity gradients, strain in the mesoscale eddy field can sharpen the lateral scales of NIWs. Because NIWs have weak group velocities relative to other types of inertia-gravity waves, their propagation speeds are low and therefore the NIW phase, θ, can evolve in a similar way to a tracer advected by a background flow such that, in regions of strain, lateral gradients in phase can experience exponential growth. The similar evolution of tracers and wave phase follows from (5.2), which, when isolating the straining term and using the definition of the wavevector as the gradient of wave phase, $\mathbf{k}_h \equiv \nabla_h \theta$, yields

$$\frac{D_g}{Dt} \nabla_h \theta = -\nabla_h \overline{\mathbf{u}}_h \cdot \nabla_h \theta \tag{5.3}$$

and is identical to the equation for the gradient of a tracer, C,

$$\frac{D}{Dt} \nabla_h C = -\nabla_h \overline{\mathbf{u}}_h \cdot \nabla_h C \tag{5.4}$$

when $D_g/Dt \equiv D/Dt + \hat{\mathbf{c}}_g \cdot \nabla = D/Dt$, that is, when the intrinsic group velocity $\hat{\mathbf{c}}_g = \nabla_k \hat{\Omega}$ is zero (Bühler and McIntyre, 2005). The exponential growth in the wavevector that can result from (5.3) when $\mathbf{k}_h$ is aligned with the extension axis of

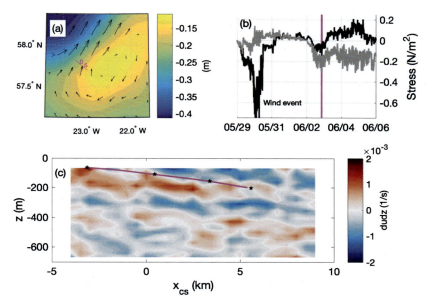

FIGURE 5.4 NIW beam observed radiating from the axial jet of a dipole vortex in the Iceland Basin down and into the dipole's anticyclone after a strong wind event. (a) The sea-surface height anomaly (colour), surface geostrophic velocity (vectors), and ship track of the section along which the beam was observed (magenta line). (b) Time series of the zonal (black) and meridional (grey) components of the wind-stress highlighting the strong wind event on 5/30/2019. The magenta-shaded region indicates the time when the section was made, i.e., 3.5 days after the wind event. (c) Section of the vertical shear in the zonal velocity illustrating the NIW beam's structure as a function of depth and a cross-stream coordinate that is perpendicular to the axial jet (the positions on the section at cross-stream distances of 0 and 5 km are indicated in (a)). The section was completed in less than 0.15 inertial periods, and hence captured a snap-shot of the NIW field. The ray path of a wavepacket with an intrinsic frequency of $1.026f$ is indicated in magenta in (c) and the black stars are the positions of the wavepacket each day of travel. See Thomas et al. (2020) for more details on the observations and the analysis.

a strain field is only a transient response for most applications, however. The response is transient, because as the lateral scales of NIWs shrink, their group velocity and dispersion increases, such that $D_g/Dt \not\approx D/Dt$, and the waves propagate out of strain regions. This mechanism is known as *wave escape* and limits the growth of the horizontal wavevector by strain (Rocha et al., 2018).

Even without wave dispersion, the effects of straining on NIWs can be muted if the lateral scales they develop are initiated by ζ-refraction (as opposed to some other process), such as wind forcing with small-scales. In this case, NIW phase is linked with the vorticity such that under strain, gradients in phase and vorticity increase in lock-step, and the horizontal wavenumber evolves as $\mathbf{k}_h = -(t/2)\nabla_h \overline{\zeta}(\mathbf{x},t)$ (Asselin et al., 2020). Thus unless the strain leads to exponential growth in $\nabla_h \overline{\zeta}$, it will not lead to exponential growth in $\mathbf{k}_h$. For example, in a steady dipole vortex, which has both strain and vorticity, a NIW that is initially horizontally uniform will develop gradients in phase that grow linearly in time *everywhere* in contrast to a random tracer field, where the gradients grow exponentially in straining regions (Asselin et al., 2020). However, when combined with lateral density gradients, strain can significantly modify NIWs by driving a frontal vertical circulation that can efficiently interact with NIWs.

5.3.3 Interaction with frontal vertical circulations

Strain acting on a front disrupts the thermal wind balance and sets up a vertical velocity $\overline{w}$ that alternates sign across the front to restore the balance (Hoskins et al., 1978). If a NIW with vertical wavenumber m is embedded in such a front, the wave will experience a vertical Doppler shift $\overline{w}m$ that changes sign across the front, inducing a lateral phase difference. This differential vertical Doppler shifting results in the growth of $\mathbf{k}_h$ at a rate given by the last term in (5.2), a process that was explored by Thomas (2019).

Vertical Doppler shifts can be significant. For example, a frontal vertical circulation modest in strength, $\overline{w} = 10$ m day^{-1}, acting on a wave with a vertical wavelength of $\mathcal{O}(100$ m$)$, typical of a low-mode NIW, yields a Doppler shift $\overline{w}m \sim 7 \times 10^{-6}$ s^{-1}, which in mid-latitudes is nearly 10% of f. High-mode NIWs will experience even larger Doppler shifts. These adjustments to the NIWs' frequency are comparable to frequency shifts associated with ζ-refraction.

However, Doppler shifting by frontal vertical motions is distinct from ζ-refraction in several ways. If the vertical circulation is driven by strain that is frontogenetic, the development of lateral phase differences is exponential, rather than

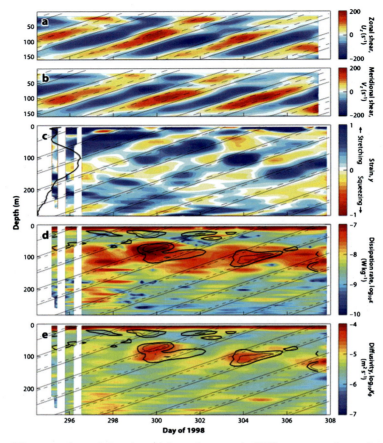

FIGURE 5.5 Depth-time series of (from top to bottom) shear (u_z, v_z), internal wave strain (IWS), a measure of a wave's stratification anomaly, dissipation rate ϵ, and diapycnal diffusivity K_ρ, associated with a downward propagating NIW observed in the Banda Sea. A profile of the mean stratification is plotted on the left of the third panel. From Alford and Gregg (2001).

linear in time, and leads to the rapid radiation of NIW beams away from fronts and down into the stratified interior. In addition, the development of lateral phase differences by vertical Doppler shifting depends on the vertical wavenumber m of the NIWs, unlike refraction. As a result, the downward flux of NIW energy out of the upper ocean should scale as m^{-1} with Doppler shifting, versus m^{-3} with refraction. Therefore high-mode NIWs should be much more effective at radiating energy in an eddy field when frontal vertical circulations are present.

5.4 Interactions of near-inertial waves with variable stratification, other internal waves, and mean flows in the interior

When NIWs leave the mixed layer and propagate into the pycnocline and beyond, they encounter variations in stratification, other internal waves, and mean flows. In this section, we will summarise how these interactions can lead to modulations in the NIW's amplitude, wavevector, and frequency and provide some observational examples of such interactions.

5.4.1 Variable stratification

As NIWs enter regions with stronger stratification, such as the pycnocline, they experience an intensification in shear, because both their vertical wavenumber and energy density increases. The amplification can be large enough to trigger turbulence and mixing, as in the exemplary observations of a downward propagating NIW in the Banda Sea described by Alford and Gregg (2001). Depth/time series from these observations demonstrate how the NIW shear strengthens as the wave approaches the peak in stratification near 100 m, resulting in a modulation of the turbulent dissipation and diffusivity over a wave period (Fig. 5.5).

The amplification in NIW shear with stratification illustrated in these observations can be understood in terms of WKB scalings for inertia-gravity waves propagating through a vertically varying stratification $N^2(z)$ (e.g. Gill (1982)). As the waves encounter higher stratification, their vertical wavenumber increases to conserve the intrinsic frequency, more specifically, $m(z) \propto N(z)$. Furthermore, the waves slow down since the increase in m lowers their group velocity, e.g., (5.1). The reduction in group velocity leads to an amplification in their energy density, $E(z) \propto N$, so that their downward energy flux, $c_{g,z} E \propto E/N$, remains constant. For NIWs, the energy density scales with their velocity squared, and hence their vertical shear scales as $S \propto N^{3/2}$, which implies that the Froude number $Fr = S/N$ increases with stratification. Consequently, vertically propagating NIWs are more likely to break and generate turbulence in regions of higher stratification, a prediction that is played out in the coherent NIW beam observed by Alford and Gregg (2001). Similar dynamics have also been observed in the deep ocean near the Mendocino Escarpment (Alford, 2010).

5.4.2 Interactions with other internal waves

Once in the ocean interior, NIWs can encounter other internal waves and through non-linear wave-wave interactions transfer energy to other frequencies and wavenumbers or, alternatively, acquire energy. Such interactions are not limited to NIWs and have been studied extensively in the context of the internal tides, but here we discuss their implications for the NIW field. We refer the reader to Chapter 6 of the monograph for a more thorough review of wave-wave interactions. In their weakly non-linear form, these interactions involve resonant triads that flux energy from low to higher vertical wavenumbers and energise the internal wave continuum while balancing the energy lost to turbulence by dissipation. There are thought to be two dominant triad interactions, parametric subharmonic instability (PSI) and induced diffusion, both of which act to transfer energy to lower frequencies (McComas and Müller, 1981; Polzin and Lvov, 2011). For NIWs that have propagated equatorward of their generation region to a latitude where their frequency is twice the local f, PSI could feed off the waves and act as a sink of their energy (Nagasawa et al., 2000). Locally however, PSI is predicted to strengthen the inertial peak and result in a vertical wavenumber kinetic energy spectrum that follows a m^{-1} power law (up to a cutoff wavenumber), yielding a blue shear spectrum in line with the canonical Garrett–Munk spectrum (McComas and Müller, 1981). The occurrence of PSI is one explanation for why shear spectra are dominated by NIWs and points to their importance in inducing shear-driven turbulence in the ocean interior. However, Monte Carlo eikonal (i.e., ray-tracing) calculations that model the interaction of internal waves, but are not restricted by the weak interaction assumption (yet are subject to the limitations of ray-tracing, which assumes scale separations) predict that there is a flux of energy from low to high frequencies, which could reduce the strength of the inertial peak (Henyey et al., 1986).

Whereas PSI and induced diffusion can produce spectra representative of what is observed, they require sources of energy at high frequencies to be sustained, which is inconsistent with our current understanding that the dominant sources of energy for the internal wave field are at near-inertial and tidal frequencies (Polzin and Lvov, 2011). Recent observational analyses suggest that NIWs can in fact energise the internal wave continuum. Le Boyer and Alford (2021) analysed kinetic energy (KE) spectra from over a thousand moorings distributed around the world's oceans and found that the KE contained in the continuum followed a seasonal cycle at most locations, suggesting a wind-driven, near-inertial source. Comparing time series of the KE contained in the near-inertial peak, the tides, and low-frequency presumably balanced, motions, Le Boyer and Alford (2021) further found that the KE in the continuum had a higher correlation with the near-inertial peak than these other two potential sources of energy for the internal wave field. Frankignoul and Joyce (1979) noted similar correlations in the records that they analysed, as well as Polzin and Lvov (2011), who commented on temporal lags between the KE in the continuum and the near-inertial peak, which they speculated were influenced by the presence or absence of mesoscale eddies.

Numerical simulations also suggest that a near-inertial energy source is essential for maintaining the internal wave continuum. Sugiyama et al. (2009) performed high-resolution, two-dimensional simulations that were able to reproduce a Garrett–Munk-like spectrum when forced at semi-diurnal tidal and near-inertial frequencies. The simulations were driven at the lowest vertical wavenumbers permissible in the model, yet generated a cascade of energy to small scales, thus filling out the continuum. This excitation of the continuum occurred only when both the semi-diurnal and near-inertial energy sources were active, whereas forcing at the semi-diurnal frequency alone did not produce a Garrett–Munk-like spectrum.

The spectral analyses described above that evidence a link between the internal wave continuum and the near-inertial peak were made in an Eulerian frame, and hence their interpretation could be confounded by Doppler shifting. For example, Pinkel (2008) demonstrated that observed frequency spectra of fine-scale shear can have a well-defined near-inertial peak in an isopycnal-following coordinate, whereas in an Eulerian frame the energy is spread into the continuum. Using a kinematic model that Doppler shifts near-inertial frequencies through vertical advection of the near-inertial shear by the internal tides and random motions, Pinkel (2008) could reproduce the observed Eulerian spectrum, raising doubts on how the internal-wave continuum should be interpreted. Alford (2001) found a similar result for near-inertial shear advected by the tides from the record described by Alford and Gregg (2001), discussed above. However, it is not clear why this Doppler shifting should be dismissed as an artifact of the coordinate system, as though it were devoid of physics. The Doppler shifting is a clear signature of advection of one wave by another, which is the essential ingredient for all wave-wave interactions. Whether the vertical advection observed in these records resulted in resonant interactions where energy is permanently transferred between waves is another question altogether. This question was addressed by Mihaly et al. (1998), who on a mooring in the northeast Pacific observed a prominent peak at the sum of the inertial and semidiurnal frequencies, which can arise from Doppler shifting of inertial motions by the semidiurnal tides. The peak had rotary spectra with properties of a freely propagating wave, suggesting that energy had been permanently transferred from inertial to higher frequencies.

5.4.3 Interactions with mean flows

Wave-mean flow interactions can Doppler shift NIW frequencies, such as wave-wave interactions, but they can additionally lead to significant refraction, as the vorticity and baroclinicity of mean flows affect the intrinsic frequency of NIWs. This refraction was discussed in Section 5.3 in the context of NIW propagation out of the mixed layer. Here, the emphasis will be on how wave-mean flow interactions can result in NIW trapping, amplification, breaking, and energy exchanges between the two types of flows in the ocean interior.

5.4.3.1 Near inertial wave trapping and amplification in anticyclones and fronts

Vorticity and baroclinicity of a background flow modify the dispersion relation of NIWs. These properties of the mean flow therefore determine the regions where NIWs are evanescent or freely propagating, and hence where the waves can be trapped. We will illustrate this physics using a zonal jet that varies in the y and z directions, $\overline{u}(y, z)$, and that satisfies the thermal wind balance: $M^2 = f \partial \overline{u}/\partial z = -\partial \overline{b}/\partial y$, where $\overline{b}$ is the buoyancy associated with the balanced flow and M^2 is a measure of its baroclinicity. The dispersion relation for a two-dimensional wave with meridional and vertical wavenumbers l and m in this flow is

$$\omega = \sqrt{f(f + \overline{\zeta}) + 2M^2 \frac{l}{m} + N^2 \frac{l^2}{m^2}}, \tag{5.5}$$

where $\overline{\zeta} = -\partial \overline{u}/\partial y$ is the vertical vorticity of the jet, N the buoyancy frequency, and the waves are assumed to be hydrostatic and satisfy the WKB approximation (Mooers, 1975; Whitt and Thomas, 2013). The frequency (5.5) is minimum for waves with phase lines that run parallel to isopycnals, i.e., $(l/m)|_{min} = -s_\rho$, where $s_\rho = M^2/N^2$ is the isopycnal slope. When this condition is met, the waves attain a minimum frequency

$$\omega_{min} = f\sqrt{1 + Ro - 1/Ri_b}, \tag{5.6}$$

where $Ro = \overline{\zeta}/f$ and $Ri_b = N^2 f^2/M^4$ are the Rossby and Richardson numbers of the balanced jet (Whitt and Thomas, 2013). For waves of a given frequency ω, the isosurface, where $\omega = \omega_{min}$, known as the separatrix, sets the boundary between where the waves can freely propagate ($\omega > \omega_{min}$) and where they are evanescent ($\omega < \omega_{min}$). In a barotropic jet ($M^2 = 0$) for example, subinertial waves are permitted, yet are trapped in regions with anticyclonic vorticity, where $\omega_{min} = f_{eff} < \omega < f$, which is the classic result of Kunze (1985). Baroclinicity expands these trapping regions since it always lowers the minimum frequency (i.e., $\omega_{min} < f_{eff}$ when $M^2 \neq 0$).

When a NIW approaches the separatrix it experiences intensification, and when it reflects off the separatrix it can be further amplified under certain conditions. This amplification follows from the behaviour of the group velocity

$$\mathbf{c}_g = \frac{N^2}{\omega m}\left[s_\rho + \left(\frac{l}{m}\right)\right]\left(1, -\frac{l}{m}\right) \tag{5.7}$$

in the proximity of the separatrix (Whitt and Thomas, 2013). At the separatrix, $\omega = \omega_{min}$, $l/m = -s_\rho$, and hence $\mathbf{c}_g = 0$. Thus as a wavepacket approaches this boundary, its energy flux vanishes and the energy density strengthens since the flux is convergent. After reflection off the separatrix and for certain background flows and wave frequencies, wave rays can converge and come to a halt, setting up a critical layer. This condition is met when the separatrix parallels isopycnals. In weakly baroclinic anticyclones, for example, critical layers are nearly flat and form at the base of the vortices, where the vorticity increases with depth (Kunze, 1985; Kunze et al., 1995). At fronts in contrast, NIW critical layers tilt with isopycnals and tend to be found on the cyclonic side of their frontal jets, where lateral density gradients are strongest (Whitt et al., 2018). A striking example of such a slantwise critical layer that formed in the Gulf Stream under strong

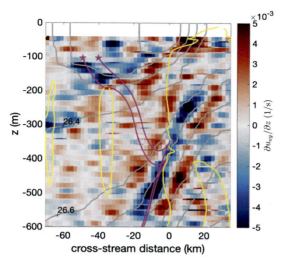

FIGURE 5.6 Cross-section of ageostrophic shear in the Gulf Stream illustrating NIW amplification in a slantwise critical layer. The NIWs are evidenced by shear bands that run nearly parallel to isopycnals (grey contours plotted at an interval of 0.2 kg m^{-3}) in the centre of the current. A slantwise critical layer for NIWs with frequency $\omega = f$ forms where the separatrix for these waves (yellow contours) parallels isopycnals. Rays for waves of this frequency (magenta lines), initialised at the locations indicated by the stars, converge after they reflect off the separatrix and run nearly tangent to isopycnals, which should result in wave amplification in this region, and is where the shear is enhanced on the section.

wintertime wind forcing was described by Whitt and Thomas (2013) and is shown in Fig. 5.6. The bands of shear that parallel isopycnals in the centre of the Gulf Stream evidence NIW amplification in the front. An analysis of the rotary nature of the shear with depth provides further evidence that the feature is associated with a downward propagating NIW (e.g., Inoue et al. (2010) and Fig. 5.8d in this chapter).

Ray-tracing demonstrates that the shear bands are associated with a slantwise critical layer. Ray paths for two wavepackets with inertial frequencies, i.e., $\omega = f$, were calculated by integrating the group velocity (5.7) evaluated using the observed distributions of M^2, N^2, and $\overline{\zeta}$ on the section (magenta lines in Fig. 5.6). The rays were initialised on the anticyclonic side of the current and at a depth of 100 m, and were directed downwards and towards the front. When the wavepackets reflect off the separatrix (yellow contours) their rays converge and run nearly parallel isopycnals, thus satisfying the conditions for a slantwise critical layer. It is in this region where the shear is intensified, resulting in elevated dissipation rates (see Section 5.55.5.2 for more details).

Enhanced near-inertial shear with similar features to the shear bands seen in the Gulf Stream have been observed in fronts throughout the world's oceans. Examples have been found in the North Pacific Subtropical Front (Kunze and Sanford, 1984; Alford et al., 2013), the Subpolar Front of the Japan/East Sea (Shcherbina et al., 2003; Thomas et al., 2010), the Kuroshio (Rainville and Pinkel, 2004; Nagai et al., 2015), the frontal boundaries of a warm-core ring (Joyce et al., 2013) and a cold-core ring (Mied et al., 1987), and on the Oregon Shelf in an upwelling front (Avicola et al., 2007). The phenomenon appears to be widespread, and thus could have important implications for mixing.

5.4.3.2 Energy exchange with mean flows

Wave-mean flow interactions are by nature two-way. The emphasis in the previous section was on how mean flows affect NIWs through refraction. In this section, we give a brief overview of how NIWs can affect mean flows through the exchange of energy.

NIWs are characterised by nearly horizontal motions and weak buoyancy anomalies, therefore their main means for exchanging energy with mean flows is through shear production,

$$SP = -\overline{u_w u_w}\frac{\partial \overline{u}}{\partial x} - \overline{u_w v_w}\frac{\partial \overline{u}}{\partial y} - \overline{v_w v_w}\frac{\partial \overline{v}}{\partial y} - \overline{v_w u_w}\frac{\partial \overline{v}}{\partial x}, \qquad (5.8)$$

where the subscript "w" denotes a wave quantity, overbars an average over wave phase (hence $(\overline{u}, \overline{v})$ is by definition the mean flow), and velocities are projected onto a Cartesian coordinate. Positive values of SP indicate a transfer of energy from the mean flow to the waves. The formula (5.8) can be simplified by switching to a coordinate system that is locally

106 Ocean Mixing

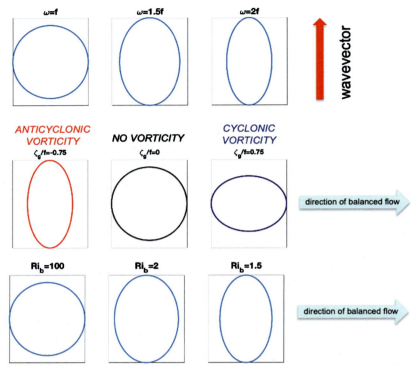

FIGURE 5.7 Schematic illustrating the factors that influence the eccentricity of NIW current ellipses that trace the wave's horizontal velocity vector over a wave period. In the absence of a background flow (top panels), the eccentricity increases with the intrinsic frequency (indicated at the top of each panel), e.g., Gill (1982), with the major axis of the ellipse pointed in the direction of the horizontal wavevector (red arrow). In a barotropic, balanced flow (middle panels), the eccentricity and orientation of the ellipse for waves with $\omega = \omega_{min}$ varies with the Rossby number $\overline{\zeta}/f$ (indicated at the top of each panel), e.g., Whitt and Thomas (2015). In a baroclinic, balanced flow without vorticity (bottom panels), the eccentricity and orientation of the ellipse for waves with $\omega = \omega_{min}$ varies with the Richardson number $Ri_b = N^2 f^2/M^4$ (indicated at the top of each panel), e.g., Thomas (2017). The eccentricity of the ellipse influences the efficacy of the energy transfer between NIWs and mean flows, e.g., (5.9).

aligned with the principle axes of the mean flow's strain tensor. In this coordinate system,

$$SP = -\frac{1}{2}\left(\overline{\hat{v}_w\hat{v}_w} - \overline{\hat{u}_w\hat{u}_w}\right)\alpha, \tag{5.9}$$

where $\hat{u}_w$ ($\hat{v}_w$) is the component of the wave's velocity parallel (perpendicular) to the extensional axis of the strain tensor, and $\alpha = \sqrt{(\partial\overline{u}/\partial y + \partial\overline{v}/\partial x)^2 + (\partial\overline{u}/\partial x - \partial\overline{v}/\partial y)^2}$ is the strain rate (Jing et al., 2017). Written in this form, it is clear that NIWs exchange energy with mean flows via SP only when the variances of their velocity parallel and normal to the extensional strain axis are different, that is, when their current ellipses are not circular. Thus processes that accentuate the eccentricity of NIW current ellipses can enhance energy transfers with mean flows.

The eccentricity of NIW current ellipses is dependent on the vorticity and baroclinicity of the background flow and the degree to which their intrinsic frequency exceeds f, which is a function of their horizontal wavenumber (e.g., Fig. 5.7). These properties of the background flow and the waves can be modified by strain, allowing for scenarios where straining adjusts the properties and strengthens the SP by changing the eccentricity of the ellipse. For example, waves travelling perpendicular to the extensional strain axis experience a growth in wavevector through (5.3), which strengthens $\overline{\hat{v}_w\hat{v}_w}$ over $\overline{\hat{u}_w\hat{u}_w}$ (given the relation between the orientation of the current ellipse and the wavevector, e.g., Fig. 5.7, top panel) and results in extraction of mean flow kinetic energy via (5.9).

Rocha et al. (2018) demonstrated that in a mean flow characterised by two-dimensional turbulence, this mechanism dominates *stimulated generation*—the transfer of energy from balanced flows to existing internal waves, such as wind-driven NIWs. They further showed that shear production by straining is self-limiting since wave escape (e.g., Section 5.3) hastens wave propagation away from strain regions. The kinetic energy removed from the balanced flow by this mechanism is manifest as an increase in NIW potential energy, and could represent a significant sink of kinetic energy for mesoscale flows in the ocean (Xie and Vanneste, 2015). Indeed, there is observational evidence that internal waves remove kinetic energy from the mesoscale following (5.9) (Polzin, 2010). However, the shear production term (5.9) can take either sign, and high resolution simulations of NIWs in the Kuroshio show that waves can be absorbed by the current and lose kinetic energy via SP as well (Nagai et al., 2015).

NIWs in mean flows with *both* strain and baroclinicity can experience yet more complex behaviour and energetics. For example, the presence of thermal wind shear can couple the evolution of the wave's vertical and horizontal wavenumbers, such that in regions of straining both components grow exponentially. In this scenario, Bühler and McIntyre (2005) posited that the wave's group velocity would go to zero and its amplitude would increase by conservation of wave action, leading to wave breaking and dissipation, a process they termed *wave capture*. Bühler and McIntyre (2005) did not however account for frontogenesis—the intensification of baroclinic currents by straining—which complicates the wave-mean flow interaction even further by introducing a time-evolving medium and an ageostrophic secondary circulation. Thomas (2012) developed a theoretical model for NIWs in currents undergoing frontogenesis and found that the waves become trapped as their phase aligns with isopycnals over time (i.e., $l/m \to -s_\rho$ and $|\mathbf{c}_g| \to 0$, e.g. (5.7)). But rather than losing energy to dissipation, the waves transfer all their energy to the ageostrophic secondary circulation via shear production terms involving vertical momentum fluxes. While the waves are being damped, however, they extract kinetic energy from the straining flow at a rate given by (5.9) and by a total, integrated amount equal to the initial kinetic energy in the wave field. Given the vast amount of kinetic energy in the near-inertial wave field, and the ubiquity of fronts in the ocean, Thomas (2012) speculated that this mechanism could be a significant sink of energy for both the NIW and mesoscale flow fields.

5.5 Dissipation of near-inertial waves

Damping of NIWs by mean flows can mark the end of their lifecycle, but not of the energy that they carried. Actual irreversible energy loss requires dissipation and/or diapycnal mixing. In this section, we summarise the ways NIWs can lose energy to dissipation in various parts of the water column, from the mixed layer to the sea floor and the stratified interior in between.

5.5.1 Near-surface dissipation

Not all the NIWs excited at the surface are able to propagate into the ocean interior; a significant fraction of their energy is lost to mixed-layer turbulence. In fact, shear instability caused by near-inertial motions at the base of the surface mixed layer is believed to be a major mechanism for deepening of the mixed layer via entrainment of fluid from the stratified layer below (e.g., Pollard and Millard, 1970; Pollard et al., 1973; Price et al., 1986). Therefore at least part of the near-inertial energy generated at the surface goes into mechanical production of turbulent kinetic energy and some of this turbulence erodes stratification, leading to an increase in the potential energy of the water column. Crawford and Large (1996) analysed results from a large ensemble of one-dimensional KPP model simulations (Large et al., 1994) and showed that the ratio between wind-induced near-inertial energy and wind work on near-inertial motions is approximately given by $0.532[1 - e^{-2.8\Gamma}]$, where Γ is a measure of how inertial the wind forcing is, and varies from 1 for pure inertial wind forcing to 0 for wind forcing with no spectral energy at the inertial frequency. For resonant inertial wind forcing ($\Gamma = 1$), a maximal fraction (~50%) of near-inertial wind work resides in kinetic energy of near-inertial motions, while the rest is lost to turbulent motions. As the wind forcing becomes increasingly less inertial ($\Gamma < 1$), the vertical velocity shear in the mixed layer increases and consequently a larger fraction of near-inertial wind work goes into turbulent kinetic energy production.

Dissipation of near-inertial energy due to shear instability at the base of the mixed layer, which occurs on a short time scale (less than a fraction of the inertial period), poses a problem for the widely-used damped slab model, which employs linear Rayleigh damping (Pollard and Millard, 1970). For example, Plueddemann and Farrar (2006) showed that the observed upper-ocean energy balance for strong resonant wind events is primarily between near-inertial wind work and energy dissipation due to shear instability processes, whereas the slab model balances near-inertial wind work with increases in near-inertial energy in the mixed layer, because it lacks a damping mechanism that operates on short time scales. As a result, the slab model can overestimate wind work on near-inertial motions by as much as a factor of 4. Alford (2020b) found that the overestimation of the wind work by slab models is most pronounced in shallow mixed layers and is less of an issue for deeper mixed layers.

The importance of accounting for near-surface dissipation when estimating near-inertial energy available for deep ocean mixing is further highlighted in recent modelling studies. For example, using a global ocean model forced by reanalysis winds but with no background flow, Furuichi et al. (2008) inferred that 75–85% of the near-inertial energy input at the sea surface was dissipated within the upper 150 m. Zhai et al. (2009) directly calculated viscous dissipation of near-inertial energy in a high-resolution model of the North Atlantic that simulates a realistic mesoscale eddy field, and found that about

70% of wind-induced near-inertial energy was lost to turbulent mixing in the upper 200 m, with the rest propagating into the ocean interior primarily through the inertial chimney effect of anticyclonic eddies. More recently, Alford (2020b) estimated the potential energy increase due to mixed-layer deepening and found that, globally, this potential energy increase is about 11% of near-inertial wind work, implying that over 50% of wind work on surface inertial motions goes to turbulence, and thus not into propagating near-inertial waves. There is also the additional ~20% damping of surface near-inertial motions by the relative wind stress (e.g., Rath et al., 2013; Zhai, 2017; Liu et al., 2019; Alford, 2020b). The implication of these studies is that the global total of near-inertial wind work that is available for generating diapycnal mixing in the deep ocean is more likely to be in the range of 0.1–0.3 TW, i.e., much less than the 0.3–1.5 TW estimated for wind work on surface inertial motions.

5.5.2 Interior dissipation

There have been several direct measurements of turbulence associated with NIW breaking in the pycnocline. These microstructure measurements can be categorised into two types: ones that targeted mesoscale eddies or fronts, where NIW-mean flow interactions can catalyse wave breaking, and others that were made irrespective of the background flow. We will start by summarising the latter.

Examples of enhanced dissipation in the pycnocline triggered by the passage of NIWs have been observed in the low, mid, and high latitudes. In the Banda Sea, at 6.5°S, for example, Alford and Gregg (2001) observed modulations in the dissipation that were correlated with the shear, internal wave strain (IWS), which is a measure of the thickening and thinning of isopycnal layers by internal waves, and Froude number of the NIW (e.g., Fig. 5.5). A similar coherency between the dissipation in the pycnocline and the shear and IWS associated with storm-driven NIWs has been seen in the Arctic (Fer, 2014). In the mid-latitudes off the Southern California coast, Gregg et al. (1986) documented persistent patches of enhanced dissipation that were correlated with a NIW in the pycnocline. The average dissipation rate was strong enough to damp the wave in 4–8 days, given the amount of energy that it carried. Further to the north off the Northern California coast, Hebert and Moum (1994) observed the decay of a NIW with enhanced dissipation rates that yielded slightly longer, yet similar wave damping times (i.e., 9 days). Similar damping times, of order 10 days, were reported by Alford and Gregg (2001) for the NIW that they observed in the Banda Sea. The breakdown of NIWs to turbulence not only damps the waves, but it increases the background potential energy of the ocean and could play an important role in mixing the pycnocline, which will be discussed further in Section 5.6.

Elevated levels of turbulence in the pycnocline have also been observed in the proximity of anticyclones and fronts, where NIW critical layers can form. Kunze et al. (1995) described a field campaign to a warm-core ring that was designed to quantify the energy balance of trapped, subinertial NIWs entering a critical layer at the base of the anticyclone. They calculated that the convergence of the inferred downward wave energy flux was approximately balanced by the observed, elevated dissipation rates in the anticyclone (which were of order 1×10^{-8} W kg^{-1}), and that energy losses to the mean flow via shear production or to higher frequency waves via wave-wave interactions were negligible. Even larger values of dissipation were measured in the slantwise critical layer within the Gulf stream front shown in Fig. 5.6 (Inoue et al., 2010). In the core of the current and over the depths, where the NIW shear bands were most prominent (200–500 m), the dissipation rates greatly exceeded levels associated with the background internal-wave field with maximum values approaching 1.0×10^{-7} W kg^{-1} (e.g. Fig. 5.8). With these dissipation values, Inoue et al. (2010) estimated that the NIW trapped in the Gulf stream could be damped in 11 days, given its kinetic energy. The Lofoten Basin eddy provides another potential example of a NIW critical layer both at its base and frontal edge. In these locations, Fer et al. (2018) reported dissipation values that were up to two orders of magnitude larger than those measured at a reference station outside of the eddy. Such NIW critical layers appear not to be limited to the open ocean. For example, on the continental shelf off of Oregon, near-inertial shear bands with elevated levels of turbulence have been observed in upwelling fronts (Avicola et al., 2007).

These examples of microstructure measurements suggest that NIW-driven enhanced dissipation is prevalent in the pycnocline, but they are far from comprehensive. However, recent global estimates of dissipation, based on a finescale strain parameterisation applied to hydrography from the Argo float array, support this inference. Using this methodology, Whalen et al. (2018) found that in regions of high mesoscale eddy kinetic energy, the parameterised dissipation was enhanced and it tracked the seasonal cycle in the wind-work on near-inertial motions. Based on these correlations, they concluded that on a global scale mesoscale flows strengthen turbulent mixing in the pycnocline caused by wind-generated NIWs. Whereas these findings are intriguing, they should be interpreted with some caution for a couple of reasons. First, the finescale parameterisation is based on IWS rather than shear, yet NIWs have high shear to IWS ratios (for example, the NIW in the Gulf stream seen in Figs. 5.6 and 5.8 had a shear to IWS ratio greater 20 (Inoue et al., 2010)), so it is not clear how

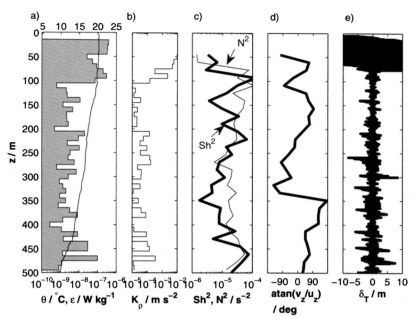

FIGURE 5.8 Fine- and microstructure profiles made in the Gulf stream on the section shown in Fig. 5.6 in the centre of the current (i.e., at a cross-stream distance of 0 km), where the prominent NIW shear bands were observed. (a) Dissipation rate averaged over 10 m bins and the potential temperature (thin line), (b) diapycnal diffusivity, (c) shear and buoyancy frequency squared, (d) the angle that the shear vector makes with the horizontal illustrating clockwise rotary motions with depth from 200–500 m (the jump above 350 m is an artifact of limiting the angle to $\pm 180°$), and (d) Thorpe displacements showing evidence of overturns in density. From Inoue et al. (2010).

well the parameterisation, as implemented, can capture NIW-driven turbulence (Polzin et al., 1995; Chinn et al., 2016). Second, non-internal-wave processes could contaminate the Argo IWS estimates, and if they share a similar seasonal cycle to the wind-work, they could confound the causal relation behind the correlation. One such non-internal-wave process is submesoscale turbulence, which is strongest in mesoscale eddy rich regions and has been observed to have a seasonal cycle that peaks in the winter when the wind-work is largest (Callies et al., 2015). Submesoscale turbulence could contaminate IWS by generating anomalies in potential vorticity that, once subducted into the pycnocline, could mimic internal wave isopycnal thickness anomalies, which could prove challenging to filter from density profiles.

5.5.3 Near-bottom dissipation

Observational studies of near-bottom dissipation are few and far between, and those that have explicitly targeted turbulent processes driven by surface-generated NIWs are even more rare. To our knowledge, to date only one study has attempted to do this. It consisted of mooring observations made just south of the Mendocino Escarpment off of California (Alford, 2010). The velocity records near the bottom revealed a dominance of near-inertial shear over the tides, even though the escarpment is a strong internal tide generator. The near-inertial shear appeared to play a role in generating turbulence, which was enhanced in the bottom 1000 m. This link was especially clear for one downgoing NIW that showed a coherency between the dissipation rate and the IWS and shear similar to what was observed in the Banda Sea (e.g., Fig. 5.5). Whereas near-inertial shear was dominant over shear at tidal frequencies, it explained only a portion of the total shear at the depths, where mixing was enhanced. This discrepancy led Alford (2010) to speculate that the internal wave continuum, energised by NIWs and the internal tides, contributed to the shear and turbulence at these levels.

Upward propagating NIWs were also observed at the moorings and are a prominent feature in other mooring records (e.g., Alford et al., 2017). If these were surface-generated NIWs that had reflected off the bottom, Alford (2010) estimated that they had longevities of over 100 days. This finding implies that bottom reflections are not necessarily the death knell for NIWs, and that NIWs can apparently persist even after running the gauntlet of wave-wave interactions, absorption into mean flows, and turbulent dissipation in the interior. Closing the lifecycle of a surface-generated NIW may involve journeys of hundreds to thousands of kilometres, the details of which remain largely unknown.

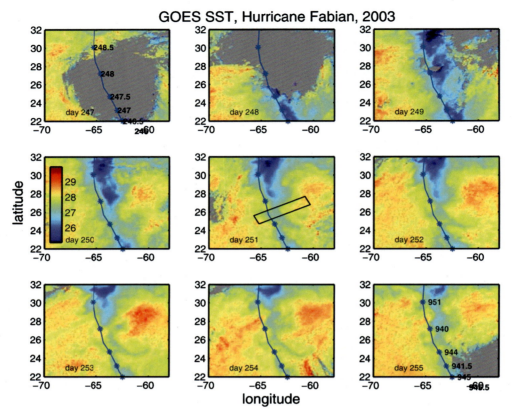

FIGURE 5.9 A sequence of SST images during and after the passage of Hurricane Fabian (2003), illustrating how a cold wake forms to the right of the storm's centre (black line). The formation of the cold wake is attributable to entrainment driven by near-inertial shear at the base of the mixed layer. The shear is stronger to the right of the storm, because there the winds at fixed point in the ocean rotate clockwise in time as the storm passes overhead, and can thus resonantly force near-inertial motions. From Price et al. (2008).

5.6 Discussion

Given the subject matter of this book, we end the chapter with a discussion on vertical and lateral mixing associated with surface-generated NIWs and their broader implications.

5.6.1 Vertical mixing

Vertical mixing driven by near-inertial motions and waves can be split into two categories: entrainment at the base of the mixed layer and mixing in the stratified interior.

Entrainment caused by near-inertial shear is thought to be a major contributor to mixed-layer deepening (e.g., Pollard and Millard, 1970; Pollard et al., 1973; Price et al., 1986) and can be especially strong under resonant wind forcing. Using large eddy simulations, Skyllingstad et al. (2000) contrasted the energetics of boundary-layer turbulence and mixed-layer deepening under resonant and non-resonant winds. Resonant winds generated much stronger inertial currents and entrainment than non-resonant winds, resulting in mixed layers that deepened twice as much for the same amplitude wind forcing. Under resonant conditions, Skyllingstad et al. (2000) found that the increase in background potential energy by entrainment was 7% of the integrated wind-work on inertial motions. This value is consistent with the recent global estimate of 11% reported by Alford (2020b). The increase in potential energy can be significant, especially under storms and hurricanes. In the latter for example, such entrainment is responsible for the creation of cold wakes (e.g., Fig. 5.9), regions of cooler sea-surface temperatures (SST) that form to the right (left) of a hurricane's path in the Northern (Southern) Hemisphere, where the winds approach resonance (Price, 1981). The ocean's response to Hurricane Frances (2004), as documented by Sanford et al. (2011) using EM-APEX floats, provides an excellent example of the formation of a cold wake. The floats revealed that, in the wake of the storm, the mixed layer deepened by ∼80 m and the SST dropped by 2.2 °C 55 km to the right of the hurricane, in contrast to a 1.2 °C cooling observed along its track (see Fig. 5.3).

Though much of the wind-work on inertial motions is thought to be lost to dissipation and entrainment in the upper ocean (see Section 5.5), the fraction that goes into propagating NIWs can drive vertical mixing in the interior. Where NIWs break and enhance dissipation, the turbulence increases the potential energy of the fluid following a mixing efficiency of $\sim$0.2 (see discussion in Section 5.5). This relation can be used to infer diapycnal diffusivities given estimates of dissipation from microstructure. From the examples of NIW wave breaking in the pycnocline discussed above (e.g., Figs. 5.5 and 5.8), this method yields diffusivities that are around one order of magnitude larger than the canonical value of 1×10^{-5} m^2 s^{-1} for the pycnocline (e.g., Ledwell et al., 1998). However, it is clear that mixing by NIWs is a highly intermittent process, since the generation of NIWs by the winds is irregular, and it can take a confluence of conditions to trigger breaking. Therefore, on average, NIWs may not elevate the diffusivity in the pycnocline over the canonical value, but they surely must contribute to its maintenance either directly or by energising the internal wave continuum.

Vertical mixing by near-inertial motions could impact the atmosphere-ocean climate system. Jochum et al. (2013) used a coupled climate model to assess these impacts. The model was adapted to include parameterisations for entrainment and interior mixing by near-inertial motions. Though mixing in the pycnocline minimally affected the ocean-atmosphere climate system, deepening of the mixed layer by entrainment did have a noticeable effect. The direct consequences of entrainment were largest in the tropics, where mixed-layer deepening modified precipitation. But the influence of the mixing extended past the tropics as the change in precipitation weakened the midlatitude westerlies through atmospheric teleconnections. On a related topic, Emanuel (2001) advocated for the importance of entrainment on the climate system through the potential impacts of hurricane cold wakes, and hence near-inertial shear-driven mixing, on the thermohaline circulation. He argued that if the temperature anomalies in cold wakes are restored by a combination of surface-heating and lateral advection, this restoration would require, for a typical hurricane season, a meridional heat transport of order 1×10^{15} W, approximately half of the observed peak poleward oceanic heat transport.

5.6.2 Lateral mixing

NIWs can also drive lateral mixing through shear dispersion. Horizontal advection of a tracer by an oscillatory, vertically sheared flow combined with vertical mixing results in an irreversible, lateral dispersal of the tracer. This shear dispersion can be modelled by a lateral diffusivity that scales with the vertical diffusivity amplified by the ratio of the shear to oscillation frequency squared (Young et al., 1982). NIWs are thus predicted to be the most effective type of internal wave for laterally mixing tracers by this mechanism, given their strong shears and low frequencies. Direct measurements of lateral dispersion made using the release and tracking of a tracer in the pycnocline reveal, however, that shear dispersion by NIWs is too weak to fully explain the inferred dispersal of the tracer over scales of $\mathcal{O}(1$ km$)$, the lengths over which shear dispersion should be most effective (Ledwell et al., 1998; Sundermeyer et al., 2020).

In contrast, shear dispersion by near-inertial motions can be quite effective in the surface boundary layer, especially in the proximity of ocean fronts. Wenegrat et al. (2020) describe the dispersion of fluorescein dye released in the surface boundary layer of the Gulf Stream front under strong wintertime forcing that illustrates this phenomenon. The dye was observed to spread over 5 km in half a day, yielding an effective lateral diffusivity of $\sim$50 m^2 s^{-1}. Large eddy simulations configured with flow and forcing parameters representative of the observations illuminated how sheared inertial oscillations, generated by the front as it was thrown out of balance by submesoscale instabilities, could rapidly mix the tracer laterally via shear dispersion. Given the strong gradients in salinity, temperature, and biogeochemical tracers at the Gulf Stream front, the process could play an important role in the cross-gyre exchange of tracers, with implications for nutrient, heat, and salinity budgets in the subtropical and subpolar gyres. As upper-ocean fronts are widespread, and can easily be driven unstable by atmospheric forcing, shear dispersion by near-inertial oscillations in the surface boundary layer is likely common.

5.7 Conclusions and outstanding questions

Many questions still remain regarding the lifecycle of surface-generated waves. Throughout this article we have assumed that wind-driven NIWs dominate over other forms of surface-generated internal waves, but is this correct? Though the wind-work on inertial motions is large, the energy input to surface gravity waves is nearly an order of magnitude larger (Huang, 2004), therefore even if a small fraction of that energy were transmitted to internal waves, say through the interaction of inertial oscillations and Langmuir turbulence (e.g., Polton et al., 2008), it could be significant. The details of processes in the surface boundary layer are undoubtedly important for quantifying the wind-work on near-inertial motions, therefore understanding how Langmuir and submesoscale turbulence (which we now realise are ubiquitous features of the surface boundary layer, e.g., Chapters 4 and 8) affect near-inertial motions and vice versa, should be explored. Preliminary studies suggest that inertial motions can energise submesoscale instabilities (Thomas and Taylor, 2014; Thomas et al., 2016), a process that has been termed *stimulated imbalance* by Barkan et al. (2017). Moreover, the interaction of submesoscale motions and NIWs can generate an internal wave continuum with a Garrett–Munk-like spectrum even in the absence of tides (Barkan et al., 2017). Clearly, the NIW field and the balanced circulation and its turbulence are intimately linked. Their interactions yield complex phenomena, many of which have yet to be discovered, with implications for the energetics of both types of motion and mixing in the ocean, and certainly merit further study.

Acknowledgements

L.N.T. was supported by ONR grant N00014-18-1-2798 under the Near-Inertial Shear and Kinetic Energy in the North Atlantic experiment (NISKINe) Departmental Research Initiative. X.Z. acknowledges support by a Royal Society International Exchanges Award (IES/R1/180037). We would like to thank Matthew Alford and an anonymous reviewer for their insightful and constructive suggestions for improving this chapter.

References

Alford, M., 2003a. Redistribution of energy available for ocean mixing by long-range propagation of internal waves. Nature 423, 159–163.

Alford, M., Gregg, M., 2001. Near-inertial mixing: modulation of shear, strain and microstructure at low latitude. J. Geophys. Res. 106, 16947–16958.

Alford, M., Shcherbina, A.Y., Gregg, M.C., 2013. Observations of near-inertial internal gravity waves radiating from a frontal jet. J. Phys. Oceanogr. 43, 1225–1239.

Alford, M.H., 2001. Fine-structure contamination: observations and a model of a simple two-wave case. J. Phys. Oceanogr. 31, 2645–2649.

Alford, M.H., 2003b. Improved global maps and 54-years history of wind-work on ocean inertial motions. Geophys. Res. Lett. 30, 1424.

Alford, M.H., 2010. Sustained, full-water-column observations of internal waves and mixing near Mendocino Escarpment. J. Phys. Oceanogr. 40, 2643–2660.

Alford, M.H., 2020a. Global calculations of local and remote near-inertial-wave dissipation. J. Phys. Oceanogr. 50, 3157–3164.

Alford, M.H., 2020b. Revisiting near-inertial wind work: slab models, relative stress and mixed-layer deepening. J. Phys. Oceanogr. 50, 3141–3156.

Alford, M.H., MacKinnon, J.A., Pinkel, R., Klymak, J.M., 2017. Space–time scales of shear in the North Pacific. J. Phys. Oceanogr. 47, 2455–2478.

Alford, M.H., MacKinnon, J.A., Simmons, H.L., Nash, J.D., 2016. Near-inertial internal gravity waves in the ocean. Annu. Rev. Mar. Sci. 8 (8), 95–123.

Alford, M.H., Whitmont, M., 2007. Seasonal and spatial variability of near-inertial kinetic energy from historical moored velocity records. J. Phys. Oceanogr. 37, 2022–2037.

Ansong, J.K., Sutherland, B.R., 2010. Internal gravity waves generated by convective plumes. J. Fluid Mech. 648, 405–434.

Asselin, O., Thomas, L.N., Young, W.R., Rainville, L., 2020. Refraction and straining of near-inertial waves by barotropic eddies. J. Phys. Oceanogr. 50 (12), 3439–3454.

Asselin, O., Young, W.R., 2020. Penetration of wind-generated near-inertial waves into a turbulent ocean. J. Phys. Oceanogr. 50 (6), 1699–1716.

Avicola, G.S., Moum, J.N., Perlin, A., Levine, M.D., 2007. Enhanced turbulence due to the superposition of internal gravity waves and a coastal upwelling jet. J. Geophys. Res. 112, C06024.

Balmforth, N.J., Young, W.R., 1999. Radiative damping of near-inertial oscillations in the mixed layer. J. Mar. Res. 57, 561–584.

Barkan, R., Winters, K.B., McWilliams, J.C., 2017. Stimulated imbalance and the enhancement of eddy kinetic energy dissipation by internal waves. J. Phys. Oceanogr. 47, 181–198.

Bühler, O., McIntyre, M., 2005. Wave capture and wave-vortex duality. J. Fluid Mech. 534, 67–95.

Callies, J., Ferrari, R., Klymak, J.M., Gula, J., 2015. Seasonality in submesoscale turbulence. Nat. Commun. 6, 6862.

Chaigneau, A., Pizarro, O., Rojas, W., 2008. Global climatology of near-inertial current characteristics from Lagrangian observations. Geophys. Res. Lett. 35.

Chang, E.K.M., Guo, Y., Xia, X., 2012. CMIP5 multimodel ensemble projection of storm track change under global warming. J. Geophys. Res. 117.

Chinn, B.S., Girton, J.B., Alford, M.H., 2016. The impact of observed variations in the shear-to-strain ratio of internal waves on inferred turbulent diffusivities. J. Phys. Oceanogr. 46, 3299–3320.

Crawford, G.B., Large, W.G., 1996. A numerical investigation of resonant inertial response of the ocean to wind forcing. J. Phys. Oceanogr. 26, 873–891.

Czeschel, L., Eden, C., 2019. Internal wave radiation through surface mixed layer turbulence. J. Phys. Oceanogr. 49, 1827–1844.

D'Asaro, E.A., 1985. The energy flux from the wind to near-inertial motions in the surface mixed layer. J. Phys. Oceanogr. 15, 1043–1059.

D'Asaro, E.A., 1989. The decay of wind-forced mixed layer inertial oscillations due to the β effect. J. Geophys. Res. 94, 2045–2056.

D'Asaro, E.A., Eriksen, C.C., Levine, M.D., Niiler, P., Paulson, C.A., Meurs, P.V., 1995. Upper-ocean inertial currents forced by a strong storm. Part I: data and comparisons with linear theory. J. Phys. Oceanogr. 25, 2909–2936.

Dippe, T., Zhai, X., Greatbatch, R.J., Rath, W., 2015. Interannual variability of wind power input to near-inertial motions in the North Atlantic. Ocean Dyn. 65, 859–875.

Duhaut, T.H., Straub, D.N., 2006. Wind stress dependence on ocean surface velocity: implications for mechanical energy input to ocean circulation. J. Phys. Oceanogr. 36, 202–211.

Egbert, R.D., Ray, R.D., 2000. Significant dissipation of tidal energy in the deep ocean inferred from satellite altimeter data. Nature 405, 775–778.

Elipot, S., Lumpkin, R., Prieto, G., 2010. Modification of inertial oscillations by the mesoscale eddy field. J. Geophys. Res. 115, C09010.

Emanuel, K.A., 2001. The contribution of tropical cyclones to the oceans' meridional heat transport. J. Geophys. Res. 58, 1427–1445.

Fer, I., 2014. Near-inertial mixing in the central Arctic Ocean. J. Phys. Oceanogr. 44, 2031–2049.

Fer, I., Bosse, A., Ferron, B., Bouruet-Aubertot, P., 2018. The dissipation of kinetic energy in the Lofoten Basin eddy. J. Phys. Oceanogr. 48, 1299–1316.
Ferrari, R., Wunsch, C., 2009. Ocean circulation kinetic energy: reservoirs, sources, and sinks. Annu. Rev. Fluid Mech. 41, 253–282.
Frankignoul, C., Joyce, T.M., 1979. On the internal wave variability during the internal wave experiment (IWEX). J. Geophys. Res. 84, 769–776.
Fu, L.L., 1981. Observations and models of inertial waves in the deep ocean. Rev. Geophys. Space Phys. 19, 141–170.
Furuichi, N., Hibiya, T., Niwa, Y., 2008. Model-predicted distribution of wind-induced internal wave energy in the world's oceans. J. Geophys. Res. 113, C09034.
Gill, A.E., 1982. Gravity waves in a rotating fluid. In: Atmosphere-Ocean Dynamics. Academic Press, Inc., pp. 247–315.
Gill, A.E., 1984. On the behavior of internal waves in the wakes of storms. J. Phys. Oceanogr. 14, 1129–1151.
Gille, S.T., 2005. Statistical characterization of zonal and meridional ocean wind stress. J. Atmos. Ocean. Technol. 22, 1353–1372.
Gregg, M., D'Asaro, E., Shay, T., Larson, N., 1986. Observations of persistent mixing and near-inertial internal waves. J. Phys. Oceanogr. 16, 856–885.
Hebert, H., Moum, J., 1994. Decay of a near-inertial wave. J. Phys. Oceanogr. 24, 2334–2351.
Henyey, F.S., Wright, J., Flatté, S.M., 1986. Energy and action flow through the internal wave field: an eikonal approach. J. Geophys. Res. 91, 8487–8495.
Hoskins, B.J., Draghici, I., Davies, H.C., 1978. A new look at the ω-equation. Q. J. R. Meteorol. Soc. 104, 31–38.
Huang, R.X., 2004. Energy flows in the ocean. In: Encyclopedia of Energy. Elsevier, Cleveland.
Hurrell, J.W., 1995. Decadal trends in the North Atlantic Oscillation and relationships to regional temperatures and precipitation. Science 269, 676–679.
Inoue, R., Gregg, M.C., Harcourt, R.R., 2010. Mixing rates across the Gulf Stream, Part 1: on the formation of eighteen degree water. J. Mar. Res. 68, 643–671.
Jiang, C., Cronin, M., Kelly, K., Thompson, L., 2005. Evaluation of a hybrid Satellite NWP based on turbulent heat flux product using TAO buoys. J. Geophys. Res. 110, C09007.
Jing, Z., Wu, L., Ma, X., 2017. Energy exchange between the mesoscale oceanic eddies and wind-forced near-inertial oscillations. J. Phys. Oceanogr. 47, 721–733.
Jochum, M., Briegleb, B., Danabasoglu, G., Large, W., Norton, N., Jayne, S., Alford, M., Bryan, F., 2013. The impact of oceanic near-inertial waves on climate. J. Climate 26, 2833–2844.
Joyce, T., Toole, J., Klein, P., Thomas, L., 2013. A near-inertial mode observed within a Gulf Stream warm core ring. J. Geophys. Res. 118, 1–10. https://doi.org/10.1002/jgrc.20141.
Klein, P., Lapeyre, G., Large, W.G., 2004a. Wind ringing of the ocean in presence of mesoscale eddies. Geophys. Res. Lett. 31.
Klein, P., Llewellyn-Smith, S., Lapeyre, G., 2004b. Organization of near-inertial energy by an eddy field. Q. J. R. Meteorol. Soc. 130, 1153–1166.
Kunze, E., 1985. Near-inertial wave propagation in geostrophic shear. J. Phys. Oceanogr. 15, 544–565.
Kunze, E., 1986. The mean and near-inertial velocity fields in a warm-core ring. J. Phys. Oceanogr. 16, 1444–1461.
Kunze, E., Sanford, T., 1984. Observations of near-inertial waves in a front. J. Phys. Oceanogr. 14, 566–581.
Kunze, E., Schmidt, R.W., Toole, J.M., 1995. The energy balance in a warm-core ring's near-inertial critical layer. J. Phys. Oceanogr. 25, 942–957.
Large, W.G., McWilliams, J.C., Doney, S.C., 1994. Oceanic vertical mixing: a review and a model with a nonlocal boundary layer parameterization. Rev. Geophys. 32, 363–403.
Le Boyer, A., Alford, M.H., 2021. Variability and sources of the internal wave continuum examined from global moored velocity records. J. Phys. Oceanogr. https://doi.org/10.1175/JPO-D-20-0155.1.
Ledwell, J., Watson, A., Law, C., 1998. Mixing of a tracer in the pycnocline. J. Geophys. Res. 103, 21499–21529.
Lee, D.K., Niiler, P.P., 1998. The inertial chimney: the near-inertial energy drainage from the ocean surface to the deep layer. J. Geophys. Res. 103, 7579–7591.
Liu, Y., Jing, Z., Wu, L., 2019. Wind power on oceanic near-inertial oscillations in the global ocean estimated from surface drifters. Geophys. Res. Lett. 46, 2647–2653.
McComas, C.H., Müller, P., 1981. The dynamic balance of internal waves. J. Phys. Oceanogr. 11, 970–986.
Mied, R.P., Lindemann, G.J., Trump, C.L., 1987. Inertial wave dynamics in the North Atlantic Subtropical Zone. J. Geophys. Res. 92 (C12), 13063–13074.
Mihaly, S.F., Thomson, R.E., Rabinovich, A.B., 1998. Evidence for nonlinear interaction between internal waves of inertial and semidiurnal frequency. Geophys. Res. Lett. 25, 1205–1208.
Mooers, C.N.K., 1975. Several effects of a baroclinic current on the cross-stream propagation of inertial-internal waves. Geophys. Fluid Dyn. 6, 245–275.
Nagai, T., Tandon, A., Kunze, E., Mahadevan, A., 2015. Spontaneous generation of near-inertial waves by the Kuroshio front. J. Phys. Oceanogr. 45, 2381–2406.
Nagasawa, M., Niwa, Y., Hibiya, T., 2000. Spatial and temporal distribution of the wind-induced internal wave energy available for deep water mixing in the North Pacific. J. Geophys. Res. 105, 13933–13943.
Ou, H., 1984. Geostrophic adjustment: a mechanism for frontogenesis? J. Phys. Oceanogr. 14, 994–1000.
Pinkel, R., 1985. A wavenumber-frequency spectrum of upper ocean shear. J. Phys. Oceanogr. 15, 1453–1469.
Pinkel, R., 2008. Advection, phase distortion, and the frequency spectrum of finescale fields in the sea. J. Phys. Oceanogr. 38, 291–313.
Plueddemann, A.J., Farrar, J.T., 2006. Observations and models of the energy flux from the wind to mixed-layer inertial currents. Deep-Sea Res. II (53), 5–30.
Pollard, R.T., Millard, R.C., 1970. Comparison between observed and simulated wind-generated inertial oscillations. Deep-Sea Res. 17, 153–175.
Pollard, R.T., Rhines, P.B., Thompson, R.O., 1973. The deepening of the wind-mixed layer. Geophys. Astrophys. Fluid Dyn. 4, 381–404.
Polton, J.A., Smith, J.A., MacKinnon, J.A., Tejada-Martinez, A.E., 2008. Rapid generation of high-frequency internal waves beneath a wind and wave forced oceanic surface mixed layer. Geophys. Res. Lett. 35, L13602.
Polzin, K.L., 2010. Mesoscale eddy-internal wave coupling. Part II: energetics and results from PolyMode. J. Phys. Oceanogr. 40, 789–801.

Polzin, K.L., Lvov, Y.V., 2011. Toward regional characterizations of the oceanic internal wavefield. Rev. Geophys. 49, RG4003. https://doi.org/10.1029/2010RG000329.

Polzin, K.L., Toole, J.M., Schmitt, R.W., 1995. Finescale parameterizations of turbulent dissipation. J. Phys. Oceanogr. 25, 306–328.

Price, J.F., 1981. Upper ocean response to a hurricane. J. Phys. Oceanogr. 11, 153–175.

Price, J.F., 1983. Internal wave wake of a moving storm. Part I. Scales, energy budget and observations. J. Phys. Oceanogr. 13, 949–965.

Price, J.F., Morzel, J., Niiler, P.P., 2008. Warming of SST in the cool wake of a moving hurricane. J. Geophys. Res. 113, C07010.

Price, J.F., Weller, R.A., Pinkel, R., 1986. Diurnal cycling: observations and models of the upper ocean response to diurnal heating, cooling, and wind mixing. J. Geophys. Res. 91, 8411–8427.

Rainville, L., Pinkel, R., 2004. Observations of energetic high-wavenumber internal waves in the Kuroshio. J. Phys. Oceanogr. 34, 1495–1505.

Rath, W., Greatbatch, R.J., Zhai, X., 2013. Reduction of near-inertial energy through the dependence of wind stress on the ocean-surface velocity. J. Geophys. Res. 118, 2761–2773.

Rath, W., Greatbatch, R.J., Zhai, X., 2014. On the spatial and temporal distribution of near-inertial energy in the Southern Ocean. J. Geophys. Res. 119, 359–376.

Rimac, A., von Storch, J.S., Eden, C., Haak, H., 2013. The influence of high-resolution wind stress field on the power input to near-inertial motions in the ocean. Geophys. Res. Lett. 40, 4882–4886.

Rocha, C.B., Wagner, G.L., Young, W.R., 2018. Stimulated generation: extraction of energy from balanced flow by near-inertial waves. J. Fluid Mech. 847, 417–451.

Rogers, J.C., 1997. North Atlantic storm track variability and its association to the North Atlantic Oscillation and climate variability of northern Europe. J. Climate 10, 1635–1647.

Sanford, T.B., Price, J.F., Girton, J.B., 2011. Upper-ocean response to Hurricane Frances (2004) observed by profiling EM-APEX floats. J. Phys. Oceanogr. 41, 1041–1056.

Shakespeare, C., Taylor, J., 2014. The spontaneous generation of inertia–gravity waves during frontogenesis forced by large strain: theory. J. Fluid Mech. 757, 817–853.

Shcherbina, A.Y., Talley, L.D., Firing, E., Hacker, P., 2003. Near-surface frontal zone trapping and deep upward propagation of internal wave energy in the Japan East Sea. J. Phys. Oceanogr. 33, 900–912.

Silverthorne, K.E., Toole, J.M., 2009. Seasonal kinetic energy variability of near-inertial motions. J. Phys. Oceanogr. 39, 1035–1049.

Simmons, H.L., Alford, M., 2012. Simulating the long-range swell of internal waves generated by ocean storms. Oceanography 25, 30–41.

Skyllingstad, E.D., Smyth, W.D., Crawford, G.B., 2000. Resonant wind-driven mixing in the ocean boundary layer. J. Phys. Oceanogr. 30, 1866–1890.

Sugiyama, Y., Niwa, Y., Hibiya, T., 2009. Numerically reproduced internal wave spectra in the deep ocean. Geophys. Res. Lett. 36, L07601. https://doi.org/10.1029/2008GL036825.

Sundermeyer, M.A., Birch, D.A., Ledwell, J.R., Levine, M.D., Pierce, S.D., Cervantes, B.T.K., 2020. Dispersion in the open ocean seasonal pycnocline at scales of 1-10 km and 1-6 days. J. Phys. Oceanogr. 50, 415–437.

Thomas, L.N., 2012. On the effects of frontogenetic strain on symmetric instability and inertia-gravity waves. J. Fluid Mech., 620–640.

Thomas, L.N., 2017. On the modifications of near-inertial waves at fronts: implications for energy transfer across scales. Ocean Dyn. 67, 1335–1350.

Thomas, L.N., 2019. Enhanced radiation of near-inertial energy by frontal vertical circulations. J. Phys. Oceanogr. 49, 2407–2421.

Thomas, L.N., Lee, C.M., Yoshikawa, Y., 2010. The subpolar front of the Japan/East Sea II: inverse method for determining the frontal vertical circulation. J. Phys. Oceanogr. 40, 3–25.

Thomas, L.N., Rainville, L., Asselin, O., Young, W.R., Girton, J., Whalen, C.B., Centurioni, L., Hormann, V., 2020. Direct observations of near-inertial wave ζ-refraction in a dipole vortex. Geophys. Res. Lett. 47 (21), e2020GL090375.

Thomas, L.N., Taylor, J.R., 2014. Damping of inertial motions by parametric subharmonic instability in baroclinic currents. J. Fluid Mech. 743, 280–294.

Thomas, L.N., Taylor, J.R., D'Asaro, E.A., Lee, C.M., Klymak, J.M., Shcherbina, A.Y., 2016. Symmetric instability, inertial oscillations, and turbulence at the Gulf Stream front. J. Phys. Oceanogr. 46, 197–217.

van Meurs, P., 1998. Interactions between near-inertial mixed layer currents and the mesoscale: the importance of spatial variabilities in the vorticity field. J. Phys. Oceanogr. 28, 1363–1388.

Vanneste, J., 2013. Balance and spontaneous wave generation in geophysical flows. Annu. Rev. Fluid Mech. 45, 147–172.

Weller, R.A., 1982. The relation of near-inertial motions observed in the mixed layer during the JASIN (1978) experiment to the local wind stress and to the quasi-geostrophic flow field. J. Phys. Oceanogr. 12, 1122–1136.

Wenegrat, J.O., Thomas, L., Sundermeyer, M., Taylor, J., D'Asaro, E., Klymak, J., Shearman, R., Lee, C., 2020. Enhanced mixing across the gyre boundary at the Gulf Stream front. Proc. Natl. Acad. Sci. 117, 17607–17614.

Whalen, C.B., MacKinnon, J., Talley, L., 2018. Large-scale impacts of the mesoscale environment on mixing from wind-driven internal waves. Nat. Geosci. 11, 842–847.

Whitt, D.B., Thomas, L.N., 2013. Near-inertial waves in strongly baroclinic currents. J. Phys. Oceanogr. 43, 706–725.

Whitt, D.B., Thomas, L.N., 2015. Resonant generation and energetics of wind-forced near-inertial motions in a geostrophic flow. J. Phys. Oceanogr. 45, 181–208.

Whitt, D.B., Thomas, L.N., Klymak, J.M., Lee, C.M., D'Asaro, E.A., 2018. Interaction of superinertial waves with submesoscale cyclonic filaments in the North Wall of the Gulf Stream. J. Phys. Oceanogr. 48, 81–99.

Wunsch, C., 1998. The work done by the wind on the oceanic general circulation. J. Phys. Oceanogr. 28, 2332–2340.

Wunsch, C., Ferrari, R., 2004. Vertical mixing, energy, and the general circulation of the oceans. Annu. Rev. Fluid Mech. 36, 281–314.

Xie, J.H., Vanneste, J., 2015. A generalised-Lagrangian-mean model of the interactions between near-inertial waves and mean flow. J. Fluid Mech. 774, 143–169.

Xu, C., Zhai, X., Shang, X., 2016. Work done by atmospheric winds on mesoscale ocean eddies. Geophys. Res. Lett. 43, 12174–12180.

Young, W.R., Ben-Jelloul, M., 1997. Propagation of near-inertial oscillations through a geostrophic flow. J. Mar. Res. 55, 735–766.

Young, W.R., Rhines, P., Garrett, C.J.R., 1982. Shear-flow dispersion, internal waves and horizontal mixing in the ocean. J. Phys. Oceanogr. 12, 515–527.

Zhai, X., 2015. Latitudinal dependence of wind-induced near-inertial energy. J. Phys. Oceanogr. 45, 3025–3032.

Zhai, X., 2017. Dependence of energy flux from the wind to surface inertial currents on the scale of atmospheric motions. J. Phys. Oceanogr. 47, 2711–2719.

Zhai, X., Greatbatch, R.J., Eden, C., Hibiya, T., 2009. On the loss of wind-induced near-inertial energy to turbulent mixing in the upper ocean. J. Phys. Oceanogr. 39, 3040–3045.

Zhai, X., Greatbatch, R.J., Zhao, J., 2005. Enhanced vertical propagation of storm-induced near-inertial energy in an eddying ocean channel model. Geophys. Res. Lett. 32.

Zhai, X., Johnson, H.L., Marshall, D.P., Wunsch, C., 2012. On the wind power input to the ocean general circulation. J. Phys. Oceanogr. 42, 1357–1365.

Chapter 6

The lifecycle of topographically-generated internal waves

Ruth Musgrave[a], Friederike Pollmann[b], Samuel Kelly[c] and Maxim Nikurashin[d]

[a]Department of Oceanography, Dalhousie University, Halifax, NS, Canada, [b]Universität Hamburg, Center for Earth System Research and Sustainability, Hamburg, Germany, [c]Large Lakes Observatory, Physics & Astronomy Department, University of Minnesota Duluth, Duluth, MN, United States, [d]Institute of Marine and Antarctic Studies, University of Tasmania, Hobart, TAS, Australia

6.1 Introduction

The interaction of currents with the ocean's varied bathymetry generates internal waves at a range of scales and frequencies. At the lowest frequencies, bottom trapped topographic waves transfer energy large distances along coastlines and submerged features, whereas at higher frequencies freely propagating internal waves redistribute energy both vertically and horizontally across ocean basins. Of particular interest for ocean mixing are internal tides and lee waves, as they form major components of the energy pathways that connect the large scales at which the ocean is forced (by astronomical gravitational forces and the winds) to the small scales at which dissipation occurs (Wunsch and Ferrari, 2004). The generation of these internal waves at topography represents a down-scale transfer of energy; a first step towards ocean mixing. Subsequent wave processes, described in this chapter, transfer wave energy to smaller scales at which instabilities form and waves break, generating stratified turbulence. A fraction of turbulent energy arising from wave breaking is viscously dissipated to heat, but the non-negligible remainder is transformed to potential energy by irreversibly mixing together different waters of different densities.

The importance of turbulence and mixing associated with internal tides and lee waves in the global ocean can be quantified by examining the rate at which energy is transferred into such waves. Assuming the energy contained in the global internal wave field is largely steady, this determines the rate at which the waves lose energy to turbulence and mixing. Internal tides are generated when the barotropic tide drives the stratified ocean interior across bathymetric contours, transferring energy to internal waves at tidal frequencies. Globally, the barotropic tide loses energy at a rate of 3.5 TW, with around 2.5 TW dissipated in shallow seas and along the continental shelf, and 1 TW dissipated in the deep ocean. Whereas the tidal energy dissipated in shallow seas has traditionally been ascribed to bottom boundary layers, the deep ocean dissipation is associated with the generation of internal tides, constituting an important source of mechanical energy in this largely quiescent region (Egbert and Ray, 2003). Lee waves are formed by the interaction of steady flows with topography, with global estimates of generation rates ranging from 0.2–0.75 TW (Nikurashin and Ferrari, 2011; Scott et al., 2011; Wright et al., 2014)—a significant component of the estimated 1 TW total rate of wind work on general circulation of the ocean (Wunsch, 1998). Taken together, internal tides and lee waves can generate turbulence and mixing in the ocean's interior at a rate of 1–2 TW, a value that is comparable to that required to sustain the ocean's stratification in the face of surface buoyancy input (Wunsch and Ferrari, 2004). The implication is that these internal waves play a critical role in basin-scale ocean circulation.

As such, the lifecycle of topographically generated internal waves, from generation to dissipation, forms an essential component of ocean physics, determining when, where and how vigorously the ocean is mixed. Understanding the spatial and temporal distribution of internal wave mixing in the global ocean is a key research goal in the field as its impacts are wide-ranging, with influence on climate (Chapter 2), ocean circulation (Chapters 3, 10, 11, 12) and biogeochemistry (Chapter 13). At a pragmatic level, research is focused on developing parameterisations for climate models, which have established sensitivities to this fundamental ocean process (Jayne and St Laurent, 2001; Simmons et al., 2004; Melet et al., 2013b).

In this chapter, we provide an overview of the key processes that influence the internal wave lifecycle, schematically illustrated in Fig. 6.1. We start with their generation at topography (Section 6.2) and a discussion of their propagation, with an integral estimate of global decay scales (Section 6.3). Possible subsequent stages in the internal wave lifecycle

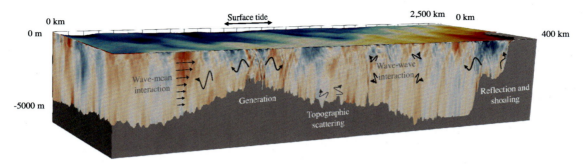

FIGURE 6.1 A schematic illustration of the key stages in the lifecycle of topographically generated internal waves. Internal waves are generated at topography and radiate into the surrounding ocean. There, they are subject to interactions with other internal waves (wave-wave interactions), mean flows (wave-mean interactions) and remote topography (wave-topography interactions). At each stage, the internal waves can transfer energy to smaller scales, ultimately leading to turbulence and ocean mixing. Colours are zonal currents (front and side face) and sea-surface height (surface) in a zonal section of the North Atlantic from the global ECCO LLC4320 simulation (see, e.g., Rocha et al., 2016, for model details).

are characterised as wave-wave interactions (Section 6.4), wave-mean flow interactions (Section 6.5) and wave-topography interactions (Section 6.6). During each of these stages, wave energy may be transferred to other waves or mean flows, or dissipated, leading to local mixing. Each section covers material that is foundational to the topic, recent key findings and the current research direction in this area. At the end of the chapter, we highlight some of the ongoing challenges and outstanding questions in this active field of research (Section 6.7).

Our primary focus throughout is on internal tides and lee waves, as these feature prominently in our understanding of internal wave ocean mixing. Other topographically generated internal waves include subinertial wind- and tide-driven topographically trapped waves, which can be regionally important for ocean mixing (e.g., Padman et al., 1992), but are likely to be of lesser global importance compared to superinertial internal tides and lee waves (Falahat and Nycander, 2015).

6.2 Generation

Interactions between stratified ocean currents and irregular bathymetry lead to the vertical displacement of isopycnals, a motion that projects onto the free modes that constitute internal waves. The nature of the generated internal wave depends on the frequency and amplitude of the forcing current, as well as the latitude, stratification and topography. The frequency ω of internal tides, compared to the local Coriolis frequency f is particularly important: subinertial internal tides, with $\omega < f$, are topographically trapped, and constrained to flux energy along isobaths, whereas superinertial internal tides, with $\omega > f$ can radiate away from topography and across ocean basins, with this latter class contributing most energetically to internal tides in the global ocean. The interaction of steady currents with irregular bathymetry leads to the formation of lee waves, which are steady in a frame of reference fixed to the generating topography, but have frequencies greater than f in the frame of reference of the current. Quasi-steady lee waves may also be generated when the currents are time-varying, for example, forming transiently above steep topography at the same time as subinertial and superinertial internal tides. The distinct generation mechanisms and energy sources for internal tides and lee waves are discussed in this section, as their impact on ocean mixing is well established (see also Sarkar and Scotti, 2017; MacKinnon et al., 2017; Legg, 2021, for recent reviews).

6.2.1 Internal tides

Frequency spectra constructed from timeseries of ocean currents usually include peaks at tidal periods, with the principle lunar semi-diurnal constituent *M2*, having the largest tidal amplitude in most parts of the ocean (Wunsch, 2015). Of the 1 TW abyssal barotropic tide loss from all tidal constituents, 0.8 TW is from *M2* (Egbert and Ray, 2003), where it may directly impact water mass transformations of the ocean's deepest waters, and the global overturning circulation (e.g., Nikurashin and Ferrari, 2013). With a period of 12.42 hours, this constituent is superinertial everywhere equatorwards of 74.5°, motivating the extensive research on the generation and propagation of superinertial internal tides that we focus on in this subsection (see box on *Inertia-gravity wave properties*). These internal tides are characterised by their vertical shears in horizontal velocity, and have amplitudes $\mathcal{O}(0.01 - 0.1)$ m s^{-1} in deep water (Garrett and Kunze, 2007).

Inertia-gravity wave properties

Both superinertial internal tides and lee waves fall into the broad category of inertia-gravity waves, occurring with frequencies greater than f (the Coriolis frequency) and lower than N (the buoyancy frequency, or stratification). The properties of these waves are determined by the dispersion relation, which relates the horizontal and vertical wavenumbers k, m to the wave frequency ω for a given stratification N and latitude. Under the traditional approximation, where only the vertical component of the Coriolis force is considered, the dispersion relation is written as

$$\omega^2 = N^2 \cos^2 \Phi + f^2 \sin^2 \Phi. \tag{6.1}$$

Lines of constant phase, forming an angle $\Phi = \operatorname{atan}(m/k)$ with the vertical, become more and more horizontal as the frequency decreases toward f (see also the textbox in Section 6.4). Both the generation and reflection properties of internal tides at bathymetry are sensitive to the slope criticality, $\epsilon = k_T h_0 / \tan \Phi$, which defines the ratio of the topographic slope to the wave slope, $\tan \Phi$.

Away from their generation site, where the ocean floor is relatively flat, inertia-gravity waves are often represented as flat bottom modes, for which the governing equations for a steady wave field can be separated in the horizontal and vertical directions. Assuming propagating periodic solutions in the horizontal, the equations reduce to a single (vertical) dimension, with solutions taking the form of vertical modes—pairs of waves that propagate diagonally upwards and diagonally downwards. Such modes are constructed to satisfy the conditions of no vertical flow at the bottom boundary, and conditions at a fixed upper boundary of either no vertical flow (the rigid lid approximation) or vertical velocity proportional to surface height (a linear free surface). Vertical modes form a complete set of basis functions that describe vertically standing waves, so they are very efficient for representing weakly dissipative waves that propagate long distances, repeatedly reflecting off the surface and bottom. Conversely, vertical modes are inefficient for representing highly dissipative and localised waves that do not propagate far enough to reflect off the surface or bottom. The superposition of many modes that often occur close to generation sites leads to beams—illustrated in Fig. 6.2—that are compact regions of large shear and isopycnal displacement extending away from their generation sites at topography with angles set by the wave frequency (e.g, Althaus et al., 2003; Cole et al., 2009). Internal tide beams are often associated with elevated turbulence and mixing (e.g., Lien and Gregg, 2001).

The baroclinicity of inertia-gravity waves over flat bathymetry makes it simple to separate internal tides from the astronomically forced barotropic tides. Equations that separately describe barotropic and baroclinic motions can be derived by decomposing the flow into these components (Baines, 1982; Carter et al., 2008; Kang, 2010; Kang and Fringer, 2012). The energy equations associated with each component include a common exchange term, the "conversion," which, by convention, takes the form of a sink in the barotropic energy equation, and a source in the baroclinic energy equation. Thus the conversion describes the rate at which energy is converted from barotropic to baroclinic motions, some of which radiates away as internal waves, and some of which dissipates locally through non-linear processes (e.g., Merrifield and Holloway, 2002; Carter et al., 2008). The energetic importance of local dissipation is represented by the ratio of depth-integrated dissipation to conversion q, which typically takes values between 0 (no local dissipation) to 1 (all local dissipation), but could be greater than 1 in regions where remotely generated waves are incident and dissipate.

The qualitative behaviour of internal tide generation is sensitive to the topographic slope at which it is generated. Topographic slope can be considered both as a bulk parameter defined by the rms height (h_0) and dominant horizontal lengthscale (k_T) of the bathymetry, and also locally, with individual topographic features having regions of subcritical, critical and supercritical slope. When an oscillatory stratified flow interacts with a topographic slope that is shallower than the internal wave slope at that frequency, $\epsilon < 1$ and the slope is termed *subcritical*, with generated wave energy only propagating upwards. In contrast, slopes that are steeper than the wave slope have $\epsilon > 1$ and are *supercritical*, with internal waves generated at the region of critical slope, and propagating both upwards and downwards away from the topography. When $\epsilon = 1$, the slope is *critical* and the solution is singular, with a resonant wave response leading to growth in time and vigorous turbulence and mixing (Gayen and Sarkar, 2011; Scotti, 2011).

Three further parameters play important roles in internal tide generation. The first is the tidal excursion length $l_{exc} = U_0 k_T / \omega$, which compares the maximum horizontal displacement of a fluid parcel in an oscillatory current with amplitude U_0, to the dominant lengthscale of the topography k_T^{-1}. In the "acoustic limit", small excursion length flows generate waves that are dominantly at the forcing frequency, whereas away from this limit, large excursion length flows generate energy both at the forcing frequency as well as at higher harmonics. In this regime the superposition of fundamental and harmonic waves leads to the formation of transient lee waves close to the generating topography (Bell, 1975a), whereas for infinite excursion length, the solutions revert to those for steady lee waves. The second parameter is the topographic Froude number $Fr = N h_0 / U_0$, essentially a ratio of the maximum vertical displacement of the fluid to the topographic height h_0. When $Fr > 1$, fluid below a depth of U_0/N is blocked and unable to surmount the topography, with the result that tidal transport is significantly enhanced in a layer at the topographic crest, leading to locally large tidal excursion lengths and non-linear processes, such as transient lee-wave generation and hydraulic jumps (Klymak et al., 2008; Legg and Klymak, 2008; Winters and Armi, 2013; Musgrave et al., 2016). Finally, the ratio of the topographic height to ocean depth $\delta = h_0/D$,

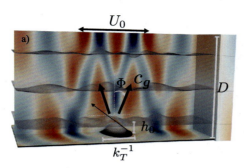 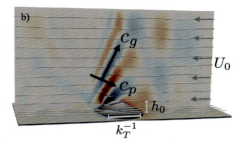

FIGURE 6.2 (a) Idealised internal tide generation (baroclinic cross-ridge velocity in colour, isopycnal surfaces contoured in grey); (b) Idealised lee wave generation (vertical velocity in colour, particle paths contoured) over 3-dimensional Gaussian topography. Group and phase velocities indicated with thick black arrows; in (b) these are marked relative to the moving background flow. Topographic slope is marked in (a) with thin black arrow. In both cases, the topographic slope is subcritical with respect to the generated waves, whose energy radiates upwards at angles Φ with respect to the vertical. Lengthscales discussed in the text are marked in light grey.

determines the validity of assuming a horizontally constant barotropic tide, and plays an important role in determining both the total conversion as well as its modal content (St Laurent et al., 2003; Garrett and Kunze, 2007).

The four parameters described above give rise to a range of dynamically distinct parameter regimes. In the case of weak topography, where the bottom boundary condition is applied at a fixed depth $z = D$, the availability of analytic solutions allows general statements to be made about parameter dependence. In contrast, where topography is tall and steep, analytic solutions are limited, and it is hard to make as much progress. In what follows we discuss each regime in turn.

6.2.1.1 Small amplitude topography

Dwarfed by comparison to dramatic topographic features such as the Hawaiian ridge, the gently sloped hills and ranges that make up the vast majority of the deep ocean floor play an energetically important role in generating internal tides. These features have $\delta \ll 1$, and include abyssal hills, a well studied subset of small amplitude topographies that are poorly resolved in current bathymetric products. Their horizontal lengthscales range from 1–10 km, and their heights are of order 50 m, with slopes spanning sub- to super-critical regimes (Goff and Jordan, 1988).

Analytic solutions in this regime were first developed in Bell (1975b) and Bell (1975a). These linear solutions provide predictions of conversion and radiated energy flux that are formally valid over small amplitude topography with subcritical slopes; the "weak topography approximation," where $\epsilon \ll 1$. This theory has since been developed and generalised (e.g., Llewellyn Smith and Young, 2002; St Laurent and Garrett, 2002; Khatiwala, 2003; Nycander, 2005), and current global estimates of tidal generation are based on this approach (e.g., Falahat et al., 2014; Vic et al., 2019). Extensions to the weak topography approximation have demonstrated that flux estimates using expressions based on this method are reasonable beyond their anticipated validity for $\epsilon < 1$, with radiated flux increasing with bathymetric height h_0^2 in this regime (St Laurent and Garrett, 2002; Balmforth et al., 2002). For supercritical slopes internal wave generation is thought to saturate, as the conversion of kinetic to potential energy at topography becomes less efficient than in the subcritical case (Nycander, 2006), but general analytic solutions are not available. As such, global estimates of conversion based on the weak topography assumption include ad hoc corrections over regions of supercritical slope that attempt to account for this, for example, by dividing the linear estimate of conversion by ϵ^2 (Melet et al., 2013a).

The wavenumbers of generated internal tide modes are restricted by the finite depth of the ocean, with individual modes preferentially generated when horizontal topographic scales match the horizontal wavelength of the mode (Khatiwala, 2003). An important corollary to this is that topographic scales that are larger than a mode one wave do not contribute to internal tide generation (St Laurent and Garrett, 2002; Llewellyn Smith and Young, 2002). Over deep, rough topography, estimates of the modal composition of generated waves show more conversion into modes greater than one (Falahat et al., 2014; Vic et al., 2018), with calculations by St Laurent and Garrett (2002) having the most conversion into modes 5 and 8 for the Mid-Atlantic ridge and East-Pacific rise, respectively.

6.2.1.2 Finite amplitude topography

Characterised by $Fr > 1$, $\epsilon > 1$ and finite values of δ, submerged mountain ranges and steep continental slopes generate internal tides, whose surface displacements can be visible from space (Ray et al., 1996; Zhao et al., 2016; Zaron, 2019). Such features are important generators of the ubiquitous low-mode internal tide energy throughout the global ocean, but the problem of generation in this parameter regime is analytically challenging, not only because the topography is usually

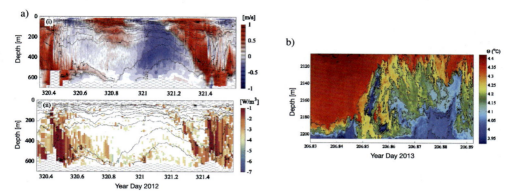

FIGURE 6.3 (a) Hydraulic jumps and lee waves form once per day at the Mendocino ridge when the diurnal and semi-diurnal tidal constituents generate strong across-ridge flow. These observations at the crest of the ridge show the advection of a turbulent transient lee wave (forming the slanted structure at yearday 321.2) and deeper hydraulic jump (at yeardays 320.4 and 321.4) past the station. (i) Cross-ridge currents (colour) with isopycnals contoured, (ii) TKE dissipation rate estimated from Thorpe scales. Adapted from Musgrave et al. (2016). (b) Depth-timeseries of high resolution temperature in the deepest 100 m above the strongly supercritical slopes of Monte Josephine. The observed breaking in these tidally driven non-linear bores is associated with turbulent diffusivities of 10^{-2} m^2 s^{-1}. Adapted from van Haren et al. (2015).

supercritical, but also because the generating barotropic flow is non-uniform over finite amplitude topography (Baines, 1973). As such, it is difficult to present general statements about how barotropic-to-baroclinic conversion depends on parameters such as topographic steepness, or tidal amplitude.

Foundational work in this area was undertaken by Baines (1973), who calculated the generation of internal tides at arbitrary 2-D topographies using the currents associated with unstratified flow solutions to derive a forcing term to the stratified equations. Using this method, he estimated that around 14 GW of tidal energy is transferred to internal tides at the world's continental shelves (Baines, 1982). Subsequent solutions for specific bathymetries suggest a general result that for tall, supercritical topography having $0.5 < \delta < 1$, most energy is generated in mode one (St Laurent et al., 2003; Llewellyn Smith and Young, 2003; Pétrélis et al., 2006), in contrast to solutions over small amplitude topography, where generation can be dominated by higher modes. For small δ, linear conversion rates at 2-D topographies can be corrected to account for saturation effects, however, for large δ, linear conversion estimates may be a significant underestimate (St Laurent et al., 2003). Solutions for arbitrary 3-D topographies are not available, but in a census of internal tide generation at idealised "pillbox" seamounts taller than 1 km, Baines (2007) estimated that these types of tall isolated features may account for around 5 GW of internal tide generation.

While important for generating radiating internal tides, tall ridges and seamounts are also responsible for the significant dissipation of tidal energy very close to topography, with observed diapycnal diffusivities many orders of magnitude larger than in the open ocean (e.g., Alford et al., 2011). Several distinct processes can be at play. The relatively large Froude numbers that occur over these topographies lead to the formation of transient hydraulic jumps within hundreds of metres of the crest on the flanks of the bathymetry, and non-linear lee waves above (Klymak et al., 2008; Alford et al., 2011; Musgrave et al., 2016), observations of which are shown in Fig. 6.3a. In the formation and dissipation of these features, each tidal cycle provide dramatic examples of mixing generated by internal waves, as their steepening and breaking leads to density overturns that can be 10–100's of metres tall, driving enhanced near-topographic turbulence and mixing. Turbulence and mixing is also enhanced at near-critical and critical slopes, where internal tide energy is concentrated in narrow beams, leading to non-linearities and elevated dissipation (Lien and Gregg, 2001; Gayen and Sarkar, 2011). Deeper, on supercritical slopes, upslope propagating non-linear bores have been associated with significant turbulence and mixing (Fig. 6.3b, van Haren et al., 2015). The fraction of energy dissipated through processes such as these, compared to that which radiates away can range from estimates of $q = 0.15$ near Hawaii, to $q = 0.4$ at the Luzon strait and Mendocino escarpment (Klymak et al., 2006; Alford et al., 2015; Musgrave et al., 2017).

Observations of internal tide beams have been made close to their generation site, associated with the energetic generation of internal tides at critical and supercritical slopes (e.g., Cole et al., 2009; Pickering and Alford, 2012). The preferential decorrelation and decay of small scales due to background shears and inhomogeneities in the density field mean that the high modes that are critical to the structure of beams tend to dissipate with distance from the source, and as such beams are not expected to persist beyond their first surface reflection (Lamb, 2014).

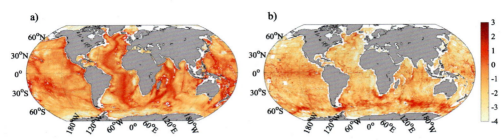

FIGURE 6.4 Theoretical estimates of the global distribution of generation for internal tides and lee waves. (a) Barotropic-baroclinic conversion for six tidal constituents from Vic et al. (2019); (b) Lee-wave generation from Nikurashin and Ferrari (2011), $\log_{10}[\text{mW/m}^2]$.

6.2.1.3 Global distribution

Derived from satellite altimetry, maps of energy loss from the barotropic tide were the first to indicate major regions of internal tide generation, with enhanced loss over deep topographic relief, such as the Mid-Atlantic and Western Indian ridges, and tall features, such as the Hawaiian ridge and Tuamoto archipelago (Ray and Mitchum, 1997; Egbert and Ray, 2003). Fully non-linear numerical simulations driven by the barotropic tide corroborated these locations, demonstrating the generation and radiation of a global field of internal tides from regions of rough topography, with calculations of global conversion that were comparable to observational estimates (Simmons et al., 2004; Niwa and Hibiya, 2011; Arbic et al., 2018).

The distribution of internal tide generation by mode plays an important role in determining the transfer of energy from the internal tide to turbulence and mixing, as high modes tend to break more readily than low. Over the Mid-Atlantic ridge, a region of deep rough topography, Vic et al. (2018) estimate that 81% of the energy lost from the barotropic tide is converted to modes greater than one, causing the fraction of local internal tide energy that is transferred to turbulence via processes, such as wave-wave interactions, to be relatively high ($q \sim 0.5$). On the other hand, mode one internal tides are generated at regional hotspots, including continental shelf breaks, ridges and seamounts, where they can account for more than 50% of local conversion (Vic et al., 2019). In these regions, local dissipation can be much less (see also Lefauve et al., 2015). The importance of internal tide generation at abyssal hills is estimated by Melet et al. (2013a) and Timko et al. (2017), who demonstrate that the inclusion of such bathymetry in global models increases global conversion estimates by around 30 GW. Abyssal hills mostly occur over mid-ocean spreading centres, such as the Mid-Atlantic ridge and the East Pacific rise, where regional increases in internal tide generation can be up to 100%. Due to their small scales, most of the increased conversion by abyssal hills is accounted for by an increase in high mode generation. However, these topographic features are too small to be observed globally, and direct measurements of bathymetry based on single- and multi-beam soundings are sparse. As a result, current global calculations of conversion are based on empirically constructed maps of abyssal hill statistics, contributing some uncertainty to their estimates.

Calculations of global internal tide generation made using semi-analytic models based on the weak topography approximation can be tuned to be brought into agreement with satellite observations for barotropic tide energy loss in many regions (Nycander, 2005; Falahat et al., 2014; Vic et al., 2019, see also Fig. 6.4a). Difficulties remain over steep topography, where the models are not formally valid; or shallower topography, where the application of linear theory can lead to unreasonably large conversions of both positive and negative sign (Falahat et al., 2014; Vic et al., 2019), an issue that has been addressed by either applying an arbitrary cap to the conversion estimates, or by discarding conversions shallower than a set depth of several hundred metres. Though negative conversion is not unphysical per se (it arises for certain phase differences between the baroclinic pressure perturbation and barotropic vertical velocity at the seafloor that have been attributed to remotely generated internal tides, e.g., Zilberman, 2009; Kurapov et al., 2003; Kelly and Nash, 2010; Gong et al., 2019), the large negative values obtained in shallower waters emphasise that the applicability of linear theory is limited to subcritical topography in the deep ocean.

Forward numerical models currently provide the best global estimates of internal tide generation at tall steep topography. Some of these models only include tidal forcing (Simmons et al., 2004; Niwa and Hibiya, 2011), whereas others also include the (wind driven) general circulation (e.g., Arbic and Garrett, 2010; Müller et al., 2012). Such models are invaluable for examining internal wave generation (and subsequent wave-wave, wave-mean flow, and wave-topography interactions), but are difficult to "tune" for optimal state estimates. For example, wave drag is required to obtain a reasonable surface tide, but this drag subsequently alters the internal wave field. In addition, these models require sub-grid scale parameterisations to extract baroclinic energy in lieu of explicitly resolved finescale dynamics. Despite these complications, global models consistently produce global estimates of internal tide generation that compare well with satellite and in situ observations (Arbic et al., 2018).

6.2.2 Quasi-steady lee waves

When a steady, stratified current flows over topography, isopycnals are lifted upstream and lowered downstream of the obstacle. In a frame of reference that moves with the background current at speed U_0, this appears as an oscillation with frequency $\omega_L = U_0 k_T$, as the wavelength is set by the topographic scale k_T^{-1} (the subscript L indicates a Lagrangian frame of reference) (Long, 1953). Upward radiating internal waves are triggered if $f < \omega_L < N$, extracting energy and momentum from the underlying flow. In this case, the horizontal phase speed of the wave is equal and opposite to the background flow such that the wave is arrested and held stationary. Thus for deep stratifications of $N = \mathcal{O}(10^{-3})$ s^{-1} and currents $U_0 = \mathcal{O}(1)$ to $\mathcal{O}(10)$ cm s^{-1}, topography with horizontal scales from $\mathcal{O}(100)$ m to $\mathcal{O}(10)$ km will generate radiating internal waves. Relative to the topography, linear lee-wave energy propagates upwards and downstream of the generating topography, with phase lines tilted backwards into the flow (Fig. 6.2b). The slope of the phase lines is given by k_T/m, where, in the hydrostatic non-rotating regime, the lee-wave dispersion relation sets the vertical wavenumber to be $m = N/U_0$. As for oscillatory flows, the ratio of the characteristic slope of the waves to the topographic slope is an important parameter in steady flows, but for lee waves this ratio coincides with the Froude number condition for deep blocking. Names for this parameter vary widely throughout the literature, including being called "Nhu" by Baines (1997), to emphasise that it is not a Froude number in the traditional sense. Here we follow Mayer and Fringer (2017) and call it the lee-wave Froude number

$$Fr_{lee} = \frac{Nh_0}{U_0}. \tag{6.2}$$

Broadly speaking, low lee-wave Froude number flows obey linear dynamics, generating steady lee-waves aloft. High lee-wave Froude number flows generate time-dependent responses due to lee wave breaking, and in addition may exhibit horizontal flow splitting at depth and the formation of a turbulent wake (Nikurashin and Ferrari, 2010b; Nikurashin et al., 2014; Perfect et al., 2020b). Finite values of the ratio of topographic height to ocean depth ($\delta = h_0/D$) lead to increased flow speeds above the topography, and the formation of relatively higher-frequency lee waves that interact with the surface-forming vertical modes (Klymak et al., 2010).

As for internal tides, we discuss lee-wave generation first at small amplitude topography, then at finite amplitude topography, making the dynamic distinction between each regime by the absence or presence of horizontally split flow.

6.2.2.1 Small amplitude topography

For regions having $Fr_{lee}, \delta \ll 1$, the bottom boundary condition can be linearised and the weak-topography approximation employed. Solutions for monochromatic topography in an infinite depth ocean yield a vertical wave energy flux $E = \frac{1}{2}\rho_0 h_0^2 k_T N U_0^2$ (Gill, 1982), but this is modified for multichromatic topography as the power law relation with U_0 is sensitive to the topographic spectrum (Bell, 1975a; Nikurashin and Ferrari, 2010a).

The assumption of infinite depth is, perhaps, better justified for lee waves than internal tides as the wavenumbers of lee waves are restricted such that $f/U_0 < k < N/U_0$, tending to generate waves with relatively small horizontal and vertical scales that are susceptible to changes in the vertical profile of background currents and are less likely to reach the ocean surface. For waves generated by bottom intensified currents the reduction in current speed aloft can bring the frequency of the wave towards f forming a critical layer, wherein the vertical group velocity tends towards zero and the shear intensifies. As such it is anticipated that these waves dissipate before reaching the surface, though some caveats remain (see Section 6.5).

For finite values of Fr_{lee} that are still less than a critical value $Fr_c \approx 0.3$, nonlinear simulations and theory indicate that steady lee wave solutions are no longer possible, as the relatively larger amplitude lee wave response steepens and breaks aloft, depositing momentum into inertial oscillations. The shear associated with the inertial oscillations lead to further breaking, resulting in a resonant interaction that causes the flow to be both time dependent and highly dissipative, with predictions of around 50% of generated wave energy dissipating in the deepest 1000 m (Nikurashin and Ferrari, 2010b,a). Despite the nonlinear processes above the topography, lee wave generation for Fr_{lee} close to Fr_c is well predicted by linear theory, increasing quadratically with Fr_{lee}, but for much larger values generation saturates and ad hoc corrections are employed in a similar manner to internal tides (Nikurashin and Ferrari, 2010a; Scott et al., 2011).

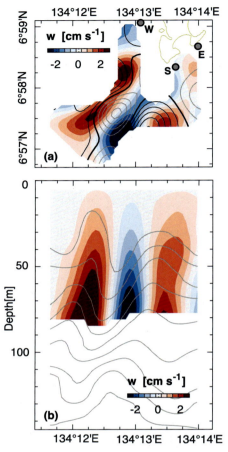

FIGURE 6.5 Quasi-steady lee waves form over a tall ridge in the equatorial Pacific. These objectively mapped surveys show the vertical velocities associated with the wave in colour at (a) 77 m depth, and (b) in a depth-longitude section over Peleliu ridge. Density anomalies are contoured in grey every 0.15 kg m^{-3} in (a), with the mean density contoured in black. Isotherms are contoured in grey every 2 °C in (b). Reproduced with permission from Johnston et al. (2019).

6.2.2.2 Finite amplitude topography

Tall topography is associated with high lee-wave Froude numbers, wherein the deep flow becomes decoupled from the lee-wave generation processes aloft. In two dimensions, the deep flow becomes blocked and stagnant as the conversion of kinetic to potential energy at depths below U_0/N of the crest of the topography is insufficient for fluid parcels to surmount the peak (Klymak et al., 2010). For three-dimensional obstacles, however, the deep flow splits laterally around the topography, forming wake eddies in the lee, which themselves can contribute to dissipation and mixing in the flow (Johnston et al., 2019; Perfect et al., 2020a), or could even generate unsteady wake lee waves (Perfect et al., 2020b).

The blocking or lateral deflection of deep flows means that the upwards wave energy flux at supercritical topography is expected to saturate with increasing topographic steepness as the deep fluid does not contribute to wave generation (Nikurashin and Ferrari, 2010b). Energy conversion from the mean flow is expected to continue to increase in this regime, with energy converted to wake eddies and turbulence rather than radiating lee waves. Numerical simulations indicate that the Fr_{lee} at which conversion saturates is sensitive to anisotropies in the bathymetry, with purely two-dimensional bathymetry (i.e., an infinite ridge) exhibiting saturation for values of $Fr_{lee} \gtrsim 0.7$, compared to three-dimensional isotropic bathymetry, which saturates at lower $Fr_{lee} \approx 0.4$ (Nikurashin et al., 2014). Observations of depth-integrated dissipation over rough topography in the Southern ocean are more consistent with theoretical predictions, where lee-wave generation saturates at lower Fr_{lee} (St. Laurent et al., 2012; Waterman et al., 2013). Discrepancies, however, remain with direct measurements of high Fr_{lee} lee waves by Cusack et al. (2017) at a deep ridge in Drake Passage, and by Johnston et al. (2019) at a shallow ridge in the equatorial Pacific (see Fig. 6.5), each finding vertical wave fluxes $\mathcal{O}(1)$ W m^{-2}, compared to the saturation value of $\mathcal{O}$ 10 mW m^{-2} found in three-dimensional simulations by Nikurashin et al. (2014). It is unclear whether these discrepancies reflect significantly different generation physics, or uncertainties in near-bottom parameters and subsequent lee-wave propagation (Kunze and Lien, 2019).

When topographic lengthscales are very large, $U_0 k_T < f$ and the linear solution is evanescent away from the bottom. Nevertheless, as Fr_{lee} also tends to be large at these scales, non-linear effects, such as blocking and hydraulic jumps may form. In this regime, numerical simulations indicate that the work done by non-propagating form drag leads to turbulence and mixing close to the bathymetry (Klymak, 2018).

6.2.2.3 Global distribution

Lee waves are expected to form in regions where bottom currents, horizontal topographic lengthscales and near-bottom stratifications satisfy the conditions of $f < k_T U_0 < N$, with larger wave fluxes associated with stronger stratification, currents and topographic steepness. As a result, global estimates of lee-wave generation are sensitive to bathymetric resolution, as well as uncertainties in near-bottom currents and stratification (Scott et al., 2011; Wright et al., 2014). These uncertainties are reflected in estimates of global conversion, which range from 0.15–0.75 TW (Nikurashin and Ferrari, 2011; Scott et al., 2011; Wright et al., 2014; Nikurashin et al., 2014; Melet et al., 2015). Unlike for internal tides, where an upper bound for their generation rate is well constrained by astronomical inferences (Munk, 1997), the upper bound for lee-wave generation is given by an estimate of the wind-work on the ocean's general circulation of around 1 TW, a quantity that itself is a tiny residual of the total wind-work on the ocean, and thus subject to considerable uncertainty (Ferrari and Wunsch, 2009).

Maps of lee-wave generation, such as in Fig. 6.4b, closely track bottom kinetic energy, with estimates consistently predicting that the Southern ocean accounts for at least 50% of global lee-wave generation due to the strong deep currents and elevated roughness (defined as the topographic height variance over the lengthscales that generate lee waves). The influence of these waves on diabatic mixing in the Southern ocean is discussed in Chapter 12. Conversion estimates in these areas are $\mathcal{O}(100)$ mW m^{-2}, comparable to the generation of internal tide energy in other parts of the ocean (Nikurashin and Ferrari, 2011; Scott et al., 2011). In contrast, the Mid-Atlantic ridge, despite being rough, is not an important source of lee waves as bottom currents over the ridge are relatively weak. Generation in the equatorial regions is expected to be enhanced as the lee-wave band extends to lower frequencies, enabling lee-wave generation at a wider range of topographic scales. Lee waves have also been observed at the western boundary of the North Atlantic, generated by the interaction of westward propagating mesoscale eddies with the continental slope (Clément et al., 2016).

Observations in the Southern ocean support theoretical predictions of energetic lee-wave generation, with both elevated internal wave fields and turbulent dissipation rates over regions of rough topography and strong bottom currents (St. Laurent et al., 2012; Waterman et al., 2013; Meyer et al., 2015). However, measurements of turbulent dissipation are considerably smaller than would be expected, raising the possibilities that either generation is over-estimated, or that lee-wave energy does not dissipate close to the generating topography, but propagates away or is reabsorbed by the mean flow (Sheen et al., 2013; Waterman et al., 2013; Kunze and Lien, 2019, see also Section 6.5).

6.3 Internal tide propagation and an integral estimate of decay

Once generated, internal waves are subject to a range of specific processes (see Sections 6.4, 6.5, 6.6) that cause them to transfer energy to smaller and smaller scales, where they eventually break leading to turbulence and mixing. The rates at which waves lose energy depend on their individual properties, with high-mode, small-scale internal waves tending to dissipate more rapidly than low-modes. In particular, low-mode superinertial internal tides propagate long distances through the open ocean, as observed in the North Pacific via acoustic tomography (Dushaw et al., 1995), satellite altimetry (Ray et al., 1996), and in situ measurements (Alford and Zhao, 2007). This propagation is clearly illustrated in the high-resolution global maps of internal tide SSH amplitude derived from satellite altimetry, which show horizontal beams (caused by the superposition of waves from multiple sources) of internal tide SSH radiating outwards from generation sites (Fig. 6.6; Zaron, 2019). Though these beams decay with distance, their decay scales in these maps is not indicative of wave dissipation as the waves become nonstationary when they propagate through time-dependent mean flow, causing their amplitudes to decrease in satellite-derived SSH maps that rely on harmonic fits over years of data.

Rough estimates of the globally averaged length and time scales of internal-tide decay can be made by combining observational data (e.g., Wunsch, 1975). Here we estimate the decay lengthscale in the following manner: (1) Estimate global generation for each vertical mode C_n; (2) Estimate global mode-1 energy E_1, from satellite observations; (3) Estimate the residence time $\tau = E_1/C_1$ from mode-1 observations; and (4) Estimate the decay lengthscale $d_n = \tau c_{g,n}$ for each mode (where $c_{g,n}$ is the group speed), assuming a uniform residence time across all modes. Falahat et al. (2014) calculated that modes 1 and 2 account for 34% and 25% of total M_2 generation (with 41% in higher modes). Since these fractions are unlikely to change at neighbouring semi-diurnal constituents, we multiply them by total semi-diurnal deep-ocean surface

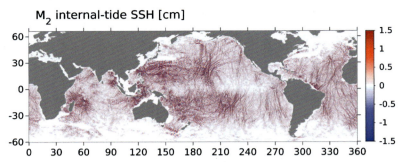

FIGURE 6.6 Satellite-derived map of M_2 internal-tide sea surface height. Only the real component of the complex amplitude is plotted. Data are courtesy of E. Zaron (Zaron, 2019).

tide loss, which is about 1 TW (Egbert and Ray, 2003), to obtain $C_1 = 0.34$ TW, $C_2 = 0.25$ TW, and $C_{3-\infty} = 0.41$ TW. Zhao et al. (2016) used satellite altimetry to estimate 36 PJ of coherent mode-1 M_2 tidal energy. We first multiply this number by 1.25, because 20% of semi-diurnal generation occurs outside the M_2 constituent (Egbert and Ray, 2003), and then multiply by 2.5 to account for underestimated zonally-propagating internal tides, which are poorly resolved by the satellite tracks (de Lavergne et al., 2019). Finally, we multiply by 1.8, because Zaron (2017) found that stationary satellite SSH variance was only 56% of total semi-diurnal variance. The corrected estimate of mode-1 semidiurnal energy is $E_1 \approx 200$ PJ, corresponding to an average energy density of about 550 J m^{-2} and a residence time of $\tau \approx 7$ days. As this is an approximate estimate, we use fixed group speeds appropriate to latitudes equatorwards of around 35° of $c_{g,1} = 2.5$ m s^{-1} and $c_{g,2} = 1.25$ m s^{-1} (Alford and Zhao, 2007), to find that, on average, mode-1 power (0.34 TW) is lost within $d_1 = 1500$ km of its generation sites, whereas mode-2 power (0.25 TW) is lost within $d_2 = 750$ km of its generation sites (see also de Lavergne et al., 2019).

These estimated decay lengthscales for modes 1 and 2 internal tides anticipates the findings from the more detailed calculations presented in the sections that follow, specifically that low-mode internal-tides decay slowly over hundreds of kilometres. It is further supported by in situ observations of propagating internal-tide decay (e.g., Alford and Zhao, 2007; Waterhouse et al., 2018), which found energy losses of $\mathcal{O}(10^{-3})$ W m^{-2} (compared to $\mathcal{O}(1)$ W m^{-2} of energy gain at generation "hot spots"). This is roughly consistent with simply dividing 0.34 TW of total mode-1 dissipation by the area of the ocean (which yields 1 mW m^{-2}). Thus internal-tides decay slowly over large spatial scales through an unknown combination of higher-order dynamics, such as small-scale topographic scattering, wave-wave interactions, wave-mean flow interactions and breaking on distant continental slopes. To accurately map the sinks of internal-tide energy, each of these processes must be carefully cataloged and quantified.

6.4 Wave-wave interactions

The non-linear interactions amongst internal gravity waves are thought to form the most important link between the large scales at which internal waves are generated and the small scales, at which they dissipate (Müller and Olbers, 1975; McComas and Müller, 1981; Winters and D'Asaro, 1997; Polzin, 2004b). By successively fluxing energy through the internal wave spectrum to where the waves become unstable via shear or gravitational instability and break, the non-linear interactions essentially sustain wave-induced turbulent mixing, and thus are a pivotal mechanism in the global ocean's energy balance.

6.4.1 Theoretical background

In a system of equations with quadratic non-linear terms, such as the equations of motion that describe internal gravity waves, two waves can combine (as a sum or difference) to force a third wave (Phillips, 1960). The phase $\mathbf{K}_3 \cdot \mathbf{x} - \omega_3 t$ of the third wave—described by the three-dimensional wavenumber and position vectors, $\mathbf{K}$ and $\mathbf{x}$, respectively, and frequency ω—is determined by the system's dispersion relation (Eq. (6.1)). If it corresponds to that of the forcing term, that is, the sum or difference combination of the two waves with phases $\mathbf{K}_1 \cdot \mathbf{x} - \omega_1 t$ and $\mathbf{K}_2 \cdot \mathbf{x} - \omega_2 t$, the natural frequency corresponds to the driving frequency, and resonance occurs as for a harmonic oscillator. The resonance condition for wavenumbers and frequencies is hence

$$\omega_3 = \omega_1 \pm \omega_2 \qquad \mathbf{K}_3 = \mathbf{K}_1 \pm \mathbf{K}_2. \tag{6.3}$$

If it is satisfied, energy is continuously transferred from the waves ($\mathbf{K_1}, \omega_1$) and ($\mathbf{K_2}, \omega_2$) to the third wave ($\mathbf{K_3}, \omega_3$) (e.g. Ball, 1964). The energy of the individual waves thus changes, but the total energy of the three waves (triad) is conserved. There are many solutions to these resonance conditions that follow the internal gravity wave dispersion relation, so that in the ocean's gravity wave field, energy can be transferred via resonant triad interactions (Olbers, 1976; Müller et al., 1986). For surface gravity waves, for example, this is not the case; instead, resonant four-wave (quadruplet) interactions dominate the non-linear energy transfer.

For the internal gravity wave field of the ocean with its continuous spectrum, the energy transfer at a given wavenumber and frequency is caused by infinitely many interacting triads. Their effect can be quantified using transport theory, assuming that the non-linear interactions are weak and that the individual waves are statistically independent (Hasselmann, 1962, 1967a). The result is called a scattering integral or kinetic equation, and it has been evaluated in slightly different formats in a number of studies (e.g., Olbers, 1976; McComas and Bretherton, 1977; Pomphrey et al., 1980; Caillol and Zeitlin, 2000; Lvov et al., 2012; Eden et al., 2019). The spectral information required for such an evaluation is typically taken from the Garrett–Munk (GM) model (Garrett and Munk, 1972, 1975; Cairns and Williams, 1976; Munk, 1981), a heuristic model of the internal gravity wave's wavenumber-frequency spectrum based on linear theory and observations. It describes a vertically symmetric and horizontally isotropic energy spectrum with a spectral slope of -2 in both frequency and wavenumber space and, although it does not describe near-inertial waves or internal tides specifically, is generally considered a good representation of the background internal wave field in the deep global ocean (McComas and Bretherton, 1977; Müller et al., 1978). Much of what is known about non-linear interactions among internal gravity waves relies on this model spectrum.

Three interaction classes

Parametric subharmonic instability (PSI) is characterised by the decay of a low-wavenumber wave $\mathbf{K_2}$ into two waves of high wavenumbers and approximately half the frequency $\mathbf{K_1}, \mathbf{K_3}$, that is, the frequency resonance condition (Eq. (6.3)) takes the form $\omega_2 \approx 2\omega_1 \approx 2\omega_3$. Successive PSI interactions transfer energy step by step toward low frequencies and high wavenumbers, where the likelihood of shear-induced wave breaking is higher (e.g., Olbers, 1983; Staquet and Sommeria, 2002). The rate of this energy transfer is determined by the energy content of the low-wavenumber wave, so that for GM-type spectra, this interaction mechanism is most efficient at near-inertial frequencies, where the energy levels are highest (McComas and Bretherton, 1977; Eden et al., 2019). *Elastic scattering* (ES) involves two high-frequency waves, one propagating energy upward ($\mathbf{K_1}$), the other downward ($\mathbf{K_3}$), and a third, low-frequency wave of almost twice the vertical wavenumber of the other two ($\mathbf{K_2}$). As these three waves interact, energy is transferred from the more energetic of the two high-frequency waves to the less energetic one until their energies are equal. This process removes vertical asymmetries in the spectrum, except at near-inertial frequencies, and hence contributes little to energy transfers in the vertically symmetric GM spectrum (McComas and Müller, 1981). *Induced diffusion* (ID) describes the interaction of a low-frequency, low-wavenumber wave ($\mathbf{K_2}$) with two waves of much higher frequencies and wavenumbers ($\mathbf{K_1}, \mathbf{K_3}$). The effect is a diffusion of wave action (energy divided by frequency) mainly in vertical wavenumber space, whose strength is determined by the shear level of the low-frequency wave. The ID mechanism dominates interactions in the high-frequency, high-wavenumber part of the spectrum and acts to fill in sharp cutoffs in the wave spectrum at high wavenumbers (Munk, 1981).

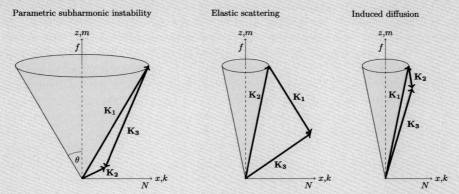

For two-dimensional internal gravity waves, a fixed frequency corresponds to a fixed ratio of zonal and vertical wavenumbers, k/m. In three dimensions, surfaces of constant frequency hence take the shape of a cone, as illustrated here for one of the interacting waves (the y-axis points into the page), opening upwards for positive m and downwards for negative m. The cone's opening angle θ is $\theta = 90° - \Phi$, where Φ is the angle between the wavenumber vector $\mathbf{K}$ (the direction of phase propagation) and the horizontal.

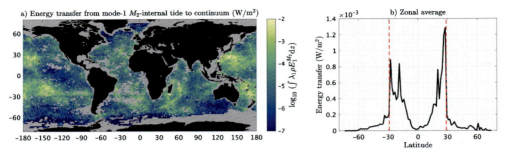

FIGURE 6.7 Energy transfer from the mode-1 M_2 tide to the higher mode gravity wave continuum via non-linear wave-wave interactions. The decay rates λ are obtained from an evaluation of the kinetic equation with climatological stratification data (Gouretski, 2018) and continuum energy levels estimated from Argo profiles in the upper ocean via an extended finestructure method (Pollmann et al., 2017). The internal M_2 tide energy E_{M_2} is taken from the Stormtide simulation (Müller et al., 2012), which resolves the first two modes (Li et al., 2015), by assuming the same modal spreading, as de Lavergne et al. (2019). The zonal average (b) shows a clear peak at the critical latitude for PSI interactions (red dashed line). Equatorward of this latitude, both sum and difference interactions (plus and minus sign in resonance condition in Eq. (6.3)) contribute to the energy transfer, whereas poleward of it freely propagating waves can only be produced by difference interactions. Right at the critical latitude, the only possible sum interaction is PSI with $\omega_3 = \omega_{M_2} = f + f = \omega_1 + \omega_2$. Adapted from Olbers et al. (2020).

6.4.2 Parametric subharmonic instability of the internal tide

McComas and Bretherton (1977) showed that for GM-type spectra, the complicated non-linear energy transfers can be understood in terms of three classes of resonant interactions that dominate different parts of the wavenumber-frequency domain (see highlighted box *Three interaction classes*): elastic scattering (ES), induced diffusion (ID), and parametric subharmonic instability (PSI). The interaction mechanism that has probably received most attention is PSI. This instability arises in many physical systems when the parameters that determine the natural frequency are varied in time, for example, when the support of a pendulum is moved, or when children alternately squat and stand up on a playground swing (see also Onuki and Hibiya, 2015). In the context of internal wave energetics and ocean mixing, PSI is of particular relevance when it transfers energy toward frequencies near f, where group velocities tend to zero and there is potential for energy build-up and breaking. It is also of particular relevance when the parent wave has a high energy content, which is why research has mainly focused on internal tides, in particular, the energetic semi-diurnal M_2 tide.

The latitude where the M_2-tide decays via PSI into (near-)inertial waves, that is, where $\omega_{M_2} = 2f$, is at 28.8°. In idealised numerical simulations, MacKinnon and Winters (2005) observed a substantial loss of M_2-tide energy to subharmonic motions near this critical latitude, combined with an increase in upper ocean dissipation rates. Further arguments for such a dramatic effect of PSI on internal tide energetics and induced turbulent mixing had been provided by other numerical analyses (Hibiya et al., 2002; Furuichi et al., 2005) and observational studies using satellite altimetry (Kantha and Tierney, 1997; Ray and Cartwright, 2001) or current profiler (Nagasawa et al., 2002; Hibiya and Nagasawa, 2004) data. Subsequent studies, however, could not confirm this dominant role of PSI for internal-tide energetics: shipboard and moored observations between 25°N and 37°N north of Hawaii revealed that PSI could notably modify the near-inertial shear equatorward of the critical latitude, but that internal-tide energy fluxes were not affected as the critical latitude was crossed (Alford and Zhao, 2007). MacKinnon et al. (2013) also report evidence of PSI near the critical latitude without any dramatic effects on internal-tide energy levels. The global analysis of M_2-tide energy levels and fluxes from twenty years of satellite altimeter observations of Zhao et al. (2016), too, showed no enhanced energy loss at the critical latitude, in agreement with their earlier investigations in the North Pacific (Zhao and Alford, 2009; Zhao et al., 2010). Instead, internal tides originating just south of the critical latitude at the Hawaiian ridge were observed to propagate thousands of kilometres to the Alaskan shelf.

Like plane gravity waves, tidal beams can loose energy to subharmonic waves by resonant triad interactions (Tabaei and Akylas, 2003; Korobov and Lamb, 2008), but this is limited by the finite width of the beam: energy can only be transferred as long as the subharmonic waves do not leave the beam, which happens sooner for narrower beams (Bourget et al., 2013; Karimi and Akylas, 2014). Maurer et al. (2016) find that the finite beam width also affects the influence of ambient rotation, showing that instability is enhanced when f is about half the beam frequency (see also Dauxois et al., 2018, and references therein). This is in agreement with the peaks seen at f = $\omega_{M_2}/2$ in Fig. 6.7b as well as other numerical (Hazewinkel and Winters, 2011; Nikurashin and Legg, 2011) and observational (Carter et al., 2006; Xie et al., 2011; Liao et al., 2012) analyses, indicating that whereas PSI can occur everywhere equatorward of the critical latitude, it is most efficient as the subharmonic frequencies approach f, albeit not catastrophically so. Stratification variations (Sutherland and Jefferson, 2020) as well as dynamic phenomena, such as mesoscale eddies (Li et al., 2018) and constant (Richet et al.,

2017) or evolving (Yang et al., 2020) currents can further affect the efficiency of PSI energy transfers and the associated dissipation.

6.4.3 Wave-wave interactions in finestructure methods, mixing parameterisations, and numerical simulations

The enhanced mixing observed throughout the water column over rough topography, e.g., in the Brazil basin (Polzin et al., 1997; Ledwell et al., 2000), has been argued to result from non-linear triad interactions transferring energy from the stable internal tides to smaller-scale internal waves that are more prone to instability (Polzin, 2004b; Nikurashin and Legg, 2011). Wave-wave interaction theory hence forms an important part of energetically consistent parameterisations of wave-induced turbulent mixing in ocean models, and is also essential to some observational turbulence estimates. The underlying concept is that, in steady state, internal wave energy dissipation is balanced by the downscale spectral energy flux. Therefore, the turbulent mixing that is driven by the dissipated internal wave energy is described in terms of the energy flux induced by non-linear wave-wave interactions. This energy flux has been estimated, e.g., by Olbers (1976), McComas and Müller (1981), or Henyey et al. (1986), showing that the non-linear energy transports in GM wave fields depend quadratically on the energy level and the buoyancy frequency. The associated dissipation rates were found to agree well with a range of fine- and microstructure observations (Gregg, 1989; Polzin et al., 1995), underlining that these theoretically derived functional dependencies permit a first-order description of wave-induced turbulent mixing. This is exploited in the so-called "finestructure method," which estimates turbulent kinetic energy dissipation induced by wave breaking from the finescale ($\mathcal{O}(10-100)$ m) shear and strain variability (Wijesekera et al., 1993; Kunze et al., 2006; Polzin et al., 2014). Vertical shear and strain at these scales can be related to the internal wave energy level needed for the energy flux calculations and, contrary to turbulence-resolving measurements, are easily observed with standard instrumentation. They are, however, typically representative of the entire internal gravity wave spectrum, and hence do not only represent topographically generated internal waves.

In parameterisations of the turbulent mixing induced by internal tides, typically a distinction between near-field and far-field tidal mixing is made. For near-field tidal mixing, a commonly implemented parameterisation (St Laurent and Garrett, 2002; Simmons et al., 2004; Jayne, 2009, see also Chapter 2) uses internal tide generation estimates and does not explicitly involve wave-wave interaction theory. Polzin (2004a), however, invoked dimensional arguments to derive semi-empirical flux laws that follow the same functional dependencies as the energy flux calculations described above and are consistent with finestructure observations collected in the Brazil basin. Replacing the uniform exponential vertical decay of internal tide energy in the approach of St Laurent and Garrett (2002) by a vertical profile derived from these semi-empirical flux laws (Polzin, 2009) provoked modest but robust differences in global numerical simulations (Melet et al., 2013b). For far-field tidal mixing, a description accounting for wave-wave interactions was recently provided by de Lavergne et al. (2019, 2020) in the form of two- and three-dimensional static maps constructed by Lagrangian tracking of tidal beams. Energy loss via wave-wave interactions is represented by PSI energy transfers and accounts for about 60% of the total energy dissipation of the first 10 internal tide modes.

The most extensive representation of how wave-wave interactions shape internal wave energetics are spectral models based on the radiative transfer equation (Müller and Natarov, 2003; Olbers and Eden, 2013), which describes the change of wave energy by propagation, refraction, wave-wave interactions, forcing, and dissipation processes (e.g., Hasselmann, 1967b). The IDEMIX concept of Olbers and Eden (2013) accounts for the vertical symmetrisation by ES and, in a follow-up version, also for PSI interactions between low-mode internal tides and the high-mode continuum (Eden and Olbers, 2014). IDEMIX reproduces global finestructure estimates of internal wave energy levels and turbulent kinetic energy dissipation rates (Pollmann et al., 2017) and, used as a mixing parameterisation in ocean general circulation models via the Osborn model (Osborn, 1980), leads to some improvements over energetically inconsistent setups (Eden et al., 2014; Nielsen et al., 2018). A recent extension includes a separate compartment for lee waves, which describes their generation at the sea floor, their interaction with the mean flow, and the transfer of lee-wave energy to the higher-mode internal wave continuum, where it is then available for mixing (Eden et al., in revision). These energy transfers are represented by a wave-wave interaction closure that follows a quadratic dependence on energy similar to the scattering integral and parameterisations described above, and integrated to 0.34 TW in a global 1/10° simulation. All these results underline the central role of topographically generated internal waves and the non-linear wave-wave interactions they are involved in for turbulent mixing in the ocean interior and the necessity to consistently account for it in ocean models.

Whereas wave breaking and turbulent mixing will presumably always need parameterisation in global general circulation models, the increase in computer power allows state-of-the-art ocean models to resolve low-mode internal tides and the non-linear wave-wave interactions they experience: Müller et al. (2015) find evidence of non-linear triad interactions in the wavenumber-frequency energy spectra modelled by the hybrid coordinate ocean model (HYCOM) north of Hawaii, consistent with the spectral peaks seen in moored current profiler observations. Based on the same model, Ansong et al. (2018) apply bispectral analysis to investigate PSI interactions of diurnal and semi-diurnal internal tides in the global ocean. The energy transfers are largest near the respective critical latitudes in agreement with previous analyses (Simmons, 2008), but are relatively weak, because positive (from low-mode tides to subharmonic waves) and negative (reverse) transfers almost balance. Since the model only fully resolves the first two internal tide modes, these results might be sensitive to model resolution (as also suggested by Hazewinkel and Winters, 2011; Müller et al., 2015), seeing that higher modes have been observed to dominate PSI energy transfers at some locations (MacKinnon et al., 2013).

6.4.4 Global perspective

The global calculation of energy transfers by wave-wave interactions requires knowledge of the energy level and the local spectral characteristics. The former involves not only the generation (see Section 6.1), but also propagation and interaction processes (see Sections 6.2–6.5), and has so far only been estimated for the energetic internal tides, for which independent observations can be combined with numerical simulations and linear theory (de Lavergne et al., 2020; Olbers et al., 2020). The latter is typically determined in reference to the GM spectral model, which represents a wave field far away from the influence of boundaries or forcing mechanisms, and hence cannot represent internal waves near the ocean's surface (Roth et al., 1981) or in the vicinity of continental shelves (MacKinnon and Gregg, 2003), submarine canyons (Kunze et al., 2002), or strong mean flows (Ruddick and Joyce, 1979). In addition, regional (Polzin and Lvov, 2011) and global (Pollmann, 2020; Le Boyer and Alford, 2021) analyses of the internal wave field's spectral parameters underline that they deviate spatially and temporally from the canonical GM values. But not only the ingredients, also the energy flux recipe itself contains a fair amount of uncertainty. It is unlikely that underlying assumptions, such as the weakness of non-linear processes and off-resonant interactions, suitably describe the ocean's wave field everywhere and at all times, and assuming exactly that, does, in the words of Pomphrey et al. (1980), "require an act of faith". In spite of these caveats, such global-scale estimates of the energy transfers by wave-wave interactions (see also Fig. 6.7) are indispensable to better understand this important part of the internal tide lifecycle and to quantify and consistently parameterise tidally driven turbulent mixing.

6.5 Wave-mean flow interactions

As topographically generated internal waves propagate throughout the ocean they encounter low-frequency flows, such as those associated with mesoscale eddies and currents. On an f-plane, linear rotating flows comprise a steady geostrophic mean-flow and a field of non-interacting inertia-gravity waves. However, the non-linear advection of momentum and buoyancy couples the waves to the mean flow (and the waves to one another), both altering their propagation and enabling energy transfers as described in this section.

6.5.1 Theoretical background

A framework for examining wave-mean flow interactions may be developed from the Boussinesq, hydrostatic, f-plane equations of motion in Cartesian coordinates. A simple wave-mean flow decomposition assumes horizontal velocity $\mathbf{u}$, pressure (normalised by reference density) p, and buoyancy b can be expanded in powers of $\mathrm{Ro} \equiv U/(fL) \ll 1$, which is the Rossby number for quasi-geostrophic flows, and a measure of non-linearity for waves with a similar lengthscale L (since then $fL \approx \omega/k$, so Ro can be interpreted as U/c, where c is the phase speed). The expansion of pressure is

$$p = \mathrm{Ro}\, p_1 + \mathrm{Ro}^2 p_2 + \ldots \tag{6.4}$$

and the expansions of $\mathbf{u}$ and b are similar. The $\mathcal{O}(\mathrm{Ro})$ equations are linear, so averaging them over several wave periods yields a geostrophic flow that can be subtracted from the total flow to isolate the conventional linear wave balance. The next-order equations include advective terms constructed from products of first-order fields. Time averaging these equations produces the standard quasi-geostrophic equation, which evolves without interacting with the wave field (Wagner et al., 2017). Finally, subtracting the mean-flow balance from the second-order equations yields the second-order wave balance. Recombining the first and second-order wave balances produces

$$\frac{\partial \mathbf{u}'}{\partial t} + \left(\overline{\mathbf{u}}_g \cdot \nabla\right) \mathbf{u}' + \left(\mathbf{u}' \cdot \nabla\right) \overline{\mathbf{u}}_g + w' \frac{\partial \overline{\mathbf{u}}_g}{\partial z} + f \hat{\mathbf{k}} \times \mathbf{u}' = -\nabla p', \tag{6.5a}$$

$$\frac{\partial b'}{\partial t} + \overline{\mathbf{u}}_g \cdot \nabla b' + \mathbf{u}' \cdot \nabla \overline{b}_g + w' \left(N_0^2 + N_g^2\right) = 0, \tag{6.5b}$$

where $\nabla = \partial/\partial x + \partial/\partial y$. Primes indicate wave terms, overlines indicate mean-flow (time-averaged) terms, and the g subscripts emphasise that the first-order mean-flow is in geostrophic balance, which requires $w_g = 0$. Wave-wave terms, relevant in Section 6.4, are omitted here, because they force a 2nd harmonic flow that is not directly involved in wave-mean interactions at this order. The small Rossby number assumption ensures that mean-flow perturbations to stratification are small, i.e., $\partial \overline{b}_g/\partial z \equiv N_g^2 \ll N_0^2$, where N_0 is the unperturbed background stratification. Kunze (1985) also noted that stability constraints limit the magnitudes of the horizontal and vertical mean-flow shear, and derived an approximate dispersion relation for (6.5) assuming a local plane wave solution. The result has a real part, which affects wave propagation, and an imaginary part, which represents wave-mean energy exchange.

In an alternative framework, a wave equation can be derived without any Rossby number restrictions by assuming the mean flow has large amplitude and waves are small perturbations (Bühler, 2009). This scaling has the advantages of allowing arbitrary mean-flows (i.e., there is no small Rossby number assumption) and automatically eliminating wave-wave interactions in the leading-order wave equations. It is appropriate for describing weak waves in energetic mesoscale or sub-mesoscale turbulence. However, if the waves become too energetic in a large Rossby number mean-flow, the wave- and mean-flows may be too entangled and distorted to define a wave-mean decomposition. In this situation, a more sensible approach may be to examine the scales of motion and energy transfers between them (e.g., Barkan et al., 2017).

Eq. (6.5) is a reasonable model throughout much of the ocean, because internal tides and lee waves interact with small-Rossby-number mean flows, and the waves and mean flows often have similar characteristic velocities. However, the system of equations is greatly simplified where the waves and mean flow have different horizontal scales. In the "refractive" limit (sometimes called the "inertial" limit), one neglects terms involving wave shear [e.g., $(\overline{\mathbf{u}}_g \cdot \nabla)\mathbf{u}'$], yielding, to first order, an effective inertial frequency of $f_{\text{eff}} = f + \zeta/2$, where $\zeta = \partial \overline{v}_g/\partial x - \partial \overline{u}_g/\partial y$ is the mean flow relative vorticity (Kunze, 1985; Zaron and Egbert, 2014). The refractive limit is most relevant to wind-driven near-inertial waves, which are often generated at very large horizontal scales.

In the horizontal "geometric" limit (a term borrowed from optics), one neglects terms with mean-flow horizontal shear [e.g., $(\mathbf{u}' \cdot \nabla)\overline{\mathbf{u}}_g$ and $\mathbf{u}' \cdot \nabla \overline{b}_g$]. This approximation also neglects mean-flow vertical shear, because it is related to the horizontal buoyancy gradient via the thermal wind. The only mean-flow gradient retained is N_g^2. Note that the geometric limit requires smooth, but not spatially uniform, mean flows. This regime allows ray-tracing solutions, which are consistent with the Wentzel Kramers Brillouin (WKB) approximation. Wave packets move like particles along trajectories defined by the local group velocity, while amplitude and wavenumber evolve slowly (relative to a wave period). The analogy with particles leads to additional conservation laws, such as conservation of wave action, via Hamiltonian mechanics (Bühler, 2009). Ray-tracing has been used as a primary tool for predicting topographically generated wave propagation through mesoscale currents (e.g., Rainville and Pinkel, 2006) and as a diagnostic tool for understanding wave-mean flow phenomenology in fully-non-linear numerical simulations (e.g., Zaron and Egbert, 2014; Duda et al., 2018).

6.5.2 Mean-flow effects on wave propagation

Early in situ measurements revealed that internal tides are typically non-stationary (i.e., intermittent; Wunsch, 1975), but more recently Ray et al. (1996) provided clear satellite observations of stationary internal tides propagating thousands of kilometres from Hawaii. Subsequent investigators sought to identify and quantify the wave-mean flow interactions that produced non-stationary internal tides. Chiswell (2002) found that perturbations in stratification by mesoscale eddies altered internal-tide propagation speeds and explained phase variability in the internal tide at Station Aloha. Rainville and Pinkel (2006) used 2-D ray-tracing through realistic barotropic mesoscale currents and found that wave-mean interactions could explain the slow decay of the stationary internal tide with distance from Hawaii. Zaron and Egbert (2014) used a regional circulation model to show that internal tides generated at the Hawaiian ridge primarily become non-stationary while propagating through mesoscale currents, instead of through temporally variable generation over the ridge. They did not make the geometric approximation, but their scale analysis suggested this approximation was appropriate, i.e., they identified wave refraction by mean-flow stratification and wave advection by mean-flow currents as the dominant forms of wave mean interaction. Wave refraction by mean-flow vorticity, as expressed through f_{eff}, was an order of magnitude weaker.

Mean-flow effects on the internal tide have also been examined using several idealised numerical models (e.g., Ward and Dewar, 2010; Ponte and Klein, 2015; Dunphy et al., 2017; Wagner et al., 2017). A recurring finding in these studies, regardless of their precise model formulation or setup, is that incoming low-mode plane waves rapidly become incoherent as they interact with quasi-geostrophic turbulence (Fig. 6.8). These dynamics have been further demonstrated numerous times in oceanic settings, with two recent examples, including ray-tracing through the (baroclinic) Gulf stream (Duda et al., 2018) and a detailed examination of internal tides propagating through equatorial jets in a general circulation model (Buijsman et al., 2017).

132 Ocean Mixing

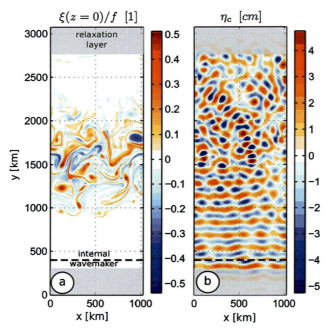

FIGURE 6.8 Numerical simulation of a low-mode internal tide propagating through mesoscale turbulence. (a) Detided surface relative vorticity normalised by the Coriolis frequency. (b) internal-tide sea surface height. The internal-tide wavemaker and relaxation layer are shown. The internal tides are scattered and refracted by the mesoscale turbulence. Reproduced with permission from Ponte and Klein (2015).

Mean-flow effects on lee-wave propagation are difficult to identify, as these waves tend not to propagate as far as low-mode internal tides. Linear lee waves are stationary in the mean flow that generates them, but time variability can lead to advection of the wave field. Secondary waves, which are generated by non-linear interactions between lee waves, are not stationary in the generating current, and are advected downstream, even in a steady current. In a numerical simulation inspired by the Antarctic circumpolar current, Zheng and Nikurashin (2019) found that 50% of wave-generated turbulence occurs downstream of the generation region, with an *e*-folding scale of 20-30 km.

6.5.3 Mean-flow effects on wave energy

Mean flows can both exchange energy with waves and catalyse the transfer of wave energy from large to small scales (i.e., scatter waves to higher modes), where breaking and mixing occurs. Within the geometric regime, horizontal wavenumbers can grow exponentially in (non-divergent) quasi-geostrophic mean flows if the mean flow rate of strain is large enough to exceed the mean-flow vorticity, typically expressed as the Okubo–Weiss criterion $S_s^2 + S_n^2 > \zeta^2$, where S_s and S_n are the shear and normal horizontal strain rates (Polzin, 2010). Under some circumstances a wave can be theoretically "captured" and grow by extracting energy from the mean flow (Bühler, 2009). Wave capture is typically not observed for internal tides, because they are dominated by low vertical modes and are sufficiently dispersive so that their group speed increases as their horizontal wavenumber grows, allowing them to escape (Rocha et al., 2018). However, von Storch et al. (2019) find indications in a 0.1° numerical simulation that wave trains with vertical structures comparable to that of low-mode internal waves propagate along axial flows as if they were captured. At present, global energy transfers between topographically generated internal waves and mean flows are poorly constrained, in part because it is difficult to interpret the output of high-resolution general circulation models, resolving a broad range of motions. Idealised simulations have found negligible energy exchanges between internal tides and mean flows across a range of mesoscale conditions (Dunphy et al., 2017; Wagner et al., 2017).

The rate at which mean flows scatter topographically generated internal waves from high to low modes is also poorly constrained. Dunphy and Lamb (2014) methodically simulated mode-1 internal-tide scattering by barotropic and baroclinic eddies with order-one Rossby numbers. They found that mode-1 waves could scatter up to 13% of their energy to higher modes under optimal conditions. However, Dunphy et al. (2017) extended these results to internal tides propagating across a 500 km wide turbulent quasi-geostrophic front and found only a 1% decrease in mode-1 energy. Regional simulations of the Gulf stream (Kelly et al., 2016) and Tasman sea (Savage et al., 2020) also produced weak mode-1 scattering (i.e.,

$\mathcal{O}(10^{-3})$ W m^{-2} depth-integrated energy losses), despite energetic internal tides. Li et al. (2019) examined mode-1 tides normally incident on steady geostrophic fronts and found that about 10% of the incident energy could be scattered to higher modes. All of these numerical results indicate that mean flows can scatter low-mode energy to higher modes at modest rates, although this scattering has not yet been explicitly observed.

Within the geometric regime, vertically propagating lee waves conserve wave action ($E/|kU|$), so that waves lose (gain) energy to (from) the mean flow if it weakens (strengthens) above topography. Guided by this fact, Kunze and Lien (2019) calculated that about 50% of lee-wave energy in the (bottom-trapped) Atlantic circumpolar current is re-absorbed by the mean flow as these waves propagate upward, implying that the fraction of lee-wave energy lost to turbulent dissipation and mixing may be reduced compared to earlier estimates (Scott et al., 2011; Nikurashin and Ferrari, 2011; Wright et al., 2014; Melet et al., 2015; Trossman et al., 2016; Yang et al., 2018).

6.5.4 Global perspective

There is overwhelming theoretical, numerical, and observational evidence that mean flows alter the horizontal propagation of low-mode internal tides via (i) refraction by stratification anomalies, (ii) advection, and (iii) refraction by mesoscale currents (listed by their order of importance in Zaron and Egbert, 2014). The details of these interactions are critical for determining how internal tides become non-stationary and correctly interpreting global satellite observations of internal tides.

At this time, relatively little can be said about the global contribution of wave-mean interactions in dissipating internal tides. In general, internal-tide energy exchanges with mean flows (Wagner et al., 2017) and between vertical modes (Dunphy et al., 2017) have been modelled and found to be weak. Regional simulations indicate that eddies can scatter mode-1 tidal energy to higher modes at a rate of $\mathcal{O}(1 \text{ mW m}^{-2})$, where internal tides are energetic (Kelly et al., 2016; Savage et al., 2020), however, an inventory of inter-modal scattering in a global circulation model is needed to establish the relative importance of this process. Quantitative in situ observations of (i) internal-tide energy exchange with a mean flow and (ii) inter-modal scattering have not been published (to our knowledge), but may be possible following the methods of Polzin (2010) and Savage et al. (2020), respectively.

The global importance of lee-wave mean-flow interactions is also unclear. Theoretical results from Kunze and Lien (2019) suggest 50% of lee-wave energy may be re-absorbed by the mean flow as a consequence of conservation of wave action. A first attempt at modelling this interaction globally was recently made by Eden et al. (in revision) with a closure for lee-wave drag based on the IDEMIX concept (see Section 6.4.3), showing that the lee-wave bottom stress was locally comparable to or even larger than the surface wind stress (e.g., in the Southern ocean, in the Gulf stream, or over the Mid-Atlantic ridge) and that the vertical wave-induced momentum flux could reach the ocean's surface layer. Further investigation and especially observations are needed to better quantify lee-wave mean-flow interactions and their variability in the global ocean. The horizontal advection of secondary waves and turbulence (generated by lee-wave interactions) is small (i.e., 20–30 km; Zheng and Nikurashin, 2019), but likely important for closing local energy balances (Meyer et al., 2015).

6.6 Wave-topography interaction

Low-mode internal tides have been shown to propagate across ocean basins thousands of kilometres away from their generation site (Dushaw et al., 1995; Ray et al., 1996; Zhao et al., 2016). As they propagate, however, internal tides encounter topographic features, such as abyssal hills, mid-ocean ridges, and continental margins, which can effectively facilitate the transfer of the low-mode internal-tide energy to smaller-scale internal waves and ultimately to a scale at which waves break. Lee waves, in contrast, are believed to dissipate most of their energy near their generation site as they radiate upward away from the bottom of the ocean. Recent observational (Meyer et al., 2015) and idealised numerical (Zheng and Nikurashin, 2019) studies have explored the remote dissipation of lee waves, finding that lee waves are advected by mean flows downstream of their generation site, where they can potentially interact with topography. However, lee waves decay quite rapidly, within $O(10 \text{ km})$, downstream of the generation site, and hence their dissipation remains fairly local in comparison to the propagation and dissipation of low-mode internal tides. As such, much attention has been given to the interaction of low-mode internal tides with topography. Theories and numerical simulations have primarily focused on two types of wave-topography interaction: (*i*) scattering of low-mode internal tides at small-scale *abyssal hills* and (*ii*) reflection of low-mode internal tides at large-scale topographic features, in particular, *continental margins*.

6.6.1 Theoretical background & observational estimates

6.6.1.1 Abyssal hills

Poorly resolved in global bathymetric maps, abyssal hills are thought to play a leading order role in the scattering of internal tides. As low-mode internal tides propagate throughout the oceans, abyssal hills can scatter their energy to smaller-scale internal waves, and ultimately to a scale at which internal waves break (Kelly et al., 2013; Waterhouse et al., 2018; de Lavergne et al., 2019), leading to bottom-intensified turbulence, energy dissipation, and mixing.

The mathematical description of internal wave scattering at small-scale topographic roughness, such as abyssal hills, is similar to that of non-linear wave-wave interactions, but with one of the interacting triads given by the stationary topography field (Cox and Sandstrom, 1962; Hasselmann, 1967a). The interaction between internal waves and bottom topography arises through the kinematic boundary condition of no flow across topography, which is non-linear in wave amplitude. Assuming subcritical topography, Müller and Xu (1992) applied a small perturbation expansion to the boundary condition, and hence examined a weakly-non-linear solution to the problem of internal wave interaction with topography, in which the wavenumber of the scattered wave is the sum of the topographic and incident wavenumbers based on Eq. (6.3). They found that both reflection and scattering of an incident internal wave at rough topography redistributes energy flux in wavenumber space. However, scattering was found to be more efficient than reflection at transferring energy to higher wavenumbers. Applying these results to a Garrett–Munk background of internal waves (Garrett and Munk, 1975) and typical abyssal hills in the ocean, Müller and Xu (1992) estimated that less than 10% of the incoming radiation is redistributed among wavenumbers during a scattering event.

Consistent with the approach of Müller and Xu (1992), the scattering of low-mode internal tides at rough abyssal hill topography can be intuitively understood the same way as the generation of internal tides by a barotropic tide, but with the barotropic tide amplitude replaced with that of the low-mode internal tide. Exploiting this similarity between internal-tide generation and the scattering of low-mode internal tides, St Laurent and Garrett (2002) estimated the scattering of the mode-1 internal tide for a region of the Mid-Atlantic ridge. They found that mode-1 scattering occurs with an efficiency of about 10%, comparable to the efficiency estimated for the Garrett–Munk spectrum by Müller and Xu (1992). More recently, Buhler and Holmes-Cerfon (2011) extended the wave scattering approach of Müller and Xu (1992) to study the decay of a mode-1 internal tide as it interacts with random bottom topography, both theoretically and numerically. They obtained a scaling for the horizontal decay of mode-1 internal-tide energy flux caused by the scattering of its energy into high-mode internal waves showing that the mode-1 internal-tide energy flux decays exponentially with distance, and that the decay rate depends on the statistics of the abyssal hills.

6.6.1.2 Continental margins

Continental margins are topographic features that are found in proximity to continents and generally consist of three different types of topography: the continental rise, the continental slope, and the continental shelf; each characterised by different slopes and depths. Remotely generated low-mode internal tides can reach and interact with the topography of continental margins. Depending on the local topographic slope and how it compares to the slope of the internal-tide rays, internal tides can reflect backwards into the ocean interior, reflect and reduce their vertical scale until they break and dissipate locally; or reflect forward and shoal.

The physical mechanism governing this interaction is different from the mechanism of internal wave interaction with abyssal hills and is largely described by linear theory for wave reflection (e.g., Phillips (1966); Staquet and Sommeria (2002)). Internal waves follow a simple reflection law at a sloping boundary of continental margins, wherein the reflected waves maintain the incident wave frequency, and thus preserve their phase angle with respect to the vertical (see Eq. (6.1)). The interaction of a low-mode internal tide with a critical slope, where the topographic slope matches that of the wave (see Section 6.2), will reflect energy into higher wavenumbers (Fig. 6.9b). In that limit, linear theory predicts a reflected wave of an infinite amplitude and infinitesimal vertical wavelength, which in practice leads to spatially concentrated wave flux, wave breaking, and energy dissipation (Legg and Adcroft, 2003; Nash et al., 2004; Martini et al., 2013). On the other hand, supercritical topographic slopes (Fig. 6.9c) will lead to the reflection of incident internal tide away from the continental margin back into the ocean interior (Klymak et al., 2013; Johnston et al., 2015), whereas subcritical slopes (Fig. 6.9a) will reflect them forward and lead to the internal wave shoaling (e.g., Müller and Liu (2000); Legg (2014)).

Direct observations of low-mode internal-tide reflection at near-critical continental margins support theoretical predictions of energy transfers into high modes and internal-tide dissipation. Nash et al. (2004) obtained estimates of an incident low-mode internal tide, turbulence, and mixing at the near-critical continental margin slope off the coast of Virginia, USA (Fig. 6.10). Their measurements showed elevated near-bottom turbulent energy dissipation rates of $\mathcal{O}(10^{-8})$ W kg^{-1} and diapycnal diffusivities of $\mathcal{O}(10^{-3})$ m^2 s^{-1} maintained by the reflection and subsequent dissipation of a remotely gener-

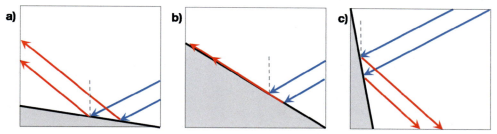

FIGURE 6.9 Reflection of an internal wave at a continental margin for (a) subcritical slope, (b) critical slope, and (c) supercritical slope. The incident internal wave is shown in (blue) and reflected wave is shown in (red). The normal direction is indicated by the dashed grey lines.

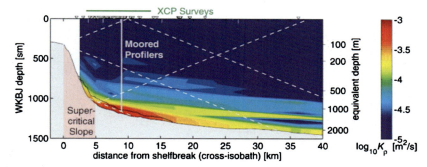

FIGURE 6.10 Turbulent mixing derived from the HRP during the period 27 May–4 Jun at the continental slope of Virginia, USA. M_2 internal-tide characteristics are shown for reference. The region of supercritical topographic slope is also indicated (pink shading). Reproduced with permission from Nash et al. (2004).

ated low-mode M_2 internal tide on the near-critical slope. The observed region of the elevated mixing was in the shadow zone for locally generated internal tides, and no local lee-wave generation by subinertial flows was found to explain the observed elevated mixing. The results suggest that the reflection of the incident low-mode internal tide at the near-critical slope produces a high-vertical wavenumber response that decays to turbulence on a time scale of $\mathcal{O}(1)$ day. In a subsequent experiment on the Oregon continental slope, Kelly et al. (2012) observed energy transfers from low to high modes and found it roughly equal to local turbulent dissipation rates, highlighting that topographic scattering at near-critical slopes is a pathway for low-mode internal wave dissipation.

The reflection of low-mode internal tides at supercritical slopes on a continental margin has been recently investigated both observationally and numerically as part of the Tasman tidal dissipation experiment (T-TIDE) off the east coast of Tasmania (Klymak et al., 2016). Numerical simulations with a realistic slope show that approximately 65% of the incident mode-1 internal-tide energy generated at the Macquarie ridge reflects as mode-1 energy from the continental slope of Tasmania, whereas 21% dissipated. The simulations showed that a simple linear incoming internal tide resulted in a rich and complex wave response as a result of its interaction with topography. These modelling results are consistent with glider (Johnston et al., 2015) and moored observations in this region (Marques et al., 2020), which show the formation of a standing wave and high reflectivity coefficients of 0.8-1.

6.6.2 Global distribution

Though all topographic features can interact with internal waves, scattering at abyssal hills and reflection at continental margins have been shown to be the most efficient at transferring internal wave energy to smaller-scales where they break. Though in principle relevant for all internal waves, these processes are particularly important for low-mode internal tides that radiate thousands of kilometres across ocean basins, and hence are exposed to and interact with a range of topographic features. In contrast, lee waves generated by subinertial flows, i.e., ocean currents and mesoscale eddies, are believed to dissipate their energy locally within $\mathcal{O}(10)$ km of their generation site.

Considering the geometric principles of internal-tide reflection at continental margins described above, and assuming a random distribution of the incident internal-tide rays, de Lavergne et al. (2019) estimated the efficiency and energy dissipation associated with the interaction of the M_2 internal tide with continental margins (Fig. 6.11). They found that shoaling produces substantial energy loss at the shelf break and shoreward, while it is negligible elsewhere. Critical slopes, primarily

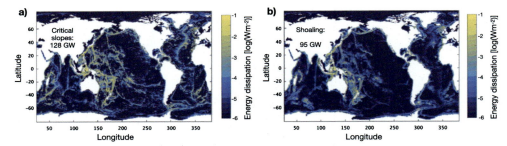

FIGURE 6.11 Estimates of internal wave energy dissipation across all tidal constituents at (a) critical and (b) subcritical slopes. Reproduced with permission from de Lavergne et al. (2019).

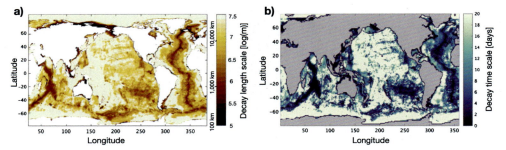

FIGURE 6.12 (a) Low-mode internal-tide decay lengthscale in due to scattering by abyssal hills. Reproduced with permission from de Lavergne et al. (2019). (b) Low-mode M_2 internal-tide decay time scale in due to scattering at topography. Updated from Eden and Olbers (2014) based on tide-continuum interaction time scales of Olbers et al. (2020, see Fig. 6.7) and the abyssal hill data of Goff and Arbic (2010).

located on the continental slopes around most of the ocean's perimeter, were associated with significant dissipation. As in similar previous studies (e.g., Kelly et al., 2013; Johnston et al., 2015; Klymak et al., 2016), backward reflection at super-critical slopes was found to dominate shoaling or dissipation, implying that much of the incident low-mode internal-tide energy reaching continental margins is reflected into the open ocean, where it can then undergo wave-wave and wave-mean interaction processes.

de Lavergne et al. (2019) applied the scaling from Buhler and Holmes-Cerfon (2011) globally using *rms* height and horizontal wavenumber statistics of abyssal hills produced by Goff and Arbic (2010). The estimated e-folding decay lengthscale (Fig. 6.12a) shows that the scattering of low-mode internal tide is most efficient at the mid-ocean ridge crests, where the abyssal hill roughness tends to be enhanced due to the formation of abyssal hills as part of the spreading of the mid-ocean ridge systems (Goff and Jordan, 1988). The horizontal decay lengthscale of several hundreds of kilometres implies that the low-mode internal tides will scatter most of their energy locally, within a few bounces, as they interact with abyssal hills at the crests of the mid-ocean ridges. On the other hand, the decay length-scale was found to be large, thousands of kilometres, over abyssal plains and region covered by sediments. In those regions, low-mode internal tides can propagate large distances nearly unaffected by bottom roughness until they reach continental margins.

Similar global estimates have been calculated in other studies. Eden and Olbers (2014) used a parameterisation of internal-tide scattering by abyssal hills based on the results of Müller and Xu (1992) in their energetically consistent global ocean model. The distribution of the associated decay time scale that they found (Fig. 6.12b) is qualitatively consistent with the estimate of de Lavergne et al. (2019), showing rapid decay over the mid-ocean ridges. However, quantitatively, their estimate of the decay lengthscale is a factor of 3 to 10 larger than the estimate of de Lavergne et al. (2019). Finally, Kelly et al. (2013) used a linear theory estimate of internal-tide generation by Nycander (2005) and bathymetry resolved by ETOPO2v2 to estimate the scattering of the mode-1 M_2 internal tide. They found scattering rates an order of magnitude weaker than those of de Lavergne et al. (2019), likely due to poor representation of abyssal hills by satellite-derived topography in ETOPO2v2.

6.7 Conclusions and outstanding questions

The interaction of ocean currents with seafloor topography creates a variety of internal gravity waves that can propagate far from their generation sites throughout the global ocean. The eventual breaking of small-scale internal waves dissipates wave energy and drives turbulence, with a fraction of the turbulent energy being used to increase potential energy by mixing the stratified ocean's interior. Turbulent mixing by internal waves is the primary mechanism by which abyssal water properties are modified, and thus it plays a critical role in determining deep stratification and global circulation.

The generation sites of the internal gravity waves discussed in this chapter cover a broad range of topographic features, from the gently sloped abyssal hills to tall, steep mountain ranges, such as Hawaii, and equally diverse is the generated wave signal. Some of it can be understood in terms of linear theory, but in other cases, numerical simulations or satellite altimeter data provide the more reliable results—and in still others, the wave response is not well understood at all. This chapter focuses mainly on internal tides and lee waves as their dynamics are important for ocean mixing, and their global signal is most easily investigated with the above-mentioned methods. Their lifecycle has been described in terms of the generation and subsequent interactions of these waves with the mean flow, bottom topography, and other internal waves.

Conceptual analogies can be drawn between the various physical processes that play a role in the internal wave lifecycle. For example, topographic scattering can be understood in the same way as resonant wave-wave interactions, when one of the interacting waves (not that which is forced) is replaced by a stationary wavenumber component of the abyssal hill topographic spectrum. Similarly, the interaction of low-mode internal tides with the large-scale slope of continental margins is analogous to internal wave propagation through mean-flow shear, in which radiating internal waves change their wavenumber to preserve their frequency (or wave action). Critical slope reflection at continental margins is thus analogous to critical layers in mean flows, where the wavenumber of incident waves reduces to zero, focusing therefore the wave energy and leading to wave breaking and energy dissipation. These analogies underline the common physics describing wave processes, and have enabled progress in understanding the mechanisms that shape the lifecycle of topographically generated internal waves, at least on a conceptual level.

When it comes to the specifics, however, even for the relatively well understood internal tides and lee waves, the list of open questions is long. Generation and interaction processes are hard to observe directly in the real ocean, expensive to model for a broad range of spatial and temporal scales, and complicated to describe in theoretical models, especially as non-linear terms begin to dominate. We here offer an overview of unresolved issues (certainly non-exhaustive and biased by the authors' interests):

- Internal-tide generation theories are only strictly valid over a subset of oceanic parameters, where topography has small amplitude and subcritical slopes. Conversion over tall, steep topography remains poorly understood, yet forms a significant source of internal-tide energy in the global ocean (Balmforth and Peacock, 2009). At supercritical locations, "weak topography" calculations likely overestimate the conversion (see Nycander, 2006). Further research on this issue is needed as about half of the global generation occurs in these regions (Nycander, 2005; Falahat et al., 2014), but a consistent method for estimating conversion over arbitrary topography has not been developed.
- The role of subinertial, topographically trapped internal waves for ocean mixing has not been fully explored. Global calculations for subinertial internal tides indicate that their role is likely negligible compared to superinertial tides (Falahat and Nycander, 2015), but the analysis is restricted to weak topographies, with limited applicability to the continental shelves, where tides are energetic. To our knowledge, dissipation and mixing by wind-driven coastal-trapped waves and low-frequency topographically trapped waves have not been considered.
- The modal partitioning of internal-tide energy plays an important role in tidal mixing parameterisations (de Lavergne et al., 2019) or decay time scale estimates (Olbers et al., 2020), but global estimates of modal composition are constrained by only a few observations at select sites (e.g., Nash et al., 2006).
- Just as for internal tides, global estimates of lee-wave generation are uncertain over tall or steep topography. Discrepancies between observed fluxes and linear generation estimates may either reflect uncertainties in wave generation, or in the subsequent evolution of the wave as it propagates upwards through vertically sheared flow (Kunze and Lien, 2019).
- In many parts of the ocean, tides co-exist with much lower frequency quasi-steady flows, and the generation of lee waves and internal tides cannot be considered to be linearly independent of one another. Wave generation, near-field dissipation and the subsequent evolution of the waves are affected, but this has only just started to be explored (Lamb and Dunphy, 2018; Voet et al., 2020; Shakespeare, 2020).
- Estimates of the energy transfers by non-linear wave-wave interactions and mixing parameterisations building on these have focused on resonant triad interactions among weakly non-linear internal tides. For an exhaustive analysis, other types of internal waves, as well as other processes, such as the generation of superharmonics by self-interaction (Sutherland and Jefferson, 2020) or the non-resonant interaction of tidal beams (Lamb, 2004) need to be taken into account.
- Eden et al. (2019) observed in their numerical evaluation of resonant and non-resonant triad interactions, that the induced energy loss depends not only on the spectral energy level and the buoyancy frequency, but also on the vertical wavenumber slope, a relation currently not taken into account in the finestructure method and mixing parameterisations. A thorough evaluation against high-resolution turbulence observations is required to assess the impact and relevance of this dependence on spectral parameters.

- In numerical simulations, pathways of lee-wave energy are strongly influenced by the waves' interplay with inertial oscillations above the sea floor (Nikurashin and Ferrari, 2010b). This remains to be confirmed by in situ observations, and is only just starting to be included in mixing parameterisations (Melet et al., 2014; Labreuche et al., 2020).
- To assess the contribution of wave-mean flow interactions to wave-energy dissipation, a global inventory of the inter-modal energy scattering by wave-eddy interactions is needed, possibly in the form of high-resolution numerical simulations combined with dedicated observational campaigns to constrain them. The limited regional numerical estimates available indicate scattering out of mode 1 at a rate of 1 mW m^{-2}, which integrates to a global value of 0.4 TW—perhaps suggesting a far more important role for wave-mean flow interactions in internal-tide dissipation than has been discussed to date, but this process is difficult to observe and validate.
- Global estimates and parameterisations of the wave-topography interaction have been developed and implemented in global ocean models (de Lavergne et al., 2020). However, large uncertainties still exist: whereas the estimates agree qualitatively, quantitatively they can differ by an order of magnitude. Reflection at continental margins relies on the knowledge of the incident low-mode internal-tide field, and hence depends on the quantitative accuracy of earlier stages of the internal wave lifecycle that govern the evolution of the low-mode internal tides while they radiate across ocean basins towards continental margins. In addition to uncertainties in incident low-mode internal-tide estimates, scattering uncertainties are also related to poor knowledge of abyssal hills and deep stratification.
- All these specific issues are overshadowed by the general problem that global estimates of internal wave generation or interaction processes are only as reliable as the input data used. The main reason for the often substantial differences between individual estimates for different generation or dissipation mechanisms arise from the use of different bathymetric products (i.e., resolving or representing abyssal hills), stratifications, and forcing (i.e., surface tide or low-frequency currents), all of which are poorly constrained by observations, particularly at small scales and over complicated topography.

Because of the central role that small-scale turbulent mixing plays for large-scale ocean and climate dynamics, it is imperative to understand and quantify the mixing driven by topographically generated internal waves. Internal-wave processes are unlikely to be resolved in global models in the near future, and hence we need to be able to parameterise their influence in ocean and climate models. This requires tackling problems, such as those outlined above, to better constrain estimates of internal-wave energy, how much is transferred to turbulence, and how and why these transfers vary spatially and temporally. Although many aspects of ocean mixing by internal gravity waves have been investigated, many more await discovery.

Acknowledgements

Stephanie Waterman and Andrew Thompson provided detailed and constructive comments that improved this chapter. Musgrave was supported by funding from the Canada Research Chairs Program; Kelly was supported by NASA (NNX16AH75G) and the National Science Foundation (NSF-OCE1434352). We thank Uriel Zajaczkovski for his help with Figs. 6.1 and 6.2, which were made using Paraview (Ahrens et al., 2005).

References

Ahrens, J., Geveci, B., Law, C., 2005. Paraview: An End-User Tool for Large Data Visualization. The Visualization Handbook, vol. 717.

Alford, M.H., MacKinnon, J.A., Nash, J.D., Simmons, H.L., Pickering, A., Klymak, J., Pinkel, R., Sun, O., Rainville, L., Musgrave, R.C., Beitzel, T., Fu, K., Lu, C., 2011. Energy flux and dissipation in Luzon Strait: two tales of two ridges. J. Phys. Oceanogr. 41 (11), 2211–2222.

Alford, M.H., Peacock, T., MacKinnon, J.A., Nash, J.D., Buijsman, M.C., Centuroni, L.R., Chao, S.-Y., Chang, M.-H., Farmer, D.M., Fringer, O.B., Fu, K.-H., Gallacher, P.C., Graber, H.C., Helfrich, K.R., Jachec, S.M., Jackson, C.R., Klymak, J.M., Ko, D.S., Jan, S., Johnston, T.M.S., Legg, S., Lee, I.-H., Lien, R.C., Mercier, M.J., Moum, J.N., Musgrave, R.C., Park, J.-H., Pickering, A.I., Pinkel, R., Rainville, L., Ramp, S.R., Rudnick, D.L., Sarkar, S., Scotti, A., Simmons, H.L., St Laurent, L.C., Venayagamoorthy, S.W., Wang, Y.-H., Wang, J., Yang, Y.J., Paluszkiewicz, T., David Tang, T.-Y., 2015. The formation and fate of internal waves in the South China Sea. Nature 521 (7550), 65–69.

Alford, M.H., Zhao, Z., 2007. Global patterns of low-mode internal-wave propagation. Part II: group velocity. J. Phys. Oceanogr. 37 (7), 1849–1858.

Althaus, A., Kunze, E., Sanford, T.B., 2003. Internal tide radiation from Mendocino Escarpment. J. Phys. Oceanogr. 33 (7), 1510–1527.

Ansong, J.K., Arbic, B.K., Simmons, H.L., Alford, M.H., Buijsman, M.C., Timko, P.G., Richman, J.G., Shriver, J.F., Wallcraft, A.J., 2018. Geographical distribution of diurnal and semidiurnal parametric subharmonic instability in a global ocean circulation model. J. Phys. Oceanogr. 48 (6), 1409–1431.

Arbic, B.K., Alford, M.H., Ansong, J.K., Buijsman, M.C., Ciotti, R.B., Farrar, J.T., Hallberg, R.W., Henze, C.E., Hill, C.N., Luecke, C.A., et al., 2018. Primer on global internal tide and internal gravity wave continuum modeling in HYCOM and MITgcm. In: New Frontiers in Operational Oceanography, pp. 307–392.

Arbic, B.K., Garrett, C.J.R., 2010. A coupled oscillator model of shelf and ocean tides. Cont. Shelf Res. 30 (6), 564–574.

Baines, P.G., 1973. The generation of internal tides by flat-bump topography. Deep-Sea Res. 20, 179–205.

Baines, P.G., 1982. On internal tide generation models. Deep-Sea Res., A, Oceanogr. Res. Pap. 29 (3), 307–338.

Baines, P.G., 1997. Topographic Effects in Stratified Flows. Cambridge University Press.

Baines, P.G., 2007. Internal tide generation by seamounts. Deep-Sea Res., Part 1 54 (9), 1486–1508.
Ball, F., 1964. Energy transfer between external and internal gravity waves. J. Fluid Mech. 19 (3), 465.
Balmforth, N.J., Ierley, G., Young, W.R., 2002. Tidal conversion by subcritical topography. J. Phys. Oceanogr. 32, 2900–2914.
Balmforth, N.J., Peacock, T., 2009. Tidal conversion by supercritical topography. J. Phys. Oceanogr. 39 (8), 1965–1974.
Barkan, R., Winters, K.B., McWilliams, J.C., 2017. Stimulated imbalance and the enhancement of eddy kinetic energy dissipation by internal waves. J. Phys. Oceanogr. 47 (1), 181–198.
Bell, T.H., 1975a. Lee waves in stratified flows with simple harmonic time dependence. J. Fluid Mech. 67 (4), 705–722.
Bell, T.H., 1975b. Topographically generated internal waves in the open ocean. J. Geophys. Res., 1–8.
Bourget, B., Dauxois, T., Joubaud, S., Odier, P., 2013. Experimental study of parametric subharmonic instability for internal waves. ArXiv preprint. arXiv:1302.0178.
Bühler, O., 2009. Waves and Mean Flows. Cambridge University Press.
Buhler, O., Holmes-Cerfon, M., 2011. Decay of an internal tide due to random topography in the ocean. J. Fluid Mech. 678, 271.
Buijsman, M.C., Arbic, B.K., Richman, J.G., Shriver, J.F., Wallcraft, A.J., Zamudio, L., 2017. Semidiurnal internal tide incoherence in the equatorial Pacific. J. Geophys. Res., Oceans 122 (7), 5286–5305.
Caillol, P., Zeitlin, V., 2000. Kinetic equations and stationary energy spectra of weakly nonlinear internal gravity waves. Dyn. Atmos. Ocean. 32 (2), 81–112.
Cairns, J.L., Williams, G.O., 1976. Internal wave observations from a midwater float, 2. J. Geophys. Res. 81 (12), 1943–1950.
Carter, G.S., Gregg, M.C., Merrifield, M.A., 2006. Flow and mixing around a small seamount on Kaena Ridge, Hawaii. J. Phys. Oceanogr. 36 (6), 1036–1052.
Carter, G.S., Merrifield, M.A., Becker, J.M., Katsumata, K., Gregg, M.C., Luther, D.S., Levine, M.D., Boyd, T.J., Firing, Y.L., 2008. Energetics of M2 barotropic-to-baroclinic tidal conversion at the Hawaiian Islands. J. Phys. Oceanogr. 38 (10), 2205–2223.
Chiswell, S.M., 2002. Energy levels, phase, and amplitude modulation of the baroclinic tide off Hawaii. J. Phys. Oceanogr. 32, 2640–2651.
Clément, L., Frajka-Williams, E., Sheen, K.L., Brearley, J.A., Garabato, A.C.N., 2016. Generation of internal waves by eddies impinging on the western boundary of the North Atlantic. J. Phys. Oceanogr. 46 (4), 1067–1079.
Cole, S.T., Rudnick, D.L., Hodges, B.A., Martin, J.P., 2009. Observations of tidal internal wave beams at Kauai Channel, Hawaii. J. Phys. Oceanogr. 39 (2), 421–436.
Cox, C.S., Sandstrom, H., 1962. Coupling of internal and surface waves in water of variable depth. J. Oceanogr. Soc. Jpn. 18, 499–513.
Cusack, J.M., Naveira Garabato, A.C., Smeed, D.A., Girton, J.B., 2017. Observation of a large lee wave in the Drake Passage. J. Phys. Oceanogr. 47 (4), 793–810.
Dauxois, T., Joubaud, S., Odier, P., Venaille, A., 2018. Instabilities of internal gravity wave beams. Annu. Rev. Fluid Mech. 50, 131–156.
de Lavergne, C., Falahat, S., Madec, G., Roquet, F., Nycander, J., Vic, C., 2019. Toward global maps of internal tide energy sinks. Ocean Model. 137, 52–75.
de Lavergne, C., Vic, C., Madec, G., Roquet, F., Waterhouse, A., Whalen, C., Cuypers, Y., Bouruet-Aubertot, P., Ferron, B., Hibiya, T., 2020. A parameterization of local and remote tidal mixing. J. Adv. Model. Earth Syst. 12 (5).
Duda, T.F., Lin, Y.-T., Buijsman, M., Newhall, A.E., 2018. Internal tidal modal ray refraction and energy ducting in baroclinic Gulf Stream currents. J. Phys. Oceanogr. 48 (9), 1969–1993.
Dunphy, M., Lamb, K.G., 2014. Focusing and vertical mode scattering of the first mode internal tide by mesoscale eddy interaction. J. Geophys. Res., Oceans 119 (1), 523–536.
Dunphy, M., Ponte, A.L., Klein, P., Le Gentil, S., 2017. Low-mode internal tide propagation in a turbulent eddy field. J. Phys. Oceanogr. 47 (3), 649–665.
Dushaw, B.D., Howe, B.M., Cornuelle, B.D., Worcester, P.F., Luther, D.S., 1995. Barotropic and baroclinic tides in the central North Pacific Ocean determined from long-range reciprocal acoustic transmissions. J. Phys. Oceanogr. 25 (4), 631–647.
Eden, C., Czeschel, L., Olbers, D., 2014. Toward energetically consistent ocean models. J. Phys. Oceanogr. 44 (12), 3160–3184.
Eden, C., Olbers, D., 2014. An energy compartment model for propagation, nonlinear interaction, and dissipation of internal gravity waves. J. Phys. Oceanogr. 44 (8), 2093–2106.
Eden, C., Olbers, D., Eriksen, T., 2020. A closure for lee wave drag on the large-scale ocean circulation, J. Phys. Oceanogr. In revision.
Eden, C., Pollmann, F., Olbers, D., 2019. Numerical evaluation of energy transfers in internal gravity wave spectra of the ocean. J. Phys. Oceanogr. 49 (3), 737–749.
Egbert, G.D., Ray, R.D., 2003. Semi-diurnal and diurnal tidal dissipation from TOPEX/Poseidon altimetry. Geophys. Res. Lett. 30 (17).
Falahat, S., Nycander, J., 2015. On the generation of bottom-trapped internal tides. J. Phys. Oceanogr. 45 (2), 526–545.
Falahat, S., Nycander, J., Roquet, F., Zarroug, M., 2014. Global calculation of tidal energy conversion into vertical normal modes. J. Phys. Oceanogr. 44 (12), 3225–3244.
Ferrari, R., Wunsch, C., 2009. Ocean circulation kinetic energy: reservoirs, sources, and sinks. Annu. Rev. Fluid Mech. 41 (1), 253–282.
Furuichi, N., Hibiya, T., Niwa, Y., 2005. Bispectral analysis of energy transfer within the two-dimensional oceanic internal wave field. J. Phys. Oceanogr. 35 (11), 2104–2109.
Garrett, C.J.R., Kunze, E., 2007. Internal tide generation in the deep ocean. Annu. Rev. Fluid Mech. 39 (1), 57–87.
Garrett, C.J.R., Munk, W.H., 1972. Space-time scales of internal waves. Geophys. Astrophys. Fluid 3 (1), 225–264.
Garrett, C.J.R., Munk, W.H., 1975. Space-time scales of internal waves: a progress report. J. Geophys. Res. 80 (3), 291–297.
Gayen, B., Sarkar, S., 2011. Direct and large-eddy simulations of internal tide generation at a near-critical slope. J. Fluid Mech. 681, 48–79.
Gill, A.E., 1982. Atmosphere-Ocean Dynamics, Vol. 30. Academic Press.

Goff, J., Arbic, B.K., 2010. Global prediction of abyssal hill roughness statistics for use in ocean models from digital maps of paleo-spreading rate, paleo-ridge orientation, and sediment thickness. Ocean Model. 32 (1–2), 36–43.

Goff, J.A., Jordan, T.H., 1988. Stochastic modeling of seafloor morphology: inversion of sea beam data for second-order statistics. J. Geophys. Res., Solid Earth 93 (B11), 13589–13608.

Gong, Y., Rayson, M.D., Jones, N.L., Ivey, G.N., 2019. The effects of remote internal tides on continental slope internal tide generation. J. Phys. Oceanogr. 49 (6), 1651–1668.

Gouretski, V., 2018. World ocean circulation experiment-Argo global hydrographic climatology. Ocean. Sci. 14 (5).

Gregg, M.C., 1989. Scaling turbulent dissipation in the thermocline. J. Geophys. Res.

Hasselmann, K., 1962. On the non-linear energy transfer in a gravity-wave spectrum Part 1. General theory. J. Fluid Mech. 12 (4), 481–500.

Hasselmann, K., 1967a. Nonlinear interactions treated by the methods of theoretical physics (with application to the generation of waves by wind). Proc. R. Soc. Lond. Ser. A, Math. Phys. Sci. 299 (1456), 77–103.

Hasselmann, K., 1967b. Weak-Interaction Theory of Ocean Waves. Schriftenreihe Schiffbau.

Hazewinkel, J., Winters, K., 2011. PSI of the internal tide on a β-plane: flux divergence and near-inertial wave propagation. J. Phys. Oceanogr. 41 (9), 1673–1682.

Henyey, F., Wright, J., Flatte, S., 1986. Energy and action flow through the internal wave field: an eikonal approach. J. Geophys. Res.

Hibiya, T., Nagasawa, M., 2004. Latitudinal dependence of diapycnal diffusivity in the thermocline estimated using a finescale parameterization. Geophys. Res. Lett. 31 (1).

Hibiya, T., Nagasawa, M., Niwa, Y., 2002. Nonlinear energy transfer within the oceanic internal wave spectrum at mid and high latitudes. J. Geophys. Res. 107 (C11), 28-1–28-8.

Jayne, S.R., 2009. The impact of abyssal mixing parameterizations in an ocean general circulation model. J. Phys. Oceanogr. 39 (7), 1756–1775.

Jayne, S.R., St Laurent, L.C., 2001. Parameterizing tidal dissipation over rough topography. Geophys. Res. Lett. 28 (5), 811–814.

Johnston, T.M.S., Rudnick, D.L., Kelly, S.M., 2015. Standing internal tides in the Tasman Sea observed by gliders. J. Phys. Oceanogr. 45 (11), 2715–2737.

Johnston, T.S., MacKinnon, J.A., Colin, P.L., Haley Jr, P.J., Lermusiaux, P.F., Lucas, A.J., Merrifield, M.A., Merrifield, S.T., Mirabito, C., Nash, J.D., Ou, C.Y., Siegelman, M., Terrill, E.J., Waterhouse, A.F., 2019. Energy and momentum lost to wake eddies and lee waves generated by the North equatorial current and tidal flows at Peleliu, Palau. Oceanography 32. https://doi.org/10.5670/oceanog.2019.417.

Kang, D., 2010. Energetics and dynamics of internal tides in Monterey Bay using numerical simulations. Ph.D. thesis. Stanford University.

Kang, D., Fringer, O.B., 2012. Energetics of barotropic and baroclinic tides in the Monterey Bay area. J. Phys. Oceanogr. 42 (2), 272–290.

Kantha, L.H., Tierney, C.C., 1997. Global baroclinic tides. Prog. Oceanogr. 40 (1), 163–178.

Karimi, H., Akylas, T., 2014. Parametric subharmonic instability of internal waves: locally confined beams versus monochromatic wavetrains. J. Fluid Mech. 757, 381.

Kelly, S.M., Jones, N.L., Nash, J.D., Waterhouse, A.F., 2013. The geography of semidiurnal mode-1 internal-tide energy loss. Geophys. Res. Lett. 40, 1–5.

Kelly, S.M., Lermusiaux, P.F., Duda, T.F., Haley Jr, P.J., 2016. A coupled-mode shallow-water model for tidal analysis: internal tide reflection and refraction by the Gulf Stream. J. Phys. Oceanogr. 46 (12), 3661–3679.

Kelly, S.M., Nash, J.D., 2010. Internal-tide generation and destruction by shoaling internal tides. Geophys. Res. Lett. 37 (23), L23611.

Kelly, S.M., Nash, J.D., Martini, K.I., Alford, M.H., Kunze, E., 2012. The cascade of tidal energy from low to high modes on a continental slope. J. Phys. Oceanogr. 42 (7), 1217–1232.

Khatiwala, S., 2003. Generation of internal tides in an ocean of finite depth: analytical and numerical calculations. Deep-Sea Res., Part 1 50, 1–19.

Klymak, J., Legg, S., Pinkel, R., 2010. High-mode stationary waves in stratified flow over large obstacles. J. Fluid Mech. 644, 321–336.

Klymak, J., Moum, J.N., Nash, J.D., Kunze, E., Girton, J.B., Carter, G.S., Lee, C., Sanford, T.B., Gregg, M.C., 2006. An estimate of tidal energy lost to turbulence at the Hawaiian Ridge. J. Phys. Oceanogr. 36, 1148–1164.

Klymak, J., Pinkel, R., Rainville, L., 2008. Direct breaking of the internal tide near topography: Kaena Ridge, Hawaii. J. Phys. Oceanogr. 38 (2), 380–399.

Klymak, J.M., 2018. Nonpropagating form drag and turbulence due to stratified flow over large-scale abyssal hill topography. J. Phys. Oceanogr. 48 (10), 2383–2395.

Klymak, J.M., Buijsman, M., Legg, S., Pinkel, R., 2013. Parameterizing surface and internal tide scattering and breaking on supercritical topography: the one- and two-ridge cases. J. Phys. Oceanogr. 43, 1380–1397.

Klymak, J.M., Simmons, H.L., Braznikov, D., Kelly, S., MacKinnon, J.A., Alford, M.H., Pinkel, R., Nash, J.D., 2016. Reflection of linear internal tides from realistic topography: the Tasman continental slope. J. Phys. Oceanogr. 46 (11), 3321–3337.

Korobov, A., Lamb, K., 2008. Interharmonics in internal gravity waves generated by tide-topography interaction. J. Fluid Mech. 611, 61–95.

Kunze, E., 1985. Near-inertial wave propagation in geostrophic shear. J. Phys. Oceanogr. 15 (5), 544–565.

Kunze, E., Firing, E., Hummon, J., 2006. Global abyssal mixing inferred from lowered ADCP shear and CTD strain profiles. J. Phys. Oceanogr.

Kunze, E., Lien, R.-C., 2019. Energy sinks for lee waves in shear flow. J. Phys. Oceanogr. 49 (11), 2851–2865.

Kunze, E., Rosenfeld, L.K., Carter, G.S., Gregg, M.C., 2002. Internal waves in Monterey Submarine Canyon. J. Phys. Oceanogr. 32, 1890–1913.

Kurapov, A.L., Egbert, G.D., Allen, J., Miller, R.N., Erofeeva, S.Y., Kosro, P., 2003. The M2 internal tide off Oregon: inferences from data assimilation. J. Phys. Oceanogr. 33 (8), 1733–1757.

Labreuche, P., Le Sommer, J., Staquet, C., 2020. Energy pathways of internal waves generated by geostrophic motions over small-scale topography. In preparation.

Lamb, K.G., 2004. Nonlinear interaction among internal wave beams generated by tidal flow over supercritical topography. Geophys. Res. Lett. 31 (9).

Lamb, K.G., 2014. Internal wave breaking and dissipation mechanisms on the continental slope/shelf. Annu. Rev. Fluid Mech. 46, 231–254.

Lamb, K.G., Dunphy, M., 2018. Internal wave generation by tidal flow over a two-dimensional ridge: energy flux asymmetries induced by a steady surface trapped current. J. Fluid Mech. 836, 192–221.

Le Boyer, A., Alford, M.H., 2021. Variability and sources of the internal wave continuum examined from global moored velocity records. J. Phys. Oceanogr.

Ledwell, J., Montgomery, E., Polzin, K.L., St Laurent, L.C., Schmitt, R., Tootle, J., 2000. Evidence for enhanced mixing over rough topography in the abyssal ocean. Nature.

Lefauve, A., Muller, C., Melet, A., 2015. A three-dimensional map of tidal dissipation over abyssal hills. J. Geophys. Res., Oceans 120 (7), 4760–4777.

Legg, S., 2014. Scattering of low-mode internal waves at finite isolated topography. J. Phys. Oceanogr. 44 (1), 359–383.

Legg, S., 2021. Mixing by oceanic lee waves. Annu. Rev. Fluid Mech. 53.

Legg, S., Adcroft, A., 2003. Internal wave breaking at concave and convex continental slopes*. J. Phys. Oceanogr. 33 (11), 2224–2246.

Legg, S., Klymak, J., 2008. Internal hydraulic jumps and overturning generated by tidal flow over a tall steep ridge. J. Phys. Oceanogr. 38 (9), 1949–1964.

Li, B., Cao, A., Lü, X., 2018. Observations of near-inertial waves induced by parametric subharmonic instability. J. Oceanol. Limnol. 36 (3), 641–650.

Li, Q., Mao, X., Huthnance, J., Cai, S., Kelly, S., 2019. On internal waves propagating across a geostrophic front. J. Phys. Oceanogr. 49 (5), 1229–1248.

Li, Z., Storch, J.-S.v., Müller, M., 2015. The M2 internal tide simulated by a 1/10° OGCM. J. Phys. Oceanogr. 45 (12), 3119–3135.

Liao, G., Yuan, Y., Yang, C., Chen, H., Wang, H., Huang, W., 2012. Current observations of internal tides and parametric subharmonic instability in Luzon Strait. Atmos.-Ocean 50 (sup1), 59–76.

Lien, R.-C., Gregg, M., 2001. Observations of turbulence in a tidal beam and across a coastal ridge. J. Geophys. Res., Oceans 106 (C3), 4575–4591.

Llewellyn Smith, S.G., Young, W.R., 2002. Conversion of the barotropic tide. J. Phys. Oceanogr. 32, 1554–1566.

Llewellyn Smith, S.G., Young, W.R., 2003. Tidal conversion at a very steep ridge. J. Fluid Mech. 495, 175–191.

Long, R.R., 1953. Some aspects of the flow of stratified fluids: I. A theoretical investigation. Tellus 5 (1), 42–58.

Lvov, Y.V., Polzin, K.L., Yokoyama, N., 2012. Resonant and near-resonant internal wave interactions. J. Phys. Oceanogr. 42 (5), 669–691.

MacKinnon, J., Alford, M.H., Sun, O., Pinkel, R., Zhao, Z., Klymak, J., 2013. Parametric subharmonic instability of the internal tide at 29 n. J. Phys. Oceanogr. 43 (1), 17–28.

MacKinnon, J., Gregg, M., 2003. Shear and baroclinic energy flux on the summer New England shelf. J. Phys. Oceanogr. 33 (7), 1462–1475.

MacKinnon, J.A., Winters, K.B., 2005. Subtropical catastrophe: significant loss of low-mode tidal energy at 28.9°. Geophys. Res. Lett. 32 (15).

MacKinnon, J.A., Zhao, Z., Whalen, C.B., Waterhouse, A.F., Trossman, D.S., Sun, O.M., St. Laurent, L.C., Simmons, H.L., Polzin, K., Pinkel, R., et al., 2017. Climate process team on internal wave–driven ocean mixing. Bull. Am. Meteorol. Soc. 98 (11), 2429–2454.

Marques, O.B., Alford, M.H., Pinkel, R., MacKinnon, J.A., Klymak, J.M., Nash, J.D., Waterhouse, A.F., Kelly, S.M., Simmons, H.L., Braznikov, D., 2020. Internal tide structure and temporal variability on the reflective continental slope of southeastern Tasmania. J. Phys. Oceanogr.

Martini, K.I., Alford, M.H., Kunze, E., Kelly, S.M., Nash, J.D., 2013. Internal bores and breaking internal tides on the Oregon continental slope. J. Phys. Oceanogr. 43 (1), 120–139.

Maurer, P., Joubaud, S., Odier, P., 2016. Generation and stability of inertia–gravity waves. J. Fluid Mech. 808, 539–561.

Mayer, F., Fringer, O., 2017. An unambiguous definition of the Froude number for lee waves in the deep ocean. J. Fluid Mech. 831.

McComas, C., Bretherton, F.P., 1977. Resonant interaction of oceanic internal waves. J. Geophys. Res. 82 (9), 1397–1412.

McComas, C., Müller, P., 1981. The dynamic balance of internal waves. J. Phys. Oceanogr.

Melet, A., Hallberg, R., Adcroft, A., Nikurashin, M., Legg, S., 2015. Energy flux into internal lee waves: sensitivity to future climate changes using linear theory and a climate model. J. Climate 28 (6), 2365–2384.

Melet, A., Hallberg, R., Legg, S., Nikurashin, M., 2014. Sensitivity of the ocean state to lee wave–driven mixing. J. Phys. Oceanogr. 44 (3), 900–921.

Melet, A., Nikurashin, M., Muller, C., Falahat, S., Nycander, J., Timko, P.G., Arbic, B.K., Goff, J.A., 2013a. Internal tide generation by abyssal hills using analytical theory. J. Geophys. Res., Oceans 118, 1–16.

Melet, A., Venayagamoorthy, S.W., Legg, S., Polzin, K.L., 2013b. Sensitivity of the ocean state to the vertical distribution of internal-tide-driven mixing. J. Phys. Oceanogr. 43 (3), 602–615.

Merrifield, M.A., Holloway, P., 2002. Model estimates of M2 internal tide energetics at the Hawaiian Ridge. J. Geophys. Res. 107 (C8).

Meyer, A., Polzin, K.L., Sloyan, B.M., Phillips, H.E., 2015. Internal waves and mixing near the Kerguelen Plateau. J. Phys. Oceanogr. 46 (2), 417–437.

Müller, M., Arbic, B.K., Richman, J.G., Shriver, J.F., Kunze, E.L., Scott, R.B., Wallcraft, A.J., Zamudio, L., 2015. Toward an internal gravity wave spectrum in global ocean models. Geophys. Res. Lett. 42 (9), 3474–3481.

Müller, M., Cherniawsky, J.Y., Foreman, M., Storch, J.S., 2012. Global M2 internal tide and its seasonal variability from high resolution ocean circulation and tide modeling. Geophys. Res. Lett. 39 (19).

Müller, P., Holloway, G., Henyey, F., Pomphrey, N., 1986. Nonlinear interactions among internal gravity waves. Rev. Geophys. 24 (3), 493–536.

Müller, P., Liu, X., 2000. Scattering of internal waves at finite topography in two dimensions. Part I: theory and case studies. J. Phys. Oceanogr. 30 (3), 532–549.

Müller, P., Natarov, A., 2003. The internal wave action model (IWAM). In: Proceedings, Aha Huliko'a Hawaiian Winter Workshop, School of Ocean and Earth Science and Technology, Special Publication. In: Citeseer, pp. 95–105.

Müller, P., Olbers, D.J., 1975. On the dynamics of internal waves in the deep ocean. J. Geophys. Res. 80 (27), 3848–3860.

Müller, P., Olbers, D.J., Willebrand, J., 1978. The IWEX spectrum. J. Geophys. Res., Oceans 83 (C1), 479–500.

Müller, P., Xu, N., 1992. Scattering of oceanic internal gravity waves off random bottom topography. J. Phys. Oceanogr. 22 (5), 474–488.

Munk, W., 1981. Internal waves and small-scale processes. In: Evolution of Physical Oceanography, pp. 264–291.

Munk, W.H., 1997. Once again: once again–tidal friction. Prog. Oceanogr.

Musgrave, R.C., MacKinnon, J.A., Pinkel, R., Waterhouse, A.F., Nash, J., Kelly, S.M., 2017. The influence of subinertial internal tides on near-topographic turbulence at the Mendocino Ridge: observations and modeling. J. Phys. Oceanogr. 47, 2139–2154.

Musgrave, R.C., MacKinnon, J.A., Pinkel, R., Waterhouse, A.F., Nash, J.D., 2016. Tidally driven processes leading to near-field turbulence in a channel at the crest of the Mendocino Escarpment. J. Phys. Oceanogr. 46, 1137–1155.

Nagasawa, M., Hibiya, T., Niwa, Y., Watanabe, M., Isoda, Y., Takagi, S., Kamei, Y., 2002. Distribution of fine-scale shear in the deep waters of the North Pacific obtained using expendable current profilers. J. Geophys. Res., Oceans 107 (C12), 14-1.

Nash, J.D., Kunze, E., Lee, C.M., Sanford, T.B., 2006. Structure of the baroclinic tide generated at Kaena Ridge, Hawaii. J. Phys. Oceanogr. 36, 1123–1135.

Nash, J.D., Kunze, E., Toole, J.M., Schmitt, R.W., 2004. Internal tide reflection and turbulent mixing on the continental slope. J. Phys. Oceanogr. 34 (5), 1117–1134.

Nielsen, S.B., Jochum, M., Eden, C., Nuterman, R., 2018. An energetically consistent vertical mixing parameterization in CCSM4. Ocean Model. 127, 46–54.

Nikurashin, M., Ferrari, R., 2010a. Radiation and dissipation of internal waves generated by geostrophic motions impinging on small-scale topography: application to the Southern Ocean. J. Phys. Oceanogr. 40 (9), 2025–2042.

Nikurashin, M., Ferrari, R., 2010b. Radiation and dissipation of internal waves generated by geostrophic motions impinging on small-scale topography: theory. J. Phys. Oceanogr. 40 (5), 1055–1074.

Nikurashin, M., Ferrari, R., 2011. Global energy conversion rate from geostrophic flows into internal lee waves in the deep ocean. Geophys. Res. Lett. 38 (8).

Nikurashin, M., Ferrari, R., 2013. Overturning circulation driven by breaking internal waves in the deep ocean. Geophys. Res. Lett. 40.

Nikurashin, M., Ferrari, R., Grisouard, N., Polzin, K.L., 2014. The impact of finite-amplitude bottom topography on internal wave generation in the Southern Ocean. J. Phys. Oceanogr. 44 (11), 2938–2950.

Nikurashin, M., Legg, S., 2011. A mechanism for local dissipation of internal tides generated at rough topography. J. Phys. Oceanogr. 41 (2), 378–395.

Niwa, Y., Hibiya, T., 2011. Estimation of baroclinic tide energy available for deep ocean mixing based on three-dimensional global numerical simulations. J. Oceanogr. 67 (4), 493.

Nycander, J., 2005. Generation of internal waves in the deep ocean by tides. J. Geophys. Res. 110 (C10), C10028.

Nycander, J., 2006. Tidal generation of internal waves from a periodic array of steep ridges. J. Fluid Mech. 567, 415–432.

Olbers, D., Eden, C., 2013. A global model for the diapycnal diffusivity induced by internal gravity waves. J. Phys. Oceanogr. 43 (8), 1759–1779.

Olbers, D., Pollmann, F., Eden, C., 2020. On PSI interactions in internal gravity wave fields and the decay of baroclinic tides. J. Phys. Oceanogr. 50 (3), 751–771.

Olbers, D.J., 1976. Nonlinear energy transfer and the energy balance of the internal wave field in the deep ocean. J. Fluid Mech. 74 (2), 375–399.

Olbers, D.J., 1983. Models of the oceanic internal wave field. Rev. Geophys. 21 (7), 1567–1606.

Onuki, Y., Hibiya, T., 2015. Excitation mechanism of near-inertial waves in baroclinic tidal flow caused by parametric subharmonic instability. Ocean Dyn. 65 (1), 107–113.

Osborn, T.R., 1980. Estimates of the local rate of vertical diffusion from dissipation measurements. J. Phys. Oceanogr.

Padman, L., Plueddemann, A.J., Muench, R.D., Pinkel, R., 1992. Diurnal tides near the Yermak Plateau. J. Geophys. Res. 97 (C8), 12639–12652.

Perfect, B., Kumar, N., Riley, J., 2020a. Energetics of seamount wakes. Part I: energy exchange. J. Phys. Oceanogr. 50 (5), 1365–1382.

Perfect, B., Kumar, N., Riley, J., 2020b. Energetics of seamount wakes. Part II: wave fluxes. J. Phys. Oceanogr. 50 (5), 1383–1398.

Pétrélis, F., Llewellyn Smith, S.G., Young, W.R., 2006. Tidal conversion at a submarine ridge. J. Phys. Oceanogr.

Phillips, O., 1960. On the dynamics of unsteady gravity waves of finite amplitude Part 1. The elementary interactions. J. Fluid Mech. 9 (2), 193–217.

Phillips, O., 1966. The Dynamics of the Upper Ocean. Cambridge Univ. Press, New York, États-Unis.

Pickering, A., Alford, M.H., 2012. Velocity structure of internal tide beams emanating from Kaena Ridge, Hawaii. J. Phys. Oceanogr. 42 (6), 1039–1044.

Pollmann, F., 2020. Global characterization of the ocean's internal wave spectrum. J. Phys. Oceanogr. 50 (7), 1871–1891.

Pollmann, F., Eden, C., Olbers, D., 2017. Evaluating the global internal wave model IDEMIX using finestructure methods. J. Phys. Oceanogr. 47 (9), 2267–2289.

Polzin, K., 2004a. A heuristic description of internal wave dynamics. J. Phys. Oceanogr. 34 (1), 214–230.

Polzin, K.L., 2004b. Idealized solutions for the energy balance of the finescale internal wave field. J. Phys. Oceanogr. 34, 231–246.

Polzin, K.L., 2009. An abyssal recipe. Ocean Model. 30 (4), 298–309.

Polzin, K.L., 2010. Mesoscale eddy–internal wave coupling. Part II: energetics and results from PolyMode. J. Phys. Oceanogr. 40 (4), 789–801.

Polzin, K.L., Lvov, Y.V., 2011. Towards regional characterizations of the oceanic internal wavefield. Rev. Geophys. 49 (4).

Polzin, K.L., Naveira Garabato, A.C., Huussen, T.N., Sloyan, B.M., Waterman, S., 2014. Finescale parameterizations of turbulent dissipation. J. Geophys. Res., Oceans 119 (2), 1383–1419.

Polzin, K.L., Toole, J.M., Ledwell, J.R., Schmitt, R.W., 1997. Spatial variability of turbulent mixing in the abyssal ocean. Science 276 (5309), 93–96.

Polzin, K.L., Toole, J.M., Schmitt, R.W., 1995. Finescale parameterizations of turbulent dissipation. J. Phys. Oceanogr. 25 (3), 306–328.

Pomphrey, N., Meiss, J.D., Watson, K.M., 1980. Description of nonlinear internal wave interactions using Langevin methods. J. Geophys. Res., Oceans 85 (C2), 1085–1094.

Ponte, A.L., Klein, P., 2015. Incoherent signature of internal tides on sea level in idealized numerical simulations. Geophys. Res. Lett. 42, 1520–1526.

Rainville, L., Pinkel, R., 2006. Propagation of low-mode internal waves through the ocean. J. Phys. Oceanogr. 36 (6), 1220–1236.

Ray, R.D., Cartwright, D.E., 2001. Estimates of internal tide energy fluxes from Topex/Poseidon altimetry: Central North Pacific. Geophys. Res. Lett. 28 (7), 1259–1262.

Ray, R.D., Eanes, R.J., Chao, B.F., 1996. Detection of tidal dissipation in the solid Earth by satellite tracking and altimetry. Nature 381, 595–597.

Ray, R.D., Mitchum, G.T., 1997. Surface manifestation of internal tides in the deep ocean: observations from altimetry and island gauges. Prog. Oceanogr. 40, 135–162.

Richet, O., Muller, C., Chomaz, J.-M., 2017. Impact of a mean current on the internal tide energy dissipation at the critical latitude. J. Phys. Oceanogr. 47 (6), 1457–1472.

Rocha, C.B., Chereskin, T.K., Gille, S.T., Menemenlis, D., 2016. Mesoscale to submesoscale wavenumber spectra in Drake Passage. J. Phys. Oceanogr. 46 (2), 601–620.

Rocha, C.B., Wagner, G.L., Young, W.R., 2018. Stimulated generation: extraction of energy from balanced flow by near-inertial waves. J. Fluid Mech. 847, 417–451.

Roth, M., Briscoe, M.G., McComas III, C., 1981. Internal waves in the upper ocean. J. Phys. Oceanogr. 11 (9), 1234–1247.

Ruddick, B.R., Joyce, T.M., 1979. Observations of interaction between the internal wavefield and low-frequency flows in the North Atlantic. J. Phys. Oceanogr. 9 (3), 498–517.

Sarkar, S., Scotti, A., 2017. From topographic internal gravity waves to turbulence. Annu. Rev. Fluid Mech. 49, 195–220.

Savage, A.C., Waterhouse, A.F., Kelly, S.M., 2020. Internal tide nonstationarity and wave-mesoscale interactions in the Tasman Sea. J. Phys. Oceanogr. 50, 2931–2951.

Scott, R.B., Goff, J.A., Naveira Garabato, A., Nurser, A., 2011. Global rate and spectral characteristics of internal gravity wave generation by geostrophic flow over topography. J. Geophys. Res., Oceans 116 (C9).

Scotti, A., 2011. Inviscid critical and near-critical reflection of internal waves in the time domain. J. Fluid Mech. 674, 464.

Shakespeare, C.J., 2020. Interdependence of internal tide and lee wave generation at abyssal hills: global calculations. J. Phys. Oceanogr. 50 (3), 655–677.

Sheen, K., Brearley, J., Naveira Garabato, A.C., Smeed, D., Waterman, S., Ledwell, J.R., Meredith, M.P., St. Laurent, L., Thurnherr, A.M., Toole, J.M., et al., 2013. Rates and mechanisms of turbulent dissipation and mixing in the Southern Ocean: results from the diapycnal and isopycnal mixing experiment in the Southern Ocean (DIMES). J. Geophys. Res., Oceans 118 (6), 2774–2792.

Simmons, H.L., 2008. Spectral modification and geographic redistribution of the semi-diurnal internal tide. Ocean Model. 21 (3–4), 126–138.

Simmons, H.L., Jayne, S.R., St Laurent, L.C., Weaver, A.J., 2004. Tidally driven mixing in a numerical model of the ocean general circulation. Ocean Model. 6 (3–4), 245–263.

St Laurent, L.C., Garrett, C.J.R., 2002. The role of internal tides in mixing the deep ocean. J. Phys. Oceanogr. 32, 2882–2899.

St Laurent, L.C., Stringer, S., Garrett, C.J.R., Perrault-Joncas, D., 2003. The generation of internal tides at abrupt topography. Deep-Sea Res., Part 1 50 (8), 987–1003.

St. Laurent, L., Naveira Garabato, A.C., Ledwell, J.R., Thurnherr, A.M., Toole, J.M., Watson, A.J., 2012. Turbulence and diapycnal mixing in Drake Passage. J. Phys. Oceanogr. 42 (12), 2143–2152.

Staquet, C., Sommeria, J., 2002. From instabilities to turbulence. Annu. Rev. Fluid Mech. 34, 559–593.

Sutherland, B.R., Jefferson, R., 2020. Triad resonant instability of horizontally periodic internal modes. Phys. Rev. Fluids 5 (3), 034801.

Tabaei, A., Akylas, T., 2003. Nonlinear internal gravity wave beams. J. Fluid Mech. 482, 141.

Timko, P.G., Arbic, B.K., Goff, J.A., Ansong, J.K., Smith, W.H., Melet, A., Wallcraft, A.J., 2017. Impact of synthetic abyssal hill roughness on resolved motions in numerical global ocean tide models. Ocean Model. 112, 1–16.

Trossman, D.S., Arbic, B.K., Richman, J.G., Garner, S.T., Jayne, S.R., Wallcraft, A.J., 2016. Impact of topographic internal lee wave drag on an eddying global ocean model. Ocean Model. 97, 109–128.

van Haren, H., Cimatoribus, A., Gostiaux, L., 2015. Where large deep-ocean waves break. Geophys. Res. Lett. 42, 2351–2357.

Vic, C., Garabato, A.C.N., Green, J.M., Waterhouse, A.F., Zhao, Z., Mélet, A., De Lavergne, C., Buijsman, M.C., Stephenson, G.R., 2019. Deep-ocean mixing driven by small-scale internal tides. Nat. Commun. 10 (1), 1–9.

Vic, C., Naveira Garabato, A.C., Green, J.A.M., Spingys, C., Forryan, A., Zhao, Z., Sharples, J., 2018. The lifecycle of semidiurnal internal tides over the Northern Mid-Atlantic Ridge. J. Phys. Oceanogr. 48 (1), 61–80.

Voet, G., Alford, M.H., MacKinnon, J.A., Nash, J.D., 2020. Topographic form drag on tides and low-frequency flow: observations of nonlinear lee waves over a Tall Submarine Ridge near Palau. J. Phys. Oceanogr. 50, 1489–1507.

von Storch, J.-S., Badin, G., Oliver, M., 2019. The interior energy pathway: inertia-gravity wave emission by oceanic flows. In: Eden, C., Iske, A. (Eds.), Energy Transfers in Atmosphere and Ocean. Springer, pp. 53–85.

Wagner, G.L., Ferrando, G., Young, W.R., 2017. An asymptotic model for the propagation of oceanic internal tides through quasi-geostrophic flow. J. Fluid Mech. 828, 779–811.

Ward, M.L., Dewar, W.K., 2010. Scattering of gravity waves by potential vorticity in a shallow-water fluid. J. Fluid Mech. 663, 478.

Waterhouse, A.F., Kelly, S.M., Zhao, Z., MacKinnon, J.A., Nash, J.D., Simmons, H., Brahznikov, D., Rainville, L., Alford, M., Pinkel, R., 2018. Observations of the Tasman Sea internal tide beam. J. Phys. Oceanogr. 48 (6), 1283–1297.

Waterman, S., Naveira Garabato, A.C., Polzin, K.L., 2013. Internal waves and turbulence in the Antarctic circumpolar current. J. Phys. Oceanogr. 43 (2), 259–282.

Wijesekera, H., Padman, L., Dillon, T., Levine, M., Paulson, C., Pinkel, R., 1993. The application of internal-wave dissipation models to a region of strong mixing. J. Phys. Oceanogr. 23.

Winters, K.B., Armi, L., 2013. The response of a continuously stratified fluid to an oscillating flow past an obstacle. J. Fluid Mech. 727, 83–118.

Winters, K.B., D'Asaro, E.A., 1997. Direct simulation of internal wave energy transfer. J. Phys. Oceanogr. 27 (9), 1937–1945.

Wright, C.J., Scott, R.B., Ailliot, P., Furnival, D., 2014. Lee wave generation rates in the deep ocean. Geophys. Res. Lett. 41 (7), 2434–2440.

Wunsch, C., 1975. Internal tides in the ocean. Rev. Geophys. 13 (1), 167–182.

Wunsch, C., 1998. The work done by the wind on the oceanic general circulation. J. Phys. Oceanogr. 28 (11), 2332–2340.
Wunsch, C., 2015. Modern Observational Physical Oceanography: Understanding the Global Ocean. Princeton University Press.
Wunsch, C., Ferrari, R., 2004. Vertical mixing, energy and the general circulation of the oceans. Annu. Rev. Fluid Mech. 36 (1), 281–314.
Xie, X.-H., Shang, X.-D., van Haren, H., Chen, G.-Y., Zhang, Y.-Z., 2011. Observations of parametric subharmonic instability-induced near-inertial waves equatorward of the critical diurnal latitude. Geophys. Res. Lett. 38 (5).
Yang, L., Nikurashin, M., Hogg, A.M., Sloyan, B.M., 2018. Energy loss from transient eddies due to lee wave generation in the Southern Ocean. J. Phys. Oceanogr. 48 (12), 2867–2885.
Yang, W., Wei, H., Zhao, L., 2020. Parametric subharmonic instability of the semidiurnal internal tides at the East China Sea shelf slope. J. Phys. Oceanogr. 50 (4), 907–920.
Zaron, E.D., 2017. Mapping the nonstationary internal tide with satellite altimetry. J. Geophys. Res., Oceans 122 (1), 539–554.
Zaron, E.D., 2019. Baroclinic tidal sea level from exact-repeat mission altimetry. J. Phys. Oceanogr. 49 (1), 193–210.
Zaron, E.D., Egbert, G.D., 2014. Time-variable refraction of the internal tide at the Hawaiian Ridge. J. Phys. Oceanogr. 44 (2), 538–557.
Zhao, Z., Alford, M.H., 2009. New altimetric estimates of mode-1 M2 internal tides in the Central North Pacific Ocean. J. Phys. Oceanogr. 39 (7), 1669–1684.
Zhao, Z., Alford, M.H., Girton, J.B., Rainville, L., Simmons, H.L., 2016. Global observations of open-ocean mode-1 M2 internal tides. J. Phys. Oceanogr. 46 (6), 1657–1684.
Zhao, Z., Alford, M.H., MacKinnon, J.A., Pinkel, R., 2010. Long-range propagation of the semidiurnal internal tide from the Hawaiian Ridge. J. Phys. Oceanogr. 40 (4), 713–736.
Zheng, K., Nikurashin, M., 2019. Downstream propagation and remote dissipation of internal waves in the Southern Ocean. J. Phys. Oceanogr. 49 (7), 1873–1887.
Zilberman, N., 2009. Internal tide genermion over deep-ocean topography. Ph.D. thesis. University of Hawai'i.

Chapter 7

Mixing at the ocean's bottom boundary

Kurt L. Polzin[a,c] and Trevor J. McDougall[b,c]

[a]*Department of Physical Oceanography, Woods Hole Oceanographic Institution, Woods Hole, MA, United States,* [b]*School of Mathematics and Statistics, University of New South Wales, Sydney, NSW, Australia*

7.1 Introduction

Once upon a time, at the dawn of modern oceanography, there was a great influx of money and the creation of many near-continuously measuring instruments. These included instruments, such as the Neil Brown Mark III CTD (Brown, 1974), which permitted <1 m resolution of temperature, salinity and pressure; Sanford's electro-magnetic velocity profiler (Sanford, 1975), which provided, for the first time, direct observations of oceanic horizontal velocity at ten-metre vertical resolution; long-term moored observations of horizontal currents and temperature that led to a quantitative understanding of the spatial distribution of mean flows and eddy variability (via the mid-ocean dynamics experiment (MODE), Group et al. (1978); PolyMODE, Kamenkovich et al. (1986); etc.). These new observations opened a multitude of avenues to explore the question, 'How does water rise from the abyss to the bottom of the thermocline?' posed, for example, by Munk (1966), but this was by no means the origin of that question (e.g., Wyrtki, 1961, 1962).

This takes us to what we call Armi vs. Garrett, after their 1979 exchange (Armi, 1979b; Garrett, 1979). Armi and Millard (1976) and Armi (1978) were attempting to interpret steps in abyssal temperature and salinity traces returned by the new fangled Neil Brown instrument as detached mixed layers, and linking those to the issue of complex topography, Fig. 7.1. Garrett, on the other hand, promotes a seductive figure of a well-mixed boundary layer over a planar sloping bottom, Fig. 7.2. The crux of this 'conversation' can be posed as a question: 'At what rate are you mixing in boundary layers that are well mixed?' with the backdrop of, 'How important are boundary processes relative to interior mixing associated with an intermittent internal wave breaking process in the interior?' Our understanding has grown significantly since that exchange. With four decades worth of perspective, Armi was likely speaking to mixing as the end product of dynamical processes, indirectly coupled to a frictional bottom stress, and an actual theoretical paradigm has been created for the figure sketched by Garrett. The reason for *this* preamble is that the dichotomy in that exchange not only is still in play today; it structures this review. The core issue is that, however reasonable a representation of the bottom boundary layer Fig. 7.2 might seem to be, when you start to look at the ocean, one finds great sympathy for Fig. 7.1.

Munk (1966) suggests several avenues by which the global upwelling of deep waters might be accomplished. One of these is that of mixing resulting from interactions between internal waves, akin to how one might consider white capping in a surface wavefield to be the end process of interactions between surface waves, for which the theoretical underpinnings (Phillips, 1960; Hasselmann, 1962) were just beginning to be articulated. With this option one might argue internal wave

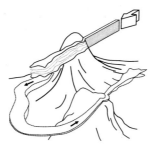

FIGURE 7.1 A schematic characterisation of flow-topography interaction, boundary layer production and flow separation. Note, amongst other things, that the ocean's bottom is not depicted as a planar sloping boundary. Figure redrawn from Armi (1979a).

c. These authors contributed equally to this work.

146 Ocean Mixing

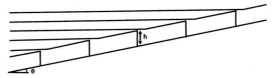

FIGURE 7.2 A schematic representation of the bottom boundary layer over a planar sloping boundary. Figure redrawn from Garrett (1979). This schematic is a prototype for more sophisticated discussions below.

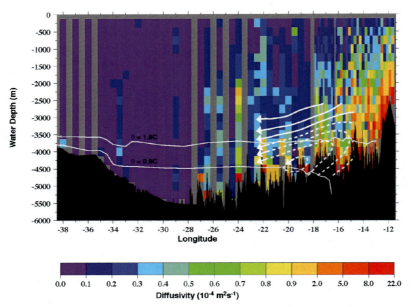

FIGURE 7.3 Depth-longitude section of diapycnal diffusivity K_ρ in the Brazil basin. The figure appears in Mauritzen et al. (2002), with data originally reported in Polzin et al. (1997) and Ledwell et al. (2000). Note the non-uniform colour map. The thin white lines mark the observed depth of the 0.8 and 1.8 C potential isotherms. The thicker white lines with arrows are a schematic representation of the stream function, estimated from an inverse calculation (St. Laurent et al., 2001), and are intended to portray the zonal overturning circulation and modification of bottom water in the Brazil basin.

breaking and mixing to be broadly distributed in space. This path takes us past the articulation of a two-dimensional internal wave spectrum (Garrett and Munk, 1972b), from which the space-time variations of mixing events can be estimated (Garrett and Munk, 1972a) and concerted efforts to provide a theoretical paradigm for non-linear interactions between internal waves (Olbers, 1976; McComas and Bretherton, 1977; Pomphrey et al., 1980). Coupled with a growing ability to measure oceanic turbulence using specialised sensors (Nasmyth, 1970; Gregg and Cox, 1971; Osborn and Siddon, 1973), theory, kinematic mixing models employing finestructure observations and direct turbulence measurements provided an evolving appreciation (Garrett, 1984) that the background internal wavefield supports a diffusivity of $1 \times 10^{-6} - 1 \times 10^{-5}$ m^2 s^{-1} in the main thermocline, an order of magnitude smaller than Munk's abyssal mixing metric of $K_\rho = 1 \times 10^{-4}$ m^2 s^{-1} required to permit the net upwelling of some 30×10^6 m^3 s^{-1}. A second avenue mentioned in Munk (1966) consists simply of a suggestion of strong mixing along the ocean boundaries due to current shear and internal wave breaking being efficiently communicated into the interior along isopycnal surfaces (Schiff, 1966). This is the paradigm that underpins Armi vs. Garrett. The viability of this boundary-mixing scenario was increasingly underscored by simple heat budgets (made possible by long-term deployments of current metre arrays (Hogg et al., 1982; Whitehead and Worthington, 1982) for abyssal basins that implied the deep ocean hosted significantly larger diapycnal diffusivities.

The box on this problem was fully opened in the context of a large multi-national field program that produced a control volume for dense water entering and exiting the deep Brazil basin (Morris et al., 2001). Many of the key issues are manifested in the iconic section of diapycnal diffusivity K_ρ estimates from the Brazil basin tracer release experiment, Fig. 7.3. Over the western side of the Brazil Basin, turbulent mixing is small ($K_\rho < 1 \times 10^{-5}$ m^2 s^{-1}) and nearly independent of depth. On the eastern side of the basin, K_ρ is an order of magnitude larger at mid-depth and increases by another order of magnitude as the bottom is approached. The key to this pattern lies in the bottom topography: On the western boundary the bottom is weakly sloping. In the east, over the Mid-Atlantic ridge, the bottom topography is steep and rough. Associated with the pattern of increasing turbulent diffusivity with depth is an increase in turbulent dissipation, which, if taken at face value, implies downwelling.[1] This inference of downwelling contradicts the net watermass budgets for the deep Brazil basin, in which more dense water enters than leaves, demanding net upwelling across isopycnals.

The focus of this review is the maintenance of the deep overturning cell: the processes and mechanisms which support the 3-D turbulence responsible for the flow across dense isopycnals and the representation of such processes in general circulation models (GCMs). There are closely allied questions of tracer dispersion and watermass transformation, horizontal circulations related to the mixing distributions and transport of mixed water from the boundary that we touch upon. We start this review by covering the things we can agree on in Section 7.2. These are equations and the importance of non-dimensional parameters in organising our thoughts. Section 7.3 reviews idealised modelling and GCM behaviour. Section 7.4 then tends to issues of observational inference directed to elucidating the idealised nature of the preceding section. After summarising in Section 7.5, we will attempt to describe pathways to answer open questions. We are aided by substantial reviews concerning the forcing of turbulence by topographic internal gravity waves (Sarkar and Scotti, 2017), the details of buoyancy fluxes in relation to turbulent kinetic energy production and dissipation in stratified flows (Gregg et al., 2018) and bottom boundary layer physics (Trowbridge and Lentz, 2018) that gratefully limit our discussion.

7.2 Common ground

The plot line for writing a story about global ocean upwelling is deceptively simple. In steady state, the velocity across an isopycnal is proportional to the divergence of the turbulent density flux, and one wants the net transformation to be equal to the rate at which dense water is formed. With boundary mixing, though, the plot line immediately branches, because we need to consider both the magnitude of that turbulent buoyancy flux and the relevant spatial scales. In this section, we gather together seemingly disparate threads to create a common language for a discussion about "boundary mixing" and elucidate ambiguities of interpretation that have arisen.

7.2.1 Equations

We start from a Reynolds decomposition of the equations of motion, in which variables ϕ have been decomposed into a mean $\overline{\phi}$ and perturbation ϕ'. This act itself carries with it a certain ambiguity: The mean could represent time invariant conditions or slowly varying subinertial times scales, or even internal waveband processes. The perturbation could represent subinertial times scales, internal waves (superinertial) or 2-dimensional and or 3-dimensional turbulence. In Section 7.4, we discuss a variety of phenomena, which imply the locus of turbulent fluxes is within the internal waveband over steep and rough topography. In the following, the three-dimensional velocity is $\mathbf{u} = (u, v, w)$, ρ is density, g is gravity, f is the Coriolis parameter relating to the local pendulum day and p is pressure. A bold font is used to indicate a vector quantity. The Reynolds averaged equations are

$$\frac{\partial \overline{u}}{\partial t} + \overline{\mathbf{u}} \cdot \nabla \overline{u} + \nabla \cdot \overline{u'\mathbf{u}'} - f\overline{v} = -\frac{1}{\rho_0} \frac{\partial \overline{p}}{\partial x}, \tag{7.1a}$$

$$\frac{\partial \overline{v}}{\partial t} + \overline{\mathbf{u}} \cdot \nabla \overline{v} + \nabla \cdot \overline{v'\mathbf{u}'} + f\overline{u} = -\frac{1}{\rho_0} \frac{\partial \overline{p}}{\partial y}, \tag{7.1b}$$

$$\frac{\partial \overline{w}}{\partial t} + \overline{\mathbf{u}} \cdot \nabla \overline{w} + \nabla \cdot \overline{w'\mathbf{u}'} = -\frac{1}{\rho_0} \frac{\partial \overline{p}}{\partial z} - \overline{b}, \tag{7.1c}$$

$$\frac{\partial \overline{\rho}}{\partial t} + \overline{\mathbf{u}} \cdot \nabla \overline{\rho} + \nabla \cdot \overline{\rho'\mathbf{u}'} = 0, \tag{7.1d}$$

in which variations of density in the momentum equations are retained only when they appear in terms multiplied by gravity g; buoyancy is $b = -g(\rho - \rho_0)/\rho_0$; and ρ_0 is a constant reference density. Eqs. (7.1) are complemented by a continuity equation, $\nabla \cdot \mathbf{u} = 0$. The direct impacts of molecular processes are assumed to be small with respect to the effects of turbulent transports: there is an inner scale of turbulent microstructure, where gradient variances are dissipated by molecular processes. This scale is much smaller than the 'outer' scales that characterise the turbulent fluxes.

A common device is to describe the net effects of turbulent fluxes $\mathbf{F}_\phi$ using flux-gradient relations, which in turn has an intellectual grounding in mixing length theory. In one dimension,

$$F_\phi \equiv \overline{u'_\perp \phi'} = -K_\phi \nabla_\perp \overline{\phi} \qquad \text{flux-gradient relation,} \tag{7.2a}$$

---

[1]. Not that this is obvious. If turbulent intensity varies in space, a tracer's centre of mass will be drawn in the direction of more intense turbulence. See Section 7.2.3 for discussions about the flows across density surfaces driven by variations in turbulent intensity.

$$\phi' = \ell_{\text{mix}} \nabla_\perp \overline{\phi} \qquad \text{mixing length,} \tag{7.2b}$$

in which $u'_\perp$ is normal to mean isopleths of $\overline{\phi}$; $\nabla_\perp \overline{\phi}$ is the normal gradient and ℓ_{mix} the 'mixing length'. This comes with important caveats: it is a 'local' closure relating K_ϕ to the gradient at that point. As such it is unlikely to be an accurate summary of convection set up in a downwelling Ekman layer (Section 7.4.3) or near-boundary wave overturning relating to linear internal wave kinematics (Section 7.4.1) rather than dynamical instabilities, or situations in which turbulent transport terms contribute to a scalar variance budget (Section 7.2.5.2).[2]

A further equation describes the evolution of a passive tracer, whose concentration we will denote by C[3]:

$$\frac{\partial \overline{C}}{\partial t} + \overline{\mathbf{u}} \cdot \nabla \overline{C} + \nabla \cdot \overline{C' \mathbf{u}'} = 0. \tag{7.3}$$

7.2.2 Boundary conditions

One of the more bedeviling issues with boundary mixing is that there is not simply one process or one relevant boundary layer height scale. Rather, there is a hierarchy as one considers larger and longer space-time scales and the dynamical balances that dictate the boundary layer structure change. Molecular viscosity acts to bring the velocity at the boundary to zero, i.e., $\mathbf{u}(x, y, z = h) = 0$, however, this happens within a viscous sublayer on scales smaller than one centimetre embedded within a log-layer, in which the only dynamically relevant lengthscale is the distance from the boundary, (e.g., Tennekes and Lumley, 1972). In the far-field of this log-layer, in excess of 1000 times wall unit lengths of ν/u_* with molecular viscosity ν and 'friction velocity' $u_* = \sqrt{\tau/\rho}$ dependent upon the bottom stress τ, one has a situation in which the velocity profile joins the 'free-stream'. The object of turbulence modelling with full-blown closure schemes (e.g., Weatherly and Martin, 1978; Taylor and Sarkar, 2008) is to estimate the evolution of turbulent kinetic energy and turbulent lengthscales as a function of height above bottom, from which the momentum and buoyancy fluxes, and thus eddy viscosity/diffusivity (7.2) follow. This turbulence modelling paradigm permits, in particular, the *prediction* of eddy viscosity / diffusivity profiles in response to stratification, rotation, or pressure gradients as a function of height above bottom.

The theoretical work discussed here is of different intent. In this work, one generally *assumes* vertical profiles of eddy diffusivity and, from a constant ratio of eddy viscosity A to eddy diffusivity K (i.e., a Prandtl number), solves for the structure of the boundary layer; and then relates that structure to global ocean behaviour. This requires some fortitude. From the perspective of the planetary boundary layer, 1000 wall units is quite small. At a free stream velocity of 0.20 m s^{-1}, 1000 wall units is less than 0.2 m. There are two things to keep in mind. First, this theoretical work will use a no-flow condition on the velocity at the boundary without attempting a self-consistent structure in the vertical for the turbulent dynamics. There is a rich history in this problem of *ad hoc* assumptions about the spatial structure of eddy diffusivity and viscosity which, in combination with no-flux and no-flow boundary conditions provide a mathematically well-posed problem, from which analytic or simple numerical solutions can provide significant insight. We report on these in considerable detail in Section 7.3. There are issues, though. These assumed profiles, Prandtl numbers and boundary conditions do not correspond to the output of any realistic turbulence closure scheme. We try to cover some of this ground in Section 7.4.3. Second, there are precious few near-bottom observations that serve as constraints. With this intent, the boundary conditions for the Reynolds averaged equations (7.1) are

$$\mathbf{u}(z = h) = 0 \quad vis \quad \overline{\mathbf{u}' u'_\perp} = \frac{1}{2} C_d \overline{\mathbf{u}} |\overline{u}|, \tag{7.4a}$$

$$\mathbf{u} \cdot \nabla h(x, y) = 0, \tag{7.4b}$$

$$\nabla_\perp C(z = h) = 0 \quad vis \quad \overline{\mathbf{u}' C'} \cdot \nabla h = 0. \tag{7.4c}$$

The first equation (7.4a) represents both the molecular condition of no-flow at the bottom (where $z = h(x, y)$) and a quadratic drag formulation for the bottom normal stress, nominally $C_d = 2 - 3 \times 10^{-3}$. The quadratic drag formulation assumes that the afore-mentioned hierarchy of boundary layers is turbulent, which is generally, but not always the case (Ruan et al., 2019; Kaiser and Pratt, 2020), to be employed when one is not resolving the log layer. The bottom normal

---

2. This flux-gradient relation is applicable only to variables that possess the "potential" property, whereby the variable is unchanged by adiabatic and isohaline changes in pressure. Examples of variables that are not "potential" variables are in situ temperature, in situ density, specific volume anomaly, and enthalpy (IOC, 2010).

3. This conservative form of an evolution equation for a variable is only applicable to variables that possess the "conservative" property, whereby the total amount of the property remains the same when two parcels are brought together and mixed to completion. Examples of variables that are not "conservative" variables are potential temperature, potential density and entropy (IOC, 2010).

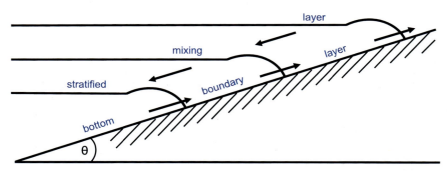

FIGURE 7.4 The basic one-dimensional model, featuring bidirectional flow as arrows crossing isopycnals that results from vertical structure in the mixing. The weakly stratified region near the boundary is dynamically distinct from a more strongly stratified region above. The near boundary region supports upwelling and is termed the bottom boundary layer (BBL), the more strongly stratified region supports downwelling and is termed the stratified mixing layer (SML). Adapted from Garrett (1991).

velocity is represented as $\overline{u}_\perp$, and h represents the equation of the boundary. The second (7.4b) is a statement that the bottom is a material surface and that fluid parcels on that surface remain there. The third (7.4c) represents both the molecular condition of no-flux, in which $\nabla_\perp C$ is the boundary normal gradient, and a statement that the turbulent flux of a quantity C normal to the boundary is zero.

Subtly hidden within the kinematic statement of no-flow through the boundary (7.4b) is that the pressure field arranges itself to accommodate (7.4b), and thereby represents a potential drag. This can be more clearly seen by integrating the Reynolds averaged equations (7.1) in the vertical and using Leibniz's integration rule:

$$\int_{h(x,y)}^{z_0} dz \nabla_{x,y} p(x,y,z) = \nabla_{x,y} \int_{h(x,y)}^{z_0} dz\, p(x,y,z) + p(x,y,z=h(x,y))\nabla_{x,y}h(x,y), \tag{7.5}$$

in which $\nabla_{x,y}$ is the horizontal gradient operator, and z_0 represents a reference height. The correlation of pressure with height gradients along the bottom boundary ($p\nabla h$) is termed 'form drag'. Form drag represents the conversion of energy between mean and perturbation fields, (with total energy conservation following from $p\mathbf{u}\cdot\nabla h = 0$), placing energy into the perturbation fields, where it is more prone to mixing.

Both form drag (7.5) and bottom drag (7.4a) represent sources for turbulent mixing and can be of comparable size for finite amplitude topography, Section 7.4.2. There is an essential difference, though. The stress-driven turbulence acts on scales of 10 s of metres, whereas form drag acts over height scales of 100 s of metres in the abyssal ocean, Section 7.4.2.

7.2.3 Coordinate transformations and the one-dimensional model

The equations of motion are cast in a Cartesian (x, y, z) coordinate system. Though this emphasises the fact that gravity defines the vertical coordinate and Earth's rotation impacts the horizontal velocities, distinctions which are fundamental in geophysical fluid dynamics, significant simplification and intuition can be attained via analytic solutions to a one-dimensional representation by representing the bottom as an infinite sloping plane, Fig. 7.4:

$$z = -h_0 + \alpha x,$$

with h_0 being an arbitrary constant and $\alpha = \tan(\theta)$ is the topographic slope.

Rotation to a semi-infinite plane

Here one assumes the mean can be further decomposed into $\overline{\phi} = \phi_i + \hat{\phi}$ with interior flow v_i and interior density profile ρ_i, and formulates equations for near-boundary deviations. Using a small-angle approximation ($\sin(\theta) \cong \alpha$, $\cos(\theta) \cong 1$), the equations for the boundary variables in a boundary normal coordinate $\hat{z}$ are (e.g., Brink and Lentz, 2010):

$$\frac{\partial \hat{u}}{\partial t} - f\hat{v} = \hat{b}\alpha + \frac{\partial}{\partial \hat{z}} A \frac{\partial \hat{u}}{\partial \hat{z}}, \tag{7.6a}$$

$$\frac{\partial \hat{v}}{\partial t} + f\hat{u} = \frac{\partial}{\partial \hat{z}} A \frac{\partial \hat{v}}{\partial \hat{z}}, \tag{7.6b}$$

$$\frac{\partial \hat{b}}{\partial t} + \hat{u}\, \alpha\, N^2 = \frac{\partial}{\partial \hat{z}} K \frac{\partial \hat{b}}{\partial \hat{z}}, \qquad (7.6c)$$

in which the interior buoyancy gradient $N^2 = -g\rho_0^{-1}\rho_{iz}$ has been assumed constant. The fluxes $\overline{\hat{\mathbf{u}}'\hat{\phi}'}$ have been represented by flux-gradient closure schemes. An important caveat is that the fluxes in (x, y, z) need not be isotropic (i.e., 3-dimensional). Though this assumption may characterise grid generated turbulence that is the subject of the Phillips (1970) and Phillips et al. (1986) laboratory studies of boundary layers over sloping topography, motions characterising turbulent fluxes over flat topography in geophysical systems are anisotropic (e.g., Kaimal et al., 1972): 3-d isotropic motions in boundary layers are inefficient at communicating the boundary conditions. Rather, such 3-d motions are part of the inertial subrange. The presence of a sloping boundary compounds the issue: the sloping boundary represents an opportunity to couple with a resonant internal waveband process, in which the internal waveband flux has an aspect ratio characteristic of the bottom slope, Sections 7.4.1 & 7.4.3. The authors agree that the one-dimensional model is a perfectly reasonable paradigm. The issues orbit around the interpretation of A and K in terms of representing physical processes, the possible conflation of Cartesian (x, y, z), bottom normal $(x, y, \hat{z})$ and isopycnal coordinates, and relevant lengthscales of boundary processes. The implications of such distinctions are regarded as a fundamental open question, Section 7.5.

A focal point for the one-dimensional model is the question encapsulated in Fig. 7.2: 'Are the turbulent processes that give rise to the boundary layer structure simply mixing unstratified water?' and, more generally, 'What are the processes that might ventilate the bottom mixed layer and renew the stratification?' Steady solutions to (7.6) hold some answers to this question. One can view this problem from the perspective in which there is no interior mean flow and the boundary layer structure is a response to turbulent mixing imposed as an enhanced diffusivity. Working from this perspective, Chris Garrett (Garrett, 1990, 1991; Garrett et al., 1993; Garrett, 2001) built on the prior theoretical work of Phillips (1970); Wunsch (1970); Thorpe (1987) to derive theoretical expressions and understanding of the steady-state flows that occur near a sloping boundary when the slope of the boundary is constant, the interior far-field stratification is constant, and for small-scale diffusivities and viscosities that are functions of only the slope-normal distance $\hat{z}$. Because of the one-dimensional nature of the imposed geometry and stratification, a mean diapycnal flow occurred only if the diffusivity, well away from the boundary, was non-zero, but Garrett was able to shed considerable light on the largely self-cancelling transports up and down the slope.

These steady solutions have a natural height scale q^{-1} (Garrett, 1990):

$$q^4 = \frac{1}{4}\left[\frac{N^2 \sin^2(\theta)}{A(\hat{z}=0)K(\hat{z}=0)} + \frac{f^2}{A^2(\hat{z}=0)}\right]. \qquad (7.7)$$

For an unstratified ocean (7.7) returns the familiar Ekman result $q^{-1} \to \sqrt{2A/f}$. Eq. (7.7) also provides insight into the relative roles of buoyancy mixing and friction in dictating the structure of the boundary layer: buoyancy effects become more important as $P_r s^2 \to 1$, in which $P_r = A/K$ is the turbulent Prandtl number and

$$s = \frac{N \sin(\theta)}{f} \qquad (7.8)$$

is the slope Burger number, with $s \sim O(1)$ providing a meaningful quantification of what constitutes 'steeply sloping topography'.

Isopycnals dip into the boundary to meet the no-flux boundary condition on this height scale with a reduction of buoyancy, with the implication that buoyancy fluxes are reduced from what might be estimated from the farfield stratification. The stratification is *not* zero, as might be imagined for a steady-state solution: these dipping isopycnals imply pressure gradients that support an upslope flow near the boundary, Fig. 7.4, that continually 'restratifies'. If the imposed diffusivity decreases with height, pressure gradients in the opposite direction arise and a bidirectional flow results aloft. This cross-slope flow serves to maintain the stratification, and there is no intrinsic need to transport 'mixed water' off-slope. On the other hand, intrusions of mixed fluid into the interior associated with a convergence in cross-slope transport resulting from changes in slope and / or interior stratification are possible (e.g., Phillips et al., 1986; Kunze et al., 2012).

If the height scale over which $K(\hat{z})$ varies is larger than q^{-1}, this model sets an expectation that the turbulent buoyancy fluxes near the boundary are smaller than those aloft. This limit is so commonly invoked that the weakly stratified near-boundary region has been labelled the bottom boundary layer (BBL) and turbulent regions aloft referred to as the stratified mixing layer (SML), Fig. 7.4, which also delineates regions having upslope and downslope transports. Ferrari et al. (2016) carry this argument to an extreme and assert that the no normal flux boundary condition (7.4c) requires the diapycnal flux to be zero, $\overline{w'\rho'}(\hat{z}=0) = 0$. The upslope/downslope dichotomy is central to theoretical developments in Section 7.3.3–7.3.2. In Section 7.5, we forward the suggestion that Ferrari et al. (2016)'s assertion that $\overline{w'\rho'}(\hat{z}=0) = 0$ results from assuming the fluxes giving rise to the turbulence are nearly isotropic and that the notion that the BBL hosts generally smaller buoyancy flux magnitudes than in the SML is in jeopardy.

A second view of this problem is one in which the mixing is driven by an overlying geostrophically balanced along slope flow (Trowbridge and Lentz, 1991; Lentz and Trowbridge, 1991; MacCready and Rhines, 1991, 1993; Garrett et al., 1993). The intent of this class of problems is slightly different: cross-slope transport results from a combination of mixing and cross-slope advection in the Ekman layer, defined as the height scale over which the bottom drag is deposited. The focal point here has been the character of the steady-state solution, known as the arrested Ekman layer, in which the boundary layer grows to sufficient height that the geostrophic shear associated with mean isopycnal tilt decreases the bottom velocity, reduces the bottom stress, reducing mixing until the bottom velocity approaches zero. Early work on this problem is focused upon the slow time evolution ($\partial_t \ll f$) to the arrested state. This knowledge is acquired using a plethora of *ad hoc* assumptions about turbulent momentum and buoyancy fluxes via A and K, with limited attempts to model the system with full-on closure schemes, (e.g., Weatherly and Martin, 1978). The work of Brink and Lentz (2010) and high-performance computing (Umlauf et al., 2015; Ruan et al., 2019) opens up the examination of the time-dependent problem using full-on turbulence closure schemes that address the potential for multiple height scales in the boundary layer and convective instabilities avoided in this early work. We attempt to summarise such efforts in Section 7.4.

Density coordinates

The transformation to a density coordinate system (McDougall, 1984, 1989; Marshall et al., 1999) provides a tremendous intellectual simplification: boundary mixing concerns the evolution of the buoyancy field in response to mixing, and transforming the equations of motion (7.1) into a density coordinate system (e.g, Cushman-Roisin and Beckers, 2011) simplifies the density equation to a balance between time dependence, the divergence of the density flux and advection across isopycnals. It focuses attention directly upon diabatic processes and discards sub-inertial fluctuations that are quite efficient at dispersion along isopycnals, but typically are not directly connected to 3-d turbulence, and thus cause limited dispersion across isopycnals (e.g., Ferrari and Polzin, 2005). The mean diapycnal velocity is given by

$$\tilde{\mathbf{e}} = -\frac{\nabla \cdot \mathbf{F}_\rho}{\|\nabla \rho\|} \mathbf{n}_\rho, \tag{7.9}$$

and note that this expression applies even when the Eulerian fields are unsteady. This intellectual simplicity, though, comes at the cost of the analytic complexity of needing to include extra terms to allow for the curvature of isopycnal surfaces, and the boundary conditions become much more complicated Garrett (2001).[4]

7.2.4 Integration

Integration is a vehicle for simplifying complex problems. This tool appears in three significant contexts.

Walin style control volume budgets

Using two surfaces of constant density as boundaries for a volume integral of the density equation leads to an expression for the time rate of change of volume V contained within those two isopycnals (Walin, 1982), Fig. 7.5:

$$\frac{\partial V}{\partial t} = -\frac{\partial G}{\partial \rho} + \Psi_{\text{advective}}, \tag{7.10a}$$

$$G = \iint \tilde{\mathbf{e}} \, dA, \tag{7.10b}$$

where G is the volume flux *across* an isopycnal maintained by turbulent processes. Ignoring inputs from geothermal sources and surface inputs, the transformation rate G is the area-integrated diapycnal velocity from (7.9), dA is an infinitesimal area element and $\Psi_{\text{advective}}$ represents the advective fluxes of *intermediate* density into and out of the volume, such as those estimated by current metre arrays or geostrophic transport sections.

---

4. The inquisitive might note that the transformation from Cartesian into density coordinates assumes a one-to-one mapping between z and ρ and this is problematic in the case of convection, such as presented by a downwelling Ekman layer, Section 7.4.3.

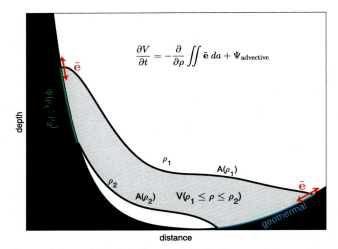

FIGURE 7.5 A schematic depiction of terms relating to the Walin budget (7.10) for the time rate of change of the volume V between two density horizons ρ_1 and ρ_2. The volume is indicated by the grey shaded area; mass and density Ψ is supplied on the left hand boundary via a gap in the side wall indicated by the green segment; diapycnal velocities $\tilde{e}$, largest at the boundary, are indicated in red. The presence of geothermal heating, neglected in (7.10), is represented in cyan.

Hypsometry

The issue of integration exposes one to the fact that, along surfaces of constant depth, the area

$$A = \iint dA \qquad (7.11)$$

is monotonically increasing as depth *decreases*. For example, $A(\rho_1) > A(\rho_2)$ in Fig. 7.5. This introduces a subtle interplay between increasing area with *decreasing* depth and the observed tendency of turbulent dissipation profiles to increase with *increasing* depth (Fig. 7.3 and Polzin et al. (1997)), implying a *local* (7.9) downwelling. The two trends compete in the Walin budget (7.10), and the net result, apart from a basin-wide and global requirement for net upwelling, is not obvious. Upwelling requires that the area of isopycnals increases with height in such a way that even though the buoyancy flux decreases with height on each cast, the area-integrated buoyancy flux increases with height (Klocker and McDougall, 2010). Specific proposals have been put forward in Polzin (2009) for single grid cells of a climate class GCM and for canyons incising larger-scale sloping topography (Kunze et al., 2012; Thurnherr et al., 2020); see Sections 7.3 and 7.5.

The moment method

The moment method is a generic method for analysing stochastic systems in a wide variety of contexts (Gardiner et al., 1985; Polzin and Lvov, 2017) involving dispersion. Given an equation for tracer concentration (7.3) and a closure, such as (7.2), one multiplies the equation by a spatial coordinate, integrates over space and applies appropriate boundary conditions. For example, a delta function initial condition in a homogeneous ocean, where the tracer never reaches the boundary, leads to

$$\langle z^2 \rangle = 2 K t, \qquad (7.12)$$

in which $\langle z^2 \rangle$ is the mean squared dispersion about the centre of mass. In mixing studies, such metrics are useful comparisons to estimates of tracer dispersion across isopycnal surfaces (Ledwell et al., 2000; Holmes et al., 2019). See Section 7.3 for further discussion concerning boundary conditions.

7.2.5 Energetics and mixing

The ability of turbulent production to result in a buoyancy flux is a significant part of the lexicon concerning ocean turbulence. How this plays out in the context of boundary mixing is central to some of our disagreements, and to address this we need to draw together basic (e.g., Tennekes and Lumley, 1972) statements about energy budgets and how measurements intersect with those budgets.

7.2.5.1 Energetics

We start from a Reynolds decomposition, multiply the equations of motion by their perturbations and time average.

TKE equation

The equation for the time-averaged turbulent kinetic energy $\overline{u_i' u_i'}/2 = TKE$ is

$$\frac{1}{2}\left[\frac{\partial \overline{u_i' u_i'}}{\partial t} + \overline{u}_j \frac{\partial \overline{u_i' u_i'}}{\partial x_j}\right] + \overline{u_i' u_j' \frac{\partial \overline{u}_i + u_i'}{\partial x_j}} \cong -\rho_0^{-1} \frac{\partial}{\partial x_i}\overline{u_i' p'} + \overline{b'w'} - \frac{\nu}{2}\overline{\left(\frac{\partial u_i}{\partial x_j} + \frac{\partial u_j}{\partial x_i}\right)^2}$$

$$\frac{d}{dt}TKE \quad - \quad \mathcal{P}_k \quad + \quad N.L. \cong \quad P_{\text{work}} \quad + \quad \mathcal{B} \quad - \quad \epsilon. \tag{7.13}$$

Time rate of change of TKE following the mean flow $\overline{u}_j$ is balanced by turbulent production $\mathcal{P}_k$ and nonlinear transports, pressure work, buoyancy flux $\mathcal{B}$ and the kinetic energy dissipation rate ϵ. Terms of $O(\ell/L)^2 \ll 1$, in which ℓ is a metric of the strain (dissipation) field and L is the outer, energy-containing scale of the turbulent velocity, have been neglected in comparison to the dissipation rate ϵ (Tennekes and Lumley, 1972). Here u_i are velocity vector components with vertical component $u_3 = w$, ν is kinematic viscosity and repeated indices imply summation.

Scalar variance equation

A similar slowly varying construct emerges for the scalar (here density ρ) variance equation:

$$\frac{1}{2}\left[\frac{\partial \overline{\rho_i' \rho_i'}}{\partial t} + \overline{u}_j \frac{\partial \overline{\rho_i' \rho_i'}}{\partial x_j}\right] + \overline{\rho' u_i' \frac{\partial \overline{\rho} + \rho'}{\partial x_i}} \cong -\frac{1}{2}\kappa \overline{\left(\frac{\partial \rho'}{\partial x_i}\right)\left(\frac{\partial \rho'}{\partial x_i}\right)}$$

$$\frac{d}{dt}SV/2 \quad - \quad \mathcal{P}_\rho \quad + \quad N.L. \cong \quad - \quad \chi_\rho/2. \tag{7.14}$$

Time rate of change of the scalar variance SV is balanced by turbulent production $\mathcal{P}_\rho$, non-linear transports and dissipation χ_ρ. Terms $O(\ell_\rho/L) < \ell/L \ll 1$, in which ℓ_ρ is a metric of the turbulent density gradient (dissipation) field, have been neglected in comparison to the rate of dissipation of density variance, as in (7.13). The *vertical* component of the density flux in (7.14) appears as a source of potential energy in (7.13). This distinction is a key issue in boundary mixing, Section 7.5.[5]

7.2.5.2 Metrics of mixing

The holy grail in the study of 'mixing' is, well, mixing, by which we mean the turbulent transport of a substance C, $\overline{\mathbf{u}'C'}$, across mean isopleths, $\overline{C}$, in (7.3) and (7.14). Direct measurement of the fluxes is quite difficult, hence the usual course of events is to argue from restricted balances of (7.13) and (7.14) and relate the fluxes, which happen at the overturning scales, to dissipation, which is the smallest turbulent scale.

This is a convoluted process. First, one assumes steady conditions, then asserts that integration of the triple correlations $\overline{u_i' u_j' \frac{\partial u_i'}{\partial x_j}}$ and $\overline{\rho' u_i' \frac{\partial \rho'}{\partial x_i}}$ over a volume and application of Green's theorem renders the triple correlations as the net turbulent transport across the volumetric boundaries, small for large volumes and negligible for spatially homogeneous conditions, but yet potentially problematic near boundaries (Shaw et al., 2001), and finally pleads ignorance about the role of turbulent pressure-velocity correlations, fervently hoping that volumetric integration smooths away any issues. Second, one acknowledges the role of stratification in differentiating stirring along isopycnals by largely adiabatic motions from mixing across isopycnals by their diabatic brethren. This brings one to the turbulent production: dissipation balances advocated by Osborn (1980):

$$-\mathcal{P}_k \cong \mathcal{B} - \epsilon, \tag{7.15}$$

and Osborn and Cox (Osborn and Cox, 1972):

$$\overline{\rho' w'}\,\overline{\rho}_z = -\chi_\rho/2, \tag{7.16}$$

---

5. We set aside questions concerning density being related to both temperature and salinity, considering the fluid to be temperature-dominated and boundary mixing to be a turbulent stress or form drag-driven paradigm.

which further assumes production is dominated by vertical terms. The working hypothesis is that one can define a flux Richardson number as the ratio between the buoyancy flux and turbulent production:

$$R_f = -\frac{\mathcal{B}}{\mathcal{P}_k}. \tag{7.17}$$

With (7.15) and (7.16) one has two indirect routes to empirical knowledge of $\overline{\mathbf{u}'\rho'}$: via ϵ and via χ. Simultaneous knowledge of both provides a dissipation ratio:

$$\Gamma_\mu = \frac{\chi_\rho}{\epsilon}\frac{N^2}{2\overline{\rho_z^2}} = \frac{R_f}{1 - R_f}. \tag{7.18}$$

Empirical constraints on Γ_μ are provided by microstructure sensors resolving the dissipation scales. A second route is possible for sufficiently energetic turbulence. Energetic turbulence is often characterised by a range of wavenumbers k that are much larger than those characterising the large scales of turbulence and much smaller than those at which molecular dissipation acts. Within this range it is reasonable to assume that the downscale transports of energy ϵ (relating to triple correlations $\overline{u_i'u_j'\frac{\partial u_i'}{\partial x_j}}$) and scalar variance χ_ρ (relating to $\overline{\rho' u_i'\frac{\partial \rho'}{\partial x_i}}$) are the only dynamically relevant parameters. Dimensional analysis relates the kinetic energy and scalar spectra:

$$E_k(k) \cong C_k\, \epsilon^{2/3}\, k^{-5/3}$$

and

$$E_\rho(k) \cong C_\rho\, \chi_\rho\, \epsilon^{-1/3}\, k^{-5/3}.$$

Knowledge of the spectral density E_k and E_ρ thus provides mid-scale estimates of mixing given the constants of proportionality (Kaimal et al., 1972; Sreenivasan, 1995, 1996) and can be summarised as a spectral transport ratio Γ_{trans}:

$$\Gamma_{\text{trans}} = \frac{C_k}{C_\rho}\frac{E_\rho(k)N^2}{2\overline{\rho_z^2}E_K(k)}. \tag{7.19}$$

Such estimates are referred to as 'inertial subrange' estimates and appear in (7.13) and (7.14) as triple correlations in the primed variables.

If we turn our eyes towards the issue of production in the TKE equation, there are at least three phenomenological end members: shear production (e.g., $-\overline{u'w'}\,\overline{u}_z$), for which Γ_μ and Γ_{trans} are less than 0.2; convection, implicit in (7.13) as either forcing or mean conditions, such as a downwelling Ekman layer that maintain a condition of static instability and result in the conversion of gravitation potential energy $\overline{w'b'}$, and for which $\Gamma = -1.0$; and form drag, which appears as a boundary condition (Section 7.2.2) and results in pressure gradients that force flow accelerations and hydraulic transitions with an indeterminate Γ assignment. See Gregg et al. (2018) for a review of lab measurements, direct numerical simulations of shear instabilities and oceanic microstructure data.

The presence of near-boundary upwelling in the one-dimensional model (Fig. 7.4) occurs in association with a relative minimum in the buoyancy flux as a function of $\hat{z}$. This minimum is part of the model construction: one assumes $K(\hat{z})$ and $A(\hat{z})$, and then solves for the buoyancy field. The two, in combination as (7.2), return the mid-depth minimum in buoyancy flux. There are substantive issues in support of this that can be expressed as structure in the mixing metrics Γ_μ and Γ_{trans}, with small values of Γ below the mid-depth minimum. These are issues of lengthscale suppression, time dependence and tertiary dependencies of Γ upon turbulent intensity and gradient Richardson number.

Lengthscale suppression

Neither dissipation range nor inertial subrange estimates of ϵ or χ are what we want, which is an estimate of the fluxes $(\overline{\mathbf{u}'\rho'})$ at the overturning scales. A tool for addressing links between dissipation scales and overturning scales can be developed by assuming that the important balance is one between the force of gravity, ensconced in the stratification N, and the inertial forces of turbulence, encapsulated in the downscale transfer rate in an inertial subrange, ϵ. Dimensional analysis leads to the Ozmidov length L_o:

$$L_o = \sqrt{\epsilon/N^3},$$

which can be compared to a direct estimate of overturning quantified as the root-mean-squared vertical displacement z' of parcels in an overturn (Thorpe, 1977) after sorting to be statically stable:

$$L_T = \overline{z'^2}^{1/2}.$$

The unbounded shear flow, for which $\Gamma \cong 0.2$, is a situation in which the overturning scales are free to evolve such that $L_T \cong L_o$. The distance from the boundary represents a fundamental imposition of an external lengthscale that plausibly impacts mixing efficiency. It is directly expressed as wall layer scalings in turbulent closure schemes (e.g., Mellor and Yamada, 1974) with the suppression of turbulent lengthscales, implying smaller gradient Richardson number R_i and smaller flux Richardson number R_f in association with smaller R_i. A reduction in mixing efficiency associated with lengthscale constraints is both fundamental and intuitive, but yet competes with increasing stress-driven dissipation as the boundary is approached and is fundamentally grounded in the paradigm of steady flow over a flat bottom.

Time dependence

Internal wave breaking has a lifecycle between overturning and dissipation (Smyth et al., 2001; Chalamalla and Sarkar, 2015) that is expressed in L_T and L_o and impacts the interpretation of data. There is the potential for wave breaking to emphasise 'younger' parts of the lifecycle and support more efficient mixing.

Tertiary dynamics

An ongoing source of uncertainty (e.g., Mashayek et al., 2017; Ijichi et al., 2020) is the potential of a tertiary dependence of Γ_μ upon either turbulent intensity:

$$R_{eb} = \epsilon/\nu N^2, \tag{7.20}$$

which is proportional to the ratio of overturning scales L_o to dissipation scales $L_k = (\epsilon/\nu^3)^{-1/4}$ and serves as a bandwidth for the inertial subrange, or upon gradient Richardson number R_i on scales relevant to overturning.

Having noted these details concerning mixing metrics, we remind ourselves about the source of knowledge for the mid-depth minimum in buoyancy flux: we have arrived at this via a mixing length closure that is appropriate for small perturbations in a steady flow that are related to the local gradients. The phenomenology of boundary mixing presented in Section 7.4 is that of convection in an Ekman layer coupled with a resonant internal wave band process, episodic convection associated with an internal wave-breaking process and O(1) variations in stratification within the boundary layer on the internal wave time scale. None of this meshes well with a mixing length closure or the discard of transport terms in (7.13) and (7.14). Skepticism is required both here and in the interpretation of mixing metrics Γ via extrapolation of results from regions without boundaries to the issue of boundary mixing. Moreover, the internal wave band processes being emphasised have motions that parallel the slope. We remind ourselves that what we want in a discussion of boundary mixing are estimates of $\overline{\mathbf{u}'C'}$ in (7.3). This is *not* identical to the buoyancy flux $\overline{w'b'}$ in the TKE equation (7.13), where the buoyancy flux enters only in the direction of gravity. With isopleths of C dipping into the boundary, these cross-slope fluxes project across those mean isopleths. The propensity to ignore this distinction leads us to articulate an open question in Section 7.5, where we call for more and better data to address these concerns.

7.3 Implications of the bottom intensification of ocean mixing for upwelling: buoyancy budgets for bottom-intensified mixing

In this section, we are mainly concerned with understanding the area-integrated volume transports of water moving both downwards and upwards across density surfaces. The crucial statements are the diapycnal advection: diffusion balance (7.9) and corresponding budget for the volume of water contained between two isopycnals (7.10). Because there is no buoyancy flux through the sea floor, it is hard to avoid upwelling very close to a sloping ocean floor. The global response, though, is complicated by hypsometric considerations (7.11). In Section 7.3.1, we discuss the impact that including such near-boundary mixing structure has for the global stratification in numerical models. In Sections 7.3.2 through 7.3.5, we discuss and then dissect the details of solutions to the one-dimensional boundary layer problem posed in (7.6) that relate to this structure. We discuss aspects of the ventilation of the boundary by adiabatic processes (eddy-stirring) in Section 7.3.2 and touch upon the interpretation of tracer releases in light of such boundary layer structure in Section 7.3.6. We return to comment upon the implications for the horizontal circulation in numerical models (Section 7.3.7).

7.3.1 Abyssal ocean circulation models are sensitive to bottom topography

Ocean models have traditionally used a constant value of the diapycnal diffusivity, or at least a vertical profile that is essentially constant below a depth of 2000 m (Bryan and Lewis, 1979). With the stratification, $N^2 = b_z$, being an increasing function of height; a depth-independent diapycnal diffusivity implies that the dissipation of turbulent kinetic energy ϵ, increases with height, in contrast with observations in the abyss (Fig. 7.3). General circulation models run with parameterisations of bottom-enhanced mixing via increasing diffusivity with depth (e.g., Jayne, 2009; Melet et al., 2013) host vigorous overturning cells, implying a complex interaction between downwelling implied by increasing diffusivity with depth and changing stratification to permit greater net upwelling.

This complex interaction was explored in an idealised setting by specifying the buoyancy flux profile (Ferrari et al., 2016). The buoyancy flux profile was constructed so as to have a relative minimum at a height above bottom of 80 m, and two variants of model topography were considered. One such model was constructed with a flat bottom and vertical walls. A steady state was reached, with sufficient "upwelling" along a thin boundary layer along the flat ocean bottom to overcome the diapycnal downwelling in the ocean interior. However, the model's interior stratification was much weaker than in the real ocean. When the model topography was reconfigured with sloping side walls, a much more realistic stratification and meridional overturning circulation was obtained, again with diapycnal downwelling in the ocean interior and relatively fast upwelling in thin bottom boundary layers on the sloping boundaries. The diapycnal upwelling transport in these boundary layers was about 2.5 times the net diapycnal upwelling, with the balance (−1.5) sinking across isopycnals in the ocean interior. The pattern of near-boundary upwelling and downwelling relates directly to that in the one-dimensional model presented in Section 7.2.3. The net (global) upwelling (7.10) then relates to basin geometry through hypsometric arguments (7.11).

Using World Ocean Circulation Experiment data, conventional topography and mixing specified through a diffusivity profile, de Lavergne et al. (2016b) diagnose negative diapycnal transports in the ocean interior and emphasise the substantial upwards vertical transport in the bottom boundary layer, with the latter resulting from the convergence of the turbulent flux of buoyancy resulting from adjustments (weakening) of the BBL stratification and due to geothermal heating. de Lavergne et al. (2017) find that the maximum in the meridional overturning streamfunction is insensitive to how abyssal mixing is specified in terms of various geophysical parameters (e.g., topographic slope, abyssal height variance, etc.) and in so doing make a reasonably strong statement that global hypsometry controls the maximum in the meridional overturning stream function.

The study of Ferrari et al. (2016) showed that when the buoyancy flux of small-scale mixing is forced to decrease with height in an ocean model, the deep circulation is very sensitive to the bottom topography, with the deep circulation being completely different when the topography consisted of vertical walls and a flat bottom, versus having more realistic non-vertical and non-horizontal topography. This sensitivity is not apparent when the magnitude of the diapycnal buoyancy flux of small-scale mixing is allowed to increase with height, thus allowing upwelling in the ocean interior. This sensitivity of the abyssal circulation to the bottom topography (the ocean's hypsometry) will be taken up again in Section 7.3.4 below.

7.3.2 One-dimensional solutions for flow near a sloping bottom boundary

Callies (2018) has shown that for oceanographically relevant parameters of the earth's rotation, abyssal boundary slopes, and the small-scale diffusivity and viscosity, the vertical stratification that is predicted by these one-dimensional near-boundary solutions is very weak, and he suggested that the sloping isopycnals that are typical of these solutions are unstable to baroclinic instability, and that including the resultant adiabatic stirring of the thickness between closely spaced density surfaces (Gent et al., 1995; McDougall and McIntosh, 2001) makes a material difference to the stratification. Naveira Garabato et al. (2019) have observed this type of adiabatic submesoscale eddy exchange in an abyssal boundary current in the Orkney passage, finding evidence of both symmetric and centrifugal instabilities along with enhanced diapycnal mixing. One way of parameterising this bolus transport is as a greatly increased vertical viscosity and Prandtl number, following the work of Greatbatch and Lamb (1990). With this modification, the one-dimensional near-boundary flow model of Garrett (1990, 1991, 2001) gives the steady-state bottom boundary layer (BBL), being the region closest to the sloping wall, where the flow is upslope, to be a few tens of metres thick. The upward flow is concentrated in the lowermost third of this distance, where the stratification is weak (see Fig. 7.6), whereas the upper half of the BBL is well stratified (Holmes and McDougall, 2020).

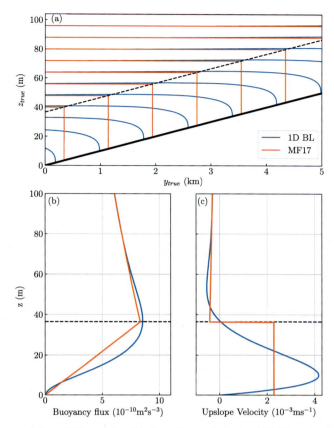

FIGURE 7.6 (a) Cross-section of isopycnals (smooth blue contours) in Garrett's one-dimensional (1D) boundary layer solution in the vertical plane, with the horizontal axis being up-slope distance. The dashed black line separates the downward-flowing stratified mixing layer SML from the upward-flowing bottom boundary layer (BBL). (b) The magnitude of the diffusive buoyancy flux (smooth blue line) of the 1D solution, which has a maximum value at the top of the BBL. (c) The upslope velocity as a function of distance normal to the boundary. In these panels the full blue lines are for the 1D boundary layer solution, while the full yellow lines are for McDougall and Ferrari (2017)'s assumed geometry in which the interior has a constant stratification, the BBL is well mixed, and the turbulent buoyancy flux in the BBL varies linearly with height.

7.3.3 Expressions for the upwelling in the BBL and downwelling in the SML

Consider a one-dimensional situation with the ocean floor sloping at an angle with respect to the isopycnals lying above the boundary layer region, and with the far-field stratification $N^2 = b_z$ being depth-independent (where $b = -g(\rho - \rho_{\text{bottom}})/\rho_{\text{bottom}}$ is the buoyancy and ρ_{bottom} is the density at the densest part of the ocean); Fig. 7.6. The slope-normal coordinate is designated $\hat{z}$ and the magnitude of the small-scale turbulent diapycnal buoyancy flux per unit area is written as a function of $\hat{z}$ as $\mathcal{B}(\hat{z})$. Considering bottom-intensified mixing, and ignoring for now the background diapycnal diffusion in the ocean interior, we take $\mathcal{B}(\hat{z})$ to decay to zero at large distance from the sea floor, while right at the sea floor, the component of the buoyancy flux normal to the boundary must be zero (apart from the geothermal heat flux coming through the sea floor). As $\mathcal{B}(\hat{z})$ increases downwards towards the sea floor, it reaches a maximum value, $\mathcal{B}_0$, before decreasing towards zero at the sea floor.

At the height of the maximum magnitude of the turbulent buoyancy flux, $\mathcal{B}_{\hat{z}}$ is zero, and so the diapycnal velocity is zero. Between the sea floor and the height at which $\mathcal{B}(\hat{z})$ reaches its maximum value $\mathcal{B}_0$, the diapycnal velocity is positive, moving up-slope (in a steady-state) towards less dense water, whereas above this height the diapycnal velocity is negative and the fluid flows parallel to the sea floor in the direction towards denser fluid. Following Ferrari et al. (2016) and McDougall and Ferrari (2017), we label the region close to the sea floor, where fluid upwells towards less dense seawater the bottom boundary layer (BBL) of thickness h and the fluid immediately above this, where the diapycnal velocity is negative, is called the stratified mixing layer (SML). We expect the BBL to occupy the region within a few tens of metres above the sea floor, while the e-folding vertical distance of the SML is expected to be between 200 m and 500 m (Callies, 2018).

The buoyancy budget for the fluid in the BBL shows that the (positive) diapycnal volume transport towards less dense fluid is given by (McDougall and Ferrari, 2017; Holmes and McDougall, 2020):

$$\mathcal{E}_{BBL} = \int \frac{\mathcal{B}_0}{|\nabla b|} \frac{1}{\tan(\theta + \psi)} dx, \qquad (7.21)$$

where $|\nabla b|$ is the magnitude of the spatial gradient of buoyancy at the top of the BBL, and $(\theta + \psi)$ is the angle between the sea floor and the isopycnals at the height of the top of the BBL (with θ being the angle between the horizontal and the sea floor, and ψ the angle between the horizontal and the isopycnal surfaces). Here the integral is taken horizontally along the incrop line (into the page), where the isopycnal intersects the sea floor. Taking $\mathcal{B}_0 = D_0|\nabla b|$ to be the product of $|\nabla b|$ and the diffusivity at the top of the BBL, D_0, this equation simplifies to

$$\mathcal{E}_{BBL} = \int \frac{D_0}{\tan(\theta + \psi)} dx. \qquad (7.22)$$

In deriving these expressions, the effect that the non-linearity of the equation of state of seawater has on the relationship between upwelling and the divergence of the buoyancy flux has been ignored. Also, in assuming one-dimensional geometry and constant far-field stratification, we have ignored any variation up the slope of the turbulent diffusive flux of buoyancy across successive buoyancy surfaces inside the BBL, F_{BBL}; Holmes and McDougall (2020) have derived the additional term that appears in the above equations (their equation (14)) when such upslope variation is present.

The expression (7.22) states that the upwards diapycnal transport in the BBL per unit horizontal distance along the isopycnal incrop line is $D_0/\tan(\theta + \psi)$, where $\tan(\theta + \psi)$ is the tangent of the angle that isopycnals at the top of the BBL make with respect to this upper boundary of the BBL. With a constant height h of the BBL, this is the same angle that these isopycnals at the top of the BBL make with respect to the sloping sea floor. The need for this transport to be proportional to the diffusivity at the top of the BBL, D_0, is understandable, and the reason for the division by $\tan(\theta + \psi)$ arises, because (i) the up-slope buoyancy gradient along the top of the BBL is smaller than $|\nabla b|$ by the factor $\sin(\theta + \psi)$, and (ii) the component of the diapycnal buoyancy flux per unit area normal to the top of the BBL is smaller than $\mathcal{B}_0 = D_0|\nabla b|$ by the factor $\cos(\theta + \psi)$.

Applying the Walin (1982) type buoyancy budget analysis (7.10) to the whole volume between a pair of closely spaced buoyancy surfaces, and continuing to ignore the geothermal buoyancy flux, the area-integrated, or net, volume transport upwelling through a buoyancy surface is given by

$$\mathcal{E}_{net} = \mathcal{E}_{BBL} + \mathcal{E}_{SML} = \frac{dF}{db}, \qquad (7.23)$$

where the net upwelling $\mathcal{E}_{net}$ (in m^3 s^{-1}) is the sum of the upwelling in the BBL, $\mathcal{E}_{BBL}$, and that in the SML, $\mathcal{E}_{SML}$, which is negative. Here the turbulent buoyancy flux across the whole area of an isopycnal is here defined as

$$F = F_{BBL} + F_{SML} = \int\int \mathcal{B}(b, x, y) dx dy, \qquad (7.24)$$

where it is recognised that there are contributions to F from both the SML and the BBL.

To develop an expression for the sinking volume transport through buoyancy surfaces in the SML, $\mathcal{E}_{SML}$, the magnitude of the vertical component of the turbulent diapycnal buoyancy flux is taken to be an exponentially decreasing function of buoyancy, $\mathcal{B}(b, x, y) = \mathcal{B}_0(x, y) e^{-(b-b_0)/\Delta b}$, with the e-folding buoyancy scale being Δb, and $\mathcal{B}_0(x, y)$ being the turbulent buoyancy flux at the height of the boundary between the BBL and the SML. The exactly vertical velocity through buoyancy surfaces (often denoted with the symbol e) is $\partial \mathcal{B}/\partial b|_{x,y} = -\mathcal{B}(x, y, b)/\Delta b$, and when this is area-integrated along an isopycnal over the full area of the SML, we find that

$$\mathcal{E}_{SML} = -\frac{1}{\Delta b} \int\int_{SML} \mathcal{B}(b, x, y) dx dy = -\frac{F_{SML}}{\Delta b} = -\frac{F_{SML}}{N^2 d}, \qquad (7.25)$$

where we have assumed that the e-folding buoyancy scale Δb does not vary with horizontal location along the buoyancy surface in the SML region, and in the last part of Eq. (7.25), we have replaced the buoyancy scale Δb with an e-folding vertical distance d (which is often taken to be 500 m) and the square of a buoyancy frequency that is typical of the SML region along this isopycnal. When the e-folding height d of the turbulent buoyancy flux is much larger than the

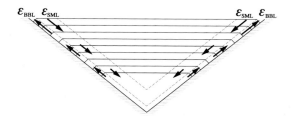

FIGURE 7.7 An ocean with conical bottom topography and a constant value of the diffusive buoyancy flux at the top of the bottom boundary layer (BBL). The magnitudes of the diapycnal transports in both the BBL and the SML increase with height, whereas the diapycnal velocities in both the BBL and the SML are independent of height (if the vertical interior stratification is depth-independent).

thickness of the BBL h, the contribution of the BBL to the area-integrated buoyancy flux F is small, so that in this case $F = F_{BBL} + F_{SML} \approx F_{SML}$. From Holmes and McDougall (2020), we expect F_{BBL}/F to be 0.05.

Just as the diapycnal upwelling in the BBL increases as the tangent of the angle between isopycnals and the sea floor, $\tan(\theta + \psi)$, decreases (as we noted from Eq. (7.22)), so too does the magnitude of the downwelling in the SML, because for smaller values of $\tan(\theta + \psi)$, the area of the integral where $\mathcal{B}(b, x, y)$ is significant expands.

7.3.4 How much larger is the upwelling in the BBL than the net upwelling?

The horizontally uniform upwelling paradigm of Munk (1966) and Munk and Wunsch (1998) has held sway for so long, and with the dynamical implications of bottom-intensified ocean mixing only now beginning to be understood; it is perhaps too early to expect that a simple paradigm might emerge to replace the horizontally uniform view. If such a simple paradigm does emerge, it will likely rely on the implications of a Walin (1982) type buoyancy budget. The example of a conical ocean with the magnitude of the turbulent buoyancy flux at the top of the BBL $\mathcal{B}_0$ being constant, might serve as an example of an alternative paradigm to the one-dimensional upwelling-diffusion model; not that we think that a conical ocean is very realistic. A conical ocean has a constant slope $\tan\theta$ with the radius R of the ocean increasing with height according to $R_z = 1/\tan\theta$. Applying the overall buoyancy budget, Eq. (7.23), shows that for heights well above the deepest part of the ocean, the net diapycnal upwelling transport is (from Section 7 of McDougall and Ferrari (2017))

$$\mathcal{E}_{net} = \frac{dF}{db} = D_0 \frac{2\pi d}{(\tan\theta)^2} \qquad \text{(conical ocean)}, \tag{7.26}$$

where the e-folding height scale d of the magnitude of the turbulent buoyancy flux has been assumed to be much larger than the thickness of the BBL h, and the isopycnals have been assumed to be horizontal outside the BBL. This expression can be understood by noting that the area of the SML scales as $R2\pi d/\tan\theta$, and when this is vertically differentiated, the second negative power of $\tan\theta$ enters since $R_z = 1/\tan\theta$. The ratio of the upwelling volume flux in the BBL to $\mathcal{E}_{net}$ in this conical ocean is given by (using Eq. (7.22))

$$\frac{\mathcal{E}_{BBL}}{\mathcal{E}_{net}} = \frac{R\tan\theta}{d} \qquad \text{(conical ocean)}. \tag{7.27}$$

From these relations, we see that (i) the net upwelling of bottom water $\mathcal{E}_{net}$ is independent of the radius R of the cone, so that the same net volume flux $\mathcal{E}_{net}$ upwells through all height levels of the conical ocean, and (ii), both $\mathcal{E}_{BBL}$ and $|\mathcal{E}_{SML}|$ are significantly larger than $\mathcal{E}_{net}$, and they both increase linearly with R, that is, they increase linearly with height (see Fig. 7.7). Holmes et al. (2018) have explored the implications for upwelling in the BBL and downwelling in the SML of many idealised situations with variable topography and variable stratification.

More generally, we can gain some insight into the ratio of $\mathcal{E}_{BBL}$ to $\mathcal{E}_{net}$ by essentially regarding the variation of $\mathcal{E}_{net}$ with buoyancy $\mathcal{E}_{net}(b)$, as known, and by integrating Eq. (7.23), we find that

$$F = F_{BBL} + F_{SML} = \int_{b_{min}}^{b} \mathcal{E}_{net} db', \tag{7.28}$$

where the integral is performed from the very densest water with buoyancy b_{min}. From Eqs. (7.25) and (7.28), we find that

$$\mathcal{E}_{SML} = -\frac{F_{SML}}{F} \frac{1}{N^2 d} \int_{b_{min}}^{b} \mathcal{E}_{net} db'. \tag{7.29}$$

McDougall and Ferrari (2017) took the BBL to be a well-mixed boundary layer so that F_{BBL} was very small and the ratio F_{SML}/F was very close to unity. More realistically, Holmes and McDougall (2020) derived an expression for F_{BBL} in the one-dimensional near-boundary formulation, finding that F_{BBL}/F is approximately $1.4/\pi$ times the ratio of the relevant vertical lengthscales, that is, $F_{BBL}/F \approx 0.45h/d$, where h is the thickness of the BBL, and taking h/d to be 0.1, we expect F_{SML}/F to be approximately 0.95. The expression for the ratio of $\mathcal{E}_{BBL}$ to $\mathcal{E}_{net}$ is found from Eq. (7.29) together with the additive relation, $\mathcal{E}_{net} = \mathcal{E}_{BBL} + \mathcal{E}_{SML}$, obtaining

$$\frac{\mathcal{E}_{BBL}}{\mathcal{E}_{net}} = 1 + \frac{F_{SML}}{F} \frac{1}{\mathcal{E}_{net}} \frac{1}{N^2 d} \int_{b_{\min}}^{b} \mathcal{E}_{net} db'. \tag{7.30}$$

This ratio of the upwelling in the BBL to the net diapycnal transport might be called an amplification factor, and this equation is a handy relationship for this amplification factor in terms of a vertical integral of the net upwelling transport. For example, Eq. (7.30) shows that in the very densest parts of the world ocean, the diapycnal transport occurs mainly in the BBL, with negligible sinking in the SML (from Eq. (7.30) with $b \approx b_{\min}$). If the net upwelling transport $\mathcal{E}_{net}$ is assumed to be independent of buoyancy (or height), then for a typical value of N^2 of 1×10^{-6} s^{-2} at a depth of 2500 m, where $b - b_{\min} \approx 2.5 \times 10^{-3}$ m s^{-2}, and with $d = 475$ m, Eq. (7.30) requires the upwelling in the BBL, $\mathcal{E}_{BBL}$, to be $6\mathcal{E}_{net}$ and the sinking diapycnal transport in the SML to be $-5\mathcal{E}_{net}$. For example, if the net upwelling transport at this depth is 25 Sv, then the sinking transport in the SML is 125 Sv, while 150 Sv is upwelled in the BBL.

It seems more realistic to assume that the net upwelling transport is a linear function of buoyancy, and under this assumption, Eq. (7.30) requires that at a depth of 2500 m $\mathcal{E}_{BBL} = 3.5\mathcal{E}_{net}$ and $\mathcal{E}_{SML} = -2.5\mathcal{E}_{net}$. For example, if the net upwelling transport at this depth is 25 Sv, then the sinking transport in the SML is 62 Sv, while 87 Sv is upwelled in the BBL. Using the Nikurashin and Ferrari (2013) parameterisation of breaking internal waves in realistic topography, Ferrari et al. (2016) evaluated $\mathcal{E}_{BBL}$, $\mathcal{E}_{SML}$ and $\mathcal{E}_{net}$ as functions of buoyancy, finding $\mathcal{E}_{BBL}/\mathcal{E}_{net} \approx 3.8$ at a neutral density of 28.1 kg m^3, whereas McDougall and Ferrari (2017) applied Eq. (7.30) to the net upwelling transport $\mathcal{E}_{net}(b)$ of the Lumpkin and Speer (2007) inverse study, finding $\mathcal{E}_{BBL}/\mathcal{E}_{net} \approx 6$ at the same neutral density value. Yet another estimate of $\mathcal{E}_{BBL}/\mathcal{E}_{net}$ can be found from the work of de Lavergne et al. (2017), where the integral in Eq. (7.30) can be estimated from Fig. 5c of that paper, yielding $\mathcal{E}_{BBL}/\mathcal{E}_{net} \approx 2$ at a neutral density of 28.11 kg m^3, so that, taking the net upwelling transport at this depth to be 25 Sv, the sinking transport in the SML is 25 Sv, while 50 Sv is upwelled in the BBL.

The above estimates of the ratio $\mathcal{E}_{BBL}/\mathcal{E}_{net}$ vary between 2 and 6 in the abyssal ocean, and in our view, values between 2 and 4 are realistic estimates. Using this range of values for $\mathcal{E}_{BBL}/\mathcal{E}_{net}$, and with the net upwelling transport at this depth estimated at 25 Sv, the sinking transport in the SML is estimated to be between 25 Sv and 75 Sv, while between 50 Sv and 100 Sv is upwelled in the BBL.

In the present paradigm, much of the ocean interior is close to being adiabatic, while there is strong diapycnal upwelling occurring hard against the sloping bottom in a BBL that is only tens of metres thick. Taking an average bottom slope of 1/400, the BBL extends only $\sim$20 km horizontally, and then there is a broader region of substantial diapycnal sinking in the SML, whose thickness is about 500 m, and which has a horizontal width of $\sim$200 km. This view of the diapycnal transport, where all of the upwelling occurs in narrow BBLs within $\sim$20 km horizontally of rough topography, stands in stark contrast to the previously assumed horizontally uniform upwelling.

In the above discussion of bottom-intensified mixing, we have ignored (i) the geothermal buoyancy flux, (ii) a small constant background diffusivity of 1×10^{-5} m^2 s^{-1}, and (iii) the effects of the non-linearities of the equation of state. The geothermal buoyancy flux arrives directly in the BBL and is estimated to cause an additional $\sim$4-8 Sv of diapycnal upwelling in the BBL (de Lavergne et al., 2016a).

A background diapycnal diffusivity of 1×10^{-5} m^2 s^{-1} causes an upwelling diapycnal volume flux of $\sim$2.5 Sv, with only a small part being in the BBL (McDougall and Ferrari, 2017), whereas the cabbeling and thermobaric dianeutral advection processes (caused by lateral mixing and the nonlinear nature of the equation of state) are thought to cause a diapycnal sinking transport of approximately this same amount ($\sim$2.5 Sv) in the ocean interior (Iudicone et al., 2008) and (Klocker and McDougall, 2010). We conclude that the net diapycnal volume transport caused by a background diapycnal diffusivity, while small, is probably counteracted by the sinking through isopycnals caused by cabbeling and thermobaricity.

A separate type of non-linear equation of state effect arises in the parameterisation of the diapycnal buoyancy flux and its divergence. Since density and buoyancy are not conservative variables, the material derivative of buoyancy is not only caused by the divergence of the turbulent diffusive flux of buoyancy (as we have assumed in this article), but there are additional terms due to the spatial variation of the thermal expansion and saline contraction coefficients. This issue has been discussed in McDougall (1984); Klocker and McDougall (2010); IOC (2010); de Lavergne et al. (2016b), where it was found to be of leading order in the upper 1500 m of the ocean, but we do not mention it further here.

7.3.5 Net upwelling in the abyss depends mainly on the shape of the ocean floor

For there to be net upwelling of water through isopycnals, the area-integrated buoyancy flux F needs to increase with buoyancy (since from Eq. (7.23), $\mathcal{E}_{net} = dF/db$). For bottom-intensified mixing characterised by a vertical decay scale d, the area of active mixing on each isopycnal A_{mix} scales as the horizontal width $d/\tan\theta$ times the perimeter L of the isopycnal along the topography. Hence, $F \sim \mathcal{B}_0 A_{mix} \sim \mathcal{B}_0 L d/\tan\theta$ so that $F^{-1} dF/db$ scales as

$$\frac{1}{F}\frac{dF}{db} \sim \frac{(\mathcal{B}_0)_b}{\mathcal{B}_0} + \frac{(A_{mix})_b}{A_{mix}} \sim \frac{(\mathcal{B}_0)_b}{\mathcal{B}_0} + \frac{(L)_b}{L} - \frac{(\tan\theta)_b}{\tan\theta} + \frac{d_b}{d}. \qquad (7.31)$$

This suggests that there are four different ways that net upwelling can be enabled, namely (i) if the magnitude of the buoyancy flux at the top of the BBL $\mathcal{B}_0$ is an increasing function of buoyancy, (ii) if the length (perimeter) L is an increasing function of buoyancy, (iii) if the slope of the sea floor $\tan\theta$ is a decreasing function of buoyancy, and (iv) if the vertical length scale d is an increasing function of buoyancy. de Lavergne et al. (2017) have implemented a wide variety of mixing scenarios and with each have evaluated the area-integrated buoyancy flux F along neutral density surfaces of a hydrographic atlas, paying attention to the ocean floor topography. From F as a function of buoyancy, the net upwelling rate $\mathcal{E}_{net} = dF/db$ was calculated as a function of buoyancy, and de Lavergne et al. (2017) found that this upwelling $\mathcal{E}_{net}(b)$ in the abyss was quite insensitive to the choice of mixing scenario. Similar shapes for $\mathcal{E}_{net}(b)$ were found for bottom-intensified mixing as for geothermal heating (which has the same implications for diapycnal transports as a vertical profile of constant turbulent buoyancy flux, which goes to zero abruptly at the sea floor).

The reason for this behaviour is the dominance of the topography of the sea floor in the calculation of $\mathcal{E}_{net}(b)$. In the case of bottom intensified mixing, the second and third terms on the right-hand side of Eq. (7.31) together represent the effect of sea floor geometry (with respect to density coordinates), and these terms were found to dominate Eq. (7.31) in the abyss. With bottom-intensified mixing, an isopycnal that has a larger area close to the sea floor is exposed to a larger area-integrated turbulent buoyancy flux than one that has a smaller area close to the bottom. This idea is encapsulated in the term "incrop area", which is the area of the sea floor that is seen by fluid within a small range of density between one density value and the next, so that "incrop area" has dimensions of area per unit buoyancy difference. de Lavergne et al. (2017) found that the "incrop area", $(A_{mix})_b$, was well-correlated with the net upwelling transport $\mathcal{E}_{net} = dF/db$ in the abyss, and this relationship held for a wide range of parameterised mixing scenarios. This finding was confirmed in the study by Holmes et al. (2018), where it was found that below a depth of 4500 m the change with buoyancy of the perimeter, where neutral density surfaces intersect the sea floor dominates, the calculation of the net diapycnal upwelling, though shallower than this 4500 m depth, the incrop area decreases with height, implying that in this depth range, the only way to achieve net upwelling is by $\mathcal{B}_0$ increasing with buoyancy (the first term in Eq. (7.31)), or by geothermal heating (which has the same transport implications as an interior buoyancy flux profile that does not change with height above the BBL).

This study by de Lavergne et al. (2017) also strongly points to the need to think of $\mathcal{E}_{net}(b)$ increasing strongly with buoyancy in the abyss rather than being a constant net upwelling transport. This, plus the strong influence of the sea floor topography ("incrop area" $(A_{mix})_b$) on the shape of $\mathcal{E}_{net}(b)$, and the relative insensitivity to the mixing scenario, represent strong contrasts to the one-dimensional, horizontally uniform, upwelling-diffusion paradigm.

7.3.6 What can be learned from purposefully released tracers?

In the above, we have concentrated on understanding the implications of the bottom intensification of ocean mixing for driving the net diapycnal upwelling $\mathcal{E}_{net}$ in the deep ocean, as well as on estimating how much larger than $\mathcal{E}_{net}$ is the upwelling $\mathcal{E}_{BBL}$ that occurs in the thin bottom boundary layers. Given the diapycnal upwelling and downwelling that occurs near the sloping sea floor, one wonders what imprint this motion might have on the distribution of tracers in the ocean. Holmes et al. (2019) have taken some initial steps in understanding this problem, concentrating their investigations on a tracer that is purposely released either in the BBL or just above it in the SML, and using the velocity field of the one-dimensional boundary layer theory of Garrett (1990, 1991) and Callies and Ferrari (2018).

The tool here is conceptually straightforward. Given observations of a tracer $C(x, y, \rho, t)$ released on an isopycnal surface $\rho(t_0)$, the effort is to interpret the time-evolving dispersion and mean drift of the tracer concentration about $\rho(t_0)$ in terms of the diapycnal mixing K. The simplest situation in which K is spatially uniform is given by (7.12). If $K = K(\rho)$ and is independent of (x, y), the first moment is proportional to $\partial K/\partial \rho$ (Ledwell et al., 1998). With boundary mixing, there is the potential for the transfer of tracer into and out of boundary layers along isopycnals as well as the complications of boundary conditions (7.4) to consider. A conceptually straightforward tool has attained a potentially complicated interpretation.

When their tracer was released in the BBL, its centre of mass moves strongly upwards to greater buoyancies while simultaneously, its diapycnal spread, as estimated from the rate of increase of tracer variance in buoyancy space, is much less than would be suggested by simply averaging the diapycnal diffusivity over the tracer patch. The reduction in the diapycnal spreading of the tracer occurs, because the tracer is constrained to not diffuse through the sea floor, and also by the dipole of upwelling in the BBL and downwelling in the SML. When the tracer is injected in the SML its centre of gravity initially moves with the fluid downwards to greater densities.

The diapycnal diffusion of tracers in this near-boundary situation is also sensitive to the presence of along-isopycnal diffusion, or equivalently, to the intermittent exchange of fluid out of and into the near-boundary region, as emphasised by Armi (1979b). Suffice it to say that the diapycnal movement and diffusion of a tracer through the upwelling and downwelling regimes of the near-boundary flows, is very complicated, especially in the presence of lateral exchanges along isopycnals. What is clear from the initial work in this area by Holmes et al. (2019) is that, for a purposefully released tracer in the BBL, it is not uncommon for the actual tracer-weighted diapycnal diffusivity to be three times as large as would be naively estimated by examining the rate of change of the tracer variance in buoyancy space.

7.3.7 Implications for the circulation of the abyssal ocean

Stommel and Arons (1959) posited a spatially uniform vertical velocity at mid-depth as the cause of vertical stretching of water columns in the deep-ocean interior that drives the abyssal circulation through the linear vorticity equation. At first blush, there would seem to be a problem with the sign of the interior Stommel–Arons abyssal circulation, simply by realising that as the dense plumes of Antarctic bottom water sink into the abyss, they are traditionally assumed to entrain, and thus increase their volume flux as they descend to the deepest part of the ocean. As pointed out by McDougall (1989) and by Rhines (1993), because of this increasing volume transport with depth, and more importantly, also due to the decreasing ocean area as a function of depth, the average vertical velocity across successive heights, required simply by continuity, is a decreasing function of height, which is the opposite of the assumed stretching of water columns in the Stommel–Arons theory. The resolution of this conundrum will likely rely on the work of Baines (2001, 2005) and Hogg et al. (2017), which shows that dense gravity currents do not always entrain fluid from the environment as they make their way to the sea floor. Rather, on sufficiently weak slopes, descending plumes detrain fluid into the environment.

In thinking about diapycnal motion across isopycnals in the abyssal ocean, it is useful to first imagine an isopycnal to be essentially at constant height. This then leaves a small area at this height that is not of the same density as the isopycnal; this being the very small area through which the very dense bottom water is descending at the rate $\mathcal{E}_{net}$ (because, assuming the abyssal ocean to be in a steady state implies that the sinking volume flux in this dense plume is equal to the net upwelling across the remainder of the surface), and we call this the first area at this height. The second area is the large approximately adiabatic interior area, where the mixing activity is small. The third and fourth areas are the regions we have identified as the SML and BBL, through which fluid sinks and rises respectively, in both cases caused by the divergence of bottom-intensified turbulent diapycnal buoyancy fluxes.

We now ask the question, 'In which of the last three regions would we expect there to be a linear vorticity balance that might support a horizontal circulation in the abyss in the sense of Stommel–Arons?' The rather adiabatic interior region would be sufficiently linear as to obey the linear vorticity equation, but with very small diapycnal motion and almost horizontal density surfaces, there is no appreciable vertical vortex stretching to drive a significant horizontal circulation. This leaves the BBL and the SML. The BBL contains much non-linear dynamics and would not be a candidate for application of the linear vorticity balance. This leaves the SML where the linear vorticity balance might be thought to apply, especially as the vertical derivative of the diapycnal velocity in the SML is of the same sign as the Stommel & Arons vortex stretching; the difference being that in the SML the vertical velocity is negative and increasing in magnitude downwards! However, the realisation following Callies and Ferrari (2018), that the adiabatic bolus advection is important in the SML to achieve the observed near-bottom stratification gives serious pause for concern, because in the context of the Eulerian average velocity, the effect of the bolus transport is equivalent to a large vertical viscosity, which will likely render a linear vorticity balance unjustified. It seems then that we must conclude that with bottom-intensified diapycnal mixing there is no region in the abyssal ocean where the linear vorticity balance is appropriate, except in the almost adiabatic interior, which does not seem capable of driving an interesting circulation.

Callies (2018) has modelled the abyssal circulation using planetary geostrophic dynamics with Rayleigh friction, in a study with bottom-intensified turbulence and sloping bottom topography. The strong and narrow upwelling in the BBLs on both the western and eastern sides of the bathtub basin was clearly evident. Unlike the Stommel–Arons case, where the poleward flow is spread throughout the zonal width of the basin and any equatorward flow occurs in the deep western boundary current, in their study poleward flow occurred near both western and eastern boundaries. Drake et al. (2020) have run this type of planetary geostrophic ocean model, but now with strong bottom-intensified mixing at a mid-ocean ridge, and with a realistic vertical stratification to which the model was restored in its southern-most region, as well as some runs with relaxation towards a constant vertical stratification. This study confirms that the BBL upwelling, and particularly the SML downwelling, are found to depend on the detailed nature of the topography and of the imposed vertical stratification. The meridional component of the horizontal circulation was concentrated in the regions of intense diapycnal mixing on the flanks of the mid-ocean ridge, with poleward flow on the eastern side of the ridge and equatorward flow on the western side of the ridge. These are regions in which the diapycnal mixing and the Rayleigh friction are both significant, and their horizontal circulation makes a striking contrast to that of the Stommel–Arons circulation.

7.3.8 Summary remarks

The realisation, from microstructure observations, that much of the diapycnal mixing in the deep ocean is concentrated near the sea floor originally came from the Brazil basin experiment more than twenty years ago. This bottom intensification of ocean mixing in the deep ocean has now been confirmed in many studies and in compilations of turbulence measurements (Waterhouse et al., 2014). Away from the rough topography at the ocean edges the background turbulent diffusivity is thought to be approximately 1×10^{-5} m^2 s^{-1}, a value which drives rather small diapycnal volume transports, which are likely cancelled by the downwards diapycnal motion arising from cabbeling and thermobaricity.

With the early one-dimensional boundary layer theoretical studies of Garrett (1990, 1991) leading the way, we now have confidence that bottom-intensified mixing leads to strong upwards diapycnal volume transport hard up against the sloping sea floor in what we now call the bottom boundary layer (BBL). Immediately above this BBL, where the magnitude of the turbulent buoyancy flux decreases with height, is the stratified mixing layer (SML) in which fluid parcels become denser and move through isopycnals towards denser (deeper) water. The Walin-type buoyancy budget approach yields rather simple relations between the diapycnal volume transports in the BBL and the SML, as a function of how the net upwelling varies in the vertical as a function of buoyancy (Eqs. (7.29) and (7.30)). These relations indicate that the upwelling transport in the BBL is often several times as large as the net upward volume transport, and this has been confirmed in general circulation model studies. However, it must be said that to date, the diapycnal transport in the BBL has not been directly observed.

These observational and theoretical insights have direct implications for how we should parameterise diapycnal mixing in numerical ocean models, and also on the importance of a proper representation of bottom topography. If a diapycnal diffusivity is used to parameterise the small-scale turbulent mixing, then care should be taken to ensure that it varies sufficiently fast in the vertical, so as to outweigh the increase in stratification in the vertical, thus ensuring that the magnitude of the turbulent buoyancy flux decreases with height. Note that this can only be checked after running the model, because the stratification changes during the model run. In addition, the ocean model needs to explicitly resolve or parameterise the sub-mesoscale instabilities that are thought to transport mass adiabatically along isopycnals when they slope significantly near ocean boundaries (Callies and Ferrari, 2018).

Several studies have shown that the net diapycnal upwelling transport in the abyss is a strongly increasing function of buoyancy (rather than being a fixed volume transport), and this has implications for the way that the dense plume-like sources of bottom water sink to the deepest parts of the ocean, and then spread out at depth. The concept of continual entrainment into these plumes is due for wholesale revision.

Many studies have pointed to the importance of the topographic shape of the sea floor in regulating the upwelling of bottom water in the abyss, and this has been convincingly demonstrated by de Lavergne et al. (2017), where a variety of different mixing parameterisations all led to essentially the same net upwelling transport across isopycnals in the abyss.

Though we have some understanding, based on buoyancy budget analyses, of the net diapycnal transport and the larger upwelling transport in the BBL, we do not yet have a clear view on the importance of these transports on the movement and diffusion of tracers in the ocean, nor of the dynamical consequences of these up- and downwelling motions in driving the horizontal abyssal circulation. The horizontal circulation patterns that can be seen in Callies (2018) and Drake et al. (2020) bear no resemblance to that of the Stommel–Arons circulation, and moreover, the bottom intensification of ocean mixing and the necessity of adiabatic bolus transport to achieve realistic vertical stratifications means that the linear vorticity balance on which the Stommel–Arons circulation rests is probably not applicable in any interesting part of the abyssal ocean. Since it is only a few years since the dominant paradigm has shifted from the horizontally uniform upwelling-diffusion balance of Munk (1966) to the paradigm of counter-flowing up- and downwelling transports near ocean boundaries, we can be hopeful that our understanding will develop rapidly in this area of oceanography.

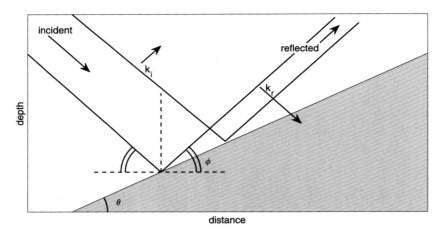

FIGURE 7.8 Ambient wave reflection. The key to understanding how the kinematic bottom boundary condition of no-normal flow changes the internal wave field over variable topography is contained in the internal wave dispersion relation. The internal wave frequency is determined not by wavelength, but rather by the angle of the wave vector, $\phi = \tan^{-1}(\sqrt{(k^2 + l^2)/m^2})$ (7.32), and this determines the trajectories of water parcels. A simple visualisation can be achieved by drawing the crests and troughs of a wave incident at the boundary and requiring that the reflected wave matches the propagation of these crests and troughs along the boundary (black solid lines). Depicted are incident and reflected wavenumbers $\mathbf{k}_i$ and $\mathbf{k}_r$, with phase propagating in the direction of the wavevectors. Wave frequency, and hence ϕ, are preserved under reflection, independent of the orientation of the slope α. Transformation of incident energy to very high wavenumber $\mathbf{k}_r$ occurs at a frequency for which $\phi \cong \theta$ (and thus $\vartheta \cong \alpha$), leading to enhanced wave-breaking and mixing (e.g., Chalamalla et al., 2013). Figure from Bracco et al. (2020), after Garrett (1991).

7.4 Production mechanisms for boundary mixing

The dynamics governing the physical evolution of stratified rotating flows over complex topography represent an especially difficult problem. A simple but key notion concerns the internal wave dispersion relation:

$$\vartheta = \sqrt{\frac{k^2 + l^2}{m^2}} = \sqrt{\frac{(\omega^2 - f^2)}{(N^2 - \omega^2)}} \qquad (7.32)$$

for plane waves with wavenumber $\mathbf{k} = (k, l, m)$ and frequency ω. Wave frequency ω is simply related to the wave aspect ratio ϑ and very efficient coupling into small vertical scales occurs when this aspect ratio meets the topographic slope, $\alpha = \tan \theta$. In this section, we present some of the basic turbulent mixing mechanisms, with an eye upon the one-dimensional model and its underpinning assumptions.

7.4.1 Internal wave reflection / internal tide generation

The prototype problem emphasises conditions in which the large-scale inclination of the topography approximately matches that of ray trajectories associated with the energy-containing frequencies of the internal wavefield (Fig. 7.8). Reflection and/or generation of internal waves under critical conditions for which wave aspect ratio ϑ (7.32) equals the topographic slope α are understood to result in significant velocity gradients and mixing (e.g., Eriksen, 1985; Garrett and Gilbert, 1988). Imprinted upon this class are the spatial inhomogeneity and temporal intermittency of the forcing parameters (i.e., tides, the background wavefield, topography). We note several caveats:

- Internal wave reflection is a less productive mixer than the idealised semi-infinite problem might have you believe. The wave reflection process was the target of a process study (Eriksen, 1998) on the steeply sloping flanks of Fieberling Guyot, a tall seamount in the eastern Pacific. While turbulent dissipation was significantly enhanced toward the bottom (Toole et al., 1997), the vertical shear responsible for that dissipation was enhanced in the along-slope direction, rather inconsistent with the linear internal wave kinematics that underlies Fig. 7.8. The reason is unresolved. However, there are ridges and ravines running up and down the flanks of the Guyot, like that depicted in Fig. 7.10 that could give rise to the along-slope shear. One finds that the linear reflection problem does not have a solution for super-critical incident waves *if* the bottom boundary condition (7.4b) requires the reflected wave to have an along-slope component (Eriksen, 1982). More speculatively, unlike the semi-infinite plane, the ocean does have a free surface.
- There are similar issues with internal tide generation. The kinematic bottom boundary condition (7.4b) stating that flow parallels the bottom is nominally applied at bottom, $z = h(x, y)$. In a small amplitude limit, for which $\vartheta \gg \alpha$, one

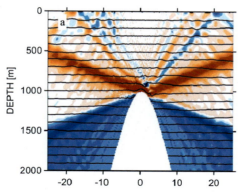

FIGURE 7.9 Internal tide generation for a super-critical ridge, in which internal waves of the tidal frequency have smaller aspect ratio (7.32) than the topographic slope α. The baroclinic velocity field from a model simulation by Klymak et al. (2010) is depicted. There are 4 main beams emanating from the ridge crest associated with the forcing and super harmonics at greater angle to the horizontal.

approximates this by applying (7.4b) at $z = 0$. The benefit of this approximation is that one can take Fourier transforms of the topographic roughness fields in Fig. 7.3 and, in a simple manner, translate those transforms into the internal wave response and ultimately to turbulent dissipation associated with wave breaking (Polzin, 2004b). The draw-back is that one loses connection with the enhanced shear and non-linear phenomena implied by the contraction of ray tubes in Fig. 7.8.

Expectations of a strong enhancement of vertical shear associated with the contraction of ray-tubes are not supported by observations from the Brazil basin (Polzin, 2004b). The manner in which this was inferred is that a Fourier representation of topographic roughness was routed through a linear model of internal tide generation, and this served as the boundary condition for a non-linear internal wave propagation model that, in turn, provided an estimate of turbulent dissipation as a function of height above boundary. The essence of that non-linear propagation model was a direct relationship between internal wave shear and dissipation (Polzin, 2004a). Both predicted shear and predicted dissipation using the small amplitude topography approximation were in agreement with observations. Qualitative assessments (Polzin, 2004b, see their appendix) using a finite topography model (Baines, 1973) suggested an order of magnitude enhancement of shear variance associated with near-critical topographic slopes that was not observed.

The following rationale was put forward concerning why the mathematically more sophisticated theoretical treatment was not consistent with the observations: that theoretical treatment of finite amplitude topography is essentially one-dimensional. This one-dimensional assumption may be germane to a continental slope—shelf break—continental shelf planform or an extended feature, such as the Hawaiian ocean ridge, but the abyssal hill roughness situated atop the Mid-Atlantic ridge in Fig. 7.3 is two-dimensional. In one-dimension, the flow is forced to be up and over topography. In two dimensions, there is the potential for steady flow to go around an obstacle in instances in which it might otherwise be 'blocked' from going up and over (e.g., Epifanio and Durran, 2001).

In summary, basic linear internal wave kinematics points to near-critical slopes, for which $\vartheta \cong \alpha$ as being places that should host enhanced internal wave shear and mixing. However, when one gets to the point of being quantitative, there are some essential details about steep, rough topography that we do not understand. In particular, these inconsistencies point to expectations based upon idealised one-dimensional representations as being naive.

Idealised simulations of critical and near-critically reflected internal waves / internal tide generation (Slinn and Riley, 1998; Gayen and Sarkar, 2010; Chalamalla et al., 2013; Sarkar and Scotti, 2017) reveal episodic convective near-boundary overturning associated with the kinematics of the internal waves. The picture is one of the wave tripping over itself in an 'internal swash zone'. The efficiency of convective mixing near the boundary rather argues against a generic lengthscale suppression of Γ to support a mid-depth minimum in buoyancy flux. Significant ($O(1)$) perturbations of density gradients that cycle with the wave period are inconsistent with the underpinnings of a mixing length closure (7.2) and suggest skepticism.

The supercritical limit

The supercritical ($\vartheta \ll \alpha$) limit of the internal tide generation problem (Fig. 7.9) is one that can also host strong mixing. Here, the topography is sufficiently steep that the contraction of ray-tubes plays little role in enhanced shear to drive mixing. Rather, enhanced near-boundary mixing is associated with a downslope jet and overturning wave in the lee of the ridge.

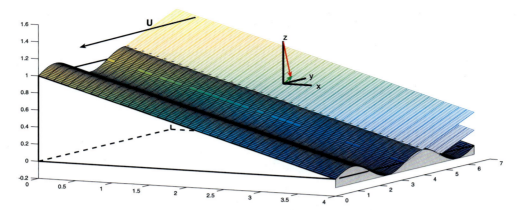

FIGURE 7.10 Lee waves: Mean flow over irregular topography can induce radiating internal waves if the topographic irregularities superimposed on the slope have a 'small' amplitude (figure from Bracco et al. (2020), after Thorpe (1992)). In this case, one assumes a quasi-stationary internal wave response that varies on the time scale of the background flow. Mean flow direction is indicated by the arrow in association with U. A right-hand x-y-z coordinate system defines the orientation of the wave vector (red) normal to the wave phase planes. On an inclined plane, the projection of the wavevector onto the horizontal (green) lies in the upslope direction. Without the inclined plane, the dynamical transition between waves and hydraulics occurs when the topographic perturbations have an amplitude similar to that of the lee wave; flow blocking initiates and vertical shear in the wave response is approximately equal to the buoyancy frequency N. The dynamical impact of the inclined plane as concerns blocking/splitting is largely unexplored.

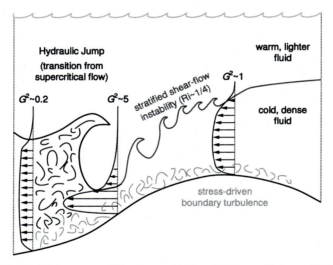

FIGURE 7.11 Internal hydraulics: Intense mixing can occur if the topographic irregularities are 'finite amplitude' and a parameter regime of internal hydraulics is entered (figure from Nash and Moum (2001)). Turbulence and mixing occur in conjunction with bottom stress in an accelerating downslope jet, high shear on an interface with overlying fluid and within the hydraulic transition (jump).

These features are reminiscent of the internal hydraulic response to a steady flow, Fig. 7.11. Here, however, the downslope jet and overturning wave cycle from one side of the ridge crest to the other in sync with the barotropic tidal flow. The foundations of a mixing length theory, i.e., small perturbations on a steady flow, are likely challenged in this environment.

7.4.2 Sub-inertial flow and topography

Rough topography on a slope

A second class of turbulence forcing mechanisms emphasises subinertial flows (i.e., eddies and mean flows) in combination with topographic variability. When the amplitude of the topographic variability is small and has length scales permitting the radiation of internal waves, then internal lee waves (Bretherton, 1969; Thorpe, 1992) are generated (Fig. 7.10). When the topographic variability is large, however, near-boundary wave breaking, hydraulic-like effects, flow separation and vortex shedding arise, and dominate the radiating response, e.g., Baines (1997); Fig. 7.11. Boundary mixing associated with these finite amplitude phenomena can be quite intense and typically occurs in stratified waters apart from the boundary.

There is a wealth of knowledge within the atmospheric science community concerning stratified flows over topography. Such studies are usually performed for isolated topography (with lengthscale ℓ) in a non-rotating limit, specifically $U/f\ell \gg 1$, where U is the typical lateral flow velocity and f the Coriolis parameter. Assuming vertically uniform background flow and constant stratification, over one-dimensional topography with height h and length ℓ scales, the steepness parameter $Nh/U : 1$ separates a linear wave regime from a high drag state featuring hydraulic effects. In two dimensions this boundary is somewhat smaller, $Nh/U : 1/2$ (Nikurashin et al., 2014), with the transition between 1-D and 2-D requiring elongated ridges of length-to-width aspect ratios of 1:10 (Epifanio and Durran, 2001). One can argue for this dynamical boundary on the basis of energetics: flow below the topographic crests lacks sufficient kinetic energy to trade for potential energy to traverse the ridge crest and is either blocked, or in 2-D, splits and flows around the topographic obstacle. The second way to argue for this Nh/U parameter boundary uses linear wave kinematics: assuming a steady response, the stationary lee wave has an apparent frequency of kU with horizontal wavenumber $k = 1/\ell$, a vertical wavelength of U/N, and propagates at a slope with respect to the horizontal of k/m that must exceed the topographic slope h/ℓ for the linear response to be valid.

Extrapolating from the atmospheric literature is hazardous for the following reasons:

- The oceanic paradigm is one in which topography is not isolated and the scales of topography are such that rotation cannot be neglected [$U/f\ell \sim O(1)$]; with rotation, the linear kinematics that links wave steepness to energetics are fundamentally altered.
- The presence of a large-scale slope alters the kinematics of the linear problem so that the lowest frequency internal wave has a wavenumber that points upslope (Thorpe, 1992, see Fig. 7.10) rather than being parallel to the roughness wavenumber. This fundamental change in the steady linear solution modifies the blocking criteria in uncertain ways.
- Filtering of topography to address lee-wave generation (Scott et al., 2011; Nikurashin and Ferrari, 2011; Kunze and Lien, 2019) is problematic, because the bottom boundary condition that flow has to be parallel to the boundary (7.4b) is non-linear in topographic height. Claims (Kunze and Lien, 2019) that filtered topography typically lands in a small Nh/U parameter regime seem to miss the point: a simple filtering of the topography is implicitly equivalent to a linearisation, and the bulk of the topographic height variance lies in the subinertial [$U/f\ell < O(1)$] regime. Fig. 7.10 presents an idealised representation of multi-scale topography, in which the linear wave generation problem is understood to differ. The dynamical impact of the inclined plane on flow blocking is largely unexplored.

Global estimates of lee-wave transformation rates (Scott et al., 2011; Nikurashin and Ferrari, 2011) have value in that they provide order of magnitude constraints on the process. They depend upon a hypothesis of a quasi-stationary response, either lee-wave or hydraulic, to quasi-stationary flow. Verification of this is difficult. Moored ADCPs embedded in current metre arrays about a large-scale [$U/f\ell < O(1)$] seamount in Drake passage (Brearley et al., 2013; Cusack et al., 2020) and along a smaller scale [$U/f\ell > O(1)$] ridge on the western boundary off the Bahamas (Clément et al., 2016; Evans et al., 2020) document that the vertical shear variance is modulated on eddy time scales, but find this enhanced vertical shear appears at near- and super-inertial Eulerian frequencies, rather than being a quasi-stationary lee-wave response. Whether this time dependence relates to the forcing of super harmonics in a wave problem (Nikurashin and Ferrari, 2010) or relates to transients apparent in models of the boundary layer on a sloping bottom (Section 7.4.3) are open questions.

Constrictions, gaps and passages

One perspective concerning the global upwelling of abyssal waters is an abrupt departure from the conical basin pictured as a prototype of hypsometric issues in Section 7.3. This perspective considers the abyssal circulation as a series of interconnected basins. These basins are related through topographic constrictions, gaps and passages featuring enhanced mean flows and intense mixing related to hydraulic phenomena with the pressure head for that flow set up by external mixing processes (tides, etc.) in the interior basin. The importance of mixing related to hydraulic processes is made manifest in major passages by abrupt spatial changes in bottom potential temperature (e.g., Mercier and Morin, 1997; Alford et al., 2013).

The deep basin experiment (Morris et al., 2001) was a multi-national effort to quantify the input and output of dense water from the Brazil basin, relating to control volume studies in the early 1980s (Hogg et al., 1982; Whitehead and Worthington, 1982). From these early studies, CTD transects across the southern entrance to the basin revealed near-vertical isopycnals banked upon the eastern side of the channel. This was interpreted by Hogg (1983) as a product of rotating hydraulic withdrawal in the face of slowly varying topography, with those steep isopycnal slopes geostrophically balanced rather than implying much about mixing. In the following deep basin experiment the northeastern exit, represented by the Romanche and Chain fracture zones, was surveyed with fine and microstructure instrumentation (Polzin et al., 1996). Even though this exit is extensive, the passages are at low latitudes (within one degree of the equator) such that intense mixing associated with hydraulically controlled flows and internal hydraulic transitions was identified as being responsible for a 0.41 °C increase in bottom potential temperature.

With reference to Fig. 7.11, mixing in an internal hydraulic scheme can occur in conjunction with intense shear on an interface and subsequent hydraulic transitions (jumps) or in conjunction with bottom drag in a downslope jet. In the Romanche Fracture Zone, the intense shear occurred on the (relatively) strongly stratified boundary between North Atlantic Deep Water and Antarctic Bottom Water, and the relative maximum of dissipation there implied a downwelling of deep water and upwelling of bottom water. The control volume estimates of Mercier and Speer (1998) and turbulence estimates in Polzin et al. (1996) provided fodder for a back-of-the-envelope estimate presented in Ferron et al. (1998), which suggested that, despite the hydraulic processes leading to mean dissipation rates orders of magnitude larger than those associated with tidal mixing above roughness in the Brazil Basin (Polzin et al., 1997), both external tidal and local hydraulic processes were required to explain the control volume constraints for mass and buoyancy (upwelling and mixing).

The robustness of that result and the spatial separation between localised hydraulic processes within the Romanche vs tidal mixing in the eastern Atlantic provides the following insight: The hydraulic flow is associated with a pressure head between the western and eastern equatorial Atlantic: the densest water is in the west. The reason for that pressure head is the broadly distributed mixing in the farfield: without complete mixing of buoyancy by hydraulic processes, in the absence of a far-field (tidal) mixing process, dense water will slowly fill up the interior basin and diminish the pressure head. Thus such hydraulic transitions are not going to be the rate controlling process for abyssal upwelling.

Alford et al. (2013) present a similar example of intense mixing associated with hydraulically controlled flows within the Samoan passage. The Samoan passage is a 40 km wide notch in the South Pacific bathymetry, through which flows most of the water supplying the North Pacific abyssal circulation. Similar to the Romanche fracture zone, it is relatively low-latitude (8°S), and the issue of hydraulic control appears in conjunction with the smaller-scale topographic features within the Passage rather than the broader scale topography. These locations constitute 'hot spots' of mixing that dominate the Passage average (Voet et al., 2015). Cusack et al. (2019) argue that hydraulic processes in a supercritical flow regime in the hot spots dominate over tidally induced mixing processes and are very persistent in time. Application of an analytic model for internal hydraulic jumps in a stratified flow (Thorpe et al., 2018) revealed a remarkable ability to predict jump locations, changes in the baroclinic structure of the flow and turbulent dissipation. The model has a continuously stratified interface between two homogeneous layers. Transitions involving this interface were identified and could provide a rationale for upwelling and downstream warming trends of bottom potential temperature, without directly invoking the one-dimensional model or its boundary layer physics.

MacKinnon et al. (2008) describe observations from the Atlantis II fracture zone, which presents a pathway for Antarctic Bottom Water and Circumpolar Deep Water from the abyssal Southern Ocean into the deep Indian ocean. Enhanced mixing along the centre line of the fracture zone was hypothesised to be associated with the interaction of internal waves, possibly of tidal origin, with the vertical shear of an $O(0.2 \text{ m s}^{-1})$ mean current. The along channel evolution of watermass properties suggested greater diapycnal transformations than documented by the centre line measurements. The issue of mixing associated with internal hydraulics was discounted. The measurements do not appear to speak to the potential of mixing along the sloping sidewalls of the fracture zone.

A microcosm of these concerns exists within rift valleys and smaller fracture zones. Within the Brazil basin, a proposal (Thurnherr et al., 2005) that basin-wide mixing was largely sustained by mean flows in conjunction with occasional sills cutting across fracture zones and internal waves interacting with these mean flows (Clément and Thurnherr, 2018) seems not to have carried the day (Thurnherr et al., 2020). The circuitous route to this conclusion seems to originate in geographic groupings of microstructure profiles into valley, slope and crest bins (St. Laurent et al., 2001) that are visually inconsistent with data presented in Fig. 3 of Polzin et al. (1997). What is called for are dynamically grounded models (Polzin, 2004b, 2009; Thorpe et al., 2018) to aid understanding of the vertical structure and patterns of dissipation profiles, rather than geographic groupings unrelated to the underlying physics. From this dynamical perspective, the singular role of this fracture zone valley is that it acts as a conduit for mass transport, pursued further in Section 7.5. Different conclusions might be applicable to the rift valleys (Thurnherr, 2006; St Laurent and Thurnherr, 2007; Tippenhauer et al., 2015) of slow spreading ridges, such as the Mid-Atlantic ridge, in which the rift valley side-walls are significantly steeper than adjacent abyssal hills and the interesting internal tide generation physics occur at the topographic crests, e.g., Fig. 7.9.

7.4.3 Friction and sub-inertial flows

A third class of forcing mechanisms is provided by turbulent stresses associated with drag at the bottom boundary (Eq. (7.4a)), and an insulating condition on temperature (buoyancy, Eq. (7.4c)). Over sloping topography the drag triggers cross-slope mass and buoyancy transport (Fig. 7.12), ultimately leading to a reduction in the near-bottom flow and

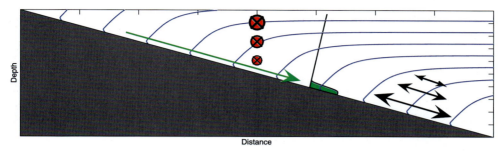

FIGURE 7.12 Mean flows and friction. This schematic depicts an Ekman layer over a sloping boundary in its downwelling configuration, rendered for a southern hemisphere. Drag exerted on the along-slope flow (red bullets and crosses) results in a downslope Ekman transport (green arrow and green near-boundary profile), advecting light water downslope and tilting isopycnals (blue contours). Statically unstable conditions are attained, driving boundary layer turbulence similar to that of a convectively unstable atmosphere. Note that the isopycnals are not normal to the bottom on the scale of the figure. The system has a resonance at frequencies for which $\vartheta = \alpha$, and the boundary forcing is carried aloft by mechanisms related to these cross-slope motions, depicted as black arrows. The schematic depicts the convectively unstable region having a smaller height scale than the penetration depth of the turbulent momentum flux (bottom drag), which in turn has a smaller height scale than the cross-slope process.

a decrease in drag as the near-boundary buoyancy anomalies come into geostrophic balance. The steady-state limit, with zero bottom velocity, is known as the arrested Ekman layer (Garrett et al., 1993). In the case in which the flow is oriented in the direction of Kelvin wave propagation, Ekman transports are downslope, near-boundary stratification is reduced, and the situation is referred to as a downwelling Ekman layer. For retrograde motion, upslope transport leads to increased stratification at the boundary. Below we focus upon the downwelling case and avoid the arrested state.

There are two key points. First, this downwelling boundary layer system is driven by downslope transports of light water, which results in statically unstable conditions on a height scale determined by the penetration of the turbulent momentum flux into the fluid. This places the oceanic boundary layer into a convectively unstable parameter regime, similar to the daytime atmospheric boundary when heated from below (Stull, 2012). Second, this convective process can resonantly couple with internal wave band motions, for which $\vartheta = \alpha$, Fig. 7.13.

Using a large eddy simulation (LES) of the downwelling Ekman layer, Ruan et al. (2019) point out that their model behaviour parallels that of the prototypical 'night-time' atmospheric boundary layer, in which the surface is cooler than the overlying air, in which the boundary provides a source for buoyancy and provides for stratification. Taken at face value, this is bizarre: the downwelling Ekman layer exhibits convection, which is the prototypical *daytime* atmospheric boundary layer when solar insolation results in a surface that is warmer than the air above. A potential reconciliation is to note that the insulating bottom boundary condition requires positive stratification and the question becomes, 'What is the lengthscale over which this pertains?'.

In the LES model, there are two modes of behaviour. If frictional effects are large, the boundary layer is turbulent. If buoyancy forcing is large, the boundary layer 'laminarises', i.e., it becomes non-turbulent. In the atmospheric boundary layer, the metric to judge whether friction or buoyancy forcing is 'large' lies in the size of the Obukov lengthscale relative to about 100 wall units (Stull, 2012). If the Obukov lengthscale is greater than approximately 100 wall units, the boundary layer is turbulent. Ruan et al. (2019) find that the downwelling Ekman layer exhibits a similar transition when the buoyancy forcing is expressed as the depth-integrated downslope Ekman buoyancy transport.

Within a molecular sublayer, the flux-gradient relation (7.2) holds, and within this molecular sublayer, the insulating bottom boundary condition (7.4c) provides $\partial_{\hat{z}} b = 0$. This translates into positive stratification at the boundary, $\partial_z b > 0$. The question apparently becomes, 'When is the flow strong enough that this molecular condition can't project into/past the viscous sublayer and trigger a laminarisation of the boundary layer?' The answer, from Ruan et al. (2019), is that their modified Obukov lengthscale needs to exceed roughly 100 wall units for the flow to laminarise. Whether such a transition exists for a hydrodynamically rough surface, one in which surface texture (e.g., gravel) extends into the viscous sublayer, is an open question, but the important point here is that the discussion concerning the structure of the 'BBL' and 'SML' has been moved from a $\hat{z}$ of approximately 100 m (Ferrari et al., 2016) to a $\hat{z}$ of 1 cm.

Observations that document the essential convective physics are rare, with prominent exceptions presented in Moum et al. (2004) and Kolås and Fer (2018). Many additional studies of observational inference (Trowbridge and Lentz (2018) and references therein) similarly come from shelf and upper slope regions. Our understanding of the downwelling Ekman boundary layer in the deep ocean is thus grounded in idealised modelling studies executed in the slope-rotated coordinate system (7.6). The most relevant of these implement turbulent closure schemes underpinned by the concept of an unstratified wall layer or the resulting quadratic drag (7.4a).

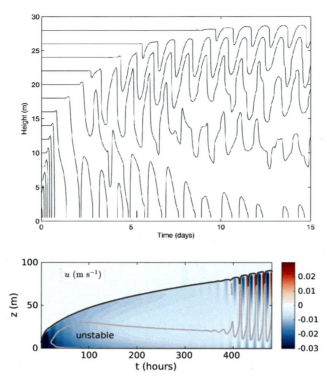

FIGURE 7.13 A key issue for the Ekman layer process is the limited penetration depth of the turbulent drag relative to the arrest height scale for typical abyssal stratifications. In the one-dimensional models of the sloping Ekman layer, mixing is carried aloft by the agency of cross-slope motions at the resonance frequency, for which $\vartheta = \alpha$. The upper panel depicts isopycnal contours from Brink and Lentz (2010) for an $s \sim O(1)$ simulation, and the lower panel depicts the cross-slope velocity from Umlauf et al. (2015) at a somewhat smaller Burger number ($s = 0.22$). The onset of cross-slope motions is much delayed in the lower Burger number simulations.

The salient features of these modelling studies are, first, that the slope Burger number $N \sin \theta / f$ (Eq. (7.8)) is arguably the key nondimensional parameter determining the structure and long-time evolution of the boundary layer. The effects of gravity relative to rotation ensconced in N/f is a close second, and, apart from the near-bottom region, the bulk of the boundary layer is stratified with a height scale proportional to U/N. Two figures serve to illustrate our points. The first example is from Brink and Lentz (2010) with $(s, N/f, U/N) = (0.98, 195, 20.5 \text{ m})$, in which time-dependent cross-slope oscillations at the resonant frequency $\omega(\alpha)$ rapidly set in. The second example is from Umlauf et al. (2015) with $(s, N/f, U/N) = (0.22, 44.7, 44.7 \text{ m})$, in which the onset of cross-slope oscillations is relatively delayed.

The cross-slope variability is crucial, but our current understanding of that variability is limited. In one-dimension the cross-slope variability is forced to map onto internal wave kinematics. In two dimensions, it is understood (Allen and Newberger, 1998) that the arrested state of the 1-d model is unstable to symmetric instabilities if the Ertel potential vorticity has the opposite sign from its interior value. Ruan et al. (2019) argue from 3-d large eddy simulations that the oscillatory motions are triggered by an episodic molecular laminarisation within the log-layer prior to the onset of arrest. Regardless of the physics, the cross-slope variability serves to ventilate the boundary layer at heights greater than the turbulent drag is able to penetrate, further bending isopycnals toward the bottom and, through the agency of thermal wind, diminishing the bottom velocity to the arrested state.

However limited our understanding may be, there is a *clear* association of fine- and microstructure vis-a-vis rough steep topography and subinertial near-bottom velocity, Naveira Garabato et al. (2004). The potential relevance of bottom friction to such measurements is provided by dedicated observations about a relatively non-descript seamount in Drake passage, in which the internal waveband shear variance is modulated in response to the subinertial flow (Brearley et al., 2013), and that modulated internal wave variability does not agree with linear internal wave kinematics (Cusack et al., 2020). With environmental parameters of $U = 0.1 \text{ m s}^{-1}$ and $N = 1 \times 10^{-3} \text{ s}^{-1}$, that non-descript hill (cf. Fig. 7.1) maps onto $(s, N/f, U/N) = (0.6, 10, 100 \text{ m})$ and has an advective timescale ($f/kU \cong 0.3$) that is too large to efficiently generate quasi-stationary lee waves. Yet that timescale is large enough to consider the initiation of boundary layer coupling with cross-slope motions without being so large that the arrest stage is of concern.

This boundary layer process is more clearly, yet imperfectly, seen in measurements from an abyssal passageway. Naveira Garabato et al. (2019) present preliminary results from a process study documenting the mixing and transformation of newly formed Antarctic Bottom Water as it transits one of the major pathways by which Antarctic Bottom Water is exported into the World Ocean. They document intense mixing of weakly stratified ($N/f \sim 5$) water over steep complex topography ($s \sim O(1)$), in which bottom-enhanced turbulence extends multiples of $U/N \sim 400$ m above the bottom. They interpret their measurements in terms of submesoscale instabilities, resulting from a sign reversal in Ertel potential vorticity. Microstructure data reported in Spingys et al. (2021) document a highly efficient mixing process with $O(1)$ dissipation ratios Γ_μ (7.18).

Finally, as concerns the Ferrari et al. (2016) proposal that the one-dimensional model be characterised by a mid-depth minimum in the buoyancy flux, the essential physics in the downwelling Ekman layer concerns convection and convection is understood to host an efficient dissipation ratio (Section 7.2.5.2, Gregg et al. (2018)) that is essentially non-local and not captured by a mixing length closure.

7.5 Discussion

We began this review by invoking the starting lines from many a fairytale. A well-honed fairytale is not necessarily complete fiction; often there is allegorical truth. That preamble contrasts with the more serious tone of scientific writing, in which the epistemology of science is empirical ratification. Upon this point, what we have confidence in, is rather meager. There is a dire need for observations to constrain the problem, and theoretical efforts would benefit from turbulence modelling to address the myriad of assumptions and approximations.

The key piece that underpins our understanding of upwelling is the diapycnal advection-diffusion balance (7.9), which makes direct reference to the divergence of the density flux. Direct estimates of this flux are quite difficult, so what we have done is to cobble together bits and pieces from our knowledge of turbulence and idealised models (Section 7.2.5.2, 7.3 etc.). Turbulence modelling studies will fundamentally improve our understanding and constrain the uncertainty. There is an associated question of representing boundary mixing in state-of-the-art regional simulations (e.g., Bracco et al., 2020) that will similarly require significant effort.

Substantial progress could be achieved with direct estimates of density and momentum fluxes, and in particular in conjunction with the spatial distribution of time-varying form drag. We require an extension of bottom lander estimates (Shaw et al., 2001) in the vertical (to $\hat{z} > U/N$) in steeply sloping ($s \sim O(1)$) environments and internal wave band estimates (van Haren et al., 1994; Gemmrich and van Haren, 2002) into turbulent frequency bands. Having a direct estimate of bed stress resulting from turbulence (Perlin et al., 2005) and estimates of reduced mixing efficiency with decreasing height above the boundary (Holleman et al., 2016) present technical issues, but our ignorance largely lies within the internal wave and submesoscale frequency bands, and centres upon the questions of how system resonance and rotation alter the behaviour of the system. These frequencies and questions could be tackled with focused numerical modelling efforts, clarifying the possible coupling between friction and form drag. Important phenomena will arise in 3-dimensions relating to non-propagating form drag, and these will be difficult to address numerically.

We do understand that there is an association between mixing and steep rough topography. The parameters in play are fairly clear: a non-dimensionalised topographic slope, especially in relation to the critical slope of internal wave dynamics, for which $\vartheta = \alpha$; external forcing about that resonance; a metric of stratification relative to rotation N/f and the height scale U/N, in which U/N appears in many physical contexts. Disentangling the effects of non-propagating form drag from boundary layer physics over sloping topography will require significant effort. A coordinated survey of this extended parameter space would be a significant community undertaking, but would amount to more than incremental progress.

Beyond that broad summary, we close by trying to frame open questions with the hope of clarifying pathways to progress. Unsurprisingly, this amounts to a dance around putting an observation into a numerical or theoretical construct.

Open Question #1: representing rough topography

The first, and foremost, question relates to Fig. 7.3, in which the strongly enhanced mixing over the MAR is supported by rough and steep topography. All these glorious mountains are bulldozed in the construction of a general circulation model (GCM), which may even lack a coarse representation of the fracture zone valleys that serve as conduits of mass transport to the MAR crest (Thurnherr et al., 2005)). Figs. 7.14, 7.15 and 7.16 attempt to depict the physical situation. Fig. 7.14 is a 3-d representation of multibeam bathymetry along a fracture zone valley with up-valley flow from west to east. Downwelling across isopycnals is depicted above the topography in association with an increase in dissipation toward the bottom; *upwelling* is depicted as fluid peels off from the valley axis and enters the area between the abyssal hills. The rationale for this is a hypsometric argument contained in Fig. 7.15: denser isopycnals occupy less surface area, and thus the

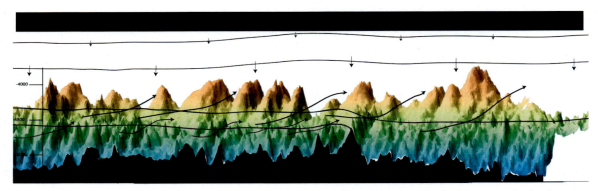

FIGURE 7.14 A schematic depiction of the secondary circulation proposed in Polzin (2009) along the axis of a fracture zone leading from abyssal plain to ridge crest. Lines without arrows represent isopycnals. Lines with arrows represent particle trajectories with net upwelling across isopycnals. Upwelling in this schematic is associated with the near-boundary decay of the internal tide generated/scattered in association with abyssal hills atop the ridge, along with hypsometric arguments. The black foreground is the bathymetry profile along the canyon axis. The peaks atop the ridge are abyssal hills. The canyon axis is significantly smoother than the adjacent ridge, but is occasionally cut by abyssal hills extending from the ridge. These features form sills that block deep flow up the canyon axis. The light grey shading depicts the face of a medium resolution GCM grid cell, $3/4° \times 3/4° \times 250$ m.

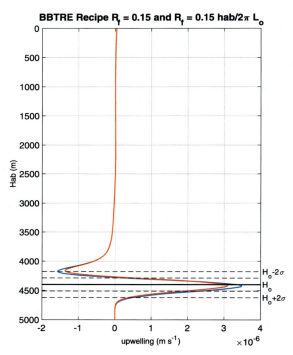

FIGURE 7.15 Diapycnal velocity profiles after invoking hypsometric arguments (7.33) and (7.34). The mean depth is assumed to be $H_o = 4400$ m and the root-mean-square topography $\sigma = 110$ m. Hypsometric effects extend several standard deviations about the mean. The blue trace appears in Fig. 6 of Polzin (2009). The red trace alters that estimate by reducing the observed dissipation with height above boundary to account for potential mixing metric limitations conjectured in Ferrari et al. (2016). The difference is minor and implies that the most important issue is the hypsometric constraints, rather than the details of the near-bottom buoyancy flux profile.

area-integrated diapycnal fluxes are smaller. The upwelling profile in Fig. 7.15 (Polzin, 2009) was estimated as

$$\epsilon(Z) = \int_0^\infty \epsilon(z) P(z \mid Z) dz, \tag{7.33}$$

in which $\epsilon(z)$ is a parameterised estimate of the observed profiles and is monotonically increasing in z over the bottom most 1500 m. The factor $P(z \mid Z)$ represents the probability density of having a station height z given the model coordinate

Mixing at the ocean's bottom boundary Chapter | 7 **173**

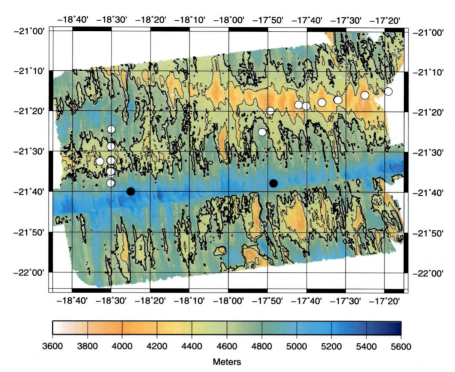

FIGURE 7.16 A 2-d representation of bathymetry appearing in Figs. 7.3 & 7.14. Two depth contours are drawn: a thick one representing 4400 m depth and a thin one at 4200 m depth. These contours are highly convoluted, and the contour length depends upon the effective resolution of the underlying data. Higher and higher resolution will return longer and longer contour lengths. The figure domain is of similar size to the grid cell of a climate resolution GCM. Circles represent station positions for a subset of the profiles appearing in (Polzin et al., 1997; Ledwell et al., 2000) and Fig. 7.3. The blue shading represents the 'canyon', which is the subject of Thurnherr et al. (2005).

(depth) Z. The vertical component of the diapycnal velocity w^* then follows as (McDougall, 1989):

$$w^* N^2 = \frac{\partial \Gamma \epsilon}{\partial Z}, \tag{7.34}$$

in which the mixing metric Γ was assumed to be 0.2. In this calculation, the abyssal hill topography was assumed to follow a Gaussian distribution with a root-mean-squared height of 110 m based upon a parametric spectral characterisation of abyssal hills (see Polzin (2009) for details). The underlying dissipation curve $\epsilon(z)$ was prescribed as the dynamically motivated characterisation presented in Polzin (2004b). The resulting diapycnal velocity supports a pattern of upwelling and downwelling reminiscent of Ferrari et al. (2016), Section 7.3: the net represents upwelling and is a small residual of contributions above and below the mean depth. This result has been attained *without* invoking *ad hoc* arguments about the near-boundary structure of the buoyancy flux, *a la* Ferrari et al. (2016). The relevance of abyssal hills to this conversation can alternately be posed as in Ferrari et al. (2016) by reducing the hypsometric arguments concerning surface area in (7.11) to the length of the boundary in a line integral. Though what one wants in that representation is the length of the line along a grounding isopycnal, contours of constant depth using multibeam data, Fig. 7.16, are highly convoluted, and the inferred line lengths are more than an order of magnitude greater than the meridional extent of the figure. The abyssal hills are consistent with a fractal process (Goff and Jordan, 1988), which in this context means that as one increases the effective resolution of the underlying topographic data set, the contour length increases without bound (e.g., Mandelbrot and Mandelbrot, 1982). Figs. 7.14–7.16 provide a strong argument that the locus for the rate controlling dynamics of abyssal upwelling reside in the abyssal hills. With that statement, though, comes a question: How does one properly represent the boundary conditions (Eqs. (7.4)) in a GCM when such radical changes have been made to the bottom boundary?

How much cause for concern is there in bulldozing the mountains to create the gently sloping topography of a GCM? Take the example of the Brazil basin and say that one eliminates the abyssal hills to create this flat landscape, and by implication the upwelling relating to their hypsometric variability in Fig. 7.15. One is left with offset fractures (valleys) that transport dense water to the ridge crest, cut by occasional sills (Thurnherr et al., 2005). If one represents the mixing associated with the abyssal hills using $\epsilon(z)$ without (7.33), the in-cropping of isopycnals into the valleys creates its own hypsometric variability and upwelling results (Thurnherr et al., 2020). However, these valleys are also eliminated in the creation of the flat landscape of a climate GCM (Figs. 7.14 & 7.16). A message from the study of de Lavergne et al. (2017) is the following: if one has a reasonable representation of the overall *energetics* within a basin interior, changing the spatial distribution of mixing within the GCM will produce a consistent representation of globally averaged upwelling that reflects global hypsometry.

The consequences of not acknowledging that the abyssal hills have been bull-dozed are at least two fold. First, one is put in a position of invoking vertical structure in the buoyancy flux (Ferrari et al., 2016) to promote upwelling, and in so doing inviting commentary about the lack of physics underpinning that assertion. Second, weakly stratified water will tend to accumulate near the boundaries, and one is put in a position of invoking a tertiary process for such a baroclinic instability to ventilate the boundary. In Callies (2018), submesoscale eddies of similar scale and aspect ratio of abyssal hills are proposed to address the ventilation of the bottom boundary, a need that is arguably a result of the creation of the flat landscape in a GCM. We remind the reader of a similar conversation involving Figs. 7.1 & 7.2 detailed in the Introduction.

One perspective in which such details matter is the context climate change predictions. Global ocean upwelling underpins the millennial timescale of climate trends, and one can view such trends as small imbalances in budgets that reflect the detailed differences discussed herein. In this context, a goal should be for GCMs to acknowledge the underlying physics.

Open Question #2a: specification of mixing efficiency and its near-boundary structure

Within this GCM world, one can ask how to configure the model to support upwelling. Two direct answers have appeared. One approach (de Lavergne et al., 2016b) relies upon a characterisation of direct numerical simulations of turbulence that contend the mixing efficiency decreases to small values as turbulent intensity quantified by R_{eb} (7.20) increases beyond a value of 400 (Shih et al., 2005) and play off weakened abyssal stratification to attain upwelling. Recent assessments of microstructure (Ijichi et al., 2020; Spingys et al., 2021) in this high R_{eb}, low N parameter regime have demonstrated that mixing efficiency estimates do not support a general decrease in Γ_μ with increasing R_{eb} for intense turbulence. A second approach assumes a well-mixed BBL, in which the vertical buoyancy flux diminishes linearly to zero, as the sea floor is approached (McDougall and Ferrari, 2017)). Whether it is important that the mixing efficiency reduces as the bottom boundary is approached, or whether it is the stratification that goes to zero just above the boundary, by one means or another, or both, the turbulent buoyancy flux is zero at the sea floor, and this is the feature that causes the diapycnal upwelling in the BBL right near the boundary. This is an unavoidable consequence of the buoyancy budget, and applies no matter the nature of the many physical processes at play in the near-boundary region (Holmes and McDougall, 2020). The inevitability of the upwards BBL diapycnal transport is captured in (7.21) and (7.22). These remarks are predicated on the idea that in the SML region, which on a vertical water column resides above the top of the BBL, the complications caused by density reversals are largely absent.

Open Question #2b

Ferrari et al. (2016) argue for a generic near-boundary diminishment of mixing efficiency, presumably in conjunction with a suppression of lengthscales with decreasing distance from the boundary. Their argument is to assume that all three components of the density flux vector in (7.1d) scale as the vertical component. This may well be untrue: over sloping boundaries, turbulent density fluxes are fed by near-critical wave processes with horizontal fluxes significantly larger than the vertical. We demonstrate this by writing the density flux vector in its Cartesian format:

$$\mathbf{F}_\rho = F_x \hat{i} + F_z \hat{k}.$$

After rotating through the angle θ into slope normal and slope parallel coordinate system, the no-flux bottom boundary condition is

$$F_\perp|_{(z=h(x,y))} = F_z|_{(z=h(x,y))} \cos\theta - F_x|_{(z=h(x,y))} \sin\theta = 0.$$

The Ferrari et al. argument is that

$$\text{if} \quad F_x \cong F_z, \tag{7.35a}$$

then, in a small angle approximation,

$$F_\perp|_{(z=h(x,y))} \cong F_z|_{(z=h(x,y))} = 0. \tag{7.35b}$$

The observational constraint that turbulent dissipation ϵ increases with decreasing height above boundary and the assumption (7.35a) thus lead Ferrari et al. (2016) to assume that the mixing metric Γ decreases more quickly than ϵ increases such that F_z attains a relative minimum off the bottom, with the consequence of downwelling above the relative minimum and upwelling below via (7.9).

The relation (7.35a) is buttressed by an intuitive notion of a lengthscale suppression grounded in the paradigm of steady flows, stable stratification and a wealth of atmospheric boundary layer lore obtained over flat bottoms. The issue of mixing above a sloping boundary could be qualitatively different. Here, turbulence is linked to internal wave processes, in which the issue of a critical slope figures prominently. *If*, for example, we choose to scale the horizontal fluxes using basic internal wave kinematics (7.32),

$$\frac{F_x}{F_z} \equiv \frac{\overline{u'\theta'}}{\overline{w'\theta'}} \sim \frac{u'_{rms}\theta'_{rms}}{w'_{rms}\theta'_{rms}} \cong \frac{N\sqrt{\omega(\alpha)^2 + f^2}}{\omega(\alpha)\sqrt{\omega(\alpha)^2 - f^2}}, \tag{7.36}$$

the bottom boundary condition becomes

$$F_\perp|_{(z=h(x,y))} \cong F_z|_{(z=h(x,y))} \left[\cos\theta - \frac{N\sqrt{\omega(\alpha)^2 + f^2}}{\omega(\alpha)\sqrt{\omega(\alpha)^2 - f^2}} \sin(\theta)\right] \cong 0$$

such that *no* constraint upon $F_z|_{(z=h(x,y))}$ is implied due to the subtractive cancellation of the geometric terms in the square brackets. The scaling (7.36) is motivated by internal wave band temperature fluxes reported by van Haren and collaborators (van Haren et al., 1994; Gemmrich and van Haren, 2002).

With such a proposed scaling, it also should be appreciated that there are phenomenological differences between a critical frequency process and steady shear flow. A key feature of critical frequency processes is not only that of reversals in cross-slope flow, it is the episodic occurrence of statically unstable conditions related to the differential advection of density associated with a near-critical nature of the internal wave field forming large scale overturning events (Sarkar and Scotti, 2017). Similar issues arise in the downwelling Ekman layer, in which the buoyancy flux is accomplished via the agency of the critical frequency wave rather than in conjunction with the bottom stress.

Our orthogonality of opinion is captured by reactions to Fig. 7.2. When you see the weak stratification near the boundary, do you think about the mixing of mixed water and decreasing buoyancy flux as the bottom is approached, or do you see the potential for a convective process and imagine a highly efficient mixing process with large near-bottom buoyancy fluxes captured in Fig. 7.12? The former opinion is aided and abetted by models that simply assume a diffusivity profile and solve for the buoyancy and velocity fields, rather than engaging with turbulence modelling.

Near-boundary microstructure estimates that might resolve this disparate point of view are difficult to come by, especially within an abyssal setting. Within the context of a highly stratified estuary, Holleman et al. (2016) use inertial subrange estimates of ϵ and χ to document a 'boundary influenced' regime of decreased Γ over a height scale $h(x,y) < 2\pi L_o$. If we translate this to the Brazil basin, near-bottom estimates of $\epsilon = 1 \times 10^{-8}$ W/kg and $N^2 = 1 \times 10^{-3}$ s^{-2} returns $2\pi L_o = 20$ m. There is little observational constraint: regions closer that 20 m to the bottom were unsampled by the free-fall instrumentation. Though an observational constraint is lacking, one can certainly pose the question of whether the root of the Ferrari et al. (2016) conjecture concerning the structure of Γ is the issue of GCM behaviour when the GCM lacks an adequate representation of rough topography. To address this, the mixing metric Γ in (7.34) was changed from a constant 0.2 to decrease linearly to zero over the bottom-most 20 m, and the hypsometric arguments (7.33) and (7.34) repeated. The end result appears alongside the original upwelling assessment in Fig. 7.15: an $O(10\%)$ change is noted in the diapycnal upwelling profile. Hypsometry dominates. Holmes and McDougall (2020) did consider the fact that while the buoyancy flux through the sloping sea floor is zero, the turbulent buoyancy flux parallel to the boundary is non-zero, as discussed in this section. However, they found that the lateral convergence of these buoyancy fluxes parallel to the sloping sea floor was unlikely to be large enough to reverse the upward flow in the BBL, even at individual locations, let alone over a significant area of the sea floor.

Open Question #3: communication with the interior

One of the issues posed with Fig. 7.2 is how one avoids the state in which water near the boundary becomes weakly stratified. The proposal in Fig. 7.15 might be applicable for the Mid-Atlantic ridge, but is likely not a generic solution to that problem. In Callies (2018), we find baroclinic instability to be an elegant (simple) way of restratifying the interior. The closure that is offered suggests a rather uniform diffusive process.

This contrasts with the apparent time history of a tracer release in the Gulf of Mexico reported in Ledwell et al. (2016), in which near-boundary tracer was swept off-slope in single, semi-deterministic event. One can also create a list of possible ventilation processes: internal hydraulic transitions, internal bores and flow separation represent efficient ways to expel mass from the boundary and accomplish a wholesale reset of near-boundary stratification that compete with restratification via baroclinic instability.

Open Question #4: whither entrainment/detrainment arguments in abyssal plumes?

It has traditionally been assumed that sinking plumes of bottom water continually entrain water into them, thus increasing their volume transport as they sink towards the abyss. This, however, leads to contradictions as outlined in Section 7.3. Do we overcome these issues by appealing to the modern realisation that sinking plumes can detrain as well as entrain, or do we need to consider multiple sources of bottom water as another way out of these contradictions? We note that both the Brazil basin and the Orkney passage measurement programs have been made in these descending plumes of dense bottom water, so these locations may not be typical of more general rough sea floor boundary locations.

Acknowledgements

General. We thank Dr. Ilker Fer and a second anonymous reviewer for their indulgence and constructive criticism that significantly improved our initial efforts. Dr. Ryan Holmes provided extensive comments on a draft of this chapter and provided Fig. 7.7. Dr. Bryan Kaiser provided key insight into issues of turbulence modelling for the one-dimensional model. Thank you.

Author contributions. All authors contributed equally to the writing of the manuscript.

Funding. We gratefully acknowledge support from the National Science Foundation support through grant OCE-1536779 (K.P.), and from the Australian Research Council through grant FL150100090 (T. McD).

Conflicts of interest. The authors declare that there is no conflict of interest regarding the publication of this chapter.

Data availability. No data were created as part of this review.

References

Alford, M.H., Girton, J.B., Voet, G., Carter, G.S., Mickett, J.B., Klymak, J.M., 2013. Turbulent mixing and hydraulic control of abyssal water in the Samoan passage. Geophys. Res. Lett. 40, 4668–4674.
Allen, J., Newberger, P., 1998. On symmetric instabilities in oceanic bottom boundary layers. J. Phys. Oceanogr. 28, 1131–1151.
Armi, L., 1978. Some evidence for boundary mixing in the deep ocean. J. Geophys. Res., Oceans 83, 1971–1979.
Armi, L., 1979a. Effects of variations in eddy diffusivity on property distributions in the oceans. J. Mar. Res. 37, 515–530.
Armi, L., 1979b. Reply to comments by C. Garrett. J. Geophys. Res., Oceans 84, 5097–5098.
Armi, L., Millard Jr., R.C., 1976. The bottom boundary layer of the deep ocean. J. Geophys. Res. 81, 4983–4990.
Baines, P., 1973. The generation of internal tides by flat-bump topography. In: Deep Sea Research and Oceanographic Abstracts. Elsevier, pp. 179–205.
Baines, P.G., 1997. Topographic Effects in Stratified Flows. Cambridge University Press.
Baines, P.G., 2001. Mixing in flows down gentle slopes into stratified environments. J. Fluid Mech. 443, 237.
Baines, P.G., 2005. Mixing regimes for the flow of dense fluid down slopes into stratified environments. J. Fluid Mech. 538, 245.
Bracco, A., Paris, C.B., Esbaugh, A.J., Frasier, K., Joye, S., Liu, G., Polzin, K., Vaz, A.C., 2020. Transport, fate and impacts of the deep plume of petroleum hydrocarbons formed during the macondo blowout. Front. Mar. Sci. 7, 764.
Brearley, J.A., Sheen, K.L., Naveira Garabato, A.C., Smeed, D.A., Waterman, S., 2013. Eddy-induced modulation of turbulent dissipation over rough topography in the Southern Ocean. J. Phys. Oceanogr. 43, 2288–2308.
Bretherton, F.P., 1969. Momentum transport by gravity waves. Q. J. R. Meteorol. Soc. 95, 213–243.
Brink, K.H., Lentz, S.J., 2010. Buoyancy arrest and bottom Ekman transport. Part I: steady flow. J. Phys. Oceanogr. 40, 621–635.
Brown, N., 1974. A precision ctd microprofiler. In: Ocean'74-IEEE International Conference on Engineering in the Ocean Environment. IEEE, pp. 270–278.
Bryan, K., Lewis, L., 1979. A water mass model of the world ocean. J. Geophys. Res., Oceans 84, 2503–2517.
Callies, J., 2018. Restratification of abyssal mixing layers by submesoscale baroclinic eddies. J. Phys. Oceanogr. 48, 1995–2010.
Callies, J., Ferrari, R., 2018. Dynamics of an abyssal circulation driven by bottom-intensified mixing on slopes. J. Phys. Oceanogr. 48, 1257–1282.
Chalamalla, V.K., Gayen, B., Scotti, A., Sarkar, S., 2013. Turbulence during the reflection of internal gravity waves at critical and near-critical slopes. J. Fluid Mech. 729, 47.
Chalamalla, V.K., Sarkar, S., 2015. Mixing, dissipation rate, and their overturn-based estimates in a near-bottom turbulent flow driven by internal tides. J. Phys. Oceanogr. 45, 1969–1987.
Clément, L., Frajka-Williams, E., Sheen, K., Brearley, J., Garabato, A.N., 2016. Generation of internal waves by eddies impinging on the western boundary of the North Atlantic. J. Phys. Oceanogr. 46, 1067–1079.
Clément, L., Thurnherr, A.M., 2018. Abyssal upwelling in mid-ocean ridge fracture zones. Geophys. Res. Lett. 45, 2424–2432.
Cusack, J.M., Brearley, J.A., Naveira Garabato, A.C., Smeed, D.A., Polzin, K.L., Velzeboer, N., Shakespeare, C.J., 2020. Observed eddy–internal wave interactions in the Southern Ocean. J. Phys. Oceanogr. 50, 3043–3062.
Cusack, J.M., Voet, G., Alford, M.H., Girton, J.B., Carter, G.S., Pratt, L.J., Pearson-Potts, K.A., Tan, S., 2019. Persistent turbulence in the Samoan passage. J. Phys. Oceanogr. 49, 3179–3197.

Cushman-Roisin, B., Beckers, J.M., 2011. Introduction to Geophysical Fluid Dynamics: Physical and Numerical Aspects. Academic Press.
Drake, H.F., Ferrari, R., Callies, J., 2020. Abyssal circulation driven by near-boundary mixing: water mass transformations and interior stratification. J. Phys. Oceanogr. 50, 2203–2226.
Epifanio, C.C., Durran, D.R., 2001. Three-dimensional effects in high-drag-state flows over long ridges. J. Atmos. Sci. 58, 1051–1065.
Eriksen, C.C., 1982. Observations of internal wave reflection off sloping bottoms. J. Geophys. Res., Oceans 87, 525–538.
Eriksen, C.C., 1985. Implications of ocean bottom reflection for internal wave spectra and mixing. J. Phys. Oceanogr. 15, 1145–1156.
Eriksen, C.C., 1998. Internal wave reflection and mixing at Fieberling Guyot. J. Geophys. Res. 103, 2977–2994.
Evans, D.G., Frajka-Williams, E., Garabato, A.N., Polzin, K., 2020. Mesoscale eddy dissipation by a 'zoo' of submesoscale processes at a western boundary. J. Geophys. Res., Oceans 125.
Ferrari, R., Mashayek, A., McDougall, T.J., Nikurashin, M., Campin, J.M., 2016. Turning ocean mixing upside down. J. Phys. Oceanogr. 46, 2239–2261.
Ferrari, R., Polzin, K.L., 2005. Finescale structure of the t–s relation in the eastern North Atlantic. J. Phys. Oceanogr. 35, 1437–1454.
Ferron, B., Mercier, H., Speer, K., Gargett, A., Polzin, K., 1998. Mixing in the Romanche Fracture Zone. J. Phys. Oceanogr. 28, 1929–1945.
Gardiner, C.W., et al., 1985. Handbook of Stochastic Methods, Vol. 3. Springer, Berlin.
Garrett, C., 1979. Comment on 'some evidence for boundary mixing in the deep ocean' by Laurence armi. J. Geophys. Res., Oceans 84, 5095.
Garrett, C., 1984. Turning points in universal speculation on internal waves. In: A Celebration in Geophysics and Oceanography, Vol. 84, pp. 38–46.
Garrett, C., 1990. The role of secondary circulation in boundary mixing. J. Geophys. Res., Oceans 95, 3181–3188.
Garrett, C., 1991. Marginal mixing theories. Atmos.-Ocean 29, 313–339.
Garrett, C., 2001. An isopycnal view of near-boundary mixing and associated flows. J. Phys. Oceanogr. 31, 138–142.
Garrett, C., Gilbert, D., 1988. Estimates of Vertical Mixing by Internal Waves Reflected Off a Sloping Bottom. Elsevier Oceanography Series, vol. 46. Elsevier, pp. 405–423.
Garrett, C., MacCready, P., Rhines, P., 1993. Boundary mixing and arrested Ekman layers: rotating stratified flow near a sloping boundary. Annu. Rev. Fluid Mech. 25, 291–323.
Garrett, C., Munk, W., 1972a. Oceanic mixing by breaking internal waves. In: Deep Sea Research and Oceanographic Abstracts. Elsevier, pp. 823–832.
Garrett, C., Munk, W., 1972b. Space-time scales of internal waves. Geophys. Astrophys. Fluid Dyn. 3, 225–264.
Gayen, B., Sarkar, S., 2010. Turbulence during the generation of internal tide on a critical slope. Phys. Rev. Lett. 104, 218502.
Gemmrich, J.R., van Haren, H., 2002. Internal wave band eddy fluxes above a continental slope. J. Mar. Res. 60, 227–253. https://doi.org/10.1357/00222400260497471.
Gent, P.R., Willebrand, J., McDougall, T.J., McWilliams, J.C., 1995. Parameterizing eddy-induced tracer transports in ocean circulation models. J. Phys. Oceanogr. 25, 463–474.
Goff, J.A., Jordan, T.H., 1988. Stochastic modelling of seafloor morphology: inversion of Sea Beam data for second-order statistics. J. Geophys. Res. 93, 13,589–13,608.
Greatbatch, R.J., Lamb, K.G., 1990. On parameterizing vertical mixing of momentum in non-eddy resolving ocean models. J. Phys. Oceanogr. 20, 1634–1637.
Gregg, M., Cox, C., 1971. Measurements of the oceanic microstructure of temperature and electrical conductivity. In: Deep Sea Research and Oceanographic Abstracts. Elsevier, pp. 925–934.
Gregg, M., D'Asaro, E., Riley, J., Kunze, E., 2018. Mixing efficiency in the ocean. Annu. Rev. Mar. Sci. 10, 443–473. https://doi.org/10.1146/annurev-marine-121916-063643.
Group, M., et al., 1978. The mid-ocean dynamics experiment. Deep-Sea Res. 25, 859–910.
van Haren, H., Oakey, N., Garrett, C., 1994. Measurements of internal wave band eddy fluxes above a sloping bottom. J. Mar. Res. 52, 909–946.
Hasselmann, K., 1962. On the non-linear energy transfer in a gravity-wave spectrum Part 1. General theory. J. Fluid Mech. 12, 481–500.
Hogg, C.A., Dalziel, S.B., Huppert, H.E., Imberger, J., 2017. Inclined gravity currents filling basins: the impact of peeling detrainment on transport and vertical structure. J. Fluid Mech. 820, 400.
Hogg, N., Biscaye, P., Gardner, W., Schmitz Jr, W.J., 1982. On the transport and modification of Antarctic bottom water in the vema channel. J. Mar. Res. 40, 231–263.
Hogg, N.G., 1983. Hydraulic control and flow separation in a multi-layered fluid with applications to the vema channel. J. Phys. Oceanogr. 13, 695–708.
Holleman, R.C., Geyer, W.R., Ralston, D.K., 2016. Stratified turbulence and mixing efficiency in a salt wedge estuary. J. Phys. Oceanogr. 46, 1769–1783. https://doi.org/10.1175/JPO-D-15-0193.1.
Holmes, R., de Lavergne, C., McDougall, T.J., 2019. Tracer transport within abyssal mixing layers. J. Phys. Oceanogr. 49, 2669–2695.
Holmes, R., McDougall, T.J., 2020. Diapycnal transport near a sloping bottom boundary. J. Phys. Oceanogr. 50, 3252–3266.
Holmes, R.M., de Lavergne, C., McDougall, T.J., 2018. Ridges, seamounts, troughs, and bowls: topographic control of the dianeutral circulation in the abyssal ocean. J. Phys. Oceanogr. 48, 861–882.
Ijichi, T., St. Laurent, L., Polzin, K.L., Toole, J.M., 2020. How variable is mixing efficiency in the abyss? Geophys. Res. Lett. 47, e2019GL086813.
IOC, S., 2010. Iapso 2010: the international thermodynamic equation of seawater—2010: calculation and use of thermodynamic properties. In: Intergovernmental Oceanographic Commission, Manuals and Guides, Vol. 56, p. 220.
Iudicone, D., Madec, G., McDougall, T.J., 2008. Water-mass transformations in a neutral density framework and the key role of light penetration. J. Phys. Oceanogr. 38, 1357–1376.
Jayne, S.R., 2009. The impact of abyssal mixing parameterizations in an ocean general circulation model. J. Phys. Oceanogr. 39, 1756–1775.
Kaimal, J., Wyngaard, J., Izumi, Y., Cote, O., 1972. Spectral characteristics of surface-layer turbulence. Q. J. R. Meteorol. Soc. 98, 563–589.

Kaiser, B., Pratt, L.J., 2020. The transition to turbulence within internal tide boundary layers in the abyssal ocean. In: Ocean Sciences Meeting 2020. AGU.

Kamenkovich, V., Monin, A., Voorhis, A., 1986. The Polymode Atlas. Woods Hole Oceanographic Institution.

Klocker, A., McDougall, T.J., 2010. Influence of the nonlinear equation of state on global estimates of dianeutral advection and diffusion. J. Phys. Oceanogr. 40, 1690–1709.

Klymak, J.M., Legg, S., Pinkel, R., 2010. A simple parameterization of turbulent mixing near supercritical topography. J. Phys. Oceanogr. 40, 2059–2074.

Kolås, E., Fer, I., 2018. Hydrography, transport and mixing of the West Spitsbergen current: the Svalbard branch in summer 2015. Ocean Sci. 14, 1603–1618.

Kunze, E., Lien, R.C., 2019. Energy sinks for lee waves in shear flow. J. Phys. Oceanogr. 49, 2851–2865.

Kunze, E., MacKay, C., McPhee-Shaw, E.E., Morrice, K., Girton, J.B., Terker, S.R., 2012. Turbulent mixing and exchange with interior waters on sloping boundaries. J. Phys. Oceanogr. 42, 910–927.

de Lavergne, C., Madec, G., Le Sommer, J., Nurser, A.G., Naveira Garabato, A.C., 2016a. On the consumption of Antarctic bottom water in the abyssal ocean. J. Phys. Oceanogr. 46, 635–661.

de Lavergne, C., Madec, G., Le Sommer, J., Nurser, A.J.G., Naveira Garabato, A.C., 2016b. The impact of a variable mixing efficiency on the abyssal overturning. J. Phys. Oceanogr. 46, 663–681. https://doi.org/10.1175/JPO-D-14-0259.1.

de Lavergne, C., Madec, G., Roquet, F., Holmes, R., McDougall, T.J., 2017. Abyssal ocean overturning shaped by seafloor distribution. Nature 551, 181–186.

Ledwell, J., Montgomery, E., Polzin, K., Laurent, L.S., Schmitt, R., Toole, J., 2000. Evidence for enhanced mixing over rough topography in the abyssal ocean. Nature 403, 179–182.

Ledwell, J.R., He, R., Xue, Z., DiMarco, S.F., Spencer, L.J., Chapman, P., 2016. Dispersion of a tracer in the deep Gulf of Mexico. J. Geophys. Res., Oceans 121, 1110–1132.

Ledwell, J.R., Watson, A.J., Law, C.S., 1998. Mixing of a tracer in the pycnocline. J. Geophys. Res., Oceans 103, 21499–21529.

Lentz, S.J., Trowbridge, J.H., 1991. The bottom boundary layer over the northern California shelf. J. Phys. Oceanogr. 21, 1186–1201.

Lumpkin, R., Speer, K., 2007. Global ocean meridional overturning. J. Phys. Oceanogr. 37, 2550–2562.

MacCready, P., Rhines, P.B., 1991. Buoyant inhibition of Ekman transport on a slope and its effect on stratified spin-up. J. Fluid Mech. 223, 631–661.

MacCready, P., Rhines, P.B., 1993. Slippery bottom boundary layers on a slope. J. Phys. Oceanogr. 23, 5–22.

MacKinnon, J.A., Johnston, T.M.S., Pinkel, R., 2008. Strong transport and mixing of deep water through the Southwest Indian Ridge. Nat. Geosci. 1. https://doi.org/10.1037/ngeo340.

Mandelbrot, B.B., Mandelbrot, B.B., 1982. The Fractal Geometry of Nature, Vol. 1. WH Freeman, New York.

Marshall, J., Jamous, D., Nilsson, J., 1999. Reconciling thermodynamic and dynamic methods of computation of water-mass transformation rates. Deep-Sea Res., Part 1, Oceanogr. Res. Pap. 46, 545–572.

Mashayek, A., Caulfield, C.P., Peltier, W.R., 2017. Role of overturns in optimal mixing in stratified mixing layers. J. Fluid Mech. 826, 522–552. https://doi.org/10.1017/jfm.2017.374.

Mauritzen, C., Polzin, K.L., McCartney, M.S., Millard, R.C., West-Mack, D.E., 2002. Evidence in hydrography and density finestructure for enhanced vertical mixing over the Mid-Atlantic Ridge in the western Atlantic. J. Geophys. Res. 107. https://doi.org/10.1029/2001JC001114.

McComas, C.H., Bretherton, F.P., 1977. Resonant interaction of oceanic internal waves. J. Geophys. Res. 82, 1397–1412.

McDougall, T.J., 1984. The relative roles of diapycnal and isopycnal mixing on subsurface water mass conversion. J. Phys. Oceanogr. 14, 1577–1589.

McDougall, T.J., 1989. Dianeutral advection. In: Parameterization of Small-Scale Processes: Proc. 'Aha Huliko' a Hawaiian Winter Workshop, pp. 289–315.

McDougall, T.J., Ferrari, R., 2017. Abyssal upwelling and downwelling driven by near-boundary mixing. J. Phys. Oceanogr. 47, 261–283. https://doi.org/10.1175/JPO-D-16-0082.1.

McDougall, T.J., McIntosh, P.C., 2001. The temporal-residual-mean velocity. Part II: isopycnal interpretation and the tracer and momentum equations. J. Phys. Oceanogr. 31, 1222–1246.

Melet, A., Hallberg, R., Legg, S., Polzin, K., 2013. Sensitivity of the ocean state to the vertical distribution of internal-tide-driven mixing. J. Phys. Oceanogr. 43, 602–615.

Mellor, G.L., Yamada, T., 1974. A hierarchy of turbulence closure models for planetary boundary layers. J. Atmos. Sci. 31, 1791–1806.

Mercier, H., Morin, P., 1997. Hydrography of the romanche and chain fracture zones. J. Geophys. Res., Oceans 102, 10373–10389.

Mercier, H., Speer, K.G., 1998. Transport of bottom water in the romanche fracture zone and the chain fracture zone. J. Phys. Oceanogr. 28, 779–790.

Morris, M.Y., Hall, M.M., St. Laurent, L.C., Hogg, N.G., 2001. Abyssal mixing in the Brazil basin. J. Phys. Oceanogr. 31, 3331–3348.

Moum, J., Perlin, A., Klymak, J., Levine, M., Boyd, T., Kosro, P., 2004. Convectively driven mixing in the bottom boundary layer. J. Phys. Oceanogr. 34, 2189–2202.

Munk, W., Wunsch, C., 1998. Abyssal recipes ii: energetics of tidal and wind mixing. Deep-Sea Res., Part 1, Oceanogr. Res. Pap. 45, 1977–2010.

Munk, W.H., 1966. Abyssal recipes. In: Deep Sea Research and Oceanographic Abstracts. Citeseer, pp. 707–730.

Nash, J.D., Moum, J.N., 2001. Internal hydraulic flows on the continental shelf: high drag states over a small bank. J. Geophys. Res., Oceans 106, 4593–4611. https://doi.org/10.1029/1999JC000183.

Nasmyth, P.W., 1970. Oceanic turbulence. Ph.D. thesis. University of British Columbia.

Naveira Garabato, A.C., Frajka-Williams, E.E., Spingys, C.P., Legg, S., Polzin, K.L., Forryan, A., Abrahamsen, E.P., Buckingham, C.E., Griffies, S.M. McPhail, S.D., Nicholls, K.W., Thomas, L.N., Meredith, M.P., 2019. Rapid mixing and exchange of deep-ocean waters in an abyssal boundary current. Proc. Natl. Acad. Sci. 116, 13233–13238. https://doi.org/10.1073/pnas.1904087116.

Naveira Garabato, A.C., Polzin, K.L., King, B.A., Heywood, K.J., Visbeck, M., 2004. Widespread intense turbulent mixing in the Southern Ocean. Science 303, 210–213.
Nikurashin, M., Ferrari, R., 2010. Radiation and dissipation of internal waves generated by geostrophic motions impinging on small-scale topography: theory. J. Phys. Oceanogr. 40, 1055–1074.
Nikurashin, M., Ferrari, R., 2011. Global energy conversion rate from geostrophic flows into internal lee waves in the deep ocean. Geophys. Res. Lett. 38.
Nikurashin, M., Ferrari, R., 2013. Overturning circulation driven by breaking internal waves in the deep ocean. Geophys. Res. Lett. 40, 3133–3137.
Nikurashin, M., Ferrari, R., Grisouard, N., Polzin, K., 2014. The impact of finite-amplitude bottom topography on internal wave generation in the Southern Ocean. J. Phys. Oceanogr. 44, 2938–2950.
Olbers, D.J., 1976. Nonlinear energy transfer and the energy balance of the internal wave field in the deep ocean. J. Fluid Mech. 74, 375–399.
Osborn, T., 1980. Estimates of the local rate of vertical diffusion from dissipation measurements. J. Phys. Oceanogr. 10, 83–89.
Osborn, T., Siddon, T., 1973. Oceanic shear measurements using the airfoil probe.
Osborn, T.R., Cox, C.S., 1972. Oceanic fine structure. Geophys. Fluid Dyn. 3, 321–345.
Perlin, A., Moum, J.N., Klymak, J., Levine, M.D., Boyd, T., Kosro, P.M., 2005. A modified law-of-the-wall applied to oceanic bottom boundary layers. J. Geophys. Res., Oceans 110.
Phillips, O., 1960. On the dynamics of unsteady gravity waves of finite amplitude Part 1. The elementary interactions. J. Fluid Mech. 9, 193–217.
Phillips, O., 1970. On flows induced by diffusion in a stably stratified fluid. In: Deep Sea Research and Oceanographic Abstracts. Elsevier, pp. 435–443.
Phillips, O., Shyu, J.H., Salmun, H., 1986. An experiment on boundary mixing: mean circulation and transport rates. J. Fluid Mech. 173, 473–499.
Polzin, K., Toole, J., Ledwell, J., Schmitt, R., 1997. Spatial variability of turbulent mixing in the abyssal ocean. Science 276, 93–96.
Polzin, K.L., 2004a. A heuristic description of internal wave dynamics. J. Phys. Oceanogr. 34, 214–230.
Polzin, K.L., 2004b. Idealized solutions for the energy balance of the finescale internal wavefield. J. Phys. Oceanogr. 34, 231–246.
Polzin, K.L., 2009. An abyssal recipe. Ocean Model. 30, 298–309.
Polzin, K.L., Lvov, Y.V., 2017. An oceanic ultra-violet catastrophe, wave-particle duality and a strongly nonlinear concept for geophysical turbulence. Fluids 2, 36.
Polzin, K.L., Speer, K., Toole, J.M., Schmitt, R.W., 1996. Mixing in the Romanche Fracture Zone. Nature 380, 54–57.
Pomphrey, N., Meiss, J.D., Watson, K.M., 1980. Description of nonlinear internal wave interactions using Langevin methods. J. Geophys. Res., Oceans 85, 1085–1094.
Rhines, P.B., 1993. Oceanic general circulation: wave and advection dynamics. In: Modelling Oceanic Climate Interactions. Springer, pp. 67–149.
Ruan, X., Thompson, A.F., Taylor, J.R., 2019. The evolution and arrest of a turbulent stratified oceanic bottom boundary layer over a slope: downslope regime. J. Phys. Oceanogr. 49, 469–487.
Sanford, T.B., 1975. Observations of the vertical structure of internal waves. J. Geophys. Res. 80, 3861–3871.
Sarkar, S., Scotti, A., 2017. From topographic internal gravity waves to turbulence. Annu. Rev. Fluid Mech. 49, 195–220.
Schiff, L., 1966. Lateral boundary mixing in a simple model of ocean convection. In: Deep Sea Research and Oceanographic Abstracts. Elsevier, pp. 621–626.
Scott, R., Goff, J., Naveira Garabato, A., Nurser, A., 2011. Global rate and spectral characteristics of internal gravity wave generation by geostrophic flow over topography. J. Geophys. Res., Oceans 116.
Shaw, W.J., Trowbridge, J.H., Williams III, A.J., 2001. Budgets of turbulent kinetic energy and scalar variance in the continental shelf bottom boundary layer. J. Geophys. Res., Oceans 106, 9551–9564. https://doi.org/10.1029/2000JC000240.
Shih, L.H., Koseff, J.R., Ivey, G.N., Ferziger, J.H., 2005. Parameterization of turbulent fluxes and scales using homogeneous sheared stably stratified turbulence simulations. J. Fluid Mech. 525, 193–214. https://doi.org/10.1017/S0022112004002587.
Slinn, D.N., Riley, J., 1998. Turbulent dynamics of a critically reflecting internal gravity wave. Theor. Comput. Fluid Dyn. 11, 281–303.
Smyth, W., Moum, J., Caldwell, D., 2001. The efficiency of mixing in turbulent patches: inferences from direct simulations and microstructure observations. J. Phys. Oceanogr. 31, 1969–1992.
Spingys, C., Naveira Garabato, A., Legg, S., Polzin, K., Abrahamsen, P., Buckingham, C., Forryan, A., Frajka-Williams, E., 2021. Mixing and transformation in a deep western boundary current: a case study. J. Phys. Oceanogr. 51 (4), 1205–1222.
Sreenivasan, K.R., 1995. On the universality of the Kolmogorov constant. Phys. Fluids 7, 2778–2784. https://doi.org/10.1063/1.868656.
Sreenivasan, K.R., 1996. The passive scalar spectrum and the Obukhov–Corrsin constant. Phys. Fluids 8, 189–196. https://doi.org/10.1063/1.868826.
St Laurent, L.C., Thurnherr, A.M., 2007. Intense mixing of lower thermocline water on the crest of the mid-Atlantic ridge. Nature 448, 680–683.
St. Laurent, L.C., Toole, J.M., Schmitt, R.W., 2001. Buoyancy forcing by turbulence above rough topography in the abyssal Brazil basin. J. Phys. Oceanogr. 31, 3476–3495.
Stommel, H., Arons, A., 1959. On the abyssal circulation of the world ocean—I. Stationary planetary flow patterns on a sphere. Deep-Sea Res. 1953 (6), 140–154.
Stull, R.B., 2012. An Introduction to Boundary Layer Meteorology, Vol. 13. Springer Science & Business Media.
Taylor, J.R., Sarkar, S., 2008. Stratification effects in a bottom Ekman layer. J. Phys. Oceanogr. 38, 2535–2555.
Tennekes, H., Lumley, J., 1972. A first course in turbulence.
Thorpe, S., Malarkey, J., Voet, G., Alford, M., Girton, J., Carter, G., 2018. Application of a model of internal hydraulic jumps. J. Fluid Mech. 834, 125–148.
Thorpe, S.A., 1977. Turbulence and mixing in a Scottish loch. Philos. Trans. R. Soc. Lond. Ser. A, Math. Phys. Sci. 286, 125–181.
Thorpe, S.A., 1987. Current and temperature variability on the continental slope. Philos. Trans. R. Soc. Lond. Ser. A, Math. Phys. Sci. 323, 471–517.

Thorpe, S.A., 1992. The generation of internal waves by flow over the rough topography of a continental slope. Proc. R. Soc. Lond. Ser. A, Math. Phys. Sci. 439, 115–130.
Thurnherr, A., 2006. Diapycnal mixing associated with an overflow in a deep submarine canyon. Deep-Sea Res., Part 2, Top. Stud. Oceanogr. 53, 194–206.
Thurnherr, A., St. Laurent, L., Speer, K., Toole, J., Ledwell, J., 2005. Mixing associated with sills in a canyon on the midocean ridge flank. J. Phys. Oceanogr. 35, 1370–1381.
Thurnherr, A.M., Clément, L., St. Laurent, L., Ferrari, R., Ijichi, T., 2020. Transformation and upwelling of bottom water in fracture zone valleys. J. Phys. Oceanogr. 50, 715–726.
Tippenhauer, S., Dengler, M., Fischer, T., Kanzow, T., 2015. Turbulence and finestructure in a deep ocean channel with sill overflow on the mid-Atlantic ridge. Deep-Sea Res., Part 1, Oceanogr. Res. Pap. 99, 10–22.
Toole, J.M., Schmitt, R.W., Polzin, K.L., Kunze, E., 1997. Near-boundary mixing above the flanks of a midlatitude seamount. J. Geophys. Res., Oceans 102, 947–959.
Trowbridge, J., Lentz, S., 1991. Asymmetric behavior of an oceanic boundary layer above a sloping bottom. J. Phys. Oceanogr. 21, 1171–1185.
Trowbridge, J.H., Lentz, S.J., 2018. The bottom boundary layer. Annu. Rev. Mar. Sci. 10, 397–420.
Umlauf, L., Smyth, W.D., Moum, J.N., 2015. Energetics of bottom Ekman layers during buoyancy arrest. J. Phys. Oceanogr. 45, 3099–3117.
Voet, G., Girton, J.B., Alford, M.H., Carter, G.S., Klymak, J.M., Mickett, J.B., 2015. Pathways, volume transport, and mixing of abyssal water in the Samoan passage. J. Phys. Oceanogr. 45, 562–588.
Walin, G., 1982. On the relation between sea-surface heat flow and thermal circulation in the ocean. Tellus 34, 187–195.
Waterhouse, A.F., MacKinnon, J.A., Nash, J.D., Alford, M.H., Kunze, E., Simmons, H.L., Polzin, K.L., St. Laurent, L.C., Sun, O.M., Pinkel, R., et al., 2014. Global patterns of diapycnal mixing from measurements of the turbulent dissipation rate. J. Phys. Oceanogr. 44, 1854–1872.
Weatherly, G.L., Martin, P.J., 1978. On the structure and dynamics of the oceanic bottom boundary layer. J. Phys. Oceanogr. 8, 557–570.
Whitehead Jr, J., Worthington, L., 1982. The flux and mixing rates of Antarctic bottom water within the North Atlantic. J. Geophys. Res., Oceans 87, 7903–7924.
Wunsch, C., 1970. On oceanic boundary mixing. In: Deep Sea Research and Oceanographic Abstracts. Elsevier, pp. 293–301.
Wyrtki, K., 1961. The thermohaline circulation in relation to the general circulation in the oceans. Deep-Sea Res. 1953 (8), 39–64.
Wyrtki, K., 1962. The oxygen minima in relation to ocean circulation. In: Deep Sea Research and Oceanographic Abstracts. Elsevier, pp. 11–23.

Chapter 8

Submesoscale processes and mixing

Jonathan Gula[a,b], John Taylor[c], Andrey Shcherbina[d] and Amala Mahadevan[e]

[a]Univ Brest, CNRS, IRD, Ifremer, Laboratoire d'Océanographie Physique et Spatiale (LOPS), IUEM, Brest, France, [b]Institut Universitaire de France (IUF), Paris, France, [c]Department of Applied Mathematics and Theoretical Physics, University of Cambridge, Cambridge, United Kingdom, [d]Applied Physics Laboratory, University of Washington, Seattle, WA, United States, [e]Woods Hole Oceanographic Institution, Woods Hole, MA, United States

8.1 Introduction

Submesoscale currents in the ocean take the form of intense jets and vortices with horizontal scales of 100 m–10 km and timescales of hours to weeks (Thomas et al., 2008; McWilliams, 2016). Almost invariably they are associated with the corresponding hydrographic features: fronts, filaments, and spirals of matching scales. This association is so tight that dynamic and hydrographic terms are commonly used interchangeably in the literature (e.g., "jets" and "fronts"). High resolution satellite observations of sea surface temperature (SST) and ocean colour or high-resolution numerical models (Fig. 8.1) are exposing the ubiquity of such submesoscale frontal features at the surface.

The upper end of the submesoscale is historically defined by the scales of the mesoscale eddies, which vary as the local Rossby deformation radius R_d, from roughly 10 km at high-latitudes to 200 km near to the equator (Chelton et al., 1998; LaCasce and Groeskamp, 2020). The lower end of the submesoscale is typically taken as the scales corresponding to the turbulent boundary layer depth $h_{bl} = 10$ m–100 m, below which the flow becomes more isotropic. Thus submesoscale currents occupy intermediate space and timescales between quasi-geostrophic mesoscale eddies and the fully three-dimensional turbulence.

Dynamically, submesoscale processes are characterised by order one Rossby and Froude numbers (Thomas et al., 2008). The Rossby number, arising from the relative scaling of the inertial and Coriolis terms in the equations of motion, is defined as $Ro = U/fL$, where U is the characteristic horizontal velocity scale, f is the local Coriolis frequency and L is the characteristic horizontal length scale. The Froude number, characterising the balance of inertia and stratification effects, is defined as $Fr = U/NH$, where $N = \sqrt{\partial b/\partial z}$ is the buoyancy frequency, $b = -g\rho/\rho_0$ is the buoyancy, ρ is the density, ρ_0 is a constant reference density, and H is the characteristic vertical scale. The order one Rossby and Froude numbers imply that stratification, rotation, and inertia are all important to submesoscale dynamics.

The important role played by submesoscale currents has been realised over the last decade (McWilliams, 2016). Submesoscales can transfer energy from larger quasi-balanced motions to small-scale three-dimensional turbulence, thereby providing a route to dissipation (Müller et al., 2005). Submesoscale currents redistribute water properties, including momentum, buoyancy, heat, freshwater, and biogeochemical tracers (Poje et al., 2014; Shcherbina et al., 2015; Mahadevan, 2016). The strong heterogeneities they generate at the ocean surface have important implications for modulating air–sea fluxes of energy and in structuring marine ecosystems (Lévy et al., 2018) and bacterioplankton communities (Fadeev et al., 2020). The strong vertical velocities they generate drive significant irreversible vertical fluxes of mass, buoyancy, and materials that control stratification of the upper ocean as well as exchanges between the surface layer and the ocean interior (Balwada et al., 2018; Su et al., 2018; Uchida et al., 2019; Mahadevan et al., 2020; Bachman and Klocker, 2020).

Submesoscale currents are generated and influenced by a number of different processes, which are summarised in Fig. 8.2 and will be discussed in detail in the following section. The stirring induced by mesoscale eddies can generate density fronts and filaments (Hoskins and Bretherton, 1972; McWilliams, 1984). This can be described using quasi-geostrophic theories, where it corresponds to a direct cascade of tracer variance toward smaller scales (see review in Lapeyre (2017) and Chapter 9 for a more detailed description). The fronts and filaments can then further intensify due to self-sustained frontogenesis (Hoskins and Bretherton, 1972). Ageostrophic effects are more important to 'submesoscale frontogenesis' when submesoscale currents deform fronts (Shakespeare and Taylor, 2013; Barkan et al., 2019).

Another efficient mechanism to generate submesoscale currents is mixed-layer baroclinic instability (MLI) (Haine and Marshall, 1999; Boccaletti et al., 2007; Fox-Kemper et al., 2008). MLI is an extension of the classical geostrophic baroclinic instability, but with a smaller horizontal scale due to the reduced Rossby deformation radius in the mixed layer. However,

182 Ocean Mixing

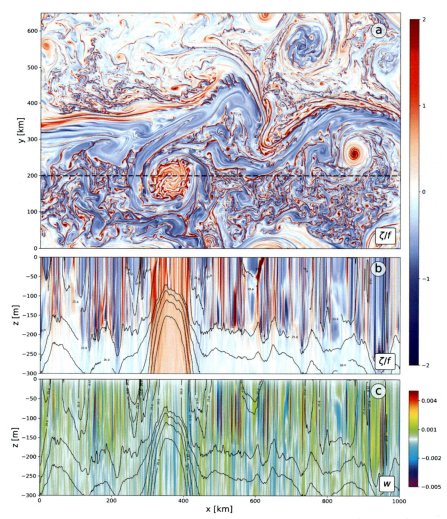

FIGURE 8.1 (a) Instantaneous relative vorticity $\zeta = v_x - u_y$ normalised by f at the surface in the wintertime Gulf Stream in a submesoscale-resolving ($\delta = 500$ m) simulation. (b) Vertical section of vertical vorticity along the dashed line in panel (a), and (c) vertical section of vertical velocity (in m s^{-1}).

since the corresponding Rossby number is typically $O(1)$, ageostrophic effects are more pronounced in MLI compared to the mesoscale baroclinic instability. Submesoscale currents can also be generated by inherently ageostrophic "symmetric" instabilities triggered by atmospheric forcing at fronts (Taylor and Ferrari, 2009; Thomas and Taylor, 2010; D'Asaro et al., 2011). The ageostrophic nature of submesoscale dynamics explains a number of profound characteristics of submesoscale features, including intense upwelling/downwelling, rapid restratification, and production of small-scale turbulence.

In most cases submesoscale currents do not directly contribute to mixing in the form of tracer homogenisation, which occurs near sub-millimetre Batchelor scales due to molecular diffusion. However, submesoscale dynamics have an important role in cascading energy and tracer variance from the largely adiabatic mesoscale down to scales where three-dimensional turbulence can take over (McWilliams et al., 2001). Submesoscale instabilities can facilitate a dynamical route to molecular diffusion by extracting energy from the geostrophically balanced flow and triggering secondary turbulent flows leading to small-scale turbulent mixing (Molemaker et al., 2010; D'Asaro et al., 2011; Brüggemann and Eden, 2015). Furthermore, submesoscale processes are typically associated with enhanced vertical velocities (Mahadevan and Tandon, 2006; Klein and Lapeyre, 2009) and vertical fluxes of heat, buoyancy, and tracers in the mixed layer (Mahadevan et al., 2012; Su et al., 2018). They can thus induce complex profiles of vertical stratification that influence vertical mixing in the surface layer and maintain elevated mixing efficiency at submesoscale fronts.

Submesoscale processes play a particularly important role in modulating the upper ocean stratification. Even though stratification changes are primarily driven by air-sea exchanges and vertical mixing, lateral stirring becomes increasingly

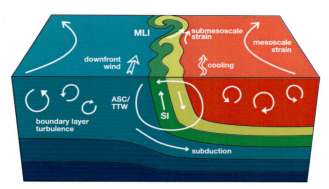

FIGURE 8.2 Idealised depiction of various submesoscale processes discussed in this chapter. Convergent mesoscale strain drives frontogenesis. Surface cooling or a down-front wind can make the front unstable to symmetric instability (SI). The frontogenetic strain and vertical mixing drive an ageostrophic secondary circulation (ASC), which, in the latter case, can be described as a turbulent thermal wind (TTW) balance. Submesoscale eddies develop through mixed-layer instability (MLI), which drives further frontogenesis and localises boundary layer turbulence and subduction of water into the thermocline.

important near density fronts (Rudnick and Ferrari, 1999; Ferrari and Rudnick, 2000). Submesoscale processes alter the horizontal and vertical density gradients in a number of interrelated and competing ways: Frontogenesis sharpens horizontal gradients of density and may also alter the vertical stratification if frontogenetic strain is depth-dependent. At the same time, submesoscale ageostrophic secondary circulation associated with frontogenesis, MLI, and symmetric instability (SI) tends to decrease isopycnal slope in the upper ocean, partially transforming horizontal density gradients into vertical, a process commonly described as submesoscale frontal slumping and restratification.

Submesoscale dynamics have been originally studied in the open ocean surface boundary layer, although they are also prevalent in coastal or estuarine environments. The local Rossby deformation radius on the continental shelf can be on the order of 1–10 km or smaller, such that structures at this scale may be classified as mesoscale in the literature. However, the Rossby and Froude numbers nearshore can reach and exceed $O(1)$ values typical of open ocean submesoscale regimes (e.g., Capet et al., 2008a; Dauhajre et al., 2017; Wang et al., 2021). Thus littoral dynamics share many processes with open ocean submesoscale dynamics, albeit modified by the strong influence of topographic, wave-driven, tidal, and freshwater runoff effects.

Energetic submesoscale currents can also be found at the bottom of the deep ocean in the presence of steep topography as on continental slopes, seamounts, and ridges. An illustration showing submesoscale currents in a numerical model at 1500 m depth around the Mid-Atlantic ridge is shown in Fig. 8.3. High-resolution numerical models have highlighted efficient mechanisms for topographic generation of submesoscale currents through interactions of geostrophic flows with steep topographic slopes (Molemaker et al., 2015; Gula et al., 2015b), ridges (Vic et al., 2018), and seamounts (Srinivasan et al., 2019). These interactions can lead to elevated local kinetic energy dissipation (Gula et al., 2016) and diapycnal mixing (Dewar et al., 2015) through triggering of centrifugal instability or intense horizontal shear instability and to the formation of submesoscale coherent vortices. Observations have recently confirmed that bottom currents interacting with sloping topography can trigger submesoscale instabilities with important implications for mixing and upwelling of deep-ocean waters (Ruan et al., 2017; Naveira Garabato et al., 2019). Theoretical process studies are beginning to highlight the full extent of the role played by submesoscale processes in the bottom boundary layer (Wenegrat et al., 2018a; Wenegrat and Thomas, 2020). Furthermore, submesoscale baroclinic instability of the bottom boundary layer might be instrumental in driving restratification in the deep ocean and exporting mixed water from the bottom boundary layer (Callies, 2018; Wenegrat et al., 2018a). These results are particularly timely, because of the recent realisation that the energetic mixing near ocean boundaries plays a fundamental role in driving the global ocean circulation, as detailed in Chapters 3 and 7 of this book. The role played by bottom submesoscale processes in this picture is still largely unknown.

Field observations of submesoscale phenomena are complicated by their small horizontal scales, rapid evolution, and spatial heterogeneity. Much of our understanding of submesoscale dynamics has been obtained from numerical modelling (Fig. 8.1), either through dedicated process studies (e.g., Boccaletti et al., 2007) or realistic high-resolution simulations (e.g., Capet et al., 2008b). Observational verification of this understanding, however, remains a challenge. Though satellite SST or ocean colour routinely reveal an abundance of submesoscale structures at the surface, they cannot be used to reconstruct the underlying velocity fields. Furthermore, the footprints of the current generation of satellite altimeters ($\approx$100 km) is unable to resolve submesoscale currents. Satellite measurements from the surface water and ocean topography mission (SWOT, Fu and Ferrari, 2008) will soon provide a global coverage of sea surface height (SSH) at better resolution (15–30 km, Morrow

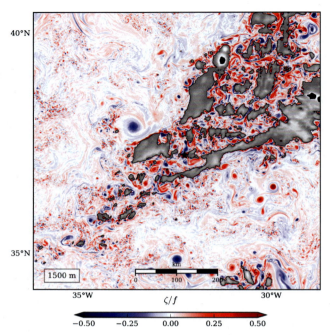

FIGURE 8.3 Instantaneous relative vorticity $\zeta = v_x - u_y$ at 1500 m depth over the Mid-Atlantic Ridge. Data from a numerical simulation with $\delta x = 750$ m horizontal resolution. Image adapted from Vic et al. (2018).

et al., 2019). However, reconstructing the surface velocity from SSH at these scales remains a challenging task due to the presence of internal waves and other unbalanced motions with similar spatial scales (Torres et al., 2018). Promising results have been obtained for the observation of intense submesoscale fronts with airborne optical and radar measurements of their surface roughness signature (Rascle et al., 2017, 2020), but they are limited to a small number of examples so far.

Only a few dedicated observational programs have been able to acquire data with enough spatial and temporal resolution to allow a quantitative characterisation of submesoscale currents in the surface layer, e.g., AESOP (D'Asaro et al., 2011; Johnson et al., 2020a), LATMIX (Shcherbina et al., 2013), OSMOSIS (Buckingham et al., 2016), ASIRI (Wijesekera et al., 2016), CARTHE-LASER (D'Asaro et al., 2020), and CALYPSO (Mahadevan et al., 2020). Such programs usually require a combination of mooring arrays, multiple ships, autonomous underwater vehicles, or swarms of surface drifters to allow for a computation of submesoscale gradients of velocity and tracers. Observations of submesoscale currents at the bottom of the ocean are even more difficult to obtain, and the examples are still rare (Naveira Garabato et al., 2019).

In the following section, we review the important mechanisms driving the formation, evolution, and decay of submesoscale fronts. These include frontogenesis in Section 8.2.1 and the instabilities that develop on submesoscale fronts in Section 8.2.2. Aspects specific to the bottom boundary are developed in Section 8.2.3. The evolution of fronts in the presence of turbulent mixing is discussed in Section 8.2.4, and finally the decay of submesoscale fronts is discussed in Section 8.2.5. In Section 8.3, we discuss the impact of the different types of submesoscale processes on the redistribution of density, focusing in particular on the effects on the restratification of the surface layer, but also on the mixing and restratification induced by submesoscale processes in the bottom boundary layer. In Section 8.4, the impact on the redistribution of passive tracers is discussed, including impacts for some biogeochemical tracers and buoyant material. Finally, we present our conclusions and perspectives in Section 8.5.

8.2 Life-cycle of submesoscale fronts

8.2.1 Frontogenesis

Open ocean fronts develop and intensify through a process known as frontogenesis. The mathematical theory of frontogenesis, which describes the intensification of cross-front density gradients in response to an imposed background flow, was originally developed in the atmospheric context (Hoskins and Bretherton, 1972; Hoskins, 1982) and later applied to the upper ocean (Macvean and Woods, 1980; Lapeyre and Klein, 2006; Capet et al., 2008c). An extensive review of oceanic

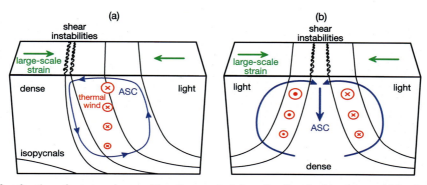

FIGURE 8.4 Sketches of surface-layer frontogenesis caused by a large-scale deformation flow for (a) a front and (b) a dense filament; adapted from McWilliams (2016).

frontogenesis has recently been published by McWilliams (2021), we briefly recall the important aspects of frontogenesis below.

A classical description of frontogenesis involves the sharpening of favourably aligned lateral density gradients by a straining flow, disruption of geostrophic balance for the along-front flow, and generation of an ageostrophic secondary circulation (ASC), see Fig. 8.2. This secondary circulation acts to restore geostrophic balance by advectively restratifying the flow, i.e., tilting the isopycnals toward the horizontal by bringing warm/light water over the top of cold/dense water. The ASC thus forms an overturning cell in the cross-front/vertical plane, with upwelling ($w' > 0$) on the light side ($\rho' < 0$ or $b' > 0$) and downwelling ($w' < 0$) on the dense side ($\rho' > 0$ or $b' < 0$), where w is the vertical velocity and primes denote deviations from a cross front average. The associated vertical buoyancy flux is thus always positive, i.e., $\overline{w'b'} > 0$, corresponding to restratification of the front.

A canonical setup to study frontogenesis considers a density front embedded in a large-scale horizontal flow with a uniform strain rate: $u_d = -\alpha x$, $v_d = \alpha y$, as illustrated in Fig. 8.4a. This configuration is well suited for mesoscale strain-induced frontogenesis, which is one of the fundamental factors leading to submesoscale currents. For a quasi-geostrophic flow, strain-induced frontogenesis causes the cross-front gradient to increase at an exponential rate (e.g., $\partial_x C \sim \exp \alpha t$ for a tracer C) (Washington, 1964).

On the dense side of a finite-width front, the ASC is convergent at the ocean surface (see Fig. 8.4). This amplifies the convergence associated with the large-scale strain flow and accelerates frontogenesis. Using the inviscid semi-geostrophic equations, Hoskins and Bretherton (1972) showed that this leads to super-exponential intensification of the density gradient and the collapse of the front into a singularity in finite time. A similar result was found in the generalised theory of Shakespeare and Taylor (2013), which included gravity waves and other ageostrophic effects. The asymptotic model derived by Barkan et al. (2019) even shows that when $Ro \sim 1$, the sharpening rate is primarily determined by the near-surface convergent motions associated with the ASC, rather than the large-scale strain flow. A bulk estimate for the vertical buoyancy fluxes of a front undergoing submesoscale frontogenesis is shown to be $\overline{w'b'} \sim -Hl(t)\delta(t)\partial_x b(t)$, with $\delta(t) < 0$ the velocity divergence due to the ASC and $l(t) \approx l(t_0)(1 + \delta(t_0)t)$ the width of the front at a time $t > t_0$. Therefore any pre-existing density front, for example, generated by the stirring of density by mesoscale eddies, may rapidly form an intense submesoscale front with order one Ro and further sharpen at a super-exponential rate.

A variant of frontogenesis applicable to dense filaments—sometimes called filamentogenesis—has been proposed by McWilliams et al. (2009). A dense filament corresponds to a surface density maximum formed by two parallel fronts with opposite-sign density gradient, as shown in Fig. 8.4b. The deformation flow acting on a favourably aligned dense filament causes an even more rapid narrowing and stronger surface convergence and downwelling at its centre than for an isolated front of similar scale and Ro. Instantaneous values of the vertical velocity at the centre of dense filaments can reach ~ 1 cm s^{-1} in realistic submesoscale-resolving simulations (Gula et al., 2014).

In general, any ASC in a finite-width front will necessarily have a zone of surface convergence, and therefore may contribute additional frontogenetic feedback. An example of this will be discussed in Section 8.2.4 dealing with the ASC induced by vertical mixing.

8.2.2 Instability of surface boundary layer fronts

Once fronts form and intensify through the process of frontogenesis, they can become susceptible to a variety of instabilities which generate additional submesoscale currents. These instabilities include mixed-layer instability (MLI); symmetric

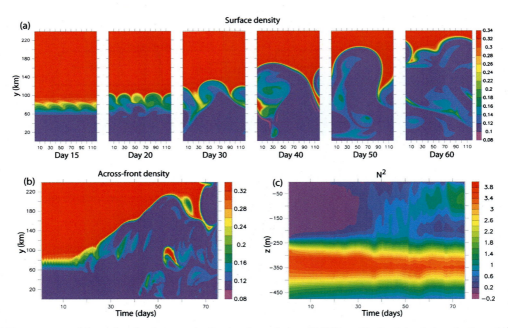

FIGURE 8.5 (a) Time sequence of the surface density anomaly showing the evolution of MLI in an idealised simulation of a front. (b) Evolution of the across front density and (c) of the vertical stratification. Adapted from Mahadevan et al. (2010).

instability (SI); ageostrophic anticyclonic instability (McWilliams et al., 2004); and horizontal shear instability (Munk et al., 2000).

In this section, we briefly describe the theory for the two most studied submesoscale instabilities: MLI and SI. Although the focus in this section will largely be on processes in the upper ocean, similar physical mechanisms are also active in the bottom boundary layer, as will be discussed in Section 8.2.3.

8.2.2.1 Mixed-layer baroclinic instability

Mixed-layer instability (MLI) is an upper ocean equivalent of the classical interior baroclinic instability (Haine and Marshall, 1999), as illustrated in Fig. 8.5. It is one of the most important sources of submesoscale currents in the surface mixed layer, and it has been extensively studied over the last decade using theory, idealised numerical simulations (Boccaletti et al., 2007; Fox-Kemper et al., 2008; Callies et al., 2016), and realistic submesoscale resolving numerical simulations (Capet et al., 2008c).

Baroclinic instabilities in the ocean interior are the primary source for the generation of mesoscale eddies at horizontal scales around or larger than the Rossby deformation radius,

$$R_{int} = \frac{N_{int} H_{tot}}{f}, \qquad (8.1)$$

where N_{int} is the stratification in the ocean interior and H_{tot} the ocean depth. MLI is a similar mechanism taking place in the weakly stratified surface layer, which has its own deformation radius,

$$R = \frac{NH}{f} \qquad (8.2)$$

where N and H are now the stratification and thickness of the weakly stratified surface layer. R is typically much smaller than R_{int}, because of the relative smallness of the mixed-layer stratification N and thickness H. Typical values of $N = 10^{-3}$ s^{-1}, $H = 100$ m and $f = 10^{-4}$ s^{-1} lead to $R = 1$ km, which is well within the submesoscale range.

Submesoscale MLI can be analysed to leading order using the Eady model (Eady, 1949). Although the ageostrophic secondary circulation can be large during submesoscale frontogenesis (Barkan et al., 2019), most theoretical studies of submesoscale instability have considered a balanced basic state. The basic state in the Eady model consists of a fluid with constant horizontal and vertical buoyancy gradients bounded from above and below by flat, rigid surfaces. The basic state velocity has a constant vertical shear in thermal wind balance with the horizontal buoyancy gradient. Specifically, if the basic state buoyancy is

$$b = N^2 z - M^2 y, \tag{8.3}$$

then the basic state geostrophic velocity is

$$u = \frac{M^2}{f} z, \tag{8.4}$$

where, in this case, z can be defined with respect to an arbitrary reference level of no motion.

As discussed in Eady (1949) and Stone (1966), there is one non-dimensional parameter in the system describing the evolution of linear, inviscid perturbations to this basic state. This parameter can be formulated as a 'balanced Richardson number',

$$Ri \equiv \frac{N^2}{u_z^2} = \frac{N^2 f^2}{M^4}, \tag{8.5}$$

where u_z is the thermal wind shear. Note that Ri is equivalent to the gradient Richardson number formed with the thermal wind shear. If we use the thermal wind to define a characteristic velocity scale, $U = M^2 H/f$, where H is the vertical distance between the top and bottom boundaries, and set a characteristic horizontal lengthscale based on the isopycnal slope, i.e., $L = N^2 H/M^2$, then we can relate the Rossby number to the balanced Richardson number,

$$Ro \equiv \frac{U}{fL} = \frac{M^4}{f^2 N^2} = \frac{1}{Ri}. \tag{8.6}$$

The analysis in Eady (1949) assumed that $Ri \gg 1$, and hence $Ro \ll 1$, whereas Stone (1966) generalised this to arbitrary Ri and Ro. When $Ri \gg 1$, the most unstable mode in the Eady model is independent of the cross-front direction (the y-direction in the basic state given above) and corresponds to the classical baroclinic instability. In this context, MLI can be viewed as baroclinic instability in the limit when $Ri, Ro \sim 1$, both characteristic of submesoscales in the mixed layer.

The wavelength of the most unstable perturbation for baroclinic instability is $L \approx 4R$, assuming that $Ri \gg 1$ as in Eady (1949). Taking into account ageostrophic effects, the wavelength of the fastest-growing MLI mode becomes $L_{MLI} \approx 4\sqrt{1 + Ri^{-1}} R$ (Stone, 1966; Nakamura, 1988), such that the instability is shifted to larger scales at small Ri if R is fixed. This scaling, with a typical value $Ri = 0.8$, has been used to estimate the surface MLI wavelength on a global scale by Dong et al. (2020b). The mixed-layer depth and stratification have been computed either directly from Argo profiles or from 1D simulations of the generalised ocean turbulence model (GOTM) embedded in the outputs of a global MITgcm simulation at submesoscale-permitting resolution (LLC4320, see Su et al. (2018)), as shown in Fig. 4.3 and 4.4 of Chapter 4. The results obtained for L_{MLI} with GOTM and a KPP parameterisation for vertical mixing are reproduced in Fig. 8.6a,b. The zonal median MLI wavelength varies from about 30 km in the tropics to 1 km at high latitudes. Given that the grid spacing of a model needs to be at least $L_{MLI}/8$ to properly resolve the eddies generated due to MLI, the requirements are still pretty high for models to be able to resolve MLI on a global scale.

MLI draws its energy from the available potential energy associated with horizontal buoyancy gradients—and converts it to kinetic energy. The energy source term in the equation of evolution for the perturbation kinetic energy is the buoyancy flux, $\overline{w'b'}$, where perturbations to the background flow are denoted $(\cdot)'$ (Capet et al., 2008c). The buoyancy flux associated with MLI is thus always positive and restratifies the mixed layer by distorting and tilting isopycnal surfaces (Fig. 8.5). Restratification by MLI represents a leading-order process in the mixed-layer buoyancy budget, as discussed in Section 8.3.

The conversion of available potential energy to kinetic energy happens near the instability scale. From there we may expect two possible non-linear energy cascades: an inverse energy cascade towards larger scales—typical of geostrophic turbulence (see review in Klein et al., 2019) – and a forward energy cascade driven by ageostrophic motions and loss of balance (Capet et al., 2008d). MLI transfers energy preferentially to larger scales (Boccaletti et al., 2007; Fox-Kemper et al., 2008), as visible in Fig. 8.5. Interactions between submesoscale eddies generated by MLI lead to an inverse cascade of energy, from the scale corresponding to the local wavelength of the fastest growing mode of MLI towards the mesoscale. This process helps to energise mesoscale eddies (Qiu et al., 2014; Sasaki et al., 2014; Schubert et al., 2020) and may explain the interannual to decadal variations of the mesoscale kinetic energy (Sasaki et al., 2020). However, the submesoscale eddy field generated as a result of MLI also develops sharp frontal features with active frontogenesis, which can lead to secondary instabilities and small-scale turbulence, and feed the forward cascade of energy (Schubert et al., 2020), as discussed in Section 8.2.5.

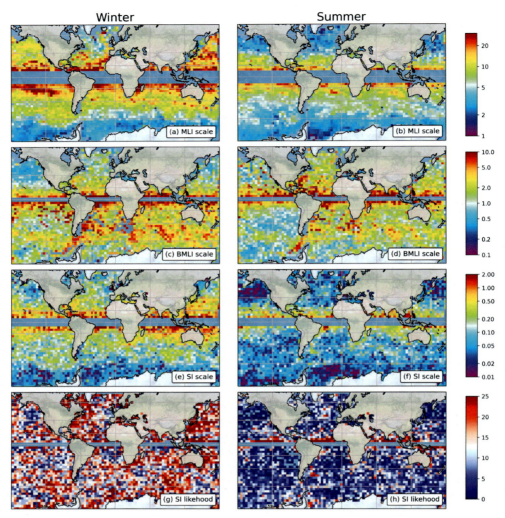

FIGURE 8.6 Global distributions of the fastest-growing wavelength (in km) for (a, b) MLI, (c, d) BMLI, and (e-f) SI, and relative likelihood (%) of SI (g-h). Results in (left) winter (February in the Northern Hemisphere and August in the Southern Hemisphere) and (right) summer (August in the Northern Hemisphere and February in the Southern Hemisphere). Figure adapted from Dong et al. (2020b) and Dong et al. (2021).

The available potential energy density at a front scales as $H^2|\nabla_h b|$. The depth of the mixed layer is thus a critical parameter controlling the energy of submesoscale eddies generated by MLI. This explains why submesoscale flows are expected to be stronger in winter, when the mixed layer deepens due to the strong negative buoyancy forcing, than in summer. This seasonality is confirmed by both the numerical simulations (Mensa et al., 2013; Sasaki et al., 2014; Brannigan et al., 2015; Ajayi et al., 2021) and in-situ observations (Callies et al., 2015). In winter, the energised submesoscale flows produce surface kinetic energy spectra that scale as approximately k^{-2} at scales below 100 km, with k the horizontal wavenumber, which is typical of active submesoscale regimes. In summer, the spectra are steeper and scale as approximately k^{-3}, which is typical of interior quasi-geostrophic turbulence. More precisely, a global analysis using outputs from the submesoscale-permitting LLC4320 simulation has highlighted two phases in this seasonal transition (Khatri et al., 2020). In late autumn the spectra first flatten as k^{-2} at scales < 50 km, but still follow a k^{-3} slope at larger scales (100–300 km). In late winter, the spectra flatten as k^{-2} also at large scales due to the inverse energy cascade initiated by MLI energising the mesoscales. There is also a time-lag – about one month in the Kuroshio region – between the mixed-layer thickness maximum and the submesoscale energy peak due to the competition between the production of eddy kinetic energy due to the vertical buoyancy fluxes and the non-linear energy cascade to larger scales (Dong et al., 2020a). The surface MLI wavelength, which scales linearly with H, also has a strong seasonality with a median value 1.6 times smaller in summer (10 km) than winter (16 km) globally (Fig. 8.6a).

Submesoscale baroclinic instability modes can sometimes extend below the mixed layer for particular stratification profiles. This can happen in the presence of a weakly stratified transition layer between the mixed layer and the thermocline, as observed by Zhang et al. (2020). The presence of a weakly stratified sublayer and an equatorward buoyancy gradient can also lead to the onset of Charney baroclinic instability, leading to an intensification of near-surface frontal activity (Capet et al., 2016).

8.2.2.2 Symmetric instability

When ageostrophic perturbations are considered in the linear stability analysis of the Eady model described previously, an additional set of instabilities are permitted. This was shown by Stone (1966, 1970, 1972) in a series of papers on non-geostrophic instability in the Eady model.

When $Ri < 1$, an ageostrophic instability develops where the most unstable perturbations are independent of x (the cross-front direction).[1] Stone (1966) and Hoskins (1974) refer to this mode as 'symmetric instability', a term that originated in studies of growing axisymmetric perturbations to a circular baroclinic vortex (Eliassen and Kleinschmidt, 1957; Ooyama, 1966).

In an unbounded, inviscid fluid, the most unstable mode of symmetric instability (hereafter SI) has motion that is aligned with isopycnals (Stone, 1966; Taylor and Ferrari, 2009). As a result, the buoyancy perturbations associated with SI modes are typically small. Whereas the baroclinic instabilities (including MLI) gain energy predominately from background potential energy, SI modes gain energy primarily at the expense of the thermal wind shear (Taylor and Ferrari, 2010).

Hoskins (1974) showed that the criterion for SI can be written $fq < 0$, where $q = (\boldsymbol{\omega} + f\hat{\mathbf{k}}) \cdot \nabla b$ is the Ertel potential vorticity (PV) and f is the Coriolis parameter.[2] This immediately presents a paradox: Since PV is materially conserved in an adiabatic, inviscid fluid, no re-arrangement of fluid parcels within a region with $fq < 0$ will change the bulk PV (Thorpe and Rotunno, 1989). A resolution of this paradox was proposed by Taylor and Ferrari (2009), who showed that the along-isopycnal motions associated with SI become unstable to a secondary Kelvin–Helmholtz shear instability. The resulting small-scale turbulence quickly increases fq by exchanging PV with a neighbouring stable region (e.g., the pycnocline) or driving a stabilising PV flux through the ocean surface.

SI can also be viewed as a hybrid of inertial and gravitational instability (Haine and Marshall, 1999). Due to the close connection with gravitational instability, SI is sometimes referred to as 'slantwise convection' (e.g., Straub and Kiladis, 2002; van Haren and Millot, 2009), although, as discussed below, it is important to note that the energetics of SI are distinct from convection. Inertial instability arises when the vertical component of the relative vorticity is anticyclonic and larger in magnitude than the Coriolis parameter, i.e., $\zeta/f < -1$. When the stratification is unstable ($N^2 < 0$) gravitational (or Rayleigh–Taylor) instability can develop.

For inviscid motions that are independent of x, the absolute momentum, $\mathcal{M} \equiv u - fy$, is conserved,

$$\frac{D\mathcal{M}}{Dt} = 0, \tag{8.7}$$

where D/Dt is the material derivative. In this case, the Ertel PV can be written

$$q = \mathcal{M}_y b_z - \mathcal{M}_z b_y. \tag{8.8}$$

For simplicity, consider the Northern Hemisphere ($f > 0$), where SI develops when $q < 0$ or

$$\frac{\mathcal{M}_y}{\mathcal{M}_z} < \frac{b_y}{b_z}. \tag{8.9}$$

In other words, SI can occur when isopycnals are steeper than surfaces of constant absolute momentum.

Given the close relationship between gravitational, inertial, and symmetric instability, a natural question is how to identify the dominant instability from a given set of conditions (e.g., in observations or an ocean model). This is particularly relevant within strong fronts, where all three instabilities can result in slantwise motion. The potential vorticity is not sufficient to distinguish between these instabilities since gravitational and inertial instability are also typically associated with $fq < 0$ (with $fN^2 < 0$ in the case of gravitational instability, and $f\zeta < 0$ in the case of inertial instability). One

---

1. Note that Stone (1966) also refers to Kelvin–Helmholtz instabilities, but Vanneste (1993) showed that these do not develop in the Eady model.
2. Note that a more general criterion for SI can be introduced for curved fronts in cyclogeostrophic or gradient wind balance. The resulting expression becomes $(1 + Cu)fq < 0$, where Cu is a non-dimensional number quantifying the curvature of the flow, as discussed in Buckingham et al. (2021a,b).

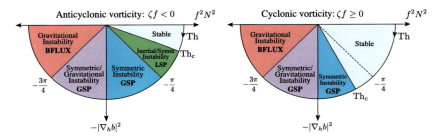

FIGURE 8.7 Regimes of gravitational instability (GI), symmetric instability (SI) and inertial instability (II) as a function of the Thomas angle (Th) when the vertical component of the relative vorticity (ζ) is anticyclonic (left) and cyclonic (right). In each region, the dominant source of energy is labelled as follows: Buoyancy flux (BFLUX), Geostrophic shear production (GSP), and lateral shear production (LSP). Adapted from Thomas et al. (2013).

approach, which was utilised by Thomas et al. (2013) is to classify the dominant instability associated with a given basic state using the energetics of the most unstable mode.

In a barotropic fluid, gravitational instability converts background potential energy into kinetic energy through the vertical buoyancy flux ($w'b' > 0$), while inertial instability grows via lateral shear production. As noted above, SI predominately extracts energy from the vertical shear associated with the thermal wind via geostrophic shear production (Taylor and Ferrari, 2010). By identifying the energy source associated with the most unstable perturbations, the dominant instability can be identified from a given basic state.

As was shown by Thomas et al. (2013), for a basic state that is in thermal wind balance, the dominant type of instability depends on the balanced Richardson number (8.5). Thomas et al. (2013) introduced what we call the 'Thomas angle', Th, where

$$\text{Th} \equiv \tan^{-1}\left(-\frac{|\nabla_h b|^2}{f^2 N^2}\right). \tag{8.10}$$

The Thomas angle is analogous to the Turner angle (often denoted Tu), which is used to distinguish different types of double-diffusive instabilities.

The Thomas angle re-maps the possible values of $Ri \in (-\infty, \infty)$ to $-\pi \leq \text{Th} \leq 0$. For a basic state consisting of uniform gradients in velocity and buoyancy, the flow is unstable when $\text{Th} < \text{Th}_c$, where

$$\text{Th}_c \equiv \tan^{-1}\left(-\frac{\zeta}{f}\right), \tag{8.11}$$

is the critical angle. Fig. 8.7 shows regions of gravitational instability (GI), symmetric instability (SI), and inertial instability (II). When the basic state vertical vorticity is anticyclonic, inertial/symmetric instability (II/SI) occurs when $-45° < \text{Th} < \text{Th}_c$. In this region the lateral shear production (LSP) is the dominant energy source for the growing perturbations, although the geostrophic shear production (GSP) also contributes. For a stably stratified barotropic fluid ($N^2 > 0$ and $|\nabla b| = 0$), the Thomas angle is $\text{Th} = 0$, and instability is only possible if $\text{Th}_c > 0$, which corresponds to $\zeta < -f$, the usual criterion for inertial instability.

Although an unstable buoyancy profile ($N^2 < 0$) corresponds to $\text{Th} < -\pi/2$, the buoyancy flux (BFLUX) is the dominant source of energy only for $\text{Th} < -3\pi/4$. When $-3\pi/4 < \text{Th} < -\pi/4$ for anticyclonic vorticity and $-3\pi/4 < \text{Th} < \text{Th}_c$ for cyclonic vorticity, the GSP is the dominant source of perturbation kinetic energy. Within these limits, Thomas et al. (2013) distinguish between SI/GI when $N^2 < 0$ and SI when $N^2 > 0$. Although the linear stability analysis was performed for a simple basic state with uniform velocity and buoyancy gradients, Thomas et al. (2013) showed that the Thomas angle was able to identify various dynamical regimes when applied *pointwise* to large-eddy simulations of the Gulf Stream front.

8.2.2.3 Forced symmetric instability

Since the criterion for SI depends on the sign of the potential vorticity, and since PV is materially conserved, the development of SI is closely tied to boundary forcing that is capable of changing the bulk PV. Further, since the growth rate of SI is relatively fast (with a characteristic timescale $\sim 1/f$), simulations and observations indicate that SI can quickly respond to de-stabilising forcing and maintain $fq \simeq 0$ (e.g., Thomas and Taylor, 2010; Thomas et al., 2013, 2016). Following Taylor and Ferrari (2010), we use 'forced symmetric instability' (or forced SI) to refer to the non-linear manifestation of symmetric instability that develops in response to destabilising surface forcing.

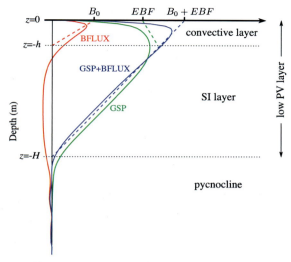

FIGURE 8.8 Schematic showing the structure of forced symmetric instability as described by Taylor and Ferrari (2010). Solid curves show the buoyancy flux (BFLUX), geostrophic shear production (GSP), and their sum as calculated from the large-eddy simulation of the response of the Gulf Stream to a storm as reported in Thomas et al. (2016). Dashed lines show the SI parameterisation further described in Bachman et al. (2017a). The surface buoyancy flux B_0, the Ekman buoyancy flux EBF, and their sum are also labelled.

Thomas (2005) discussed the conditions required for destruction of PV by surface forcing and showed that fq is reduced when the surface of the ocean is cooled by losing heat to the atmosphere, or when the surface wind stress points in the direction of the thermal wind, a so called 'downfront wind'. In the case of a downfront wind, dense water tends to be drawn over light water by the cross-front Ekman flow acting on the horizontal density gradient associated with the front. When sampled at a fixed position, this cross-front advection will result in a decrease in the surface buoyancy. The rate of change in buoyancy induced by the cross-front Ekman flow can be quantified through the Ekman buoyancy flux (EBF):

$$\text{EBF} \equiv \mathbf{M}_E \cdot \nabla_h b, \tag{8.12}$$

where

$$\mathbf{M}_E = \boldsymbol{\tau}_w \times \frac{f\hat{\mathbf{k}}}{\rho_0 f^2}, \tag{8.13}$$

is the Ekman transport, $\boldsymbol{\tau}_w$ is the wind stress, $\nabla_h b$ is the horizontal buoyancy gradient, f is the Coriolis parameter, and $\hat{\mathbf{k}}$ is the vertical unit vector. For a boundary layer of thickness H, the rate of change of PV scales with the sum of the EBF, and the surface buoyancy flux, B_0:

$$\frac{Dq}{Dt} \sim -\frac{f}{H^2}(\text{EBF} + B_0). \tag{8.14}$$

Hence, when $\text{EBF} + B_0 > 0$, fq will be reduced and conditions could become favourable for SI. It is worth emphasising that the EBF is an *advective* buoyancy flux and that advection does not change the integrated buoyancy. However, the appearance of the EBF in Eq. (8.14) shows that the EBF is generally associated with a change in the bulk PV. This is due to the fact that a downfront wind stress induces a frictional PV flux that removes PV from the ocean (Thomas, 2005).

As noted above, $fq < 0$ is a necessary but not sufficient condition for SI, and the dominant instability will depend on the Thomas angle. For example, if $\nabla_h b = 0$, surface cooling ($B_0 > 0$) will decrease fq, but the flow will become susceptible to gravitational instability (convection) rather than SI. Taylor and Ferrari (2010) used idealised simulations to study buoyancy and wind-driven convection within a frontal zone. They found that like convection, SI acts fast enough to keep the PV close to zero in a 'low PV layer'. PV conservation then allows a prediction for the deepening of the low PV layer, given the rate of PV destruction due to the surface forcing.

Taylor and Ferrari (2010) found that for surface cooling and downfront winds, the low PV layer can be further decomposed into two layers, as shown in Fig. 8.8. In a 'convective layer', near the surface, stratification is weak, the buoyancy flux is positive, and convective plumes are visible. Note that despite its name, turbulence generated by wind and/or wave breaking can dominate in the 'convective layer', and any plumes might be strongly distorted by the associated shear. Below the convective layer, SI develops, resulting in a stable stratification. In this 'forced SI layer', small-scale turbulence is generated by secondary shear instabilities that develop between the SI cells. This small-scale turbulence mixes the stable stratification, and hence the buoyancy flux is negative. In the forced SI layer, a cross-front shear (which will be described later) maintains the stable stratification.

Taylor and Ferrari (2010) derived a scaling for the depth of the convective layer h:

$$\frac{|\nabla_h b|^2}{f^2}(B_0 + \text{EBF})^{1/3} h^{4/3} = c \left[(B_0 + \text{EBF}) \left(1 - \frac{h}{H} \right) \right], \tag{8.15}$$

where $c \simeq 14$ is an empirical scaling coefficient. For a known low PV layer depth H; horizontal buoyancy gradient $|\nabla_h b|$; and surface and Ekman buoyancy flux B_0 and EBF, Eq. (8.15) can be solved numerically to find h. This provides a means for identifying when SI will be active under a given set of conditions. At a strong front and/or with weak surface forcing, $h/H \to 0$, and SI will develop throughout the low PV layer. However, when the horizontal buoyancy gradient is weak or forcing is sufficiently strong, $h/H \to 1$, and the low PV layer is expected to be weakly stratified.

Taylor and Ferrari (2010) used a turbulent Ekman balance (which is discussed below in Section 8.2.4) to show that the sum of the geostrophic shear production and buoyancy flux in a boundary layer with forced SI is a linear function of depth within the low PV layer, specifically,

$$\text{GSP} + \text{BFLUX} \simeq (B_0 + \text{EBF}) \left(\frac{z + H}{H} \right). \tag{8.16}$$

Thomas et al. (2013) proposed a model profile for the GSP by further assuming that the buoyancy flux takes the following linear profile within the convective layer:

$$\text{BFLUX} \simeq B_0 \frac{z + h}{h}. \tag{8.17}$$

These model profiles are shown in Fig. 8.8 and compared with the LES in Thomas et al. (2013). Bachman et al. (2017a) used this as a basis for a parameterisation for SI and implemented it in the MITgcm.

Observational evidence for SI typically consists of finding of signs of enhanced ventilation, fine-scale interleaving, and small-scale mixing within stratified baroclinic fronts. Early suggestions that SI (under the name of 'slantwise convection') may be partially responsible for ventilation of the Labrador current was reported in Straneo et al. (2002). D'Asaro et al. (2011) and Thomas et al. (2013) observed regions with $fq < 0$ in the Kuroshio and Gulf Stream fronts, respectively. D'Asaro et al. (2011) and Thomas et al. (2016) observed elevated dissipation rates within these regions, providing indirect evidence for geostrophic shear production associated with SI.

Several studies from the OSMOSIS project have shown evidence for SI in the subtropical northeast Atlantic. Using a full year of glider and mooring observations, Thompson et al. (2016); Buckingham et al. (2019); Yu et al. (2020) found a stratified low PV layer in the winter with scalings that were consistent with the model and theory in Taylor and Ferrari (2010). Yu et al. (2019b) found evidence for wind-forced SI at a transient front in the same region, as illustrated in Fig. 8.9. In-situ observations from the mooring array were used to compute the vertical stratification, the lateral buoyancy gradient, and the most likely instability (GI, SI, II) according to the classification described in Fig. 8.7. The mixed layer was favourable to GI triggered by atmospheric cooling ($\sim$200 W m^{-2}) between 3 and 7 April 2013, which resulted in a deepening of the mixed layer down to almost 300 m. A strong lateral buoyancy gradient, generated by the confluence of mesoscale eddies, developed between 8 and 11 April 2013. Persistent down-front winds during this period led to the generation of SI (Fig. 8.9c) and an EBF $\sim$650 W m^{-2}. The SI event was also associated with elevated glider-derived turbulent dissipation rates. Other evidence of SI were found by Ramachandran et al. (2018) in a shallow front in the Bay of Bengal, by Peng et al. (2020) in a density filament in the Benguela upwelling system, by Koenig et al. (2020) in a thermohaline front caused by sea ice melt in the Nansen basin, and by Bosse et al. (2021) around a deep convection area in the northwestern Mediterranean Sea. Together these studies suggest that SI does not require strong climatological fronts like the Gulf Stream and Kuroshio, and instead can occur in typical open ocean conditions.

Perhaps the most direct evidence for SI comes from observations of its influence on temperature and passive tracers. Savelyev et al. (2018) observed step-like patterns in the surface temperature in the southern Gulf Stream from airborne thermal imagery. These features closely resemble the feature seen in numerical simulations of SI (e.g., Stamper and Taylor, 2017), and concurrent ship-based measurements confirmed that conditions were favourable for SI. Wenegrat et al. (2020) analysed a tracer release experiment conducted in the Gulf Stream during a period with strong wind forcing and near-inertial

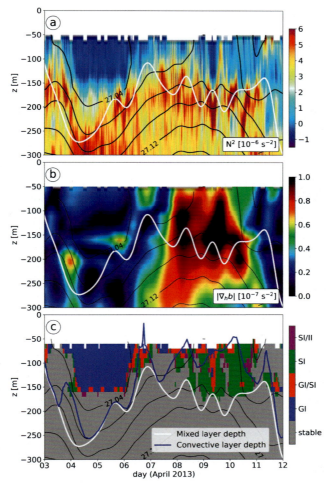

FIGURE 8.9 Time series of (a) vertical buoyancy stratification N^2, (b) lateral buoyancy gradient |$\nabla_h b$|, and (c) submesoscale instability category at the OSMOSIS central mooring site from 3 to 12 April 2013. The white line indicates the mixed-layer depth, and the blue line indicates the convective layer depth. Isopycnals are overlaid as black contours at intervals of 0.02 kg m^{-3}. Adapted from Yu et al. (2019b).

oscillations. The observations and numerical simulations suggest that circulation associated with SI mixed the dye along isopycnals. This, combined with episodic vertical mixing associated with near-inertial shear led to significant cross-front mixing of the tracer due to shear dispersion.

A global view of the scale and relative likelihood of SI, as computed by Dong et al. (2021), is provided in Fig. 8.6. The most unstable wavelength for SI (L_{SI}) is estimated from Stone (1966)'s linear prediction applied to the upper ocean state extracted from the global submesoscale-permitting LLC4320 simulation. The SI scale varies strongly with latitude, with median values of 1.5 km at the equator to 15 m at high latitudes. It also varies with seasons as the median scale roughly doubles in winter compared to summer. The SI scales are 1 or 2 orders of magnitude smaller than the MLI scale, and basin or global-scale models do not yet have sufficient resolution to capture SI.

8.2.3 Submesoscale processes at the bottom of the ocean

Fronts can also form at the bottom of the ocean. This has been widely studied for application in coastal areas. For example, over a continental shelf, the alongshore winds in the presence of a cross-shelf density gradient can create upwelling or downwelling situations. In the upwelling case, which has been the most studied (Barth, 1989a,b, 1994; Durski and Allen, 2005), the surface front leads to the same type of baroclinic instability or symmetric instability, as described previously. However, in the case of downwelling, a density front intersects the bottom, rather than the surface. This can lead to baroclinic instability as well, generating an eddy field with the same typical scale (1–10 km) (Brink, 2016).

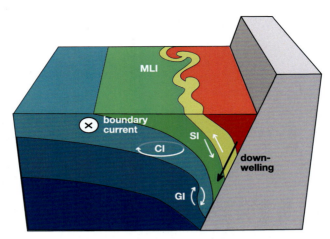

FIGURE 8.10 Idealised depiction of various submesoscale processes in the bottom boundary layer. A boundary slope current moving in the direction of Kelvin wave propagation generates a downslope Ekman current. After loss of potential vorticity due to friction and diapycnal mixing at the bottom, the front can be unstable to gravitational instability (GI), symmetric instability (SI), or centrifugal instability (CI). Submesoscale eddies develop through baroclinic instability of the bottom mixed layer (MLI). Schematic adapted from Naveira Garabato et al. (2019).

Submesoscale currents and fronts are also found in deeper parts of the ocean, especially in the presence of steep topography, such as seamounts and ridges. For example, energetic submesoscale currents are visible near the Mid-Atlantic ridge in Fig. 8.3. The corresponding kinetic energy spectra exhibit a shallow $\sim k^{-2}$ slope, contrasting with open ocean—i.e., far from topographic features—regimes of turbulence, dominated by mesoscales and a $\sim k^{-3}$ slope (Vic et al., 2018).

The bottom boundary layer (BBL) of the ocean exhibits key similarities with the surface boundary layer. Available potential energy associated with horizontal buoyancy gradients in the BBL can feed an equivalent of MLI (Wenegrat et al., 2018a). Friction and diapycnal mixing at the bottom can also inject or destroy PV in the BBL (Benthuysen and Thomas, 2012; Morel et al., 2019), which can either stabilise it or trigger a variety of unbalanced submesoscale instabilities, such as symmetric or centrifugal (inertial)[3] instabilities (Allen and Newberger, 1998; Molemaker et al., 2015; Dewar et al., 2015). The frictional drag can play the same role as the wind in the upper ocean and lead to forced instability regimes (Wenegrat and Thomas, 2020). Thus most of the submesoscale instabilities described in Section 8.2.2 in the surface layer have a counterpart at the bottom as summarised in Fig. 8.10 and described below.

8.2.3.1 Bottom boundary layer baroclinic instability

A counterpart to MLI occurs in weakly stratified boundary layers (Brink, 2012, 2013; Wenegrat et al., 2018a). The problem of baroclinic instability over a slope is a classic one (Blumsack and Gierasch, 1972; Mechoso, 1980; Solodoch et al., 2016), which has recently been revisited for buoyancy-driven flow by Hetland (2017) and for baroclinic instability over the continental shelf by Chen et al. (2020). It has also been used for low Ri regimes characteristics of the BBL by Wenegrat et al. (2018a).

The model is similar to the Eady model presented in Section 8.2.2.1, with the addition of a topographic slope θ. The ratio of the topographic slope to minus the isopycnal slope is then given by the slope parameter:

$$\alpha = \frac{N^2}{f u_z}\theta = S Ri^{\frac{1}{2}}, \quad (8.18)$$

where

$$S = \frac{N\theta}{f} \quad (8.19)$$

is the slope Burger number and Ri is the Richardson Number. As in Hetland (2017) and Wenegrat et al. (2018a), $\alpha = 1$ corresponds to an isopycnal slope equal and opposite to the topographic slope, and $\alpha = -1$ to isopycnals parallel to the topography. With uniform background vertical stratification and slope-normal shear, the effect of the topographic slope is to stabilise baroclinic instability for $\alpha < -1$ (Blumsack and Gierasch, 1972).

---

3. Here we use the term 'centrifugal instability' synonymously with 'inertial instability' (as defined in Section 8.2.2.2) since the former has become the standard convention in studies of the BBL (McWilliams, 2016).

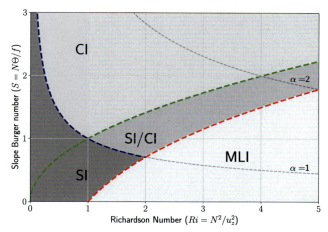

FIGURE 8.11 Regime diagram for submesoscale baroclinic instability (MLI), symmetric instability (SI), centrifugal instability (CI), and mixed instability (SI/CI) in the ocean BBL as a function of the Richardson number $Ri = N^2/u_z^2$ and the slope Burger number $S = N\theta/f$. The dashed red line shows the limit of zero PV, the dashed green line shows the limit of zero absolute vorticity $f + \zeta = 0$, and the dashed blue line is the theoretical expectation for the ratio of lateral and vertical shear production LSP/GSP = 1 in the hydrostatic limit. Adapted from Wenegrat et al. (2018a).

Realistic BBL structures have been shown to support submesoscale baroclinic instability (Wenegrat et al., 2018a). A first case corresponding to a BBL generated by bottom-intensified mixing supports a baroclinic instability that restratifies the thick outer layer of the BBL. A second case, corresponding to an interior flow in the direction of Kelvin wave propagation (meaning that the flow has the coast on its right in the Northern Hemisphere), drives a downslope Ekman flow and generates available potential energy that also feeds a submesoscale baroclinic instability. In most BBL cases, the isopycnals intersect the bottom almost perpendicularly, corresponding to a positive α, such that small positive values of α and S might be the most common configuration in the BBL (Wenegrat et al., 2018a).

The most unstable bottom MLI wavelength (L_{BMLI}, Fig. 8.6c, d) has also been estimated globally by Dong et al. (2020b), following the same methodology than for MLI in the surface layer (see Section 8.2.2.1). The median values are in the order of O(1) km, thus slightly smaller than the surface MLI scales and well into submesoscales.

8.2.3.2 Bottom injection of PV and submesoscale instabilities

Friction and diapycnal mixing at the bottom are strong sources of PV (Williams and Roussenov, 2003; Benthuysen and Thomas, 2012; Molemaker et al., 2015; Gula et al., 2015b). PV is extracted in the case of a boundary slope current moving in the direction of Kelvin wave propagation because the frictional drag reduces velocity along the slope, induces a negative horizontal velocity shear and a downslope Ekman flow that advects lighter water under denser water and drives diabatic mixing. PV is increased when the current is in the opposite direction (meaning that the flow has the coast on its left in the Northern Hemisphere) when it leads to a positive horizontal velocity shear and an upslope Ekman advection that tends to restratify the bottom boundary layer (Benthuysen and Thomas, 2012).

The PV extraction related to downslope buoyancy advection is a close analogous of the PV extraction in the surface layer in the case of a downfront wind, as described in Subsection 8.2.2.3. The advection of lighter water under denser water can generate convective mixing, which acts to increase the horizontal buoyancy gradient and decrease the vertical gradient, and can lead to a regime of forced SI or centrifugal (inertial) instability (CI) (Allen and Newberger, 1998; Molemaker et al., 2015; Dewar et al., 2015; Jiao and Dewar, 2015; Gula et al., 2016; Yankovsky and Legg, 2018; Wenegrat et al., 2018a; Wenegrat and Thomas, 2020; Naveira Garabato et al., 2019).

As in the surface layer, these additional instabilities (GI, SI, and CI/II) appear in the linear stability analysis of the Eady model when ageostrophic perturbations are included. The stability analysis for a geostrophically balanced along-slope flow in the direction of Kelvin wave propagation, with uniform stable stratification and uniform slope-normal velocity shear, is illustrated in Fig. 8.11 based on the scalings derived in Wenegrat et al. (2018a). When the slope Burger number is small, the criterion for instability is similar to that in the surface layer, namely, SI is dominant for $Ri \lesssim 1$ and MLI is dominant for $Ri \gtrsim 1$. When the slope Burger number is increased, the dominant mode becomes CI or a mixed SI/CI mode at moderate Ri, and the baroclinic instability mode is shifted to larger Ri.

The different types of instabilities occurring when the PV is negative can be identified using the same criteria as in the surface layer (Section 8.2.2.2). In particular, instabilities can be classified based on their dominant source of energy:

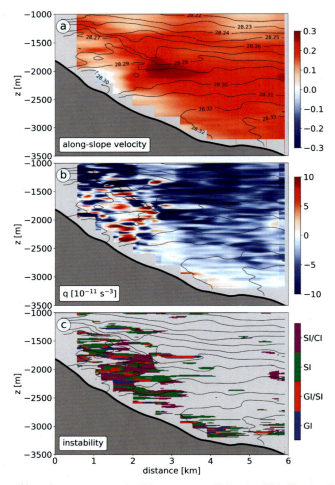

FIGURE 8.12 Transect across an abyssal boundary current near the Orkney passage sill (Section B3 in Naveira Garabato et al. (2019)). (a) Along-slope velocity (in m s^{-1}, colour) and neutral density (in kg m^{-3}, black contours). (b) PV (colour) and neutral density (black contours). (c) Instability type (colour) and neutral density (black contours). Adapted from Naveira Garabato et al. (2019).

GI extract their energy predominantly from buoyancy flux (BFLUX), SI from geostrophic shear production (GSP), and CI from lateral shear production (LSP) terms. The limit between the CI and SI regimes in the stability diagram (Fig. 8.11) corresponds to the ratio of lateral and vertical shear production being equal to 1 (LSP/GSP = 1), which gives the condition $\alpha = 1$ in the hydrostatic limit (see Eq. 33 in Wenegrat et al., 2018a).

Recent high-resolution observations taken in an abyssal boundary current flowing along steep topography in the Orkney passage highlighted the generation of GI, SI, and CI in a configuration corresponding to a downslope Ekman flow (Fig. 8.12, Naveira Garabato et al., 2019). GI is dominant in a $\approx$100 m thick layer over the bottom, whereas SI and CI happen over a larger region, which is about 1–2 km high and 500 m wide, as shown in Fig. 8.12. These instabilities generate an intensified near-boundary turbulence and drive buoyancy exchanges between the boundary layer and the interior.

8.2.3.3 Topographic wakes

A feature specific of the bottom is that, in the presence of a slope, it is possible for the highly sheared flow in the BBL to separate from the boundary and move into the stratified interior of the ocean, thus forming turbulent topographic wakes (Molemaker et al., 2015; Gula et al., 2015b, 2016), which can be even more efficient at driving diapycnal mixing. If the boundary is horizontal, which happens in the surface mixed layer or over a flat bottom, the frictional drag can only induce a vertical shear of horizontal velocity, such that the flow only acquires horizontal vorticity. It is thus unlikely to separate from the boundary as vertical motions are strongly limited by rotation and stratification, and instabilities and turbulence mostly stay inside the boundary layer (McWilliams, 2017). However, over a sloped bottom, the topographic drag induces both a vertical and a horizontal shear, and the sheared layer has the capability of separating from the boundary

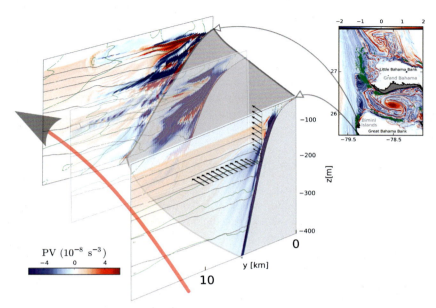

FIGURE 8.13 Snapshots of potential vorticity (PV, in 10^{-8} s^{-3}) along the Bahamas banks showing generation of negative PV (in colours) along the slope and the onset of centrifugal instability in the wake. The big red arrow indicates the direction of the Gulf Stream, and the small black arrows the velocity vectors close to the slope. Density is shown in black contours with an interval of 0.5 kg m^{-3}, and along-slope velocity is shown in green contours with an interval of 0.2 m s^{-1}. The inset on the right shows the instantaneous surface relative vorticity $\zeta = v_x - u_y$, normalised by f (in colours), with regions of high kinetic energy dissipation (depth-integrated energy dissipation $\langle \epsilon \rangle > 2 \times 10^{-4}$ W m kg^{-1}) highlighted in green. Adapted from Gula et al. (2016).

either due to the geometry of the boundary (Molemaker et al., 2015) or due to the background mesoscale straining (Vic et al., 2015). Escaping the constraint of the boundary, the sheared flow undergoes instabilities, such as CI/SI or horizontal shear instabilities, that lead to elevated local dissipation and mixing in the stratified interior (Gula et al., 2016) and to the generation of submesoscale coherent vortices (SCV) (Molemaker et al., 2015; Perfect et al., 2018; Srinivasan et al., 2019), which can further export tracers and water masses over much longer time and distances (Gula et al., 2019).

A typical case corresponds to a flow past a seamount, which may generate a turbulent topographic wake with both cyclonic and anticyclonic vorticity, leading to instabilities and formation of coherent vortices in the wake of the seamount, as seen from idealised (Perfect et al., 2018; Srinivasan et al., 2019) or realistic simulations (Srinivasan et al., 2017; Gula et al., 2019; Napolitano et al., 2021). There are similarities with the more classical island wake, which forms as a von Karman street with alternating cyclones and anticyclones (Stegner, 2014).

Vorticity is similarly generated when a current flows along sloped topography. The vorticity may be cyclonic or anticyclonic, depending on the orientation of the current relative to the topography. For example, the Gulf Stream along the U.S. seaboard generates cyclonic vorticity. There, the topographic drag amplifies the horizontal cyclonic shear of the jet until the current separates from the slope. The highly sheared flow becomes unstable to barotropic shear instability, which leads to the formation of streets of submesoscale vortices, as described in Gula et al. (2015a,b).

The anticyclonic case is particularly interesting as it can trigger CI, which has a strong impact on energy dissipation and mixing (Dewar et al., 2015; Gula et al., 2016). Following the sequence of processes described in the context of the California undercurrent (Molemaker et al., 2015) or the Gulf Stream (Gula et al., 2016), relative vorticity can locally become much less than $-f$ and result in negative PV. The slope has a stabilising effect upstream of separation, but the negative PV strip leads to intense CI and energy dissipation in the separated wake (see Fig. 8.13). In a Gulf Stream model, the vertically integrated dissipation rates of eddy kinetic energy due to parameterised turbulence (KPP), reach values up to 8×10^{-4} W m kg^{-1} instantaneously at 26°N following separation of the negative PV strip from the slope. It is of the same order as the dissipation rates observed in an intense surface front within the Kuroshio current (D'Asaro et al., 2011) integrated over the mixed layer.

Equatorial topographic wakes are particularly efficient at dissipating energy (Srinivasan et al., 2021). The background rotation is important to drive the merging and alignment of vorticity structures in the wakes. However, in the tropics, the decreasing f leads to more vertically-sheared wake structures and increasing energy dissipation.

Topographic wakes are a common feature in realistic submesoscale-resolving simulations, and most of the studies on the subject come from idealised or realistic numerical simulations (McWilliams, 2016). It remains very complicated to sample such localised events of turbulence in the ocean, especially in the deep regions. Intense small-scale turbulence and the formation of submesoscale eddies have been observed in the wake of the Palau island chain, in the western Pacific, during the program FLEAT (see articles in the special issue of Oceanography on the subject: Johnston et al., 2019; MacKinnon et al., 2019). Direct observations of turbulent wakes in the deep ocean are needed to better quantify the impact of theses processes. Deep seamounts and steep topographic features are numerous and the cumulative impact of topographic wakes on mixing might still be strongly underestimated.

8.2.4 The influence of vertical mixing on the evolution of a submesoscale front

In this section, we will discuss the influence of small-scale turbulence and vertical mixing on the evolution of submesoscale fronts, and thereby the submesoscale currents within. Study of these multi-scale interactions is at a relatively early stage, and many open questions remain. Here, we discuss some of the advances that have been made in this area. We start by discussing the ageostrophic circulation in 'generalised Ekman' (GE) or 'turbulent thermal wind' (TTW) balance, driven by vertical mixing of momentum at fronts, emphasising the connections between these balance relations. We then discuss the role that a GE/TTW-balanced flow plays in the evolution of a front. Finally, we discuss the connection between GE/TTW balance and submesoscale instabilities.

As noted in Chapter 4, when averaged over a sufficiently long period of time, the wind-driven currents in a horizontally homogeneous mixed layer sufficiently far from the equator are expected to be in a state of turbulent Ekman balance, (e.g., Cushman-Roisin and Beckers, 2011):

$$-f \langle v^a \rangle = -\frac{d}{dz} \langle u'w' \rangle = \frac{d}{dz}\left(v_T(z) \frac{d \langle u^a \rangle}{dz} \right), \qquad (8.20)$$

$$f \langle u^a \rangle = -\frac{d}{dz} \langle v'w' \rangle = \frac{d}{dz}\left(v_T(z) \frac{d \langle v^a \rangle}{dz} \right), \qquad (8.21)$$

where superscript a denotes the ageostrophic velocity; angle brackets denote a time (and/or horizontal) average; and v_T is a turbulent viscosity, which is assumed to be horizontally isotropic. The solution to these equations yields the classical Ekman spiral (Ekman, 1905), albeit modified by the depth-dependence of the turbulent viscosity. This balance has been confirmed with observations, although this requires a careful analysis and a long-time average (Price et al., 1987; Lenn and Chereskin, 2009; Polton et al., 2013; Johnson et al., 2020a).

In classical turbulent Ekman balance, the geostrophic velocity (which was implicitly cancelled with the pressure gradient in Eqs. (8.20)–(8.21)) is assumed to be depth-independent, and hence only the ageostrophic shear appears in the last term in each equation. At a front, however, the geostrophic flow also includes a vertical shear. The vertically-sheared geostrophic flow $\mathbf{u}_g(z)$ can be mixed vertically by small-scale turbulence, thereby driving an ageostrophic response. In the context of wind-driven flow at fronts in the Tropical Pacific, Cronin and Kessler (2009) derived what they termed the 'generalised Ekman' balance, which can be written

$$-f \langle v \rangle = -\frac{1}{\rho_0}\left\langle \frac{\partial p}{\partial x} \right\rangle + \frac{d}{dz}\left(v_T(z) \frac{d \langle u \rangle}{dz} \right), \qquad (8.22)$$

$$f \langle u \rangle = -\frac{1}{\rho_0}\left\langle \frac{\partial p}{\partial y} \right\rangle + \frac{d}{dz}\left(v_T(z) \frac{d \langle v \rangle}{dz} \right), \qquad (8.23)$$

$$\frac{1}{\rho_0}\left\langle \frac{\partial p}{\partial z} \right\rangle = \langle b \rangle. \qquad (8.24)$$

Note that the full velocity, $\mathbf{u} = \mathbf{u}_g + \mathbf{u}_a$, appears in the Coriolis and vertical mixing terms. Wenegrat and McPhaden (2016) found analytical solutions to Eqs. (8.22)–(8.24) and extended the model to include the influence of surface waves, whereas Taylor and Ferrari (2010) found that this balance holds during forced symmetric instability (with the vertical momentum flux terms instead of a turbulent viscosity).

Alternatively, Eqs. (8.22)–(8.24) can be viewed as a generalisation of thermal wind balance. Gula et al. (2014) note that when vertical mixing of momentum is included in a state that is otherwise in geostrophic and hydrostatic balance, the

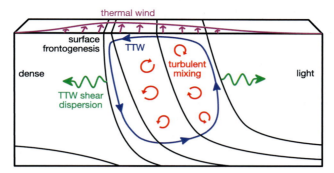

FIGURE 8.14 Schematic of the cross-front turbulent thermal wind (TTW) circulation and its influence on the evolution of submesoscale fronts. Near the surface, convergent cross-front TTW circulation leads to frontogenesis, as described in McWilliams et al. (2015); McWilliams (2017). Re-stratification by the TTW circulation and vertical mixing leads to frontolysis through shear-dispersion, as described in Crowe and Taylor (2018, 2019).

vertical shear satisfies the following equations:

$$-f\frac{d\langle v\rangle}{dz} = -\left\langle\frac{\partial b}{\partial x}\right\rangle + \frac{\partial^2}{\partial z^2}\left(\nu_T \frac{\partial \langle u\rangle}{\partial z}\right), \tag{8.25}$$

$$f\frac{d\langle u\rangle}{dz} = -\left\langle\frac{\partial b}{\partial y}\right\rangle + \frac{\partial^2}{\partial z^2}\left(\nu_T \frac{\partial \langle v\rangle}{\partial z}\right). \tag{8.26}$$

Gula et al. (2014) called this 'turbulent thermal wind' (TTW) balance and they found that this described the circulation in simulations of submesoscale cold filaments in the Gulf Stream.

Note that Eqs. (8.25)–(8.26) can be obtained by taking the vertical derivative of Eqs. (8.22) and (8.23) and using (8.24). With a constant turbulent viscosity, Charney (1973) showed that this is the leading order balance in the hydrostatic Boussinesq equations for small Rossby number, and Crowe and Taylor (2018) showed that the same analysis can be extended to include a depth-dependent turbulent viscosity. Garrett and Loder (1981) and Ponte et al. (2013) discussed the use of Eqs. (8.25)–(8.26) to diagnose the circulation for given a density field and $\nu_T(z)$ and showed that simple solutions can be obtained in the limit of small Ekman number (i.e., weak vertical mixing), where the geostrophic shear can be used in the viscous terms. Finally, the dynamical balance underlying TTW is the same as the leading order balance in the subinertial mixed layer (SML) model of Young (1994).

Unlike the inviscid thermal wind balance, TTW balance is associated with a vertically-sheared cross-front flow. A simple explanation for this cross-front flow is as follows. Consider starting with a front in thermal wind balance. Then, suppose that small-scale turbulence (e.g., convection driven by surface cooling) leads to vertical mixing, which reduces the along-front shear. Since the mixed-layer depth is typically much smaller than the width of fronts, isotropic mixing by small-scale turbulence will not have a strong direct influence on the horizontal density gradient. As a result, the hydrostatic pressure gradient will also be relatively unaffected by small-scale mixing. Therefore a reduction in the along-front shear by vertical mixing will lead to an unbalanced cross-front pressure gradient, which will, in turn, drive a cross-front flow. The cross-front flow associated with this mechanism will always be in the sense to restratify the front with light water advected over dense water. The TTW circulation is also associated with a surface PV flux at submesoscale fronts, which may dominate both wind and surface buoyancy-flux-driven PV fluxes. These effects have been shown to modify the seasonal cycle of mode water formation in the Gulf Stream (Wenegrat et al., 2018b).

The cross-front GE/TTW flow can influence the evolution of the front and the associated submesoscale instabilities. This represents a mechanism whereby vertical mixing by small-scale turbulence can influence submesoscales (Garrett and Loder, 1981; Thompson, 2000). As discussed in Garrett and Loder (1981), the unbalanced pressure gradient, and hence the cross-front flow, will tend to be largest at the location where the horizontal density gradient is maximum, and the cross-front flow will naturally tend to zero outside the front. As sketched in Fig. 8.14, the horizontal convergence/divergence of the TTW circulation will lead to downwelling on the dense side of the front and upwelling on the light side of the front. Near the surface on the dense side of the front, the convergence of the TTW flow will lead to frontogenesis. McWilliams et al. (2015) and Sullivan and McWilliams (2018) show that this leads to a rapid collapse of outcropping dense filaments (where the TTW circulation consists of two counter-rotating cells with downwelling in the centre of the dense filament). McWilliams (2017) developed a diagnostic framework, which combined the TTW circulation with the Omega equation to analyse the secondary circulation and frontogenetic tendency for fronts and filaments. Bodner et al. (2019) used a perturbation approach to investigate the influence of horizontal and vertical mixing on the evolution of a front based on the inviscid frontogenetic model, described in Shakespeare and Taylor (2013). A diurnal modulation of the small-scale turbulence, driven by the solar cycle, has been shown to drive a diurnal modulation cycle of frontogenesis and relaxation (Dauhajre and McWilliams, 2018), which can be described using transient turbulent thermal wind (T3W) equations.

As discussed in Garrett and Loder (1981), Young (1994), and Crowe and Taylor (2018), the vertically-sheared cross-front TTW circulation can cause the width of the front to increase through shear dispersion (e.g., Young and Jones, 1991). As the cross-front circulation slumps the isopycnals, small-scale turbulence mixes density vertically, and the sustained combination of these two processes leads to an increase in the frontal width and a spindown of the front. When the turbulent viscosity depends only on z, Crowe and Taylor (2018) found that surface-intensified frontogenesis occurs during a brief transient phase before the front spreads through shear dispersion. They also found that as the front spreads, it limits to a self-similar shape with a roughly constant horizontal density gradient in the interior of the front. At the edges of the front, intense narrow bands of upwelling and downwelling appear in analytical solutions (Crowe and Taylor, 2018) and numerical simulations (Crowe and Taylor, 2019). Crowe and Taylor (2020) extended this work to include the circulation driven by a surface wind stress and a surface buoyancy flux and identified the conditions when the front sharpens or spreads.

Most of the studies referenced above represent vertical mixing using a prescribed turbulent viscosity (e.g., Crowe and Taylor, 2018), a simple model for mixing (e.g., Young, 1994), or a boundary layer parameterisation (such as the KPP scheme) (e.g., Gula et al., 2014; McWilliams et al., 2015). Although large-eddy simulations have shown that turbulent mixing can be significantly modified at submesoscale fronts (e.g., Taylor, 2016; Smith et al., 2016), much less is understood about how turbulence reacts and couples with the evolution of submesoscale fronts. One notable exception is Sullivan and McWilliams (2018), who studied large-eddy simulations of a collapsing submesoscale cold filament. Small-scale turbulence was generated in their simulations by wind forcing and/or surface cooling. Turbulence triggered the collapse of the filament through GE/TTW-induced frontogenesis. However, in the later stages of the filament evolution, this frontogenesis is arrested by enhanced turbulence generated through horizontal shear instabilities, as further discussed in Section 8.2.5.

There are many open questions related to the interactions between submesoscale fronts and turbulent mixing. Such questions include the following: Does the suppression of vertical mixing following frontal restratification (Taylor and Ferrari, 2011) enhance or inhibit TTW-induced frontogenesis? How is the spreading of a front by shear-dispersion changed when turbulence is modified by frontal shear/stratification? What factors influence the arrest of frontogenesis by turbulence, and hence equilibrium frontal widths? One of the major challenges in this area is the computational cost associated with simulations that capture both changes to small-scale turbulence and the evolution of submesoscale fronts. For example, the simulations in Sullivan and McWilliams (2018) used $\sim 10^{10}$ grid points. However, advances in computing power should make these simulations more accessible, and will allow an exploration of parameter space that will likely be necessary to develop and test parameterisations of these processes.

8.2.5 Frontal arrest and routes to dissipation

The theory of frontogenesis does not explain how the sharpening of the front is ultimately stopped. As noted above, the inviscid theory predicts the formation of a discontinuity in the surface density in a finite time (Hoskins and Bretherton, 1972). In reality the sharpening of the front can be arrested due to spatio-temporal variability of the large-scale flow (McWilliams et al., 2019), dissipation by intensified small-scale turbulence or an instability and eddy-feedback equilibration process. Frontal arrest due to the effect of baroclinic instability has been shown in McWilliams and Molemaker (2011). Another likely scenario is the triggering of horizontal shear instability (Gula et al., 2014; Samelson and Skyllingstad, 2016; Sullivan and McWilliams, 2018; Verma et al., 2019), which has growth rates scaling with the cross-front velocity gradient. This scenario for frontal arrest has been confirmed using large-eddy simulations of an idealised dense filament by Sullivan and McWilliams (2018). Within less than a day, frontogenesis is arrested at a small width (≈ 100 m), primarily by an enhancement of the turbulence through a small submesoscale, horizontal shear instability of the sharpened filament, followed by a subsequent slow decay of the filament by further turbulent mixing. However, other studies have found sharpening of a front down to $O(10)$ m until the release of a gravity current (Pham and Sarkar, 2018).

Submesoscale processes can thus initiate a forward cascade of energy and tracer variance, ultimately leading to energy dissipation and diapycnal mixing in frontal regions. Evidence of such a forward cascade of energy to dissipative scales has been observed at isolated small-scale frontal features in idealised high-resolution numerical simulations of MLI (Molemaker et al., 2010; Skyllingstad and Samelson, 2012; Samelson and Skyllingstad, 2016). Quantitative assessments in submesoscale-permitting realistic model have confirmed that the forward cascade of kinetic energy at the smallest resolved scales of the models occurs mainly in frontogenetic regions (Schubert et al., 2020), and that superinertial ageostrophic motions play an important role (Ajayi et al., 2021). However, the submesoscale energy cascades have also been shown to be strongly time- and region-dependent (Yang et al., 2021).

Very high levels of turbulent kinetic energy dissipation have been observed at strong persistent fronts, such as the Kuroshio or Gulf Stream (Nagai et al., 2009; D'Asaro et al., 2011; Nagai et al., 2012). In the open ocean, on the other hand, the contribution of submesoscale processes to small-scale turbulence may seem weak compared to the contribution of winds and waves, as observed during the OSMOSIS campaign (Buckingham et al., 2019), but they still have an important impact for extracting energy from the geostrophic circulation. Observations of a front in the Baltic sea, following the passage of a storm, have also highlighted a direct route to turbulent dissipation, linked to shear instability (Carpenter et al., 2020). Another observation of a density filament in the Benguela upwelling system also confirmed high level of energy dissipation driven by forced SI (Peng et al., 2020).

8.3 Redistribution of density and restratification at the submesoscale

Submesoscale instabilities tend to reduce horizontal density gradients and enhance vertical density gradients, i.e., they are restratifying in nature. The reason is twofold: Some of these instabilities tap into the available potential energy associated with lateral density gradients directly, whereas others draw kinetic energy from currents, disrupt the thermal wind balance, and force the lateral gradients to adjust. This results in ageostrophic secondary circulations that act to restratify the flow on timescales of hours to weeks. The stratification of the upper ocean is thus controlled by the competing effects of vertical mixing and restratification. The relative importance of these two processes is important in determining the density structure of the upper ocean.

In this section, we first describe separately the effect of the different submesoscale processes presented in Section 8.2 on the restratification of the ocean. Then, we comment on the competition between the different type of processes and their overall importance in the ocean.

8.3.1 Restratification induced by submesoscale processes

Submesoscale processes can increase vertical stratification on timescales that compete with surface radiative forcing. Submesoscale restratification is usually quantified in terms of the horizontal ($u'b'$) or vertical ($w'b'$) buoyancy flux, which correspond to a redistribution of density, but are often expressed as an equivalent surface heat flux to facilitate comparison with the ubiquitous restratification due to surface heat fluxes and freshwater input.

Earlier works on submesoscale frontal slumping have highlighted the role of geostrophic adjustment (Tandon and Garrett, 1994) and of the ASC associated with surface frontogenesis (Lapeyre et al., 2006) to restratify the horizontal gradients generated by wind mixing and the straining of mesoscale eddies. In this case, the vertical buoyancy flux associated with strain-induced frontogenesis of a single front scales as $\overline{w'b'} \sim H^2|\nabla_h b|/f^2$ (McWilliams, 2016), where H is the mixed-layer depth, and $\nabla_h b$ is the horizontal buoyancy gradient at the front.

During SI, restratification occurs primarily via a horizontal buoyancy flux associated with the vertically sheared cross-front flow (Taylor and Ferrari, 2010). Bachman et al. (2017a) proposed a parameterisation for restratification by SI. Since SI is generally faster than MLI, in an initial value problem (or following a sudden forcing event) SI typically acts in the early stages of the response before transitioning to MLI (Stamper and Taylor, 2017). In the idealised experiments of Fox-Kemper et al. (2008), SI brings the front to $Ri = 1$, i.e., $N^2 = |\nabla_h b|^2/f^2$, then MLI becomes the primary instability process. The complementary roles of SI and MLI are illustrated by the simulations of frontal regions under destabilising heat forcing in Taylor and Ferrari (2011): restratification occurs very rapidly (less than a day) due to SI, whereas MLI achieves a much stronger restratification, but on much longer times (several days). In Verma et al. (2019) as well, during the spindown of a front, the initial configuration with $Ri = 0.26$ is unstable to SI that then develops secondary KH instability and three-dimensional turbulence. The MLI soon becomes the dominant mode as Ri increases to beyond 0.95 during restratification of the front.

Restratification driven by finite-amplitude MLI is an adiabatic process that can be represented through advection by an eddy-driven overturning streamfunction (Boccaletti et al., 2007; Fox-Kemper et al., 2008):

$$\Psi^* \sim -C_e \frac{H^2|\nabla_h b|}{f} \tag{8.27}$$

and an associated buoyancy flux:

$$\overline{w'b'} \sim C_e \frac{H^2|\nabla_h b|^2}{f}, \tag{8.28}$$

with $C_e \approx 0.06$ determined from idealised experiments by Fox-Kemper et al. (2008). A parameterisation of MLI restratification effects has been developed by Fox-Kemper et al. (2008); Fox-Kemper and Ferrari (2008) based on this scaling, analogously to what was done by Gent and Mcwilliams (1990) to parameterise interior baroclinic instability for non-eddying ocean models. This parameterisation contributed to improve the properties of the mixed layer in climate models (Fox-Kemper et al., 2011; Gent et al., 2011). For example, it can reduce the maximum mixed-layer depth by up to 200 m in the Southern ocean and correct systematic biases with important climatic impacts (Calvert et al., 2020). However, it was also noted by Callies and Ferrari (2018b) that this scaling only represents the initial phase of the instability, but is not valid anymore when MLI reaches the finite-amplitude stage and develop larger eddies and stronger velocities than assumed in the scalings. In the numerical experiments of Capet et al. (2008b), for example, the submesoscale vertical buoyancy flux due predominantly to MLI is ~ 100 W m^{-2}, exceeding the prediction of Eq. (8.28). Note, however, that the rate of restratification by MLI may be impacted by the parameterisation of vertical eddy viscosity in a given model (Mukherjee et al., 2016).

8.3.2 Competition between destratification and restratification of a front

The two main factors driving turbulence in the mixed layer are the surface buoyancy flux and wind. When the horizontal density gradient is sufficiently weak, surface cooling (corresponding to a surface buoyancy flux $B_0 > 0$) will drive mixing and destratification through convection. Wind blowing over a front may restratify the mixed layer if the wind is blowing upfront (and advecting light water over dense water) or destratify if the wind is blowing downfront (and advecting dense water over light water). As discussed in Subsection 8.2.2.3, the wind drives an Ekman buoyancy flux (EBF) that scales as

$$\text{EBF} \sim \tau_w^a \frac{|\nabla_h b|}{\rho f}, \tag{8.29}$$

where τ_w^a is the along-front wind stress. Note however that, even though the relative orientation of fronts and winds is nearly isotropic on a global scale, there is an asymmetry in the processes, leading to a larger upwind restratification than downfront destratification for an equal wind stress (Thomas and Ferrari, 2008).

In the presence of wind forcing, the impact of the submesoscale processes on restratification can be different than the picture drawn in the previous subsection. Depending on the relative strengths of the wind-driven Ekman flow and the frontal ASC, wind-driven destratification can overcome restratification induced by the ASC related to frontogenesis (Thomas and Ferrari, 2008). The competition between destratification by downfront winds and restratification by MLI has been studied by Mahadevan et al. (2010). The relative importance of the destratification by down-front winds compared to the restratification by MLI can be characterised by a ratio between the respective buoyancy fluxes. Using Eq. (8.29) and Eq. (8.28), this ratio is

$$R = \frac{\tau_w^a}{C_e H^2 \rho |\nabla_h b|}. \tag{8.30}$$

In the simulations of Mahadevan et al. (2010), the ratio is equal to 1 and the restratification by MLI is countered by a down-front wind stress $\tau_w \approx 0.2$ N m^{-1}. However, the amplitude of the destratifying buoyancy flux diminishes with time as the alignment between the wind and front is disrupted by the growth of the frontal instability.

MLI is relatively resilient to convective mixing and found to grow in most conditions. MLI shuts off convection in conditions of weak buoyancy forcing (Mahadevan et al., 2010). MLI is also active and capable of generating submesoscale eddies in the presence of strong convection, even if it is unable to completely restratify the flow and shut off convection (Callies and Ferrari, 2018a). The convective deepening of the mixed layer is partially balanced by the submesoscale buoyancy flux due to MLI, leading to a more heterogeneous and shallower mixed layer than would be the case in the absence of submesoscales (Couvelard et al., 2015). A similar resilience of MLI-driven restratification was observed during the passage of an intense idealised autumn storm by Whitt and Taylor (2017). Despite the strong destratifying effect due to downfront winds, the submesoscales intensified during the storm and maintained a strong stratification in local patches. Considering a balance between MLI and the effect of a fixed vertical diffusion would imply a restratification rate one order of magnitude larger than Eq. (8.28) (Bachman and Taylor, 2016). In a more realistic situation, based on a submesoscale-resolving simulation of the central part of the Baltic Sea, Chrysagi et al. (2021) confirmed that submesoscales are able to maintain shallow mixed-layer depths in local patches during the passage of storms, and that a rapid restratification is at work after the passage of the storms. This submesoscale restratifying effect has important implications, because the mixing efficiency can remain high ($\Gamma = 0.2$) at submesoscale fronts, contrary to regions where the mixed layer is already well-mixed (Chrysagi et al., 2021).

Global estimates using Argo profiling float observations highlight the importance of MLI to the restratification of the upper ocean (Johnson et al., 2016). The mechanism is particularly effective in the regions and times where the mixed layer is deep, in particular during the transition from winter to spring, when the mixed layer is at its deepest. Regions of deep convection (Haine and Marshall, 1999) and inside anticyclones (Bosse et al., 2019) are also characterised by enhanced MLI-driven restratification. A number of in-situ observations have confirmed the importance of the MLI-driven restratification around the globe: e.g., in the north-east Pacific (Hosegood et al., 2006), in the Southern ocean (Bachman et al., 2017b; du Plessis et al., 2017), or in the Antarctic marginal ice zone and under sea ice, where the equivalent heat flux can reach $\sim$240 W m^{-2} in mid-winter (Biddle and Swart, 2020). However, the few high-resolution observations at individual fronts, such as the ones taken along the rim of a Southern ocean mesoscale eddy (Adams et al., 2017) or over a California current upwelling front (Johnson et al., 2020a,b), are also exposing that the destratification/restratification of the front is often complex and involve interactions between many processes.

Observations of submesoscale fronts over long periods are very difficult to obtain, and the ocean surface mixing, ocean submesoscale interaction study (OSMOSIS), which took place in the north-east Atlantic, is one rare example. The year-long records of in-situ measurements from a glider and a mooring array highlight a strong seasonal cycle of submesoscale turbulence (Buckingham et al., 2016) with several instabilities controlling the stratification of the upper ocean: The mixed layer deepens in the fall due to GI, and is restratified by SI and MLI throughout the winter (Thompson et al., 2016; Yu et al., 2019b).

Year-long observations were also obtained with a similar mooring array in the North Pacific subtropical countercurrent during the SubMESI experiment (Zhang et al., 2020). The SubMESI mooring was located in a region more typical of subtropical gyres with a shallower mixed layer and higher eddy kinetic energy than the north-east Atlantic. Submesoscale turbulence is driven there by a combination of MLI and strain-induced frontogenesis, leading to an equivalent upward heat flux comparable in magnitude with the net surface heat fluxes during late winter and early spring.

Submesoscale processes play an important role in the surface mixed-layer, but it remains unclear how effectively they can penetrate into the underlying pycnocline and significantly contribute to vertical heat and buoyancy fluxes in the interior. Intense frontogenetic regions with ageostrophic dynamics are present in the ocean interior in the idealised simulations of Molemaker et al. (2010) and Barkan et al. (2015). However, the dynamical regimes in these simulations are characterised by a large M^2/N^2 ratio ($= 0.2$ in Molemaker et al. (2010)) compared to typical interior ocean values (see discussion in Capet et al. (2016)).

Examples of submesoscale fronts that penetrate below the mixed layer have been documented by Yu et al. (2019a); Zhang et al. (2020). Recent observations and modelling in the Southern ocean have highlighted the presence of deep submesoscale fronts reaching down to 1000 m depth. These fronts are generated by the straining associated with Southern ocean mesoscale eddies and are associated with active frontogenesis. They drive an anomalous upward heat transport from the ocean interior to the surface that acts to restratify the ocean interior (Siegelman et al., 2020; Siegelman, 2020). Similar deep-reaching fronts might be present in regions with strong horizontal gradients (e.g., western boundary currents and regions with intense mesoscale activity) and low stratification (e.g., high latitudes regions). The submesoscale vertical heat transport computed at 200 m depth by Su et al. (2018, their Fig. 3c, d) might give a good idea of where to expect such deep-reaching fronts.

8.3.3 Bottom boundary layer mixing and restratification

In the BBL, unbalanced submesoscale processes linked to negative PV (GI, CI, SI) can produce small-scale turbulence and contribute to mixing the fluid toward a state of marginal stability. Recent observations by Ruan et al. (2017) in the southern Drake passage have highlighted the presence of strong lateral buoyancy gradients along slopes and enhanced bottom turbulence associated with submesoscale instabilities, leading to thick bottom mixed layers and water-mass modifications. Observations in the Orkney passage, shown in Fig. 8.12, have also highlighted vigorous turbulent mixing associated with submesoscale instabilities, in particular CI and SI (Naveira Garabato et al., 2019). Microstructure measurements revealed turbulent dissipation and mixing rates exceeding oceanic background values by one to three orders of magnitude in the abyssal boundary current. However, to maintain efficient water-mass transformations, it is also essential that the mixed waters are exported out of the mixed layer and replaced by stratified waters from the interior. These instabilities also drive such lateral exchanges (Naveira Garabato et al., 2019), leading to a high mixing efficiency (0.6–1) compared to a typical efficiency of 0.2 (Spingys et al., 2021).

The contribution of topographic wakes to mixing may also be significant as the mixing efficiency of CI is particularly high in the stratified interior (Dewar et al., 2015; Jiao and Dewar, 2015; Gula et al., 2016). In the idealised numerical experiments of Perfect et al. (2020), large volume-averaged diapycnal diffusivities are found in the wake of a seamount.

They are shown to scale like the product of the Froude and Rossby numbers squared: $K_{rho} \sim (Fr\, Ro)^2 = (\frac{U}{NH}\frac{U}{fD})^2$, where H and D are the height and half-width of the seamount.

Similarly to the surface, bottom MLI draws energy through the vertical buoyancy flux and contributes to a restratification of the bottom mixed layer ($\overline{w'b'} > 0$). MLI is thus suspected to directly affect the strength and structure of the abyssal overturning circulation by playing a crucial role to balance the effect of bottom-intensified mixing and allow for sustained water mass transformations (Callies, 2018), see also Chapters 3 and 7. The stratification over the flanks of mid-ocean ridge systems seem to be largely maintained by submesoscale baroclinic eddies, contrary to the stratification within ridge-flank canyons, which is maintained by mixing-driven mean flows (Ruan and Callies, 2020).

8.4 Redistribution of passive tracers and particles

In this section, we describe the role of submesoscale processes in redistributing *passive* ocean tracers (those that do not affect the density of sea water). We consider two kinds of passive tracers: Firstly, those that are *conservative*, i.e., have no interior sources and sinks, and hence affected only by boundary fluxes, physical transport, and mixing. Secondly, those that are *reactive*, i.e., altered by growth and decay processes. Naturally, most biogeochemical tracers are regarded as reactive, since they are affected by biological growth or decay that occurs in the presence of another variable or property, such as photosynthesis in the sunlit region.

The same tracers may exhibit different active/passive or conservative/reactive behaviours depending on the situation. For example, temperature and salinity affect the density of sea water, and therefore are undeniably active. However, most currents in the ocean are isopycnal, even at the submesoscales. When such motions are considered, temperature or salinity anomalies act as *passive* tracers that are stirred on isopycnal surfaces by the eddying flow field. Density-compensated thermohaline variability is often the most convenient passive tracer to study, since it is found ubiquitously in the world's oceans, even where other passive tracers may be absent.

Discrete particulate matter suspended in the ocean (such as plankton, larvae, sediment, microplastics, oil droplets, etc.) as well as macroscopic objects (marine debris, kelp, ice floes) often behave similarly to passive tracers, although their redistribution is additionally complicated by the buoyancy and inertia effects.

8.4.1 Conservative tracers

Conservative passive tracers in the ocean are stirred by the eddying flow field, creating tracer variability over a range of scales. This variability can be characterised by an isopycnal spectrum of a tracer variance $P(k)$, where k is the horizontal wave number. The passive tracer spectrum is theoretically predicted to be universally related to the stirring eddy flow field and its turbulent kinetic energy spectrum $E(k)$, at least for well-developed turbulence (Vallis, 2006). For a uniformly rolling-off kinetic energy spectrum $E(k) \sim k^{-n}$, $n < 3$, the corresponding conservative passive tracer variance spectrum rolls off as

$$P(k) \sim k^{(n-5)/2}. \tag{8.31}$$

In the interior ocean, quasi-geostrophic dynamics suggests that $E(k) \sim k^{-3}$, corresponding to the tracer variance spectrum $P(k) \sim k^{-1}$. On the other hand, surface-quasigeostrophic theory applicable to near-surface submesoscale processes produces kinetic energy and passive tracer spectra with the same slope, i.e., $E(k) \sim P(k) \sim k^{-5/3}$. Relationship (8.31) breaks down for steep kinetic energy spectra ($n > 3$), where stirring becomes non-local and $P(k) \sim k^{-1}$ (Batchelor, 1959). As (8.31) suggests, when smaller-scale turbulence is more energetic (the $E(k)$ spectrum is flatter) there is more vigorous stirring of passive tracers, resulting in less tracer variance at smaller scales and a steeper tracer variance spectrum, $P(k)$ (Scott, 2006; Callies and Ferrari, 2013; Jaeger et al., 2020a).

A number of observational studies have tried to analyse the spectra of passive tracers in the pycnocline, but until recently, they lacked the horizontal resolution to conclusively measure the spectrum at submesoscales. A major oceanographic field experiment called LatMix (Scalable lateral mixing and coherent turbulence, Shcherbina et al., 2015), aimed at observing submesoscale stirring, was conducted in 2012 in the northwestern Atlantic. Direct measurements of dye dispersion highlighted stirring by submesoscale eddies (Sundermeyer et al., 2014). Spectra of salinity anomalies along isopycnals, which behave mostly like a passive tracer, pointed to a steeper variance at submesoscale than expected from (surface) quasi-geostrophic motions (Kunze et al., 2015), consistent with other sets of observations (Klymak et al., 2015). Another recent study (Jaeger et al., 2020b) that compiled 4800 km of ship track with underway-CTD profiles at spacings of 0.3–3 km in the Bay of Bengal, also showed a steepening of the tracer variance spectrum at submesoscales, suggesting that submesoscales are acting to mix tracers along isopycnal surfaces in the pycnocline. Submesoscale streamers were also observed during the

LatMix experiment. They were associated with a significant freshwater flux across the sharp north wall of the Gulf Stream (Klymak et al., 2016), which was previously considered as a barrier to lateral mixing. These exchanges across the Gulf Stream front have been further characterised using dye observations and numerical simulations as a result of a combination of strong vertical mixing events and shear dispersion induced by submesoscale instabilities (Wenegrat et al., 2020).

Submesoscale isopycnal stirring appears to vary strongly across isopycnals, creating complex vertical interleaving of water masses (Shcherbina et al., 2009; Jaeger et al., 2020a). Interleaving layers can be 1–10 m thick, tens of kilometres in horizontal extent, and tends to be tilted with respect to the isopycnals. These interleaving intrusions of water are generated by the swirling flow associated with eddies, both at mesoscales (Smith and Ferrari, 2009) and submesoscales. The submesoscale signatures suggest that the vertical shear of the horizontal velocity, which is associated with the strong lateral density gradients in the mixed layer, is critical for generating layers in water mass anomaly that are tilted with respect to the isopycnal surfaces (Jaeger et al., 2020a). The very thin (depth to length) aspect ratio of the layers creates a strong vertical gradient in the tracers, which is then smoothed through diapycnal mixing. In this way, lateral stirring by submesoscale processes creates thin layers with sharp vertical gradients that result in enhanced diapycnal mixing.

8.4.2 Mixing and transport of reactive tracers

Along-isopycnal vertical transport is associated with isopycnal stirring, which becomes more efficient with submesoscale dynamics. This is why submesoscale processes are responsible for enhancing the vertical transport of biogeochemical tracers, such as nutrients, carbon, and oxygen (Thomas et al., 2008). The net vertical flux of a biogeochemical tracer is the integral of $w'c'$, where c' is the tracer anomaly from the tracer mean at that depth. Hence, the flux depends on the covariance of the vertical velocity and tracer anomaly, and is dependent on the relative timescales for the biogeochemical reaction (or modification) and vertical transport. This is one of the reasons that submesoscale vertical velocities are efficient in transporting biogeochemical properties (Whitt et al., 2019).

Model simulations have shown that increasing horizontal grid resolution leads to a greater vertical supply of nutrients (Lévy et al., 2012; Uchida et al., 2020) and stronger tracer subduction into the pycnocline (Balwada et al., 2018). A careful analysis (Freilich and Mahadevan, 2019) shows that although the vertical uplift of isopycnals is the larger component of vertical transport, the along-isopycnal component of vertical transport is more sensitive to model resolution and accounts for a larger fraction of vertical flux when the submesoscale activity is increased.

8.4.3 Impacts on the dispersion of buoyant material

Surface submesoscale currents can produce strong local dispersion (spreading) of passive tracers down to 100 m scales with implications for the predictive modelling of oceanic pollutants (Poje et al., 2014). At the same time, frontogenetic submesoscale flows are characterised by zones of sharp convergences of horizontal currents at the ocean surface (see discussion in Section 8.2.1), which are naturally anti-dispersive. Convergence zones concentrate buoyant material, such as plankton, natural surfactants, pollutants (particularly, oil, and oil degradation products), and marine debris. Highly buoyant material remains close to the surface, therefore breaking the three-dimensional non-divergence, and allowing strong accumulation within convergence zones (D'Asaro et al., 2018).

Taylor (2018) used large-eddy simulations of MLI under various intensities of convective forcing to study the influence of submesoscales on buoyant tracers. When the buoyant rise velocity was not small compared to the *rms* vertical fluid velocity, the buoyancy tracers accumulate near the surface in a submesoscale front that wraps around the submesoscale eddy, mirroring the patterns seen in observations (D'Asaro et al., 2018). Intense downwelling at the submesoscale front pulls the buoyant tracers beneath the surface. In some cases, this results in the buoyant tracer extending deeper into the water column than it would in the absence of submesoscales.

8.4.4 Dispersion by the deep submesoscale currents

Submesoscale currents also contribute to the dispersion and transport of passive tracers at the bottom and in the interior of the ocean. Submesoscale currents (as well as tidal currents) generate significant horizontal and vertical dispersion of tracers close to the topography. For example, over the Mid-Atlantic ridge, submesoscale currents have been shown to influence dispersion of deep-sea hydrothermal vents effluents and larvae and impact connectivity between deep-sea ecosystems (Vic et al., 2018).

Deep submesoscale currents can also generate long-range anomalous transport of tracers in the ocean interior in the form of subsurface submesoscale eddies, known as submesoscale coherent vortices (SCVs) (McWilliams, 1985) or Intra

Thermocline Eddies (ITEs) (Dugan et al., 1982). SCVs are usually defined as energetic eddies with a radius smaller than the Rossby deformation radius, a structure localised in the vertical, and an interior velocity maximum (McWilliams, 1985). SCVs may be generated due to various frictional or diabatic effects, which include convection (Marshall and Schott, 1999), wind-driven destruction of PV (Thomas, 2008); geothermal forcing (Baker et al., 1987); bottom mixing (McWilliams, 1985); or friction (D'Asaro, 1988); or by inviscid instability processes of bottom currents, such as baroclinic instability (Morvan et al., 2019). Topographic wakes are suspected to be the major source of SCVs (McWilliams, 2016).

SCVs can be very long-lived (>1 year) and travel far from their origins, being primarily advected by mesoscale and mean currents. One of the most well-known types of SCVs are meddies (even if they are formally closer to mesoscale than submesoscale) formed at the exit of the Mediterranean Sea (McDowell and Rossby, 1978), which spread salty Mediterranean waters in the subtropical Atlantic ocean. SCVs that form in eastern boundary regions are essential for spreading oxygen-poor and nutrient-rich waters into the interior of gyres (Frenger et al., 2018). SCVs also form from wintertime deep convection, as observed in the Labrador sea (Clarke, 1984; Lilly and Rhines, 2002) and the northwestern Mediterranean Sea (Testor and Gascard, 2003; Bosse et al., 2016, 2017), where they are essential for spreading the newly formed deep waters within ocean basins. SCVs formed in the wake of topographic features, for example the Charleston bump along the U.S. seaboard, can transport waters from the bottom mixed layer over long distances and spread them within the subtropical gyre (Gula et al., 2019).

SCVs have always been very difficult to sample due to the sparsity of in situ observations able to resolve submesoscales in the interior of the ocean. Thanks to the coverage of Argo floats, it is now possible to make more robust statistics on the existence and properties of the SCVs (Li et al., 2017; McCoy et al., 2020). The cumulative effect of SCVs could potentially affect the large-scale transport and distribution of heat, nutrients, and biogeochemical tracers. However, a quantitative assessment of such impacts still remains to be done.

8.5 Conclusion and future directions

As an intermediary between balanced flows and small-scale turbulence, submesoscales play an important part in the story of ocean mixing. Through the instabilities described in Sections 8.2.2 and 8.2.3, submesoscales restratify the surface and bottom mixed layers (see Section 8.3), whereas turbulent mixing influences the evolution of submesoscale fronts by generating a cross-front circulation, as described in Section 8.2.4. At the same time, submesoscales drive strong vertical circulations that redistribute passive, active, reactive, and buoyant tracers while stirring tracers along isopycnals, transporting material through the ocean interior and shortening the route to small-scale mixing (Section 8.4).

The body of knowledge on submesoscale processes has grown substantially over the last 15 years. However, many theoretical and modelling studies have considered only a subset of the submesoscale processes mentioned here, either by design or by limitations in resolution. The multi-scale interactions between various processes (e.g., boundary layer turbulence, SI, MLI, internal waves, and mesoscale eddies) remains relatively unexplored. Recent advances in computational power have made it possible to begin to study at least some of these interactions. For example, Skyllingstad and Samelson (2020) used large-eddy simulations to study the interaction between boundary layer turbulence, SI, and MLI, whereas submesoscale-permitting global ocean models (e.g., Su et al., 2018, 2020) have provided insight into the influence of submesoscales on large-scale circulation and heat transport. Piecing together the links between small-scale turbulence and mixing, submesoscales, mesoscale eddies, internal waves, and the general circulation remains a grand challenge.

The effect of air-sea coupling at the submesoscale on the ocean surface layer is also relatively unexplored. Using submesoscale-resolving numerical simulations, Renault et al. (2018) found a sink of energy at the submesoscale related to induced Ekman pumping velocities, but also an increase of the injection of energy by baroclinic conversion into the submesoscale. The effect of waves and Langmuir turbulence on the submesoscale processes, described above, is also a relatively new topic. Some studies have shown that Langmuir turbulence can significantly increase small-scale turbulence and counter the restratifying effects of MLI (Hamlington et al., 2014), and that surface waves also have an impact on frontogenesis (Suzuki et al., 2016; Sullivan and McWilliams, 2019). A next step in our understanding of the oceanic submesoscale processes and their impact on the stratification of the ocean surface layer may require a fully interactive atmosphere as well as the wave feedbacks, which will modulate the turbulent exchanges of momentum, heat, and tracers between the atmosphere and ocean.

Submesoscale currents impact the large-scale ocean circulation in many counter-intuitive ways, including upscale energy transfer to mesoscale eddies or upgradient fluxes of tracers (Klein et al., 2019). Recent studies are highlighting the global impacts of submesoscales, for example, the contribution of near-surface submesoscale currents in driving vertical heat fluxes (Su et al., 2018, 2020) and restratifying the upper ocean (Johnson et al., 2016). The role played by submesoscale instabilities at the bottom, which has been discovered much more recently, may also be a very important ingredient for the generation of near-bottom mixing, the export of buoyancy and ultimately for the closure of the overturning circulation.

The current generation of OGCM is far from being able to resolve the full range of submesoscale processes in the surface and bottom boundary layers. In the ongoing CMIP6 exercise, ocean models will typically have a 1/4 degree resolution, allowing for mesoscales, but not totally resolving them (Griffies et al., 2016). One can expect that truly mesoscale-resolving ocean models will be a standard for climate studies a decade from now, but the use of submesoscale-resolving models in climate studies might still be several decades away, up to 40–50 years, based on Moore's law (Dong et al., 2020b). More precisely, according to the global estimates of Dong et al. (2020b) and Dong et al. (2021) (see Fig. 8.6), the required grid spacing to resolve MLI over 90 % of the ocean's surface is about 1 km in winter and 500 m in summer; for BMLI it falls to about 100 m, and, finally, less than 10 m for SI. The current state-of-the-art for global or basin-scale forced ocean simulations have grid spacings on the order of 1–2 km (LLC4320 at 1/48°, eNATL60 at 1/60°, or GIGATL1 at 1 km), which can only be said to safely resolve winter MLI. Thus these state-of-the-art models just scratch the surface of the submesoscale realm.

Though submesoscales are not fully resolved in many models, many submesoscale effects on mixing have yet to be parameterised. Some parameterisations are already used in ocean models to incorporate the restratifying effects of MLI (Fox-Kemper et al., 2011; Calvert et al., 2020) and the extraction of energy from the geostrophic currents and diffusive tracer mixing due to SI and associated secondary shear instabilities (Bachman et al., 2017a; Dong et al., 2020c). These efforts are encouraging, but these parameterisations still have issues (Calvert et al., 2020) and do not account yet for all the effects of submesoscales in the surface boundary layer. Furthermore, no parameterisations exist yet for bottom boundary submesoscale processes, and given their potential impact on the large-scale circulation, designing one should stand out as an important objective. The non-linear, interacting processes active at the submesoscale make the parameterisation problem particularly challenging. However, recent developments in empirical data-driven methods and machine learning offer a promising path forward.

Acknowledgements

The authors thank C. Vic, J. Dong, B. Fox-Kemper, X. Yu, J. Wenegrat, and A. Naveira Garabato for providing the original data used to plot Figs. 8.3, 8.6, 8.9, 8.11, and 8.12. J. Gula gratefully acknowledges support from the French National Agency for Research (ANR) through the projects DEEPER (ANR-19-CE01-0002-01) and ISblue "Interdisciplinary graduate school for the blue planet" (ANR-17-EURE-0015).

References

Adams, K.A., Hosegood, P., Taylor, J.R., Sallée, J.-B., Bachman, S., Torres, R., Stamper, M., 2017. Frontal circulation and submesoscale variability during the formation of a Southern Ocean mesoscale eddy. J. Phys. Oceanogr. 47 (7), 1737–1753.

Ajayi, A., Le Sommer, J., Chassignet, E.P., Molines, J.-M., Xu, X., Albert, A., Dewar, W., 2021. Diagnosing cross-scale kinetic energy exchanges from two submesoscale permitting ocean models. J. Adv. Model. Earth Syst. 13 (6), e2019MS001923.

Allen, J.S., Newberger, P.A., 1998. On symmetric instabilities in oceanic bottom boundary layers. J. Phys. Oceanogr. 28 (6), 1131–1151.

Bachman, S., Fox-Kemper, B., Taylor, J., Thomas, L., 2017a. Parameterization of frontal symmetric instabilities. I: theory for resolved fronts. Ocean Model. 109, 72–95.

Bachman, S.D., Klocker, A., 2020. Interaction of jets and submesoscale dynamics leads to rapid ocean ventilation. J. Phys. Oceanogr. 50, 2873–2883.

Bachman, S.D., Taylor, J.R., 2016. Numerical simulations of the equilibrium between eddy-induced restratification and vertical mixing. J. Phys. Oceanogr. 46 (3), 919–935.

Bachman, S.D., Taylor, J.R., Adams, K.A., Hosegood, P.J., 2017b. Mesoscale and submesoscale effects on mixed layer depth in the Southern Ocean. J. Phys. Oceanogr. 47 (9), 2173–2188.

Baker, E.T., Massoth, G.J., Feely, R.A., 1987. Cataclysmic hydrothermal venting on the Juan de Fuca Ridge. Nature 329, 149–151.

Balwada, D., Smith, K.S., Abernathey, R., 2018. Submesoscale vertical velocities enhance tracer subduction in an idealized Antarctic circumpolar current. Geophys. Res. Lett. 45 (18), 9790–9802.

Barkan, R., Molemaker, M.J., Srinivasan, K., McWilliams, J.C., D'Asaro, E.A., 2019. The role of horizontal divergence in submesoscale frontogenesis. J. Phys. Oceanogr. 49 (6), 1593–1618.

Barkan, R., Winters, K.B., Llewellyn Smith, S.G., 2015. Energy cascades and loss of balance in a reentrant channel forced by wind stress and buoyancy fluxes. J. Phys. Oceanogr. 45 (1), 272–293.

Barth, J., 1989a. Stability of a coastal upwelling front, 1. Model development and a stability theorem. J. Geophys. Res. 94, 10844–10856.

Barth, J., 1989b. Stability of a coastal upwelling front, 2. Model results and comparison with observations. J. Geophys. Res. 94, 10857–10883.

Barth, J., 1994. Short-wavelenght instabilities on coastal jets and fronts. J. Geophys. Res. 99, 16095–16115.

Batchelor, G.K., 1959. Small-scale variation of convected quantities like temperature in turbulent fluid Part 1. General discussion and the case of small conductivity. J. Fluid Mech. 5 (1), 113–133.

Benthuysen, J., Thomas, L., 2012. Friction and diapycnal mixing at a slope: boundary control of potential vorticity. J. Phys. Oceanogr. 42, 1509–1523.

Biddle, L.C., Swart, S., 2020. The observed seasonal cycle of submesoscale processes in the Antarctic marginal ice zone. J. Geophys. Res., Oceans 125 (6), e2019JC015587.

Blumsack, S.L., Gierasch, P.J., 1972. Mars: the effects of topography on baroclinic instability. J. Atmos. Sci. 29 (6), 1081–1089.

Boccaletti, G., Ferrari, R., Fox-Kemper, B., 2007. Mixed layer instabilities and restratification. J. Phys. Oceanogr. 37, 2228–2250.

Bodner, A.S., Fox-Kemper, B., Van Roekel, L.P., McWilliams, J.C., Sullivan, P.P., 2019. A perturbation approach to understanding the effects of turbulence on frontogenesis. J. Fluid Mech. 883, A25.

Bosse, A., Fer, I., Lilly, J.M., Søiland, H., 2019. Dynamical controls on the longevity of a non-linear vortex: the case of the lofoten basin eddy. Sci. Rep. 9 (1), 13448.

Bosse, A., Testor, P., Damien, P., Estournel, C., Marsaleix, P., Mortier, L., Prieur, L., Taillandier, V., 2021. Wind-forced submesoscale symmetric instability around deep convection in the northwestern Mediterranean Sea. Fluids 6 (3).

Bosse, A., Testor, P., Houpert, L., Damien, P., Prieur, L., Hayes, D., Taillandier, V., Durrieu de Madron, X., d'Ortenzio, F., Coppola, L., Karstensen, J., Mortier, L., 2016. Scales and dynamics of submesoscale coherent vortices formed by deep convection in the northwestern Mediterranean Sea. J. Geophys. Res., Oceans 121 (10), 7716–7742.

Bosse, A., Testor, P., Mayot, N., Prieur, L., D'Ortenzio, F., Mortier, L., Goff, H.L., Gourcuff, C., Coppola, L., Lavigne, H., Raimbault, P., 2017. A submesoscale coherent vortex in the Ligurian Sea: from dynamical barriers to biological implications. J. Geophys. Res., Oceans 122 (8), 6196–6217.

Brannigan, L., Marshall, D.P., Naveira-Garabato, A., Nurser, A.G., 2015. The seasonal cycle of submesoscale flows. Ocean Model. 92, 69–84.

Brink, K., 2013. Instability of a tidal mixing front in the presence of realistic tides and mixing. J. Mar. Res. 71 (3), 227–251.

Brink, K.H., 2012. Baroclinic instability of an idealized tidal mixing front. J. Mar. Res. 70 (4), 661–688.

Brink, K.H., 2016. Continental shelf baroclinic instability. Part I: relaxation from upwelling or downwelling. J. Phys. Oceanogr. 46 (2), 551–568.

Brüggemann, N., Eden, C., 2015. Routes to dissipation under different dynamical conditions. J. Phys. Oceanogr. 45 (8), 2149–2168.

Buckingham, C.E., Gula, J., Carton, X., 2021a. The role of curvature in modifying frontal instabilities. Part I: review of theory and presentation of a nondimensional instability criterion. J. Phys. Oceanogr. 51 (2), 299–315.

Buckingham, C.E., Gula, J., Carton, X., 2021b. The role of curvature in modifying frontal instabilities. Part II: application of the criterion to curved density fronts at low Richardson numbers. J. Phys. Oceanogr. 51 (2), 317–341.

Buckingham, C.E., Lucas, N.S., Belcher, S.E., Rippeth, T.P., Grant, A.L.M., Le Sommer, J., Ajayi, A.O., Naveira Garabato, A.C., 2019. The contribution of surface and submesoscale processes to turbulence in the open ocean surface boundary layer. J. Adv. Model. Earth Syst. 11 (12), 4066–4094.

Buckingham, C.E., Naveira Garabato, A.C., Thompson, A.F., Brannigan, L., Lazar, A., Marshall, D.P., George Nurser, A.J., Damerell, G., Heywood, K.J., Belcher, S.E., 2016. Seasonality of submesoscale flows in the ocean surface boundary layer. Geophys. Res. Lett. 43 (5), 2118–2126.

Callies, J., 2018. Restratification of abyssal mixing layers by submesoscale baroclinic eddies. J. Phys. Oceanogr. 48 (9), 1995–2010.

Callies, J., Ferrari, R., 2013. Interpreting energy and tracer spectra of upper-ocean turbulence in the submesoscale range (1-200 km). J. Phys. Oceanogr. 43, 2456–2575.

Callies, J., Ferrari, R., 2018a. Baroclinic instability in the presence of convection. J. Phys. Oceanogr. 48 (1), 45–60.

Callies, J., Ferrari, R., 2018b. Note on the rate of restratification in the baroclinic spindown of fronts. J. Phys. Oceanogr. 48 (7), 1543–1553.

Callies, J., Ferrari, R., Klymak, J., Gula, J., 2015. Seasonality in submesoscale turbulence. Nat. Commun. 6, 6862.

Callies, J., Flierl, G., Ferrari, R., Fox-Kemper, B., 2016. The role of mixed-layer instabilities in submesoscale turbulence. J. Fluid Mech. 788, 5–41.

Calvert, D., Nurser, G., Bell, M.J., Fox-Kemper, B., 2020. The impact of a parameterisation of submesoscale mixed layer eddies on mixed layer depths in the nemo ocean model. Ocean Model. 154, 101678.

Capet, X., Campos, E.J., Paiva, A.M., 2008a. Submesoscale activity over the Argentinian shelf. Geophys. Res. Lett. 35 (15).

Capet, X., McWilliams, J.C., Molemaker, M., Shchepetkin, A., 2008b. Mesoscale to submesoscale transition in the California Current System. Part I: flow structure, eddy flux, and observational tests. J. Phys. Oceanogr. 38, 29–43.

Capet, X., McWilliams, J.C., Molemaker, M., Shchepetkin, A., 2008c. Mesoscale to submesoscale transition in the California Current System. Part II: frontal processes. J. Phys. Oceanogr. 38, 44–64.

Capet, X., McWilliams, J.C., Molemaker, M.J., Shchepetkin, A.F., 2008d. Mesoscale to submesoscale transition in the California Current System. Part III: energy balance and flux. J. Phys. Oceanogr. 38 (10), 2256–2269.

Capet, X., Roullet, G., Klein, P., Maze, G., 2016. Intensification of upper-ocean submesoscale turbulence through Charney baroclinic instability. J. Phys. Oceanogr. 46 (11), 3365–3384.

Carpenter, J.R., Rodrigues, A., Schultze, L.K.P., Merckelbach, L.M., Suzuki, N., Baschek, B., Umlauf, L., 2020. Shear instability and turbulence within a submesoscale front following a storm. Geophys. Res. Lett. 47 (23), e2020GL090365. https://doi.org/10.1029/2020GL090365.

Charney, J.G., 1973. Symmetric circulations in idealized models. In: Planetary Fluid Dynamics. D. Reidel Publishing Company, pp. 128–141.

Chelton, D.B., deSzoeke, R.A., Schlax, M.G., El Naggar, K., Siwertz, N., 1998. Geographical variability of the first baroclinic Rossby radius of deformation. J. Phys. Oceanogr. 28 (3), 433–460.

Chen, S.-N., Chen, C.-J., Lerczak, J.A., 2020. On baroclinic instability over continental shelves: testing the utility of Eady-type models. J. Phys. Oceanogr. 50 (1), 3–33.

Chrysagi, E., Umlauf, L., Holtermann, P., Klingbeil, K., Burchard, H., 2021. High-resolution simulations of submesoscale processes in the Baltic Sea: the role of storm events. J. Geophys. Res., Oceans 126 (3), e2020JC016411.

Clarke, R., 1984. Transport through the Cape Farewell-Flemish Cap section. In: Rapports et procès-verbaux des réunions - Conseil international pour l'exploration de la mer, vol. 185, pp. 120–130.

Couvelard, X., Dumas, F., Garnier, V., Ponte, A., Talandier, C., Treguier, A., 2015. Mixed layer formation and restratification in presence of mesoscale and submesoscale turbulence. Ocean Model. 96, 243–253.

Cronin, M.F., Kessler, W.S., 2009. Near-surface shear flow in the tropical Pacific cold tongue front. J. Phys. Oceanogr. 39 (5), 1200–1215.
Crowe, M., Taylor, J., 2020. The effects of surface wind stress and buoyancy flux on the evolution of a front in a turbulent thermal wind balance. Fluids 5 (2), 87.
Crowe, M.N., Taylor, J.R., 2018. The evolution of a front in turbulent thermal wind balance. Part 1. Theory. J. Fluid Mech. 850, 179–211.
Crowe, M.N., Taylor, J.R., 2019. The evolution of a front in turbulent thermal wind balance. Part 2. Numerical simulations. J. Fluid Mech. 880, 326–352.
Cushman-Roisin, B., Beckers, J.-M., 2011. Introduction to Geophysical Fluid Dynamics: Physical and Numerical Aspects. Academic Press.
D'Asaro, E., 1988. Generation of submesoscale vortices: a new mechanism. J. Geophys. Res. 93 (C6), 6685–6693.
D'Asaro, E., Lee, C., Rainville, L., Harcourt, R., Thomas, L., 2011. Enhanced turbulence and energy dissipation at ocean fronts. Science 332 (318).
D'Asaro, E.A., Carlson, D.F., Chamecki, M., Harcourt, R.R., Haus, B.K., Fox-Kemper, B., Molemaker, M.J., Poje, A.C., Yang, D., 2020. Advances in observing and understanding small-scale open ocean circulation during the Gulf of Mexico research initiative era. Front. Mar. Sci. 7, 349.
D'Asaro, E.A., Shcherbina, A.Y., Klymak, J.M., Molemaker, J., Novelli, G., Guigand, C.M., Haza, A.C., Haus, B.K., Ryan, E.H., Jacobs, G.A., Huntley, H.S., Laxague, N.J.M., Chen, S., Judt, F., McWilliams, J.C., Barkan, R., Kirwan, A.D., Poje, A.C., Özgökmen, T.M., 2018. Ocean convergence and the dispersion of flotsam. Proc. Natl. Acad. Sci. 115 (6), 1162–1167.
Dauhajre, D.P., McWilliams, J.C., 2018. Diurnal evolution of submesoscale front and filament circulations. J. Phys. Oceanogr. 48 (10), 2343–2361.
Dauhajre, D.P., McWilliams, J.C., Uchiyama, Y., 2017. Submesoscale coherent structures on the continental shelf. J. Phys. Oceanogr. 47 (12), 2949–2976.
Dewar, W., Molemaker, M., McWilliams, J.C., 2015. Centrifugal instability and mixing in the California Undercurrent. J. Phys. Oceanogr. 45, 1224–1241.
Dong, J., Fox-Kemper, B., Zhang, H., Dong, C., 2020a. The seasonality of submesoscale energy production, content, and cascade. Geophys. Res. Lett. 47 (6), e2020GL087388.
Dong, J., Fox-Kemper, B., Zhang, H., Dong, C., 2020b. The scale of submesoscale baroclinic instability globally. J. Phys. Oceanogr.
Dong, J., Fox-Kemper, B., Zhang, H., Dong, C., 2021. The scale and activity of symmetric instability globally. J. Phys. Oceanogr.
Dong, J., Fox-Kemper, B., Zhu, J., Dong, C., 2020c. Application of symmetric instability parameterization in the coastal and regional ocean community model (croco). Earth Space Sci. Open Arch., 30.
du Plessis, M., Swart, S., Ansorge, I.J., Mahadevan, A., 2017. Submesoscale processes promote seasonal restratification in the Subantarctic Ocean. J. Geophys. Res., Oceans 122 (4), 2960–2975.
Dugan, J.P., Mied, R.P., Mignerey, P.C., Schuetz, A.F., 1982. Compact, intrathermocline eddies in the Sargasso Sea. J. Geophys. Res., Oceans 87 (C1), 385–393.
Durski, S., Allen, J., 2005. Finite-amplitude evolution of instabilities associated with the coastal upwelling front. J. Phys. Oceanogr. 35, 1606–1628.
Eady, E., 1949. Long waves and cyclone waves. Tellus 1, 33–52.
Ekman, V.W., 1905. On the influence of the Earth's rotation on ocean-currents.
Eliassen, A., Kleinschmidt, E., 1957. Dynamic meteorology. In: Geophysik II/Geophysics II. Springer, pp. 1–154.
Fadeev, E., Wietz, M., von Appen, W.-J., Nöthig, E.-M., Engel, A., Grosse, J., Graeve, M., Boetius, A., 2020. Submesoscale dynamics directly shape bacterioplankton community structure in space and time. bioRxiv.
Ferrari, R., Rudnick, D.L., 2000. Thermohaline variability in the upper ocean. J. Geophys. Res., Oceans 105 (C7), 16857–16883.
Fox-Kemper, B., Danabasoglu, G., Ferrari, R., Griffies, S., Hallberg, R., Holland, M., Maltrud, M., Peacock, S., Samuels, B., 2011. Parameterization of mixed layer eddies. III: implementation and impact in global ocean climate simulations. In: Modelling and Understanding the Ocean Mesoscale and Submesoscale. Ocean Model. 39 (1), 61–78.
Fox-Kemper, B., Ferrari, R., 2008. Parameterization of mixed layer eddies. Part II: prognosis and impact. J. Phys. Oceanogr. 38 (6), 1166–1179.
Fox-Kemper, B., Ferrari, R., Hallberg, R., 2008. Parameterization of mixed layer eddies. Part I: theory and diagnosis. J. Phys. Oceanogr. 38, 1145–1165.
Freilich, M.A., Mahadevan, A., 2019. Decomposition of vertical velocity for nutrient transport in the upper ocean. J. Phys. Oceanogr. 49 (6), 1561–1575.
Frenger, I., Bianchi, D., Stührenberg, C., Oschlies, A., Dunne, J., Deutsch, C., Galbraith, E., Schütte, F., 2018. Biogeochemical role of subsurface coherent eddies in the ocean: tracer cannonballs, hypoxic storms, and microbial stewpots? Glob. Biogeochem. Cycles 32 (2), 226–249.
Fu, L.-L., Ferrari, R., 2008. Observing oceanic submesoscale processes from space. Eos Trans. AGU 89 (48), 488.
Garrett, C., Loder, J., 1981. Dynamical aspects of shallow sea fronts. Philos. Trans. R. Soc. Lond. 302, 563–581.
Gent, P.R., Danabasoglu, G., Donner, L.J., Holland, M.M., Hunke, E.C., Jayne, S.R., Lawrence, D.M., Neale, R.B., Rasch, P.J., Vertenstein, M., Worley, P.H., Yang, Z.-L., Zhang, M., 2011. The community climate system model version 4. J. Climate 24 (19), 4973–4991.
Gent, P.R., Mcwilliams, J.C., 1990. Isopycnal mixing in ocean circulation models. J. Phys. Oceanogr. 20 (1), 150–155.
Griffies, S.M., Danabasoglu, G., Durack, P.J., Adcroft, A.J., Balaji, V., Böning, C.W., Chassignet, E.P., Curchitser, E., Deshayes, J., Drange, H., Fox-Kemper, B., Gleckler, P.J., Gregory, J.M., Haak, H., Hallberg, R.W., Heimbach, P., Hewitt, H.T., Holland, D.M., Ilyina, T., Jungclaus, J.H., Komuro, Y., Krasting, J.P., Large, W.G., Marsland, S.J., Masina, S., McDougall, T.J., Nurser, A.J.G., Orr, J.C., Pirani, A., Qiao, F., Stouffer, R.J., Taylor, K.E., Treguier, A.M., Tsujino, H., Uotila, P., Valdivieso, M., Wang, Q., Winton, M., Yeager, S.G., 2016. Omip contribution to cmip6: experimental and diagnostic protocol for the physical component of the ocean model intercomparison project. Geosci. Model Dev. 9 (9), 3231–3296.
Gula, J., Blacic, T.M., Todd, R.E., 2019. Submesoscale coherent vortices in the Gulf Stream. Geophys. Res. Lett. 46.
Gula, J., Molemaker, M., McWilliams, J.C., 2014. Submesoscale cold filaments in the Gulf Stream. J. Phys. Oceanogr. 44 (10), 2617–2643.
Gula, J., Molemaker, M., McWilliams, J.C., 2015a. Gulf Stream dynamics along the Southeastern U.S. Seaboard. J. Phys. Oceanogr. 45 (3), 690–715.
Gula, J., Molemaker, M., McWilliams, J.C., 2015b. Topographic vorticity generation, submesoscale instability and vortex street formation in the Gulf Stream. Geophys. Res. Lett. 42, 4054–4062.
Gula, J., Molemaker, M., McWilliams, J.C., 2016. Topographic generation of submesoscale centrifugal instability and energy dissipation. Nat. Commun. 7, 12811.
Haine, T., Marshall, J., 1999. Gravitational, symmetric, and baroclinic instability of the ocean mixed layer. J. Phys. Oceanogr. 28, 634–658.

Hamlington, P.E., Van Roekel, L.P., Fox-Kemper, B., Julien, K., Chini, G.P., 2014. Langmuir–submesoscale interactions: descriptive analysis of multiscale frontal spindown simulations. J. Phys. Oceanogr. 44 (9), 2249–2272.

Hetland, R.D., 2017. Suppression of baroclinic instabilities in buoyancy-driven flow over sloping bathymetry. J. Phys. Oceanogr. 47 (1), 49–68.

Hosegood, P., Gregg, M.C., Alford, M.H., 2006. Sub-mesoscale lateral density structure in the oceanic surface mixed layer. Geophys. Res. Lett. 33 (22).

Hoskins, B.J., 1974. The role of potential vorticity in symmetric stability and instability. Q. J. R. Meteorol. Soc. 100, 480–482.

Hoskins, B.J., 1982. The mathematical theory of frontogenesis. Annu. Rev. Fluid Mech. 14, 131–151.

Hoskins, B.J., Bretherton, F.P., 1972. Atmospheric frontogenesis models: mathematical formulation and solution. J. Atmos. Sci. 29, 11–37.

Jaeger, G.S., Lucas, A.J., Mahadevan, A., 2020a. Formation of interleaving layers in the Bay of Bengal. In: Atmosphere-Ocean Dynamics of Bay of Bengal - Volume 2. Deep-Sea Res., Part 2, Top. Stud. Oceanogr. 172, 104717.

Jaeger, G.S., MacKinnon, J.A., Lucas, A.J., Shroyer, E., Nash, J., Tandon, A., Farrar, J.T., Mahadevan, A., 2020b. How spice is stirred in the Bay of Bengal. J. Phys. Oceanogr. 50 (9), 2669–2688.

Jiao, W., Dewar, W., 2015. The energetics of centrifugal instability. J. Phys. Oceanogr. 45, 1554–1573.

Johnson, L., Lee, C.M., D'Asaro, E.A., 2016. Global estimates of lateral springtime restratification. J. Phys. Oceanogr. 46 (5), 1555–1573.

Johnson, L., Lee, C.M., D'Asaro, E.A., Thomas, L., Shcherbina, A., 2020a. Restratification at a California current upwelling front. Part I: observations. J. Phys. Oceanogr. 50 (5), 1455–1472.

Johnson, L., Lee, C.M., D'Asaro, E.A., Wenegrat, J.O., Thomas, L.N., 2020b. Restratification at a California current upwelling front. Part II: dynamics. J. Phys. Oceanogr. 50 (5), 1473–1487.

Johnston, T.S., Schönau, M.C., Paluszkiewicz, T., MacKinnon, J.A., Arbic, B.K., Colin, P.L., Alford, M.H., Andres, M., Centurioni, L., Graber, H.C., Helfrich, K.R., Hormann, V., Lermusiaux, P.F., Musgrave, R.C., Powell, B.S., Qiu, B., Rudnick, D.L., Simmons, H.L., Laurent, L.S., Terrill, E.J., Trossman, D.S., Voet, G., Wijesekera, H.W., Zeiden, K.L., 2019. Flow encountering abrupt topography (fleat): a multiscale observational and modeling program to understand how topography affects flows in the western North Pacific. Oceanography 32 (4), 10–21.

Khatri, H., Griffies, S.M., Uchida, T., Wang, H., Menemenlis, D., 2020. A synthesis of upper ocean geostrophic kinetic energy spectra from a global submesoscale permitting simulation. Earth Space Sci. Open Arch., 16.

Klein, P., Lapeyre, G., 2009. The oceanic vertical pump induced by mesoscale and submesoscale turbulence. Annu. Rev. Mar. Sci. 1, 351–375.

Klein, P., Lapeyre, G., Siegelman, L., Qiu, B., Fu, L.-L., Torres, H., Su, Z., Menemenlis, D., Le Gentil, S., 2019. Ocean-scale interactions from space. Earth Space Sci. 6 (5), 795–817.

Klymak, J., Shearman, R., Gula, J., Lee, C., D'Asaro, E., Thomas, L., Harcourt, R., Shcherbina, A., Sundermeyer, M., Molemaker, M., McWilliams, J.C., 2016. Submesoscale streamers exchange water on the North Wall of the Gulf Stream. Geophys. Res. Lett. 43, 1226–1233.

Klymak, J.M., Crawford, W., Alford, M.H., MacKinnon, J.A., Pinkel, R., 2015. Along-isopycnal variability of spice in the North Pacific. J. Geophys. Res., Oceans 120 (3), 2287–2307.

Koenig, Z., Fer, I., Kolås, E., Fossum, T.O., Norgren, P., Ludvigsen, M., 2020. Observations of turbulence at a near-surface temperature front in the Arctic Ocean. J. Geophys. Res., Oceans 125 (4), e2019JC015526. https://doi.org/10.1029/2019JC015526.

Kunze, E., Klymak, J.M., Lien, R.-C., Ferrari, R., Lee, C.M., Sundermeyer, M.A., Goodman, L., 2015. Submesoscale water-mass spectra in the Sargasso Sea. J. Phys. Oceanogr. 45 (5), 1325–1338.

LaCasce, J.H., Groeskamp, S., 2020. Baroclinic modes over rough bathymetry and the surface deformation radius. J. Phys. Oceanogr., 1–40.

Lapeyre, G., 2017. Surface quasi-geostrophy. Fluids 2 (1).

Lapeyre, G., Klein, P., 2006. Impact of the small-scale elongated filaments on the oceanic vertical pump. J. Mar. Res. 64, 835–851.

Lapeyre, G., Klein, P., Hua, B.L., 2006. Oceanic restratification forced by surface frontogenesis. J. Phys. Oceanogr. 36 (8), 1577–1590.

Lenn, Y.-D., Chereskin, T.K., 2009. Observations of Ekman currents in the Southern Ocean. J. Phys. Oceanogr. 39 (3), 768–779.

Lévy, M., Ferrari, R., Franks, P.J.S., Martin, A.P., Rivière, P., 2012. Bringing physics to life at the submesoscale. Geophys. Res. Lett. 39 (14).

Lévy, M., Franks, P.J.S., Smith, K.S., 2018. The role of submesoscale currents in structuring marine ecosystems. Nat. Commun. 9 (1), 4758.

Li, C., Zhang, Z., Zhao, W., Tian, J., 2017. A statistical study on the subthermocline submesoscale eddies in the northwestern Pacific Ocean based on Argo data. J. Geophys. Res., Oceans 122 (5), 3586–3598.

Lilly, J.M., Rhines, P.B., 2002. Coherent eddies in the Labrador Sea observed from a mooring. J. Phys. Oceanogr. 32 (2), 585–598.

MacKinnon, J.A., Alford, M.H., Voet, G., Zeiden, K.L., Shaun Johnston, T.M., Siegelman, M., Merrifield, S., Merrifield, M., 2019. Eddy wake generation from broadband currents near Palau. J. Geophys. Res., Oceans 124 (7), 4891–4903.

Macvean, M.K., Woods, J.D., 1980. Redistribution of scalars during upper ocean frontogenesis: a numerical model. Q. J. R. Meteorol. Soc. 106 (448), 293–311.

Mahadevan, A., 2016. The impact of submesoscale physics on primary productivity of plankton. Annu. Rev. Mar. Sci. 8, 161–184.

Mahadevan, A., D'Asaro, E., Lee, C., Perry, M.J., 2012. Eddy-driven stratification initiates North Atlantic spring phytoplankton blooms. Science 337 (6090), 54–58.

Mahadevan, A., Pascual, A., Rudnick, D.L., Ruiz, S., Tintoré, J., D'Asaro, E., 2020. Coherent pathways for vertical transport from the surface ocean to interior. Bull. Am. Meteorol. Soc. 101 (11), E1996–E2004.

Mahadevan, A., Tandon, A., 2006. An analysis of mechanisms for submesoscale vertical motion at ocean fronts. Ocean Model. 14, 241–256.

Mahadevan, A., Tandon, A., Ferrari, R., 2010. Rapid changes in mixed layer stratification driven by submesoscale instabilities and winds. J. Geophys. Res., Oceans 115 (C3).

Marshall, J., Schott, F., 1999. Open-ocean convection: observations, theory, and models. Rev. Geophys. 37, 1–64.

McCoy, D., Bianchi, D., Stewart, A.L., 2020. Global observations of submesoscale coherent vortices in the ocean. Prog. Oceanogr. 189, 102452.

McDowell, S., Rossby, H., 1978. Mediterranean water: an intense mesoscale eddy off the Bahamas. Science 202, 1085–1087.

McWilliams, J., Colas, F., Molemaker, M., 2009. Cold filamentary intensification and oceanic surface convergence lines. Geophys. Res. Lett. 36, L18602.
McWilliams, J.C., 1984. The emergence of isolated, coherent vortices in turbulent flow. AIP Conf. Proc. 106 (1), 205–221.
McWilliams, J.C., 1985. Submesoscale, coherent vortices in the ocean. Rev. Geophys. 23, 165–182.
McWilliams, J.C., 2016. Submesoscale currents in the ocean. Proc. R. Soc. A 472 (2189).
McWilliams, J.C., 2017. Submesoscale surface fronts and filaments: secondary circulation, buoyancy flux, and frontogenesis. J. Fluid Mech. 823, 391–432.
McWilliams, J.C., 2021. Oceanic frontogenesis. Annu. Rev. Mar. Sci. 13 (1), 227–253. PMID: 33395349.
McWilliams, J.C., Gula, J., Molemaker, M., Renault, L., Shchepetkin, A., 2015. Filament frontogenesis by boundary layer turbulence. J. Phys. Oceanogr. 45, 1988–2005.
McWilliams, J.C., Gula, J., Molemaker, M.J., 2019. The Gulf Stream North wall: ageostrophic circulation and frontogenesis. J. Phys. Oceanogr. 49 (4), 893–916.
McWilliams, J.C., Molemaker, M., 2011. Baroclinic frontal arrest: a sequel to unstable frontogenesis. J. Phys. Oceanogr. 41, 601–619.
McWilliams, J.C., Molemaker, M., Yavneh, I., 2001. From stirring to mixing of momentum: cascades from balanced flows to dissipation in the oceanic interior. In: Aha Huliko'a Proceedings, pp. 59–66.
McWilliams, J.C., Molemaker, M.J., Yavneh, I., 2004. Ageostrophic, anticyclonic instability of a geostrophic, barotropic boundary current. Phys. Fluids 16 (10), 3720–3725.
Mechoso, C.R., 1980. Baroclinic instability of flows along sloping boundaries. J. Atmos. Sci. 37 (6), 1393–1399.
Mensa, J., Garraffo, Z., Griffa, A., Ozgokmen, T., Haza, A., Veneziani, M., 2013. Seasonality of the submesoscale dynamics in the Gulf Stream region. Ocean Dyn., 53–85.
Molemaker, M., McWilliams, J.C., Capet, X., 2010. Balanced and unbalanced routes to dissipation in an equilibrated Eady flow. J. Fluid Mech. 654, 35–63.
Molemaker, M., McWilliams, J.C., Dewar, W., 2015. Submesoscale instability and generation of mesoscale anticyclones near a separation of the California Undercurrent. J. Phys. Oceanogr. 45, 613–629.
Morel, Y., Gula, J., Ponte, A., 2019. Potential vorticity diagnostics based on balances between volume integral and boundary conditions. Ocean Model. 138, 23–35.
Morrow, R., Fu, L.-L., Ardhuin, F., Benkiran, M., Chapron, B., Cosme, E., d'Ovidio, F., Farrar, J.T., Gille, S.T., Lapeyre, G., Le Traon, P.-Y., Pascual, A., Ponte, A., Qiu, B., Rascle, N., Ubelmann, C., Wang, J., Zaron, E.D., 2019. Global observations of fine-scale ocean surface topography with the surface water and ocean topography (swot) mission. Front. Mar. Sci. 6, 232.
Morvan, M., L'Hégaret, P., Carton, X., Gula, J., Vic, C., Sokolovskiy, M., Koshel, K., 2019. The life cycle of submesoscale eddies generated by topographic interactions. Ocean Sci. Discuss. 2019, 1–20.
Mukherjee, S., Ramachandran, S., Tandon, A., Mahadevan, A., 2016. Production and destruction of eddy kinetic energy in forced submesoscale eddy-resolving simulations. Ocean Model. 105, 44–59.
Müller, P., McWilliams, J.C., Molemaker, M., 2005. Routes to dissipation in the ocean: the 2d/3d turbulence conundrum. In: Baumert, H., Simpson, J., Sundermann, J. (Eds.), Marine Turbulence: Theories, Observations and Models. Cambridge Press.
Munk, W., Armi, L., Fischer, K., Zachariasen, F., 2000. Spirals on the sea. Proc. R. Soc. Lond., Ser. A, Math. Phys. Eng. Sci. 456 (1997), 1217–1280.
Nagai, T., Tandon, A., Yamazaki, H., Doubell, M., 2009. Evidence of enhanced turbulent dissipation in the frontogenetic Kuroshio Front thermocline. J. Atmos. Sci. 50 (14), 2159–2179.
Nagai, T., Tandon, A., Yamazaki, H., Doubell, M.J., Gallager, S., 2012. Direct observations of microscale turbulence and thermohaline structure in the Kuroshio front. J. Geophys. Res., Oceans 117 (C8).
Nakamura, N., 1988. Scale selection of baroclinic instability - effects of stratification and nongeostrophy. J. Atmos. Sci. 45 (21), 3253–3267.
Napolitano, D.C., da Silveira, I.C.A., Tandon, A., Calil, P.H.R., 2021. Submesoscale phenomena due to the Brazil current crossing of the Vitória-Trindade ridge. J. Geophys. Res., Oceans 126 (1), e2020JC016731.
Naveira Garabato, A.C., Frajka-Williams, E.E., Spingys, C.P., Legg, S., Polzin, K.L., Forryan, A., Abrahamsen, E.P., Buckingham, C.E., Griffies, S.M., McPhail, S.D., Nicholls, K.W., Thomas, L.N., Meredith, M.P., 2019. Rapid mixing and exchange of deep-ocean waters in an abyssal boundary current. Proc. Natl. Acad. Sci. 116 (27), 13233–13238.
Ooyama, K., 1966. On the stability of the baroclinic circular vortex: a sufficient criterion for instability. J. Atmos. Sci. 23 (1), 43–53.
Peng, J.-P., Holtermann, P., Umlauf, L., 2020. Frontal instability and energy dissipation in a submesoscale upwelling filament. J. Phys. Oceanogr.
Perfect, B., Kumar, N., Riley, J.J., 2018. Vortex structures in the wake of an idealized seamount in rotating, stratified flow. Geophys. Res. Lett. 45 (17), 9098–9105.
Perfect, B., Kumar, N., Riley, J.J., 2020. Energetics of seamount wakes. Part I: energy exchange. J. Phys. Oceanogr. 50 (5), 1365–1382.
Pham, H.T., Sarkar, S., 2018. Ageostrophic secondary circulation at a submesoscale front and the formation of gravity currents. J. Phys. Oceanogr. 48 (10), 2507–2529.
Poje, A.C., Özgökmen, T.M., Lipphardt, B.L., Haus, B.K., Ryan, E.H., Haza, A.C., Jacobs, G.A., Reniers, A.J.H.M., Olascoaga, M.J., Novelli, G., Griffa, A., Beron-Vera, F.J., Chen, S.S., Coelho, E., Hogan, P.J., Kirwan, A.D., Huntley, H.S., Mariano, A.J., 2014. Submesoscale dispersion in the vicinity of the deepwater horizon spill. Proc. Natl. Acad. Sci. 111 (35), 12693–12698.
Polton, J.A., Lenn, Y.-D., Elipot, S., Chereskin, T.K., Sprintall, J., 2013. Can drake passage observations match Ekman's classic theory? J. Phys. Oceanogr. 43 (8), 1733–1740.
Ponte, A., Klein, P., Capet, X., Le Traon, P.-Y., Chapron, B., Lherminier, P., 2013. Diagnosing surface mixed layer dynamics from high-resolution satellite observations: numerical insights. J. Phys. Oceanogr. 43, 1345–1355.
Price, J.F., Weller, R.A., Schudlich, R.R., 1987. Wind-driven ocean currents and Ekman transport. Science 238 (4833), 1534–1538.

Qiu, B., Chen, S., Klein, P., Sasaki, H., Sasai, Y., 2014. Seasonal mesoscale and submesoscale eddy variability along the North Pacific subtropical countercurrent. J. Phys. Oceanogr. 44 (12), 3079–3098.

Ramachandran, S., Tandon, A., Mackinnon, J., Lucas, A.J., Pinkel, R., Waterhouse, A.F., Nash, J., Shroyer, E., Mahadevan, A., Weller, R.A., Farrar, J.T., 2018. Submesoscale processes at shallow salinity fronts in the Bay of Bengal: observations during the winter monsoon. J. Phys. Oceanogr. 48 (3), 479–509.

Rascle, N., Chapron, B., Molemaker, J., Nouguier, F., Ocampo-Torres, F.J., Osuna Cañedo, J.P., Marié, L., Lund, B., Horstmann, J., 2020. Monitoring intense oceanic fronts using sea surface roughness: satellite, airplane, and in situ comparison. J. Geophys. Res., Oceans 125 (8), e2019JC015704. https://doi.org/10.1029/2019JC015704.

Rascle, N., Molemaker, J., Marié, L., Nouguier, F., Chapron, B., Lund, B., Mouche, A., 2017. Intense deformation field at oceanic front inferred from directional sea surface roughness observations. Geophys. Res. Lett. 44 (11), 5599–5608.

Renault, L., McWilliams, J.C., Gula, J., 2018. Dampening of submesoscale currents by air-sea stress coupling in the Californian upwelling system. Sci. Rep. 8 (1), 13388.

Ruan, X., Callies, J., 2020. Mixing-driven mean flows and submesoscale eddies over mid-ocean ridge flanks and fracture zone canyons. J. Phys. Oceanogr. 50 (1), 175–195.

Ruan, X., Thompson, A.F., Flexas, M.M., Sprintall, J., 2017. Contribution of topographically generated submesoscale turbulence to Southern Ocean overturning. Nat. Geosci. 10, 840.

Rudnick, D.L., Ferrari, R., 1999. Compensation of horizontal temperature and salinity gradients in the ocean mixed layer. Science 283 (5401), 526–529.

Samelson, R.M., Skyllingstad, E.D., 2016. Frontogenesis and turbulence: a numerical simulation. J. Atmos. Sci. 73 (12), 5025–5040.

Sasaki, H., Klein, P., Qiu, B., Sasa, Y., 2014. Impact of oceanic-scale interactions on the seasonal modulation of ocean dynamics by the atmosphere. Nat. Commun. 5, 5636.

Sasaki, H., Qiu, B., Klein, P., Sasai, Y., Nonaka, M., 2020. Interannual to decadal variations of submesoscale motions around the North Pacific subtropical countercurrent. Fluids 5 (3), 116.

Savelyev, I., Thomas, L.N., Smith, G.B., Wang, Q., Shearman, R.K., Haack, T., Christman, A.J., Blomquist, B., Sletten, M., Miller, W.D., Fernando, H.J.S., 2018. Aerial observations of symmetric instability at the North wall of the Gulf Stream. Geophys. Res. Lett. 45 (1), 236–244.

Schubert, R., Gula, J., Greatbatch, R.J., Baschek, B., Biastoch, A., 2020. The submesoscale kinetic energy cascade: mesoscale absorption of submesoscale mixed layer eddies and frontal downscale fluxes. J. Phys. Oceanogr. 50 (9), 2573–2589.

Scott, R.K., 2006. Local and nonlocal advection of a passive scalar. Phys. Fluids 18 (116601).

Shakespeare, C.J., Taylor, J.R., 2013. A generalized mathematical model of geostrophic adjustment and frontogenesis: uniform potential vorticity. J. Fluid Mech. 736, 366–413.

Shcherbina, A., D'Asaro, E., Lee, C., Klymak, J., Molemaker, M., McWilliams, J., 2013. Statistics of vertical vorticity, divergence, and strain in a developed submesoscale turbulence field. Geophys. Res. Lett. 40.

Shcherbina, A.Y., Gregg, M.C., Alford, M.H., Harcourt, R.R., 2009. Characterizing thermohaline intrusions in the North Pacific subtropical frontal zone. J. Phys. Oceanogr. 39 (11), 2735–2756.

Shcherbina, A.Y., Sundermeyer, M.A., Kunze, E., D'Asaro, E., Badin, G., Birch, D., Brunner-Suzuki, A.-M.E.G., Callies, J., Kuebel Cervantes, B.T., Claret, M., Concannon, B., Early, J., Ferrari, R., Goodman, L., Harcourt, R.R., Klymak, J.M., Lee, C.M., Lelong, M.-P., Levine, M.D., Lien, R.-C., Mahadevan, A., McWilliams, J.C., Molemaker, M.J., Mukherjee, S., Nash, J.D., Özgökmen, T., Pierce, S.D., Ramachandran, S., Samelson, R.M., Sanford, T.B., Shearman, R.K., Skyllingstad, E.D., Smith, K.S., Tandon, A., Taylor, J.R., Terray, E.A., Thomas, L.N., Ledwell, J.R., 2015. The latmix summer campaign: submesoscale stirring in the upper ocean. Bull. Am. Meteorol. Soc. 96 (8), 1257–1279.

Siegelman, L., 2020. Energetic submesoscale dynamics in the ocean interior. J. Phys. Oceanogr. 50 (3), 727–749.

Siegelman, L., Klein, P., Rivière, P., Thompson, A.F., Torres, H.S., Flexas, M., Menemenlis, D., 2020. Enhanced upward heat transport at deep submesoscale ocean fronts. Nat. Geosci. 13 (1), 50–55.

Skyllingstad, E.D., Samelson, R.M., 2012. Baroclinic frontal instabilities and turbulent mixing in the surface boundary layer. Part I: unforced simulations. J. Phys. Oceanogr. 42 (10), 1701–1716.

Skyllingstad, E.D., Samelson, R.M., 2020. Instability processes in simulated finite-width ocean fronts. J. Phys. Oceanogr.

Smith, K.M., Hamlington, P.E., Fox-Kemper, B., 2016. Effects of submesoscale turbulence on ocean tracers. J. Geophys. Res., Oceans 121 (1), 908–933.

Smith, K.S., Ferrari, R., 2009. The production and dissipation of compensated thermohaline variance by mesoscale stirring. J. Phys. Oceanogr. 39 (10), 2477–2501.

Solodoch, A., Stewart, A.L., McWilliams, J.C., 2016. Baroclinic instability of axially symmetric flow over sloping bathymetry. J. Fluid Mech. 799, 265–296.

Spingys, C.P., Garabato, A.C.N., Legg, S., Polzin, K.L., Abrahamsen, E.P., Buckingham, C.E., Forryan, A., Frajka-Williams, E.E., 2021. Mixing and transformation in a deep western boundary current: a case study. J. Phys. Oceanogr.

Srinivasan, K., McWilliams, J.C., Jagannathan, A., 2021. High vertical shear and dissipation in equatorial topographic wakes. J. Phys. Oceanogr.

Srinivasan, K., McWilliams, J.C., Molemaker, M.J., Barkan, R., 2019. Submesoscale vortical wakes in the Lee of topography. J. Phys. Oceanogr. 49 (7), 1949–1971.

Srinivasan, K., McWilliams, J.C., Renault, L., Hristova, H.G., Molemaker, J., Kessler, W.S., 2017. Topographic and mixed layer submesoscale currents in the near-surface southwestern tropical Pacific. J. Phys. Oceanogr. 47 (6), 1221–1242.

Stamper, M.A., Taylor, J.R., 2017. The transition from symmetric to baroclinic instability in the Eady model. Ocean Dyn. 67 (1), 65–80.

Stegner, A., 2014. Oceanic Island Wake Flows in the Laboratory. American Geophysical Union (AGU), pp. 265–276. Chapter 14.

Stone, P., 1966. On non-geostrophic baroclinic instability. J. Atmos. Sci. 23, 390–400.

Stone, P., 1970. On non-geostrophic baroclinic instability: Part II. J. Atmos. Sci. 27, 721–726.
Stone, P.H., 1972. On non-geostrophic baroclinic stability: Part III. The momentum and heat transports. J. Atmos. Sci. 29 (3), 419–426.
Straneo, F., Kawase, M., Pickart, R.S., 2002. Effects of wind on convection in strongly and weakly baroclinic flows with application to the Labrador sea. J. Phys. Oceanogr. 32 (9), 2603–2618.
Straub, K., Kiladis, G., 2002. Observations of a convectively coupled Kelvin wave in the Eastern Pacific ITCZ. J. Atmos. Sci. 59, 30.
Su, Z., Torres, H., Klein, P., Thompson, A.F., Siegelman, L., Wang, J., Menemenlis, D., Hill, C., 2020. High-frequency submesoscale motions enhance the upward vertical heat transport in the global ocean. J. Geophys. Res., Oceans 125 (9), e2020JC016544.
Su, Z., Wang, J., Klein, P., Thompson, A.F., Menemenlis, D., 2018. Ocean submesoscales as a key component of the global heat budget. Nat. Commun. 9, 775.
Sullivan, P.P., McWilliams, J.C., 2018. Frontogenesis and frontal arrest of a dense filament in the oceanic surface boundary layer. J. Fluid Mech. 837, 341–380.
Sullivan, P.P., McWilliams, J.C., 2019. Langmuir turbulence and filament frontogenesis in the oceanic surface boundary layer. J. Fluid Mech. 879, 512–553.
Sundermeyer, M.A., Skyllingstad, E., Ledwell, J.R., Concannon, B., Terray, E.A., Birch, D., Pierce, S.D., Cervantes, B., 2014. Observations and numerical simulations of large-eddy circulation in the ocean surface mixed layer. Geophys. Res. Lett. 41 (21), 7584–7590.
Suzuki, N., Fox-Kemper, B., Hamlington, P.E., Van Roekel, L.P., 2016. Surface waves affect frontogenesis. J. Geophys. Res., Oceans 121 (5), 3597–3624.
Tandon, A., Garrett, C., 1994. Mixed layer restratification due to a horizontal density gradient. J. Phys. Oceanogr. 24 (6), 1419–1424.
Taylor, J., Ferrari, R., 2009. The role of secondary shear instabilities in the equilibration of symmetric instability. J. Fluid Mech. 622, 103.
Taylor, J.R., 2016. Turbulent mixing, restratification, and phytoplankton growth at a submesoscale eddy. Geophys. Res. Lett. 43 (11), 5784–5792.
Taylor, J.R., 2018. Accumulation and subduction of buoyant material at submesoscale fronts. J. Phys. Oceanogr. 48 (6), 1233–1241.
Taylor, J.R., Ferrari, R., 2010. Buoyancy and wind-driven convection at mixed layer density fronts. J. Phys. Oceanogr. 40 (6), 1222–1242.
Taylor, J.R., Ferrari, R., 2011. Ocean fronts trigger high latitude phytoplankton blooms. Geophys. Res. Lett. 38 (23).
Testor, P., Gascard, J.-C., 2003. Large-scale spreading of deep waters in the western Mediterranean Sea by submesoscale coherent eddies. J. Phys. Oceanogr. 33 (1), 75–87.
Thomas, L., 2005. Destruction of potential vorticity by winds. J. Phys. Oceanogr. 35, 2457–2466.
Thomas, L., Ferrari, R., 2008. Friction, frontogenesis, and the stratification of the surface mixed layer. J. Phys. Oceanogr. 38 (11), 2501–2518.
Thomas, L., Tandon, A., Mahadevan, A., 2008. Submesoscale processes and dynamics. J. Geophys. Res.
Thomas, L., Taylor, J., Ferrari, R., Joyce, T., 2013. Symmetric instability in the Gulf Stream. Deep-Sea Res. 91, 96–110.
Thomas, L.N., 2008. Formation of intrathermocline eddies at ocean fronts by wind-driven destruction of potential vorticity. Dyn. Atmos. Ocean. 45 (3–4), 252–273.
Thomas, L.N., Taylor, J.R., 2010. Reduction of the usable wind-work on the general circulation by forced symmetric instability. Geophys. Res. Lett. 37 (18), L18606.
Thomas, L.N., Taylor, J.R., D'Asaro, E.A., Lee, C.M., Klymak, J.M., Shcherbina, A., 2016. Symmetric instability, inertial oscillations, and turbulence at the Gulf Stream front. J. Phys. Oceanogr. 46 (1), 197–217.
Thompson, A.F., Lazar, A., Buckingham, C., Naveira-Garabato, A.C., Damerell, G.M., Heywood, K.J., 2016. Open-ocean submesoscale motions: a full seasonal cycle of mixed layer instabilities from gliders. J. Phys. Oceanogr. 46, 1285–1307.
Thompson, L., 2000. Ekman layers and two-dimensional frontogenesis in the upper ocean. J. Geophys. Res., Oceans 105 (C3), 6437–6451.
Thorpe, A., Rotunno, R., 1989. Nonlinear aspects of symmetric instability. J. Atmos. Sci. 46 (9), 1285–1299.
Torres, H.S., Klein, P., Menemenlis, D., Qiu, B., Su, Z., Wang, J., Chen, S., Fu, L., 2018. Partitioning ocean motions into balanced motions and internal gravity waves: a modeling study in anticipation of future space missions. J. Geophys. Res., Oceans.
Uchida, T., Balwada, D., Abernathey, R., McKinley, G., Smith, S., Lévy, M., 2019. The contribution of submesoscale over mesoscale eddy iron transport in the open Southern Ocean. J. Adv. Model. Earth Syst. 11 (12), 3934–3958.
Uchida, T., Balwada, D., Abernathey, R.P., McKinley, G.A., Smith, S.K., Lévy, M., 2020. Vertical eddy iron fluxes support primary production in the open Southern Ocean. Nat. Commun. 11 (1), 1125.
Vallis, G.K., 2006. Atmospheric and Oceanic Fluid Dynamics. Cambridge University Press, Cambridge, U.K.
van Haren, H., Millot, C., 2009. Slantwise convection: a candidate for homogenization of deep newly formed dense waters. Geophys. Res. Lett. 36 (12).
Vanneste, J., 1993. The Kelvin-Helmoltz instability in a non-geostrophic baroclinic unstable flow. Math. Comput. Model. 17, 149–154.
Verma, V., Pham, H.T., Sarkar, S., 2019. The submesoscale, the finescale and their interaction at a mixed layer front. Ocean Model. 140, 101400.
Vic, C., Gula, J., Roullet, G., Pradillon, F., 2018. Dispersion of deep-sea hydrothermal vent effluents and larvae by submesoscale and tidal currents. Deep-Sea Res., Part 1 133, 1–18.
Vic, C., Roullet, G., Carton, X., Capet, X., Molemaker, M., Gula, J., 2015. Eddy-topography interactions and the fate of the Persian Gulf Outflow. J. Geophys. Res., Oceans 120, 6700–6717.
Wang, T., Barkan, R., McWilliams, J.C., Molemaker, M.J., 2021. Structure of submesoscale fronts of the Mississippi River plume. J. Phys. Oceanogr.
Washington, W., 1964. A note on the adjustment towards geostrophic equilibrium in a simple fluid system. Tellus 16, 530–534.
Wenegrat, J., Thomas, L., Sundermeyer, M., Taylor, J., D'Asaro, E., Klymak, J., Shearman, R., Lee, C., 2020. Enhanced mixing across the gyre boundary at the Gulf Stream front. Proc. Natl. Acad. Sci. 117 (30), 17607–17614.
Wenegrat, J.O., Callies, J., Thomas, L.N., 2018a. Submesoscale baroclinic instability in the bottom boundary layer. J. Phys. Oceanogr. 48 (11), 2571–2592.
Wenegrat, J.O., McPhaden, M.J., 2016. Wind, waves, and fronts: frictional effects in a generalized Ekman model. J. Phys. Oceanogr. 46 (2), 371–394.

Wenegrat, J.O., Thomas, L.N., 2020. Centrifugal and symmetric instability during Ekman adjustment of the bottom boundary layer. J. Phys. Oceanogr. 50 (6), 1793–1812.

Wenegrat, J.O., Thomas, L.N., Gula, J., McWilliams, J.C., 2018b. Effects of the submesoscale on the potential vorticity budget of ocean mode waters. J. Phys. Oceanogr. 48 (9), 2141–2165.

Whitt, D.B., Lévy, M., Taylor, J.R., 2019. Submesoscales enhance storm-driven vertical mixing of nutrients: insights from a biogeochemical large eddy simulation. J. Geophys. Res., Oceans 124 (11), 8140–8165.

Whitt, D.B., Taylor, J.R., 2017. Energetic submesoscales maintain strong mixed layer stratification during an autumn storm. J. Phys. Oceanogr. 47 (10), 2419–2427.

Wijesekera, H.W., Shroyer, E., Tandon, A., Ravichandran, M., Sengupta, D., Jinadasa, S., Fernando, H.J., Agrawal, N., Arulananthan, K., Bhat, G., et al., 2016. Asiri: an ocean–atmosphere initiative for Bay of Bengal. Bull. Am. Meteorol. Soc. 97 (10), 1859–1884.

Williams, R., Roussenov, V., 2003. The role of sloping sidewalls in forming potential vorticity contrasts in the ocean interior. J. Phys. Oceanogr. 33, 633–648.

Yang, Y., McWilliams, J.C., Liang, X.S., Zhang, H., Weisberg, R.H., Liu, Y., Menemenlis, D., 2021. Spatial and temporal characteristics of the submesoscale energetics in the Gulf of Mexico. J. Phys. Oceanogr. 51 (2), 475–489.

Yankovsky, E., Legg, S., 2018. Symmetric and baroclinic instability in dense shelf overflows. J. Phys. Oceanogr. 49 (1), 39–61.

Young, W., 1994. The subinertial mixed layer approximation. J. Phys. Oceanogr. 24 (8), 1812–1826.

Young, W.a., Jones, S., 1991. Shear dispersion. Phys. Fluids A, Fluid Dyn. 3 (5), 1087–1101.

Yu, X., Naveira Garabato, A.C., Martin, A.P., Buckingham, C.E., Brannigan, L., Su, Z., 2019a. An annual cycle of submesoscale vertical flow and restratification in the upper ocean. J. Phys. Oceanogr. 49 (6), 1439–1461.

Yu, X., Naveira Garabato, A.C., Martin, A.P., Gwyn Evans, D., Su, Z., 2019b. Wind-forced symmetric instability at a transient mid-ocean front. Geophys. Res. Lett. 46 (20), 11281–11291.

Yu, X., Naveira Garabato, A.C., Martin, A.P., Marshall, D.P., 2020. The annual cycle of upper-ocean potential vorticity and its relationship to submesoscale instabilities. J. Phys. Oceanogr.

Zhang, Z., Zhang, X., Qiu, B., Zhao, W., Zhou, C., Huang, X., Tian, J., 2020. Submesoscale currents in the subtropical upper ocean observed by long-term high-resolution mooring arrays. J. Phys. Oceanogr., 1–65.

Chapter 9

Isopycnal mixing

Ryan Abernathey[a], Anand Gnanadesikan[b], Marie-Aude Pradal[b] and Miles A. Sundermeyer[c]
[a]Lamont Doherty Earth Observatory of Columbia University, New York, NY, United States, [b]Johns Hopkins University, Baltimore, MD, United States, [c]University of Massachusetts Dartmouth, North Dartmouth, MA, United States

9.1 Introduction

The fact that we have a special chapter for isopycnal mixing suggests that much of the rest of this book is, in one way or another, concerned with its opposite: diapycnal mixing. Isopycnal mixing is indeed an outlier in ocean mixing research. Because diapycnal mixing irreversibly increases the potential energy of the ocean, it is directly tied to and constrained by the ocean energy budget. This creates a single over-arching framework for the study of diapycnal mixing: energy conservation. Isopycnal mixing, in contrast, is by definition energetically neutral, and therefore plays no direct role in the ocean energy budget. In the absence of a single over-arching framework, isopycnal mixing must be understood by bringing together many different viewpoints and concepts from fluid dynamics. *A central goal of this chapter is to present these different viewpoints and explain how they can be used to illuminate isopycnal mixing from different angles.*

Compared to diapycnal mixing, the space and timescales of isopycnal mixing on basin scales are massive: mesoscale eddies, the primary driver of isopycnal mixing, occupy space scales of 100 km or more and evolve on timescales of months. Even at the submesoscale, particularly in the stratified interior, isopycnal mixing occurs at horizontal scales of 100 s of m to many km and on timescales from hours to days (see also Chapter 8 for discussion of submesoscale mixing). Consequently, observational techniques for measuring isopycnal mixing differ completely from those used for diapycnal mixing. Microstructure probes are of limited use here[1]; instead, oceanographers use floats, drifters, tracer release experiments, and satellite observations to quantify isopycnal mixing rates. Even with these tools, obtaining a coherent global picture of isopycnal mixing from observations alone remains a difficult challenge. *Another main goal of this chapter is to review the methods used for observing isopycnal mixing and the results they have obtained.*

Numerical models of the ocean play a central role in the study of isopycnal mixing. On the one hand, coarse-resolution global ocean models must parameterise the effects of unresolved isopycnal mixing to properly simulate tracer distributions, biogeochemical cycles, and climate variability. Informing and refining such parameterisations is a big motivation for isopycnal mixing research. On the other hand, certain isopycnal mixing processes can be simulated directly by high-resolution models in both idealised configurations (including quasigeostrophic models) and—a more recent development—in global eddy-resolving configurations. *In this chapter, we review how isopycnal mixing is parameterised in coarse-resolution basin- and global-scale circulation models and synthesise the results from recent high-resolution modelling studies.*

In reviewing recent progress on isopycnal mixing, several cross-cutting themes emerge. These themes represent confounding factors, which complicate our understanding and prevent a simplistic description of the mixing process.

Inhomogeneity The magnitude and structure of isopycnal mixing vary extremely in space, both horizontally and vertically, and also potentially in time. Observations made in one place don't necessarily generalise to a global view. Considerable recent progress has been made, however, on characterising inhomogeneity from observations and explaining it with theory.
Anisotropy Isopycnal mixing can act very differently in different directions. Anisotropy in mixing is largely generated by anisotropy in the background flow, shear, and potential vorticity gradients. This anisotropy poses a persistent challenge for both diagnosis and parameterisation of isopycnal mixing.
Non-locality Most theoretical models for mixing assume that stirring and dissipation (molecular-level homogenisation of tracer concentration) occur in the same place, but, for isopycnal mixing, it's clear this is not the case in many regions. A related challenge is that Lagrangian measurements naturally sample a broad swath of space, whereas numerical models need to parameterise isopycnal mixing in Eulerian grid cells.

---

1. An exception to this rule can be found in Naveira Garabato et al. (2016).

Scale dependence Isopycnal mixing acts over a range of scales, and the strength and character of the mixing process depends strongly on what scale you zoom into. Research aimed at parameterisation has traditionally focused on mixing by the largest mesoscale eddies. However, as numerical models become finer in resolution, the question of how to parameterise isopycnal mixing in models with finer grids is emerging.

These themes represent much of the focus of contemporary research. We do not dedicate sections to these themes, but rather attempt to weave them through the manuscript as unifying threads.

This chapter is intended both as a general introduction to isopycnal mixing for those new to the topic and as a review of some of the important recent developments in the theory, observations, and modelling. A graduate student beginning research is our target audience, and we assume a corresponding familiarity with general physical oceanography and basic geophysical fluid dynamics. The chapter is organised as follows: Section 9.2 introduces core concepts, such as the mixing itself and the neutral plane. In Section 9.3, we provide a qualitative overview of the physical mechanisms / processes involved in isopycnal mixing. Section 9.4 describes the theoretical frameworks which must come together to give a quantitative understanding of the mixing process, including the parameterisation problem. In Section 9.5, we review observational estimates of isopycnal mixing, highlighting evidence for inhomogeneity, anisotropy, and scale dependence. Section 9.6 describes the growing body of information about isopycnal mixing that has emerged from numerical simulations. In Section 9.7, we summarise some important impacts of isopycnal mixing on ocean circulation and biogeochemistry (these topics are also addressed in Chapters 2, 3 and 14). We conclude in Section 9.8 with an outlook for future research.

9.2 Background concepts

Before we can turn to isopycnal mixing proper, we must establish a basic understanding of two things: *What is mixing?* and *What is isopycnal?*. We begin with the first question.

9.2.1 What is mixing?

> We define mixing as the combined action of stirring and molecular diffusion leading to the homogenisation of tracers.

For the purposes of this chapter, we are interested in the mixing of scalar quantities (i.e., *tracers*) which obey a conservation equation of the form

$$\frac{\partial c}{\partial t} + \mathbf{u} \cdot \nabla c = \kappa \nabla^2 c + \gamma, \tag{9.1}$$

where c is the tracer concentration, κ represents molecular diffusivity, and γ represents any other non-conservative sources or sinks. An equation of this form describes the evolution of salinity, potential temperature, biogeochemical tracers (such as nutrient concentration), and potential vorticity. (We explicitly exclude from consideration the 'mixing' of momentum or energy, which are not well modelled by (9.1).) Mixing refers to the combined action of the velocity field $\mathbf{u}$ and the molecular diffusion κ to homogenise the concentration of c. In the absence of any forcing γ, mixing would eventually lead to a completely homogeneous field of c. Most tracers in the ocean exist in a state of quasi-equilibrium between forcing, which reinforces tracer gradients, and mixing, which tries to eliminate them.

We will now develop a mathematical model of mixing, which provides definitions of the concepts of dispersion, eddy diffusion, stirring, and dissipation. Readers who feel comfortable with their intuition on these concepts should feel free to skip to the next section. We begin by taking a Reynolds average[2]—i.e., an average over many realisations of the stirring process—of (9.1). We will make several simplifying assumptions: (i) the flow is incompressible, i.e., $\nabla \cdot \mathbf{u} = 0$; (ii) the mean velocity field is zero, i.e. $\bar{\mathbf{u}} = 0$; (iii) there is no explicit forcing ($\gamma = 0$), but the mean tracer $\bar{c}$ has a fixed uniform gradient in the x-direction; i.e., $\partial \bar{c}/\partial x = const$; and (iv) the system is in a statistically homogeneous state in time, y- and z-dimensions. This final assumption requires that any time- and y- or z-derivatives of mean quantities be zero. With these

---

[2]. We denote the Reynolds average with the operator -. The average of a quantity ϕ is $\bar{\phi}$. The deviation from the average—often called the *fluctuation* or the *eddy* term—is denoted with a prime: $\phi' = \phi - \bar{\phi}$. For a more through discussion of Reynolds averaging and its properties, consult a textbook, such as Olbers et al. (2012).

assumptions, we arrive at a mean tracer equation of the form[3]

$$\frac{\partial}{\partial x}\overline{u'c'} = 0. \tag{9.2}$$

Since there is nothing to balance convergence of the eddy tracer flux $\overline{u'c'}$ in this simple, homogeneous scenario, the eddy tracer flux must be a constant. But what determines the magnitude and direction of the flux?

Mixing occurs when the velocity field **u** causes water parcels to move around in a random and/or irreversible way. This suggests that mixing is in fact a Lagrangian property of the velocity field. The spreading and exchange of water parcels from an initial reference position is generically termed *dispersion*. A key result was derived by Taylor (1921), who found a relationship between parcel dispersion statistics and the eddy tracer flux. For isotropic, homogeneous turbulent flow, the passive tracer is transported in the direction of the background gradient at the rate

$$\overline{u'c'} = -K\frac{\partial \overline{c}}{\partial x}. \tag{9.3}$$

This is the equation for Fickian diffusion. The quantity K is a *diffusivity*, sometimes called the *eddy diffusivity* to distinguish from κ, the molecular diffusivity. It relates to dispersion via the formula

$$K = \lim_{t \to \infty} \frac{1}{2}\frac{d}{dt}\langle (\mathbf{x}_i(t) - \mathbf{x}_i(0))^2 \rangle, \tag{9.4}$$

where the angle brackets represent an ensemble average over many parcels, and x_i the positions of each parcel. (9.4) states that K is proportional to the growth rate of the squared displacement of water parcels from their original positions; the faster the particles spread, the stronger the diffusion. Diffusivity is one of the primary quantitative diagnostics of mixing. The notion of eddy diffusion as linear with time increase in ensemble distance squared recurs throughout this chapter, particularly in the observational methods of Section 9.5. For most geophysical flows, $K \gg \kappa$: the diffusion induced by the velocity field far exceeds molecular diffusion.

It is also useful to distinguish between stirring and mixing. When the displacement of water parcels brings parcels of differing origins into close proximity, the motion is often called *stirring*. A hallmark of stirring is the creation of sharp, fine-scale gradients in tracer concentration, such as those seen in the satellite image in Fig. 9.1. True *mixing* however, ultimately requires molecular diffusion to homogenise water properties. To see this mathematically, we must go a step further and derive the conservation equation for tracer *variance* $(\overline{c'^2}/2)$. The variance quantifies the magnitude of the fluctuations in the tracer field. We obtain it by first subtracting (9.2) from (9.1) to arrive at a conservation equation for c', the tracer fluctuation:

$$\frac{\partial c'}{\partial t} + \mathbf{u} \cdot \nabla c' + u'\frac{\partial \overline{c}}{\partial x} - \overline{u'\frac{\partial c'}{\partial x}} = \kappa \nabla^2 c'. \tag{9.5}$$

Next we multiply this equation by c' and take the average again. After applying the assumptions of statistical isotropy and homogeneity, plus considerable algebra, we arrive at the following variance budget:

$$\overline{u'c'}\frac{\partial \overline{c}}{\partial x} = -\kappa \overline{|\nabla c'|^2}. \tag{9.6}$$

The left-hand-side term represents the production of tracer variance by the stirring of the background gradient. The right-hand-side term, $-\kappa \overline{|\nabla c'|^2}$, is associated with the *dissipation* of tracer variance by molecular diffusion; it requires κ to be non-zero, but the strength of the dissipation depends on the strength of the tracer gradients, which are determined by the stirring process. By applying Taylor's formula (9.3) for the eddy diffusivity, we obtain

$$-K\left(\frac{\partial \overline{c}}{\partial x}\right)^2 = -\kappa \overline{|\nabla c'|^2}. \tag{9.7}$$

This equation provides a quantitative relationship between the kinematic stirring of water parcels (represented by K), and the microscale dissipation of tracer variance, which is fundamental to ocean mixing, both isopycnal and diapycnal (Osborn and Cox, 1972). The fact that K is positive (i.e., that the flux is down the gradient), is required to satisfy the variance budget in (9.7).

---

3. We use the fact that, $\overline{u'\partial c'/\partial x} = \partial(\overline{u'c'})/\partial x$. This can be obtained via the product rule, together with the continuity equation $\nabla \cdot \mathbf{u'} = 0$.

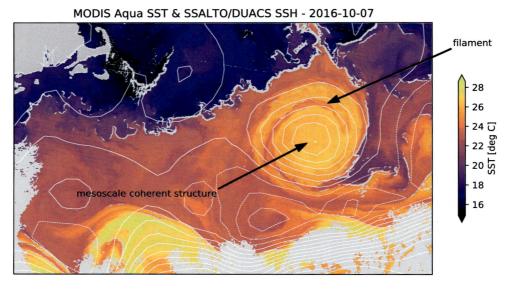

FIGURE 9.1 Satellite data illustrating stirring by a mesoscale eddy in the North Atlantic, just north of the Gulf stream. The colours show sea-surface temperature (SST) observed by the MODIS Aqua infrared imager on Oct. 7, 2016. (The MODIS L2P sea surface temperature data are sponsored by NASA.) Masked light grey areas indicate poor data quality, mostly related to cloud cover. The white contours show the sea surface height (SSH) observed by satellite altimetry. (Data are from the Copernicus marine environment monitoring service SSALTO/DUACS delayed-time Level-4 sea surface height, measured by multi-satellite altimetry observations over Global ocean). The SSH contours are approximate streamlines for the geostrophic flow, showing the direction of the surface currents. A large warm-core anticyclonic mesoscale eddy is highlighted. Within the eddy, a filament of cooler water is clearly visible. This image illustrates how stirring by the velocity field can create sharp tracer gradients, which can then be dissipated by smaller-scale mixing processes. Eventually the dissipation reaches the microscale, and molecular diffusion homogenises the temperature among adjacent water masses. Similar stirring and mixing processes are active along neutral surfaces in the ocean interior, but they cannot be directly observed by satellite.

Although this description of mixing is very generic, it contains the seeds of many of the fundamental questions in contemporary isopycnal mixing research. How do we quantify mixing if the flow is *not* isotropic and homogeneous? Can we reconcile Lagrangian descriptions of mixing (e.g. (9.4)) with Eulerian ones (e.g., (9.6))? Is tracer variance dissipated in the same location it is produced? Before getting to these questions, we must first understand what's so special about the isopycnal direction of motion.

9.2.2 What is isopycnal?

> Water parcels can move freely along neutral surfaces (isopycnals) without experiencing any restoring forces, creating a preferential direction for stirring motions.

The ocean is stratified in density. At timescales longer than a few hours, this stratification obeys the hydrostatic balance. This stratification means that when water parcels are displaced in the vertical direction (either up or down), they experience a restoring gravitational force, pulling them back to their level of neutral buoyancy (Fig. 9.2). For small displacements, the strength of this restoring force is proportional to the square of the local Brunt–Väisälä frequency N^2, which measures the strength of the stratification.

If there were no horizontal variations in ocean stratification, then the isopycnals (density surfaces) would all be flat, or more precisely, aligned with the geopotential field. Water parcels would be free to move horizontally (perpendicular to the direction of gravity) without experiencing any restoring forces. However, this is not the case; the presence of horizontal temperature and salinity gradients in the ocean means that parcels displaced in a horizontal direction will also experience a vertical restoring force, pushing them up or down towards their level of neutral buoyancy, as illustrated in Fig. 9.2.

So where can a water parcel move without experiencing *any* restoring forces? The answer is, *along the neutral plane*. The neutral plane is the plane perpendicular to the diapycnal direction $\hat{\mathbf{n}}$, defined by

$$\hat{\mathbf{n}} = \frac{\nabla \sigma}{|\nabla \sigma|} = \frac{-\alpha^\Theta \nabla \Theta + \beta^\Theta \nabla S_A}{|-\alpha^\Theta \nabla \Theta + \beta^\Theta \nabla S_A|}, \tag{9.8}$$

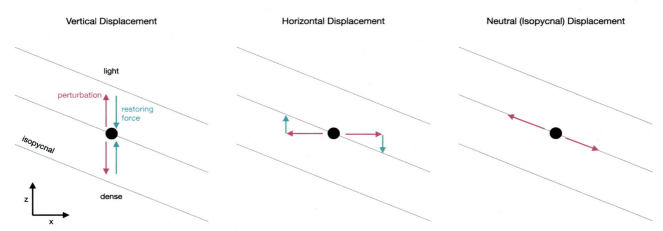

FIGURE 9.2 Illustration of how water parcels respond to displacements. The direction of displacement (perturbation) is shown in magenta, and the resulting buoyancy restoring force in cyan. The water parcel is represented as a black dot. Isopycnals are sketched in light grey. Left: vertical perturbations result in restoring forces, pulling the parcel back to its level of neutral buoyancy. Centre: in the presence of sloping isopycnals, horizontal perturbations also result in *vertical* restoring forces. Left: restoring forces are only absent if the perturbation is along the neutral (isopycnal) plane.

where σ is *locally-referenced* potential density[4]; the second equality shows this is equal to the sum of conservative temperature Θ and absolute salinity S_A gradients, multiplied by their respective thermal expansion (α^Θ), and haline contraction (β^Θ) coefficients (McDougall et al., 2009). Any sustained motion in the diapycnal direction requires work to be done, and is thus constrained by energy conservation. (Indeed, this was a primary focus of earlier chapters.) Isopycnal motion, by contrast, requires no work against gravity. This means water parcels are free to slide around on the neutral plane without any energy input. The existence of this special direction leads to extreme anisotropy in mixing; isopycnal diffusion coefficients K can be more than 10^8 times larger than turbulent diapycnal mixing coefficients! (Note that this obvious anisotropy between diapycnal and isopycnal mixing is distinct from the more subtle anisotropy that occurs within the neutral plane, which gives isopycnal mixing a preferred lateral direction—e.g., see Bachman et al., 2020—a topic we take up later.)

If a water parcel's temperature or salinity is actively changing, the restoring-force arguments made above do not apply. For this reason, the concept of isopycnal mixing is most relevant in circumstances when temperature and salinity are conserved following the flow, such as in the ocean interior. (These conditions are conventionally termed "adiabatic", although that is not a thermodynamically correct use of the term; *adiabatic* formally applies only to heat conservation.) Within the surface and bottom boundary layers, or in coastal regions, the presence of strong diabatic forcing and other mixing processes means that water parcels are less constrained by the neutral plane, and mixing can become more isotropic. This chapter focuses primarily on the conservative, interior regime of isopycnal mixing. However, we do briefly consider the transition to lateral mixing in the surface diabatic layer in Section 9.2.4.

It is mentally convenient, and standard practice, to imagine an "isopycnal" as a smooth two-dimensional surface spanning ocean basins. However, this simplistic view can be misleading. The neutral plane defined in (9.8) defines the plane on which isopycnal mixing occurs. For simplified models of the ocean, which assume a linear equation of state (constant α^Θ and β^S), the neutral plane is parallel everywhere to isosurfaces of buoyancy, defined as

$$b = g(\alpha^\Theta \Theta - \beta^S S) , \tag{9.9}$$

or equivalently, to potential density. These are proper global "neutral surfaces", because water parcels can travel along them conservatively without experiencing any gravitational restoring forces.[5]

---

4. *Potential density* σ is defined as $\sigma(S, \Theta) = \rho(S, \Theta, p_{ref})$, where ρ is the *in-situ* density, and p_{ref} is a constant *reference pressure*. In (9.8), which uses infinitesimal gradients of temperature and salinity, p_{ref} is taken as the pressure at the point where the gradients are measured. However, potential density is also commonly used as a water mass tracer due to its conservative nature (it depends only on temperature and salinity). In this case, it is common to denote the reference pressure with a subscript; σ_0 is potential density referenced to the surface; σ_2 is referenced to 2000 dbar, etc. But note that the first equality in (9.8) is not necessarily accurate if the *in-situ* pressure is far from the reference pressure.
5. We note that much of our theoretical understanding of isopycnal mixing derives from idealised quasigeostrophic (QG) models, in which advection occurs in the horizontal plane (Vallis, 2006). Key to translating these insights to the real ocean is to note that QG theory assumes a horizontally constant background stratification N^2, and consequently horizontal motion in QG is asymptotically analogous to isopycnal motion in primitive equations. The "neutral surfaces" in z-coordinate QG models are simply surfaces of constant depth.

The leap from a local neutral plane to a global neutral surface is not possible for real seawater, due to the non-linear "helical" nature of seawater's equation of state. Global isosurfaces that are everywhere parallel to the neutral plane simply don't exist (Klocker and McDougall, 2010). This unfortunate fact complicates the analysis of isopycnal mixing and frustrates attempts to model the ocean in density coordinates. Though locally referenced potential density provides a good approximation of the neutral angle near the reference pressure level, it does not do so over the whole water column. Most isopycnal-coordinate ocean models use σ_2, potential density referenced to 2000 dbar, as their vertical coordinate, and assume that the σ_2 isopycnals are the neutral surfaces relevant for mixing; the latest version of the National atmospheric and oceanic association's (NOAA) modular ocean model (MOM6) is the only model, to our knowledge, which attempts to account for the mismatch between the isopycnal coordinate and the true neutral plane (Shao et al., 2020).

The desire for convenient global neutral surfaces on which to analyse ocean transport and water masses motivated the development of the concept of neutral density γ_n (Jackett and McDougall, 1997). Neutral density is calculated in an iterative process designed to minimise the misfit between γ_n isosurfaces and the local neutral plane. Though analysis of data in neutral-density coordinates has become widespread, it is not a silver bullet to eliminate the fundamental problem of helicity. Crucially, γ_n is not a conservative variable (i.e., a function of only Θ and S), but depends also on pressure, latitude and longitude. This complicates its interpretation and use in water-mass-transformation studies (Iudicone et al., 2008). An alternative, thermodynamic neutral density γ_T was recently proposed by Tailleux (2016). We also note that the very concept of mixing along the neutral angle, broadly accepted as conventional wisdom by most of the field, remains a topic of some controversy among specialists (Tailleux, 2017; McDougall et al., 2017).

Isopycnal mixing cannot not directly transport buoyancy, since buoyancy is homogeneous along the neutral plane. However, it can transport heat and salt, since temperature and salinity can both have neutral gradients. These gradients are mutually compensating across a broad range of scales (Rudnick and Ferrari, 1999; Spiro Jaeger and Mahadevan, 2018); as a result, neutral surfaces are significantly smoother than either isotherms or isohalines. This is evident in Fig. 9.3, which shows the depth of an isopycnal together with Θ on that isopycnal.

Actual data-based calculations of the neutral angle and neutral gradients, whether in numerical models or in data analysis, require numerical methods appropriate for discrete data. This turns out to be a surprisingly subtle problem. Groeskamp et al. (2019a) recently assessed the errors associated with common numerical methods (e.g., Griffies, 1998) and proposed a new, more accurate algorithm.

9.2.3 What then is isopycnal mixing?

> Isopycnal mixing is the action by a velocity field oriented in the neutral plane, together with small-scale dissipation, to homogenise tracer concentration along the neutral plane.

The above energetic constraints establish a preferred direction for water parcel motion in the ocean interior. Bringing this together with our generic definition of mixing, we arrive finally at a working definition of isopycnal mixing: *action by a velocity field* **u** *oriented in the neutral plane, together with molecular diffusion, to homogenise tracer concentration* c *along the neutral plane.*[6] In this context, we can interpret (9.4) as describing displacements occurring along the neutral plane, and K becomes an isopycnal diffusion coefficient. (In Section 9.4.1, we place this hand-waving assertion on more rigorous mathematical grounds.) Adiabatic flow in the ocean interior is, to-first order, two-dimensional and non-divergent on the neutral plane. So ideas from two-dimensional turbulence theory are frequently invoked in the study of isopycnal mixing.

Here again it is important to note that many of Taylor's assumptions from Section 9.2.1—isotropic turbulence, uniform background gradient—do not hold in the real ocean. The breakdown of these assumptions and the resulting complexity in characterising isopycnal mixing is a major theme of this chapter. Much research has aimed at generalising Taylor's concept and measuring K with floats and drifters (see Section 9.5.2.) For isopycnal mixing, the strength of the mixing depends on the magnitude and structure of the stirring process, with either submesoscale or microscale 3D turbulent processes mediating the route to molecular dissipation. In Section 9.3, we review the processes that drive isopycnal stirring.

---

6. We avoid the term 'along-isopycnal mixing', which appears sometimes in the literature, because it is redundant. The term 'isopycnal' already means 'occurring at constant density', so 'isopycnal mixing' means 'mixing occurring at constant density'.

9.2.4 Lateral mixing near boundaries

> Near the surface, mixing occurs in the horizontal plane, acquiring a diapycnal component. Many of the processes driving isopycnal mixing in the interior are the same that drive near-surface diapycnal horizontal mixing.

As noted above, proper isopycnal mixing occurs only in the quasi-conservative regime of the interior ocean. Near the boundaries, strong boundary-layer turbulent mixing and fluxes from air-sea interaction break the constraint on water parcels to stay on the neutral plane (see also Chapter 8 for more discussion submesoscale mixing in the surface boundary layer). Additionally, the presence of the ocean surface enforces a geometric constraint, inducing parcels to move predominantly along the geopotential (i.e., horizontal) plane. As a consequence of both these conditions, the mixing angle transitions from the neutral plane in the ocean interior to the horizontal plane at the surface, acquiring a diapycnal component (Treguier, 1999; Badin et al., 2011; Abernathey et al., 2013). Many of the processes that drive interior isopycnal mixing (e.g., mesoscale eddies) are the same that drive near-surface diapycnal horizontal mixing. This is evident in Fig. 9.3, which compares sea-surface temperature to potential temperature on an interior isopycnal; most of the same coherent structures are visible in both fields. This assumption proves convenient for studying isopycnal mixing, since surface data (e.g., drifters, satellite observations) are far more abundant than interior data.

9.3 Mechanisms of isopycnal stirring and dissipation

Before delving into more theoretical detail on isopycnal mixing, we next give a qualitative overview of the physical processes involved, including stirring mechanisms, which generate isopycnal tracer variance, and dissipation mechanisms, which dissipate this variance.

9.3.1 Mesoscale turbulence

> Baroclinic instability drives mesoscale turbulence in most of the ocean. Mesoscale turbulence is one of the primary stirring processes for isopycnal mixing.

The leading mechanism driving isopycnal stirring on basin scales is widely accepted to be mesoscale turbulence. Mesoscale turbulence describes the highly energetic, non-linear flows that occur on scales at or above the ocean baroclinic deformation radius (roughly 10–200 km). These flows are characterised by low Rossby number and generally obey quasigeostrophic dynamics; as a result, in the adiabatic interior, mesoscale velocities are oriented along the neutral plane. The dominant source of energy for these motions is baroclinic instability of the large-scale ocean circulation, which is nearly ubiquitous yet varies considerably in its local growth rate (Gill, 1973; Tulloch et al., 2009). This variability is likely a root cause of the inhomogeneity in isopycnal mixing (Vollmer and Eden, 2013). However, other energy sources may also produce mesoscale turbulence, including barotropic instability (particularly relevant in boundary currents, e.g., Qiu et al., 2009), wind forcing (Kersalé et al., 2011), and flow interactions with topography (MacFadyen and Hickey, 2010).

Regardless of the energy source, once mesoscale flows reach sufficient strength, the dynamics become dominated by non-linear interactions, resulting in the characteristic phenomenology of geostrophic turbulence: an inverse cascade of energy to horizontal scales larger than the baroclinic deformation radius and graver vertical scales; and the formation of mesoscale coherent structures, such as eddies, fronts, and filaments (e.g., McWilliams, 1990, more on this later).

Spatial variability in the intensity of mesoscale turbulence is correlated with the strength of isopycnal stirring, although the quantitative relationship between the two processes is an active topic of research (Vollmer and Eden, 2013; Klocker and Abernathey, 2014; Busecke and Abernathey, 2019; Bachman et al., 2020). Mesoscale turbulence is generally most intense in magnitude near the surface and decays with depth, corresponding to the ocean's inherent vertical modes of variability (Wunsch, 1977; LaCasce, 2016; Bachman et al., 2020). However, particularly in the Southern ocean, mesoscale eddy variability can span the whole water column, driving isopycnal mixing even in the abyssal ocean (Mashayek et al., 2017). Complicating matters is that the mesoscale eddy diffusivity does not always correspond directly to mesoscale eddy kinetic energy, but also depends on the so-called mixing length; we take this up in Section 9.4.2.

Some qualitative views of isopycnal mixing by mesoscale turbulence are shown in Fig. 9.1, the stirring of sea-surface temperature within a Gulf stream ring as observed by satellite, and in Fig. 9.3, which plots potential temperature on a potential-density surface from a high-resolution numerical model.

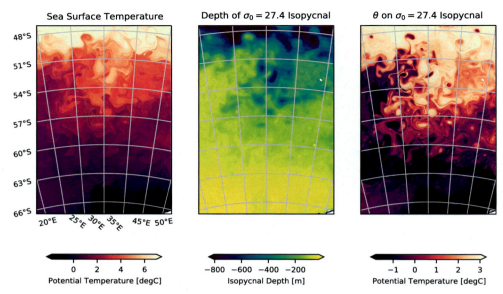

FIGURE 9.3 Qualitative illustration of the isopycnal mixing of temperature using outputs from the MITgcm LLC2160 simulation (Rocha et al., 2016). The region selected is a sector of the Southern ocean directly south of Africa. Left panel shows sea-surface temperature (SST). Middle panel shows the depth of the $\sigma_0 = 1027.4$ kg m^{-3} isopycnal surface, which is over 800 m deep near the northern boundary but shoals moving south. (Here we use σ_0 as an approximation to the neutral plane, due to the computational complexity of calculating γ_n for high-resolution model data.) Right panel shows the potential temperature field θ on that same isopycnal surface, revealing both the large-scale background meridional temperature gradient and the filamentary structures that are a signature of isopycnal stirring. Not shown is salinity; however, we know its gradient must be equal and opposite to the temperature gradient in such a way to maintain constant σ_0. Also note how many of the same stirring patterns visible in the subsurface temperature are also visible in SST, suggesting a link between isopycnal mixing and near-surface lateral mixing.

9.3.2 Transport by coherent structures

> All types of flow structures, including eddies, fronts, and tracer filaments, contribute to isopycnal mixing. The qualitative description of the relationships between coherent structures and measures of isopycnal mixing remains an area of active and controversial research.

Mesoscale turbulence refers to the whole gamut of velocity fluctuations in the flow. However, different types of fluctuations stir differently. Linear waves (e.g., Rossby waves) produce purely reversible deformations of tracer contours; this is not really stirring and doesn't lead to mixing under conservative conditions (Plumb, 1979; Nakamura, 2001). (Note, however, that linear wavelike motion combined with tracer sources / sinks can lead to irreversible transport, for example, when nutrients are consumed in the mixed layer; Freilich and Mahadevan, 2019.) Thus regions where mesoscale variability is more wavelike (predominantly low latitudes) are typically associated with relatively lower mesoscale isopycnal mixing rates compared to highly nonlinear regions with the same eddy kinetic energy (Klocker and Abernathey, 2014).

Turbulent geophysical flows commonly produce coherent structures, such as eddies, fronts, and filaments (Haller, 2015). In the ocean, the terms 'mesoscale turbulence' and 'mesoscale eddies' are often used interchangeably. However, qualitatively, it is clear that all types of coherent structures, not just eddies, contribute to isopycnal mixing. Tracer filaments, for example, are a manifestation of a stirring process that produces fine-scale tracer variance, en-route to molecular mixing. Such filaments are produced by the underlying 'Lagrangian Coherent Structures' of the velocity field (e.g., Mathur et al., 2007; Haller, 2015). The term Lagrangian coherent structures refers to special material surfaces within the fluid that maintain coherence (i.e., stay together) over a given time interval. The fingerprint of mesoscale coherent structures is clearly visible in both the SST satellite image (Fig. 9.1) and the simulated isopycnal temperature distribution shown in Fig. 9.3. Whereas these particular examples show filamentation by mesoscale eddies, higher resolution data and simulations that resolve submesoscale processes also reveal increasingly finer-scale filamentation as grid resolution increases (e.g., Molemaker et al., 2015), making clear that eddies, fronts, and filaments all exist at a variety of scales (see also Section 9.3.7). Quantitative relationships between coherent structure properties and more conventional measures of isopycnal mixing, such as eddy diffusivity, have yet to be firmly established. Based on Eulerian eddy tracking methods, some authors have argued that a large fraction of the transport in ocean mesoscale turbulence can be attributed to the action of a few discrete coherent

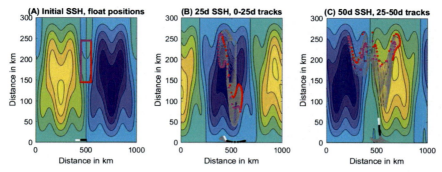

FIGURE 9.4 Qualitative illustration of chaotic advection following Pierrehumbert (1991). A freely propagating Rossby wave with one zonal cycle and half a meridional cycle in the domain (producing the yellow highs and blue lows), has superimposed on it a second much smaller-amplitude freely propagating wave with five zonal and one meridional cycles (resulting in the rich small-scale spatial structure). (A) Initial conditions, with a quadrilateral array of floats in the centre of the domain and a linear array along the southern boundary. (B) Situation after 25 days, with float tracks during the first 25 days shown by the grey lines. (C) Situation after 50 days with the float tracks from days 25–50 shown by the grey lines.

eddies (Dong et al., 2014; Zhang et al., 2014). Other recent studies, mostly using the methods of Lagrangian coherent structures, have argued the opposite; that mixing outside of coherent eddies accounts for most of the transport (Wang et al., 2015; Abernathey and Haller, 2018; Liu et al., 2019; Zhang et al., 2020). The sensitivity of the conclusion to the method of defining and tracking coherent structures highlights the importance of precise and careful definitions.

There is a rich literature on the regional generation and structure of coherent eddies. Certain current systems, such as the Agulhas retroflection and the Gulf stream, periodically shed highly coherent 'rings', which may persist for months or years, trapping fluid inside and transporting it over a long range. During the coherent phase, this type of transport is more akin to advection than to diffusion / mixing. However, such rings inevitably break apart or dissipate, mixing their trapped contents with the ambient water. Agulhas and Gulf-stream rings are surface intensified, but other regions produce sub-surface coherent eddies that are nearly invisible at the surface; prominent examples include 'Meddies,' salty boluses which originate at the Mediterranean outflow (Tychensky and Carton, 1998; Colin de Verdiere, 1992; Oliveira et al., 2010; Richardson and Tychensky, 1998); "Cuddies," emitted from the California undercurrent (Garfield et al., 1999; Collins et al., 2013; Pelland et al., 2013); and similar structures off the Peru-Chile undercurrent (Frenger et al., 2018). Incorporating the roles of these structures into a broader description of isopycnal mixing and parameterising their effects in coarse-resolution numerical models remains an open problem.

9.3.3 Chaotic advection

> Non-turbulent but periodically varying velocity fields can create the highly filamented tracer fields, upon which small-scale mixing can act to produce diffusion via a process known as chaotic advection. The tools of chaotic advection theory represent a powerful way to characterise anisotropy in ocean mixing.

Up to this point, we have implied that the velocity that does the stirring is fundamentally turbulent, i.e., unsteady in time and inherently chaotic. However, a large body of research has shown that even deterministic, smooth velocity fields with a bit of periodic temporal variability can produce chaotic trajectories, a phenomenon known as 'chaotic advection' (Aref, 1984; Ottino, 1989; Pierrehumbert, 1991). The mental model for chaotic advection is a baker folding pastry dough; the regular, repeated stretching, and folding of the dough leads to exponentially fine layers. Fig. 9.4 illustrates how chaotic advection can be produced by a system as simple as two propagating finite-amplitude Rossby waves of different wavelengths. An initial quadrilateral of floats (points in Fig. 9.4a) is stretched by the meridional flow associated with the long-mode Rossby wave. Superimposed on this stretching, a smaller-scale, less energetic mode introduces a twist at either end of the quadrilateral, producing the distribution at 25 days (Fig. 9.4b). Between 25 and 50 days (Fig. 9.4c) the gravest mode then moves the southern cluster back to the north, completing the "fold", and intermingling the purple floats that started on the northwest of the quadrilateral with the red ones that started on the southeast. Note however, that a linear array of floats that began at the southern edge of the domain shows no such folding (black and white floats are still in a line after 50 days) so that the mixing accomplished by this field is spatially non-uniform. The resulting spread of the floats is much greater in the zonal direction than in the meridional direction; the mixing is also anisotropic. Qualitatively similar behaviour can be found with eddies that 'pulse' and off, for example as a result of vortex shedding in an unsteady flow (Rypina et al., 2012),

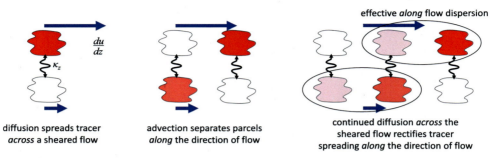

FIGURE 9.5 Schematic of shear dispersion showing diffusion *across* a sheared flow, leading to effective dispersion *along* the direction of flow. The example shown here is for a vertically sheared horizontal flow du/dz, combined with vertical diffusion κ_z, but the same mechanism can also apply to horizontally sheared flow, e.g., du/dy combined with cross-stream diffusion κ_y.

with some regions showing active stretching and folding, and thus acting as 'blenders', while others show relatively little exchange, potentially acting as barriers to mixing.

One way of distinguishing regions where chaotic advection produces 'blenders' as opposed to 'barriers' is that trajectories in the former tend to separate exponentially in time. A primary diagnostic of such behaviour is the finite-time Lyapunov exponent (FTLE), which characterises the rate at which a small patch of fluid is stretched. Waugh et al. (2012) computed FTLEs using geostrophic surface velocities derived from altimetry. They found a reasonable relationship between dispersion and FTLE when the timescales involved were comparable to the eddy turnover time. FTLEs are also widely used in the study of the Lagrangian coherent structures, described in the previous section.

Chaotic advection theory has been applied extensively to the study of isentropic mixing in the atmosphere (Pierrehumbert and Yang, 1993), a problem conceptually similar to ocean isopycnal mixing. Those authors argued that much of the eddy diffusive tracer transport in the atmosphere could be explained as chaotic advection by weakly non-linear propagating Rossby waves. The evidence for chaotic advection in isopycnal mixing processes in the ocean is less clear. In one study, Rypina et al. (2010) demonstrated that mixing associated with localised vortex dipoles found in a high-resolution model of the Philippine sea could be well described in terms of this theory. MacGilchrist et al. (2017) argued that chaotic advection plays a major role in ocean ventilation via isopycnal mixing. Bower (1991) showed that a similar mechanism associated with propagating meanders changing in amplitude could explain patterns of cross-stream mixing in the Gulf stream. Though it is unclear how important chaotic advection is relative to turbulent dispersion over most of the ocean, the tools of chaotic advection theory remain a powerful way to characterise anisotropy in ocean mixing.

9.3.4 Shear-driven mixing

> The combined action of large-scale shear and small-scale diffusion can lead to isopycnal mixing on a range of different scales.

The shearing and straining associated with the eddies, fronts, and filaments of geostrophic turbulence are in part responsible for the irreversible spreading of tracers in the ocean through the generation of small-scale tracer variance. However, the same shears and strains also contribute importantly to the dissipation of tracer variance at small scales (the right-hand-side term in (9.7)) in part through the mechanism of shear dispersion. The basic mechanism of shear dispersion is shown schematically in Fig. 9.5, where diffusion *across* a sheared flow leads to spreading of a tracer *along* the direction of flow. As tracer diffuses *across* the flow, fluid parcels are separated *along* the flow by the shear. Continued diffusion *across* the flow further spreads the tracer *across* the flow, including to regions up- and down-stream of previously diffused fluid parcels, thus effectively spreading tracer *along* the direction of flow.

The theoretical underpinnings of shear dispersion originated with Taylor (1953) in the context of steady capillary pipe flow, and were soon expanded to oceanographic contexts under a variety of flow conditions, including both spatially and temporally varying flows (e.g., Saffman, 1962; Bowden, 1965; Csanady, 1966; Carter and Okubo, 1965; Young et al., 1982). Though the principal ingredients for shear dispersion are relatively straightforward—a sheared flow and a diffusion acting across the direction of flow—the details of both the flow and the cross-stream diffusion can have important, and counter-intuitive, consequences. For example, if the diffusive timescale acting across the flow is small compared to the advective timescale of the flow, then the spreading of a tracer *along* the direction of flow will be *inversely* proportional to the *cross*-flow diffusion: effectively the cross-flow diffusion is so fast that the fluid does not have time to build up cross-stream tracer gradients even amid the shear. Alternatively, if the diffusive timescale is long compared to the advective timescale, the *along*-flow spreading instead becomes *proportional* to the *cross*-flow diffusion: here the shear is able to differentially advect fluid parcels large distances before diffusion spreads the tracers they carry across the flow. Finally, depending on whether the shear is time-dependent (e.g., oscillatory) or steady, the rate of along-flow spreading can range anywhere from linear to cubic in time.

Applying the concepts of shear dispersion to isopycnal mixing in the ocean interior, Young et al. (1982) showed that the interaction between diapycnal diffusivity and sinusoidally varying velocities $u(z, t)$, caused by internal waves will result in an effective isopycnal diffusivity that is proportional to the diapycnal diffusivity, and with tracers spreading linearly in time. Applying the same principles to laterally sheared currents, shear dispersion can also occur in mesoscale flows, such as boundary currents or jets, where both vertical and lateral shears combine with submesoscale diffusivities acting across the direction of flow. For example, examining 2D β-plane turbulence (β being the meridional gradient in the Coriolis parameter), Smith (2005) showed that shear-dispersion acts to enhance mixing *along* the β-plane jet axis, and derived scaling relations for the strength of the effective *along*-axis diffusivity due to shear dispersion, showing that it followed the *inverse* proportionality to *cross*-stream diffusivity. This mechanism is frequently invoked to explain anisotropy in isopycnal mixing, as measured by floats and tracers.

9.3.5 Additional submesoscale isopycnal mixing processes

> Stirring by internal-wave Stokes drift and vortical modes can produce isopycnal mixing of $O(1)$ m^2 s^{-1}.

As noted above, shear dispersion can occur on a variety of scales. Focusing for the moment on the submesoscale, internal wave shear dispersion, as described by Young et al. (1982), was long believed to be a significant contributor to isopycnal dispersion at the submesoscale. However, observations of tracer dispersion on scales of $O(1-10)$ km in a variety of environments have repeatedly exceeded theoretical predictions for classic internal-wave shear dispersion (Ledwell et al., 1998; Sundermeyer and Ledwell, 2001; Shcherbina et al., 2015). Kunze and Sundermeyer (2015) have argued that since turbulent mixing and internal-wave shear magnitude are intermittent, correlated, and log-normally distributed, the effect of internal-wave shear dispersion in the ocean may be larger than Young et al. (1982) originally anticipated. However, further work is required to confirm this result.

There are a number of other mechanisms that can also drive isopycnal mixing at these scales. A more comprehensive discussion of submesoscale processes and mixing is provided in Chapter 8, particularly in the context of boundary layer processes (i.e., near the sea surface, the bottom, or along lateral boundaries). As noted in Section 9.2.4, near boundaries, isopycnal mixing processes also tend to involve diapycnal processes. However, in the stratified interior, two additional purely isopycnal submesoscale mixing processes are of note besides internal-wave shear dispersion: random stirring by internal-wave Stokes drift and vortical mode stirring.

Stirring by internal-wave Stokes drift occurs as the result of non-linear, second-order corrections to the velocity field associated with internal waves. Early work by Herterich and Hasselmann (1982) studied similar dispersion effects for surface waves. Relative to internal waves, a more recent series of papers by Bühler and colleagues (Bühler and Holmes-Cerfon, 2009; Holmes-Cerfon et al., 2011; Bühler et al., 2013; Bühler and Guo, 2016) showed that, when combined with dissipation and non-linear interactions associated with the forward energy cascade of a typical open ocean internal-wave field, internal-wave Stokes drift can lead to submesoscale isopycnal stirring of order 0.25 m^2 s^{-1} (Bühler and Guo, 2016). This value is conspicuously small compared to isopycnal dispersion by mesoscale eddies, which readily exceed 1000 m^2 s^{-1}. However, at the submesoscale, where observed isopycnal dispersion rates are only observed to be only of order $O(1)$ m^2 s^{-1}, stirring by non-linear internal waves may be a significant contributor.

A third mechanism of submesoscale isopycnal dispersion is vortical mode stirring. In rotating stratified fluids, the vortical mode is the zero frequency solution associated with the geostrophically balanced portion of the flow, complementing the faster non-balanced internal-wave solutions. The vortical mode is often (though not exclusively) associated with the potential vorticity of the fluid (e.g., Kunze and Sanford, 2019). At large scales, this is represented by features, such as mesoscale eddies, whereas at submesoscales the vortical mode consists of smaller geostrophically balanced motions, including submesoscale eddies. Since the vortical mode is by definition geostrophically balanced, motions associated with it are isopycnal. A field of vortical modes, or vortical mode eddies, can thus drive isopycnal dispersion, very much in the same way that mesoscale eddies drive dispersion on larger scales. A number of studies have shown that vortical mode stirring is able to generate submesoscale diffusivities in the ocean pycnocline of order $O(1)$ m^2 s^{-1}, consistent with observed estimates (e.g., Polzin and Ferrari, 2004; Sundermeyer et al., 2005; Sundermeyer and Lelong, 2005). A significant challenge that remains, however, is that the submesoscale vortical mode is extremely difficult to observe due to its low energy levels compared to internal waves. Notable successes in this regard include detailed analysis of shear and strain spectra by Pinkel (2014), and the pioneering electromagnetic float measurements of Lien and Sanford (2019).

The generation of the submesoscale vortical mode is believed to stem from the relaxation of mixed regions generated by internal-wave breaking (Garrett and Munk, 1972), and also via eddy shedding of flow past topography (Kunze and Sanford, 1993). As noted previously, the mesoscale vortical mode associated with QG motions classically exhibits an inverse energy cascade, in contrast to internal waves, which characteristically cascade energy to smaller scales. Theoretical and numerical studies of non-linear interactions between internal waves and vortical mode have shown that the vortical mode can act as a catalyst, facilitating the forward cascade of energy within the internal-wave field (e.g., Riley and Lelong, 2000). Additionally, however, there is growing evidence that non-linear interactions between wave and vortical motions can also lead to forward cascades in both internal waves and the vortical mode. For example, Lelong et al. (2020) and Early et al. (2021) showed that inertial oscillations superimposed on an anticyclonic mesoscale eddy can lead to the generation of higher frequency waves without significantly affecting the eddy itself. Barkan et al. (2017) and Thomas and Arun (2020) further showed that steady and variable winds together produced a strong energy cascade in geostrophic turbulence, especially within cyclonic eddies.

Many modern ocean models now resolve mesoscales, but not submesoscales; therefore parameterisation of these submesoscale dispersion processes is becoming increasingly important. Additionally, both internal waves and the vortical mode may be important in climate models due to their potential role in modulating energy transfer from the mesoscale ultimately down to dissipative scales (e.g., Ferrari and Wunsch, 2009). Lastly, though the submesoscale vortical mode typically contains orders of magnitude less energy than its internal wave counterpart, it is many times more efficient at isopycnal dispersion, and hence still generates comparable dispersion to internal-wave-driven processes.

9.3.6 Diapycnal dissipation of isopycnal tracer variance

> Tracer variance generated by mesoscale mixing along isopycnals is not dissipated only in this 2D neutral plane. Other processes described in this book can dissipate tracer variance created by mesoscale stirring, resulting in a smoothing of the tracer filament variance in the diapycnal direction as well.

As noted above, mesoscale stirring processes drive the formation of fine-scale tracer gradients. From a turbulence perspective, this represents a forward cascade of tracer variance to small scales (e.g., Smith et al., 2002). Eventually, at the Batchelor scale, this variance can be dissipated by molecular diffusion. But en-route to this scale, many other mixing processes, described in other chapters of this book, may become involved. For instance, interior temperature and salinity variability generated by mesoscale isopycnal stirring may be dissipated by the diapycnal mixing associated with internal wave breaking.

Although isopycnal stirring itself is quasi-2D, occurring on the neutral plane, the dissipation process need not be. As described by (Smith and Ferrari, 2009), mesoscale straining increases tracer variance in both the horizontal and vertical directions, efficiently filamenting the tracer into streaks. Vertical shears associated with the same geostrophic flow simultaneously tilt the tracer filaments, maintaining a scale-independent aspect ratio of N/f, which when combined with the action of diapycnal diffusion, acts to smooth the filaments, removing small-scale variance and increasing their scale. The analogous phenomenon in the atmosphere was studied by Haynes and Anglade (1997).

The competing tendency of mesoscale strain to increase tracer variance by elongating and narrowing tracer filaments vs. that of submesoscale diffusive processes (e.g., shear dispersion or other diffusive processes acting between the mesoscale and 3D turbulent scale) to remove tracer variance and arrest the thinning of streaks was cleverly exploited by Ledwell et al. (1998). Specifically, by estimating the total mass of tracer observed 5–6 months after injection and the length of the observed tracer streaks, they estimated the mean mesoscale strain rate (i.e., the gradient of the velocity along the direction of flow) acting on the filaments, which, combined with estimates of the filament widths, allowed them to infer the effective submesoscale diffusivity acting on the tracer. Similar approaches taking into account the large-scale deformation of a tracer patch by shearing/straining and diffusive processes at scales smaller than the deformation field have been used by numerous studies to infer both submesoscale (e.g., Sundermeyer and Ledwell, 2001; Sundermeyer et al., 2020; Wenegrat et al., 2020) and mesoscale (Naveira Garabato et al., 2016; Zika et al., 2020) diffusion.

9.3.7 Frontogenesis and loss of balance

> Mesoscale turbulence generates fronts. As a front strengthens, it can become unstable and develops submesoscale eddies that then mix the fluid across the front.

In addition to generating tracer variance, the horizontal straining and filamentation associated with mesoscale turbulence also leads to the generation of mesoscale, and eventually submesoscale, fronts. As described by Hoskins and Bretherton (1972), large-scale geostrophic motions enhance both active and passive tracer gradients, eventually leading to loss of balance and ageostrophic motions that then further enhance mesoscale fronts. As a front strengthens, it eventually becomes baroclinically unstable, developing meanders and eddies that then mix fluid across the front, countering the isopycnal steepening by the strain field that produces frontogenesis. This eddy generation is itself another mechanism of removing small-scale isopycnal variance, effectively arresting the filamentation process caused by the mesoscale field. The exact scale at which tracer filamentation and frontogenesis are arrested depends on the details of the instabilities. For example, Spall (1997) showed that under certain conditions, instabilities along the front may be suppressed by the mesoscale strain; however, McWilliams and Molemaker (2011) showed that as the cross-frontal scale decreases, the growth rate of the most unstable mode increases, hastening the arrest. A better understanding of the underlying dynamics of fronts and their associated instabilities is an ongoing area of research, as are a myriad of other mixing mechanisms that are active at the submesoscale (see Chapter 8).

9.4 Frameworks for thinking about isopycnal mixing

Because isopycnal mixing is a multi-faceted problem, several different theoretical and analytical frameworks must come together to provide a complete picture. These frameworks draw from geophysical fluid dynamics, turbulence theory, and numerical modelling. In this section, we review some of these frameworks, providing context to the subsequent discussions of observations and models.

9.4.1 Reynolds-averaged tracer equations

> The thickness-weighted-average framework provides a solid theoretical foundation for analysing isopycnal mixing.

A simple, practical, and widespread way of thinking about isopycnal mixing is by considering the Reynolds-averaged tracer conservation equation. Reynolds averaging aims to analyse the bulk statistical effects of eddies in a turbulent flow without describing the details or scales of the turbulent fluctuations. Reynolds-averaged equations also form the basis for the numerical simulation of the ocean (see Section 9.4.4).

We already saw a simple Reynolds-averaged model in Section 9.2.1, which described the relationship between eddy diffusion and microscale dissipation of variance; here we specialise this model to isopycnal mixing. When considering Reynolds-averaged equations, many choices exist regarding the nature of the averaging operator itself. We assert that the most useful framework for reasoning about isopycnal mixing is the *thickness-weighted average* (TWA), as detailed in Young (2012). In this framework, the original governing equations are averaged at levels of constant buoyancy (a.k.a. isopycnal averaging) and weighted by a factor inversely proportional to the stratification, i.e., the continuous version of isopycnal layer 'thickness'. By averaging over many turbulent fluctuations, we capture the statistical effects of eddies on the mean state.

The TWA equation for a passive scalar c is (Young, 2012, Eq. 89)

$$\frac{D^{\#}\hat{c}}{Dt} = -\nabla \cdot \mathbf{J}^c + \hat{\gamma}. \tag{9.10}$$

Young's notation is complex yet necessary to address the subtleties of thickness-weighted averaging. The ˆ symbol represents a thickness-weighted average,[7] and γ is any non-advective source / sink of the tracer (e.g., molecular diffusion.) The

---

[7]. The TWA operator is defined as $\hat{\phi} = \overline{\phi\sigma}/\overline{\sigma}$, where ϕ is any quantity; $\sigma = (\partial b/\partial z)^{-1}$ is a continuous analog of isopycnal layer 'thickness' (b is buoyancy, proportional to potential density), and the overbar is a standard Reynolds average operator. In this context, σ is *not to be confused with potential density!* (The notational conflict exists in the literature and cannot be avoided here.) Crucially, the averaging is performed with buoyancy (equivalently, potential density), rather than z, as the independent vertical coordinate.

operator $D^\#/Dt$ is the material derivative following the thickness-weighted residual flow $\mathbf{u}^\#$. Although this term involves eddy variability, its transport is advective in nature, and therefore it is not considered part of isopycnal mixing; in numerical models, $\mathbf{u}^\#$ comprises both the resolved Eulerian-mean velocity and the eddy bolus transport parameterised by the scheme of Gent and McWilliams (1996). The eddy tracer flux is contained in the vector

$$\mathbf{J}^c = (\widehat{u''c''}, \widehat{v''c''}, \widehat{\varpi''c''}), \tag{9.11}$$

where the double-prime represents an anomaly from the TWA quantity; the first two components of the vector are in the neutral direction, whereas the third component is in the vertical direction. (See Young (2012) for detailed explanations of the TWA basis vectors.) The quantity ϖ represents any non-advective sources / sinks of buoyancy; when $\varpi = 0$, the flow is often dubbed "adiabatic"; under these conditions, the TWA equations show that the tracer flux $\mathbf{J}^c$ indeed *must* lie along the neutral plane.

To reach a theoretical justification for the concept of isopycnal diffusion, we examine the TWA tracer variance equation (Young, 2012, Eq. 90), which can be derived in a manner similar to Section 9.2.1. We allow the tracer to be dissipated by a diffusive term $\kappa \nabla^2 c$ (representing molecular diffusion), assume a steady state, and integrate over the entire domain $\mathcal{V}$ to obtain

$$\int_\mathcal{V} \mathbf{J}^c \cdot \nabla \hat{c} \, dV = -\kappa \int_\mathcal{V} \widehat{|\nabla c''|^2} dV, \tag{9.12}$$

an isopycnal version of the classic relation of Osborn and Cox (1972), and formally very similar to (9.7). The quantity on the RHS is the dissipation of tracer variance, and is negative definite. The equation shows that, on average, the tracer flux $\mathbf{J}^c$ must be directed *down* the mean gradient of $\hat{c}$. Under adiabatic conditions ($\varpi = 0$), this constraint can be satisfied by isotropic isopycnal diffusion with diffusion coefficient K, such that

$$(\widehat{u''c''}, \widehat{v''c''}) = -K \left(\frac{\partial \hat{c}}{\partial \tilde{x}}, \frac{\partial \hat{x}}{\partial \tilde{y}} \right), \tag{9.13}$$

where the fluxes on the LHS are understood to be in the neutral plane and the derivatives on the RHS are taken at constant buoyancy (i.e., the isopycnal tracer gradient). Note that if K is allowed to vary in space or become an anisotropic tensor (see Section 9.4.4) as expected, there is no requirement that it be always positive; only that (9.12) is satisfied in an integral sense. Furthermore, when $\varpi \neq 0$ (non-adiabatic conditions, such as near the surface), the vertical tracer flux term $\widehat{\varpi''c''}$ comes into play, and the notion of two-dimensional eddy mixing on the neutral plane is no longer valid. Finally, we note that the presence of non-diffusive sources/sinks of tracer may further break the expectation of down-isopycnal-gradient fluxes. This is particularly relevant for biogeochemical tracers, which undergo significant biological production/consumption.

9.4.2 Mixing-length theory

> Mixing-length theory expresses eddy diffusivity as the product of a mixing-length scale, an eddy velocity, and a mixing efficiency parameter. This framework is useful for interpreting observations and building theoretical models of isopycnal mixing.

From the Reynolds-average perspective, the primary challenge is to specify K. Prandtl (1925) proposed that turbulent mixing rates could be estimated as

$$K = \alpha V_e L_{mix}, \tag{9.14}$$

where V_e is a characteristic eddy velocity at the energy-containing scale; L_{mix} is the 'mixing length', i.e., the distance over which water parcels are displaced before mixing with their environment during the stirring process; and α is a dimensionless 'mixing efficiency' parameter.[8] An alternative formulation is in terms of timescales. If a fluid parcel maintains its properties

---

8. The mixing efficiency in diapycnal mixing is commonly defined as the ratio of the change in background potential energy to the total energy expended in the mixing process (Gregg et al., 2018). This is not the same thing as α in (9.14). For one thing, discussed in Section 9.2.2, isopycnal mixing is energetically neutral, producing no change in potential energy. Instead α quantifies the effectiveness of a velocity field of given kinetic energy V_e at producing an eddy tracer flux. It is related to the persistence of correlations between the velocity anomalies and the tracer anomalies (Taylor, 1915). Values reported in the literature range from 0.1 to 0.5 (Klocker and Abernathey, 2014). This spread is somewhat related to the ambiguity of precisely defining L_{mix}, but there is no doubt further research on mixing efficiency, specifically in isopycnal mixing, would be valuable.

for a time interval T_{mix} before mixing with its environment, that implies $L_{mix} = V_e T_{mix}$, and thus

$$K = \alpha V_e^2 T_{mix}. \tag{9.15}$$

Plugging in representative values for the ocean mesoscale of $V_e = 0.1$ m s^{-1}, $L_{mix} = 100$ km, and $\alpha = 0.1$, one obtains the classical value of $K = 1000$ m^2 s^{-1}. This is indeed the canonical constant value of isopycnal diffusivity used in many ocean models (see Section 9.4.4).

Beyond such back-of-the-envelope estimates, mixing-length theory provides a framework for theory-based closures for K, where the goal is to estimate isopycnal mixing purely from the large-scale environment or forcing. Such theories can be used to explain and predict inhomogeneity in isopycnal mixing. Using this framework, a large body of theoretical work has sought universal scaling laws for various flavours of turbulent diffusion in idealised geophysical flows. Most studies in this vein have used two-dimensional or quasigeostrophic models. V_e can typically be estimated based on energetic arguments, assuming a balance between the injection of energy by large-scale forcing and its dissipation by drag. L_{mix} instead, generally requires some knowledge of the 'halting scale' of the inverse energy cascade (more on this in the next subsection).

In homogeneous two-dimensional f-plane turbulence, both V_e and L_{mix} ultimately depend on the strength of bottom drag, which regulates both the eddy kinetic energy and the maximum size of eddies; (Smith et al., 2002; Grianik et al., 2004). However, the presence of the β-effect fundamentally changes the picture. For strong values of β, the energy from the inverse cascade is redirected into zonal jets, limiting both the size and kinetic energy of eddies (Vallis and Maltrud, 1993; Smith et al., 2002; Srinivasan and Young, 2014). A particularly comprehensive perspective on eddy diffusivity in two-dimensional flows was recently published by Kong and Jansen (2017), who provided a unified theory spanning multiple flow regimes.

For baroclinic flows, the net energy source for the turbulence is not fixed, as in the models above, but instead is part of the solution; the conversion from available potential energy to eddy kinetic energy by baroclinic instability itself depends on the strength of eddy mixing. Held and Larichev (1996) proposed a widely used scaling theory for diffusivity in baroclinically unstable flow on a β-plane. Other scaling theories for baroclinic energy conversion have been posed based on linear QG stability analysis (e.g., Visbeck et al., 1997; Smith and Marshall, 2009; Vollmer and Eden, 2013).

An important kinematic insight into the role of wavelike behaviour on turbulent mixing was introduced by Ferrari and Nikurashin (2010, henceforth FN10). Partly in an attempt to explain the findings of Marshall et al. (2006) of unexpectedly low diffusivity in the core of the ACC, FN10 developed a weakly non-linear stochastic model for eddy diffusivity in the presence of a potential vorticity gradient. Their model showed how eddy propagation relative to the background mean flow can significantly suppress L_{mix}. Subsequent studies (e.g., Klocker et al., 2012a,b) explored the implications of this concept for Lagrangian diffusivities, arguing that suppressed mixing length was associated with a negative lobe in the Lagrangian autocorrelation function that was ignored by prior authors (see Section 9.5.2 for more detail). Many later studies have used this insight to interpret diagnostics of isopycnal mixing, explain patterns of spatial variability, and develop predictive theories. Kong and Jansen (2017) later connected this kinematic approach to the earlier f-plane and β-plane scaling theories.

Mixing length theory can also be applied in an observational context by using spatially variable observations to estimate V_e and L_{mix}, as pioneered by Armi and Stommel (1983). V_e is readily estimated from the velocity variance observed by current metres, repeat hydrography, or satellite altimeters. The mixing length can be estimated from the variance of a quasi-conservative tracer c (e.g., temperature, salinity) as

$$L_{mix} = \sqrt{\langle c'^2 \rangle}/|\nabla \langle c \rangle|, \tag{9.16}$$

where the numerator is the standard deviation of repeated concentration measurements and the denominator is the large-scale background tracer gradient. The method assumes that the variance is generated by stirring of the tracer gradient along isopycnals. Applications of this approach are discussed in Section 9.5.1.3.

9.4.3 Spectral-space view of turbulence and mixing

> Energy and variance cascades in wavenumber space provide an important framework for understanding multi-scale interactions in turbulent flows, helping explain the formation of coherent structures, the small-scale dissipation of variance, and the statistics of float dispersion.

The phenomena and dynamics described in Section 9.3 are all facets of the same governing equations that describe stratified geostrophic flow. Likewise, the Reynolds-average approach averages over a wide spectrum of different modes of variability with a broad brush. Different processes, such as baroclinic instability, which occurs predominantly at the scale of the baroclinic deformation radius, and transport by coherent structures, can be viewed from a spectral perspective within a continuum of wavenumber space.[9] At the core of turbulence theory is the concept of energy transfer among different scales in the ocean via non-linear (a.k.a. triad) interactions. In the classic turbulence theory of Kolmogorov (1941, 1962) and Kraichnan (1967), the conservation of energy and enstrophy (the square of vorticity) dictate a preferred direction of energy transfer as a function of wavenumber (a.k.a. a turbulent 'cascade'). For three-dimensional turbulence, conservation of energy requires a downscale cascade from the forcing scale to the molecular scale, where energy dissipation can occur. For geostrophic turbulence (most relevant to isopycnal mixing), the combined constraints of energy and enstrophy conservation dictate an *upscale* cascade of energy and a downscale cascade of tracer variance and enstrophy. The upscale energy cascade is associated with the formation of coherent structures, such as eddies, whereas the downscale tracer variance cascade is associated with microscale variance dissipation. Thus both the stirring process and the dissipation process of mesoscale isopycnal mixing are inherent in the spectral view.

Spectral turbulence theory makes specific predictions about the shape of the wavenumber power spectra of conserved quantities which are germane to isopycnal mixing. For homogeneous isotropic turbulence, the steepness of the energy spectrum in wavenumber space determines whether dispersion of tracers is governed locally as a function of scale (i.e., by scales of motion similar to the scale of the tracer or particles in question), or whether it is dominated by energy at larger scales (Bennett, 1984). Specifically, if the kinetic energy wavenumber spectrum of the turbulence has a slope less than k^{-3}, then tracer dispersion at any given scale within the spectrum is controlled locally in wavenumber space by turbulence at that scale. Alternatively, if the energy spectrum of the turbulence has a slope greater than k^{-3}, then large-scale rather than small-scale eddies dominate the turbulence, and hence the rate of tracer dispersion, and the dispersion is non-local.

The question of local vs. non-local dispersion is important to understanding submesoscale isopycnal mixing processes. The dominant energy-containing scale in the ocean is the mesoscale, i.e., typically slightly larger scale than the first baroclinic deformation radius. It is thus mesoscale turbulence that predominantly stirs tracers in the ocean, from active tracers (such as temperature and potential vorticity) to passive tracers, such as chlorophyll-a and other biogeochemical constituents (see Figs. 9.1 and 9.3). An important implication of this is that, if we release a tracer or cluster of Lagrangian particles at a separation distance much less than the dominant energy-containing scale, the tracer patch or particle cluster will initially follow approximately the same streamlines, with only minor differences in direction and speed. As the tracer or particles further spread, however, they increasingly span more of the large-scale turbulent flow field, leading to more rapid separation. This type of dispersion process, in which the squared separation of fluid parcels increases faster than time to the first power is termed 'anomalous' dispersion. Anomalous dispersion continues until the velocities experienced by tracer parcels or Lagrangian particles are no longer correlated with one another. At that point the diffusion is governed by the dominant energy-containing motions of the mesoscale eddy field (e.g. Garrett, 1983; Ledwell et al., 1998; Sundermeyer and Price, 1998; Balwada et al., 2016a, 2021).

Combining the ideas of mixing-length theory with the idea that as tracers and particles span larger scales they encounter larger eddy variance scales, a natural outcome is that the effective diffusivity enacted by the turbulence increases with scale. This was first noted empirically by Richardson (1926) in the context of atmospheric turbulence, and later by Richardson and Stommel (1948) in the context of the ocean on scales of 0.2–100 m. In both cases the observed dispersion grew according to a 4/3 power law, i.e., $K \propto L^{4/3}$, consistent with predictions from similarity theories by Kolmogorov (1941) and Batchelor (1950). Similar scale-dependent diffusion was shown by Okubo (1971) for scales ranging from 100 m up to 100 km, but with a slightly modified power law, $K \propto L^{1.1}$. (Notable in the case of Okubo's results, however, is that it is unlikely that the same mechanisms are responsible for the observed dispersion over the entire range of scales (roughly 3 decades) spanned by his analysis.) The key point is that, whether by homogeneous isotropic turbulence or by a variety of mechanisms ranging from internal-wave shear dispersion to submesoscale frontal instability to mesoscale stirring, the observational result being that dispersion in the ocean increases with scale appears to be largely robust, at least up to the mesoscale. Once the separation between parcels is larger than the largest eddies, the 'asymptotic regime', diffusivity should no longer be scale dependent. Though scale-dependence in diffusion was once studied primarily for academic reasons, it is increasingly a practical concern for high-resolution ocean models, which must model sub-gridscale fluxes on scales of 100 m to 10 km. We review some contemporary observational evidence for scale-dependent relative dispersion in Section 9.5.2.2

---

9. "Wavenumber" refers to the coordinate produced by taking a Fourier transform of the governing equations over the spatial dimensions. Wavenumber has units of m^{-1}. High wavenumbers are equivalent to small spatial scales.

The idea of scale-dependent dispersion in the ocean necessarily breaks down when scales become too large, and this breakdown can induce anisotropy in the isopycnal mixing process. As noted in Section 9.4.2, dynamical constraints can halt the inverse cascade of geostrophic turbulence before energy reaches the basin scale. One is bottom drag acting on the mesoscale flow field, which is required to remove energy accumulated by the QG inverse cascade at those large scales. The Rhines arrest scale, given by $L_\beta \sim \sqrt{U/\beta}$, where U is a typical mesoscale eddy velocity, and β is the linear gradient of planetary vorticity (Rhines, 1975) is another important factor. The Rhines scale arises, because parcels exchanging across meridional distances L_β must change their planetary vorticity by $\beta * L_\beta$. For the total vorticity to be conserved, U/L_β must be of comparable size. As a result, planetary β leads to the formation of zonal jets, limiting effective mixing lengthscales in the meridional vs. zonal direction. Smith (2005) showed that such jets can also enhance zonal dispersion over meridional dispersion via the process of shear dispersion (Note the same process can also lead to local anisotropy within non-zonal jets or currents). Additionally, on a β-plane, coherent structures propagate predominantly westward, with minimal meridional translation, again, enhancing zonal exchange over meridional exchanges. We have already seen an example of this in the context of chaotic advection in Fig. 9.4. In the context of mesoscale dispersion, the presence of β thus manifests as an anisotropy of the mesoscale dispersion coefficient, with zonal dispersion typically found to exceed meridional dispersion by a factor of two or more (see also Section 9.5.1).

9.4.4 Isopycnal mixing in numerical models

- Global and ocean climate models (resolution of 1° or less) do not resolve mesoscale turbulence and rely on parameterisations of isopycnal mixing.
- Eddy transport is typically represented via a diffusion tensor in the context of non-TWA Reynolds averaging at constant height.
- The diffusion tensor contains an antisymmetric (Gent McWilliams) and a symmetric (Redi) component. The latter is associated with isopycnal mixing.
- Implementing these equations in numerical ocean models is numerically and conceptually challenging, with many open questions.

The evolution of the physical, chemical, and biological properties of ocean water can be represented using numerical models of the ocean circulation. Some high-resolution global models are capable of partially or fully resolving mesoscale turbulence; such models can thus be used as tools for studying isopycnal mixing (see Section 9.6.2). However, typical global and ocean climate models, with resolutions of 1° and coarser, do not resolve these processes and must parameterise isopycnal mixing. The need to parameterise isopycnal mixing is both an important practical concern and a useful theoretical framing: if we can truly understand isopycnal mixing, parameterising it should be feasible. A central complication here is that isopycnal mixing is closely related to the process of eddy-induced advection; both are part of the broader problem of *mesoscale eddy parameterisation* (which also includes sub-grid eddy momentum fluxes); however, they have distinct physics and impacts.

One starting point for understanding how ocean models represent sub-grid transport is the Reynolds-averaged TWA tracer equation (Eq. 89, Young, 2012, also reviewed in Section 9.4.1). Under adiabatic conditions, the TWA transport of a tracer in the isopycnal x-direction is given by

$$\widehat{uc} = \frac{\overline{uch}}{\overline{h}} = \hat{u}\hat{c} + \widehat{u''c''}, \tag{9.17}$$

$$= \left(\overline{u} + \frac{\overline{u'h'}}{\overline{h}}\right)\hat{c} + \widehat{u''c''}, \tag{9.18}$$

where the overbar denotes a simple time or ensemble average. Whereas Young advocates making $\hat{u}$, the TWA residual velocity, the prognostic variable for ocean models, in practice nearly all ocean models solve for $\overline{u}$, the unweighted averaged quantity. In this case, two separate transport terms must be parameterised: $\widehat{u''c''}$, the isopycnal Reynolds tracer transport (discussed in Section 9.4.1); and $\overline{u'h'}/\overline{h}$, the eddy "bolus transport", which appears as an additional advective velocity next to $\overline{u}$.

The bolus transport is associated with conversion of available potential energy to eddy kinetic energy in baroclinic instability. It is the target of the parameterisation of Gent and McWilliams (1990) (see also Gent and McWilliams, 1996;

Gent, 2016). A simple form of their closure is to write

$$\overline{u'h'} = -K_{GM}\frac{\partial \overline{h}}{\partial x}, \qquad (9.19)$$

a model sometimes referred to as 'thickness diffusion'. K_{GM} has units of diffusivity and is often called the Gent–McWilliams or GM coefficient. Parameterisation of the bolus transport is crucial for ocean models to accurately reproduce the ocean mean state. However, as (9.18) and (9.6) show, it does not result in the production of tracer variance, and therefore is not part of mixing. Isopycnal mixing pertains to the final term in (9.18). A diffusive parameterisation for isopycnal mixing is essentially just (9.13).

Building on the insights of Solomon (1971), isopycnal mixing in coarse-resolution models was first implemented by Redi (1982), who framed it instead in terms of a diffusivity tensor. The diffusion-tensor framework (Griffies, 1998), which is primarily deployed in the context of standard (non-TWA) Reynolds averaging at constant height, provides an alternative route to understanding how isopycnal mixing is parameterised in ocean models. A diffusion tensor is a form of turbulent closure, which relates the eddy tracer flux vector to the mean gradients in all directions, i.e.,

$$\overline{\mathbf{u}'c'} = -\mathbf{K}\nabla \overline{c}, \qquad (9.20)$$

where $\mathbf{K}$ is a rank-2 tensor (3 × 3 in three dimensions). It is convenient to decompose $\mathbf{K}$ into the sum of a symmetric component $\mathbf{S} = (\mathbf{K} + \mathbf{K}^T)/2$ and an antisymmetric component $\mathbf{A} = \mathbf{K} - \mathbf{S}$. As shown by Griffies (1998), the impacts of $\mathbf{A}$ are fundamentally advective in nature and does not impact the tracer variance budget: the fact that the diffusion tensor also parameterises advection can be a persistent source of confusion. The mixing tensor approach is not only useful for implementing mixing in numerical models; as described in Section 9.6.2.2, it can also be turned into a tool for diagnosing mixing from high-resolution numerical simulations.

In the standard Gent and McWilliams (1990) parameterisation, this component is expressed as

$$\mathbf{A} = K_{GM}\begin{bmatrix} 0 & 0 & -S_x \\ 0 & 0 & -S_y \\ S_x & S_y & 0 \end{bmatrix}, \qquad (9.21)$$

where S_x and S_y are the isopycnal slopes in the x and y directions. (Note we are using the 'small-angle' approximation, appropriate to isopycnal slopes $\ll 1$, viz. Redi, 1982; Griffies et al., 1998.)

The symmetric component $\mathbf{S}$ contains the isopycnal mixing. Redi (1982) was the first to implement the insight of Solomon (1971) that isopycnal diffusion could be accomplished by combining a horizontal diffusion operator with a rotation operator into the neutral plane. Under the small-angle approximation, the combined operator takes the form

$$\mathbf{S} = K_{Redi}\begin{bmatrix} 1 & 0 & S_x \\ 0 & 1 & S_y \\ S_x & S_y & S^2 \end{bmatrix}, \qquad (9.22)$$

where $S^2 = S_x^2 + S_y^2$ and K_{Redi} is an isotropic isopycnal diffusivity. Since it diffuses tracers down their isopycnal gradient and reduces tracer variance, K_{Redi} is presumably equivalent to the generic isopycnal diffusivity K discussed throughout the rest of this paper.

Implementing isopycnal mixing in numerical models poses a number of challenges.

- *Defining the neutral plane and rotating this plane relative to the coordinate system.* In level/pressure coordinate models (which make up the bulk of climate models) this is commonly implemented via the symmetric mixing tensor of (9.22). The isopycnal slopes S_x and S_y are computed relative to the horizontal plane of the model coordinate system (Griffies et al. (1998)). In the special case where the coordinate system is isopycnal, $S = 0$ and the diffusion is just horizontal diffusion along the isopycnal coordinates. Note that even in nominally 'isopycnal' coordinate models where the vertical coordinate is the density referenced to a specific depth, neutral surfaces will generally not follow this coordinate exactly, so that for strict accuracy rotation may still be required.

Despite the simplicity of (9.22), actually implementing it in models requires dealing with some subtle numerical issues. For example, if density is only a function of potential temperature or salinity, then by (9.8), the slopes of density and temperature/salinity surfaces are identical, and the flux of both salinity and temperature should be zero. Ensuring that this is the case requires carefully computing the density slope by linearising the density equation. If this is not done, the model can become numerically unstable. Additionally, the slope S is a ratio of horizontal and vertical gradients, which are not defined across the same face of a grid cell, and so when the density field has high local curvature care must be taken to properly capture the effect of grid-scale density gradients on the local slope. These and other issues are discussed in detail in Griffies et al. (1998).

- *Defining what happens as the system enters the mixed layer, where $S \to \infty$.* As noted in Section 9.2.4, there must be a transition between the isopycnal diffusion in the interior and lateral diffusion in the surface diabatic layer. Various schemes have been proposed for this transition (Gerdes et al., 1991; Danabasoglu and McWilliams, 1995; Ferrari et al., 2010), but most of the analysis of their impact has focused on their effect of limiting K_{GM} (Gnanadesikan et al., 2007; Farneti and Gent, 2011). The physics and impact of this sort of lateral diapycnal mixing remain relatively understudied (Badin et al., 2011).

- *The mixing coefficient K must be defined.* As discussed in Gnanadesikan et al. (2015b) models used in the Fifth coupled model intercomparison project (CMIP5) varied greatly in how they do this. Some models used a constant coefficient, with values varying between 500 m^2 s^{-1} (the HadGEM model reported in Johns et al., 2006) to 2000 m^2 s^{-1} (the INGV CMCC model in Fogli et al., 2009). Others used a spatially-varying K that scaled with the spacing of the grid boxes (Marsland et al., 2003); and still others use methods developed for estimating K_{GM}. This usually involves letting the timescale be given by the growth rate of baroclinic instability (Visbeck et al., 1997) so that $1/T \sim f^2 * S_\rho/N$, where f is the Coriolis parameter and S_ρ may be averaged over some part of the water column. The mixing lengthscale is then either the width of the baroclinic zone (GFDL ESM2G, Dunne et al., 2012) or some combination of the Rossby radius of deformation and Rhines scale (Eden and Greatbatch, 2008a, implemented in Bentsen et al., 2013).

- *Scale dependence and mixing in the 'grey zone'.* Modelling isopycnal mixing as a subgridscale diffusion implicitly assumes that there is a scale separation between the eddies doing the mixing and the flow field. This assumption is violated even for models with a resolution of 1° in the equatorial zone (Hallberg, 2013), where tropical instability waves have scales of hundreds of km. As one moves to 1/2° or 1/4°, eddies become 'permitted' over more of the domain and more of the simulated ocean falls into a 'grey zone', where some fraction of the processes involved in isopycnal mixing are directly simulated. 'Scale-aware' parameterisations, which can transition smoothly from regimes where the eddies are too small to be resolved to regimes where the main energy-containing scale is well resolved, are currently being developed (Pearson et al., 2017; Zanna et al., 2017), but these parameterisations currently focus primarily on momentum transport, rather than on isopycnal mixing. Additionally, recent work (Nummelin et al., 2021) has argued that as one moves to coarser and coarser scales, the diffusion coefficient increases in part due to having to capture more shear dispersion.

A primary point of interaction between the modelling and observational communities involves the specification of K, since K can presumably be measured. Below we review some of the many estimates of K from different observational techniques. Though detailed observational knowledge of the spatial patterns of K provides a useful reference point for models, simply prescribing a fixed coefficient is unsatisfactory. The "holy grail" is a closure theory for both K_{GM} and K, validated by observations, which is numerically stable and applicable to the full range of scales, regions, and climate states.

9.5 Observational estimates of isopycnal mixing

Given the above theoretical underpinnings of the physical processes governing isopycnal mixing, the challenge remains of how to measure and parameterise the rates of such mixing as a function of time, space, and scale throughout the world oceans. From an observational perspective, two ways of doing this are with either natural or anthropogenic tracers, or alternatively, using water-following drifters or floats to mimic Lagrangian particles in the ocean. In broad brushstroke, both tracers and floats are Lagrangian in the sense that they naturally integrate in both space and time following fluid parcels. They differ, however, in that tracers are continuous, allowing concentration and gradients to be measured at any/all scales, whereas floats are discrete, requiring large numbers of drifters/floats to represent 'concentration'. An advantage of drifters/floats is that in practice they represent unique particles with unique IDs, such that if two drifters trade places, this is readily detected, whereas if two parcels of fluid with the same tracer concentration trade places, it is not. There are of course many other differences between the two approaches: suffice it to say, both are useful tools with advantages as well as shortcomings.

9.5.1 Tracer-based methods

- Different methods exist to estimate isopycnal mixing from natural and anthropogenic tracers.
- Natural tracers (temperature, salinity, potential vorticity, biogeochemical tracers) are most often used in the context of inverse methods or mixing-length theory.
- Anthropogenic tracers, especially deliberate tracer release experiments, usually have sources that are more localised in space and/or time and can be used to estimate dispersion via the spreading of a tracer patch.

9.5.1.1 Natural vs. anthropogenic tracers

Tracer-based methods of diagnosing rates of isopycnal dispersion in the ocean fall broadly into two categories: natural tracers and anthropogenic tracers. For both, the ultimate goal from an observational perspective is to determine what causes tracers to spread irreversibly along isopycnals. By inferring mixing rates and the processes that control them, these processes can then be parameterised in ocean models to forecast and/or hindcast the effects of mixing on global circulation and climate.

The use of natural tracers to infer global circulation and mixing rates is rooted in traditional watermass analysis (e.g., Worthington, 1981; Broecker and Peng, 1982). Natural tracers that fill the ocean basins have sources and sinks that vary spatially, at once both making them useful for diagnosing both mean transport and diffusivity, but also requiring these sources and sinks to be sufficiently constrained to be useful. Examples of natural tracers include temperature, salt, and potential vorticity, as well as a variety of biogeochemical tracers that are modified through air-sea exchange, or internally in the ocean (e.g., Broecker and Peng, 1982). Often simply examining sections of tracers reveals much about isopycnal mixing. For example, Bower et al. (1985) analysed sections of oxygen and other tracers across the Gulf stream and concluded that the upper part of the Gulf stream front served as a barrier to isopycnal mixing but that, at greater depths, there was very active cross-frontal exchange. These insights were borne out by later float-based observations (Bower and Lozier, 1994).

A common approach for inferring isopycnal mixing rates from ocean tracers is to use inverse methods to infer the combined effects of mean advection, diapycnal, and isopycnal mixing from hydrographic and/or other tracer distribution information, with various assumed dynamical constraints (e.g., Wunsch, 1987; Killworth, 1986; Zika et al., 2010; Groeskamp et al., 2019b). The inverse problem is simplified if the ocean is presumed to be in steady state, since in that case even relatively sparse observations (e.g., hydrographic cruises sometimes separated by many years) can be combined and used to constrain the inversion.

Anthropogenic ocean tracers typically have sources that are more localised in space and/or time; examples include ^{14}C, ^{3}H, and ^{90}Sr produced during nuclear testing of the mid-20th century (Broecker and Peng, 1982), chlorofluorocarbons (CFC's), and sulfur hexafluoride (SF$_6$) used industrially from the mid-to-late 20th century (Fine, 2011), as well as more catastrophic events, such as the release of radionuclides (e.g., ^{134}Cs and ^{137}Cs) during the Chernobyl and Fukushima Daiichi power plant disasters in 1986 and 2011, respectively (Buesseler et al., 1987, 2012). An advantage of tracers whose sources are more localised in space and/or time (e.g., relative to the 100–500 year meridional overturning circulation timescale) is that they provide clear initial conditions and evolution timescales. Conversely, if the combined space-time evolution of a particular transient tracer cannot be sufficiently determined, the inversion problem can also become poorly constrained.

9.5.1.2 Tracer release experiments

Intentional tracer release experiments offer the advantage of providing a means of computing mass budgets, and hence using quantitative tracer conservation arguments to help determine the rates of spreading both through increased variance and decreasing tracer concentration. A particularly powerful aspect of tracer release techniques is that they can be used to infer diffusion over a range of spatial scales (Garrett, 1983), from the submesoscale stirring that dissipates filaments (and might thus be relevant within an eddy-resolving model) to the mesoscale stirring appropriate for parameterising a coarse-resolution climate model.

The widespread use of intentional tracer release experiments to measure ocean mixing began in earnest in the 1960s with fluorescent tracers. Early experiments using rhodamine and fluorescein were initially used to measure dispersion in near-surface waters in coastal environments (Pritchard and Carpenter, 1960), and later in stratified environments to measure both isopycnal and diapycnal mixing rates (e.g., Kullenberg, 1971; Wanninkhof et al., 1997; Houghton, 1997; Sundermeyer and Ledwell, 2001; Sundermeyer et al., 2020). A common approach in purposeful tracer experiments is to infer diffusivity either in the horizontal, or along isopycnal or neutral density surfaces from the time evolution of the 2nd moment of the tracer,

$$\sigma_s^2 = \frac{\int_{-\infty}^{\infty} (s - \bar{s})^2 C \, ds}{\int_{-\infty}^{\infty} C \, ds}, \quad (9.23)$$

where $\bar{s}$ is the centre of mass of the tracer in the relevant coordinate system. In the event that the patch is Gaussian, the horizontal or isopycnal diffusivity is then exactly given by

$$K_s = \frac{1}{2} \frac{\partial \sigma_s^2}{\partial t}. \quad (9.24)$$

The use of purposeful non-fluorescent chemical tracers to measure dispersion on larger time and space scales (months to years and hundreds of km) was pioneered through early research of Ledwell et al. (1986) using SF_6 in the Santa Monica basin, and later expanded to a variety of open ocean environments using both SF_6 and later SF_5CF_3 (Ho et al., 2008). The advantage of these chemical tracers over fluorescent dyes lies in their ability to be detected over background in the ocean at levels of order 10^{-17} parts by mass in the case of the former, compared to 10^{-10} parts by mass in the case of the latter. Major tracer release experiments have subsequently been performed in the subtropical North Atlantic (NATRE, SaltFinger, GUTRE), Greenland sea, deep Brazil basin (BBTRE) and in the Southern ocean (DIMES). Results from several such experiments are summarised in Table 9.1. Ledwell et al. (1998) reported results from the North Atlantic tracer release experiment, finding zonal and meridional diffusion coefficients of 1500 and 600 $m^2\ s^{-1}$, respectively, more than 2.5 years after deploying the tracer. Banyte et al. (2013) report similar values in the Guinea dome, with zonal and meridional diffusion coefficients of 1000–1200 $m^2\ s^{-1}$ and 500 $m^2\ s^{-1}$, respectively. Ledwell and Watson (1991) found a lateral diffusion coefficient in the Santa Monica basin off the coast of California on scales of about 10 km of 20 $m^2\ s^{-1}$. A deep tracer release in the Brazil basin yielded a diffusion coefficient of order 100 $m^2\ s^{-1}$ (Rye et al., 2012) on timescales of several months.

Like any method, purposeful tracer experiments also have limitations. First, the spread of the tracer patch typically only gives a bulk estimate of diffusivity, with at best minimal information about its temporal and/or spatial variation, e.g., from early to late evolution of the tracer patch. An example of the latter is given by Zika et al. (2020) who analysed the DIMES tracer experiment in the Southern ocean sampled over a period of four years, finding a relatively slow rate of spreading during the first two years after release while the patch was in the Pacific, followed by more rapid spreading as it passed through the Drake passage and the Scotian sea to the east. Second, since such experiments are by design local in space and time, construction of global diffusivity maps would require multiple tracer releases over large areas. Third, tracer spreading in the ocean results from the cumulative action of the turbulent eddy field and mean currents alike (e.g., through shear dispersion and shearing/straining on scales larger than the tracer patch), which can be difficult to disentangle observationally, even in the best of circumstances. This is problematic if one is interested in estimating K for the purpose of parameterisation of turbulent diffusion.

9.5.1.3 Tracer variance and mixing length

Making use of natural tracers to explain the salinity budget of the North Atlantic, Armi and Stommel (1983) introduced a new method for inferring isopycnal diffusivities from standard hydrographic measurements. They used salinity variance inferred from repeat hydrographic sections in (9.16) to estimate L_{mix}, the mixing length in Prandtl's formula, together with dynamic height variability to estimate V_e. Together with a nominal mixing efficiency of $\alpha = 0.25$, these resulted in profiles of K ranging between 500 and 1500 $m^2\ s^{-1}$.

This method was revived by Ferrari and Polzin (2005), who used data from the North Atlantic tracer release experiment (NATRE) to calculate a vertical profile of K, which decayed from $\sim$1000 $m^2\ s^{-1}$ near the surface to $\sim$200 $m^2\ s^{-1}$ at depth. Using additional microstructure dissipation estimates, they also compared the relative roles of diapycnal and isopycnal mixing in temperature variance production. Naveira Garabato et al. (2016) made similar use of tracer and microstructure data from the diapycnal and isopycnal mixing experiment in the Southern ocean (DIMES). Naveira-Garabato et al. (2011) used repeat hydrography and altimetry to make a mixing-length estimate of K across the fronts of the Antarctic circumpolar current (ACC); they found regions of enhanced K between and below the ACC jets and suppressed values in the jet cores, as anticipated by Marshall et al. (2006) and Ferrari and Nikurashin (2010).

Cole et al. (2015) recently made the first global, three-dimensional estimate of K by using salinity variance measured by ARGO floats to estimate L_{mix}. The results from this estimate were published openly, providing a broadly useful dataset for many different applications Cole et al. (2017). In Box 9.1 and Fig. 9.6, we use Cole's dataset to summarise the contemporary consensus about the global spatial distribution of isopycnal mixing.

TABLE 9.1 Observed lateral diffusivities from different studies: drifter (D), float (F), mixed drifter plus float (DF), and tracer (T), targeting specific regions and depths. For (D, F & DF), the (Methods) column distinguishes between single-particle diffusivity (spd; and zc and asym when first zero-crossing or asymptotic integration of the auto-correlation was done, respectively) or relative particle diffusivity (rpd) and whether the value is a basin average (basin), spatially-resolving using binning (bin) or clustering (cluster) or bicubic splines (bss), or specific to groups of floats (group); for (T), the case when tracer spreading was corrected using altimetry is marked. Studies: (D1–10 are Colony and Thorndike (1985); Brink et al. (2000); Falco et al. (2000); Lumpkin and Flament (2001); Lumpkin et al. (2002); Bauer et al. (2002); Koszalka et al. (2009); Lumpkin and Elipot (2010); Falco and Zambianchi (2011); Koszalka et al. (2013)), (F1-5, Zhang et al. (2001); Ollitrault and Colin de Verdière (2002); LaCasce et al. (2014); Balwada et al. (2016b, 2021)), (DF1,2: Rupolo (2007) and Zhurbas et al. (2014)), (T1-5 are Ledwell and Watson (1991); Ledwell et al. (1998); Banyte et al. (2013); Rye et al. (2012); Tulloch et al. (2014)). Error bounds are not given, but all numbers are approximate.

Study	Region	Depth	Method	K [1000 m^2 s^{-1}]
D1	Arctic	Surface	rpd, basin	1.1
D2	Cal. current	15 m	spd, basin	3.5/3.1
D3	Adriatic	15 m	spd, basin, bss	1.0/.0
D4	Pacific NEC	15 m	spd, 6 groups, zc	2.3
D4	N. Pac subtrop.	15 m	spd, 6 groups, zc	12.4
D5	N Atl.	15 m	spd, zc	1–23
D5	N. Atl.	z < 1500 m	spd, zc	0.07–1.8
D6	trop. Pacific	15 m	spd, bins, bss	5-74/2–9
D7	Norw. sea	15 m	rpd, basin	2.9
D8	Gulf stream	15 m	rpd	14
D8	E. subtrop. Atl.	15 m	rpd	1.7
D9	ACC	15 m	spd, basin, bss	9.0/2.0
D10	E. Nordic seas	15 m	spd, cluster & bin	0.7–2.4
F1	Newf. basin	$\sigma = 27.2$	spd, basin	5.8/5.6
F1	Newf. basin	$\sigma = 27.5$	spd, basin	3.6/4.1
F2	Canary basin	700 m	spd, basin	1.5
F2	Newf. basin	700 m	spd, basin	3.5
F2	Corner rise	700 m	spd, basin	4.1
F3	ACC basin	$\sigma \sim 27.7$	spd, asymp	0.8
F4	ACC & Scotia sea	500–1500 m	spd, asymp	0.7–2.8
F5	ACC (S. Pac.)	500–1500 m	rpd	N/A
DF1	global map	15+400–1000 m	spd, zc	0–40
DF2	global map	15 m	spd, zc vs. asymp	0–10(asymp)
T1	Cal. current	4000 m	Spread	0.02
T2	Subtr. N, Atl.	300 m	Spread	1.5/0.6
T3	Guinea dome	300 m	Spread	1–1.2/0.5
T4	Brazil basin	4000 m	Spread	0.1
T5	Drake passage	1500 m	Corrected spread.	0.7

Box 9.1 Spatial variability of isopycnal mixing rates

Most tracer and float experiments yield just a few values in a particular region, making it hard to generalise to the global scale. Here we summarise some of the main conclusions from Cole et al. (2015), one of the only studies to estimate isopycnal mixing globally. We note that these results are broadly consistent with many of the more localised studies reviewed in this section, and thus provide a good overview of the current observational knowledge about isopycnal mixing rates. Cole's open data (Cole et al., 2017), shown in Fig. 9.6, permit some broad conclusions about the spatial distribution of isopycnal mixing to be drawn:

- The spatial variability of K spans at least two orders of magnitude, between 100 and 10,000 m^2 s^{-1} with a global median near 1000 m^2 s^{-1}.
- Regions of high mesoscale eddy activity have higher K than more quiescent regions (e.g., eastern subpolar gyres).
- Despite high mesoscale activity, diffusivity is somewhat weaker in the cores of strong currents (e.g., Gulf stream, ACC), supporting the mean-flow-suppression theory of Ferrari and Nikurashin (2010).
- Near-surface values are generally higher than those at depth, except for in the ACC, consistent with the predictions of Smith and Marshall (2009); Abernathey et al. (2010).

One caveat about Cole's estimate, however, is that the V_e values were drawn from an eddy-permitting numerical model, not from observations.

The mixing-length approach to estimating K is appealing, because it is based on hydrographic temperature and salinity measurements, which are widespread. Its Eulerian character means that the results can easily be translated to numerical

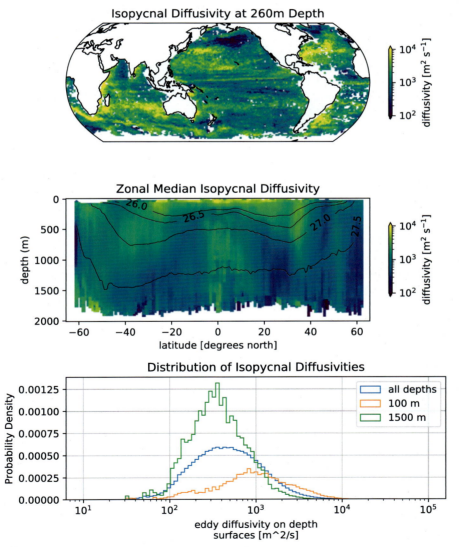

FIGURE 9.6 ARGO-based diffusivity estimate of Cole et al. (2015). Original figure generated from published data (Cole et al., 2017). *Top panel:* A horizontal slice of diffusivity at 260 m depth. *Middle panel:* The zonal median isopycnal diffusivity. *Bottom panel:* Histograms of isopycnal diffusivity near the surface (100 m depth), at 1500 m, and for all depths.

models. Its downsides include the dependence on the mixing efficiency parameter α, which has mostly been estimated from laboratory experiments, its assumption of isotropic mixing, and its applicability only to regions of the ocean where the production of T/S variance is dominated by mesoscale stirring (which excludes the near-surface layer and other regions of strong diapycnal mixing.)

9.5.2 Drifter and float-based methods

> Statistics from drifters and floats can be used to measure quantities related to stirring. Single-particle statistics measure diffusivity via Lagrangian autocorrelation in velocity fields. Relative dispersion measures spreading relative to a centre of mass of particles.

The link between the turbulent diffusion and dispersion has motivated kinematic approaches of estimating K via experiments quantifying the dispersion of floating instruments as proxies of water particles: surface drifters drogued at 15 m (Lumpkin and Pazos, 2007), and subsurface floats that can be ballasted to follow particular isopycnal or isobaric surfaces and tracked acoustically (Rossby, 2007). We present an outline of the theoretical underpinnings for interpreting such data below; more detailed descriptions can be found in LaCasce (2008).

9.5.2.1 Single-particle dispersion and diffusivity

Single-particle dispersion directly uses Taylor's diffusivity formula (9.4) and requires following the dispersion of individual particles up to the asymptotic diffusive regime, in which their ensemble average squared separation grows linearly. Note that this approach reflects both the spread about the centre of mass and movement relative to the starting location. To isolate K and quantify the turbulent dispersion, one needs to separate out the movement by the 'mean' large-scale circulation. There are several equivalent formulations of (9.4) (e.g., LaCasce et al., 2014). One commonly used approach capitalises on the relationship between particle dispersion and Lagrangian velocity noted by Taylor (1921). As a particle's trajectory is the integral of its velocity,

$$x_i(t) - x_i(0) = \int_{t_o}^{t} u(x_i, t) \, dt, \tag{9.25}$$

the diffusivity in (9.4) can be expanded as

$$K = \lim_{t \to \infty} \langle u_i(t)(x_i(t) - x_i(0)) \rangle = \lim_{t \to \infty} \langle u_i(t) \int_0^t u_i(t') dt' \rangle = \lim_{t \to \infty} u_{rms}^2 \int_0^t R(\tau) \, d\tau, \tag{9.26}$$

where u_{rms} is the root-mean square turbulent Lagrangian velocity, R is the ensemble-mean velocity autocorrelation, and $\tau = t - t'$ is the time lag. This is the so-called 'Taylor formula'. The notion of the 'asymptotic regime' ($t \to \infty$) can be recast in terms of the Lagrangian integral time,

$$T_L = \lim_{t \to \infty} \int_0^t R(\tau) d\tau, \tag{9.27}$$

as a measure of the characteristic eddy timescale and $L_{int} = u_{rms} T_L$ as a corresponding eddy lengthscale. In the asymptotic regime the integration period in (9.26) should be long compared to the integral timescale, $t \gg T_L$, although in practice, integration to the first zero crossing is sometimes used (see also Section 9.4.2). Note that the long integrations required for this method mean that the diffusion coefficients estimated capture the impacts of mesoscale eddies, and are thus most appropriate for non-eddy-resolving models, and cannot generally be used to assess submesoscale diffusion.

It is worth noting that though Taylor's approach formally assumes the flow to be statistically stationary and isotropic, this is generally not the case in a real ocean. Davis (1991) generalised Taylor's approach, allowing eddy velocities and characteristic scales to vary slowly in space and time with respect to particle displacement and time lag. The Davis recipe applies (9.26) locally to estimate $K(\mathbf{x})$, and in practical terms can be summarised as follows:

1. Divide the data into subsets that best satisfy the statistical stationarity condition (e.g., seasonal or interannual).
2. Group particle trajectories in geographical regions so that the statistics (variance, length, and timescales) are homogeneous locally and normally distributed.
3. In these regions, average the particle velocities to estimate the "pseudo–Eulerian" mean velocity; interpolate this onto particle trajectories, and subtract this from velocity time series to obtain residual velocities.

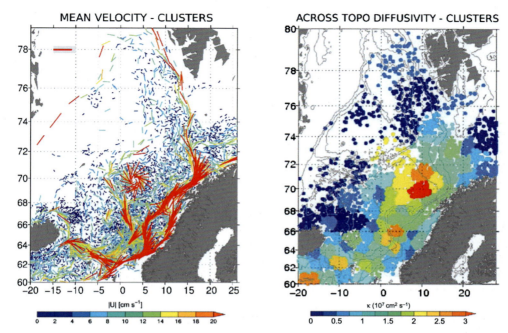

FIGURE 9.7 Surface mean velocity field (left, a) and eddy diffusivity, across-topography (right, b) in the Nordic seas, derived from surface drifters using a clustering algorithm and trajectory segmentation for better convergence of the mean velocities and diffusivities, from Koszalka et al. (2011).

4. Estimate the autocorrelation, eddy scales, and a geographical map of diffusivity, $K(\mathbf{x})$, from the residual velocities of trajectory segments in the different regions.

Note that the Davis approach is inherently iterative, i.e., it involves a search of a definition (grid) of regions that are large enough to collect sufficient data for averaging and to ensure the convergence of the estimates in time, but small enough to resolve the lengthscales of the eddy and the mean flow. Further variations and enhancements of the Davis method were made by Oh et al. (2000) and Rypina et al. (2012) in introducing local projections with respect to directions of maximum variance or the mean flow direction/topographic gradient (e.g., Koszalka et al., 2011).

Regional variation of eddy activity has led to efforts to construct diffusivity maps $K(\mathbf{x})$, motivating several studies using (9.4) and a myriad of technical interpretations of the Davis approach: a regional grid of bins, overlapping bins, circular bins (Lumpkin et al., 2002), using bicubic splines (Bauer et al., 2002), or Gaussian–Markov decomposition (Lumpkin and Johnson, 2013) to estimate the variable mean flow and residual velocities, and clustering of trajectory segments for better autocorrelation convergence (Koszalka et al., 2011). These studies produce an extremely wide range of values, from 70 $m^2\ s^{-1}$ in parts of the North Atlantic (Lumpkin et al., 2002) to 1500 $m^2\ s^{-1}$ up to 35,000 $m^2\ s^{-1}$ in the equatorial regions and close to boundary currents (Zhurbas et al., 2014).

The DIMES experiment produced several studies of single-particle-based diffusivity based on subsurface RAFOS floats deployed in the ACC (LaCasce et al., 2014; Balwada et al., 2016b). Together with tracer release experiments and numerical simulations, these studies weakly confirmed the hypothesis of FN10 of enhanced cross-stream mixing (1000–2000 $m^2\ s^{-1}$) at mid-depth and suppressed mixing ($<1000\ m^2\ s^{-1}$) at shallower depths. Balwada et al. (2016b) also revealed significant along-stream variations in mixing rates between the South Pacific sector and the Scotia sea.

On a global scale, Zhurbas et al. (2014) produced a global map of $K(\mathbf{x})$ estimated from surface drifters, whereas Roach et al. (2016) used ARGO float trajectories at 1000 m depth to assess cross-ACC diffusivity in the Southern ocean. Table 9.1 is our attempt to coarsely summarise many of the results of different tracer, drifter, and float estimates of eddy diffusivity.

A focus of research on particle statistics is to eliminate some of the *ad hoc* choices in the original Davis recipe that add subjectivity to the results. For example, Koszalka et al. (2011) applied k-means clustering, a form of unsupervised machine learning, to automatically define clusters so as to minimise the intra-cluster variance in position. The results, shown in Fig. 9.7, reveal a diffusivity that is spatially variable, with the highest values found in the Lofoten basin, near (but not in) the boundary Norwegian Atlantic current. Note also that there are substantial blank parts in the domain due to data sparsity. Another promising methodological innovation is the Bayesian approach of Ying et al. (2019); by fitting a stochastic trajectory model to float data, they estimate an anisotropic diffusivity and accompanying uncertainty from relatively sparse observations. (That model was tested with numerical data but has not yet been applied to real observations.)

9.5.2.2 Relative particle dispersion and diffusivity

Relative particle dispersion relies on tracking particles relative to a centre of mass. The central statistic used here is the mean square separation of particle pairs (Richardson, 1926; Batchelor, 1952; LaCasce, 2008):

$$D_R(t)^2 = \frac{1}{2N(N-1)} \sum_{i \neq j} |x_i(t) - x_j(t)|^2, \tag{9.28}$$

which represents the ensemble distance between particle pairs, or, equivalently, the spread of particles relative to their centre of mass. The diffusivity associated with relative dispersion is

$$K = \frac{1}{2} \frac{\partial}{\partial t} D_R(t)^2 . \tag{9.29}$$

Relative dispersion is a particularly useful method for probing the scale dependence in mixing described in Section 9.4.3, as different regimes of stirring and turbulence produce measurably different dispersion statistics. At short and intermediate timescales (relative to the Lagrangian decorrelation timescale), relative dispersion generally increases rapidly with time, with particles possibly experiencing a ballistic, non-local, and local dispersion regime; and the diffusivity in increases continuously with scale. A particular emphasis of relative-dispersion field research has been to determine whether submesoscale dispersion is local or non-local (see Section 9.4.3). At large space and time scales, relative dispersion of particle pairs is expected grow linearly with time and (9.29) will asymptote to a constant value (the diffusive regime). Below, and in Table 9.1, we summarise results from the large-scale asymptotic diffusive regime. We refer readers to Balwada et al. (2021) and references therein for a state-of-the-art tour of the intricacies of relative dispersion at small and intermediate scales.

The challenge of making observational estimates of relative dispersion is that they require large numbers of floats released together, and to date there have been only a few such experiments, some of which are listed in the second section of Table 9.1. Surface drifters, being relatively cheap and easy to deploy, have been the focus of most studies (e.g., LaCasce and Ohlmann, 2003; Koszalka et al., 2009; Lumpkin and Elipot, 2010). Lumpkin and Elipot (2010) obtained values as high as 5800 m^2 s^{-1} from the surface drifters deployed in the Gulf stream region and about 1700 m^2 s^{-1} in the eastern subtropical North Atlantic. Subsurface relative dispersion has been analysed from RAFOS floats by LaCasce and Bower (2000); Ollitrault et al. (2005) and Balwada et al. (2021). The first two used isobaric floats; for the latter (DIMES experiment), the floats were intended to be isopycnal, but, as reported by the authors, in practice were closer to isobaric. Ollitrault et al. (2005) found values about 5000 m^2 s^{-1} at 700 dbar in the North Atlantic, which is unexpectedly high for an eddy flow at depth.

9.6 Simulation-based estimates

9.6.1 Inverse methods

> Inverse methods can infer isopycnal mixing by figuring out the optimal diffusion coefficients needed to satisfy conservation laws. Modern inverse models are essentially data-assimilating GCMs.

For many decades oceanographers have used hydrographic observations, together with conservation laws for mass, heat, salt, and other tracers, to infer transport in an inverse framework (Wunsch, 1977). In an early attempt, Fiadeiro and Craig (1978) used gradients in salinity within the deep Pacific ocean to constrain the lateral diffusion coefficient below 1000 m to lie somewhere between 100 and 1000 m^2 s^{-1}. Similarly, Bower et al. (1985) used an oxygen budget to estimate the isopycnal diffusivity in the Gulf stream region pycnocline to be around 250 m^2 s^{-1}. Such methods were limited by poor knowledge of the structure of the deep flow and sparse data. Modern practitioners of hydrographic inverse methods have refined bulk estimates of diffusion coefficients at the ocean basin scale, generally finding values less than 1000 m^2 s^{-1} (e.g., Naveira Garabato et al., 2003; Zika et al., 2010).

Contemporary inverse methods often involve data assimilation within the context of general circulation models. The use of data assimilation to estimate ocean mixing coefficients was pioneered by Ferreira et al. (2005), who used a GCM and its adjoint to infer K_{GM}. Both Liu et al. (2012) and Forget et al. (2015) expanded this approach to include K and diapycnal diffusion, systematically adjusting the coefficients to reduce model misfit relative to a wide range of satellite and *in-situ* observations (including ARGO floats). In both cases, however, K was found to have a relatively minor impact on the solution compared to K_{GM}, and was not adjusted dramatically away from the prior value of 1000 m^2 s^{-2}. This is likely, because the models only assimilated temperature and salinity data, which are less sensitive to K than passive chemical tracers with stronger isopycnal gradients. Using a novel inverse method in thermohaline coordinates, Groeskamp et al. (2017) inferred that K in the deep ocean could be as low as 20 m^2 s^{-1} (see also Groeskamp et al., 2020). In all of these studies, the take away message is that such assimilation methods represent a powerful and promising way forward for inferring isopycnal mixing rates.

9.6.2 Direct simulation

> High-resolution models, which directly resolve mesoscale turbulence, can simulate the isopycnal mixing process. Estimating mixing coefficients from such models is surprisingly challenging, but provides uniquely comprehensive data.

As reviewed in Section 9.3, mesoscale turbulence is the dominant process driving isopycnal mixing. As computer power has increased over the past decades, it has become increasingly feasible to simulate mesoscale turbulence and the resulting isopycnal mixing processes directly. Models can claim to be 'mesoscale-resolving' if they resolve the first deformation radius with several grid points (Hallberg, 2013), a criterion which varies strongly with latitude. In mid-latitudes, models with resolution of approximately 10 km and higher are likely sufficient for studying mesoscale mixing and transport, but polar regions require much finer resolution. Though questions will always remain about the numerical convergence and parameter sensitivity of such models, they nevertheless provide an important viewpoint on isopycnal mixing. Griesel et al. (2019) provide an excellent review of this subject.

9.6.2.1 Lagrangian trajectory simulation

As described in Section 9.5.2, Lagrangian trajectories contain important information about isopycnal dispersion and mixing. Statistics, such as absolute and relative dispersion of Lagrangian parcels, can be used to quantify the spatial variability in isopycnal mixing. Since real isopycnal Lagrangian observations are expensive to collect and highly sparse, many researchers have sought to use simulated isopycnal trajectories from mesoscale-resolving models as a diagnostic of isopycnal mixing.

Griesel et al. (2010) deployed Lagrangian particles at various depths around the Antarctic circumpolar current in a 0.1°-resolution global ocean model. Using single-particle diffusivity (Section 9.5.2.1) with dispersion projected along and across the ACC streamlines, they identified strong anisotropy in isopycnal mixing. Even after attempting to remove the effects of shear dispersion, they found that along-stream diffusivity was several times stronger than in the cross-stream direction, and that diffusivities generally decayed with depth. Chen et al. (2014) and Chen et al. (2017) applied similar methods in the Kuroshio extension, finding anisotropic diffusivity (elevated in the along-stream direction) and some limited evidence for critical-layer enhancement at depth (Smith and Marshall, 2009). These studies found qualitative evidence supporting the mixing-length suppression model of Ferrari and Nikurashin (2010).

Although simulated float experiments can overcome the observational challenge of sparse sampling, they cannot overcome other more fundamental limitations of Lagrangian methods. Since the methods for measuring diffusivity all involve an integral over time, the particles inevitably sample a broad region of space before the integrals converge. Griesel et al. (2019) observed how this non-locality in space limits the applicability of Lagrangian estimates of K to numerical models, which require values specific to each grid cell.

9.6.2.2 Tracer-based methods

Deploying Eulerian tracers in ocean models offers an alternative way to study isopycnal mixing. Because models simulate tracers on their native Eulerian grid, the results of such experiments may be more useful for parameterisation than Lagrangian methods. In contrast to observations, where dyes can only be deployed as instantaneous point sources, numerical tracers can be initialised and forced in any way imaginable. Despite this apparently infinite possibility, recovering estimates of isopycnal diffusivity from numerical tracer experiments is subtle and computationally difficult.

For idealised homogeneous 2D or QG simulations, the interest is often a single-valued bulk diffusivity, averaged over space and time, for the entire simulation with a given set of parameters (e.g., Smith et al., 2002; Thompson and Young, 2006, 2007; Kong and Jansen, 2017; Zhang et al., 2020). In this case, the diffusion coefficient is defined simply as $K = -\langle \overline{v'q'} \rangle / Q_y$, where $\langle \overline{v'q'} \rangle$ is the eddy flux of potential vorticity (or any other conserved scalar) averaged over the domain, and Q_y is a constant imposed background gradient. Smith (2005) developed a clever numerical method to recover both zonal and meridional diffusivities from such models, revealing the anisotropy of mixing induced by shear dispersion.

For more realistic primitive equation simulations in homogeneous domains, the goal is often to recover a spatially variable mixing coefficient. One powerful method, first developed for atmospheric applications by Plumb and Mahlman (1987), is to solve an inverse problem to determine **K**, the mixing tensor of (9.20), at every point in space based on multiple tracers with orthogonal gradients. For a single tracer, (9.20) is underdetermined; given a single value of $\nabla \overline{c}$, many different **K** can produce the same $\overline{\mathbf{u}'c}$. However, by adding more and more tracers with different gradients, **K** becomes overdetermined and (9.20) can be solved approximately via a least-squares approach. The diffusive and advective components of eddy transport can then be decomposed via the symmetric and anti-symmetric components of the tensor, as described in Section 9.4.4. By diagonalising the symmetric component, the principal axes of mixing and the corresponding diffusivity coefficients can be recovered. When the mixing axis is the neutral plane, these coefficients can be interpreted as the isopycnal diffusivity.

When a zonal mean is included in the averaging operator, **K** reduces to a 2 × 2 matrix, which is simpler to compute and analyse. The multi-tracer inversion method was applied to zonally symmetric primitive equation ocean simulations by Bachman and Fox-Kemper (2013), who used it to characterise K and K_{GM} in idealised Eady-model setups; and by Abernathey et al. (2013), who compared the results with other diagnostic methods in the context of an ACC channel simulation. These two studies reached incompatible conclusions regarding the relationship of K and K_{GM}; Abernathey et al. (2013) verified the QG theoretical relationship proposed by Smith and Marshall (2009) between the two coefficients, with K being larger and experiencing a mid-depth maximum, whereas Bachman and Fox-Kemper (2013) concluded that K and K_{GM} were the same at all depths. These differences may be attributable to the different background states and forcing between the two studies.

Applying the multi-tracer method on a global scale in 3D is computationally expensive, involving the simulation of at least nine different numerical tracers in a high-resolution global model. The only known instance is the recent study of Bachman et al. (2020), an update to an earlier implementation by the same team (Fox-Kemper et al., 2012). The emphasis of this comprehensive study is the horizontal anisotropy and vertical structure of the diffusivity coefficients. They found strong anisotropy in isopycnal diffusivity, with a major-axis diffusivity generally 1–20 times larger than the minor axis. The background potential-vorticity gradient was shown to predict the orientation of mixing and drive anisotropy, with strong PV gradients acting to suppress mixing in the cross-gradient direction (Ferrari and Nikurashin, 2010; Srinivasan and Young, 2014). The vertical structure of both coefficients was shown to decay with depth, in agreement with the predictions of Marshall et al. (2012). By comparing the antisymmetric and symmetric components of **K**, Bachman et al. (2020) concluded that K and K_{GM} were practically indistinguishable.

A key drawback to the multi-tracer method is the fact that the diagnosed **K** may contain contributions from processes besides irreversible mixing, such as the advection of tracer variance (Nakamura, 2001; Wilson and Williams, 2004). This sort of non-locality is particularly severe in locations of strong, meandering currents, such as western boundary currents and meanders of the ACC. A related issue is that **K** is diagnosed from the raw eddy fluxes, but only the divergence of these fluxes is relevant for tracer budgets, leading to a gauge freedom in the specification of the coefficients (Marshall and Shutts, 1981; Eden et al., 2007). To address these challenges, some authors have diagnosed isopycnal mixing using techniques based explicitly on tracer variance dissipation. One of these methods is the 'effective diffusivity' framework of Nakamura (1995), which analyses transport in a tracer-based coordinate system. Effective diffusivity has been applied to the Southern ocean (Marshall et al., 2006; Shuckburgh et al., 2009a,b; Abernathey et al., 2010; Boland et al., 2015) and the subtropical salinity maxima (Busecke et al., 2017). Though these studies have revealed interesting spatial patterns of mixing, highlighting the importance of mixing-length suppression by jets, their applicability to Eulerian eddy parameterisation is limited by the semi-Lagrangian nature of the tracer-based coordinate system.

9.7 Impacts of isopycnal mixing

Regardless of how it is measured, be it from natural or purposeful tracers in the ocean, from Lagrangian or quasi-Lagrangian particles, or with the help of inverse methods using numerical models, the importance of isopycnal mixing in the ocean is widespread, affecting a variety of physical, chemical, and biological properties and dynamics locally, regionally, and globally. Other chapters in this book describe in depth the role of ocean mixing in ocean circulation (Chapter 3) and the climate system (Chapter 2). Here we highlight briefly some specific roles of mesoscale isopycnal mixing in the world oceans.

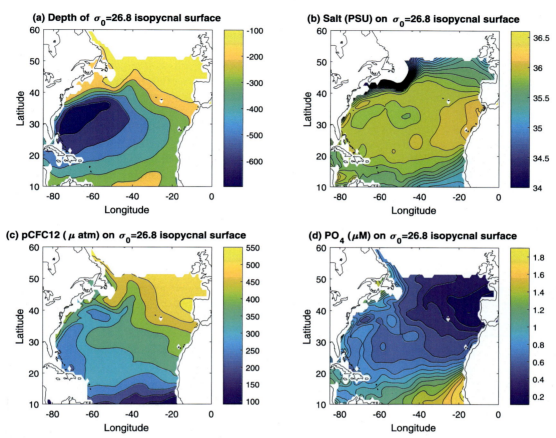

FIGURE 9.8 Illustration of mixing along an isopycnal surface in the North Atlantic. (a) Depth of $\sigma_0 = 26.8$ surface. Flow would be expected to be largely along contours of constant depth. (b) Salinity in PSU along this surface. (c) Partial pressure of CFC-12 along this surface (from Key et al. (2004)). (d) Phosphate along this surface.

9.7.1 Physical circulation

> Isopycnal mixing influences water mass transformation (via cabbeling), subpolar convection, and tropical climate variability.

The impact of isopycnal mixing on the physical circulation of the ocean can be illustrated by considering its impacts on a single isopycnal surface. As shown in Fig. 9.8a, the $\sigma_0 = 26.8$ isopycnal has a bowl-shaped structure in the North Atlantic, reflecting the wind-driven circulation of the subtropical gyre. As is the case in all ocean basins, salinity in the centre of the subtropical gyre is higher than on its poleward flank, reflecting the downward mixing of warm, salty tropical waters. Note that because salt and heat have compensating impacts on density, the centre of the gyre is also warmer. As a result, isopycnal mixing transports both salt and heat into subpolar regions. In the North Atlantic, the subtropical gyre is also saltier than waters on its equatorward side, reflecting the injection of Mediterranean water through the straits of Gibraltar, so that in the Atlantic there is also a transport of heat and salt towards the equator along this isopycnal. This behaviour is also found in shallower isopycnals across ocean basins.

Three big impacts of isopycnal mixing have been identified in the literature. The first (also discussed in Chapter 3) stems from the fact that the equation of state is non-linear, so that mixing two parcels with the same density produces a parcel with greater density. This process, known as *cabbeling*, can drive a significant fraction of water mass transformation, particularly in the Southern ocean (Urakawa and Hasumi, 2012; Groeskamp et al., 2016). Additionally, by making the ocean as a whole denser, cabbeling serves as a sink of energy of around 0.4 TW, making it a noticeable player in the global energy budget (Palter et al., 2014).

The second big impact of isopycnal mixing comes from its influence on subpolar convection. Because heat transported into the high-latitude mixed layer equilibrates relatively quickly with the atmosphere, but salinity does not (Zhang, 2017), the effect of increasing isopycnal mixing is to make the wintertime subpolar gyres more unstable, resulting in an increase in convection (Pradal and Gnanadesikan, 2014). Properly simulating this is difficult, as it requires capturing the suppression of mixing by strong currents. As the locations of these currents in models are often offset slightly relative to their true ones (Bahl et al., 2019), one cannot simply import the observed mixing coefficients into a model and expect to get realistic results.

Finally, changing isopycnal mixing can have significant impacts on climate variability. Gnanadesikan et al. (2017) found that although higher lateral mixing in the mixed layer would be expected to damp the variability associated with El Niño, the net warming of the cold tongue driven by greater mixing of heat into upwelling waters caused convection to move eastward. As this enhances the coupling between the ocean and atmosphere, the result is that higher mixing was associated with a stronger El Niño. By contrast, in the Southern ocean smaller values of K in a coarse-resolution version of the GFDL model are associated with stronger, more regular multidecadal variability in deep convection (Thomas et al., 2018). Gnanadesikan et al. (2020) suggest that intermodel variation in K may contribute to the variation in Southern ocean convective variability seen across the range of CMIP5 models (Reintges et al., 2017).

9.7.2 Passive tracers

> Isopycnal mixing significantly influences the distribution and uptake of passive tracers, such as CFC, in the ocean interior.

The impact of isopycnal mixing can also be seen on dissolved trace gasses. Of particular interest to physical oceanographers are the chlorofluorocarbons that have been rising in the atmosphere for many decades. High concentrations of these gasses indicate that water has recently been in contact with the atmosphere, whereas lower concentrations indicate that it has been out of direct contact with the atmosphere. The pCFC-12 distribution (Fig. 9.8c) shows the impact of the gyre circulation with higher concentrations on the eastern side of the basin and lower concentrations on the western side (Key et al., 2004). Note however, as we approach the outcrop within the Gulf stream, pCFC-12 concentrations *rise*. This is only possible because water parcels have multiple ways of reaching a given point within the Gulf stream, a long advective journey around the gyre, and a short more geometrically direct diffusive journey from the outcrop. This has potentially important implications for how the "age" of water calculated from CFCs will react to a decrease in the advective circulation. In regions where the relatively short diffusive pathway dominates, reducing the advective circulation will actually decrease the apparent age, whereas in regions where the advective pathway dominates, it will increase the age. Understanding and quantifying the diffusion is thus key to finding the fingerprint of changing ocean circulation within the tracer field.

The relative importance of advective vs. diffusive ventilation along isopycnal surfaces has also been considered for anthropogenic carbon dioxide. The ocean plays an important role in mitigating the rise of carbon dioxide, currently removing between 1/4 and 1/3 of annual emissions. As discussed in Gnanadesikan et al. (2015a) the rate at which this happens is dependent on K, with increase from a constant value 400 to 2400 m^2 s^{-1} producing a roughly 30% increase in uptake. A realistic distribution of eddy mixing (Abernathey and Marshall, 2013) produces an uptake between the low and high values of K, despite having much larger values in the gyre interiors, indicating that these large values are not important for uptake.

9.7.3 Isopycnal mixing and ocean biogeochemical cycles

> Isopycnal mixing transports oxygen and nutrients, with significant regional effects on ocean ecosystems.

Isopycnal mixing also has an impact on biologically cycled tracers. Nutrients (e.g., phosphate and nitrate) are taken up by phytoplankton in the surface, incorporated into sinking particles and remineralised at depth. This leads to a buildup of nutrient and drawdown of oxygen at depth, particularly within the 'shadow zones' (Luyten et al., 1983), where there are no direct advective pathways to the mixed layer. Isopycnal mixing is extremely important for bringing oxygen into the shadow zones, and for explaining how this supply changes under global warming (see Chapter 13 for more discussion of this). But this means that isopycnal mixing is also important for removing nutrient from the shadow zones. This can be clearly seen in Fig. 9.8d, as waters flowing past the shadow zones pick up phosphate on their way into the Caribbean. When brought back near the surface as the isopycnal shallows, this nutrient can play an important role in driving surface productivity.

There is broad evidence that mesoscale eddies play an important role in transporting nutrient to phytoplankton; however, studies of this transport have mostly focused on the mechanisms of passive isopycnal uplift and lateral transport (McGillicuddy et al., 1998; Lévy et al., 2001; Lévy, 2008; Resplandy et al., 2011; McGillicuddy, 2016; Letscher et al., 2016). The specific role of isopycnal mixing in nutrient transport has been less studied. Freilich and Mahadevan (2019) developed a novel decomposition of eddy vertical transport that isolated the isopycnal flux component. This method was applied to idealised simulations, where it was shown that isopycnal transport occurred mostly at the horizontal scales of less than 10 km. Uchida et al. (2019) recently demonstrated that accounting for isopycnal mixing was critical for capturing the supply of iron to the surface of the Southern ocean. Setting K to a value of around 1000 m^2 s^{-1} within a 100 km-resolution non-eddy resolving model of the Southern ocean yielded a reasonably good match with biomass and productivity predicted by a 2 km resolution model that resolved meso- and submesoscale eddies. Setting K to zero, by contrast, produced a factor of two decline in iron flux and phytoplankton biomass.

The role of isopycnal mixing on nutrient supply likely has strong regional variation. Eddy intrusions onto shelves have been shown to supply nitrogen and phosphorus to otherwise nutrient-limited coastal regions (Lee et al., 2007). However, in upwelling regions eddies may actually remove nutrients from the coastal zone (Whitney and Robert, 2002; Gruber et al., 2011).

On the global scale, the influence of K on the magnitude of biological cycling is much weaker. Increasing K is associated with decreases in the along-isopycnal gradient in nutrients, leaving the total flux relatively constant in tropical and subtropical regions (Bahl et al., 2019). However, because it changes the transport of salt into the subpolar gyre, changing isopycnal mixing can change intensity of convection in subpolar gyres. This indirect effect of changing K in turn can lead to large differences in productivity within the subpolar gyres and to differing sensitivities of this biological productivity to global warming (Bahl et al., 2019, 2020).

In addition to providing nutrients that drive productivity at the base of the food web, mesoscale eddies may also play an important role in transporting organisms higher in the food web, both by coherent transport in the cores of mesoscale eddies (Lobel and Robinson, 1986; Lehahn et al., 2011) or simply when eddy currents sweep larvae from a spawning site to a settling site. Siegel et al. (2003) explored the latter idea by developing a 'dispersion kernel', in which the spread of larvae was modelled as an advective-diffusive process, and proposed that the relevant mixing length was determined by the rms velocity times the time spent by an organism in its free-swimming stage of life. Further work has shown that numerical simulations of dispersion can be used to construct more realistic maps of connectivity, including the effects of large-scale currents in advecting parcels of water, but also suppressing diffusion across the current. In one example of this, White et al. (2010) demonstrate that the connectivity predicted by such a model reflected the observed pattern of genetic variability in a species of whelk across sites far better than did the straight-line distance between sites. An important application of such work is for the design of marine reserves, as spawning sites that are connected to many other sites by the mesoscale eddy field may be more important to preserve marine species and biodiversity. However, providing such information requires addressing many of the same problems that are faced by physical oceanographers in calculating diffusion coefficients, namely characterising the relevant velocities and mixing lengths (for more discussion of this issue see Hilário et al., 2015).

The dispersion associated with mesoscale mixing also has implications for measuring biogeochemical fluxes within the ocean. Biogenic particles sink at rates of O(10–100) metres/day, speeds far slower than the horizontal velocities associated with mesoscale eddies, which can advect particles tens of km in a day. This means that particles collected at a single site (whether on the ocean floor or from particle traps on mooring) represent an average over some broader area. The characteristics of the resulting "statistical funnel" (Siegel and Deuser, 1997) are tightly linked to the eddy diffusion coefficient.

9.8 Summary and future directions

If the real ocean were as simple as the isotropic, homogeneous turbulence considered by Taylor (1921), we could summarise isopycnal mixing with a single number: the eddy diffusivity. This number would suffice to parameterise sub-grid isopycnal mixing in ocean models, provided care is taken to apply the diffusion along the neutral plane. Different methods of measuring mixing, whether from floats, from natural tracers, or from inverse methods, would all give the same result. Isopycnal mixing would be as simple, and as boring, as molecular diffusion.

The real ocean violates these simplifying assumptions in many ways. First, the observational studies reviewed above show that isopycnal mixing is strongly *inhomogeneous* in space. This inhomogeneity can be explained from the perspective of linear baroclinic instability, which varies strongly by region. It can also be analysed through mixing-length theory, which describes mixing rates via the product between a characteristic eddy velocity and a mixing-length scale. Though, traditionally, mixing-length theory has equated mixing length with eddy size, recent theoretical progress has been made by considering the ways in which mixing length is modulated in the presence of mean flows and eddy propagation, which can suppress exchange in the cross-frontal direction (Ferrari and Nikurashin, 2010; Srinivasan and Young, 2014; Kong and Jansen, 2017).

The existence of jets, fronts, and potential vorticity gradients also leads to significant anisotropy in mixing. Mixing across jets / fronts is likely weaker by several times, or even a factor of 10, than mixing in the along-stream direction. The anisotropy is due to both the aforementioned mixing-length suppression in the cross-stream direction, combined with the action of shear dispersion in the along-stream direction. To our knowledge, no numerical models currently implement anisotropic isopycnal mixing, so assessing the importance of anisotropy on global-scale phenomena remains an open question.

Both inhomogeneity and anisotropy are closely tied to the problem of non-locality. Non-locality manifests in several ways. Simple diffusive closures make sense when tracer variance is both generated and dissipated in the same location. However, the existence of strong jets and other inhomogeneity in the background flow means that variance can be advected very far from its generation location before it is dissipated. This complicates the interpretation of inferences of mixing based on observed tracer variance and empirical mixing length. A related challenge arises when using floats or point-source tracers to infer mixing rates; because these Lagrangian observations integrate over broad regions of space and time, drawing conclusions about mixing rates at a specific point becomes difficult.

Finally, scale dependence, once a matter of theoretical curiosity, is emerging as a major practical concern for ocean modelling. Oceanographers since Richardson and Stommel (1948) have described how ocean mixing rates increase with spatial scale. However, as long as the grid cells of ocean models were larger than the largest mesoscale eddies, this scale dependence was not especially relevant for parameterisation. Now that global models are commonly run at 1/4° or 1/10° resolution, we must consider what are the appropriate sub-grid isopycnal mixing rates to use. It seems that ocean modellers have a deep body of observational knowledge to draw upon here.

Given all these difficulties, it is somewhat surprising that eddy diffusivity remains, in one way or another, the central diagnostic of isopycnal mixing. We speculate that this is driven by the needs of ocean models. Nearly all proposals to study isopycnal mixing cite the desire to inform or constrain climate models as a central motivation, and climate models universally parameterise isopycnal mixing as diffusion. So, to remain relevant to the modelling enterprise, observational and diagnostic studies must report a diffusivity.

Yet there is no shortage of novel and interesting frameworks, beyond diffusivity, for analysing observations. For instance, the "transition matrix" approach, in which probabilities of transition from one location to another are estimated from drifters, has been applied extensively to the study of ocean debris and pollution transport (Maximenko et al., 2012; Van Sebille et al., 2012). The connection of this approach to dynamical systems theory opens up many new modes of data analysis aimed at unveiling the underlying structures of transport (Froyland et al., 2007; Froyland and Padberg, 2009; Haller, 2015). Yet these approaches have mostly remained confined to the applied mathematics community and have not yet penetrated deeply in mainstream isopycnal mixing research.

For progress to occur, it seems that innovation is required on the modelling side. We offer two suggestions:

1. *Reconcile scale-dependent dissipation schemes from large-eddy simulation (LES) with observed scale-dependent dispersion.* Modern high-resolution ocean models employ scale-aware LES closures based on turbulence theory to determine grid-scale dissipation at scales of 1–10 km (Fox-Kemper and Menemenlis, 2008; Pearson et al., 2017; Zanna et al., 2017; Bachman, 2019). However, the observational literature on relative dispersion provides ample and detailed evidence for how dispersion varies over the same scale range. Can these viewpoints be brought into closer dialog? Kämpf and Cox (2016) offer one path forward.

2. *Explicitly model sub-grid tracer variance.* Mesoscale eddy parameterisation is currently undergoing a major restructuring with the advent of models that explicitly simulate sub-grid eddy energy (Eden and Greatbatch, 2008b; Marshall et al., 2012; Jansen et al., 2015). Explicitly tracking unresolved energy, from its generation to its dissipation, is both theoretically satisfying and practically useful. Although this is very useful for the advective Gent and McWilliams (1990) part of mesoscale transport (which is directly linked to baroclinic instability), it is less useful for isopycnal mixing. This is because isopycnal mixing is, by definition, energetically neutral, neither generating nor dissipating energy. It does, however, generate an analogous quadratic quantity: tracer variance. Explicitly modelling the generation, transport, and dissipation of tracer variance may offer a way forward, particularly on the issue of non-locality. It also offers a new point of comparison between models and observations; rather than compare diffusivities, models could compare their simulated isopycnal tracer variance with observations.

3. *Explore non-local transport parameterisations.* Diffusivity is a local closure for mixing; it assumes that unresolved transport can be represented in terms of the large-scale gradients at each point in space. Non-local mixing closures, in contrast, allow mass and tracers to be exchanged over larger distances. Non-local transport schemes have been used to

diagnose and parameterise atmospheric convection and boundary-layer mixing (Stull, 1993; Romps and Kuang, 2011). Such schemes usually rely on the notion of a transfer matrix, which redistributes the contents of one grid cell to many others. This framework can potentially better represent the effect of persistent coherent structures, such as eddies and plumes. Oceanographers should investigate the applicability of such schemes to the mesoscale transport problem.

Finally, we observe that the machine learning revolution sweeping through science is likely to impact the study of isopycnal mixing. A few pioneering studies have used tools from machine learning to improve how mixing is diagnosed from observations (Koszalka et al., 2011; Ying et al., 2019). Deep neural networks have been applied to learn parameterisations for mesoscale transports of momentum (Bolton and Zanna, 2019) and heat (George et al., 2021). Applying such approaches outside of the idealised world of quasigeostrophy to the isopycnal mixing problem is not without technical challenges, but will likely be tried anyway. Combining such methods with the ever-increasing volume of data from autonomous observing systems, satellites, and high-resolution numerical models may yield new breakthroughs, and new surprises, in the field of isopycnal mixing research over the next decade. At the same time, we should not forget the key role that basic theory and physical insight have played in illuminating the mixing process.

Acknowledgements

Inga Koszalka contributed significantly to an earlier draft of this manuscript, writing Section 9.5.2, creating Fig. 9.7, and compiling Table 9.1. We are deeply grateful for these contributions, which were incorporated here with her permission. Dhruv Balwada provided useful suggestions and references. Abernathey acknowledges support from NSF award OCE 1553595.

References

Abernathey, R., Ferreira, D., Klocker, A., 2013. Diagnostics of isopycnal mixing in a circumpolar channel. Ocean Model. 72, 1–16. https://doi.org/10.1016/j.ocemod.2013.07.004.

Abernathey, R., Haller, G., 2018. Transport by Lagrangian vortices in the Eastern Pacific. J. Phys. Oceanogr. 48 (3), 667–685. https://doi.org/10.1175/JPO-D-17-0102.1. https://journals.ametsoc.org/view/journals/phoc/48/3/jpo-d-17-0102.1.xml.

Abernathey, R., Marshall, J., Shuckburgh, E., Mazloff, M., 2010. Enhancement of mesoscale eddy stirring at steering levels in the Southern Ocean. J. Phys. Oceanogr. 40, 170–185. https://doi.org/10.1175/2009JPO4201.1.

Abernathey, R., Marshall, J.C., 2013. Global surface eddy diffusivities derived from satellite altimetry. J. Geophys. Res. 118, 901–916. https://doi.org/10.1002/jgrc.20066.

Aref, H., 1984. Stirring by chaotic advection. J. Fluid Mech. 143, 1–21. https://doi.org/10.1017/S0022112084001233.

Armi, L., Stommel, H., 1983. Four views of a portion of the North Atlantic subtropical gyre. J. Phys. Oceanogr. 13 (5), 828–857. https://doi.org/10.1175/1520-0485(1983)013<0828:FVOAPO>2.0.CO;2.

Bachman, S., Fox-Kemper, B., 2013. Eddy parameterization challenge suite I: Eady spindown. Ocean Model. 64, 12–18. https://doi.org/10.1016/j.ocemod.2012.12.003.

Bachman, S.D., 2019. The GM-E closure: a framework for coupling backscatter with the Gent and McWilliams parameterization. Ocean Model. (ISSN 1463-5003) 136, 85–106. https://doi.org/10.1016/j.ocemod.2019.02.006. https://www.sciencedirect.com/science/article/pii/S1463500318301975.

Bachman, S.D., Fox-Kemper, B., Bryan, F.O., 2020. A diagnosis of anisotropic eddy diffusion from a high-resolution global ocean model. J. Adv. Model. Earth Syst. 12 (2), e2019MS001904. https://doi.org/10.1029/2019MS001904.

Badin, G., Tandon, A., Mahadevan, A., 2011. Lateral mixing in the pycnocline by baroclinic mixed layer eddies. J. Phys. Oceanogr. 41 (11), 2080–2101. https://doi.org/10.1175/JPO-D-11-05.1.

Bahl, A., Gnanadesikan, A., Pradal, M.-A., 2019. Variations in ocean deoxygenation across Earth system models: isolating the role of parameterized lateral mixing. Glob. Biogeochem. Cycles 33 (6), 703–724. https://doi.org/10.1029/2018GB006121.

Bahl, A., Gnanadesikan, A., Pradal, M.-A., 2020. Scaling global warming impacts on ocean ecosystems: results from a suite of Earth System Models. Front. Mar. Sci. 7, 698. https://doi.org/10.3389/fmars.2020.00698.

Balwada, D., LaCasce, J.H., Speer, K.G., 2016a. Scale-dependent distribution of kinetic energy from surface drifters in the Gulf of Mexico. Geophys. Res. Lett. 43 (20), 10–856. https://doi.org/10.1002/2016GL069405.

Balwada, D., LaCasce, J.H., Speer, K.G., Ferrari, R., 2021. Relative dispersion in the Antarctic circumpolar current. J. Phys. Oceanogr. 51 (2), 553–574. https://doi.org/10.1175/JPO-D-19-0243.1. https://journals.ametsoc.org/view/journals/phoc/51/2/jpo-d-19-0243.1.xml.

Balwada, D., Speer, K.G., LaCasce, J.H., Owens, W.B., Marshall, J., Ferrari, R., 2016b. Circulation and stirring in the southeast Pacific Ocean and the Scotia Sea sectors of the Antarctic circumpolar current. J. Phys. Oceanogr. 46 (7), 2005–2027. https://doi.org/10.1175/JPO-D-15-0207.1. https://journals.ametsoc.org/view/journals/phoc/46/7/jpo-d-15-0207.1.xml.

Banyte, D., Visbeck, M., Tanhua, T., Fischer, T., Krahmann, G., Karstensen, J., 2013. Lateral diffusivity from tracer release experiments in the tropical North Atlantic thermocline. J. Geophys. Res., Oceans 118 (5), 2719–2733. https://doi.org/10.1002/jgrc.20211.

Barkan, R., Winters, K.B., McWilliams, J.C., 2017. Stimulated imbalance and the enhancement of eddy kinetic energy dissipation by internal waves. J. Phys. Oceanogr. (ISSN 0022-3670) 47 (1), 181–198. https://doi.org/10.1175/jpo-d-16-0117.1.

Batchelor, G.K., 1950. The application of the similarity theory of turbulence to atmospheric diffusion. Q. J. R. Meteorol. Soc. 76, 133–146. https://doi.org/10.1002/qj.49707632804.

Batchelor, G.K., 1952. Diffusion in a field of homogeneous turbulence, II: the relative motion of the particles. Proc. Philos. Soc. Camb. 48, 345–362. https://doi.org/10.1017/S0305004100027687.

Bauer, S., Swenson, M.S., Griffa, A., 2002. Eddy mean flow decomposition and eddy diffusivity estimates in the tropical Pacific Ocean: 2. Results. J. Geophys. Res. 107 (C10), 3154. https://doi.org/10.1029/2000JC000613.

Bennett, A.F., 1984. Relative dispersion: local and nonlocal dynamics. J. Atmos. Sci. 41, 1881–1886. https://doi.org/10.1175/1520-0469(1984)041<1881:RDLAND>2.0.CO;2.

Bentsen, M., Bethke, I., Debernard, J.B., Iversen, T., Kirkevåg, A., Seland, Ø., Drange, H., Roelandt, C., Seierstad, I.A., Hoose, C., et al., 2013. The Norwegian Earth system model, NorESM1-M—Part 1: description and basic evaluation of the physical climate. Geosci. Model Dev. 6 (3), 687–720. https://doi.org/10.5194/gmd-6-687-2013.

Boland, E.J., Shuckburgh, E., Haynes, P.H., Ledwell, J.R., Messias, M.-J., Watson, A.J., 2015. Estimating a submesoscale diffusivity using a roughness measure applied to a tracer release experiment in the Southern Ocean. J. Phys. Oceanogr. 45 (6), 1610–1631. https://doi.org/10.1175/JPO-D-14-0047.1.

Bolton, T., Zanna, L., 2019. Applications of deep learning to ocean data inference and subgrid parameterization. J. Adv. Model. Earth Syst. 11 (1), 376–399. https://doi.org/10.1029/2018MS001472.

Bowden, K.F., 1965. Horizontal mixing in the sea due to a shearing current. J. Fluid Mech. 21, 83–95. https://doi.org/10.1017/S0022112065000058.

Bower, A.S., 1991. A simple kinematic mechanism for mixing fluid parcels across a meandering jet. J. Phys. Oceanogr. 21, 173–182. https://doi.org/10.1175/1520-0485(1991)021<0173:ASKMFM>2.0.CO;2.

Bower, A.S., Lozier, M.S., 1994. A closer look at particle exchange in the Gulf Stream. J. Phys. Oceanogr. 24 (6), 1399–1418. https://doi.org/10.1175/1520-0485(1994)024<1399:ACLAPE>2.0.CO;2.

Bower, A.S., Rossby, H.T., Lillibridge, J.L., 1985. The Gulf Stream–barrier or blender. J. Phys. Oceanogr. 15, 24–33. https://doi.org/10.1175/1520-0485(1985)015<0024:TGSOB>2.0.CO;2.

Brink, K.H., Breadsley, R.C., Paduan, J., Limeburner, R., Caruso, M., Sires, J.G., 2000. A view of the 1993-1994 California Current based on surface drifters, floats, and remotely sensed data. J. Geophys. Res. 105 (C4), 8575–8604. https://doi.org/10.1029/1999JC900327.

Broecker, W.S., Peng, T.-H., 1982. Tracers in the Sea. Eldigio Press, Lamont-Doherty Geological Observatory, Columbia University Palisades, New York.

Buesseler, K., Livingston, H., Honjo, S., Hay, B., Manganini, S., Degens, E., Ittekkot, V., Izdar, E., Konuk, T., 1987. Chernobyl radionuclides in a Black Sea sediment trap. Nature 329 (6142), 825–828. https://doi.org/10.1038/329825a0.

Buesseler, K.O., Jayne, S.R., Fisher, N.S., Rypina, I.I., Baumann, H., Baumann, Z., Breier, C.F., Douglass, E.M., George, J., Macdonald, A.M., et al., 2012. Fukushima-derived radionuclides in the ocean and biota off Japan. Proc. Natl. Acad. Sci. 109 (16), 5984–5988. https://doi.org/10.1073/pnas.1120794109.

Bühler, O., Grisouard, N., Holmes-Cerfon, M., 2013. Strong particle dispersion by weakly dissipative random internal waves. J. Fluid Mech. 719, R4. https://doi.org/10.1017/jfm.2013.71.

Bühler, O., Guo, Y., 2016. Particle dispersion by nonlinearly damped random waves. J. Fluid Mech. 786, 332–347. https://doi.org/10.1017/jfm.2015.605.

Bühler, O., Holmes-Cerfon, M., 2009. Particle dispersion by random waves in rotating shallow water. J. Fluid Mech. 638, 5. https://doi.org/10.1017/S0022112009991091.

Busecke, J., Abernathey, R.P., Gordon, A.L., 2017. Lateral Eddy Mixing in the subtropical salinity maxima of the global Ocean. J. Phys. Oceanogr. 47 (4), 737–754. https://doi.org/10.1175/JPO-D-16-0215.1.

Busecke, J.J., Abernathey, R.P., 2019. Ocean mesoscale mixing linked to climate variability. Sci. Adv. 5 (1), eaav5014. https://doi.org/10.1126/sciadv.aav5014.

Carter, H., Okubo, A., 1965. A study of the physical processes of movement and dispersion in the Cape Kennedy area. Chesapeake Bay Institute, The Johns Hopkins University. Rep. Contract No. AT (30-1)-2973, Report Number NYO-2973-1, Pritchard, DW (Director).

Chen, R., Gille, S.T., McClean, J.L., 2017. Isopycnal eddy mixing across the Kuroshio Extension: stable versus unstable states in an eddying model. J. Geophys. Res., Oceans 122 (5), 4329–4345. https://doi.org/10.1002/2016JC012164.

Chen, R., McClean, J.L., Gille, S.T., Griesel, A., 2014. Isopycnal eddy diffusivities and critical layers in the Kuroshio Extension from an eddying ocean model. J. Phys. Oceanogr. 44 (8), 2191–2211. https://doi.org/10.1175/JPO-D-13-0258.1.

Cole, S.T., Wortham, C., Kunze, E., Owens, W.B., 2015. Eddy stirring and horizontal diffusivity from Argo float observations: geographic and depth variability. Geophys. Res. Lett. 42 (10), 3989–3997. https://doi.org/10.1002/2015GL063827.

Cole, S.T., Wortham, C.J.L., Owens, W.B., Kunze, E., 2017. Eddy diffusivity from Argo temperature and salinity profiles.

Colin de Verdiere, A., 1992. On the southward motion of Mediterranean salt lenses. J. Phys. Oceanogr. 22, 413–420. https://doi.org/10.1175/1520-0485(1992)022<0413:OTSMOM>2.0.CO;2.

Collins, C.A., Margolina, T., Rago, T.A., Ivanov, L., 2013. Looping RAFOS floats in the California Current System. Deep-Sea Res., Part 2 85, 42–61. https://doi.org/10.1016/j.dsr2.2012.07.027.

Colony, R., Thorndike, A., 1985. Sea ice motion as a drunkard's walk. J. Geophys. Res., Oceans 90 (C1), 965–974. https://doi.org/10.1029/JC090iC01p00965.

Csanady, G.T., 1966. Accelerated diffusion in the skewed shear flow of lake currents. J. Geophys. Res. 71, 411–420. https://doi.org/10.1029/JZ071i002p00411.

Danabasoglu, G., McWilliams, J.C., 1995. Sensitivity of the global ocean circulation to parameterizations of mesoscale tracer transports. J. Climate 8, 2967–2987. https://doi.org/10.1175/1520-0442(1995)008<2967:SOTGOC>2.0.CO;2.

Davis, R.E., 1991. Observing the general circulation with floats. Deep-Sea Res. Suppl. 38, S531–S571. https://doi.org/10.1016/S0198-0149(12)80023-9.

Dong, C., McWilliams, J.C., Liu, Y., Chen, D., 2014. Global heat and salt transports by eddy movement. Nat. Commun. 5 (3294). https://doi.org/10.1038/ncomms4294.
Dunne, J.P., John, J.G., Adcroft, A.J., Griffies, S.M., Hallberg, R.W., Shevliakova, E., Stouffer, R.J., Cooke, W., Dunne, K.A., Harrison, M.J., et al., 2012. GFDL's ESM2 global coupled climate–carbon Earth system models. Part I: physical formulation and baseline simulation characteristics. J. Climate 25 (19), 6646–6665. https://doi.org/10.1175/JCLI-D-11-00560.1.
Early, J.J., Lelong, M., Sundermeyer, M., 2021. A generalized wave-vortex decomposition for rotating Boussinesq flows with arbitrary stratification. J. Fluid Mech. 912, A32. https://doi.org/10.1017/jfm.2020.995.
Eden, C., Greatbatch, R., 2008a. Diapycnal mixing by meso-scale eddies. Ocean Model. 23, 113–120. https://doi.org/10.1016/j.ocemod.2008.04.006.
Eden, C., Greatbatch, R.J., 2008b. Towards a mesoscale eddy closure. Ocean Model. 20 (3), 223–239. https://doi.org/10.1016/j.ocemod.2007.09.002.
Eden, C., Greatbatch, R.J., Olbers, D., 2007. Interpreting eddy fluxes. J. Phys. Oceanogr. 37 (5), 1282–1296. https://doi.org/10.1175/JPO3050.1.
Falco, P., Griffa, A., Poulain, P.-M., Zambianchi, E., 2000. Transport properties in the Adriatic Sea as deduced from drifter data. J. Phys. Oceanogr. 30, 2055–2071. https://doi.org/10.1175/1520-0485(2000)030<2055:TPITAS>2.0.CO;2.
Falco, P., Zambianchi, E., 2011. Near-surface structure of the Antarctic Circumpolar Current derived from World Ocean Circulation Experiment drifter data. J. Geophys. Res., Oceans 116 (C5), C05003. https://doi.org/10.1029/2010JC006349.
Farneti, R., Gent, P.R., 2011. The effects of the eddy-induced advection coefficient in a coarse-resolution coupled climate model. Ocean Model. (ISSN 1463-5003) 39 (1), 135–145. https://doi.org/10.1016/j.ocemod.2011.02.005.
Ferrari, R., Griffies, S.M., Nurser, A.G., Vallis, G.K., 2010. A boundary-value problem for the parameterized mesoscale eddy transport. Ocean Model. 32 (3), 143–156. https://doi.org/10.1016/j.ocemod.2010.01.004.
Ferrari, R., Nikurashin, M., 2010. Suppression of eddy diffusivity across jets in the Southern Ocean. J. Phys. Oceanogr. 40, 1501–1519. https://doi.org/10.1175/2010JPO4278.1.
Ferrari, R., Polzin, K.L., 2005. Finescale structure of the T–S relation in the eastern North Atlantic. J. Phys. Oceanogr. 35 (8), 1437–1454. https://doi.org/10.1175/JPO2763.1.
Ferrari, R., Wunsch, C., 2009. Ocean circulation kinetic energy: reservoirs, sources, and sinks. Annu. Rev. Fluid Mech. 41. https://doi.org/10.1146/annurev.fluid.40.111406.102139.
Ferreira, D., Marshall, J., Heimbach, P., 2005. Estimating eddy stresses by fitting dynamics to observations using a residual-mean ocean circulation model and its adjoint. J. Phys. Oceanogr. 35, 1891–1910. https://doi.org/10.1175/JPO2785.1.
Fiadeiro, M., Craig, H., 1978. Three-dimensional modeling of tracers in the deep Pacific Ocean: I. Salinity and oxygen. J. Mar. Res. 36, 323–346.
Fine, R.A., 2011. Observations of CFCs and SF6 as ocean tracers. Annu. Rev. Mar. Sci. 3, 173–195. https://doi.org/10.1146/annurev.marine.010908.163933.
Fogli, P.G., Manzini, E., Vichi, M., Alessandri, A., Patara, L., Gualdi, S., Scoccimarro, E., Masina, S., Navarra, A., 2009. INGV-CMCC carbon (ICC): a carbon cycle Earth system model. CMCC Res. Pap. 61. https://doi.org/10.2139/ssrn.1517282.
Forget, G., Ferreira, D., Liang, X., 2015. On the observability of turbulent transport rates by Argo: supporting evidence from an inversion experiment. Ocean Sci. 11 (5), 839. https://doi.org/10.5194/os-11-839-2015.
Fox-Kemper, B., Lumpkin, R., Bryan, F.O., 2012. Lateral transport in the ocean interior. In: Siedler, G., Church, J., Gould, J., Griffies, S. (Eds.), Ocean Circulation and Climate – Observing and Modelling the Global Ocean. Elsevier. https://dx.doi.org/10.1016/B978-0-12-391851-2.00008-8.
Fox-Kemper, B., Menemenlis, D., 2008. Can large eddy simulation techniques improve mesoscale rich ocean models? In: Hecht, M.W., Hasumi, H. (Eds.), Ocean Modeling in an Eddying Regime. American Geophysical Union. https://dx.doi.org/10.1029/177GM19.
Freilich, M.A., Mahadevan, A., 2019. Decomposition of vertical velocity for nutrient transport in the upper ocean. J. Phys. Oceanogr. 49 (6), 1561–1575. https://doi.org/10.1175/JPO-D-19-0002.1.
Frenger, I., Bianchi, D., Stührenberg, C., Oschlies, A., Dunne, J., Deutsch, C., Galbraith, E., Schütte, F., 2018. Biogeochemical role of subsurface coherent eddies in the ocean: tracer cannonballs, hypoxic storms, and microbial stewpots? Glob. Biogeochem. Cycles 32 (2), 226–249. https://doi.org/10.1002/2017GB005743.
Froyland, G., Padberg, K., 2009. Almost-invariant sets and invariant manifolds—connecting probabilistic and geometric descriptions of coherent structures in flows. Phys. D, Nonlinear Phenom. 238 (16), 1507–1523. https://doi.org/10.1016/j.physd.2009.03.002.
Froyland, G., Padberg, K., England, M.H., Treguier, A.M., 2007. Detection of coherent oceanic structures via transfer operators. Phys. Rev. Lett. 98 (22), 224503. https://doi.org/10.1103/PhysRevLett.98.224503.
Garfield, N.C.A.C., Paquette, R.G., Carter, E., 1999. Lagrangian exploration of the California Undercurrent, 1992-95. J. Phys. Oceanogr. 29, 560–583. https://doi.org/10.1175/1520-0485(1999)029<0560:LEOTCU>2.0.CO;2.
Garrett, C., 1983. On the initial streakiness of a dispersing tracer in two- and three-dimensional turbulence. Dyn. Atmos. Ocean. 7, 265–277. https://doi.org/10.1016/0377-0265(83)90008-8.
Garrett, C., Munk, W., 1972. Oceanic mixing by breaking internal waves. Deep-Sea Res. 19, 823–932. https://doi.org/10.1016/0011-7471(72)90001-0.
Gent, P., 2016. Effects of Southern Hemisphere Wind Changes on the Meridional Overturning Circulation in Ocean Models. Annu. Rev. Mar. Sci. 8, 79–94. https://doi.org/10.1146/annurev-marine-122414-033929.
Gent, P., McWilliams, J., 1990. Isopycnal mixing in ocean circulation models. J. Phys. Oceanogr. 20, 150–155. https://doi.org/10.1175/1520-0485(1990)020<0150:IMIOCM>2.0.CO;2.
Gent, P.R., McWilliams, J.C., 1996. Eliassen-palm fluxes and the momentum equations in non-eddy-resolving ocean circulation models. J. Phys. Oceanogr. 26, 2539–2547. https://doi.org/10.1175/1520-0485(1996)026<2539:EFATME>2.0.CO;2.
George, TM.., Manucharyan, G.E., Thompson, A.F., 2021. Deep learning to infer eddy heat fluxes from sea surface height patterns of mesoscale turbulence. Nat. Commun. 12, 800. https://doi.org/10.1038/s41467-020-20779-9.

Gerdes, R., Köberle, C., Willebrand, J., 1991. The influence of numerical advection schemes on the results of ocean general circulation models. Clim. Dyn. 5. https://doi.org/10.1007/BF00210006.

Gill, A.E., 1973. Circulation and bottom water production in the Weddell Sea. Deep-Sea Res. 20, 111–140. https://doi.org/10.1016/0011-7471(73)90048-X.

Gnanadesikan, A., Griffies, S.M., Samuels, B.L., 2007. Effects in a climate model of slope tapering in neutral physics schemes. Ocean Model. (ISSN 1463-5003) 16 (1), 1–16. https://doi.org/10.1016/j.ocemod.2006.06.004.

Gnanadesikan, A., Pradal, M.-A., Abernathey, R., 2015a. Isopycnal mixing by mesoscale eddies significantly impacts oceanic anthropogenic carbon uptake. Geophys. Res. Lett. (ISSN 1944-8007) 42 (11), 4249–4255. https://doi.org/10.1002/2015GL064100.

Gnanadesikan, A., Pradal, M.-A., Abernathey, R., 2015b. Exploring the isopycnal mixing and helium–heat paradoxes in a suite of Earth system models. Ocean Sci. 11, 591–605. https://doi.org/10.5194/os-11-591-2015.

Gnanadesikan, A., Russell, A., Pradal, M.-A., Abernathey, R., 2017. Impact of lateral mixing in the ocean on El Nino in a suite of fully coupled climate models. J. Adv. Model. Earth Syst. 9 (7), 2493–2513. https://doi.org/10.1002/2017MS000917.

Gnanadesikan, A., Speller, C.M., Ringlein, G., Soucie, J.S., Thomas, J., Pradal, M.-A., 2020. Feedbacks driving interdecadal variability in Southern Ocean convection in climate models: a coupled oscillator mechanism. J. Phys. Oceanogr. 50 (8), 2227–2249. https://doi.org/10.1175/JPO-D-20-0037.1. https://journals.ametsoc.org/view/journals/phoc/50/8/jpoD200037.xml.

Gregg, M., D'Asaro, E., Riley, J., Kunze, E., 2018. Mixing efficiency in the ocean. Annu. Rev. Mar. Sci. 10 (1), 443–473. https://doi.org/10.1146/annurev-marine-121916-063643.

Grianik, N., Held, I., Smith, K., Vallis, G., 2004. The effects of quadratic drag on the inverse cascade of two-dimensional turbulence. Phys. Fluids 16 (1), 73–78. https://doi.org/10.1063/1.1630054.

Griesel, A., Dräger-Dietel, J., Jochumsen, K., 2019. Diagnosing and parameterizing the effects of oceanic eddies. In: Energy Transfers in Atmosphere and Ocean. Springer, pp. 193–224. https://dx.doi.org/10.1007/978-3-030-05704-6_6.

Griesel, A., Gille, S., Sprintall, J., McClean, J.L., Lacasce, J.H., Maltrud, M.E., 2010. Isopycnal diffusivities in the Antarctic circumpolar current inferred from Lagrangian floats in an eddying model. J. Geophys. Res. 115 (C6), C06006.

Griffies, S.M., 1998. The Gent-McWilliams skew flux. J. Phys. Oceanogr. 28, 831–841. https://doi.org/10.1175/1520-0485(1998)028<0831:TGMSF>2.0.CO;2.

Griffies, S.M., Gnanadesikan, A., Pacanowski, R.C., Larichev, V.D., Dukowicz, J.K., Smith, R.D., 1998. Isoneutral diffusion in a z-coordinate ocean model. J. Phys. Oceanogr. 28 (5), 805–830. https://doi.org/10.1175/1520-0485(1998)028<0805:IDIAZC>2.0.CO;2.

Groeskamp, S., Abernathey, R.P., Klocker, A., 2016. Water mass transformation by cabbeling and thermobaricity. Geophys. Res. Lett. (ISSN 1944-8007) 43 (20), 10,835–10,845. https://doi.org/10.1002/2016GL070860.

Groeskamp, S., Barker, P.M., McDougall, T.J., Abernathey, R.P., Griffies, S.M., 2019a. VENM: an algorithm to accurately calculate neutral slopes and gradients. J. Adv. Model. Earth Syst. 11 (7), 1917–1939. https://doi.org/10.1029/2019MS001613.

Groeskamp, S., Griffies, S.M., Iudicone, D., Marsh, R., Nurser, A.G., Zika, J.D., 2019b. The water mass transformation framework for ocean physics and biogeochemistry. Annu. Rev. Mar. Sci. 11, 271–305. https://doi.org/10.1146/annurev-marine-010318-095421.

Groeskamp, S., LaCasce, J.H., McDougall, T.J., Rogé, M., 2020. Full-depth global estimates of ocean mesoscale eddy mixing from observations and theory. Geophys. Res. Lett. 47 (18), e2020GL089425. https://doi.org/10.1029/2020GL089425. https://agupubs.onlinelibrary.wiley.com/doi/abs/10.1029/2020GL089425.

Groeskamp, S., Sloyan, B.M., Zika, J.D., McDougall, T.J., 2017. Mixing inferred from an ocean climatology and surface fluxes. J. Phys. Oceanogr. 47 (3), 667–687. https://doi.org/10.1175/JPO-D-16-0125.1.

Gruber, N., Lachkar, Z., Frenzel, H., Marchesiello, P., Münnich, M., McWilliams, J.C., Nagai, T., Plattner, G.-K., 2011. Eddy-induced reduction of biological production in eastern boundary upwelling systems. Nat. Geosci. 4 (11), 787–792. https://doi.org/10.1038/ngeo1273.

Hallberg, R., 2013. Using a resolution function to regulate parameterizations of oceanic mesoscale eddy effects. Ocean Model. 72, 92–103. https://doi.org/10.1016/j.ocemod.2013.08.007.

Haller, G., 2015. Lagrangian coherent structures. Annu. Rev. Fluid Mech. 47, 137–162. https://doi.org/10.1146/annurev-fluid-010313-141322.

Haynes, P.H., Anglade, J., 1997. The vertical-scale cascade of atmospheric tracers due to large-scale differential advection. J. Atmos. Sci. 54, 2317–2354. https://doi.org/10.1175/1520-0469(1997)054<1121:TVSCIA>2.0.CO;2.

Held, I.M., Larichev, V.D., 1996. A scaling theory for horizontally homogeneous, baroclinically unstable flow on a Beta plane. J. Atmos. Sci. 53 (7), 946–953. https://doi.org/10.1175/1520-0469(1996)053<0946:ASTFHH>2.0.CO;2.

Herterich, K., Hasselmann, K., 1982. The horizontal diffusion of tracers by surface waves. J. Phys. Oceanogr. 12 (7), 704–711. https://doi.org/10.1175/1520-0485(1982)012<0704:THDOTB>2.0.CO;2.

Hilário, A., Metaxas, A., Gaudron, S.M., Howell, K.L., Mercier, A., Mestre, N.C., Ross, R.E., Thurnherr, A.M., Young, C., 2015. Estimating dispersal distance in the deep sea: challenges and applications to marine reserves. Front. Mar. Sci. 2, 6. https://doi.org/10.3389/fmars.2015.00006. https://www.frontiersin.org/article/10.3389/fmars.2015.00006.

Ho, D.T., Ledwell, J.R., Smethie, W.M., 2008. Use of SF5CF3 for ocean tracer release experiments. Geophys. Res. Lett. 35 (4). https://doi.org/10.1029/2007GL032799.

Holmes-Cerfon, M., Bühler, O., Ferrari, R., 2011. Particle dispersion by random waves in the rotating Boussinesq system. J. Fluid Mech. 670, 150–175. https://doi.org/10.1017/S0022112010005240.

Hoskins, B.J., Bretherton, F.P., 1972. Atmospheric frontogenesis models: mathematical formulation and solution. J. Atmos. Sci. 29 (1), 11–37. https://doi.org/10.1175/1520-0469(1972)029<0011:AFMMFA>2.0.CO;2.

Houghton, R.W., 1997. Lagrangian flow at the foot of a shelfbreak front using a dye tracer injected into the bottom boundary layer. Geophys. Res. Lett. 24, 2035–2038. https://doi.org/10.1029/97GL02000.

Iudicone, D., Madec, G., McDougall, T.J., 2008. Water-mass transformations in a neutral density framework and the key role of light penetration. J. Phys. Oceanogr. 38, 1357–1376. https://doi.org/10.1175/2007JPO3464.1.

Jackett, D.R., McDougall, T.J., 1997. A neutral density variable for the world's oceans. J. Phys. Oceanogr. 27 (2), 237–263. https://doi.org/10.1175/1520-0485(1997)027<0237:ANDVFT>2.0.CO;2.

Jansen, M.F., Held, I.M., Adcroft, A., Hallberg, R., 2015. Energy budget-based backscatter in an eddy permitting primitive equation model. Ocean Model. 94, 15–26. https://doi.org/10.1016/j.ocemod.2015.07.015.

Johns, T.C., Durman, C.F., Banks, H.T., Roberts, M.J., McLaren, A.J., Ridley, J.K., Senior, C.A., Williams, K., Jones, A., Rickard, G., et al., 2006. The new Hadley Centre climate model (HadGEM1): evaluation of coupled simulations. J. Climate 19 (7), 1327–1353. https://doi.org/10.1175/JCLI3712.1.

Kämpf, J., Cox, D., 2016. Towards improved numerical schemes of turbulent lateral dispersion. Ocean Model. (ISSN 1463-5003) 106, 1–11. https://doi.org/10.1016/j.ocemod.2016.08.003. https://www.sciencedirect.com/science/article/pii/S1463500316300798.

Kersalé, M., Doglioli, A.M., Petrenko, A.A., 2011. Sensitivity study of the generation of mesoscale eddies in a numerical model of Hawaii islands. Ocean Sci. 7 (3), 277–291. https://doi.org/10.5194/os-7-277-2011. https://os.copernicus.org/articles/7/277/2011/.

Key, R.M., Kozyr, A., Sabine, C.L., Lee, K., Wanninkhof, R., Bullister, J.L., Feely, R.A., Millero, F.J., Mordy, C., Peng, T.-H., 2004. A global ocean carbon climatology: results from global data analysis project (GLODAP). Glob. Biogeochem. Cycles 18 (4). https://doi.org/10.1029/2004GB002247.

Killworth, P.D., 1986. A Bernoulli inverse method for determining the ocean circulation. J. Phys. Oceanogr. 16 (12), 2031–2051. https://doi.org/10.1175/1520-0485(1986)016<2031:ABIMFD>2.0.CO;2.

Klocker, A., Abernathey, R., 2014. Global patterns of mesoscale eddy properties and diffusivities. J. Phys. Oceanogr. 44, 1030–1047. https://doi.org/10.1175/JPO-D-13-0159.1.

Klocker, A., Ferrari, R., LaCasce, J.H., 2012a. Estimating suppression of eddy mixing by mean flows. J. Phys. Oceanogr. 42, 1566–1576. https://doi.org/10.1175/JPO-D-11-0205.1.

Klocker, A., Ferrari, R., LaCasce, J.H., Merrifield, S.T., 2012b. Reconciling float-based and tracer-based estimates of lateral diffusivities. J. Mar. Res. 70, 569–602. https://doi.org/10.1357/002224012805262743.

Klocker, A., McDougall, T.J., 2010. Quantifying the consequences of the ill-defined nature of neutral surfaces. J. Phys. Oceanogr. 40 (8), 1866–1880. https://doi.org/10.1175/2009JPO4212.1.

Kolmogorov, A.N., 1941. The local structure of turbulence in incompressible viscous fluid for very large Reynolds' numbers. Dokl. Akad. Nauk USSR 30, 301–305. https://doi.org/10.1098/rspa.1991.0075.

Kolmogorov, A.N., 1962. A refinement of previous hypotheses concerning the local structure of turbulence in a viscous incompressible fluid at high Reynolds number. J. Fluid Mech. 13 (1), 82–85. https://doi.org/10.1017/S0022112062000518.

Kong, H., Jansen, M.F., 2017. The eddy diffusivity in barotropic β-plane turbulence. Fluids 2 (4), 54. https://doi.org/10.3390/fluids2040054.

Koszalka, I., LaCasce, J.H., Andersson, M., Orvik, K.A., Mauritzen, C., 2011. Surface circulation in the Nordic Seas from clustered drifters. Deep-Sea Res., Part 1 58 (4), 468–485. https://doi.org/10.1016/j.dsr.2011.01.007.

Koszalka, I., LaCasce, J.H., Mauritzen, C., 2013. In pursuit of anomalies - analyzing the poleward transport of Atlantic Water with surface drifters. Deep-Sea Res., Part 2 85, 96–108. https://doi.org/10.1016/j.dsr2.2012.07.035.

Koszalka, I., LaCasce, J.H., Orvik, K.A., 2009. Relative dispersion in the Nordic Seas. J. Mar. Res. 67, 411–433. https://doi.org/10.1357/002224009790741102.

Kraichnan, R.H., 1967. Inertial ranges in two-dimensional turbulence. Phys. Fluids 10 (7), 1417–1423. https://doi.org/10.1063/1.1762301.

Kullenberg, G., 1971. Vertical diffusion in shallow waters. Tellus 23, 129–135. https://doi.org/10.1111/j.2153-3490.1971.tb00555.x.

Kunze, E., Sanford, T.B., 1993. Submesoscale dynamics near a seamount. Part I: measurements of Ertel vorticity. J. Phys. Oceanogr. 23 (12), 2567–2588. https://doi.org/10.1175/1520-0485(1993)023<2567:SDNASP>2.0.CO;2.

Kunze, E., Sanford, T.B., 2019. Vortical motion. In: Encyclopedia of Ocean Sciences, 3rd ed. Elsevier, pp. 702–706.

Kunze, E., Sundermeyer, M.A., 2015. The role of intermittency in internal-wave shear dispersion. J. Phys. Oceanogr. 45 (12), 2979–2990. https://doi.org/10.1175/JPO-D-14-0134.1.

LaCasce, J., 2008. Statistics from Lagrangian observations. Prog. Oceanogr. 77 (1), 1–29. https://doi.org/10.1016/j.pocean.2008.02.002.

LaCasce, J.H., 2016. Estimating Eulerian energy spectra from drifters. Fluids 1 (33). https://doi.org/10.3390/fluids1040033.

LaCasce, J.H., Bower, A., 2000. Relative dispersion in the subsurface North Atlantic. J. Mar. Res. 58, 863–894. https://doi.org/10.1357/002224000763485737.

LaCasce, J.H., Ferrari, R., Marshall, J., Tulloch, R., Balwada, D., Speer, K., 2014. Float-derived isopycnal diffusivities in the DIMES experiment. J. Phys. Oceanogr. 44, 764–780. https://doi.org/10.1175/JPO-D-13-0175.1.

LaCasce, J.H., Ohlmann, C., 2003. Relative dispersion at the surface of the Gulf of Mexico. J. Mar. Res. 61 (3), 285–312. https://doi.org/10.1357/002224003322201205.

Ledwell, J.R., Watson, A.J., 1991. The Santa Monica Basin tracer experiment: a study of diapycnal and isopycnal mixing. J. Geophys. Res., Oceans 96 (C5), 8695–8718. https://doi.org/10.1029/91JC00102.

Ledwell, J.R., Watson, A.J., Broecker, W.S., 1986. A deliberate tracer experiment in Santa Monica Basin. Nature 323 (6086), 322–324. https://doi.org/10.1038/323322a0.

Ledwell, J.R., Watson, A.J., Law, C.S., 1998. Mixing of a tracer in the pycnocline. J. Geophys. Res. 103 (C10), 21,499–21,592. https://doi.org/10.1029/98JC01738.

Lee, R.S., Pritchard, T.R., Ajani, K.P., Black, P.A., 2007. The influence of the East Australian Current eddy field on phytoplankton dynamics in the coastal zone. J. Coast. Res. 50, 576–584.

Lehahn, Y., d'Ovidio, F., Lévy, M., Amitai, Y., Heifetz, E., 2011. Long range transport of a quasi isolated chlorophyll patch by an Agulhas ring. Geophys. Res. Lett. 38 (16). https://doi.org/10.1029/2011GL048588.

Lelong, M.P., Cuypers, Y., Bouruet-Aubertot, P., 2020. Near-inertial energy propagation inside a Mediterranean anticyclonic eddy. J. Phys. Oceanogr. 50 (8), 2271–2288. https://doi.org/10.1175/jpo-d-19-0211.1.

Letscher, R.T., Primeau, F., Moore, J.K., 2016. Nutrient budgets in the subtropical ocean gyres dominated by lateral transport. Nat. Geosci. 9 (11), 815–819. https://doi.org/10.1038/ngeo2812.

Lévy, M., 2008. The modulation of biological production by oceanic mesoscale turbulence. In: Transport and Mixing in Geophysical Flows. Springer, pp. 219–261.

Lévy, M., Klein, P., Treguier, A.-M., 2001. Impact of sub-mesoscale physics on production and subduction of phytoplankton in an oligotrophic regime. J. Mar. Res. 59 (4), 535–565. https://doi.org/10.1357/002224001762842181.

Lien, R.-C., Sanford, T.B., 2019. Small-scale potential vorticity in the upper-ocean thermocline. J. Phys. Oceanogr. 49 (7), 1845–1872. https://doi.org/10.1175/JPO-D-18-0052.1.

Liu, C., Köhl, A., Stammer, D., 2012. Adjoint based estimation of eddy induced tracer mixing parameters in the global ocean. J. Phys. Oceanogr. 42, 1186–1206. https://doi.org/10.1175/JPO-D-11-0162.1.

Liu, T., Abernathey, R., Sinha, A., Chen, D., 2019. Quantifying Eulerian eddy leakiness in an idealized model. J. Geophys. Res., Oceans 124 (12), 8869–8886. https://doi.org/10.1029/2019JC015576.

Lobel, P., Robinson, A., 1986. Transport and entrapment of fish larvae by ocean mesoscale eddies and currents in Hawaiian Waters. Deep-Sea Res. A 33, 483–500. https://doi.org/10.1016/0198-0149(86)90127-5.

Lumpkin, R., Elipot, S., 2010. Surface drifter pair spreading in the North Atlantic. J. Geophys. Res. 115, C12017. https://doi.org/10.1029/2010JC006338.

Lumpkin, R., Flament, P., 2001. Lagrangian statistics in the central North Pacific. J. Mar. Syst. 29, 141–155. https://doi.org/10.1016/S0924-7963(01)00014-8.

Lumpkin, R., Johnson, G., 2013. Global ocean surface velocities from drifters: mean, variance, El Nino–Southern Oscillation response, and seasonal cycle. J. Geophys. Res. 118, 2992–3006. https://doi.org/10.1002/jgrc.20210.

Lumpkin, R., Pazos, M., 2007. Measuring surface currents with Surface Velocity Program drifters: the instrument, its data, and some recent results. In: Lagrangian Analysis and Prediction of Coastal and Ocean Dynamics, p. 39. Chapter 2.

Lumpkin, R., Treguier, A.-M., Speer, K., 2002. Lagrangian eddy scales in the Northern Atlantic Ocean. J. Phys. Oceanogr. 32, 2425–2440. https://doi.org/10.1175/1520-0485(2002)032<2425:LESITN>2.0.CO;2.

Luyten, J.R., Pedlosky, J., Stommel, H., 1983. The ventilated thermocline. J. Phys. Oceanogr. 13, 292–309. https://doi.org/10.1175/1520-0485(1983)013<0292:TVT>2.0.CO;2.

MacFadyen, A., Hickey, B.M., 2010. Generation and evolution of a topographically linked, mesoscale eddy under steady and variable wind-forcing. Cont. Shelf Res. (ISSN 0278-4343) 30 (13), 1387–1402. https://doi.org/10.1016/j.csr.2010.04.001. https://www.sciencedirect.com/science/article/pii/S0278434310001111.

MacGilchrist, G.A., Marshall, D.P., Johnson, H.L., Lique, C., Thomas, M., 2017. Characterizing the chaotic nature of ocean ventilation. J. Geophys. Res., Oceans 122 (9), 7577–7594. https://doi.org/10.1002/2017JC012875.

Marshall, D.P., Maddison, J.R., Berloff, P.S., 2012. A framework for parameterizing eddy potential vorticity fluxes. J. Phys. Oceanogr. 42 (4), 539–557. https://doi.org/10.1175/JPO-D-11-048.1.

Marshall, J., Shuckburgh, E., Jones, H., Hill, C., 2006. Estimates and implications of surface eddy diffusivity in the Southern Ocean derived from tracer transport. J. Phys. Oceanogr. 36, 1806–1821. https://doi.org/10.1175/JPO2949.1.

Marshall, J., Shutts, G., 1981. A note on rotational and divergent eddy fluxes. J. Phys. Oceanogr. 21, 1677–1681. https://doi.org/10.1175/1520-0485(1981)011<1677:ANORAD>2.0.CO;2.

Marsland, S.J., Haak, H., Jungclaus, J.H., Latif, M., Röske, F., 2003. The Max-Planck-Institute global ocean/sea ice model with orthogonal curvilinear coordinates. Ocean Model. 5 (2), 91–127. https://doi.org/10.1016/S1463-5003(02)00015-X.

Mashayek, A., Ferrari, R., Merrifield, S., Ledwell, J., St Laurent, L., Garabato, A.N., 2017. Topographic enhancement of vertical turbulent mixing in the Southern Ocean. Nat. Commun. 8 (1), 1–12. https://doi.org/10.1038/ncomms14197.

Mathur, M., Haller, G., Peacock, T., Ruppert-Felsot, J.E., Swinney, H.L., 2007. Uncovering the Lagrangian skeleton of turbulence. Phys. Rev. Lett. 98 (14), 144502. https://doi.org/10.1103/PhysRevLett.98.144502.

Maximenko, N., Hafner, J., Niiler, P., 2012. Pathways of marine debris derived from trajectories of Lagrangian drifters. Mar. Pollut. Bull. 65 (1–3), 51–62. https://doi.org/10.1016/j.marpolbul.2011.04.016.

McDougall, T., Feistel, R., Millero, F., Jackett, D., Wright, D., King, B., Marion, G., Chen, C., Spitzer, P., Seitz, S., 2009. The International Thermodynamic Equation of Seawater 2010 (TEOS-10): Calculation and Use of Thermodynamic Properties. Global Ship-based Repeat Hydrography Manual, IOCCP Report No. 14.

McDougall, T.J., Groeskamp, S., Griffies, S.M., 2017. Comment on Tailleux, R. Neutrality versus materiality: a thermodynamic theory of neutral surfaces. Fluids 2016, 1, 32. Fluids (ISSN 2311-5521) 2 (2). https://doi.org/10.3390/fluids2020019. https://www.mdpi.com/2311-5521/2/2/19.

McGillicuddy, D., 2016. Mechanisms of physical-biological-biogeochemical interaction at the oceanic mesoscale. Annu. Rev. Mar. Sci. 8, 125–159. https://doi.org/10.1146/annurev-marine-010814-015606.

McGillicuddy, D., Robinson, A., Siegel, D., Jannasch, H., Johnson, R., Dickey, T., McNeil, J., Michaels, A., Knap, A., 1998. Influence of mesoscale eddies on new production in the Sargasso Sea. Nature 394 (6690), 263–266. https://doi.org/10.1038/28367.

McWilliams, J.C., 1990. The vortices of two-dimensional turbulence. J. Fluid Mech. 219, 361–385. https://doi.org/10.1017/S0022112090002981.

McWilliams, J.C., Molemaker, M.J., 2011. Baroclinic frontal arrest: a sequel to unstable frontogenesis. J. Phys. Oceanogr. 41 (3), 601–619. https://doi.org/10.1175/2010JPO4493.1.

Molemaker, M.J., McWilliams, J.C., Dewar, W.K., 2015. Submesoscale instability and generation of mesoscale anticyclones near a separation of the California Undercurrent. J. Phys. Oceanogr. 45 (3), 613–629. https://doi.org/10.1175/JPO-D-13-0225.1.

Nakamura, N., 1995. Modified Lagrangian-mean diagnostics of the stratospheric polar vortices Part 1: formulation and analysis of GFDL SKYHI GCM. J. Atmos. Sci. 52 (11), 2096–2109. https://doi.org/10.1175/1520-0469(1995)052<2096:MLMDOT>2.0.CO;2.

Nakamura, N., 2001. A new look at eddy diffusivity as a mixing diagnostic. J. Atmos. Sci. 58, 3685–3702. https://doi.org/10.1175/1520-0469(2001)058<3685:ANLAED>2.0.CO;2.

Naveira Garabato, A.C., Polzin, K.L., Ferrari, R., Zika, J.D., Forryan, A., 2016. A microscale view of mixing and overturning across the Antarctic Circumpolar Current. J. Phys. Oceanogr. 46 (1), 233–254. https://doi.org/10.1175/JPO-D-15-0025.1.

Naveira Garabato, A.C., Stevens, D.P., Heywood, K.J., 2003. Water mass conversion, fluxes, and mixing in the Scotia Sea diagnosed by an inverse model. J. Phys. Oceanogr. 33 (12), 2565–2587. https://doi.org/10.1175/1520-0485(2003)033<2565:WMCFAM>2.0.CO;2.

Naveira-Garabato, A.R., Ferrari, R., Polzin, K., 2011. Eddy stirring in the Southern Ocean. J. Geophys. Res. 116, C09019. https://doi.org/10.1029/2010JC006818.

Nummelin, A., Busecke, J.J.M., Haine, T.W.N., Abernathey, R.P., 2021. Diagnosing the scale- and space-dependent horizontal eddy diffusivity at the global surface ocean. J. Phys. Oceanogr. 51 (2), 279–297. https://doi.org/10.1175/JPO-D-19-0256.1. https://journals.ametsoc.org/view/journals/phoc/51/2/jpo-d-19-0256.1.xml.

Oh, I.S., Zhurbas, V., Park, W., 2000. Estimating horizontal diffusivity in the East Sea (Sea of Japan) and the northwest Pacific from satellite-tracked drifter data. J. Geophys. Res., Oceans 105 (C3), 6483–6492. https://doi.org/10.1029/2000JC900002.

Okubo, A., 1971. Oceanic diffusion diagrams. Deep-Sea Res. 18, 789–802. https://doi.org/10.1016/0011-7471(71)90046-5.

Olbers, D., Willebrand, J., Eden, C., 2012. Ocean Dynamics. Springer Science & Business Media.

Oliveira, P.B., Serra, N., Fiúza, A., Ambar, I., 2010. A study of Meddies using simultaneous in situ and satellite observations. Elsevier Oceanogr. Ser. 63, 125–148. https://doi.org/10.1016/S0422-9894(00)80008-2.

Ollitrault, M., Colin de Verdière, A., 2002. SOFAR floats reveal midlatitude intermediate North Atlantic general circulation. Part I: a Lagrangian descriptive view. J. Phys. Oceanogr. 32 (7), 2020–2033. https://doi.org/10.1175/1520-0485(2002)032<2020:SFRMIN>2.0.CO;2.

Ollitrault, M., Gabillet, C., De Verdiere, A.C., 2005. Open ocean regimes of relative dispersion. J. Fluid Mech. 533, 381–407. https://doi.org/10.1017/S0022112005004556.

Osborn, T.R., Cox, C.S., 1972. Oceanic fine structure. Geophys. Astrophys. Fluid Dyn. 3 (1), 321–345. https://doi.org/10.1080/03091927208236085.

Ottino, J.M., 1989. The Kinematics of Mixing: Stretching, Chaos, and Transport. Cambridge University Press.

Palter, J.B., Griffies, S.M., Samuels, B.L., Galbraith, E.D., Gnanadesikan, A., Klocker, A., 2014. The deep ocean buoyancy budget and its temporal variability. J. Climate 27 (2), 551–573. https://doi.org/10.1175/JCLI-D-13-00016.1.

Pearson, B., Fox-Kemper, B., Bachman, S., Bryan, F., 2017. Evaluation of scale-aware subgrid mesoscale eddy models in a global eddy-rich model. Ocean Model. 115, 42–58. https://doi.org/10.1016/j.ocemod.2017.05.007.

Pelland, N.A., Eriksen, C.C., Lee, C.M., 2013. Subthermocline eddies over the Washington continental slope as observed by Seagliders, 2003–09. J. Phys. Oceanogr. 43, 2025–2053. https://doi.org/10.1175/JPO-D-12-086.1.

Pierrehumbert, R.T., 1991. Chaotic mixing of a tracer and vorticity by modulated traveling Rossby waves. Geophys. Astrophys. Fluid Dyn. 58, 285–320. https://doi.org/10.1080/03091929108227343.

Pierrehumbert, R.T., Yang, H., 1993. Global chaotic mixing on isentropic surfaces. J. Atmos. Sci. 50 (15), 2462–2480. https://doi.org/10.1175/1520-0469(1993)050<2462:GCMOIS>2.0.CO;2.

Pinkel, R., 2014. Vortical and internal wave shear and strain. J. Phys. Oceanogr. 44 (8), 2070–2092.

Plumb, R.A., 1979. Eddy fluxes of conserved quantities by small-amplitude waves. J. Atmos. Sci. 36, 1699–1705. https://doi.org/10.1175/1520-0469(1979)036<1699:EFOCQB>2.0.CO;2.

Plumb, R.A., Mahlman, J.D., 1987. The zonally-averaged transport characteristics of the GFDL general circulation/tracer model. J. Atmos. Sci. 44, 298–327. https://doi.org/10.1175/1520-0469(1987)044<0298:TZATCO>2.0.CO;2.

Polzin, K.L., Ferrari, R., 2004. Isopycnal dispersion in NATRE. J. Phys. Oceanogr. 34, 247–257. https://doi.org/10.1175/1520-0485(2004)034<0247:IDIN>2.0.CO;2.

Pradal, M.-A., Gnanadesikan, A., 2014. How does the Redi parameter for mesoscale mixing impact global climate in an Earth System Model? J. Adv. Model. Earth Syst. (ISSN 1942-2466) 6 (3), 586–601. https://doi.org/10.1002/2013MS000273.

Prandtl, L., 1925. Bericht über Untersuchungen zur ausgebildeten Turbulenz. Z. Angew. Math. Mech. 5 (2), 136–139. https://doi.org/10.1002/zamm.19250050212.

Pritchard, D.W., Carpenter, J., 1960. Measurements of turbulent diffusion in estuarine and inshore waters. Hydrol. Sci. J. 5 (4), 37–50. https://doi.org/10.1080/02626666009493189.

Qiu, B., Chen, S., Kessler, W.S., 2009. Source of the 70-day mesoscale eddy variability in the Coral Sea and the North Fiji Basin. J. Phys. Oceanogr. 39 (2), 404–420. https://doi.org/10.1175/2008JPO3988.1.

Redi, M., 1982. Oceanic isopycnal mixing by coordinate rotation. J. Phys. Oceanogr. 12, 1154–1158. https://doi.org/10.1175/1520-0485(1982)012<1154:OIMBCR>2.0.CO;2.

Reintges, A., Martin, T., Latif, M., Park, W., 2017. Physical controls of Southern Ocean deep-convection variability in CMIP5 models and the Kiel Climate Model. Geophys. Res. Lett. 44, 6951–6958. https://doi.org/10.1002/2017GL074087.

Resplandy, L., Lévy, M., Madec, G., Pous, S., Aumont, O., Kumar, D., 2011. Contribution of mesoscale processes to nutrient budgets in the Arabian Sea. J. Geophys. Res., Oceans 116 (C11), C11007. https://doi.org/10.1029/2011JC007006.

Rhines, P.B., 1975. Waves and turbulence on a beta-plane. J. Fluid Mech. 69 (3), 417–443. https://doi.org/10.1017/S0022112075001504.

Richardson, L.F., 1926. Atmospheric diffusion shown on a distance-neighbour graph. Proc. R. Soc. Lond. A 110, 709–737. https://doi.org/10.1098/rspa.1926.0043.

Richardson, L.F., Stommel, H., 1948. Note on eddy diffusion in the sea. J. Meteorol. 5, 238–240. https://doi.org/10.1175/1520-0469(1948)005<0238:NOEDIT>2.0.CO;2.

Richardson, P.L., Tychensky, A., 1998. Meddy trajectories in the Canary Basin measured during the SEMAPHORE experiment, 1993–1995. J. Geophys. Res. 103, 25,029–25,045. https://doi.org/10.1029/97JC02579.

Riley, J.J., Lelong, M.P., 2000. Fluid motions in the presence of strong stable stratification. Annu. Rev. Fluid Mech. 32 (1), 613–657. https://doi.org/10.1146/annurev.fluid.32.1.613.

Roach, C.J., Balwada, D., Speer, K., 2016. Horizontal mixing in the Southern Ocean from Argo float trajectories. J. Geophys. Res., Oceans 121 (8), 5570–5586. https://doi.org/10.1002/2015JC011440. https://agupubs.onlinelibrary.wiley.com/doi/abs/10.1002/2015JC011440.

Rocha, C.B., Gille, S.T., Chereskin, T.K., Menemenlis, D., 2016. Seasonality of submesoscale dynamics in the Kuroshio Extension. Geophys. Res. Lett. 43 (21), 11–304. https://doi.org/10.1002/2016GL071349.

Romps, D.M., Kuang, Z., 2011. A transilient matrix for moist convection. J. Atmos. Sci. 68 (9), 2009–2025. https://doi.org/10.1175/2011JAS3712.1.

Rossby, T., 2007. Evolution of Lagrangian methods in oceanography. In: Griffa, A., Kirwan, A.D., Mariano, A.J., Ozgokmen, T., Rossby, T. (Eds.), Lagrangian Analysis and Prediction of Coastal and Ocean Dynamics. Cambridge University Press, pp. 1–38. Chapter 1.

Rudnick, D.L., Ferrari, R., 1999. Compensation of horizontal temperature and salinity gradients in the ocean mixed layer. Science (ISSN 0036-8075) 283 (5401), 526–529. https://doi.org/10.1126/science.283.5401.526. https://science.sciencemag.org/content/283/5401/526.

Rupolo, V., 2007. A Lagrangian-based approach for determining trajectories taxonomy and turbulence regimes. J. Phys. Oceanogr. 37, 1584–1609. https://doi.org/10.1175/JPO3038.1.

Rye, C.D., Messias, M.-J., Ledwell, J.R., Watson, A.J., Brousseau, A., King, B.A., 2012. Diapycnal diffusivities from a tracer release experiment in the deep sea, integrated over 13 years. Geophys. Res. Lett. 39 (4). https://doi.org/10.1029/2011GL050294.

Rypina, I.I., Kamenkovich, I., Berloff, P., Pratt, L., 2012. Eddy-induced particle dispersion in the near-surface Atlantic. J. Phys. Oceanogr. 42, 2206–2228. https://doi.org/10.1175/JPO-D-11-0191.1.

Rypina, I.I., Pratt, L.J., Pullen, J., Levin, J., Gordon, A.L., 2010. Chaotic advection in an Archipelago*. J. Phys. Oceanogr. 40 (9), 1988–2006. https://doi.org/10.1175/2010JPO4336.1.

Saffman, P.G., 1962. The effect of wind shear on horizontal spread from an instantaneous ground source. Q. J. R. Meteorol. Soc. 88, 382–393. https://doi.org/10.1002/qj.49708837803.

Shao, A.E., Adcroft, A., Hallberg, R., Griffies, S.M., General-Coordinate, A., 2020. Nonlocal neutral diffusion operator. J. Adv. Model. Earth Syst. 12 (12), e2019MS001992. https://doi.org/10.1029/2019MS001992. https://agupubs.onlinelibrary.wiley.com/doi/abs/10.1029/2019MS001992.

Shcherbina, A.Y., Sundermeyer, M.A., Kunze, E., D'Asaro, E., Badin, G., Birch, D., Brunner-Suzuki, A.-M.E., Callies, J., Kuebel Cervantes, B.T., Claret, M., et al., 2015. The LatMix summer campaign: submesoscale stirring in the upper ocean. Bull. Am. Meteorol. Soc. 96 (8), 1257–1279. https://doi.org/10.1175/BAMS-D-14-00015.1.

Shuckburgh, E., Jones, H., Marshall, J., Hill, C., 2009a. Robustness of an effective diffusivity diagnostic in oceanic flows. J. Phys. Oceanogr. 39, 1993–2010. https://doi.org/10.1175/2009JPO4122.1.

Shuckburgh, E., Jones, H., Marshall, J., Hill, C., 2009b. Understanding the regional variability of eddy diffusivity in the Pacific sector of the Southern Ocean. J. Phys. Oceanogr. 39, 2011–2023. https://doi.org/10.1175/2009JPO4115.1.

Siegel, D., Deuser, W., 1997. Trajectories of sinking particles in the Sargasso Sea: modeling of statistical funnels above deep-ocean sediment traps. Deep-Sea Res., Part 1, Oceanogr. Res. Pap. 44 (9–10), 1519–1541. https://doi.org/10.1016/S0967-0637(97)00028-9.

Siegel, D., Kinlan, B., Gaylord, B., Gaines, S., 2003. Lagrangian descriptions of marine larval dispersion. Mar. Ecol. Prog. Ser. 260, 83–96. https://doi.org/10.3354/meps260083.

Smith, K.S., 2005. Tracer transport along and across coherent jets in two-dimensional turbulent flow. J. Fluid Mech. 544, 133–142. https://doi.org/10.1017/S0022112005006750.

Smith, K.S., Boccaletti, G., Henning, C.C., Marinov, I., Tam, C.Y., Held, I.M., Vallis, G.K., 2002. Turbulent diffusion in the geostrophic inverse cascade. J. Fluid Mech. 469, 13–48. https://doi.org/10.1017/S0022112002001763.

Smith, K.S., Ferrari, R., 2009. The production and dissipation of compensated thermohaline variance by mesoscale stirring. J. Phys. Oceanogr. 39 (10), 2477–2501. https://doi.org/10.1175/2009JPO4103.1. https://journals.ametsoc.org/view/journals/phoc/39/10/2009jpo4103.1.xml.

Smith, K.S., Marshall, J., 2009. Evidence for enhanced eddy mixing at mid-depth in the Southern Ocean. J. Phys. Oceanogr. 39, 50–69. https://doi.org/10.1175/2008JPO3880.1.

Solomon, H., 1971. On the representation of isentropic mixing in ocean circulation models. J. Phys. Oceanogr. 1 (3), 233–234. https://doi.org/10.1175/1520-0485(1971)001<0233:OTROIM>2.0.CO;2. https://journals.ametsoc.org/view/journals/phoc/1/3/1520-0485_1971_001_0233_otroim_2_0_co_2.xml.

Spall, M.A., 1997. Baroclinic jets in confluent flow. J. Phys. Oceanogr. 27 (6), 1054–1071. https://doi.org/10.1175/1520-0485(1997)027<1054:BJICF>2.0.CO;2.

Spiro Jaeger, G., Mahadevan, A., 2018. Submesoscale-selective compensation of fronts in a salinity-stratified ocean. Sci. Adv. 4 (2). https://doi.org/10.1126/sciadv.1701504. https://advances.sciencemag.org/content/4/2/e1701504.

Srinivasan, K., Young, W., 2014. Reynolds stress and eddy diffusivity of β-plane shear flows. J. Atmos. Sci. 71 (6), 2169–2185. https://doi.org/10.1175/JAS-D-13-0246.1.

Stull, R.B., 1993. Review of non-local mixing in turbulent atmospheres: transilient turbulence theory. Bound.-Layer Meteorol. 62 (1–4), 21–96. https://doi.org/10.1007/BF00705546.

Sundermeyer, M.A., Birch, D.A., Ledwell, J.R., Levine, M.D., Pierce, S.D., Kuebel, B.T., 2020. Dispersion in the open ocean seasonal pycnocline at scales of 1–10 km and 1–6 days. J. Phys. Oceanogr. 50 (2), 415–437. https://doi.org/10.1175/JPO-D-19-0019.1.

Sundermeyer, M.A., Ledwell, J.R., 2001. Lateral dispersion over the continental shelf: analysis of dye release experiments. J. Geophys. Res., Oceans 106 (C5), 9603–9621. https://doi.org/10.1029/2000JC900138.

Sundermeyer, M.A., Ledwell, J.R., Oakey, N.S., Greenan, B.J.W., 2005. Stirring by small-scale vortices caused by patchy mixing. J. Phys. Oceanogr. 35, 1245–1262. https://doi.org/10.1175/JPO2713.1.

Sundermeyer, M.A., Lelong, M.-P., 2005. Numerical simulations of lateral dispersion by the relaxation of diapycnal mixing events. J. Phys. Oceanogr. 35, 2368–2386. https://doi.org/10.1175/JPO2834.1.

Sundermeyer, M.A., Price, J.F., 1998. Lateral mixing and the North Atlantic Tracer Release Experiment: observations and numerical simulations of Lagrangian particles and a passive tracer. J. Geophys. Res., Oceans 103 (C10), 21481–21497. https://doi.org/10.1029/98JC01999.

Tailleux, R., 2016. Generalized patched potential density and thermodynamic neutral density: two new physically based quasi-neutral density variables for ocean water masses analyses and circulation studies. J. Phys. Oceanogr. 46 (12), 3571–3584. https://doi.org/10.1175/JPO-D-16-0072.1.

Tailleux, R., 2017. Reply to "Comment on Tailleux, R. Neutrality versus materiality: a thermodynamic theory of neutral surfaces. Fluids 2016, 1, 32". Fluids (ISSN 2311-5521) 2 (2). https://doi.org/10.3390/fluids2020020. https://www.mdpi.com/2311-5521/2/2/20.

Taylor, G.I., 1915. Eddy motion in the atmosphere. Philos. Trans. R. Soc., Ser. A 215, 1–26. https://doi.org/10.1098/rsta.1915.0001.

Taylor, G.I., 1921. Diffusion by continuous movements. Proc. Lond. Math. Soc. s2–20, 196–212. https://doi.org/10.1112/plms/s2-20.1.196.

Taylor, G.I., 1953. Dispersion of soluble matter in solvent flowing slowly through a tube. Proc. R. Soc. Lond. Ser. A 219, 186–203. https://doi.org/10.1098/rspa.1953.0139.

Thomas, J., Arun, S., 2020. Near-inertial waves and geostrophic turbulence. Phys. Rev. Fluids 5 (1), 014801. https://doi.org/10.1103/PhysRevFluids.5.014801.

Thomas, J., Waugh, D., Gnanadesikan, A., 2018. Relationship between ocean carbon and heat multidecadal variability. J. Climate 31 (4), 1467–1482. https://doi.org/10.1175/JCLI-D-17-0134.1.

Thompson, A.F., Young, W.R., 2006. Scaling baroclinic eddy fluxes: vortices and energy balance. J. Phys. Oceanogr. 36, 720–738. https://doi.org/10.1175/JPO2874.1.

Thompson, A.F., Young, W.R., 2007. Two-layer baroclinic eddy heat fluxes: zonal flows and energy balance. J. Atmos. Sci. 64, 3214–3232. https://doi.org/10.1175/JAS4000.1.

Treguier, A.M., 1999. Evaluating eddy mixing coefficients from eddy-resolving ocean models: a case study. J. Mar. Res. 57, 89–108. https://doi.org/10.1357/002224099765038571.

Tulloch, R., Ferrari, R., Jahn, O., Klocker, A., LaCasce, J., Ledwell, J.R., Marshall, J., Messias, M.-J., Speer, K., Watson, A., 2014. Direct estimate of lateral eddy diffusivity upstream of Drake Passage. J. Phys. Oceanogr. 44 (10), 2593–2616. https://doi.org/10.1175/JPO-D-13-0120.1.

Tulloch, R., Marshall, J., Smith, K.S., 2009. Interpretation of the propagation of surface altimetric observations in terms of planetary waves and geostrophic turbulence. J. Geophys. Res. 114, C02005. https://doi.org/10.1029/2008JC005055.

Tychensky, A., Carton, X., 1998. Hydrological and dynamical characterization of Meddies in the Azores region: a paradigm for baroclinic vortex dynamics. J. Geophys. Res. 103, 25,061–25,079. https://doi.org/10.1029/97JC03418.

Uchida, T., Balwada, D., Abernathey, R., McKinley, G., Smith, S., Lévy, M., 2019. The contribution of submesoscale over mesoscale eddy iron transport in the open Southern Ocean. J. Adv. Model. Earth Syst. 11 (12), 3934–3958. https://doi.org/10.1029/2019MS001805.

Urakawa, L.S., Hasumi, H., 2012. Eddy-resolving model estimate of the cabbeling effect on the water mass transformation in the Southern Ocean. J. Phys. Oceanogr. 42 (8), 1288–1302. https://doi.org/10.1175/JPO-D-11-0173.1.

Vallis, G., 2006. Atmospheric and Oceanic Fluid Dynamics. Cambridge University Press.

Vallis, G.K., Maltrud, M.E., 1993. Generation of mean flows and jets on a beta plane and over topography. J. Phys. Oceanogr. 23 (7), 1346–1362. https://doi.org/10.1175/1520-0485(1993)023<1346:GOMFAJ>2.0.CO;2.

Van Sebille, E., England, M.H., Froyland, G., 2012. Origin, dynamics and evolution of ocean garbage patches from observed surface drifters. Environ. Res. Lett. 7 (4), 044040. https://doi.org/10.1088/1748-9326/7/4/044040.

Visbeck, M., Marshall, J., Haine, T., 1997. Specification of eddy transfer coefficients in coarse-resolution ocean circulation models. J. Phys. Oceanogr. 27, 381–403. https://doi.org/10.1175/1520-0485(1997)027<0381:SOETCI>2.0.CO;2.

Vollmer, L., Eden, C., 2013. A global map of mesoscale eddy diffusivities based on linear stability analysis. Ocean Model. 72, 198–209. https://doi.org/10.1016/j.ocemod.2013.09.006.

Wang, Y., Olascoaga, M.J., Beron-Vera, F.J., 2015. Coherent water transport across the South Atlantic. Geophys. Res. Lett. 42 (10), 4072–4079. https://doi.org/10.1002/2015GL064089.

Wanninkhof, R., et al., 1997. Gas exchange, dispersion, and biological productivity on the West Florida shelf: results from a Lagrangian tracer study. Geophys. Res. Lett. 24, 1767–1770. https://doi.org/10.1029/97GL01757.

Waugh, D.W., Keating, S.R., Chen, M.-L., 2012. Diagnosing ocean stirring: comparison of relative dispersion and finite-time Lyapunov exponents. J. Phys. Oceanogr. 42 (7), 1173–1185. https://doi.org/10.1175/JPO-D-11-0215.1. https://journals.ametsoc.org/view/journals/phoc/42/7/jpo-d-11-0215.1.xml.

Wenegrat, J.O., Thomas, L.N., Sundermeyer, M.A., Taylor, J.R., D'Asaro, E.A., Klymak Jody, M., Shearman, K.R., Lee Craig, M., 2020. Enhanced mixing across the gyre boundary at the Gulf Stream front. Proc. Natl. Acad. Sci. USA 117 (30), 17607–17614. https://doi.org/10.1073/pnas.2005558117.

White, C., Selkoe, K.A., Watson, J., Siegel, D.A., Zacherl, D.C., Toonen, R.J., 2010. Ocean currents help explain population genetic structure. Proc. R. Soc. B 277, 1685–1694. https://doi.org/10.1098/rspb.2009.2214.

Whitney, F., Robert, M., 2002. Structure of Haida eddies and their transport of nutrient from coastal margins into the NE Pacific Ocean. J. Oceanogr. 58 (5), 715–723. https://doi.org/10.1023/A:1022850508403.

Wilson, C., Williams, R.G., 2004. Why are eddy fluxes of potential vorticity difficult to parameterize? J. Phys. Oceanogr. 34, 142–155. https://doi.org/10.1175/1520-0485(2004)034<0142:WAEFOP>2.0.CO;2.

Worthington, L., 1981. The water masses of the world ocean: some results of a fine-scale census. In: Warren, B.A., Wunsch, C. (Eds.), Evolution of Physical Oceanography, pp. 42–69.

Wunsch, C., 1977. Determining the general circulation of the oceans: a preliminary discussion. Science 196 (4292), 871–875. https://doi.org/10.1126/science.196.4292.871.

Wunsch, C., 1987. Using transient tracers: the regularization problem. Tellus B 39 (5), 477–492. https://doi.org/10.3402/tellusb.v39i5.15363.

Ying, Y., Maddison, J., Vanneste, J., 2019. Bayesian inference of ocean diffusivity from Lagrangian trajectory data. Ocean Model. 140, 101401. https://doi.org/10.1016/j.ocemod.2019.101401.

Young, W.R., 2012. An exact thickness-weighted average formulation of the Boussinesq equations. J. Phys. Oceanogr. 42 (5), 692–707. https://doi.org/10.1175/JPO-D-11-0102.1.

Young, W.R., Rhines, P.B., Garrett, C.J.R., 1982. Shear-flow dispersion, internal waves and horizontal mixing in the ocean. J. Phys. Oceanogr. 12, 515–527. https://doi.org/10.1175/1520-0485(1982)012<0515:SFDIWA>2.0.CO;2.

Zanna, L., Mana, P.P., Anstey, J., David, T., Bolton, T., 2017. Scale-aware deterministic and stochastic parametrizations of eddy-mean flow interaction. Ocean Model. 111, 66–80. https://doi.org/10.1016/j.ocemod.2017.01.004.

Zhang, H.-M., Prater, M.D., Rossby, T., 2001. Isopycnal Lagrangian statistics from the North Atlantic Current RAFOS float observations. J. Geophys. Res. 106 (C7), 13817–13836. https://doi.org/10.1029/1999JC000101.

Zhang, R., 2017. On the persistence and coherence of subpolar sea surface temperature and salinity anomalies associated with the Atlantic multidecadal variability. Geophys. Res. Lett. 44 (15), 7865–7875. https://doi.org/10.1002/2017GL074342.

Zhang, W., Wolfe, C.L., Abernathey, R., 2020. Role of surface-layer coherent eddies in potential vorticity transport in quasigeostrophic turbulence driven by eastward shear. Fluids 5 (1), 2. https://doi.org/10.3390/fluids5010002.

Zhang, Z., Wang, W., Qiu, B., 2014. Oceanic mass transport by mesoscale eddies. Science 345, 322–324. https://doi.org/10.1126/science.1252418.

Zhurbas, V., Lyzhkov, D., Kuzmina, N., 2014. Drifter-derived estimates of lateral eddy diffusivity in the World Ocean with emphasis on the Indian Ocean and problems of parametrization. Deep-Sea Res., Part 1 83, 1–11. https://doi.org/10.1016/j.dsr.2013.09.001.

Zika, J.D., McDougall, T.J., Sloyan, B.M., 2010. A tracer-contour inverse method for estimating ocean circulation and mixing. J. Phys. Oceanogr. 40 (1), 26–47. https://doi.org/10.1175/2009JPO4208.1.

Zika, J.D., Sallée, J.-B., Meijers, A.J., Naveira-Garabato, A.C., Watson, A.J., Messias, M.-J., King, B.A., 2020. Tracking the spread of a passive tracer through Southern Ocean water masses. Ocean Sci. 16 (2), 323–336. https://doi.org/10.5194/os-16-323-2020.

Chapter 10

Mixing in equatorial oceans

James N. Moum[a], Andrei Natarov[b], Kelvin J. Richards[b], Emily L. Shroyer[a] and William D. Smyth[a]

[a]*College of Earth, Ocean, and Atmospheric Sciences, Oregon State University, Corvallis, OR, United States,* [b]*IPRC, University of Hawaii at Manoa, Honolulu, HI, United States*

10.1 Introduction

Equatorial regions have a global footprint on Earth's climate. The equatorial current system and associated wave guide set the background for several atmosphere-ocean coupled phenomena, notably the El Niño Southern Oscillation (ENSO) sourced from the Pacific Ocean and the Madden–Julian Oscillation (MJO) that originates in the Indian Ocean, with corresponding impacts on regional climate and weather around the world. Cold tongue regions in the Atlantic, and particularly the Pacific, are known sites of intense ocean heat uptake (Holmes et al., 2019), yet are poorly represented in climate models, with persistent biases in both the Pacific (cool bias, Li and Xie, 2014; Li et al., 2019) and the Atlantic (warm bias, Dippe et al., 2018). Due to broadscale upwelling of CO_2-rich deep water, equatorial regions are the main oceanic source of carbon to the atmosphere (Murray et al., 1994). The equatorial oceans are also regions of significant air-sea oxygen exchange associated with subsurface changes that contribute to long-term global reductions in marine oxygen (Oschlies et al., 2018). These few examples highlight the importance of air-sea coupled process within the equatorial band. This coupling determines the transfer of heat, freshwater, dissolved gasses, and momentum across the sea surface. Through redistribution of subsurface momentum and scalars, ocean mixing plays a fundamental role in regulating fluxes between ocean and atmosphere.

The equatorial current system is comprised of alternating bands of zonal currents with dynamics distinguished by vanishing Coriolis frequency f. In the annual mean, easterly trade winds in the Atlantic and the Pacific drive the broad, westward South equatorial current (SEC), which spans roughly 500 km across the equator and is meridionally divergent. On its northern flank (spanning $\sim$2–10°N), the North Equatorial Countercurrent (NECC) flows eastward, counter to prevailing winds, weakening and shallowing to the east as its axis shifts northward. The zonal pressure gradient set up by surface currents is partially balanced by the eastward-flowing Equatorial Undercurrent (EUC), a thin ribbon of a subsurface current extending $\mathcal{O}(10,000)$ km zonally, $\mathcal{O}(100)$ km meridionally, and $\mathcal{O}(100)$ m vertically.

Equatorial currents are associated with strong velocities ($\mathcal{O}(1)$ m s^{-1}) with accompanying lateral and vertical shear that results in dynamic instability at a variety of scales. With opposing currents separated by $\mathcal{O}(100)$ m in depth, vertical shear (Sh) is $\mathcal{O}(0.01)$ s^{-1} above the EUC. (Here, $Sh = \sqrt{u_z^2 + v_z^2}$, where u, v are zonal and meridional currents, respectively, and the subscript z refers to differentiation in the vertical coordinate direction.) Since Sh is roughly matched by the stratification, $N = \sqrt{-g/\rho \cdot \sigma_{\theta_z}}$, where g is the local gravitational acceleration, ρ is density, σ_θ is the potential density anomaly of the fluid, the gradient Richardson number, $Ri = N^2/Sh^2$ is consistently $O(1)$ or less.

Near-surface meridional divergence driven by easterly winds in most of the equatorial Atlantic and Pacific causes upwelling of cool waters near the equator. As a result, meridional sea surface temperature (SST) gradients with pronounced fronts form within $\pm 5°$ of the equator. Tropical Instability Waves (TIWs) are clearly distinguished in SST fields (e.g., Fig. 10.1) as westward propagating oscillations (wavelength $\sim$1000 km) that perturb these off-equatorial SST fronts. Formation of TIWs is thought to be linked to instability associated with lateral and vertical shear between the NECC, SEC and EUC (Von Schuckmann et al., 2008). TIWs control SST in part through advection of temperature fields, and their signature has been noted within atmospheric boundary layer wind perturbations (e.g., see summary presented in Contreras, 2002).

In contrast to the Pacific and Atlantic Oceans, winds over the tropical Indian Ocean reverse direction with the South Asian monsoon. Likewise, equatorial currents reverse with the monsoon (Schott et al., 2009). Variability includes the development of westerly wind-driven, eastward-flowing Yoshida-Wyrtki surface jets during inter-monsoon periods and intermittent occurrence of an eastward-flowing EUC between February and June (Reppin et al., 1999). Equatorial currents also have pronounced intraseasonal variability (e.g., McPhaden and Taft, 1988; Senan et al., 2003) in association with westerly wind bursts (WWBs) accompanying the MJO (Madden and Julian, 1971, 1972) and its counterpart, the Boreal

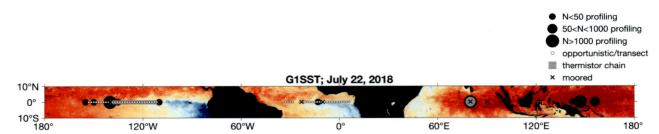

FIGURE 10.1 Locations of ocean turbulence measurements on the equator overlaid on SST.

TABLE 10.1 Summary of shipboard turbulence measurements at equatorial sites.

Location	Reference/Project	Source/Type	Data Content	Depth Range
155 W	Gregg 1976	χ_T, χ_S profiling	6 profiles	upper 700 m
150 W	Crawford 1982	ε profiling	33 profiles	upper 500 m
140 W	Tropic Heat I & II; TIWE; EQ08	ε, χ profiling	$\mathcal{O}(10000)$ profiles	upper 500 m
135 W	Herbert et al. 1992	ε profiling, thermistor chain	$\mathcal{O}(10)$ profiles, one hour	upper 125 m
137–153 W	Moum, Osborn, & Crawford 1986	ε profiling	8 profiles	upper 1000 m
110–140 W	Moum et al. 1992	ε profiling, thermistor chain	$\mathcal{O}(1000)$ profiles, 2 weeks	upper 125 m
110–140 W	Herbert, Moum, & Caldwell 1991	ε transect	$\mathcal{O}(1000)$ profiles	upper 110 m
110 W	Holmes, Moum, & Thomas 2016	CTD with χ	27 profiles	full depth
13 W	Polzin et al. 1996	ε profiling	20 profiles	full depth
28–33 W	Crawford & Osborn 1980	ε, χ transect	8 profiles	upper 300 m
24 W–6 E	Hummels, Dengler & Bourles 2013	ε, χ transect	783 profiles	upper 300 m
80 E	Pujiana, Moum & Smyth 2018	ε, χ profiling	6957 profiles	upper 500 m
80 E	Moulin, Moum & Shroyer 2018	thermistor chain	5 days	upper 10 m
142 E	Liang et al. 2019	ε, χ profiling	7 profiles	full depth
147 E	Brainerd and Gregg 1997	ε, χ profiling	$\mathcal{O}(1000)$ profiles	upper 70 m
156 E	Richards et al. 2015	ε, χ profiling	$\mathcal{O}(1000)$ profiles	upper 750 m

Summer Intraseasonal Oscillation (BSISO, Hartmann et al., 1992). These large-scale atmospheric disturbances propagate around the globe within 30–90-day period, contributing to intraseasonal variations in weather and climate on regional (Hartmann and Michelsen, 1989; Goswami et al., 2006) and global scales (Zhang, 2005, 2013).

At longer timescales, ENSO drives variability across the air-sea interface, which has recently been shown to project onto the scale of turbulent mixing (Section 10.7). Though originating in the Pacific, ENSO has far-reaching impacts, facilitated in part by the nature of the equatorial wave guide in the atmosphere. Through air-sea coupling, ocean mixing has the potential to modify the coupled response to large-scale teleconnections, such as ENSO. However, confirmation requires sampling at a vast range of time- and space scales, spanning the global equatorial wave guide. Existing turbulence measurements on the equator are limited (Fig. 10.1). The combination of a series of shipboard profiling experiments and more recent moored measurements of turbulence above the EUC core depth at 0°140°W suggest great temporal variability: from hourly (Sections 10.3, 10.4) to weekly (Section 10.6) to multi-year (Section 10.7). See Table 10.1

Our knowledge of spatial variability is rudimentary, the preponderance of measurements having been made at 0°140°W to date. Cross-equatorial gradients in stratification and currents are large; and, though we expect this to be reflected in turbulence variations, existing observations are sparse. Given measurement challenges, it is possible that improved parameterisations will help to constrain the temporal and spatial variations in ocean mixing that contribute to the net heat, carbon, oxygen and momentum transfer from the atmosphere through the upper ocean to depth (Section 10.8). A summary of known unknowns as well as anecdotal surprises is presented in Section 10.9. The two supplementary boxes highlight recently-observed phenomena of as-yet undetermined importance (Box 10.1, Box 10.2).

10.2 Ocean turbulence peaks at the equator, or does it?

Off the equator and away from special phenomena, such as fronts and large amplitude non-linear internal waves, mixed-layer turbulence beneath the zone of surface wave-breaking is forced by buoyancy and wind stress and follows relatively well-established scalings, (Gargett, 1989; Brainerd and Gregg, 1993; Thorpe, 2005). These scalings reasonably predict turbulence in mid-latitude mixed layers as forcing varies on seasonal timescales (Evans et al., 2018). In the thermoclines underlying these mid-latitude mixed layers, turbulence is strongly reduced and prediction has defied simple scalings.

The first turbulence measurements in the upper equatorial oceans, made during the Global atmospheric research program (GARP) Atlantic tropical experiment (GATE), showed enhanced turbulence isolated above the Atlantic EUC core within a narrow band $\pm 1°$ of the equator, and a direct analogue in the Pacific was made a few years later (Crawford and Osborn, 1981). Subsequent measurements constrained to the equator showed significant time variations, especially on diurnal timescales (Gregg et al., 1985; Moum and Caldwell, 1985). These studies established that the turbulent kinetic energy dissipation rate ε varied by more than an order of magnitude between measurement periods and/or locations. The relative contribution from meridional or temporal variation to this signal was unclear. Ultimately, cross-equatorial turbulence profiling transects showed the localised turbulence maximum unambiguously, particularly beneath the mixed layers and extending to the EUC core in both Pacific (Hebert et al., 1991a; Peters et al., 1989) and Atlantic cold tongues (Hummels et al., 2013). Confirming the results of the original GATE study, enhanced turbulence above the EUC core encompasses a band of about $\pm 1°$ across the equator, roughly the width of the EUC.

The intense mixing above the EUC core is the principal focus of this paper, largely because of its tremendous importance to global heat uptake from the atmosphere (Kosaka and Xie, 2013; Holmes et al., 2019), at least in the cold tongues. However, this signal is not the only one that sets the equatorial regime apart from others. Surprisingly, and in contrast to the equatorial peak above the EUC core, it is possible that mixing from breaking internal gravity waves in the deep thermocline beneath the EUC core is reduced at equatorial latitudes relative to mid-latitudes through geometrical considerations (Gregg et al., 2003), albeit with relatively few measurements. More surprising still, near-bottom mixing at very low latitudes is intensified in the absence of any topography, again in contrast to near-bottom mixing at mid-latitudes. This latter effect may be associated with the downward propagation and bottom reflection of equatorially-trapped waves formed at the sea surface (see Box 10.1).

In sum, the equator represents a local maximum in near-surface and near-bottom mixing and a local minimum in mid-water column mixing, although near-bottom and mid-water column results could stand more observational support.

10.3 Mixing in the cold tongues: diurnal forcing of turbulence below the mixed layer

Early observations of mixing in the eastern equatorial oceans showed the surprising result that, while there was a well-defined nocturnal mixed layer, turbulence extended well below and into stably stratified waters on a daily basis (Moum and Caldwell, 1985; Gregg et al., 1985, Fig. 10.2e). This deep nocturnal turbulence was eventually termed the 'deep cycle'. Since the turbulent heat flux is represented by the product of a turbulence diffusivity and the thermal gradient, the combination of high diffusivity and stratification carries a strong flux of heat from the mixed layer into the ocean interior, and thus plays an important role in the climate system (Moum et al., 2009, 2013).

The diurnal character of the deep cycle turbulence suggested a connection to the surface forcing, but the nature of that connection was not immediately clear. Observations revealed band-limited vertical motions, strongest at night, that were consistent with either shear instability (Section 10.4) or vertically propagating internal waves (Moum et al., 1992a; McPhaden and Peters, 1992). Though related, those two processes imply different mechanisms for transferring the diurnal signal downward. Shear instability accomplishes the task by creating layers of sheared turbulence that spread downward by means of entrainment, whereas internal waves carry a flux of momentum that can be deposited upon reaching a critical level. The model of Schudlich and Price (1992) included turbulent entrainment only and successfully reproduced the deep cycle. Later observations, combined with linear stability analysis, showed that the observed mean flows were indeed vulnerable to shear instability at the right times and depths (Sun et al., 1998; Moum et al., 2011; Smyth et al., 2011, 2013), tending to confirm the Schudlich–Price scenario, as described below.

A fundamental component of the deep-cycle mechanism is the shear between the EUC and the SEC (Fig. 10.2a, c). Acting against the stable stratification (Fig. 10.2d), which is maintained by the combination of surface heating and upwelling of cold water, this shear keeps Ri well below unity and often close to the critical value 1/4 (Moum et al., 1989). Stability analysis of the observed currents, together with measurements from moored turbulence sensors (Sun et al., 1998; Moum et al., 2011; Smyth et al., 2011, 2013), show that shear instability occurs, although sporadically (see Section 10.4 below).

The process that triggers the deep cycle begins at sunrise with the establishment of the daytime (or diurnal) warm layer (Price et al., 1986b). Immediately after sunrise, penetrating solar radiation begins to stratify the upper ~ 1 m of the ocean,

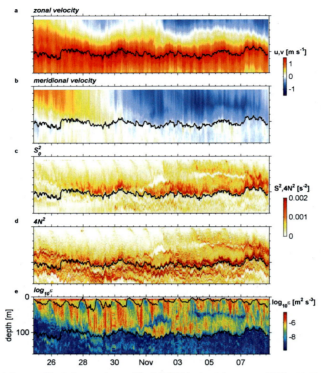

FIGURE 10.2 Measurements from 16-day profiling time series at 0°, 140°W in boreal autumn 2008. (a) Zonal velocity; (b) meridional velocity; (c) squared shear, S^2; (d) $4N^2$; (e) turbulence dissipation rate, ε. The black line in (a)–(d) shows the core (eastward velocity maximum) of the eastward-flowing EUC. The black line in (e) is the mixed-layer depth defined as the depth at which ρ deviates by 0.01 kg m^{-3} from its surface value. Adapted from Moum et al. (2009).

causing remnant nocturnal turbulence to decay rapidly (Sutherland et al., 2016; Moulin et al., 2018). In the presence of even light winds (>2 m s^{-1}), the resulting suppression of the turbulent momentum flux allows a significant surface current to develop (Hughes et al., 2020). The resultant shear is sufficient to induce mixing that slowly deepens the diurnal warm layer at a rate of ∼2 m per hour (Sutherland et al., 2016).

In mid-to-late afternoon, the surface buoyancy flux (Fig. 10.3a) weakens until it can no longer stabilise the flow against wind forcing, and the surface current becomes fully turbulent. It then begins to erode the underlying stable stratification, now descending more rapidly, ∼6 m/hour (Fig. 10.3b). Because the prevailing wind is easterly, this descending shear layer adds to the existing shear of the zonal current system, decreasing Ri to values below 1/4 (Fig. 10.3d) and triggering the onset of deep-cycle turbulence (Fig. 10.3e).

The deep cycle is a ubiquitous feature of equatorial turbulence, occurring at various longitudes and in all seasons (e.g., Pham et al., 2017). Its vertical extent is governed by the wind. In boreal spring, for example, trade winds are weak in the eastern equatorial Pacific and the deep cycle is confined to a thin layer near the surface. In the equatorial Indian ocean, even after an extended calm period, a westerly wind burst has been observed to accelerate the surface current and lead quickly to the appearance of the deep cycle (Pujiana et al., 2018).

10.4 The concepts of marginal instability and self-organised criticality and how they apply to mixing in the cold tongues

The change in sign of the Coriolis acceleration makes the equatorial oceans, particularly the cold tongues, a natural laboratory for the study of stratified, parallel shear flows. Lacking rapid inertial oscillations, winds and currents remain nearly steady and zonal over spatiotemporal intervals that are large compared to turbulent motions. (Alternatively, we might simply consider the wind and current systems to be the inertial response at the equator, cf. Section 10.5.)

The most important mechanism for turbulence production in equatorial parallel shear flows is the inflectional instability, which operates at local maxima of the mean shear profile (Smyth and Carpenter, 2019). In the presence of stable stratification, inflectional instability is damped, but it may yet grow, provided that the minimum value of Ri is less than critical.

Mixing in equatorial oceans Chapter | 10 **261**

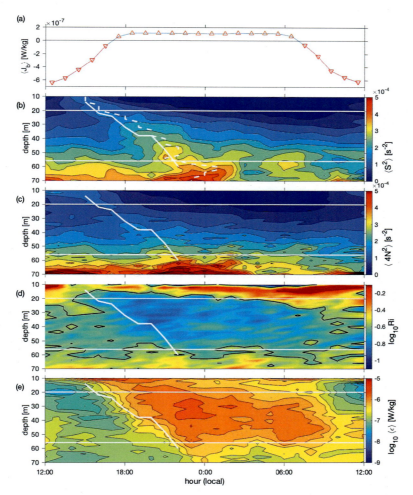

FIGURE 10.3 Diurnal composite computed from the data shown in Fig. 10.2 over 25 Oct. 2008 to 02 Nov. 2008. The abscissa is the local time, which lags UTC by nine hours. (a) surface buoyancy flux, J_b. (b) Squared shear, S^2. (c) Four times the squared buoyancy frequency N^2. (The factor 4 is chosen so that S^2 and N^2 are the same colour when Ri is near the critical value of 1/4.) (d) $Ri = N^2/S^2$. The black contour represents $Ri = 1/4$. (e) ε. In (b), solid/dashed white curves indicate different measures of the descent rate of the shear layer; the solid line is echoed in (c)–(e). Here, the descent rate is about 6 m per hour. Adapted from Smyth and Moum (2013).

In this case, the process is termed Kelvin–Helmholtz (KH) instability. In the inviscid limit, the critical value of Ri is 1/4 (Miles, 1961; Howard, 1961). In the presence of viscosity and diffusion (which may be either molecular or turbulent in origin), the critical value is usually (though not always) reduced (Thorpe and Liu, 2009; Smyth et al., 2013; Thorpe et al., 2013; Li et al., 2015). Nevertheless, in the context of observational oceanography, with its inevitable complications and measurement uncertainties, 1/4 remains a useful rule of thumb.

It has long been known that, in the turbulent upper flank of the equatorial undercurrent, Ri remains near 1/4 much of the time (Peters et al., 1988; Moum et al., 1989; Smyth et al., 2013; Pham et al., 2017), but is significantly lower than 1/4 (suggesting instability) only occasionally. This is now understood to indicate that Ri fluctuates around its critical value as the mean flow cycles between stable and unstable states (Smyth et al., 2017). In the unstable state, turbulence smooths the mean velocity and density gradients so as to increase Ri. When Ri exceeds 1/4, the inflectional mechanism stops and turbulence decays. But the forcing continues to accelerate the mean flow and, in the absence of turbulence, eventually reduces Ri to values $<1/4$, and the cycle repeats (Smyth et al., 2017, 2019). This quasi-equilibrium state has come to be known as "marginal instability" (MI), although variations of Ri from the critical value are not necessarily small.

Although it has been studied in detail at the equator, the MI state has been observed at locations around the world (Smyth, 2020). The essential requirement is a delicate balance, where the shear is forced strongly enough, relative to the

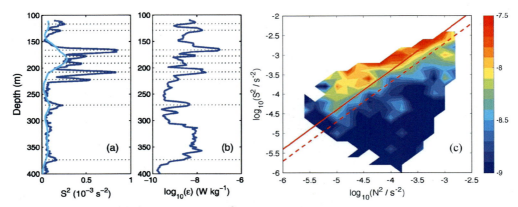

FIGURE 10.4 (a) and (b): Time-averaged vertical shear squared (Sh^2) and ε, respectively, from measurements in the WEP. The cyan curve in (a) is Sh^2 from lower resolution shear measurements. (c) ε binned averaged with respect to N^2 and Sh^2. The solid and dashed red lines are Richardson number, $Ri = 0.25$ and 0.5, respectively. (a and b used with permission from Richards et al. (2012a); c used with permission from Richards et al. (2015)).

forcing of the stratification, to maintain Ri near 1/4, but not so strongly as to obliterate the stratification entirely (as in boundary layers, e.g., Grachev et al., 2015).

MI is an example of a broader class of phenomena, denoted 'self-organised criticality' (SOC), in which a complex system regulates itself so as to remain near a critical state (Smyth et al., 2019). The canonical example is the sandpile model of Bak et al. (1987), in which sand is poured steadily onto a horizontal surface. Growing steeper at first, the sandpile eventually exceeds a critical slope, whereupon avalanches act to relieve the resulting gravitational instability. Called the 'angle of repose', the critical slope is independent of the pouring rate. Over time, sporadic avalanches occur at just the rate needed to keep the average steepness near the angle of repose. In the case of deep cycle mixing, the trade winds play the role of the sand source, whereas the critical state $Ri = 1/4$ is analogous to the angle of repose. SOC has been identified in a wide range of complex phenomena, including earthquakes, forest fires, solar flares, stock market crashes and biological extinction events (Jensen, 1998; Aschwanden, 2014, 2016). Analogous behaviour is evident in other geophysical instability phenomena, such as baroclinic instability (Stone, 1978) and salt fingering (Schmitt, 1981). This ubiquity suggests a promise of cross-fertilisation: an advance in understanding one system might furnish a clue to understanding another.

Recognition of the MI state is useful in predicting mixing characteristics of sheared, stratified turbulence. For example, the flux coefficient Γ (sometimes called the 'mixing efficiency') is the ratio of the turbulent buoyancy flux to the kinetic energy dissipation rate ε. That ratio is reliably close to 0.2 in marginally unstable flows (Smyth et al., 2019; Smyth, 2020), allowing prediction of scalar fluxes on the basis of microstructure measurements. One example is the measurement of the extraordinary turbulent heat flux that occurs when deep-cycle turbulence is amplified by tropical instability waves (Moum et al., 2009, see Box 10.2).

10.5 The importance of inertia-gravity waves and flow instabilities

Away from the hectic eastern Pacific (with its TIWs, deep cycle turbulence and shallow and strongly-sheared EUC) the western equatorial Pacific (WEP) appears at first sight to be relatively placid (less eddy activity and a deep, weakly sheared EUC). High-resolution salinity measurements suggest a more complex state (Richards and Pollard, 1991), where alternate layers of fresher and saltier water, $\mathcal{O}(10)$ m thick, extend a few hundred kilometres across the equator. Subsequent measurements from a lowered ADCP (600 kHz) (Richards et al., 2012b) reveal the vertical shear to be dominated by small-scale features that are associated with peaks in ε (Fig. 10.4a, b). In contrast, the shear at the top of the EUC, from the hull-mounted and lower frequency ADCP, is much reduced (cyan curve in Fig. 10.4a). Measurements from a number of cruises to the WEP using a similar combination of instruments show the dominance of the small vertical scale (finescale) shear to be a common feature (Richards et al., 2015). Theory and numerical studies suggest two mechanisms for the small-scale shear.

The first mechanism for small-scale shear generation is inertia-gravity waves (IGWs), long recognised as a significant source of energetic motion at small vertical scales in the ocean (Voorhis and Perkins, 1966, for example). Most research has focused on IGWs in regions sufficiently removed from the equator so that internal waves are separated from the mean flow and eddies by a measurable gap in frequency spectra. In this case, dynamical understanding of the wave field can be conveniently conceptualised in terms of forced motions at two sharp, isolated spectral peaks (inertial and tidal frequencies), plus a 'continuum': a broadband spectrum presumably maintained by non-linear wave-wave interactions redistributing near-inertial and tidal energy among other allowable frequencies (Kantha and Clayson, 2000, for example). The continuum has been viewed as having a high degree of universality (Garrett and Munk, 1972), and various mixing parameterisations have been devised based on the Garrett–Munk spectrum.

IGWs at the equator are different from the IGWs at midlatitudes in the following important ways (cf., Natarov and Richards, 2019):

- *No spectral gap.* Equatorial IGWs can have arbitrarily long periods. One important dynamic consequence is that almost any wind stress forcing will excite IGWs, in contrast to midlatitudes, where waves can only be excited by high-frequency wind forcing. Another consequence is that the IGWs overlap with mean flows and eddies in the frequency domain, leading to potential interactions between various types of motion.
- *Equatorial trapping.* Equatorial IGWs cannot propagate meridionally past their turning latitude.[1] The important consequence of this is the meridional convergence of wave energy, its accumulation and subsequent wave breaking at turning latitudes. Equatorial trapping is often described in terms of standing meridional modes: solutions of highly idealised, linear models, such as the well-known equatorial β-plane with its Hermite functions (cf., (Matsuno, 1966)). Though for low baroclinic modes, standing-wave meridional structure has been successfully identified in buoy and altimetry data (e.g., Farrar and Durland, 2012), there is little observational evidence that such structure forms for high baroclinic modes. Models suggest that, instead, a wave packet composed of high baroclinic and meridional modes forms a beam bouncing off turning latitudes at a different depth each time it encounters one, breaking and dissipating energy as it travels. Argo-based estimates show elevated levels of mixing close to the equator, with the maxima found not directly on the equator, but a few degrees off (Whalen et al., 2012). We note that though these estimates are consistent with the role of meridional convergence of wave action at turning latitudes described above, Whalen et al. estimates are based on the assumption that wave-breaking rate is governed by weakly non-linear wave-wave interactions.
- *Strong zonal equatorial currents.* Unlike midlatitude IGWs, equatorial IGWs exist within the equatorial current system and its associated strong vertical and meridional shear. The complex background current system leads to the implication of increased likelihood of propagating waves encountering critical levels (location where the phase speed of the wave matches the background velocity), where wave energy is deposited. Theory and idealised numerical experiments suggest that critical layer absorption plays an important role in shaping the spatial distributions of both mixing and wave field.

Observation of equatorial IGWs is challenging, since they are hard to separate from other types of motion in terms of frequency, and because it may be necessary to combine spatial coverage (especially in the meridional direction) with conventional time-series measurements. Altimetry measurements have provided one means of wavenumber-frequency decomposition for these signals (e.g., Farrar and Durland, 2012), but this approach in isolation cannot yield detailed information about the internal structure, particularly that associated with high vertical wavenumbers. Some IGWs are well represented in ocean general circulation models (OGCMs), but typically OGCMs do not adequately capture the small vertical-scale IGWs that produce mixing, in part because vertical resolution is still too coarse. Mixing parameterisations applied in OGCMs are also typically too diffusive and dissipate energy of small vertical-scale IGWs before they can manifest themselves dynamically.

Another mechanism that produces small-scale shear is instability of the system of intense equatorial currents (EUC, NECC, SEC, etc.). It has long been known that both the equatorial ocean and atmosphere are susceptible to instabilities that favour small vertical scales. Perhaps the best known example is the inertial instability, first extensively studied in the atmosphere (Dunkerton, 1981; Stevens, 1983). In the equatorial ocean, it has been considered a potential mechanism for deep jet formation (Hua et al., 1997), and a possible explanation for similar phenomena observed in the upper ocean (Richards and Edwards, 2003). A necessary condition for inertial instability is the existence of anomalous potential vorticity regions, where the potential vorticity associated with a flow, Q, is of opposite sign to the local Coriolis parameter f, i.e., if $fQ < 0$. The change in sign of f across $f = 0$ contributes to such regions of $fQ < 0$. Inertial instability, however, is just one type of instability that favours growth of small vertical-scale features.

Another type of instability favouring small-scale vertical structures is the Kelvin-wave instability (Boyd and Christidis, 1982; Natarov and Boyd, 2001), which does not require a condition on potential vorticity. Instead, Kelvin-wave instability relies on the existence of regions where the velocity of the background current matches the phase speed of the equatorial

---

1. Stratified, rotating fluids support freely propagating IGWs only in the range of frequencies between $f = 2\Omega|sin\theta|$ and N. Here Ω is Earth's rotation frequency and θ is latitude. A poleward propagating IGW with frequency ω less than 2Ω will therefore encounter a latitude $\theta_{turn} = \pm \sin^{-1}\left(\frac{\omega}{2\Omega}\right)$, beyond which free-wave propagation is not supported. Such latitudes are called *turning latitudes* since, here, poleward-propagating IGWs reflect and propagate back toward and past the equator, until encountering another turning latitude in the opposite hemisphere. An introductory discussion of turning latitudes and equatorial waves is found in Gill (1982).

Kelvin wave. Such instabilities are referred to as critical layer, critical latitude, or 'supersonic' instabilities (Balmforth, 1999). Unlike gravity and Rossby waves, which tend to be absorbed at critical layers, equatorial Kelvin waves are amplified (Natarov and Boyd, 2001). While low-order baroclinic Kelvin modes have phase speeds much faster than typical velocities found in the EUC or NECC, and are therefore not susceptible to Kelvin-wave instability, phase speeds of high-order baroclinic Kelvin modes can be matched by swift equatorial currents, making them unstable. A classification of such instabilities and explanation of their mechanism in terms of resonant wave interactions in a shear flow is reviewed by Taniguchi and Ishiwatari (2006).

Stability theory that determines the above criteria is, strictly speaking, only applicable to steady background flows. Equatorial currents, on the other hand, are observed to be strongly unsteady and meandering. For background flows that are periodic in time, application of Floquet analysis (cf. Bender and Orszag, 2013) reveals that they are highly susceptible to parametric subharmonic instability (PSI; cf. LeBlonde and Mysak, 1978). Specific results in the context of large-scale equatorial waves and currents can be found in d'Orgeville and Hua (2005); Natarov et al. (2008); Hua et al. (1997) and Natarov and Richards (2009). An important conclusion drawn from these studies is that a periodic background flow can generate small vertical-scale features without violating stability criteria for steady flows (Natarov and Richards, 2009).

10.6 Westerly wind bursts in the Indian Ocean and western Pacific

MJO and monsoon predictability have been limited in part by the lack of observations, in particular the feedback role that the ocean plays in modification of SST via subsurface turbulence-enhanced mixing (Zhang, 2005, 2013; Moum et al., 2016). This lack of observations led to field campaigns to study the evolution of the MJO in the western Pacific ocean (Coupled ocean-atmosphere response experiment, COARE) in 1992–93 and its origins in the central Indian ocean through the Dynamics of the MJO experiment (DYNAMO) in 2011–12.

The MJO is associated with deep convection in the atmosphere, surface WWBs and heavy precipitation. WWBs lasting a few days accelerated upper ocean currents over at least 450 km across the equator in the central equatorial Indian ocean, increasing eastward transport in the upper 100 m by a factor of 2, e.g., from 12 to 23 Sv in November 2011 during DYNAMO (Moum et al., 2014). These short bursts of wind energise upper ocean turbulence not only directly for short periods while the winds blow, but also indirectly and for more extended periods, through shear-driven mixing at the base of and within the accelerated surface current (Pujiana et al., 2018; Hoecker-Martinez et al., 2016).

The heavy precipitation creates freshwater layers that spread laterally, at least partly as buoyant gravity currents (Moulin et al., 2021) and persist following MJO events, at least under winds less than about 8 m s^{-1} (Thompson et al., 2019). The enhanced stratification caused by these rain layers effectively isolates the water beneath from surface forcing. Consequently, pre-existing subsurface turbulence (such as generated by an accompanying WWB or local squall) decays through viscous dissipation, although it is more strongly concentrated nearer the surface. Smyth et al. (1997) showed that turbulence decays exponentially with a decay timescale $\sim 2\pi/N$, much faster than laboratory measurements in unstratified fluids.

Extended turbulence profiling during DYNAMO has clearly shown that turbulence-enhanced mixing during WWBs and surface fluxes contribute equally to mixed layer cooling. Turbulence also contributes to vertical redistribution of salt, typically upwards from intermediate layers sourced from the salty Arabian sea and into the relatively fresh near-surface layers. Finally, turbulent momentum (or Reynolds) stresses are important to the deceleration of the near-surface jet initially accelerated by WWB surface wind stress (Pujiana et al., 2018). Inadequate representation of these processes in numerical models of MJO evolution may contribute to notable discrepancies between models and observations (Jensen et al., 2015), particularly through control on SST by subsurface mixing of temperature (Moum et al., 2016).

10.7 Variations on subseasonal, seasonal and interannual timescales

Technical development of χpods[2] (Moum and Nash, 2009) has led to extended time series of ocean mixing. These time series are much longer than can be obtained by shipboard measurements with traditional vertical microstructure profilers. Through redeployments at a few fixed-depth levels, a record at the NOAA TAO mooring at 0°, 140°W in the central equatorial Pacific toward the western extent of the cold tongue is presently nearly 15 years long, albeit with multiple extended gaps. This record has yielded new insights into the role that mixing plays on subseasonal and longer timescales, including ENSO.

From the first six years of the record at 0°, 140°W, a seasonal cycle in mixing is apparent (Moum et al., 2013; Xie, 2013). In the tropics, net surface heating varies semiannually, dominated by variations in incoming short wave radiation as

---

2. χpods are self-contained instruments that measure turbulence on oceanographic moorings using multiple sensors and high sampling rates (100 Hz) for periods up to 18 months.

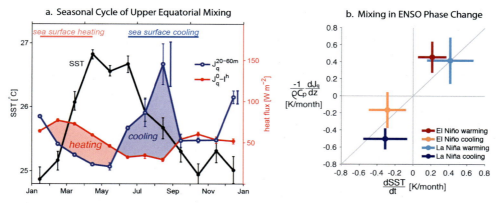

FIGURE 10.5 (a) Seasonal cycle of upper ocean turbulent heat flux at 0°, 140°W (blue) in comparison to net surface heat flux (red) and SST (black) over 2005–2011 (Moum et al., 2013). (b) Comparison of the vertical divergence of turbulent heat flux (ordinate) and time rate of change of SST (abscissa) for the periods of El Niño warming/cooling and La Niña warming/cooling at 0°, 140°W over 2005–2016 (Warner and Moum, 2019).

the sun passes overhead twice per year. In contrast, cold tongue SST has a strong annual cycle. This apparent inconsistency is resolved through assessment of the heat flux divergence across the sea surface using surface flux measurements from the mooring buoy and subsurface flux estimates from χ pod measurements. Moum et al. (2013) showed that weak mixing during the spring surface heating peak is followed by peak annual SST. Strong mixing during late boreal summers is sufficient to overcome the second semiannual surface heating peak (Fig. 10.5a). While the deep cycle of mixing persists into spring, it is likely too shallow to be detected by existing measurements or contribute significant subsurface cooling (Pham et al., 2017).

The first eleven years of the record at 0°, 140°W span several ENSO cycles. Warner and Moum (2019) conditionally sampled ENSO phase transitions to and from the neutral state and found mixing to be greatest during transition from the neutral state to La Niña, when SST cools, and weakest when the neutral state transitions to the warm SST state of El Niño. The magnitudes of air-sea heat flux divergences during these transitions over the small ensemble of events were the same sign and amplitude as sea-surface heating and cooling events during the phase transitions (Fig. 10.5b). This analysis suggests a modification of the Bjerknes feedback (Bjerknes, 1969)[3] to include mixing. In this modification, subsurface mixing reduces equatorial SST by cooling from below, thereby strengthening easterly winds, which in turn enhance surface currents and upper ocean shear, reducing Ri, and thus reinforcing mixing.

Though it seems that vertical transport by turbulence-enhanced mixing on long timescales plays a major role in the heat budget of the upper equatorial ocean, the role of turbulence in the momentum budget is less clear (Dillon et al., 1989; Hebert et al., 1991b). A root cause may be the different ways that heat and momentum are mixed. Diapycnal heat transport is effected only by mixing enhanced by turbulence. Momentum is transported vertically by IGWs and turbulence. Relatively small-scale IGWs are a ubiquitous feature of the upper equatorial ocean, at least in the cold tongues (Moum et al., 1992b; Hebert et al., 1992; Moum et al., 1992a; McPhaden and Peters, 1992). These are likely generated by shear instability above the core of the EUC (Sun et al., 1998; Moum et al., 2011; Smyth et al., 2011). We expect that their role is to transport momentum downward from generation sites above the EUC core, but the measurements required to clearly quantify this have yet to be made.

10.8 Equatorial mixing in large-scale models

An important goal of turbulence research, both intellectually and practically, is to develop simplified models, or parameterisations, that can predict the turbulent fluxes of momentum, heat and other properties based on information about the environment. This is a serious challenge. Even direct and large-eddy simulations, which are based closely on the Navier–Stokes equations and are too computationally expensive for operational applications, reproduce observations in a statistical sense at best (e.g., Skyllingstad et al., 1999; Smyth et al., 2001). Parameterising turbulent fluxes for large-scale simulations without adding excessive computational cost requires significant compromises, and the resulting inaccuracies are an important impediment to operational and climate forecasting (Dippe et al., 2018; Li et al., 2019).

At the equator, the wind drives a mostly zonal current system not entirely different from the stratified, parallel shear flows that have been studied theoretically and in the lab for over a century. Early attempts to model equatorial currents

---

3. Bjerknes first noted that changes in the zonal SST gradient in the equatorial Pacific was linked to changes in the intensity of the atmospheric Walker cell, which in turn was linked to surface winds. The feedback is both positive and negative in amplifying/diminishing ENSO events.

assumed a flux-gradient relationship with uniform eddy viscosity and diffusivity (Philander and Pacanowski, 1980). This resulted in an unrealistically thick undercurrent and thermocline. Measurements show that a strong thermocline coincides with the undercurrent core, leading to weak turbulence in this depth range. Early models, by forcing the turbulence to have uniform magnitude, applied excessive smoothing, and thereby caused both thermocline and undercurrent core to spread unrealistically.

Based on the existing understanding of parallel shear flows, Pacanowski and Philander (1981, hereafter PP) hypothesised that the combined effect of shear and stratification would be best expressed if eddy viscosity ν were a smoothly-decreasing function of Ri:

$$\nu = \nu_0 \phi(Ri) + \nu_b, \tag{10.1}$$

where ν_0 is the amplitude of the shear-driven turbulence, $\phi(Ri)$ is the functional dependence on Ri and ν_b is a constant representing background turbulence that is independent of shear. A similar format was used to represent the eddy diffusivity κ. The adjustable parameters ν_0 and ν_b were calibrated by comparing the resulting modelled flows with observations, producing a much-improved representation of the velocity and temperature fields around the equatorial undercurrent. Evidently, Ri-dependence captures the first-order differences in eddy viscosity/diffusivity between the core (weak shear, strong stratification) and flanks (strong shear, weaker stratification) of the EUC.

Although the focus on Ri is well-justified, dimensional considerations show that Ri alone cannot determine the eddy viscosity. One parameter in the PP model is the uniform amplitude ν_0, having dimensions of viscosity, which would be expected to vary based on the dominant length and timescales of the flow. The value chosen, though optimal for the measurements PP examined, is expected to vary in different stratified shear flows, or even at the equator under different conditions. For example, using direct measurement of Ri, ν and κ at $0°140°$W, Peters et al. (1988) found that, for $Ri \gtrsim \frac{1}{2}$, ν and κ decreased smoothly with Ri, but were smaller by an order of magnitude than the PP values. At $Ri < 1/2$, in contrast, ν and κ increased rapidly to values exceeding those predicted by PP. Moum et al. (1989) interpreted similar measurements as a two-state process: at large Ri, turbulence levels decrease smoothly consistent with PP and Peters et al. (1988), but when Ri is sufficiently small, turbulence levels vary widely and are essentially independent of Ri. The latter regime is what we now call MI (Section 10.4), and it represents the turbulence found in the deep cycle.

Despite its evident limitations, Ri-dependence was the basis for several subsequent parameterisations, some of which are still in use. The K-profile parameterisation (KPP; Large et al., 1994) divides the ocean into two depth ranges: the surface boundary layer and the interior. In the interior, an Ri-dependent function similar to (10.1) is used. In the surface boundary layer, a more elaborate formulation accounts directly for non-local processes, such as convection, by adding a dependence on surface fluxes of buoyancy and momentum (e.g., Smyth et al., 2002).

The limitations of the Ri-dependent interior model of KPP have been examined by Zaron and Moum (2009), who effectively make the dimensional amplitude ν_0 dependent on length and timescales derived from the local mean velocity and its shear. Although the mean velocity is superficially non-Galilean, it likely acts as a proxy for the distance from the surface (or possibly from the EUC core), which is a credible lengthscale for turbulence in the deep cycle (also see Hebert et al., 1991a). The result is a much-improved representation of interior mixing at the equator.

Whereas KPP is essentially empirical, other investigators have developed second-moment closures that are derivable from the Navier–Stokes equations. These models include Ri-dependence, but also account for other effects. The parameters ν and κ depend on Ri, but the amplitude parameter ν_0 varies, depending on, for example, the local turbulent kinetic energy and its dissipation rate, either or both of which are calculated prognostically in separate equations. These models include the well-known Mellor–Yamada hierarchy (Mellor and Yamada, 1982) together with several variants summarised by Umlauf and Burchard (2005).

A different approach is taken in the Price–Weller–Pinkel model (PWP; Price et al., 1986a; Schudlich and Price, 1992). There, shear instability is treated the way static instability is often treated: if the conditions for its growth ($Ri < 1/4$) appear at any point, the flow is mixed so as to remove the instability. This is similar in spirit to the MI scenario (Section 10.4), in which mixing is 'as strong as it needs to be' to maintain Ri near 1/4. Though that critical Ri is formally the only parameter in the PWP-mixing formulation, values of eddy viscosity and diffusivity are implicit in the grid spacing and the time interval over which the instability is mixed away.

In assessing a parameterisation scheme an often overlooked aspect is how well the observations (or model) capture the flow features generating the turbulence. This tendency is in part due to an observational limitation: typically the measured vertical resolution of Sh is significantly coarser than that for N. Due to this limitation, any relationship between Ri and ε can be lost (see Richards et al., 2015). A useful guide is how measurements of the turbulent kinetic energy dissipation rate, ε, fall in N^2, Sh^2 space, as shown in Fig. 10.4c for the WEP. In this example, when the vertical resolution is O(1 m), a clear correlation between high values of ε and low Richardson number, Ri (the solid red line is $Ri = 0.25$), emerges. This relationship gives confidence that the shear and stratification are properly resolved. The impact of unresolved motions on mixing is often tucked away in models as a 'background' diffusivity (e.g., in KPP) often taken to be a constant. Conventional vertical resolution in models is simply too coarse to capture the fine-scale features discussed in Section 10.5.

Converting ε to a vertical diffusion coefficient κ_v using the Osborn relationship, Richards et al. (2015) shows that the Ri-based parameterisation for shear-driven turbulence used in the KPP scheme of Large et al. (1994) fits the data reasonably well, provided the critical value of Ri (above which turbulence amplitude decreases rapidly) is taken between 0.25–0.3 and the maximum eddy viscosity ν_0 around 1.8×10^{-4} m^2 s^{-1} (parameters Ri_0 and ν_0 are defined in Large et al. (1994)). However, in this case, the data-constrained values are considerably smaller than those quoted by Large et al. (1994) and used in many studies. Increasing Ri_0 to account for unresolved features should not be used as a substitute when the flow features themselves have small vertical scales. It is noteworthy that ε at constant Richardson number in Fig. 10.4 increases as N^2 (or equivalently S^2) increases, while the associated diffusivity decreases.

Recently Richards et al. (2021) formulated an expression for ε in terms of a turbulent vertical lengthscale ℓ_v, the inverse timescale N and the Richardson number, with ℓ_v dependent on the characteristics of fine-scale flow features. The scaling works well with the estimated ε capturing the changes in the vertical distribution of the observed ε. Many of the changes are attributable to changes in the vertical distribution of the lengthscale ℓ_v, highlighting the need to go beyond purely Richardson-based parameterisation schemes.

10.9 Shortcomings, surprises and targets for future investigation

The upper equatorial ocean is uniquely rich in fluid dynamical phenomena. Despite its importance to global heat uptake, we are just beginning to understand and quantify the processes that mix heat (and other properties) into the deep ocean. For example, it has only been recently observed and understood that TIWs spawn buoyant gravity currents that then advect warm water equatorwards, while mixing it downward (Box 10.2). And, we have only recently recognised that enhanced mixing over *topographically smooth* seafloor occurs, let alone that it may be generated by the downward propagation of large-scale equatorially trapped waves (Box 10.1). Whether the latter is *solely* an equatorial phenomenon is likely, but as yet unknown.

The details of the cross-equatorial structure on any timescale is a matter of some importance. The few cross-equatorial transects made to date do not adequately distinguish between time and spatial variability and certainly not on any timescale longer than a few days. The implication is profound: we do not know how to integrate mixing across the equator so, for example, have only rudimentary estimates of heat transfer rates through the cold tongues, which extend roughly ±5° from the equator, more than the meridional scale of the EUC.

At midlatitudes, finescale parameterisations have undergone substantial comparison to direct observations using multiple data sets, allowing for a means of extrapolating turbulence information. In contrast, finescale parameterisations based on shear (suggest relatively low equatorial mixing, Kunze et al., 2006) and strain (suggest relatively high equatorial mixing, Whalen et al., 2012) diverge from one another near the equator (Waterhouse et al., 2014). The applicability of such approaches is generally unknown as f approaches zero, although Gregg et al. (2003) indicates agreement with shear-based estimates using the handful of direct turbulence observations available. It has also been suggested that the discrepancy between shear-based (limited to ship transects) and strain-based (include Argo profiles) estimates can be attributed to a sampling bias caused by the lack of ship crossings during high-mixing La Niña states (Kunze, 2017). However, the paucity of deep turbulence data prohibits a definitive assessment of the validity of finescale parameterisations in this region, limiting our ability to assess the apparent bias and providing additional motivation for new cross-equatorial turbulence measurements. Climate models are now incorporating spatially variable mixing (Schmittner et al., 2015), raising the need for a timely resolution of this issue.

The extent, both in time and space, of mixing induced by small vertical-scale flow features is not well known, but its impact can be large, as demonstrated through multiple model sensitivity studies. For example, a number of modelling studies have demonstrated that the magnitude and time evolution of ENSO events depend on the state of the tropical Pacific Ocean (e.g., Jin and Neelin, 1993; Timmermann et al., 1999; Meehl et al., 2001). The state of the tropical ocean, and its interaction with the atmosphere, in turn depends strongly on the vertical mixing of water properties and momentum in the equatorial thermocline (e.g., Yu and Schopf, 1997; Meehl et al., 2001; Jochum, 2009; Richards et al., 2009). For instance Meehl et al. (2001) show that a reduction in vertical diffusivity in the ocean component of their model leads to an increase in the amplitude of ENSO variability. Using a regional atmosphere–ocean coupled model, Richards et al. (2009) found that a decrease in the background vertical mixing produces a warmer cold tongue SST and decreased easterly winds, while at the same time increasing the north–south temperature difference and southerly winds. Deppenmeier et al. (2020) found that increasing vertical mixing in the models reduces Atlantic cold tongue biases.

Sensitivity in models indicates that the vertical distribution and spatial resolution of mixing are also critical. Motivated by observations of finescale shear and associated enhancement of mixing within and above the thermocline, Sasaki et al. (2012, 2013) find increased diffusivity above the thermocline results in a significant *increase* in cold tongue SST, which is further amplified by coupling with the atmosphere (through the Bjerknes effect), and an increase in the skewness of ENSO events. Both of these responses reduce coupled model biases. The sensitivity to mixing distribution is non-local. Jia et al. (2015) show that an increased diffusivity within and above the thermocline of the WEP leads to an increase in SST in the eastern equatorial Pacific through advection of the changed thermocline by swift equatorial currents. Recently, Jia et al. (2021) have investigated the impact of increasing the vertical resolution in the thermocline so that finescale flow features start to be explicitly captured. Results show remarkable similarities to the experiments of Sasaki et al. (2013) in that increased resolution also leads to enhancement of the diffusivity above the thermocline.

A better understanding is needed of the characteristics of mixing, particularly its spatial and temporal distribution, from both observational and numerical frameworks. In addition to this being a general problem to geophysical fluid dynamics, it may be particularly acute in these highly-sheared equatorial flows, where small-scale wrinkles in the shear can result in instability. For the equatorial region, mixing has a demonstrated impact on the mixed layer temperature, which in turn influences coupled atmospheric phenomena over seasonal and interannual timescales. This connection has clearly been shown in the Pacific cold tongue, with its extensive measurement record (Section 10.5). However, the role of mixing in the Atlantic is still underexplored and a clear target for investigation.

> **Box 10.1 Deep equatorial mixing: a case of intensified bottom mixing in the absence of topography**
>
> Mixing enhanced by turbulence is essential to the maintenance of the deep ocean's stratification. Experiments comparing mixing at midlatitudes in the presence and absence of steep topography (Toole et al., 1994; Ledwell et al., 2000) have shown significantly greater mixing over rough topography, and topographic mixing has been assumed to be the only source of abyssal mixing (Ferrari, 2014, for example). Observations in the equatorial Atlantic at depth have shown high-mixing rates over the mid-Atlantic ridge, a region of rough topography (Polzin et al., 1996). However, measurements in the deep equatorial Pacific ocean over a particularly smooth seafloor suggest another source for intensified mixing in the equatorial abyss. (Fig. 10.6). Fast thermistors carefully deployed on a standard ship-provided CTD showed high values of turbulence from 5 m to 700 m above the seafloor (Holmes et al., 2016). The significance of this result is in the unexpected contribution to intensified diapycnal upwelling near the equator in the abyss.
>
>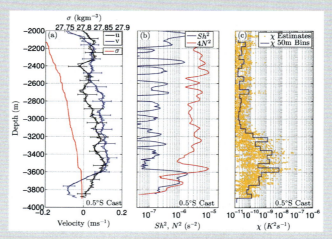
>
> **FIGURE 10.6** Profiles of (a) zonal and meridional velocity (bottom axis) and potential density (top axis), (b) S^2 and $4N^2$ and (c) temperature variance dissipation rate χ from cast 7 at 1/2 S. In Fig. 10.3a, both raw LADCP velocities with error bars (thin lines) and 100 m triangle-filtered velocities (thick lines) are shown. Adapted from Holmes et al. (2016).
>
> This intensification of mixing may be due to the vertical propagation and seafloor reflection of surface-generated and equatorially trapped waves. With consideration of the horizontal component of the Earth's rotation, near-bottom near-inertial energy is generated from these reflections (Delorme and Thomas, 2019). Subsequent vertical shear is sufficient to create low Ri and consequently enhanced mixing over broad zonal and meridional regions above topographically smooth equatorial seafloors.

Box 10.2 The role of tropical instability waves

Tropical instability waves (TIWs) propagate zonally along the thermal front of the equatorial cold tongues late in the year, at least in the absence of El Niño conditions. TIWs contribute to equatorial mixing in two recently-observed ways. TIWs are identified by significant meridional currents above the EUC. The additional current shear can nudge Ri toward critical values above the EUC, thereby enhancing turbulence (Moum et al., 2009). This additional shear extends strong upper ocean turbulence to greater depths than had previously been observed in the deep cycle beneath the mixed layer. Analysis of the long-term record of meridional currents and SST at 0°140°W indicates significant correlation between TIW kinetic energy and SST cooling. Local experiments show that the enhanced mixing has a TIW phase-dependence (Inoue et al., 2012, 2019).

Another way that TIWs contribute to equatorial mixing is by spawning buoyant gravity currents that propagate away from the main TIW front (Fig. 10.7), carrying relatively warm water equatorward. Enhanced current shear at the base of the gravity current generates turbulence. Modelling results indicate that TIW fronts release gravity currents through frontogenesis and loss of balance as the fronts approach the equator and the Coriolis force weakens (Warner et al., 2018). Further analysis shows that such fronts are ubiquitous in the eastern tropical Pacific. These features are expected to play an important role not only in transferring warm surface fluid equatorward, but also in cascading energy out of the TIW field toward smaller scales and dissipation.

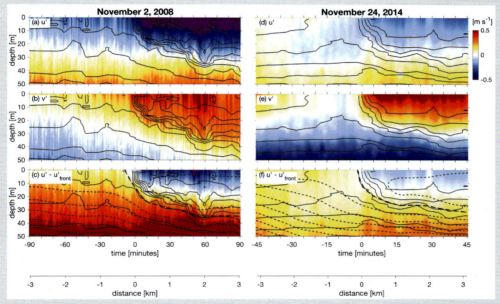

FIGURE 10.7 (a), (d): ADCP velocity in the across-front direction, and (b),(e): along-front direction. (c), (f): Across-front velocity with the frontal propagation speed removed. (a), (b), (d), (e) are in the geographic reference frame. (c), (f) are in a reference frame fixed to the front of the gravity current to highlight the flow around and within the stationary warm layer. Streamlines, calculated by integrating down from the surface, are shown with dashed lines with the zero streamline in bold. Isopycnals (solid black lines) are shown in all subplots. Adapted from Warner et al. (2018).

References

Aschwanden, M., 2014. A macroscopic description of a generalized self-organized criticality system: astrophysical applications. Astrophys. J. 782, 54.
Aschwanden, M., 2016. J. 25 years of self-organized criticality: solar and astrophysics. Space Sci. Rev. 198, 47–166.
Bak, P., Tang, C., Weisenfeld, K., 1987. Self-organized criticality: an explanation of 1/f noise. Phys. Rev. Lett. 59, 381–384.
Balmforth, N.J., 1999. Shear instability in shallow water. J. Fluid Mech. 387, 97–127.
Bender, C.M., Orszag, S.A., 2013. Advanced Mathematical Methods for Scientists and Engineers I. Asymptotic Methods and Perturbation Theory. Springer.
Bjerknes, J., 1969. Atmospheric teleconnections from the equatorial Pacific. Mon. Weather Rev. 97, 163–172.
Boyd, J.P., Christidis, Z., 1982. Low-wavenumber instability on the equatorial Beta-plane. Geophys. Res. Lett. 9 (7), 769.
Brainerd, K., Gregg, M., 1993. Diurnal restratification and turbulence in the oceanic surface mixed layer: 1. Observations. J. Geophys. Res., Oceans 89. https://doi.org/10.1029/93JC02297.
Contreras, R.F., 2002. Long-term observations of tropical instability waves. J. Phys. Oceanogr. 32, 2715–2722.
Crawford, W.R., Osborn, T.R., 1981. Control of equatorial currents by turbulent dissipation. Science 212, 539–540.
Delorme, B., Thomas, L., 2019. Abyssal mixing through critical reflection of equatorially trapped waves off smooth topography. J. Phys. Oceanogr. 49. https://doi.org/10.1175/jpo-D-18-0197.1.

Deppenmeier, A.L., Haarsma, R.J., LeSager, P., Hazeleger, W., 2020. The effect of vertical ocean mixing on the tropical Atlantic in a coupled global climate model. Clim. Dyn. 54, 5089–5109. https://doi.org/10.1007/s00382-020-05270-x.

Dillon, T.M., Moum, J.N., Chereskin, T.K., Caldwell, D.R., 1989. Zonal momentum balance at the equator. J. Phys. Oceanogr. 19, 561–570.

Dippe, T., Krebs, M., Harlaß, J., Lübbecke, J., 2018. Can climate models simulate the observed strong summer surface cooling in the equatorial Atlantic? In: Jungblut, S., Leibich, V., Bode, M. (Eds.), YOUMARES 8 – Oceans Across Boundaries: Learning from Each Other. Springer, pp. 7–23.

d'Orgeville, M., Hua, B.L., 2005. Equatorial inertial-parametric instability of zonally symmetric oscillating shear flows. J. Fluid Mech. 531, 261–291.

Dunkerton, T.J., 1981. On the inertial stability of the equatorial middle atmosphere. J. Atmos. Sci. 38, 2354–2364.

Evans, D., Lucas, N., Hemsley, V., Naveira-Garabato, C., Frajka-Williams, E., Martin, A., Painter, S., Inall, M., Palmer, M., 2018. Annual cycle of turbulent dissipation estimated from seagliders. Geophys. Res. Lett. 45 (19), 10,560–10,569. https://doi.org/10.1029/2018GL079966.

Farrar, J.T., Durland, T.S., 2012. Wavenumber–frequency spectra of inertia–gravity and mixed Rossby–gravity waves in the equatorial Pacific Ocean. J. Phys. Oceanogr. 42, 1859–1881.

Ferrari, R., 2014. What goes down must come up. Nature 513, 179–180.

Gargett, A., 1989. Ocean turbulence. Annu. Rev. Fluid Mech. 21, 419–451.

Garrett, C., Munk, W., 1972. Space-time scales of internal waves. Geophys. Fluid Dyn. 2, 225–264.

Gill, A.E., 1982. Atmosphere-Ocean Dynamics. Academic.

Goswami, B., Wu, G., Yasunari, T., 2006. The annual cycle, intraseasonal oscillations, and roadblock to seasonal predictability of the Asian summer monsoon. J. Climate 19, 5078–5099.

Grachev, A.A., Andreas, E.L., Fairall, C.W., Guest, P.S., Persson, P.O.G., 2015. Similarity theory based on the Dougherty-Ozmidov length scale. Q. J. R. Meteorol. Soc. 141, 1845–1856.

Gregg, M.C., Peters, H., Wesson, J.C., Oakey, N.S., Shay, T.J., 1985. Intensive measurements of turbulence and shear in the Equatorial Undercurrent. Nature 318, 140–144.

Gregg, M.C., Sanford, T.B., Winkel, D.P., 2003. Reduced mixing from the breaking of internal waves in equatorial waters. Nature 422, 513–515.

Hartmann, D.L., Michelsen, M.L., 1989. Intraseasonal periodicities in Indian rainfall. J. Atmos. Sci. 46, 2838–2862.

Hartmann, D.L., Michelsen, M.L., Klein, S.A., 1992. Seasonal variations of tropical intraseasonal oscillations: a 20–25-day oscillation in the western Pacific. J. Atmos. Sci. 49, 1277–1289.

Hebert, D., Moum, J.N., Caldwell, D.R., 1991a. Does ocean turbulence peak at the equator? Revisited. J. Phys. Oceanogr. 21, 1690–1698.

Hebert, D., Moum, J.N., Paulson, C.A., Caldwell, D.R., 1992. Turbulence and internal waves at the equator. Part II: details of a single event. J. Phys. Oceanogr. 22, 1346–1356.

Hebert, D., Moum, J.N., Paulson, C.A., Caldwell, D.R., Chereskin, T.K., McPhaden, M.J., 1991b. The role of the turbulent stress divergence in the equatorial Pacific zonal momentum balance. J. Geophys. Res. 96, 7127–7136.

Hoecker-Martinez, M., Smyth, W., Skyllingstad, E., 2016. Oceanic turbulent energy budget using large eddy simulation of a wind event during dynamol. J. Phys. Oceanogr. 46, 827–840.

Holmes, R., Moum, J., Thomas, L., 2016. Evidence for seafloor-intensified mixing by surface-generated equatorial waves. Geophys. Res. Lett. 43, 1202–1210.

Holmes, R.M., Zika, J., England, M., 2019. Diathermal heat transport in a global ocean model. J. Phys. Oceanogr. 49, 141–161.

Howard, L., 1961. Note on a paper of John W. Miles. J. Fluid Mech. 10, 509–512.

Hua, B.L., Moore, D.W., Gentil, S.L., 1997. Inertial nonlinear equilibration of equatorial inertial instability. J. Fluid Mech. 331, 345–371.

Hughes, K., Moum, J., Shroyer, E., 2020. Evolution of the velocity structure in the diurnal warm layer. J. Phys. Oceanogr. 50, 615–631.

Hummels, R., Dengler, M., Bourles, B., 2013. Seasonal and regional variability of upper ocean diapycnal heat flux in the Atlantic cold tongue. Prog. Oceanogr. 111, 52–74.

Inoue, R., Lien, R.C., Moum, J., 2012. Modulation of equatorial turbulence by a tropical instability wave. J. Geophys. Res. 117.

Inoue, R., Lien, R.C., Moum, J., Perez, R., Gregg, M., 2019. Variations of equatorial shear, stratification, and turbulence within a tropical instability wave cycle. J. Geophys. Res. 124, 1858–1875.

Jensen, H., 1998. Self-Organized Criticality: Emergent Complex Behavior in Physical and Biological Systems. Cambridge University Press.

Jensen, T., Shinoda, T., Chen, S., Flatau, M., 2015. Ocean response to CINDY/DYNAMO MJOs in air-sea-coupled COAMPS. J. Meteorol. Soc. Jpn. 93A, 157–178.

Jia, Y., Furue, R., McCreary, J.P., 2015. Impacts of regional mixing on the temperature structure of the equatorial Pacific Ocean. Part 2: depth-dependent vertical diffusion. Ocean Model. 91, 112–127.

Jia, Y., Richards, K.J., Annamalai, H., 2021. The impact of vertical resolution in reducing biases in sea surface temperature in a tropical Pacific Ocean model. Ocean Model. 54, 5089–5109.

Jin, F.F., Neelin, J., 1993. Modes of interannual tropical ocean-atmosphere interaction a unified view. Part I: numerical results. J. Atmos. Sci. 50, 3477–3503.

Jochum, M., 2009. Impact of latitudinal variations in vertical diffusivity on climate simulations. J. Geophys. Res. 114. https://doi.org/10.1029/2008JC005030.

Kantha, L., Clayson, C., 2000. Small Scale Processes in Geophysical Fluid Flows, 1st edition ed. Academic Preass.

Kosaka, Y., Xie, S.P., 2013. Recent global-warming hiatus tied to equatorial Pacific surface cooling. Nature 501, 403–407. https://doi.org/10.1038/nature12534.

Kunze, E., 2017. Internal-wave-driven mixing: global geography and budgets. J. Phys. Oceanogr. 47, 1325–1345. https://doi.org/10.1175/JPO-D-16-0141.1.

Kunze, E., Firing, E., Hummon, J.M., Chereskin, T.K., Thurnherr, A.M., 2006. Global abyssal mixing inferred from lowered adcp shear and ctd strain profiles. J. Phys. Oceanogr. 36, 1553–1576. https://doi.org/10.1175/JPO2926.1. https://journals.ametsoc.org/view/journals/phoc/36/8/jpo2926.1.xml.

Large, W., McWilliams, J., Doney, S., 1994. Ocean vertical mixing: a review and a model with a nonlocal boundary layer parameterization. Rev. Geophys. 32, 363–403.

LeBlonde, P.H., Mysak, L.A., 1978. Waves in the Ocean. Elsevier, Amsterdam.

Ledwell, J., Montgomery, E., Polzin, K., Laurent, L.S., Schmitt, R., Toole, J., 2000. Evidence for enhanced mixing over rough topography in the abyssal ocean. Nature 403, 179–182.

Li, G., Jian, Y., Yang, S., Du, Y., Wang, Z., Jiang, W., Huang, G., 2019. Effect of excessive equatorial Pacific cold tongue bias on the El Niño – Northwest Pacific summer monsoon relationship in CMIP5 multi-model ensemble. Clim. Dyn. 52, 6192–6212.

Li, G., Xie, S., 2014. Tropical biases in CMIP5 multimodel ensemble: the excessive equatorial Pacific cold tongue and double ITCZ problems. J. Climate 27, 1765–1780.

Li, L., Smyth, W., Thorpe, S., 2015. Destabilization of a stratified shear layer by ambient turbulence. J. Fluid Mech. 771, 1–15.

Madden, R., Julian, P., 1971. Detection of a 40-50 day oscillation in the zonal wind in the tropical Pacific. J. Atmos. Sci. 28, 702–708.

Madden, R., Julian, P., 1972. Description of global-scale circulation cells in the tropics with a 40–50 day period. J. Atmos. Sci. 29, 1109–1123.

Matsuno, T., 1966. Quasi-geostrophic motions in the equatorial area. J. Meteorol. Soc. Jpn. 44, 25–42.

McPhaden, M.J., Peters, H., 1992. Diurnal cycle of internal wave variability in the equatorial Pacific Ocean. J. Phys. Oceanogr. 22, 1317–1329.

McPhaden, M.J., Taft, B.A., 1988. Dynamics of seasonal and intraseasonal variability in the eastern equatorial Pacific. J. Phys. Oceanogr. 18, 1713–1732.

Meehl, G., Gent, P., Arblaster, J., Otto-Bliesner, B., Brady, E., Craig, A., 2001. Factors that affect the amplitude of El Niño in global coupled models. Clim. Dyn. 17, 515–527.

Mellor, G.L., Yamada, T., 1982. A hierarchy of turbulence closure models for planetary boundary layers. J. Atmos. Sci. 31, 1791–1806.

Miles, J.W., 1961. On the stability of heterogeneous shear flows. J. Fluid Mech. 10, 496–508.

Moulin, A., Moum, J., Shroyer, E., 2018. Evolution of turbulence in the diurnal warm layer. J. Phys. Oceanogr. 48, 383–396.

Moulin, A., Moum, J., Shroyer, E., Hoecker-Martinez, M., 2021. Freshwater lens fronts propagating as buoyant gravity currents in the equatorial Indian Ocean. J. Geophys. Res. Oceans. In press.

Moum, J., Nash, J., Smyth, W., 2011. Narrowband high-frequency oscillations at the equator. Part I: interpretation as shear instabilities. J. Phys. Oceanogr. 41, 397–411.

Moum, J., Perlin, A., Nash, J., McPhaden, M., 2013. Seasonal sea surface cooling in the equatorial Pacific cold tongue controlled by ocean mixing. Nature 500, 64–67.

Moum, J., Pujiana, K., Lien, R.C., Smyth, W., 2016. Ocean feedback to pulses of the Madden–Julian oscillation in the equatorial Indian ocean. Nat. Commun. 7, 13203.

Moum, J., de Szoeke, S., Smyth, W., Edson, J., DeWitt, H., Moulin, A., Thompson, E., Zappa, C., Rutledge, S., Johnson, R., Fairall, C., 2014. Air-sea interactions from westerly wind bursts during the November 2011 mjo in the Indian Ocean. Bull. Am. Meteorol. Soc. 95, 1185–1199.

Moum, J.N., Caldwell, D.R., 1985. Local influences on shear flow turbulence in the equatorial ocean. Science 230, 215–315.

Moum, J.N., Caldwell, D.R., Paulson, C.A., 1989. Mixing in the equatorial surface layer and thermocline. J. Geophys. Res. 94, 2005–2021.

Moum, J.N., Hebert, D., Paulson, C.A., Caldwell, D.R., 1992a. Turbulence and internal waves at the equator. Part 1: statistics from towed thermistors and a microstructure profiler. J. Phys. Oceanogr. 22, 1330–1345.

Moum, J.N., Lien, R.C., Perlin, A., Nash, J.D., Gregg, M.C., Wiles, P.J., 2009. Sea surface cooling at the equator by subsurface mixing in tropical instability waves. Nat. Geosci. 2, 761–765.

Moum, J.N., McPhaden, M.J., Hebert, D., Peters, H., Paulson, C.A., Caldwell, D.R., 1992b. Internal waves, dynamic instabilities and turbulence in the equatorial thermocline: an introduction to 3 papers in this issue. J. Phys. Oceanogr. 22, 1357–1359.

Moum, J.N., Nash, J.D., 2009. Mixing measurements on an equatorial ocean mooring. J. Atmos. Ocean. Technol. 26, 317–336.

Murray, J.W., Barber, R.T., Roman, M.R., Bacon, M.P., Feely, R.A., 1994. Physical and biological controls on carbon cycling in the equatorial Pacific. Science 266, 58–65. https://doi.org/10.1126/science.266.5182.58.

Natarov, A., Boyd, J.P., 2001. Beyond-all-orders instability in the equatorial Kelvin wave. Dyn. Atmos. Ocean. 33, 191–200.

Natarov, A., Richards, K.J., 2009. Three-dimensional instabilities of equatorial oscillatory flows. J. Fluid Mech. 627.

Natarov, A., Richards, K.J., 2019. Enhanced energy dissipation in the equatorial pycnocline by wind-induced internal wave activity. J. Geophys. Res. 124. https://doi.org/10.1029/2019JC015228.

Natarov, A., Richards, K.J., McCreary, J.P., 2008. Two-dimensional instabilities of time-dependent zonal flows: linear shear. J. Fluid Mech. 599, 29–50.

Oschlies, A., Brandt, P., Stramma, L., Schmidtko, S., 2018. Drivers and mechanisms of ocean deoxygenation. Nat. Geosci. 11, 467–473. https://doi.org/10.1038/s41561-018-0152-2.

Pacanowski, R., Philander, S., 1981. Parameterization of vertical mixing in numerical models of tropical oceans. J. Phys. Oceanogr. 11, 1443–1451.

Peters, H., Gregg, M., Toole, J., 1988. On the parameterization of equatorial turbulence. J. Geophys. Res. 93, 1199–1218.

Peters, H., Gregg, M.C., Toole, J.M., 1989. Meridional variability of turbulence through the equatorial undercurrent. J. Geophys. Res. 94, 18003–18009.

Pham, H., Smyth, W., Moum, J., Sarkar, S., 2017. Seasonality of deep cycle turbulence in the eastern equatorial Pacific. J. Phys. Oceanogr. 47, 2189–2209.

Philander, S.G.H., Pacanowski, R.C., 1980. The generation of equatorial currents. J. Geophys. Res. 85, 1123–1136. https://doi.org/10.1029/JC085iC02p01123.

Polzin, K.L., Speer, K.G., Toole, J.M., Schmitt, R.W., 1996. Intense mixing of Antarctic bottom water in the equatorial Atlantic Ocean. Nature 380, 54–57.

Price, J., Weller, R., Pinkel, R., 1986a. Diurnal cycling: observations and models of the upper ocean's response to diurnal heating, cooling and wind mixing. J. Geophys. Res. 91, 8411–8427.

Price, J.F., Weller, R.A., Pinkel, R., 1986b. Diurnal cycling: observations and models of the upper ocean response to diurnal heating, cooling, and wind mixing. J. Geophys. Res. 91, 8411–8427.

Pujiana, K., Moum, J.N., Smyth, W.D., 2018. The role of turbulence in redistributing upper ocean heat, freshwater and momentum in response to the MJO in the equatorial Indian Ocean. J. Phys. Oceanogr. 48, 197–220.

Reppin, J., Schott, F.A., Fischer, J., Quadfasel, D., 1999. Equatorial currents and transports in the upper central Indian Ocean: annual cycle and interannual variability. J. Geophys. Res., Oceans 104, 15495–15514.

Richards, K., Kashino, Y., Natarov, A., Firing, E., 2012a. Mixing in the western equatorial Pacific and its modulation by enso. Geophys. Res. Lett. 39. https://doi.org/10.1029/2011GL050439.

Richards, K.J., Edwards, N.R., 2003. Lateral mixing in the equatorial Pacific: the importance of inertial instability. Geophys. Res. Lett. 30, 1888. https://doi.org/10.1029/2003GL017768.

Richards, K.J., Kashino, Y., Natarov, A., Firing, E., 2012b. Mixing in the western equatorial Pacific and its modulation by ENSO. Geophys. Res. Lett. 39. https://doi.org/10.1029/2011GL050439.

Richards, K.J., Natarov, A., Carter, G.S., 2021. Scaling of shear–generated turbulence: the equatorial thermocline, a case study. J. Geophys. Res. 126 (5), e2020JC016978.

Richards, K.J., Natarov, A., Firing, E., Kashino, Y., Soares, S.M., Ishizu, M., Carter, G.S., Lee, J.H., Chang, K.I., 2015. Shear-generated turbulence in the equatorial Pacific produced by small vertical scale flow features. J. Geophys. Res. 120, 3777–3791.

Richards, K.J., Pollard, R.T., 1991. Structure of the upper ocean in the western equatorial Pacific. Nature 350, 48–50. https://doi.org/10.1038/350048a0.

Richards, K.J., Xie, S.P., Miyama, T., 2009. Vertical mixing in the ocean and its impact on the coupled ocean-atmosphere system in the eastern tropical Pacific. J. Climate 22, 3703–3719. https://doi.org/10.1175/2009JCLI2702.1.

Sasaki, W., Richards, K.J., Luo, J.J., 2012. Role of vertical mixing originating from small vertical scale structures above and within the equatorial thermocline in an OGCM. Ocean Model. 57–58, 29–42. https://doi.org/10.1016/j.ocemod.2012.09.002.

Sasaki, W., Richards, K.J., Luo, J.J., 2013. Impact of vertical mixing induced by small vertical scale structures above and within the equatorial thermocline on the tropical Pacific in a CGCM. Clim. Dyn. 41, 443–453. https://doi.org/10.1007/s00382-012-1593-8.

Schmitt, R.W., 1981. Form of the temperature-salinity relationship in the Central Water, evidence for double-diffusive mixing. J. Phys. Oceanogr. 11, 1015–1026.

Schmittner, A., Green, J., Wilmes, S.B., 2015. Glacial ocean overturning intensified by tidal mixing in a global circulation model. Geophys. Res. Lett. 42, 4014–4022.

Schott, F.A., Xie, S.P., McCreary Jr, J.P., 2009. Indian Ocean circulation and climate variability. Rev. Geophys. 47.

Schudlich, R., Price, J., 1992. Diurnal cycles of current, temperature, and turbulent dissipation in a model of the equatorial upper ocean. J. Geophys. Res. 97, 5409–5422.

Senan, R., Sengupta, D., Goswami, B., 2003. Intraseasonal "monsoon jets" in the equatorial Indian Ocean. Geophys. Res. Lett. 30.

Skyllingstad, E., Smyth, W., Moum, J., Wijesekera, H., 1999. Turbulent dissipation during a westerly wind burst: a comparison of large eddy simulation results and microstructure measurements. J. Phys. Oceanogr. 29, 5–29.

Smyth, W., 2020. Marginal instability and the efficiency of ocean mixing. J. Phys. Oceanogr. 50, 2141–2150.

Smyth, W., Carpenter, J., 2019. Instability in Geophysical Flows. Cambridge Univ. Press.

Smyth, W., Moum, J., 2013. Marginal instability and deep cycle mixing in the eastern equatorial Pacific Ocean. Geophys. Res. Lett. 40, 6181–6185.

Smyth, W., Moum, J., Li, L., Thorpe, S., 2013. Diurnal shear instability, the descent of the surface shear layer and the deep cycle of equatorial turbulence. J. Phys. Oceanogr. 43, 2432–2455.

Smyth, W., Moum, J., Nash, J., 2011. Narrowband high-frequency oscillations at the equator. Part II: properties of shear instabilities. J. Phys. Oceanogr. 41, 412–428.

Smyth, W., Pham, H., Moum, J., Sarkar, S., 2017. Pulsating stratified turbulence in the upper equatorial oceans. J. Fluid Mech. 822, 327–341.

Smyth, W., Skyllingstad, E., Crawford, G., Wijesekera, H., 2002. Nonlocal fluxes and Stokes drift effects in the K-profile parameterization. Ocean Dyn. 52, 104–115.

Smyth, W.D., Moum, J.N., Caldwell, D.R., 2001. The efficiency of mixing in turbulent patches: inferences from direct simulations and microstructure observations. J. Phys. Oceanogr. 31, 1969–1992.

Smyth, W.D., Nash, J.D., Moum, J.N., 2019. Self-organized criticality in geophysical turbulence. Nat. Sci. Rep. 3747.

Smyth, W.D., Zavialov, P.O., Moum, J.N., 1997. Decay of turbulence in the upper ocean following sudden isolation from surface forcing. J. Phys. Oceanogr. 27, 810–822.

Stevens, D., 1983. On symmetric stability and instability of zonal mean flows near the equator. J. Atmos. Sci. 40, 882–893.

Stone, P.H., 1978. Baroclinic adjustment. J. Atmos. Sci. 35, 561–571.

Sun, C., Smyth, W.D., Moum, J.N., 1998. Dynamic instability of stratified shear flow in the upper equatorial ocean. J. Geophys. Res. 103, 10,323–10,337.

Sutherland, G., Marie, L., Reverdin, G., Christensen, K., Brostrom, G., Ward, B., 2016. Enhanced turbulence associated with the diurnal jet in the ocean surface boundary layer. J. Phys. Oceanogr. 46, 3051–3067.

Taniguchi, H., Ishiwatari, M., 2006. Physical interpretation of unstable modes of a linear shear flow in shallow water on an equatorial beta-plane. J. Fluid Mech. 567, 1–26.

Thompson, E., Moum, J., Rutledge, S., Fairall, C., 2019. Wind limits on rain layers and diurnal warm layers. J. Geophys. Res., Oceans. https://doi.org/10.1029/2018JC014130.

Thorpe, S., 2005. The Turbulent Ocean. Cambridge Univ. Press.

Thorpe, S., Liu, Z., 2009. Marginal instability? J. Phys. Oceanogr. 39, 2373–2381.

Thorpe, S., Smyth, W., Li, L., 2013. The effect of small viscosity and diffusivity on the marginal stability of stably stratified shear flows. J. Fluid Mech. 731, 461–476.

Timmermann, A., Latif, M., Bacher, A., Oberhuber, J., Roeckner, E., 1999. Increased El Niño frequency in a climate model forced by future greenhouse warming. Nature 398, 694–696.

Toole, J.M., Polzin, K.L., Schmitt, R.W., 1994. Estimates of diapycnal mixing in the abyssal ocean. Science 264, 1120–1123.

Umlauf, L., Burchard, H., 2005. Second-order turbulence closure models for geophysical boundary layers: a review of recent work. Cont. Shelf Res. 25, 795–827.

Von Schuckmann, K., Brandt, P., Eden, C., 2008. Generation of tropical instability waves in the Atlantic Ocean. J. Geophys. Res., Oceans 113.

Voorhis, A., Perkins, H., 1966. The spatial spectrum of short-wave temperature fluctuations in the near-surface thermocline. Deep-Sea Res. Oceanogr. Abstr. 13, 641–654. https://doi.org/10.1016/0011-7471(66)90596-1.

Warner, S., Holmes, R., Hawkins, E.M., Hoecker-Martinez, M., Savage, A., Moum, J., 2018. Buoyant gravity currents released from tropical instability waves. J. Phys. Oceanogr. 48. https://doi.org/10.1175/JPO-D-17-0144.1.

Warner, S., Moum, J., 2019. Feedback of mixing to ENSO phase change. Geophys. Res. Lett. 46, 13920–13927.

Waterhouse, A.F., MacKinnon, J.A., Nash, J.D., Alford, M.H., Kunze, E., Simmons, H.L., Polzin, K.L., Laurent, L.C.S., Sun, O.M., Pinkel, R., Talley, L.D., Whalen, C.B., Huussen, T.N., Carter, G.S., Fer, I., Waterman, S., Garabato, A.C.N., Sanford, T.B., Lee, C.M., 2014. Global patterns of diapycnal mixing from measurements of the turbulent dissipation rate. J. Phys. Oceanogr. 44, 1854–1872. https://doi.org/10.1175/JPO-D-13-0104.1.

Whalen, C.B., Talley, L.D., MacKinnon, J.A., 2012. Spatial and temporal variability of global ocean mixing inferred from Argo profiles. Geophys. Res. Lett. 39. https://doi.org/10.1029/2012GL053196.

Xie, S.P., 2013. Unequal equinoxes. Nature 500, 33–34.

Yu, Z., Schopf, P.S., 1997. Vertical eddy mixing in the tropical upper ocean: its influence on zonal currents. J. Phys. Oceanogr. 27, 1447–1458.

Zaron, E.D., Moum, J.N., 2009. A new look at Richardson number mixing schemes for equatorial ocean modeling. J. Phys. Oceanogr. 39, 2652–2664.

Zhang, C., 2005. Madden-Julian oscillation. Rev. Geophys. 43. https://doi.org/10.1029/2004RG000158.

Zhang, C., 2013. Madden–Julian oscillation: bridging weather and climate and climate. Bull. Am. Meteorol. Soc. 94, 1849–1870.

Chapter 11

Mixing in the Arctic Ocean

Yueng-Djern Lenn[a], Ilker Fer[b], Mary-Louise Timmermans[c] and Jennifer A. MacKinnon[d]

[a]School of Ocean Sciences, Bangor University, Wales, United Kingdom, [b]Geophysical Institute, University of Bergen, Bergen, Norway, [c]Department of Earth and Planetary Sciences, Yale University, New Haven, CT, United States, [d]Scripps Institution of Oceanography, University of California San Diego, La Jolla, CA, United States

11.1 Introduction

In the far north of our planet, the Arctic is an ocean surrounded by land, unlike the Antarctic, which is a landmass surrounded by the Southern Ocean. Here in the Arctic, ocean circulation is characterised by an anti-clockwise circumpolar boundary current, fed by the Atlantic Water (AW) inflow through Fram strait and supplemented by the Pacific inflow entering through Bering Strait (Fig. 11.1). Several circulations are also present within the Arctic interior basins, of which the wind-driven clockwise Beaufort Gyre is most significant, and opposes the boundary current flowing along the North American coastline (Fig. 11.1). In this true polar ocean, the stratification associated with the distribution of Atlantic- and Pacific-origin waters, the presence of Arctic sea ice and exceptionally high river discharge set the stage for unique ocean dynamics (Fig. 11.1).

Arctic sea ice and its seasonal cycle of freeze and melt are perhaps the most well known feature of the Arctic Ocean. Now, a dramatic decline in both Arctic sea-ice extent and thickness (Stroeve et al., 2007; Kwok and Rothrock, 2009; Stroeve

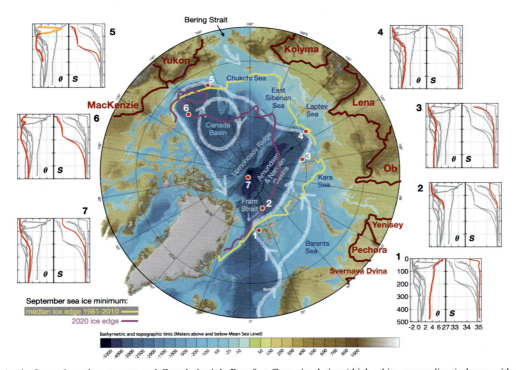

FIGURE 11.1 Arctic Ocean boundary current and Canada basin's Beaufort Gyre circulation (thick white crayon lines) shown with the MacKenzie, Kolyma, Lena, Ob, Yenisey, Pechora and Severnaya Dvina river drainage basins (outlined in red). Arctic September minima median ice edge for 1981–2010 (yellow), and for 2020 (magenta) are overlaid. Selected observational stations (1) to (7) are indicated. Associated temperature and salinity profiles (red, against the remaining 6 profiles in grey) are arrayed counter-clockwise from station (1) to (7), depicting variability in stratification, including the presence of an intra-halocline PW eddy (orange) observed at location (5). Bathymetry and topography courtesy of IBCAO (The International bathymetric chart of the Arctic ocean (Jakobsson et al., 2012)).

et al., 2012; Kwok and Cunningham, 2015; Perovich et al., 2019) is seen as the clearest sign of the polar amplification of climate change (Holland and Bitz, 2003; Bekryaev et al., 2010; Screen and Simmonds, 2010). Sea-ice formation occurs everywhere in open water during the freezing season, with up to a fifth formed in coastal polynyas in the marginal Kara Sea (Pfirman et al., 1997; Eicken, 2004). The Kara Sea is thought to play this pivotal role because of the freshwater it receives, mainly from the Ob and Yenisey rivers, which together constitute 25% of all Arctic river discharge (Pemberton et al., 2014). Riverine freshwater reduces mixed-layer salinity, increasing stratification, in a region where prevailing winds continually flux newly-formed ice away from the coasts. Indeed the Arctic Ocean as a whole is the freshest major ocean, receiving 10% of global run-off in 6% of the area covered by oceans. A lot of this freshwater ends up stored within the Beaufort Gyre halocline, in part, as a result of Ekman pumping driven by the prevailing atmospheric high.

Some of the estuarine freshwater is exported from the continental shelves into the central basins as sea ice (Pfirman et al., 1997; Eicken, 2004; Proshutinsky et al., 2019), where it is a principle source of freshwater for the surface mixed layer. Consequently, the Arctic Ocean surface mixed layer is cooler and fresher than the halocline and thermocline below it (Fig. 11.2a–c). This water column structure, so different from the warm, salty surface mixed layers of the temperate and tropical oceans that overlie cooler, fresher water, led pioneering oceanographer, Nansen (1904), to describe the Arctic as an 'upside-down' ocean. The low temperatures of the Arctic Ocean mean that, instead of temperature T, salinity S dominates seawater density. Hence, warm (up to 4 °C) and saltier AW flowing north through Fram Strait, rapidly subducts below the fresher polar surface mixed layer as it turns east to follow the continental slope as the circumpolar Arctic boundary current. On the western side, Pacific Water (PW) similarly contributes to a warm, sub-surface layer in the Canada Basin. Mixing between the AW or PW and polar surface mixed layer sets the shape and strength of the halocline that separates these water masses across most of the Arctic Ocean (Fig. 11.2a–c). The halocline itself owes its existence to the dynamical redistribution of fresh water throughout the Arctic Ocean's upper layer, and constitutes the largest oceanic freshwater reservoir in the northern hemisphere (Haine et al., 2015; Rabe et al., 2014).

Mixing in the Arctic Ocean has profound regional and global impact, as the mixing that drives water-mass transformations here is critical to functioning of the wider general Arctic circulation (see the review by Timmermans and Marshall, 2020), at the same time as influencing the fate of sea ice, and Arctic and global atmospheric circulation patterns (e.g., Kennel and Yulaeva, 2020). This chapter will explore how Arctic mixing is dominated by the interactions between the relatively warm, salty water masses of Atlantic and Pacific origin with the fresher Arctic water masses and the sea ice, and how these are all subject to wind and tidal forcing. The cold ocean temperatures and distinct T-S stratification give rise to a variety of mixing processes. Mixing occurring along the warm AW pathways (Fig. 11.1), via Fram Strait and also the Barents Sea opening to the south are of particular interest due to the potential of the subsurface AW heat to reach the polar mixed layer, where it can melt sea ice. Thinning or removal of the sea-ice cover then impacts the surface energy balance, and can have far reaching consequences and feedbacks. Other key Arctic mixing questions pertain to the storage and release of both liquid and solid (sea-ice) freshwater and how reducing sea-ice cover may be changing Arctic stratification (with feedbacks to mixing), and consequently Arctic primary productivity.

11.2 Foundations

Early Arctic research was advanced by a pioneering community, which braved the difficult environmental conditions to test theories, observe its main circulations patterns and monitor flow through key gateways and along the boundary. Arctic oceanographers observed an ice-covered ocean, in which the ice floats atop a cold fresh surface mixed layer, above a halocline transition to saltier water (e.g., Nansen, 1904; Aagaard et al., 1981; Rudels et al., 1994). Within the main halocline sub-surface temperature, maxima and minima associated with summer and winter PW were observed in the Canadian basin (e.g., Coachman and Barnes, 1961). Saltier AW was found everywhere in the Arctic (deeper than PW in the Canadian basin) (e.g., Coachman and Barnes, 1963) overlying very cold Arctic deep waters below (Meincke et al., 1997). The strongest currents were found along the continental slopes and shelf seas, but elsewhere currents were found to be weak within the basins (Rudels et al., 2000; Woodgate et al., 2001), with the exception of energetic eddies, coherent lenses rotating both cyclonically (Hunkins, 1974; D'Asaro, 1988; Padman and Dillon, 1991) or anti-cyclonically (Hunkins, 1974; Muench et al., 2000; Woodgate et al., 2001). Tides too were found to be generally modest, with shelf sea tidal currents typically $O(10 \text{ cm s}^{-1})$ except along the western Eurasian slopes (Padman et al., 1992; Kowalik and Proshutinsky, 1994). In these circumstances, internal tides were assumed to be of minimal influence.

The Surface Heat Budget of the Arctic Ocean (SHEBA) experiment mounted in the late 1990s, was the first major coordinated effort to quantify ocean-atmosphere fluxes (e.g., Persson et al., 2002), and the dynamics of both the atmospheric and oceanic boundary layers (e.g., McPhee, 2002) from a drifting year-long ice-camp. SHEBA was key to advancing our understanding of Arctic clouds (e.g., Intrieri et al., 2002), and how together with the seasonal cycle of the floating ice and

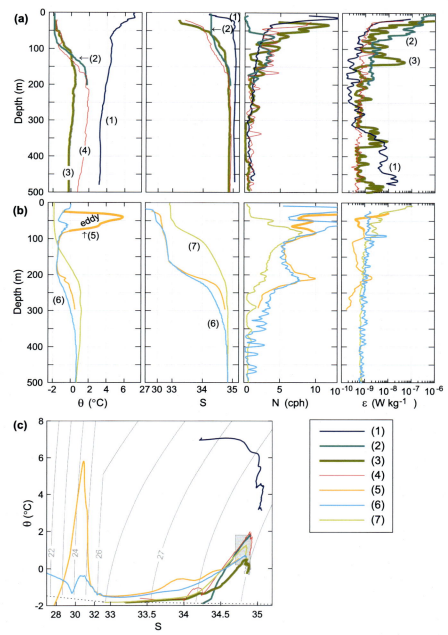

FIGURE 11.2 From left to right, profiles of potential temperature θ, practical salinity S, buoyancy frequency N (in cycles per hour (cph)), and dissipation rate of turbulent kinetic energy per unit mass (ϵ) are shown in (a) and (b) for locations 1–7 mapped out in Fig. 11.1. The θ-S relationships are shown in (c), where the grey shaded box corresponds to range depicted in Fig. 11.9. Data are derived from (1): average of 15 profiles on 19 Aug. 2015 (Kolås and Fer, 2018); (2): average of 3 profiles on 6 Feb 2015 (Meyer et al., 2017); (3): one profile on Oct. 2008 offshore from Severnaya Zemlya; (4): one profile on Oct. 2008 on the Laptev slope (Lenn et al., 2011b); (5) 51 profiles averaged between 16 and 17 Sept. 2015 in an intra-halocline eddy core on the Chukchi slope (Fine et al., 2018); (6) a background Canada Basin profile collected on 6 Sept. 2012 (Lincoln et al., 2016); and (7) 100 profiles averaged between 24 and 29 Apr. 2007 (Fer, 2009) near the North Pole. Note the change of scale at $S = 33$. Averaged profiles are shown for locations with multiple repeat observations; single profiles are shown where the data are synoptic.

snow pack, they mediate how heat derived from solar radiation is stored in the upper ocean and subsequently released, to the underside of the sea ice (McPhee, 2002). Together SHEBA and the other oceanographic observations drove the conceptual paradigm of the Arctic Ocean as a quiescent environment, where the perennial ice presented a significant barrier to the transfer of heat, freshwater, momentum and radiative fluxes (notably sunlight) between the ocean and atmosphere. In other words, an ocean with a near-permanent lid, surrounded by slightly more energetic marginal shelf seas.

Early water-column measurements under a drifting ice island in the central Arctic revealed a temperature structure consisting of a series of stacked well-mixed layers, separated by high-gradient interfaces (a staircase), indicating that vertical heat transport from the core of the AW was via double diffusion (Neshyba et al., 1971). Hydrographic observations later revealed the widespread presence of lateral thermohaline intrusions within the AW layer (e.g., Carmack et al., 1998; Rudels et al., 1999). These intrusions were seen as the primary mechanism for mixing of the relatively warm and salty AW from the boundaries into the interior basins (Ruddick and Richards, 2003; Walsh and Carmack, 2002). At the top and bottom interfaces of the intrusions, staircases characteristic of double diffusion were observed (Ruddick and Gargett, 2003; Woodgate, 2005; Carmack et al., 2015). The slow diapycnal mixing associated with these staircases was found to drive a small heat flux of at most $O(1)$ W m^{-2} from the AW to the overlying halocline (Carmack et al., 2015). Due to their ubiquity and year-round presence across the Arctic, thermohaline intrusions (and the related staircases) were considered the dominant diapycnal process occurring within the central Arctic basins. This only further reinforced the view of the Arctic as a largely quiescent environment where slow, but steady processes win the race.

Observations of dramatic summer sea-ice loss (Stroeve et al., 2007), brought a growing recognition of the rapid climate change occurring in the Arctic, and hence, renewed scientific focus. This was fueled, in part, by the misnamed International Polar Year (IPY) spanning a two-year period from 2007 to 2008, that focused international efforts on both polar regions. *In situ* and satellite observations showed that Beaufort Gyre freshwater storage has increased through the end of the century and into the 2000s as a result of increased Ekman pumping and a net freshening of the surface layer due to sea-ice melt and increased river fluxes from the shelves (Proshutinsky et al., 2002, 2009; Rabe et al., 2011; Giles et al., 2012).

Other new observations in the Barents and Kara Seas showed that these two regions did in fact have vigorous tides and well-developed internal tides (Sundfjord et al., 2007; Morozov et al., 2008; Sirevaag and Fer, 2009). Holloway and Proshutinsky (2007), extending a coarse resolution (55 km) coupled ocean/ice model to include tidal effects and a simple parameterisation for ocean mixing associated with tides, demonstrated that Arctic tides could have a profound impact on the transformation of the AW flowing in the boundary currents along the slope, and consequently Arctic Ocean stratification. Newer, more extensive turbulence observations around the Yermak Plateau showed that the energetic internal wave field previously observed (Padman and Dillon, 1991; D'Asaro and Morison, 1992) are limited to the northern reaches, whereas stronger stratification and weaker diffusivities suppress mixing to the south (Fer et al., 2010).

In the Arctic boundary current, IPY-era microstructure turbulence observations of double-diffusive heat fluxes were found to be an order of magnitude too low to explain the along-slope cooling of the AW thermocline (Lenn et al., 2009), implying that other processes must also be at work. At the same time, detailed measurements from ice-tethered profilers (ITPs, Krishfield et al., 2008; Toole et al., 2011) within the central basins showed that individual thin layers making up the staircase could be traced for 100 s of kms (Timmermans et al., 2008a), and that these were not infrequently subject to disruptions such that the layers split and merged over time and space (Radko et al., 2014).

Observations from moored instruments along the Chukchi shelf provided definitive evidence that the ice-free summer conditions resulted in more wind energy input to the oceanic near-inertial currents than during the ice-covered winter months (Rainville and Woodgate, 2009). Yet within the central Arctic basins, still largely ice-covered even in summer, currents and the internal wave field remain quiescent (Rainville and Winsor, 2008; Fer, 2014; Guthrie et al., 2013; Lincoln et al., 2016).

The most recent decade from 2010 onwards has brought even more evidence that Arctic stratification is changing along the margins, driven in part by the warming and increasing salt content of the inflowing AW (Polyakov et al., 2017; Barton et al., 2018; Polyakov et al., 2020b). These results have highlighted the AW influence on sea-ice melt in the Barents Sea and Eurasian Arctic (Årthun et al., 2012; Barton et al., 2018; Polyakov et al., 2020b). On the Pacific side, both the volume and heat content of the sub-surface Pacific summer water layer have been steadily increasing, with significant implications for accelerating ice melt in the Canada Basin (Timmermans et al., 2018).

As the heat content and stratification of the Arctic have been evolving, questions have been raised about the role of the Arctic Ocean's sea ice in regional and midlatitude climate and weather (e.g., Screen and Simmonds, 2014; Kolstad and Screen, 2019). This has only served to thrust the spotlight back onto Arctic mixing and how this ocean has been redistributing heat and freshwater on a planetary scale.

The original paradigm of a largely quiescent Arctic Ocean with mostly negligible turbulent mixing is now changing as the ice recedes and the frequency and coverage of observations increase. Double-diffusive processes and the associated thermohaline intrusions remain ubiquitous, but clearly these are only part of the puzzle of what sets Arctic stratification, and consequently its circulation. Tides appear to be more important than previously assumed, despite the high spatial inhomogeneity (Rippeth et al., 2015; Fer et al., 2020). Although no significant long-term trends have yet been observed in Arctic's internal wave field, future response to changing stratification and increasing wind energy is unknown. Meanwhile, increasing interest in mesoscale and submesoscale eddies has been driven by new technologies that allow us to begin to

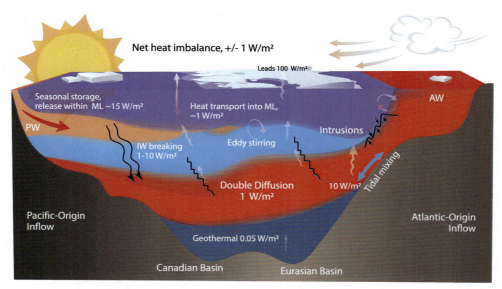

FIGURE 11.3 Schematic of Arctic mixing processes shown against an stylised cross-section of the Arctic Ocean with the Lomonosov Ridge separating the Canadian and Eurasian basins.

quantify these features in both observations and models. In the next section, we outline the key findings on all these aspects of Arctic mixing.

11.3 Key findings

Here, we review the underlying physics and major advances in our understanding of key Arctic Ocean mixing processes, schematically represented in Fig. 11.3. Recent observations set the stage for these discussions: characteristic profiles of temperature, salinity, stratification and the dissipation rate, ϵ, collected in selected stations in the Arctic ocean (Fig. 11.2, see Fig. 11.1 for station locations) show the water-mass transformations and typical turbulence levels associated with the different processes to be discussed.

A whistle-stop tour follows the boundary current east from Fram Strait: Relatively warm and saline Atlantic-origin waters observed west of Svalbard (1) subduct (2) and are transported in a boundary current along the slope of the Eurasian basin (3, 4). The changes of temperature and salinity hydrographic properties as the water flows across the Arctic can be followed in the insets of Fig. 11.1 and in the T/S diagram (Fig. 11.2c). Staircases indicative of double diffusion, plus overlying thermohaline intrusions, are seen on the outer Laptev sea slope (4). On the western side, upper layers are substantially less saline. Mixing between PW and AW and the polar surface waters sets the shape and strength of the halocline that characterise Canada basin stratification (5, 6), including eddies that trap and export waters from slope regions into the interior (5). At the central Arctic near the North Pole in the Amundsen basin, the cold halocline layer is well defined above the AW found below 200 m depth (7).

The selected profiles exemplify the turbulence levels associated with mixing processes sketched in Fig. 11.3. In the central Arctic, and away from boundaries, the dissipation rates are small, $(2-5) \times 10^{-10}$ W kg^{-1}, at the lowest detection level of the instruments. Around Svalbard, enhanced near-bottom mixing can occur in the bottom 100–200 m (1), driven by strong vertical shear of the mean current and by convection induced by Ekman advection of buoyant water across the slope (Kolås and Fer, 2018). Winter storms increased turbulence levels to order 10^{-6} W kg^{-1} under the drifting sea ice, in otherwise quiescent waters in the Nansen Basin (2) (Meyer et al., 2017). North of Severnaya Zemlya (3), 20–50 m thick layers of energetic turbulence were captured over the slope (Lenn et al., 2011b). An intrahalocline eddy observed on the Chukchi slope was turbulent near the edges (5) with modest heat fluxes, but the largest heat transport was laterally out of the eddy through thermohaline intrusions (Fine et al., 2018). Elsewhere, Canada basin stratification (6) is dominated by halocline, PW and AW with very little turbulent kinetic energy dissipation (Lincoln et al., 2016). The dissipation profile averaged in the pycnocline near the North Pole (7) following a storm event was near-inertially modulated, decaying approximately at a rate implied by the reduction of wind energy over time (Fer, 2014). The heat flux and mixing across the cold halocline layer was negligible (Fer, 2009), and dominated by fluxes associated with double diffusion above the temperature maximum (Sirevaag and Fer, 2012).

Each Arctic mixing process affects water-mass transformations and heat transport in different ways, and in a spatially inhomogeneous manner. In this section, we have loosely organised Arctic mixing processes in order of energetics and scales, from the buoyancy-driven mixing arising from ice-ocean interactions, tidally-generated and shear-driven mixing to eddies, double diffusion and lateral thermohaline intrusions.

11.3.1 Ice-ocean interactions

Arctic sea ice is almost always in motion, driven by wind forcing or ocean currents (tidally-induced ocean flows or energetic mixed-layer eddies, for example). The relative motion of the sea ice and adjacent ocean sets up a vertical shear, and a boundary layer, which controls the turbulent exchange of momentum, heat and salt. In turn, mixing in the boundary layer is strongly controlled by buoyancy fluxes produced by sea-ice growth (a destabilising buoyancy flux of brine) and melt (a stabilising buoyancy flux of fresh water). Given the critical sensitivity of sea-ice thickness and extent to ocean heat fluxes of order 1 W m^{-2} (Maykut and Untersteiner, 1971), as eluded to in Section 4.2.2.9 of Chapter 4, understanding mixing processes driving ice-ocean exchange is needed to quantify the role of oceanic heat in a changing Arctic. Air-sea-ice interactions and processes in the under-ice boundary layer have been described thoroughly in recent overviews (McPhee, 2008, 2017). The relatively stable platform offered by sea ice allows for measurements in the surface ocean boundary layer (by sensors mounted in the sea ice) that are not possible in ice-free regions; many field experiments have employed ice-mounted systems to measure turbulent fluctuations of velocity, temperature and salinity and infer the statistics of turbulent fluxes (see McPhee, 2008). In fact, the first clear observational evidence for an Ekman spiral structure of velocity in the surface ocean forced by ice-ocean stresses was from an ice-drift station (Hunkins, 1966). This velocity structure has since been studied in more detail through the use of autonomous ice-mounted systems (e.g., Cole et al., 2014).

Momentum fluxes are achieved via shear stresses at the ice-ocean interface, which also regulate the exchange of heat, salt and gases between the ocean and sea ice. In the absence of direct sampling of under-ice turbulent temperature, salinity and velocity fields, ocean-to-ice heat fluxes are typically estimated with the following parameterisation (see e.g., McPhee et al., 2003):

$$F_H = \rho c_p c_H u_{*0} \delta T, \tag{11.1}$$

where c_p is the specific heat of seawater, c_H is a heat transfer coefficient, δT is the difference between mixed-layer temperature and freezing temperature, and u_{*0} is the interface friction speed; u_{*0} may be calculated from velocity covariances (i.e., momentum fluxes) if these are measured at the ice-ocean interface (see e.g., Cole et al., 2014; Peterson et al., 2017). If direct velocity perturbations are not measured, a quadratic drag relationship is often used to infer u_{*0} from the relative velocity between the surface ocean and sea-ice (see Section 4.2.2.1 of Chapter 4). This assumes either a constant drag coefficient or one that depends upon geometric properties of the ice (see Lu et al., 2011). Another accepted parameterisation for u_{*0} assumes that a Rossby similarity relationship, which applies for neutral stratification and a laterally-homogeneous boundary layer under the sea ice, characterises the boundary layer (McPhee, 1992; Shaw et al., 2008). This parameterisation requires some knowledge of under-ice morphology (a roughness lengthscale) (see e.g., Shaw et al., 2008; Cole et al., 2017). While a multitude of factors influence these parameterised relationships (e.g., surface-ocean stratification, sea-ice ridging and keels) (McPhee, 2012), for the same oceanographic setting, the various approaches to parameterising momentum fluxes generally indicate comparable relationships between ice-ocean shear and turbulent fluctuations (Cole et al., 2014). Observations of turbulent momentum fluxes over the course of a melt season in the Beaufort Sea marginal ice zone (MIZ) indicate about a 20–30% decrease in ice-ocean drag coefficients and roughness lengths with increased winds and when sea-ice concentration decreases from 100%; further, for ice concentrations as low as 10–30%, ice-ocean drag parameterisations do not apply (Cole et al., 2017).

Ocean-to-ice heat fluxes range from a fraction of a W m^{-2} to O(100) W m^{-2}. Ice-mounted sensors drifting in the Transpolar Drift Stream indicate average summer ocean-to-ice heat fluxes in the range to 5 to 15 W m^{-2} (during the period 2002–2010), with heat fluxes derived almost entirely from surface-ocean warming by incoming solar radiation through open water areas, rather than heat entrained from below the mixed layer (Timmermans et al., 2011; Stanton et al., 2012). In the Beaufort Gyre region, summer-averaged ocean-to-ice fluxes (sampled in 1998) were similar: around 16 W m^{-2} (Shaw et al., 2009). Shaw et al. (2009) point out that outside of the summer heating period, ocean-to-ice heat fluxes were only around 0.2 to 1.2 W m^{-2}, comparable with fluxes at the base of the mixed layer (i.e., the upper halocline) estimated during the same period. During the summer heating period, fluxes at the base of the mixed layer (0.1 to 1.5 W m^{-2}) were around the same as in winter, and relatively much smaller than summer ocean-to-ice heat fluxes. Intense solar heating in low-ice summers can lead to storage of heat in the upper ocean that can slow the growth of winter sea-ice; Timmermans (2015) found that the release of solar heat stored in summer 2007 in the Canada Basin reduced 2008 sea-ice thickness by about 25%.

Heat fluxes can be highly variable in space and time in the MIZ due to strong variability in absorbed solar radiation, depending on the presence or absence of sea ice. Patterns of these fluxes further depend upon the distribution of fresh-water input, which influences the capacity of the upper ocean to store heat (e.g., Perovich et al., 2007; Gallaher et al., 2016; Perovich et al., 2011). Observations from the Beaufort Sea MIZ in 2014 indicate a 4-day event of ocean-to-ice heat fluxes reaching 200 W m^{-2} accounted for around 20% of the total summer basal ice melt (Gallaher et al., 2016).

On the other side of the Arctic basin, near the AW inflow, a wide range of heat fluxes have also been documented. For example, in 2015 in the Nansen basin north of Svalbard, Peterson et al. (2017) found average winter ocean-to-ice heat fluxes of around 1.4 W m^{-2}, with episodic local upwelling events and proximity to AW pathways increasing the heat fluxes by one order of magnitude. In spring, heat fluxes were enhanced (to O(100) W m^{-2}) from a combination of heat from the AW layer from below, solar heating, wind events driving mixing and enhanced turbulent mixing by flow-topography interactions in the vicinity of the Yermak Plateau (Peterson et al., 2017).

The buoyancy flux associated with growing sea ice (i.e., brine rejection) influences turbulence in the ice-ocean boundary layer in some regions of the Arctic. The rate of sea-ice growth, and therefore the salt flux, diminishes rapidly after the formation of new ice, which insulates the ocean from the cold atmosphere (Wettlaufer et al., 1997). Under sea-ice cover in winter, brine is not rejected in sufficient volumes to penetrate the base of the mixed layer; fluxes are too weak and the halocline is too strongly stratified. However, in open-water areas (leads) intense heat loss from the ocean drives sea-ice growth and sufficient brine production to create turbulent plumes and precondition the mixed layer for deepening by shear-driven turbulence (Morison et al., 1992). Using direct eddy-covariance measurements in the oceanic boundary layer under freezing leads, McPhee and Stanton (1996) found that even a weak destabilising surface buoyancy flux can influence the turbulence scales and efficiency of turbulent exchange compared to forcing by stress alone.

Regions of sustained sea-ice growth, such as coastal polynyas, have the highest dense-water fluxes (Cavalieri and Martin, 1994). Direct observations of energetic dense plumes are rare, and the total influence of shelf regions on setting the Arctic halocline structure is not clear. In the eastern Barents, Kara and Laptev Seas, dense flows have been observed, but these may be confined to the continental shelves (rather than contributing to the inventory of Arctic dense water) by entrainment and dilution and the confining effects of rotation (e.g., Ivanov and Golovin, 2007). In the western Barents Sea, high-salinity dense water has been observed to generate an important downslope cascade of dense plumes. For example, the plume originating from the Storfjorden polynya (on the south side of the Svalbard archipelago) has been documented to spread along the continental slope for over 600 km away from its formation area, penetrating to around 2000 m depth in Fram strait (Akimova et al., 2011), having undergone substantial vertical mixing (Fer et al., 2004).

11.3.2 Tidal mixing

In the Arctic Ocean, as in the wider global ocean (see Chapter 6), many of the strongest hotspots of vertical mixing are associated with turbulence generated by processes over topography and along margins, mainly forced by tides (Padman and Dillon, 1991; Lenn et al., 2011b; Rippeth et al., 2015; Fer et al., 2015). Over regions of steep topography turbulent heat fluxes reach 100 W m^{-2} (Meyer et al., 2017; Fer et al., 2020). These are typically areas of conversion of barotropic tidal energy to baroclinic waves and mixing.

A collection of microstructure measurements from several surveys shows that the longitudinal variation of dissipation rate averaged in the AW layer for a sequence of observational sections along the continental slope (close to locations 2, 3, 4 and 6 profiles in Fig. 11.1) correlates well ($r = 0.67$) with the tidal energy dissipation rate (Fig. 3 in Rippeth et al. (2015)). The high correlation led the authors to conclude that the tides are the source of energy to support the dissipation rates along the circum-Arctic continental slopes.

There are several pathways of energy from tides to dissipation. Bottom stress at the seabed is a sink of energy for the tidal current, driving dissipation and mixing in a bottom boundary layer. The dissipation rates can be large in continental shelf seas, where tidal currents are strong and the bottom boundary layers cover a substantial fraction of the water column. A first-order estimate can be made if we assume the work done by the bottom stress is balanced by the energy dissipation in the boundary layer. Using a quadratic drag parameterisation, the work done by bottom stress, $\tau_b = \rho_0 C_D \mathbf{u} \cdot |\mathbf{u}|$, is $\tau_b \cdot \mathbf{u}$, where $\mathbf{u}(u, v)$ is the total horizontal tidal current. Using a typical drag coefficient of 2.5×10^{-3} and an average (strong) tidal current speed of 0.15 m s^{-1}, the bottom stress is about 0.06 Pa, comparable to a surface stress from wind speeds in excess of 6 m s^{-1}, and we obtain a depth-integrated dissipation rate of 0.01 W m^{-2}. Over a water column of 100 m, this corresponds to a substantial dissipation rate, on the order of 10^{-7} W kg^{-1}. However, the regions with strong tidal currents and relatively shallow depths are limited in geographical extent. Tidal current predictions over a one-month period using

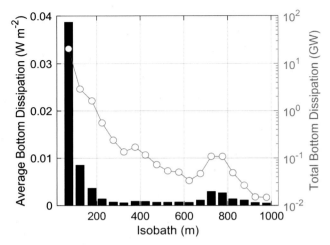

FIGURE 11.4 Average bottom dissipation (left axis, bars) from 30-day average tidal currents from Arc5km2018 using a drag coefficient of 2.5×10^{-3}. Values calculated between 50 m and 1000 m total depth above 74°N are averaged in isobath bins at 50 m intervals. Total bottom dissipation (right axis, circles) obtained by multiplying the bin averaged dissipation by the total surface area between each isobath bin interval.

all constituents from Arc5km2018 (Erofeeva and Egbert, 2020), a barotropic inverse tidal model on a 5-km grid, suggest that an average total tidal current speed $|\mathbf{u}| > 0.15$ m s^{-1} occurs over approximately 7% of the surface area between the 50 and 1000 m isobaths above 74°N. Hence, the contribution of bottom dissipation to the volume-averaged dissipation is considered to be small. In the presence of sea ice, a similar turbulent stress term can be considered, with a comparable drag coefficient, effectively doubling the dissipation rate if the ice-relative tidal current is the same.

Total dissipation from tidally-driven bottom stress is largest over the shelf areas with strong tidal currents and over topography, such as Yermak Plateau, where strong currents occur. We estimated the total dissipation using the average currents predicted from Arc5km2018, and averaged in 50-m intervals of isobaths (Fig. 11.4). Over the shelf, integrated over regions shallower than 200 m depth, total energy dissipated in the bottom boundary layer is between 1 and 20 GW (1 GW = 10^9 W), and decays with increasing total depth. This can be compared to the upper bound for local dissipation of the energy extracted from the surface tide, which includes local dissipation from both bottom stresses and high-mode internal waves, estimated to be 10 GW at the Hawaiian ridge (Rudnick et al., 2003). In the Arctic, a relatively strong dissipation is obtained between the 700 and 800 m isobaths, corresponding to energetic regions north of Svalbard and the northern flanks of the Yermak plateau. This is manifested in large dissipation rates observed in these regions (Padman and Dillon, 1991; Fer et al., 2010, 2020). The Yermak plateau is a region where diurnal tidal currents are greatly enhanced (Padman et al., 1992), consistent with resonance by trapped diurnal topographic waves circulating around the plateau.

Another mechanism by which the energy is removed from barotropic tide is the conversion of energy to baroclinic (internal) tide as the tidal flow interacts with sloping topography in a stratified environment. Linear internal waves of tidal frequency then propagate, interact and eventually break, dissipating their energy. At lower latitudes, the geography of mixing in the deep ocean is largely set by the patterns of internal tide generation, propagation and dissipation (MacKinnon et al., 2017). In contrast, much of the Arctic is above the turning latitude (also called the critical latitude ϕ_c) for both diurnal and semidiurnal internal tides. This is the latitude at which the tidal frequency matches the local inertial period (~74.5°N for M$_2$). At these latitudes, linear internal waves generated by tidal flow over sloping topography are evanescent, and their cross-slope radiation is not permitted (Vlasenko et al., 2003). However, topographically-trapped baroclinic Kelvin waves may propagate along the topographic waveguide (Hughes and Klymak, 2019; Musgrave et al., 2016). When the topographic obstruction of the stratified flow is large, the response can be non-linear. At lower latitudes, such non-linear response is typically in the form of excitation of higher harmonics, disintegration and steepening of short-wave packets over the carrying linear tidal wave. On the other hand, above the turning latitude of the forcing frequency, the non-linear internal waves have properties inherent to lee waves with relatively short wavelengths comparable to the lengthscales of the bottom topography (Vlasenko et al., 2003; Musgrave et al., 2016). Their characteristics are defined by the shape of the bottom topography, the intensity of the tidal flux and stratification.

A measure of the obstruction of the flow by the topography is the topographic Froude number, $\text{Fr}_t = U_0/Nh_0$, where N is the stratification near the topography and h_0 is the topographic height. Non-linear lee waves are generated for $\text{Fr}_t > 1$. A measure of the non-linearity of the response is the non-dimensional tidal excursion length, $l_e = U_0/(\omega L)$, where U_0 is the barotropic tidal current amplitude at the topography (e.g., shelf break, ridge etc.), L is a characteristics cross-isobath width of the topography and ω is the tidal frequency. For $l_e > 1$, non-linearity becomes important. This is achieved, for the semidiurnal frequency, for $L = 1$ km and $U_0 = 0.15$ m s^{-1}, and for $U_0 = 0.07$ m s^{-1} for a diurnal wave. Converting the tidal excursion length into a vertical excursion distance using the bottom slope s, and expressing Fr_t using this lengthscale, Legg and Klymak (2008) found that for the inverse Froude number sN/ω, larger than values of 3 (steep slopes, weak stratification and low-frequency forcing) highly non-linear jumplike lee waves occur. For a semidiurnal wave, this is expected for steep slopes of 0.1 (about 6°) and a stratification of 3 cycles per hour, corresponding to $N = 5 \times 10^{-3}$ s^{-1}.

For conditions favourable for tidal conversion above the critical latitudes of the forcing wave, the pathway for the energy from the tide to turbulence is non-linear and dominated by breaking unsteady lee waves and critical flow (Rippeth et al., 2017; Fer et al., 2020; Musgrave et al., 2016). Measurements have shown increased mixing rates linked to an unsteady lee wave forced by the M$_2$ tide (Rippeth et al., 2017). Fer et al. (2020) observed enhanced mixing associated with a trapped lee wave modulated to supercritical flow and strong turbulence by the tide. Turbulent heat fluxes exceed 100 W m^{-2} in some profiles, with an average over the observed turbulent event of 15 W m^{-2}, compared with background values of 1 W m^{-2}.

In addition to bottom boundary-layer mixing and mixing induced by nonlinear waves in the water column, tides can also enhance pycnocline mixing by an additional mechanism. In the Laptev Sea winter, barotropic tidal currents are sheared across the pycnocline as a result of the surface drag from the ice (Lenn et al., 2011b; Janout and Lenn, 2014). When this tidally-modulated pycnocline shear vector aligns with the surface stresses, the shear is enhanced in a manner consistent with an intense burst of turbulent dissipation and mixing observed by Lenn et al. (2011b). This result highlighted the potential of tides to drive pycnocline mixing even under 100% sea-ice conditions when less near-inertial energy is transferred from the atmosphere to the ocean.

11.3.3 Near-inertial motions and the internal wave continuum

In most lower latitude oceans, turbulent mixing away from boundary layers is primarily driven by breaking internal waves (see Chapter 5). Power input through wind and tides, followed by a cascade through non-linear wave-wave interactions, leads to a broadband internal wave spectrum. Within that spectrum, energy is transferred from larger-scale internal waves to smaller-scale internal waves that directly break down into turbulence. The broad-brush global patterns of resultant turbulent mixing rates are hence governed by the patterns of internal wave generation, propagation, and interaction (Olbers and Eden, 2013; Waterhouse et al., 2014; MacKinnon et al., 2017; de Lavergne et al., 2019).

The internal wave field in the Arctic Ocean was observed to be quite weak, historically. For example, observations in the SHEBA and AIWEX experiments found internal wave spectral energy levels to be an order of magnitude or more lower than those typical of lower latitudes (Levine et al., 1987; Pinkel, 2008). The weak internal wave climate has been variously attributed to a combination of low direct wind forcing on an ice-covered ocean, inability for semi-diurnal internal tides to propagate poleward of their turning latitude, and dissipation of vertically-propagating internal waves on the underside of sea ice (Morison et al., 1985). However, a closer inspection revealed that even evanescent trapped internal tides may be appreciably elevating turbulence near steep topography (Section 11.3.2).

In most of the Arctic, near-inertial internal waves (NIW) dominate the internal wave field. Time-variable wind stress leads to generation of near-inertial oscillations in the ocean mixed layer. Convergences and divergences of those motions pump the mixed-layer base and create downward propagating near-inertial internal waves (Alford et al., 2016). Enhanced dissipation associated with near-inertial internal waves have been measured in the central Arctic (Fer, 2014); however, typically the near-inertial energy under sea ice is weak (Martini et al., 2014; Dosser and Rainville, 2016). An example NIW is shown in Fig. 11.5. Here data from an ITP-V drifting with the pack ice shows presence of both inertial motions in the ML and propagating NIW below, see also (Cole et al., 2017). Mesoscale motions, such as the eddy apparent between 24 and 27 July in Fig. 11.5, modulate the effective inertial frequency and allow trapping or refraction of the waves.

With rapidly expanding seasonal open-ice conditions, there has been considerable speculation that the wind-input into the internal wave field might dramatically increase (Rainville and Woodgate, 2009), potentially significantly raising the background mixing levels in the Arctic. Some recent evidence shows clear links between wind events and enhanced upper-ocean turbulence. For example, north of Svalbard, winter storms increased the pycnocline fluxes by a factor of two (profile (2) in Fig. 11.2; Meyer et al., 2017), and the ocean-to-ice fluxes by a factor of three (Graham et al., 2019). Near 89°N, increased dissipation rates have been recorded in the pycnocline following strong winds, coherent with energised near-inertial internal waves (profile (7) in Fig. 11.2; Fer, 2014). Episodic energetic vertical mixing in the interior ocean has been observed in response to strong winds and ice-divergence. In the strongest cases, upwelling of warm AW in the central Arctic led to vertical fluxes of 100 W m^{-2} (McPhee, 1992).

However, in an integrated basin-wide sense, the increasing geographical area and seasonal extent of wind forcing onto open or mostly open water seems to be leading to a modest overall increase in the internal wave energy level and associated mixing rates (Lincoln et al., 2016). There appears to be several interesting reasons for this. Guthrie et al. (2013)

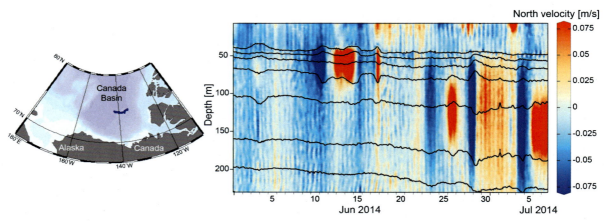

FIGURE 11.5 Depth-time section of north-south velocity from an ITP with a velocity sensor (ITP-V) that sampled in the Canada Basin in summer 2014 (right panel); profile locations are shown in the map (left panel). Potential density contours are shown in black. Data are presented in Cole et al. (2017). Several mesoscale eddies appear in the halocline alongside near-inertial internal waves.

comment on the importance of unusually thin surface mixed layers overlying strongly-stratified haloclines. Deposition of wind momentum into such thin layers may both increase the percentage of power input that is dissipated at the mixed-layer base through shear instability, and project onto very high vertical modes of internal waves. Very high-mode internal waves are more likely to dissipate near the generation region rather than propagating into the interior. Additionally, the gradient of planetary vorticity, which has been found to play an important role in creating of vertically-propagating near-inertial internal waves through convergences and divergences of mixed-layer velocities (D'Asaro et al., 1995), is very small in the Arctic. Finally, the frequency of wind events in the Arctic may not map onto the local inertial frequency as efficiently as do events in the midlatitude storm track (Dosser, 2015).

Though direct microstructure estimates of turbulent mixing in the Arctic are limited, turbulence mediated by internal wave breaking may be indirectly estimated through a finescale parameterisation based on wave-wave interaction theory and observations (Polzin et al., 2014). Use of this technique has allowed the first broadscale geographical patterns of internal waves driving mixing to emerge (Gregg et al., 2003; Kunze et al., 2006; Whalen et al., 2012, 2015). Guthrie et al. (2013) apply this technique to Arctic expendable current profiler data, and find no appreciable increase in mixing rates over the last 30 years. Chanona and Waterman (2020) apply the same technique to the comprehensive database of ice-tethered profiler data. She finds mixing rates that are low away from boundaries, with associated heat fluxes generally less than 1 W m^{-2}.

11.3.4 Eddies

Mesoscale eddies are a ubiquitous feature of the Arctic Ocean. Early field programs in the Canada Basin indicated that eddies accounted for most of the water-column kinetic energy (Hunkins, 1974). Drifting ice camps in the Beaufort sea (1975–76) as part of the Arctic ice dynamics joint experiment (AIDJEX) encountered tens of eddies (Fig. 11.6), and inferred that eddies may take up about a quarter of the Beaufort Sea area (Manley and Hunkins, 1985). A wide variety of eddy types, with a range of dynamics and water-mass origins have been documented (e.g., Hunkins, 1974; D'Asaro, 1988; Padman and Dillon, 1989; Muench et al., 2000; Woodgate et al., 2001; Pickart et al., 2005; Zhao et al., 2014). Most of the eddies sampled and studied are intra-halocline lenses (less than a couple of hundred metres thick), centred at depths around the core of Pacific or Atlantic Waters (e.g., Hunkins, 1974; Zhao et al., 2016, and represented in Fig. 11.3), rather than being surface-intensified or barotropic like the mesoscale eddies of the Southern Ocean and elsewhere (Chapters 8 and 12). A partitioning of water-column kinetic energy into dynamic modes in the Canada Basin indicates that energy tends to be concentrated in the second baroclinic mode, consistent with this prevalence of intra-halocline mesoscale eddies (Zhao et al., 2018). A relatively smaller number of taller eddies, more than 1000 m thick and centred below the main Arctic halocline, some even having a stacked double-core structure, have also been observed (e.g., Woodgate et al., 2001; Schauer et al., 2002; Aagaard et al., 2008; Carpenter and Timmermans, 2012; Zhao and Timmermans, 2015). Eddy vertical scales are inversely related to the background stratification, and horizontal scales are of the order of the first baroclinic Rossby deformation radius (see Zhao and Timmermans, 2015). The intra-halocline eddies have both warm or cold cores, radii of O(10 km), are easily distinguished by azimuthal velocities of O(10) cm s^{-1}, and are predominantly anti-cyclonic in much of the Arctic (e.g. Newton et al., 1974; Zhao et al., 2014, and Fig. 11.5), except in the Eastern Eurasian basin,

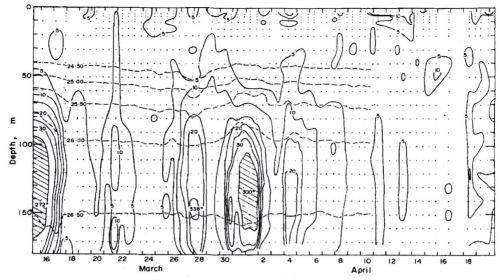

FIGURE 11.6 Current speed and sigma-t as functions of time at the AIDJEX main camp. Solid lines are contours of current speed at intervals of 5 cm s^{-1}. Dashed lines are density surfaces at intervals of 0.5 σ_t units based on hydrographic stations. Dots represent individual current measurements. Hatched areas indicate current speeds in excess of 35 cm s^{-1}. Current directions are shown at eddy cores. Figure and caption reproduced from (Hunkins, 1974).

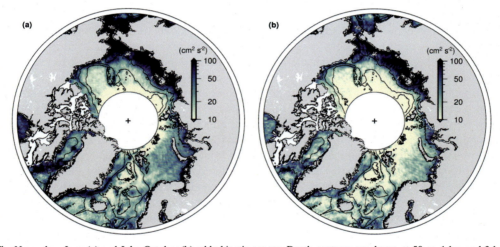

FIGURE 11.7 The November–June (a) and July–October (b) eddy kinetic energy. Depth contours are drawn at 50 m, 1 km and 3 km, taken from the ETOPO1 global bathymetry model (Amante and Eakins, 2009). Note that the Canadian Arctic archipelago has been masked out in these plots. Figure reproduced from Armitage et al. (2017).

where cyclonic eddies are just as frequently found (Pnyushkov et al., 2018). Early microstructure measurements of eddies found negligible turbulent kinetic energy dissipation rates, leading to predictions of 1–2-year lifespans (D'Asaro, 1988; Padman and Dillon, 1991). Arctic eddies are thought to be formed as a result of baroclinic instability of boundary currents, fronts, flow-topography interactions, or instability of the Beaufort Gyre (e.g., Hunkins, 1974; Spall et al., 2008; Zhao et al., 2016). Eddies contribute to the isopycnal stirring and mixing of freshwater and heat, nutrients and chemical tracers from the boundaries into the Arctic Ocean interior (e.g., Muench et al., 2000; Spall et al., 2008; Nishino et al., 2011).

The strong influence of planetary rotation in the high Arctic latitudes results in modest internal deformation radii (Zhao et al., 2014) that would be considered submesoscale in other oceans. This means that, where new remote-sensing methods have enabled it possible to infer sea surface height in sea-ice covered regions, only the largest eddies can be detected by satellite altimetry (Armitage et al., 2017, and Fig. 11.7); thermal stratification at the surface can also mask sea surface temperature signals of eddies (Porter et al., 2020). Satellite altimetry observations imply that Arctic eddy kinetic energy is at most O(100 cm^2 s^{-2}) (Armitage et al., 2017), approximately 1–2 orders of magnitude lower than eddy-rich regions, such as the Southern Ocean (e.g. Sheen et al., 2014; Lenn et al., 2011a). However, the inability to detect intrahalocline eddies

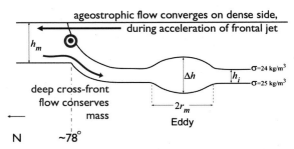

FIGURE 11.8 Schematic depicting the formation and subduction of cold-core anti-cyclonic eddies within the Arctic basins. Figure modified from Timmermans et al. (2008b), Fig. 10.

by remote sensing, contrasts with results from high-resolution numerical models that suggest these altimetric estimates are likely to be an underestimate (see Maslowski et al., 2008). The advent of ITPs, long-term moored observing systems in the US-Canadian Beaufort Gyre exploration project (BGEP, Proshutinsky et al., 2009), programs under the Arctic observing network (AON), and the international Nansen and Amundsen basin observing system (NABOS) in the early 2000s have now resolved *in situ* variability of the Arctic Ocean and boundary currents as never before. These observations show that energetic eddies are widespread in the Arctic Ocean and, as with the thermohaline intrusions (Section 11.3.6), have a potentially important dynamical role to play in stirring, mixing and the redistribution of heat and salt.

Along the Beaufort Gyre boundary, low potential vorticity anti-cyclonic eddies, with cold fresh shelf water at their cores, are formed from frontal instabilities in the shelf-break current (Spall et al., 2008). Eddy-formation hotspots also were found in the Eastern Arctic near Fram Strait, where AW enters the Arctic; and the St Anna Trough, where the Fram Strait and Barents Sea AW branches converge (Pnyushkov et al., 2018). Thus eddies formed by boundary-current instabilities populate the interiors of all Arctic basins (Zhao et al., 2014). Similarly, instabilities of the shallow fronts separating the different surface water masses associated with Canadian and Eurasian Basin waters also give rise to intra-halocline eddies; Timmermans et al. (2008b) showed that thin (12–35 m) cold-core anti-cyclonic eddies were formed during wind-forced acceleration of the frontal jet as a result of convergence of ageostrophic surface flows towards the dense side of the front, which in turn forces a deep cross-frontal flow to conserve mass (Fig. 11.8). This formation and subduction mechanism is also consistent with warm-core eddies observed in the vicinity of a meltwater front in the Chukchi sea (Lu et al., 2015). In each case, the subducted bolus gains negative relative vorticity (i.e., anti-cyclonic) as a result of the vortex squeezing experienced as it moves into the halocline on the other side of the front. This water-column squashing mechanism is also consistent with observations of thick (~1000 m) deep (below the halocline) warm-core anti-cyclonic eddies in the vicinity of topography (Woodgate et al., 2001; Aagaard et al., 2008; Carpenter and Timmermans, 2012). These appear to be formed when water parcels develop negative relative vorticity associated with squashing of flow moving over a topographic ridge; the resulting anticyclone takes on a thicker depth distribution when it moves off the ridge and the ambient weakly-stratified water column adjusts to the potential vorticity anomaly (see Carpenter and Timmermans, 2012; Zhao and Timmermans, 2015).

Numerical modelling experiments indicate that some intra-halocline anti-cyclonic eddies may be the surviving vortex of a vortex dipole (i.e., a shallow cyclone and a deep anti-cyclone, offset horizontally) (Manucharyan and Timmermans, 2013). A surface-layer cyclonic vortex, growing out of instabilities at a slumping front at the same time as the deeper anti-cyclone, would be more rapidly dissipated by under-ice friction than its deeper anti-cyclone partner. This is consistent with the predominance of anti-cyclonic eddies in the observations (Hunkins, 1974; D'Asaro, 1988; Padman and Dillon, 1989; Muench et al., 2000; Zhao et al., 2014; Woodgate et al., 2001). In the absence of under-ice friction, Manucharyan and Timmermans (2013) found that these vortex dipoles can self-propagate over long distances. Further numerical simulations by Brannigan et al. (2017) show that the separation of these Arctic eddy dipoles from the fronts depends on surface stress directed upfront, rather than other orientations.

Analysing the extensive ITP dataset (which has sufficient horizontal resolution to characterise Arctic mesoscale eddies), Zhao et al. (2016), confirmed many of these eddy-generation theories. Zhao et al. (2016) showed that in the eastern Canada Basin, eddies are typically generated by instability of surface fronts, and in the southwestern Canada Basin, or in the vicinity of ridge features of the Eurasian basin, by instability of boundary currents. Cold core eddies spawned at the shelf breaks can clearly carry fresh shelf water into the main Arctic halocline (Spall et al., 2008), whereas warm Bering seawater eddies subducting under meltwater fronts provide heat that promotes melting and delays freezing (Lu et al., 2015). Warm core eddies may also carry locally-significant heat into the main basin (Kawaguchi et al., 2012; Fine et al., 2018). Thus the question of their true impact on Arctic heat pathways and stratification depends on how these eddies dissipate their properties. In particular, there is strong interest in how warm core Arctic eddies disperse their heat.

Mesoscale eddies are generally considered the principle isopycnal mixing mechanism in the global ocean (see Section 8.3 of Chapter 8). Isopycnal mixing by eddies is carried out by propagation of anomalous bolus properties away from the source regions, and through an effective diffusive dispersion of properties through the horizontal extremities of the eddies along route. Using an inverse method and repeat hydrographic measurements between 2003 and 2015, Dosser and Timmermans (2018) estimated the along-isopycnal diffusivity in the Canada Basin to be 300–600 m^2 s^{-1}. Since Arctic eddies observed within the central basins are easily distinguished from the ambient flow by higher velocities (e.g., Timmermans et al., 2008b), there is an expectation of high geostrophic current shear on the periphery of the eddies. Hence a key question is whether this shear production is sufficient to balance turbulent kinetic energy dissipation rates (ϵ) that can drive vertical fluxes of heat?

From microstructure measurements resolving the 3D structure of an intra-halocline eddy on the Chukchi slope, Fine et al. (2018) estimated a lateral heat loss of 2000 W m^{-2} along the sides of the eddy, in contrast to much smaller doubly diffusive convection fluxes of 5 W m^{-2} out of the top (Section 11.3.5), and 0.5 W m^{-2} shear-driven turbulent fluxes out of the bottom of the eddy. These combined heat fluxes imply a 1–2-year eddy lifespan (Fine et al., 2018). These weak shear-driven turbulence observed at the base of this modern eddy (Fine et al., 2018) are consistent with the historical observations of Arctic eddies (D'Asaro, 1988; Padman and Dillon, 1991). The large lateral eddy heat fluxes, primarily responsible for the dispersion of the eddy's heat content, were instead driven by thermohaline intrusions (Section 11.3.6) propagating from the core of the eddy, which are in turn subject to double diffusion. Indeed, the role of double diffusion (Section 11.3.5) in distributing eddy heat has been documented and quantified in several other studies (Dmitrenko et al., 2008; Bebieva and Timmermans, 2016).

Finally, numerical modelling studies of the Beaufort Gyre demonstrate that eddy fluxes from the central basin towards the boundaries are critical for this basin's dynamics and freshwater storage. Lique et al. (2015) used a 2-layer idealised model to show that the balance between Ekman pumping and eddy volume flux towards the boundary mediates adjustment of the surface layer to a change in (wind-driven) forcing. The Beaufort Gyre surface-layer circulation was found to adjust on decadal timescales, in contrast to the deeper AW layer circulation, where propagating boundary-trapped waves communicate changes in boundary forcing on timescales of a month (Lique et al., 2015). Manucharyan and Spall (2016) had similar results from an MITGcm simulation that predicts basin adjustment timescales of about 6 years. Meneghello et al. (2017) further showed that Ekman pumping rates imply highly variable lateral eddy diffusivities. These diffusivities are spatially heterogeneous ranging from 50 to 400 m^2 s^{-1}, and decrease with depth. Using observations to calculate the Beaufort Gyre's mechanical energy budget, Armitage et al. (2020) demonstrated that increased surface stresses related to sea-ice decline have been offset by increased eddy activity. This increased activity results in a higher available potential energy dissipation and an apparent stabilisation of Beaufort Gyre freshwater storage from the late 2000s, and through the 2010s.

11.3.5 Double diffusion

The presence of relatively warm and salty water of Atlantic Ocean origins underlying cooler fresher water in the Arctic Ocean's central basins sets up conditions amenable for double diffusive convection, driven by the potential energy in the unstable temperature field. This is the type of double diffusion associated with both temperature and salinity increasing with depth, and is referred to as diffusive convection (see the overview by Kelley et al., 2003). The other type of double diffusion can be active when both temperature and salinity decrease with depth, and is referred to as the salt-finger regime (see the overview by Schmitt, 2003). The diffusive-convective type is most prominent in the Arctic Ocean, but the salt-finger regime also plays a role in the widespread thermohaline intrusions (see Section 11.3.6).

Throughout most of the central Arctic basin, well-mixed layers of uniform temperature and salinity separated by sharp interfaces—a manifestation of diffusive convection—have been observed at the upper boundary of the AW (Neal et al., 1969; Neshyba et al., 1971; Perkin and Lewis, 1984; Melling et al., 1984; Padman and Dillon, 1987, 1989; Timmermans et al., 2008a; Polyakov et al., 2012). Where this diffusive-convective staircase is present, heat from the AW is transported upward across the interfaces between the well-mixed layers in the staircase. The exact formation mechanism for Arctic diffusive-convective staircases remains an open area of study. Theoretical analyses consistent with observations suggest that staircases like those at the top boundary of the AW can result from interleaving motions for particular background temperature and salinity gradients, and/or that staircases may be formed by the rundown of thermohaline intrusions (Bebieva and Timmermans, 2017, 2019).

The diffusive-convective staircase can be characterised by the density ratio R_ρ, a measure of the change in density due to salinity S across the staircase to the change in density due to potential temperature θ. Diffusive convection can occur when the density ratio $R_\rho = \overline{\beta \partial S/\partial z}/\overline{\alpha \partial \theta/\partial z}$ is greater than 1 (i.e., the stabilising effect of salinity is larger than the destabilising effect of temperature). Here, α is the thermal expansion coefficient $\alpha = -\rho^{-1}\partial\rho/\partial\theta$ and β the saline contraction coefficient $\beta = \rho^{-1}\partial\rho/\partial S$; the overline denotes a mean over the staircase. Diffusive-convective staircases may be found in the ocean for density ratios up to around 10 (e.g., Kelley et al., 2003). R_ρ is lowest in the Eurasian basin, where the AW layer is warmest, and increases along the cyclonic pathway of the AW layer as it cools and propagates around the Arctic basin. In the Eurasian Basin, $3 < R_\rho < 4$, whereas in the central Canadian Basin, $5 < R_\rho < 7$ (Shibley et al., 2017).

In general, the staircase spans about 50 to 100 m at the top boundary of the AW, with homogeneous layers a few metres thick, separated by high-gradient interfaces several centimetres thick, across which temperature and salinity jumps are about 0.04 °C and 0.01, respectively (e.g., Padman and Dillon, 1987). Timmermans et al. (2008a) analysed ITP data to show that the individual mixed layers in the Canada Basin staircase are laterally coherent for at least 800 kilometres across the basin. In the Eastern Arctic, Polyakov et al. (2019) observed that the coherent layering can exceed 1000 km. Layer properties cluster along well-defined lines in $\theta - S$ space, with the slopes of these lines being attributable to a balance between along-layer advective divergence of temperature and salinity and vertical divergences of diffusive-convective temperature and salt fluxes.

Whereas well-defined layers exist throughout most of the central Arctic basin, a diffusive-convective staircase is less pronounced around Arctic Ocean continental shelf-slope regions and in the vicinity of Fram Strait (e.g., Shibley et al., 2017). The absence of a staircase is likely because of higher mixing levels in those regions (e.g., Rippeth et al., 2015); a staircase cannot form above a critical level of intermittent turbulence (Shibley and Timmermans, 2019). Using microstructure measurements, Guthrie et al. (2017) show that vertical diffusivity an order of magnitude above the molecular level is sufficient to disrupt the staircase. Where low mixing levels in the central basins allow for the persistence of a staircase, diffusive-convective fluxes are the main mechanism for vertical heat transport from the AW. Note that even in disrupted staircases, buoyancy fluxes remain the primary driver for mixing (Polyakov et al., 2019; Middleton and Taylor, 2020).

The diffusive-convective heat flux through the staircase may be estimated by employing a flux-law parameterisation that depends on the temperature, and salinity jumps across individual staircase interfaces (Kelley, 1990), or by direct microstructure measurements (Guthrie et al., 2015). Heat flux estimates range from 0.02 to 0.3 W m^{-2} in the central basins (Padman and Dillon, 1987, 1989; Timmermans et al., 2008a; Sirevaag and Fer, 2012; Guthrie et al., 2015; Shibley et al., 2017), to approximately O(1) W m^{-2} at the eastern boundary of the Eurasian basin and in the vicinity of the east Siberian continental slope (Lenn et al., 2009; Polyakov et al., 2012). These heat fluxes are only about one tenth of the mean surface-ocean heat flux to the sea ice. In general, summer solar heating of the surface-ocean layer in the absence of sea ice provides the main heat source for ocean-to-ice heat fluxes (Maykut and Untersteiner, 1971; Maykut and McPhee, 1995; Fer, 2009; Toole et al., 2010; Timmermans, 2015). Yet change is afoot, recent observations show that heat from the ocean interior in the eastern Eurasian basin has increased in the last decade, becoming comparable to summer radiative atmospheric heating for the sea-ice decay (Polyakov et al., 2020b).

The deep and bottom waters of the Arctic Ocean also have a finescale temperature and salinity structure shaped by diffusive convection. In the deep water, diffusive convection is powered by geothermal heat. A well-mixed bottom layer, about 1000 m thick from the bottom at ~3500-m depth (with the top of the layer coinciding with the approximate sill depth of the Alpha/Mendeleev Ridge), implies that convective mixing is occurring as a consequence of geothermal heating (Timmermans et al., 2003); Björk and Winsor (2006) document a similar structure in the Amundsen Basin. A staircase structure, suggestive of diffusive convection, is observed above the homogeneous bottom layer, and acts to govern the transport of heat and salt between the warm, salty bottom layer and the overlying cooler, fresher water. The essential governing physics and mechanisms for the origin and preservation of the deep staircase remain largely unknown. Both an approximate heat budget (Timmermans et al., 2003; Carmack et al., 2012) and a density overturn analysis of the interfaces between mixed layers, Timmermans et al. (2003) indicate that the heat flux through interfaces is much less than that predicted using diffusive-convective flux parameterisations (Kelley, 1990). Theory and direct numerical simulations indicate that interfaces in the deep staircase are sufficiently thick that the effects of planetary rotation may suppress heat fluxes (Kelley, 1987; Carpenter and Timmermans, 2014).

11.3.6 Thermohaline intrusions

In the early 1980s, when sea ice could be reliably expected to be encountered north of Svalbard even in summer, early-spring (March–April 1981), CTDs deployed from seaplanes landing on the ice provided some of the earliest detailed observations of thermohaline intrusions in the Arctic Ocean (Perkin and Lewis, 1984). These early observations tracking the Fram Strait AW branch (also known as the West Spitsbergen current) found that this current bifurcated into one pathway following the coastline of northern Svalbard, and another pathway the continental shelf break north of the Yermak Plateau (Perkin and

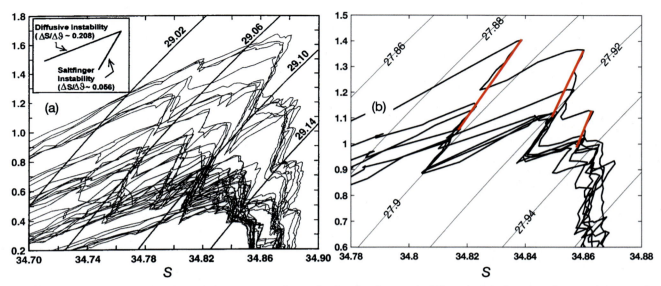

FIGURE 11.9 (a) Expanded-scale θ/S correlations near θ-maximum showing the alignment in θ/S space of the intrusions that extended across the Makarov basin observed in 1994. Inset shows the diffusive ($\Delta S/\Delta \theta \sim 0.21$) and salt-finger ($\Delta S/\Delta \theta \sim 0.056$) regimes discussed in the text; panel reproduced from (Carmack et al., 1997). (b) Expanded-scale θ/S correlations across the Arctic boundary current flowing along the East Siberian sea slope. The red lines track θ-maxima associated with warm intrusions across the boundary current front sloping downwards relative to isopycnals on the cold side of the front. For comparison, Fig. 11.2c shows the T-S plot of representative profiles from Fig. 11.1. Note that σ_{1000} is shown in (a), whereas σ_0 are plotted in (b) and Fig. 11.2(c).

Lewis, 1984). Along both these pathways, (Perkin and Lewis, 1984) reported that within the depth range associated with the AW temperature maximum, CTD profiles exhibited the now-classic zig-zag pattern when viewed in T-S diagrams (e.g., Fig. 11.9). This was taken as evidence of warmer saltier AW layers (peaks in T-S space) intruding and interleaving with the cooler fresher ambient water layers (troughs in T-S space) at the AW boundary current front (Perkin and Lewis, 1984, and Fig. 11.9).

Subsequent observations have shown that these structures are further enhanced by formation of new intrusions as the West Spitsbergen AW encounters Barents Sea Water, i.e., modified AW, joining the boundary current at the St. Anna trough (Schauer et al., 2002). Observations elsewhere in the Arctic ocean basins and over the last few decades have demonstrated not only the ubiquity of these features (e.g. Carmack et al., 1995), but also their coherence over vast distances as nested T-S peaks (Fig. 11.9a). These nested T-S peaks have been shown by Woodgate et al. (2007) to carry fingerprints of their water-mass origins all the way from Fram Strait to the Bering Strait. The extensive reach of these Arctic intrusions highlights their role in transferring properties from the AW boundary current into the central basins (Carmack et al., 1997).

Although the thermohaline intrusions transfer heat and freshwater predominantly laterally from the boundaries into the Arctic interior, the associated mixing is not purely isopycnal; intrusions cross isopycnals as they propagate, driven by buoyancy changes due to vertical divergences of vertical double diffusive heat and salt fluxes at their top and bottom boundaries. Within the intrusions, at the boundaries between cooler, fresher water and warmer, saltier water below, conditions are favourable for diffusive convection (see Section 11.3.5) (May and Kelley, 2001). Conversely, at the underside of the warm salty intrusion overlying cooler fresher water, salt fingering (see Section 11.3.5) dominates (Fig. 11.9 inset). Bebieva and Timmermans (2019) analyse observations in the Canada Basin to demonstrate that the nesting slopes (i.e., red lines in Fig. 11.9b) change with depth in response to the specific interplay between diffusive convective and salt fingering fluxes in a given layer.

The warm salty intrusive layer loses buoyancy to the overlying colder fresher layer through diffusive convection characterised by the non-dimensional flux ratio $\gamma_d^{-1} = \frac{\alpha F_\theta^d}{\beta F_S^d} > 1$, where θ is potential temperature. Conversely, the warm salty intrusive layer gains buoyancy via salt fingering fluxes through its underside as characterised by the flux ratio $\gamma_f = \frac{\alpha F_\theta^f}{\beta F_S^f} < 1$. For a steady-state warm salty intrusion, May and Kelley (2001) balance lateral advection against the salt-fingering and diffusive-convective fluxes to arrive at the following relationship:

$$-\frac{\beta F_S^d - \alpha F_\theta^d}{\beta F_S^f - \alpha F_\theta^f} = \frac{R_l - \gamma_f}{1 - \gamma_f}\left(\frac{\gamma_d^{-1} - 1}{\gamma_d^{-1} - R_l}\right), \quad (11.2)$$

where $R_l = (\alpha \partial \theta / \partial l)/(\beta \partial S / \partial l)$ is the along-intrusion (l) density ratio. Two solutions exist, in which $R_l < 1$ or $R_l > 1$. In the former case, the LHS of (11.2) is less than one, indicating that salt fingering dominates, causing the intrusions to slope upwards relative to isopycnals. In the latter, the LHS of (11.2) is greater than one, indicating that diffusive convection dominates, causing the intrusions to slope downwards relative to isopycnals.

A key result of May and Kelley's (2001) comparisons of observations, with both instability theory and steady-state model predictions, is their conclusion that the intrusions' initial development is driven by salt-fingering-dominated diffusive interleaving, with additional forcing from baroclinity of the boundary current front, but diffusive convection takes over as the dominant process for the intrusions to approach steady state. Thus well-developed AW intrusions observed north of Svalbard sloping upwards in depth, but downwards in density space on the cold side of the front, indicate that diffusive convection is the mediating process here (May and Kelley, 2001). This behaviour can also be seen farther east in cross-boundary current profiles from the East Siberian Sea slope (Fig. 11.9b).

At the cool temperatures of the Arctic Ocean, seawater is firmly within the non-linear regime of the equation of state (see Timmermans and Jayne, 2016). Walsh and Carmack (2002) proposed that large-scale spatial gradients in α modify thermohaline fluxes and the spreading of intrusions. They argued that this α-induced spatial flux divergence drives a recirculation between adjacent layers that both explains the slow vertical propagation of the intrusions (Walsh and Carmack, 2002), and the nesting of these AW intrusions across ocean basins (Walsh and Carmack, 2003).

Taken together, these key studies highlight the role of vertical diffusive-convective and salt-finger fluxes in the growth of AW thermohaline intrusions. Hence these AW intrusions were and continue to be considered an important mechanism for not just distributing boundary properties laterally into the Arctic interior thermocline, but also an important oceanic source of heat for its upper ocean (Rudels et al., 2009). Even in a changing Arctic, the expectation is that lateral intrusions will continue to be important for transferring properties from the boundaries to the interior basins.

11.4 Grand challenges

Arctic sea-ice cover variability is sensitive to small perturbations of the energy budget, e.g., 1 W m^{-2} change in oceanic-ice heat fluxes is significant (Maykut and Untersteiner, 1971). In an Arctic-in-transition, we observe notable changes in sea-ice volume, ocean heat and fresh water content, the strength and distribution of vertical stratification and the distribution of the warm AW (e.g., Polyakov et al., 2017). Overall, these changes will affect wind forcing, near-inertial wave radiation, baroclinic tide generation and radiation, double-diffusive processes, and the ultimate dissipation and vertical mixing in the water column. Indeed, oceanic heat fluxes from the warming AW reached parity with summer radiative atmospheric heating in driving Eastern Arctic sea-ice decay by 2013–2015 (Polyakov et al., 2017). More recent 2017–2018 observations indicate that oceanic heat is now the dominant driver of Eastern Arctic sea-ice loss (Polyakov et al., 2020b).

Understanding the numerous pathways for the energy from tides and winds to turbulence, the magnitude and distribution of the associated ocean mixing rates, and the role of feedbacks between forcing, mixing rates, stratification and sea ice, will help improve description and representation of the Arctic climate system. This requires a step change in the spatial and temporal coverage of our observations. Furthermore accurate measurements that are needed to identify the relative contribution of different processes face challenges. The environment is quiescent with low signal-to-noise ratios in terms of turbulent parameters. Methods must further be developed to accurately separate semi-diurnal internal tides from near-inertial wind-forced waves. The complexity of this separation is illustrated by ice-based CEAREX observations of zonal shear in the vicinity of an intra-halocline eddy (Fig. 11.10), that was itself oscillating under the ice-sheet with a 12.4 hour (M$_2$) period. Modification of the near-inertial shear is apparent at the crown of the eddy (Fig. 11.10), but distinguishing these modifications from tidally-induced eddy motion remains non-trivial.

Autonomous underwater glider technology is matured and widely used in the oceanographic community (Testor et al., 2019). Glider navigation, however, is dependent on accurate GPS fixes, which are challenging in high latitudes, particularly because the sea-ice cover hinders the glider from surfacing. Developing technology, such as sea-ice-avoidance algorithms and acoustic navigation, however, facilitate the use of gliders in ice-covered waters. This is particularly encouraging because a glider is a suitable platform for resolving the hydrographic features in eddies, across oceanic frontal systems as well as boundary currents and will be a key component of future oceanic observing systems. Gliders are also excellent platforms for ocean microstructure measurements (Fer et al., 2014; Peterson and Fer, 2014). Recent developments include on-board

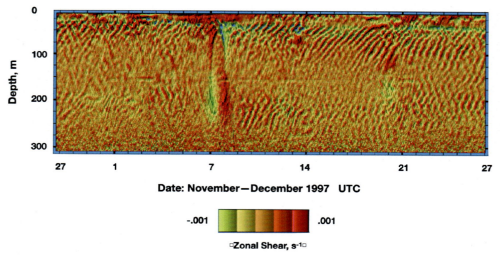

FIGURE 11.10 Observations of under-ice zonal shear resolved by data from the 1989 CEAREX experiment, 83-82 N, 13 E, as in the earlier papers (e.g. Padman et al., 1990). The data has 3.3 m vertical resolution, and the shear is calculated across 6-m segments. Figure courtesy of Rob Pinkel.

processing of microstructure data such that band-averaged spectra and estimates of dissipation rates can be relayed by Iridium messages when the glider surfaces. Similar advances in profiling float technology, building on the Argo program, are showing promise for sampling in the Arctic Ocean.

ITPs have substantially increased the spatial and temporal resolution and extent of sampling in recent years. A challenge is to equip such autonomous drifting platforms with microstructure sensors similar to those used in gliders, and together with onboard data processing, to collect near-real time dissipation profiles. Similarly there is progress in Argo float technology which, together with ice-avoidance capability, will make dissipation measurements possible (Roemmich et al., 2019).

The ultimate bottle-neck of the ocean-sea-ice turbulent exchange, the under-ice boundary layer, must be further studied to increase the knowledge and parameterisation regarding the thermodynamics and buoyancy effect from growing or melting sea ice. Application of the present exchange parameterisations lead to large uncertainties when double-diffusive effects in the boundary layer become important.

Furthermore, there are several unresolved processes that are likely to contribute to turbulent mixing which we are either just beginning to understand and parameterise in models, or have not yet grasped the full physics of. These include leads in the ice, lateral melting of marginal ice chunks, the effect of keels and wind momentum input to the ocean in variable ice cover which are also challenges for Antarctic oceanography (see Chapter 12). Of these, the impact of very deep keels from ridged multi-year ice on upper-ocean mixing is the only process than will probably become less important as the seasonal sea ice continues its decline in both extent and thickness. With the melt seasons arriving earlier and lasting longer every year, the rafted sea ice floes are breaking up sooner and more easily with the more frequent emergence of leads and ice chunks in the marginal ice zone. Sea-ice studies in the surface boundary layers indicate that the transfer of momentum from the atmosphere to the ocean is highest at about 80–90% sea ice concentration due to form drag from the multiple sea ice chunks (Castellani et al., 2014; Tsamados et al., 2014). To properly constrain these processes, we will need new robust observational techniques for the marginal ice zone and improved numerical modelling.

With ever-expanding areas of open water and its free surface, we know increased transfer of energy from the winds to the ocean will enhance the energy in the near-inertial band of the surface mixed layer (Rainville and Woodgate, 2009; Polyakov et al., 2020a), as well as result in increased generation and propagation of surface gravity waves, specifically the increased presence of ocean swell (Thomson and Rogers, 2014; Hunkins, 1962). How these will impact mixing, particularly in the marginal ice zone, is now the focus of a re-invigorated, long-dormant topic of research (e.g. Smith et al., 2018; Smith and Thomson, 2019).

Arctic freshwater storage and release has important consequences for the North Atlantic meridional overturning circulation which in turn mediates global climate. Thus we also need to improve our understanding of how the Beaufort Gyre's halocline is supplied with and loses freshwater through the action of eddies. Arctic eddy studies have, with few exceptions (e.g., Meneghello et al., 2017; Zhao et al., 2016), mainly focused on coherent vortex features (Section 11.3.4), but the eddy fluxes inferred from numerical simulations include a wider range of mesoscale variability. At present, we lack the types of *in situ* observational arrays needed to fully resolve and directly quantify these eddy effects for model validation, and to assess the eddies' true impact on basin adjustment.

Another urgent question is how the Arctic Ocean influences Northern Hemisphere weather and climate. The relationship between Arctic sea ice and regional air temperatures has gained recognition (Screen and Simmonds, 2010) as robust correlations between Barents and Kara sea ice extent with inflowing AW temperature (Barton et al., 2018; Årthun et al., 2012), and also with harsh Eurasian winters have been demonstrated (e.g., Kolstad and Screen, 2019). Yet more questions have arisen about whether the sea ice extent and resulting increase oceanic heat loss is a driver of atmospheric jet stream fluctuations or are the correlated phenomenon symptoms of a common upstream cause (Blackport et al., 2019; Warner et al., 2020). Consequently, exactly where, how and how much oceanic heat is lost to the atmosphere in the northern North Atlantic and over the Barents and Kara Seas, and the resolving the relevant ocean-ice fluxes remains a high priority.

Finally, possibly the most important question relates to when the Arctic Ocean will transition to a new state, potentially losing many of the characteristics that define a polar ocean. A likely inevitable transition is to an Arctic where ice-free summers become the norm. Ocean heat exchanges, and the mixing processes responsible, have an important role to play in such a scenario and ensuring proper representation of mixing in numerical models remains a challenge to achieving accurate predictions. Another impending transition is already evident in the Barents Sea, where a warming ocean and stratification changes are linked to changing AW inflow and sea ice variability (Årthun et al., 2012; Barton et al., 2018). Changes here hint at a future Arctic Ocean that is warm enough for temperature to play a bigger role in the dynamics (i.e., in influencing the density alongside salinity) (see Timmermans and Jayne, 2016). Indeed, Skagseth et al. (2020) report that the efficiency of AW cooling in the Barents Sea gateway to the Arctic Ocean has been declining in recent decades, though this decline may be temporarily slowed when an approaching AW fresh anomaly reaches the Barents Sea in 2020. A third potential outcome of increased turbulent fluxes from deep waters to the surface, is a transition of Arctic ecosystems from oligotrophic to eutrophic ecosystems during the ice-free summers, a possibility that may already be coming to fruition with the 3-decade-plus-long rise in primary productivity unexplained solely by increased light availability (e.g. Arrigo and van Dijken, 2015). Our ability to predict and identify major Arctic system transitions relies on continued efforts to observe and understand the full range of Arctic ocean mixing processes.

11.5 Conclusions

The Arctic Ocean is undergoing rapid changes in response to climate change. The high latitudes, warm inflows, high river discharge, high air-sea fluxes and sea ice pre-condition this ocean for the many different mixing processes that together conspire to set the Arctic's unique stratification and circulation. As climate change continues, the balance between the less energetic diffusive-convective staircases and thermohaline intrusions may be giving way to an increasing impact of wind-driven mixing, tidal mixing, and eddies. New mixing mechanisms, such as the myriad processes associated with the marginal ice zone and surface waves are emerging or gaining recognition and attracting interest. At the same time, the focus on the ocean's role in melting sea ice remains an enduring hot topic, while attention is also drawing in on the Arctic Ocean's role in driving Northern Hemisphere jet stream fluctuations. The coming years and decades are sure to bring major advances in our understanding of Arctic Ocean mixing processes and their role in sea ice and climate.

Acknowledgements

Authors acknowledge the following funding for this contribution: IF, the Research Council of Norway, AROMA project (294396); MLT, the National Science Foundation Office of Polar Programs (award 1950077); YDL, the UK-German NERC-BMBF Changing Arctic Ocean PEANUTS project (NE/R01275X/1); JM, the National Science Foundation PLR-1303791 and the Office of Naval Research N00014-16-1-2378.

The example profiles shown in Fig. 11.2 included data collected during the IPY NABOS/ASBO 2008 (Lenn et al., 2011b) and from the TEA-COSI 2012 (Lincoln et al., 2016) field seasons. NERC-funded data (ASBO and TEA-COSI) are available from the British Oceanographic Data Centre (http://www.bodc.ac.uk). Fig. 11.2 also includes modular microstructure profiler data from the NSF-funded 2015 ArcticMix experiment (Fine et al., 2018), available through the US microstructure database (https://microstructure.ucsd.edu/). The projects collecting the microstructure profiles around Svalbard and near the North Pole were supported by the Research Council of Norway (projects 229786 and 178641). The ice-tethered profiler data shown in Fig. 11.5 were collected and made available by the Ice-tethered profiler program (Toole et al., 2011; Krishfield et al., 2008) based at the Woods Hole Oceanographic Institution (http://www.whoi.edu/itp). Authors thank John Taylor and Igor Polyakov for their comments that helped to improve an earlier version of this manuscript.

References

Aagaard, K., Andersen, R., Swift, J., Johnson, J., 2008. A large eddy in the central Arctic Ocean. Geophys. Res. Lett. 35.

Aagaard, K., Coachman, L., Carmack, E., 1981. On the halocline of the Arctic Ocean. Deep-Sea Res. 28A, 529–545.

Akimova, A., Schauer, U., Danilov, S., Núẽz-Riboni, I., 2011. The role of the deep mixing in the Storfjorden shelf water plume. Deep-Sea Res., Part 1 58, 403–414. https://doi.org/10.1016/j.dsr.2011.02.001.

Alford, M.H., MacKinnon, J.A., Simmons, H.L., Nash, J.D., 2016. Near-inertial internal gravity waves in the ocean. Annu. Rev. Mar. Sci. 8, 95–123. https://doi.org/10.1146/annurev-marine-010814-015746.

Amante, C., Eakins, B.W, 2009. ETOPO1 arc-minute global relief model: procedures, data sources and analysis.

Armitage, T.W., Bacon, S., Ridout, A.L., Petty, A.A., Wolbach, S., Tsamados, M., 2017. Arctic Ocean surface geostrophic circulation 2003-2014. Cryosphere 11, 1767–1780.

Armitage, T.W., Manucharyan, G.E., Petty, A.A., Kwok, R., Thompson, A.F., 2020. Enhanced eddy activity in the Beaufort gyre in response to sea ice loss. Nat. Commun. 11, 1–8.

Arrigo, K.R., van Dijken, G.L., 2015. Continued increases in Arctic Ocean primary production. Prog. Oceanogr. 136, 60–70.

Årthun, M., Eldevik, T., Smedsrud, L., Skagseth, Ø., Ingvaldsen, R., 2012. Quantifying the influence of Atlantic heat on Barents sea ice variability and retreat. J. Climate 25, 4736–4743.

Barton, B.I., Lenn, Y.D., Lique, C., 2018. Observed atlantification of the Barents sea causes the polar front to limit the expansion of winter sea ice. J. Phys. Oceanogr. 48, 1849–1866.

Bebieva, Y., Timmermans, M.L., 2016. An examination of double-diffusive processes in a mesoscale eddy in the Arctic Ocean. J. Geophys. Res., Oceans 121, 457–475.

Bebieva, Y., Timmermans, M.L., 2017. The relationship between double-diffusive intrusions and staircases in the Arctic Ocean. J. Phys. Oceanogr. 47, 867–878.

Bebieva, Y., Timmermans, M.L., 2019. Double-diffusive layering in the Canada Basin: an explanation of along-layer temperature and salinity gradients. J. Geophys. Res., Oceans 124, 723–735.

Bekryaev, R.V., Polyakov, I.V., Alexeev, V.A., 2010. Role of polar amplification in long-term surface air temperature variations and modern Arctic warming. J. Climate 23, 3888–3906.

Björk, G., Winsor, P., 2006. The deep waters of the Eurasian basin, Arctic Ocean: geothermal heat flow, mixing and renewal. Deep-Sea Res., Part 1, Oceanogr. Res. Pap. 53, 1253–1271.

Blackport, R., Screen, J.A., van der Wiel, K., Bintanja, R., 2019. Minimal influence of reduced Arctic sea ice on coincident cold winters in mid-latitudes. Nat. Clim. Change 9, 697–704.

Brannigan, L., Johnson, H., Lique, C., Nycander, J., Nilsson, J., 2017. Generation of subsurface anticyclones at Arctic surface fronts due to a surface stress. J. Phys. Oceanogr. 47, 2653–2671.

Carmack, E., Aagaard, K., Swift, J., Perkin, R., McLaughlin, F., Macdonald, R., Jones, E., 1998. Thermohaline transitions. Coast. Estuar. Stud., 179–186.

Carmack, E., Aagard, K., Swift, J.H., MacDonald, R.W., McLaughlin, F.A., Jones, E.P., Perkin, R.G., Smith, J.N., Ellis, K.M., Killius, L.R., 1997. Change in temperature and tracer distributions within the Arctic Ocean: results from the 1994 Arctic Ocean section. Deep-Sea Res., Part 2 44, 1487–1502.

Carmack, E., Polyakov, I., Padman, L., Fer, I., Hunke, E., Hutchings, J., Jackson, J., Kelley, D., Kwok, R., Layton, C., Melling, H., Perovich, D., Persson, O., Ruddick, B., Timmermans, M.L., Toole, J., Ross, T., Vavrus, S., Winsor, P., 2015. Towards quantifying the increasing role of oceanic heat in sea ice loss in the new Arctic. Bull. Am. Meteorol. Soc. 96, 2079–2105. https://doi.org/10.1175/BAMS-D-13-00177.1.

Carmack, E.C., Macdonald, R.W., Perkin, R.G., McLaughlin, F.A., Pearson, R.J., 1995. Evidence for warming of Atlantic water in the southern Canadian basin of the Arctic Ocean: results from the larsen-93 expedition. Geophys. Res. Lett. 22, 1061–1064.

Carmack, E.C., Williams, W.J., Zimmermann, S.L., McLaughlin, F.A., 2012. The Arctic Ocean warms from below. Geophys. Res. Lett. 39.

Carpenter, J., Timmermans, M.L., 2012. Deep mesoscale eddies in the Canada Basin, Arctic Ocean. Geophys. Res. Lett. 39.

Carpenter, J., Timmermans, M.L., 2014. Does rotation influence double-diffusive fluxes in polar oceans? J. Phys. Oceanogr. 44, 289–296.

Castellani, G., Lüpkes, C., Hendricks, S., Gerdes, R., 2014. Variability of Arctic sea-ice topography and its impact on the atmospheric surface drag. J. Geophys. Res., Oceans 119, 6743–6762.

Cavalieri, D.J., Martin, S., 1994. The contribution of Alaskan, Siberian, and Canadian coastal polynyas to the cold halocline layer of the Arctic Ocean. J. Geophys. Res., Oceans 99, 18343–18362.

Chanona, M., Waterman, S., 2020. Temporal variability of internal wave-driven mixing in two distinct regions of the Arctic Ocean. J. Geophys. Res., Oceans 25 (10), e2020JC016181. https://doi.org/10.1029/2020JC016181. arXiv: https://agupubs.onlinelibrary.wiley.com/doi/pdf/10.1029/2020JC016181. https://agupubs.onlinelibrary.wiley.com/doi/abs/10.1029/2020JC016181.

Coachman, L., Barnes, C., 1961. The contribution of Bering Sea water to the Arctic Ocean. Arctic 14, 147–161.

Coachman, L., Barnes, C., 1963. The movement of Atlantic water in the Arctic Ocean. Arctic 16, 8–16.

Cole, S.T., Timmermans, M.L., Toole, J.M., Krishfield, R.A., Thwaites, F.T., 2014. Ekman veering, internal waves, and turbulence observed under Arctic sea ice. J. Phys. Oceanogr. 44, 1306–1328.

Cole, S.T., Toole, J.M., Lele, R., Timmermans, M.L., Gallaher, S.G., Stanton, T.P., Shaw, W.J., Hwang, B., Maksym, T., Wilkinson, J.P., et al., 2017. Ice and ocean velocity in the Arctic marginal ice zone: ice roughness and momentum transfer. Elem. Sci. Anthropocene 5.

D'Asaro, E., Morison, J., 1992. Internal waves and mixing in the Arctic Ocean. Deep-Sea Res., Part 1 39, 459. https://doi.org/10.1016/S0198-0149(06)80016-6.

D'Asaro, E.A., 1988. Observations of small eddies in the Beaufort sea. J. Geophys. Res., Oceans 93, 6669–6684.

D'Asaro, E.A., Eriksen, C.C., Levine, M.D., Niiler, P., Paulson, C.A., Vanmeurs, P., 1995. Upper-ocean inertial currents forced by a strong storm. Part 1: data and comparisons with linear theory. J. Phys. Oceanogr. 25, 2909–2936.

Dmitrenko, I.A., Kirillov, S.A., Ivanov, V., Woodgate, R., 2008. Mesoscale Atlantic water eddy off the Laptev Sea continental slope carries the signature of upstream interaction. J. Geophys. Res., Oceans 113.

Dosser, H.V., 2015. Internal Waves in the Western Arctic Ocean. Ph.D. thesis.

Dosser, H.V., Rainville, L., 2016. Dynamics of the changing near-inertial internal wave field in the Arctic Ocean. J. Phys. Oceanogr. 46, 395–415. https://doi.org/10.1175/jpo-d-15-0056.1.

Dosser, H.V., Timmermans, M.L., 2018. Inferring circulation and lateral eddy fluxes in the Arctic Ocean's deep Canada basin using an inverse method. J. Phys. Oceanogr. 48, 245–260. https://doi.org/10.1175/jpo-d-17-0190.1.

Eicken, H., 2004. The role of Arctic Sea Ice in transporting and cycling terrigenous organic matter. In: Stein, R., MacDonald, W. (Eds.), The Organic Carbon Cycle in the Arctic Ocean. Springer-Verlag, Heidelberg, pp. 45–53.

Erofeeva, S., Egbert, G., 2020. Arc5km2018: Arctic Ocean Inverse Tide Model on a 5 kilometer grid, 2018. Dataset.

Fer, I., 2009. Weak diffusion allows maintenance of cold halocline in the central Arctic. Atmos. Oceanic Sci. Lett. 2, 148–152.

Fer, I., 2014. Near-inertial mixing in the central Arctic Ocean. J. Phys. Oceanogr. 44, 2031–2049. https://doi.org/10.1175/JPO-D-13-0133.1.

Fer, I., Koenig, Z., Kozlov, I.E., Ostrowski, M., Rippeth, T.P., Padman, L., Bosse, A., Kolås, E., 2020. Tidally-forced lee waves drive turbulent mixing along the Arctic Ocean margins. Geophys. Res. Lett. 47, e2020GL088083. https://doi.org/10.1029/2020GL088083.

Fer, I., Müller, M., Peterson, A.K., 2015. Tidal forcing, energetics, and mixing near the Yermak Plateau. Ocean Sci. 11, 287–304. https://doi.org/10.5194/os-11-287-2015.

Fer, I., Peterson, A.K., Ullgren, J.E., 2014. Microstructure measurements from an underwater glider in the turbulent Faroe Bank Channel Overflow. J. Atmos. Ocean. Technol. 31, 1128–1150. https://doi.org/10.1175/JTECH-D-13-00221.1.

Fer, I., Skogseth, R., Geyer, F., 2010. Internal waves and mixing in the marginal ice zone near the Yermak Plateau. J. Phys. Oceanogr. 40, 1613–1630.

Fer, I., Skogseth, R., Haugan, P.M., 2004. Mixing of the Storfjorden overflow (Svalbard Archipelago) inferred from density overturns. J. Geophys. Res. 109. https://doi.org/10.1029/2003JC001968.

Fine, E.C., MacKinnon, J.A., Alford, M.H., Mickett, J.B., 2018. Microstructure observations of turbulent heat fluxes in a warm-core Canada basin eddy. J. Phys. Oceanogr. 48, 2397–2418. https://doi.org/10.1175/jpo-d-18-0028.1.

Gallaher, S.G., Stanton, T.P., Shaw, W.J., Cole, S.T., Toole, J.M., Wilkinson, J.P., Maksym, T., Hwang, B., 2016. Evolution of a Canada basin ice-ocean boundary layer and mixed layer across a developing thermodynamically forced marginal ice zone. J. Geophys. Res., Oceans 121, 6223–6250.

Giles, K.A., Laxon, S.W., Ridout, A.L., Wingham, D.J., Bacon, S., 2012. Western Arctic Ocean freshwater storage increased by wind-driven spin-up of the Beaufort gyre. Nat. Geosci. 5, 194–197.

Graham, R.M., Itkin, P., Meyer, A., Sundfjord, A., Spreen, G., Smedsrud, L.H., Liston, G.E., Cheng, B., Cohen, L., Divine, D., Fer, I., Fransson, A., Gerland, S., Haapala, J., Hudson, S.R., Johansson, A.M., King, J., Merkouriadi, I., Peterson, A.K., Provost, C., Randelhoff, A., Rinke, A., Rösel, A., Sennéchael, N., Walden, V.P., Duarte, P., Assmy, P., Steen, H., Granskog, M.A., 2019. Winter storms accelerate the demise of sea ice in the Atlantic sector of the Arctic Ocean. Sci. Rep. 9, 9222. https://doi.org/10.1038/s41598-019-45574-5.

Gregg, M.C., Sanford, T.B., Winkel, D.P., 2003. Reduced mixing from the breaking of internal waves in equatorial waters. Nature 422, 513–515.

Guthrie, J.D., Fer, I., Morison, J., 2015. Observational validation of the diffusive convection flux laws in the Amundsen Basin, Arctic Ocean. J. Geophys. Res., Oceans 120, 7880–7896.

Guthrie, J.D., Fer, I., Morison, J.H., 2017. Thermohaline staircases in the Amundsen Basin: possible disruption by shear and mixing. J. Geophys. Res., Oceans 122. https://doi.org/10.1002/2017JC012993.

Guthrie, J.D., Morison, J.H., Fer, I., 2013. Revisiting internal waves and mixing in the Arctic Ocean. J. Geophys. Res., Oceans 118, 3966–3977.

Haine, T.W., Curry, B., Gerdes, R., Hansen, E., Karcher, M., Lee, C., Rudels, B., Spreen, G., de Steur, L., Stewart, K.D., et al., 2015. Arctic freshwater export: status, mechanisms, and prospects. Glob. Planet. Change 125, 13–35.

Holland, M.M., Bitz, C.M., 2003. Polar amplification of climate change in coupled models. Clim. Dyn. 21, 221–232.

Holloway, G., Proshutinsky, A., 2007. Role of tides in Arctic Ocean/ice climate. J. Geophys. Res. 112. https://doi.org/10.1029/2006JC003643.

Hughes, K.G., Klymak, J.M., 2019. Tidal conversion and dissipation at steep topography in a channel poleward of the critical latitude. J. Phys. Oceanogr. 49, 1269–1291. https://doi.org/10.1175/jpo-d-18-0132.1.

Hunkins, K., 1962. Waves on the Arctic Ocean. J. Geophys. Res. 67, 2477–2489.

Hunkins, K., 1966. Ekman drift currents in the Arctic Ocean. In: Deep Sea Research and Oceanographic Abstracts. Elsevier, pp. 607–620.

Hunkins, K.L., 1974. Subsurface eddies in the Arctic Ocean. In: Deep Sea Research and Oceanographic Abstracts. Elsevier, pp. 1017–1033.

Intrieri, J., Fairall, C., Shupe, M., Persson, P., Andreas, E., Guest, P., Moritz, R., 2002. An annual cycle of Arctic surface cloud forcing at sheba. J. Geophys. Res., Oceans 107, SHE-13.

Ivanov, V.V., Golovin, P.N., 2007. Observations and modeling of dense water cascading from the northwestern Laptev Sea shelf. J. Geophys. Res., Oceans 112.

Jakobsson, M., Mayer, L., Coakley, B., Dowdeswell, J.A., Forbes, S., Fridman, B., Hodnesdal, H., Noormets, R., Pedersen, R., Rebesco, M., et al., 2012. The international bathymetric chart of the Arctic Ocean (ibcao) version 3.0. Geophys. Res. Lett. 39.

Janout, M.A., Lenn, Y.D., 2014. Semidiurnal tides on the Laptev Sea shelf with implications for shear and vertical mixing. J. Phys. Oceanogr. 44, 202–219.

Kawaguchi, Y., Itoh, M., Nishino, S., 2012. Detailed survey of a large baroclinic eddy with extremely high temperatures in the western Canada basin. Deep-Sea Res., Part 1, Oceanogr. Res. Pap. 66, 90–102.

Kelley, D., 1987. The influence of planetary rotation on oceanic double-diffusive fluxes. J. Mar. Res. 45, 829–841.

Kelley, D., Fernando, H., Gargett, A., Tanny, J., Özsoy, E., 2003. The diffusive regime of double-diffusive convection. Prog. Oceanogr. 56, 461–481.

Kelley, D.E., 1990. Fluxes through diffusive staircases: a new formulation. J. Geophys. Res., Oceans 95, 3365–3371.

Kennel, C.F., Yulaeva, E., 2020. Influence of Arctic sea-ice variability on Pacific trade winds. Proc. Natl. Acad. Sci. 117, 2824–2834.

Kolås, E., Fer, I., 2018. Hydrography, transport and mixing of the West Spitsbergen Current: the Svalbard Branch in summer 2015. Ocean Sci. 14, 1603–1618. https://doi.org/10.5194/os-14-1603-2018.

Kolstad, E., Screen, J., 2019. Non-stationary relationship between autumn Arctic sea ice and the winter North Atlantic oscillation. Geophys. Res. Lett.

Kowalik, Z., Proshutinsky, A.Y., 1994. The Arctic Ocean tides. In: The Polar Oceans and Their Role in Shaping the Global Environment, Vol. 85, pp. 137–158.

Krishfield, R., Toole, J., Proshutinsky, A., Timmermans, M.L., 2008. Automated ice-tethered profilers for seawater observations under pack ice in all seasons. J. Atmos. Ocean. Technol. 25, 2091–2105.

Kunze, E., Firing, E., Hummon, J.M., Chereskin, T.K., Thurnherr, A.M., 2006. Global abyssal mixing inferred from lowered ADCP shear and CTD strain profiles. J. Phys. Oceanogr. 36, 1553–1576.

Kwok, R., Cunningham, G., 2015. Variability of Arctic sea ice thickness and volume from cryosat-2. Philos. Trans. R. Soc. A, Math. Phys. Eng. Sci. 373, 20140157.

Kwok, R., Rothrock, D., 2009. Decline in Arctic sea ice thickness from submarine and icesat records: 1958–2008. Geophys. Res. Lett. 36.

de Lavergne, C., Falahat, S., Madec, G., Roquet, F., Nycander, J., Vic, C., 2019. Toward global maps of internal tide energy sinks. Ocean Model. 137, 52–75.

Legg, S., Klymak, J., 2008. Internal hydraulic jumps and overturning generated by tidal flow over a tall steep ridge. J. Phys. Oceanogr. 38, 1949–1964. https://doi.org/10.1175/2008jpo3777.1.

Lenn, Y.D., Chereskin, T.K., Sprintall, J., McClean, J.L., 2011a. Near-surface eddy heat and momentum fluxes in the Antarctic circumpolar current in drake passage. J. Phys. Oceanogr. 41, 1385–1407.

Lenn, Y.D., Rippeth, T.P., Old, C.P., Bacon, S., Polyakov, I., Ivanov, V., Hölemann, J., 2011b. Intermittent intense turbulent mixing under ice in the Laptev Sea continental shelf. J. Phys. Oceanogr. 41, 531–547.

Lenn, Y.D., Wiles, P.J., Torres-Valdez, S., Abrahamsen, P., Rippeth, T.P., Simpson, J.H., Bacon, S., Laxon, S., Polyakov, I., Ivanov, V., Kirillov, S., 2009. Vertical mixing at intermediate depths in the Arctic boundary current. Geophys. Res. Lett. 36. https://doi.org/10.1029/2008GL036792.

Levine, M.D., Clayton, A.P., Morison, J.H., 1987. Observations of internal gravity waves under the Arctic pack ice. J. Geophys. Res. 92, 779–782.

Lincoln, B.J., Rippeth, T.P., Lenn, Y.D., Timmermans, M.L., Williams, W.J., Bacon, S., 2016. Wind-driven mixing at intermediate depths in an ice-free Arctic Ocean. Geophys. Res. Lett. 43, 9749–9756. https://doi.org/10.1002/2016GL070454.

Lique, C., Johnson, H.L., Davis, P.E., 2015. On the interplay between the circulation in the surface and the intermediate layers of the Arctic Ocean. J. Phys. Oceanogr. 45, 1393–1409.

Lu, K., Weingartner, T., Danielson, S., Winsor, P., Dobbins, E., Martini, K., Statscewich, H., 2015. Lateral mixing across ice meltwater fronts of the Chukchi sea shelf. Geophys. Res. Lett. 42, 6754–6761.

Lu, P., Li, Z., Cheng, B., Leppäranta, M., 2011. A parameterization of the ice-ocean drag coefficient. J. Geophys. Res., Oceans 116.

MacKinnon, J.A., Zhao, Z., Whalen, C.B., Waterhouse, A.F., Trossman, D.S., Sun, O.M., St. Laurent, L.C., Simmons, H.L., Polzin, K., Pinkel, R., Pickering, A., Norton, N.J., Nash, J.D., Musgrave, R., Merchant, L.M., Melet, A.V., Mater, B., Legg, S., Large, W.G., Kunze, E., Klymak, J.M., Jochum, M., Jayne, S.R., Hallberg, R.W., Griffies, S.M., Diggs, S., Danabasoglu, G., Chassignet, E.P., Buijsman, M.C., Bryan, F.O., Briegleb, B.P., Barna, A., Arbic, B.K., Ansong, J.K., Alford, M.H., 2017. Climate process team on internal wave–driven ocean mixing. Bull. Am. Meteorol. Soc. 98, 2429–2454. https://doi.org/10.1175/bams-d-16-0030.1.

Manley, T., Hunkins, K., 1985. Mesoscale eddies of the Arctic Ocean. J. Geophys. Res., Oceans 90, 4911–4930.

Manucharyan, G.E., Spall, M.A., 2016. Wind-driven freshwater buildup and release in the Beaufort gyre constrained by mesoscale eddies. Geophys. Res. Lett. 43, 273–282.

Manucharyan, G.E., Timmermans, M.L., 2013. Generation and separation of mesoscale eddies from surface ocean fronts. J. Phys. Oceanogr. 43, 2545–2562.

Martini, K.I., Simmons, H.L., Stoudt, C.A., Hutchings, J.K., 2014. Near-inertial internal waves and sea ice in the Beaufort sea. J. Phys. Oceanogr. 44, 2212–2234. https://doi.org/10.1175/JPO-D-13-0160.1.

Maslowski, W., Kinney, J.C., Marble, D.C., Jakacki, J., 2008. Towards Eddy-Resolving Models of the Arctic Ocean. Washington DC American Geophysical Union Geophysical Monograph Series, vol. 177, pp. 241–264.

May, B.D., Kelley, D.E., 2001. Growth and steady state stages of the thermohaline intrusion in the Arctic Ocean. J. Geophys. Res. 106, 16783–16794.

Maykut, G., McPhee, M.G., 1995. Solar heating of the Arctic mixed layer. J. Geophys. Res., Oceans 100, 24691–24703.

Maykut, G.A., Untersteiner, N., 1971. Some results from a time-dependent thermodynamic model of sea ice. J. Geophys. Res. 76, 1550–1575.

McPhee, M., 2008. Air-Ice-Ocean Interaction: Turbulent Ocean Boundary Layer Exchange Processes. Springer Science & Business Media.

McPhee, M.G., 1992. Turbulent heat flux in the upper ocean under sea ice. J. Geophys. Res. 97, 5365–5379.

McPhee, M.G., 2002. Turbulent stress at the ice/ocean interface and bottom surface hydraulic roughness during the SHEBA drift. J. Geophys. Res. 107, 8037. https://doi.org/10.1029/2000JC000633.

McPhee, M.G., 2012. Advances in understanding ice–ocean stress during and since aidjex. Cold Reg. Sci. Technol. 76, 24–36.

McPhee, M.G., 2017. The sea ice–ocean boundary layer. In: Thomas, D. (Ed.), Sea Ice. Wiley, pp. 138–159. Chapter 5.

McPhee, M.G., Kikuchi, T., Morison, J.H., Stanton, T.P., 2003. Ocean-to-ice heat flux at the North Pole environmental observatory. Geophys. Res. Lett. 30, 2274. https://doi.org/10.1029/2003GL018580.

McPhee, M.G., Stanton, T.P., 1996. Turbulence in the statistically unstable oceanic boundary layer under Arctic leads. J. Geophys. Res. 101, 6409–6428.

Meincke, J., Rudels, B., Friedrich, H.J., 1997. The Arctic Ocean-Nordic Seas thermohaline system. ICES. J. Mar. Sci. 54, 283–299.

Melling, H., Lake, R., Topham, D., Fissel, D., 1984. Oceanic thermal structure in the western Canadian Arctic. Cont. Shelf Res. 3, 233–258.

Meneghello, G., Marshall, J., Cole, S.T., Timmermans, M.L., 2017. Observational inferences of lateral eddy diffusivity in the halocline of the Beaufort gyre. Geophys. Res. Lett. 44, 12–331.

Meyer, A., Fer, I., Sundfjord, A., Peterson, A.K., 2017. Mixing rates and vertical heat fluxes north of Svalbard from Arctic winter to spring. J. Geophys. Res., Oceans 122, 4569–4586. https://doi.org/10.1002/2016JC012441.

Middleton, L., Taylor, J.R., 2020. A general criterion for the release of background potential energy through double diffusion. J. Fluid Mech. 893.

Morison, J., McPhee, M., Curtin, T., Paulson, C., 1992. The oceanography of winter leads. J. Geophys. Res., Oceans 97, 11199–11218.

Morison, J.H., Long, C.E., Levine, M.D., 1985. Internal wave dissipation under sea ice. J. Geophys. Res. 90, 11,959–11,966.

Morozov, E.G., Paka, V.T., Bakhanov, V.V., 2008. Strong internal tides in the Kara Gates Strait. Geophys. Res. Lett. 35. https://doi.org/10.1029/2008GL033804.

Muench, R.D., Gunn, J.T., Whitledge, T.E., Schlosser, P., Smethie Jr, W., 2000. An Arctic Ocean cold core eddy. J. Geophys. Res., Oceans 105, 23997–24006.

Musgrave, R.C., Pinkel, R., MacKinnon, J.A., Mazloff, M.R., Young, W.R., 2016. Stratified tidal flow over a tall ridge above and below the turning latitude. J. Fluid Mech. 793, 933–957. https://doi.org/10.1017/jfm.2016.150.

Nansen, F., 1904. Farthest North: Being the Record of a Voyage of Exploration of the Ship "Fram" 1893–96 and of a Fifteen Months' Sleigh Journey by Dr. Nansen and Lieut. Johansen (Complete), Vol. 1. Library of Alexandria.

Neal, V.T., Neshyba, S., Denner, W., 1969. Thermal stratification in the Arctic Ocean. Science 166, 373–374.

Neshyba, S., Neal, V.T., Denner, W., 1971. Temperature and conductivity measurements under ice island t-3. J. Geophys. Res. 76, 8107–8120.

Newton, J., Aagaard, K., Coachman, L.K., 1974. Baroclinic eddies in the Arctic Ocean. In: Deep Sea Research and Oceanographic Abstracts. Elsevier, pp. 707–719.

Nishino, S., Itoh, M., Kawaguchi, Y., Kikuchi, T., Aoyama, M., 2011. Impact of an unusually large warm-core eddy on distributions of nutrients and phytoplankton in the southwestern Canada basin during late summer/early fall 2010. Geophys. Res. Lett. 38.

Olbers, D., Eden, C., 2013. A global model for the diapycnal diffusivity induced by internal gravity waves. J. Phys. Oceanogr. 43, 1759–1779.

Padman, L., Dillon, T.M., 1987. Vertical heat fluxes through the Beaufort Sea thermohaline staircase. J. Geophys. Res., Oceans 92, 10799–10806.

Padman, L., Dillon, T.M., 1989. Thermal microstructure and internal waves in the Canada Basin diffusive staircase. Deep-Sea Res., A, Oceanogr. Res. Pap. 36, 531–542.

Padman, L., Dillon, T.M., 1991. Turbulent mixing near the Yermak Plateau during the Coordinated Eastern Arctic Experiment. J. Geophys. Res., Oceans 96, 4769–4782.

Padman, L., Levine, M.D., Dillon, T.M., Morison, J., Pinkel, R., 1990. Hydrography and microstructure of an Arctic cyclonic eddy. J. Geophys. Res., Oceans 95, 9411–9420.

Padman, L., Plueddemann, A., Muench, R., Pinkel, R., 1992. Diurnal tides near the Yermak Plateau. J. Geophys. Res. 97, 12639–12652.

Pemberton, P., Nilsson, J., Meier, H.E.M., 2014. Arctic Ocean freshwater composition, pathways and transformations from a passive tracer simulation. Tellus, Ser. A Dyn. Meteorol. Oceanogr. 66, 23988. https://doi.org/10.3402/tellusa.v66.23988. https://www.tandfonline.com/doi/full/10.3402/tellusa.v66.23988.

Perkin, R.G., Lewis, E.L., 1984. Mixing in the West Spitsbergen Current. J. Phys. Oceanogr. 14, 1315–1325.

Perovich, D., Jones, K., Light, B., Eicken, H., Markus, T., Stroeve, J., Lindsay, R., 2011. Solar partitioning in a changing Arctic sea-ice cover. Ann. Glaciol. 52, 192–196.

Perovich, D., Meier, W., Tschudi, M., Farrell, S., Hendricks, S., Gerland, S., Haas, S., Krumpen, T., Polashenski, C., Ricker, R., Webster, M., 2019. [the Arctic] sea ice cover [in "State of the Climate in 2018"]. Bull. Am. Meteorol. Soc. 100, S146–S150. https://doi.org/10.1175/2019BAMSStateoftheClimate.1.

Perovich, D.K., Light, B., Eicken, H., Jones, K.F., Runciman, K., Nghiem, S.V., 2007. Increasing solar heating of the Arctic Ocean and adjacent seas, 1979–2005: attribution and role in the ice-albedo feedback. Geophys. Res. Lett. 34.

Persson, P.O.G., Fairall, C.W., Andreas, E.L., Guest, P.S., Perovich, D.K., 2002. Measurements near the atmospheric surface flux group tower at sheba: near-surface conditions and surface energy budget. J. Geophys. Res., Oceans 107, SHE-21.

Peterson, A.K., Fer, I., 2014. Dissipation measurements using temperature microstructure from an underwater glider. Methods Oceanogr. 10, 44–69. https://doi.org/10.1016/j.mio.2014.05.002.

Peterson, A.K., Fer, I., McPhee, M.G., Randelhoff, A., 2017. Turbulent heat and momentum fluxes in the upper ocean under Arctic sea ice. J. Geophys. Res., Oceans 122, 1439–1456.

Pfirman, S., Colony, R., Nürnberg, D., Eicken, H., Rigor, I., 1997. Reconstructing the origin and trajectory of drifting Arctic sea ice. J. Geophys. Res., Oceans 102, 12575–12586.

Pickart, R.S., Weingartner, T.J., Pratt, L.J., Zimmermann, S., Torres, D.J., 2005. Flow of winter-transformed Pacific water into the western Arctic. Deep-Sea Res., Part 2, Top. Stud. Oceanogr. 52, 3175–3198.

Pinkel, R., 2008. The wavenumber-frequency spectrum of vortical and internal-wave shear in the western Arctic Ocean. J. Phys. Oceanogr. 38, 277–290. https://doi.org/10.1175/2006jpo3558.1.

Pnyushkov, A., Polyakov, I.V., Padman, L., Nguyen, A.T., 2018. Structure and dynamics of mesoscale eddies over the Laptev Sea continental slope in the Arctic Ocean. Ocean Sci. 14.

Polyakov, I., Rippeth, T., Fer, I., Baumann, T., Carmack, E., Ivanov, V., Janout, M., Padman, L., Pnyushkov, A., Rember, R., 2020a. Intensification of near-surface currents and shear in the eastern Arctic Ocean. Geophys. Res. Lett. 47, e2020GL089469. https://doi.org/10.1029/2020GL089469.

Polyakov, I.V., Padman, L., Lenn, Y.D., Pnyushkov, A., Rember, R., Ivanov, V.V., 2019. Eastern Arctic Ocean diapycnal heat fluxes through large double-diffusive steps. J. Phys. Oceanogr. 49, 227–246.

Polyakov, I.V., Pnyushkov, A.V., Alkire, M.B., Ashik, I.M., Baumann, T.M., Carmack, E.C., Goszczko, I., Guthrie, J., Ivanov, V.V., Kanzow, T., et al., 2017. Greater role for Atlantic inflows on sea-ice loss in the Eurasian basin of the Arctic Ocean. Science 356, 285–291.

Polyakov, I.V., Pnyushkov, A.V., Rember, R., Ivanov, V.V., Lenn, Y.D., Padman, L., Carmack, E.C., 2012. Mooring-based observations of double-diffusive staircases over the Laptev Sea slope. J. Phys. Oceanogr. 42, 95–109.

Polyakov, I.V., Rippeth, T.P., Fer, I., Alkire, M.B., Baumann, T.M., Carmack, E.C., Ingvaldsen, R., Ivanov, V.V., Janout, M., Lind, S., et al., 2020b. Weakening of cold halocline layer exposes sea ice to oceanic heat in the eastern Arctic Ocean. J. Climate 33, 8107–8123. https://doi.org/10.1175/jcli-d-19-0976.1.

Polzin, K.L., Naveira Garabato, A.C., Huussen, T.N., Sloyan, B.M., Waterman, S., 2014. Finescale parameterizations of turbulent dissipation. J. Geophys. Res. 119, 1383–1419. https://doi.org/10.1002/2013JC008979.

Porter, M., Henley, S., Orkney, A., Bouman, H., Hwang, B., Dumont, E., Venables, E., Cottier, F., 2020. A polar surface eddy obscured by thermal stratification. Geophys. Res. Lett. 47, e2019GL086281.

Proshutinsky, A., Bourke, R., McLaughlin, F., 2002. The role of the Beaufort Gyre in Arctic climate variability: seasonal to decadal climate scales. Geophys. Res. Lett. 29, 15-1.

Proshutinsky, A., Krishfield, R., Timmermans, M.l., Toole, J., Carmack, E., Mclaughlin, F., Williams, W.J., Zimmermann, S., Itoh, M., Shimada, K., 2009. Beaufort Gyre freshwater reservoir: state and variability from observations. J. Geophys. Res. 114, 1–25. https://doi.org/10.1029/2008JC005104.

Proshutinsky, A., Krishfield, R., Toole, J., Timmermans, M.L., Williams, W., Zimmerman, S., Yamamoto-Kawai, M., Armitage, T., Dukhovskoy, D., Golubeva, E., et al., 2019. Analysis of the Beaufort gyre freshwater content in 2003-2018. J. Geophys. Res., Oceans.

Rabe, B., Karcher, M., Kauker, F., Schauer, U., Toole, J., Krishfield, R., Pisarev, S., Kikuchi, T., Su, J., 2014. Arctic Ocean basin liquid freshwater storage trend 1992–2012. Geophys. Res. Lett. 41, 961–968.

Rabe, B., Karcher, M., Schauer, U., Toole, J.M., Krishfield, R.A., Pisarev, S., Kauker, F., Gerdes, R., Kikuchi, T., 2011. An assessment of Arctic Ocean freshwater content changes from the 1990s to the 2006–2008 period. Deep-Sea Res., Part 1, Oceanogr. Res. Pap. 58, 173–185.

Radko, T., Flanagan, J., Stellmach, S., Timmermans, M.L., 2014. Double-diffusive recipes. Part II: layer-merging events. J. Phys. Oceanogr. 44, 1285–1305.

Rainville, L., Winsor, P., 2008. Mixing across the Arctic Ocean: microstructure observations during the Beringia 2005 expedition. Geophys. Res. Lett. 35. https://doi.org/10.1029/2008GL033532.

Rainville, L., Woodgate, R., 2009. Observations of internal wave generation in the seasonally ice-free Arctic. Geophys. Res. Lett. 36. https://doi.org/10.1029/2009GL041291.

Rippeth, T.P., Lincoln, B.J., Lenn, Y.D., Green, J.M., Sundfjord, A., Bacon, S., 2015. Tide-mediated warming of Arctic halocline by Atlantic heat fluxes over rough topography. Nat. Geosci. 8, 191–194.

Rippeth, T.P., Vlasenko, V., Stashchuk, N., Scannell, B.D., Green, J.A.M., Lincoln, B.J., Bacon, S., 2017. Tidal conversion and mixing poleward of the critical latitude (an Arctic case study). Geophys. Res. Lett. 44, 12349–12357. https://doi.org/10.1002/2017gl075310.

Roemmich, D., Alford, M.H., Claustre, H., Johnson, K., King, B., Moum, J., Oke, P., Owens, W.B., Pouliquen, S., Purkey, S., Scanderbeg, M., Suga, T., Wijffels, S., Zilberman, N., Bakker, D., Baringer, M., Belbeoch, M., Bittig, H.C., Boss, E., Calil, P., Carse, F., Carval, T., Chai, F., Conchubhair, D.O., d'Ortenzio, F., Dall'Olmo, G., Desbruyeres, D., Fennel, K., Fer, I., Ferrari, R., Forget, G., Freeland, H., Fujiki, T., Gehlen, M., Greenan, B., Hallberg, R., Hibiya, T., Hosoda, S., Jayne, S., Jochum, M., Johnson, G.C., Kang, K., Kolodziejczyk, N., Körtzinger, A., Traon, P.Y.L., Lenn, Y.D., Maze, G., Mork, K.A., Morris, T., Nagai, T., Nash, J., Garabato, A.N., Olsen, A., Pattabhi, R.R., Prakash, S., Riser, S., Schmechtig, C., Schmid, C., Shroyer, E., Sterl, A., Sutton, P., Talley, L., Tanhua, T., Thierry, V., Thomalla, S., Toole, J., Troisi, A., Trull, T.W., Turton, J., Velez-Belchi, P.J., Walczowski, W., Wang, H., Wanninkhof, R., Waterhouse, A.F., Waterman, S., Watson, A., Wilson, C., Wong, A.P.S., Xu, J., Yasuda, I., 2019. On the future of Argo: a global, full-depth, multi-disciplinary array. Front. Mar. Sci. 6. https://doi.org/10.3389/fmars.2019.00439.

Ruddick, B., Gargett, A.E., 2003. Oceanic double diffusion: introduction. Prog. Oceanogr. 56, 381–393.

Ruddick, B., Richards, K., 2003. Oceanic thermohaline intrusions: observations. Prog. Oceanogr. 56, 449–527.

Rudels, B., Björk, G., Muench, R., Schauer, U., 1999. Double-diffusive layering in the Eurasian Basin of the Arctic Ocean. J. Mar. Syst. 21, 3–27.

Rudels, B., Jones, E.P., Anderson, L.G., Kattner, G., 1994. On the intermediate depth waters of the Arctic Ocean. In: Johannessen, O., Muench, R., Overland, J. (Eds.), The Polar Oceans and Their Role in Shaping the Global Environment. In: Geophysics Monogr. Ser., vol. 85. AGU, Washington, D.C., pp. 33–46.

Rudels, B., Kuzmina, N., Schauer, U., Stipa, T., Zhurbas, V., 2009. Double-diffusive convection and interleaving in the Arctic Ocean–distribution and importance. Geophysica 45, 199–213.

Rudels, B., Muench, R.D., Gunn, J., Schauer, U., Freidrich, H.J., 2000. Evolution of the Arctic boundary current North of the Siberian shelves. J. Mar. Syst. 25, 77–99.

Rudnick, D.L., Boyd, T.J., Brainard, R.E., Carter, G.S., Egbert, G.D., Gregg, M.C., Holloway, P.E., Klymak, J.M., Kunze, E., Lee, C.M., Levine, M.D., Luther, D.S., Martin, J.P., Merrifield, M.A., Moum, J.N., Nash, J.D., Pinkel, R., Rainville, L., Sanford, T.B., 2003. From tides to mixing along the Hawaiian Ridge. Science 301, 355–357.

Schauer, U., Rudels, B., Jones, E., Anderson, L., Muench, R.D., Björk, G., Swift, J., Ivanov, V., Larsson, A., 2002. Confluence and redistribution of Atlantic water in the Nansen, Amundsen and Makarov basins. Ann. Geophys. 20, 257–273.

Schmitt, R.W., 2003. Observational and laboratory insights into salt finger convection. Prog. Oceanogr. 56, 419–433.

Screen, J.A., Simmonds, I., 2010. The central role of diminishing sea ice in recent Arctic temperature amplification. Nature 464, 1334–1337.

Screen, J.A., Simmonds, I., 2014. Amplified mid-latitude planetary waves favour particular regional weather extremes. Nat. Clim. Change 4, 704.

Shaw, W., Stanton, T., McPhee, M., Kikuchi, T., 2008. Estimates of surface roughness length in heterogeneous under-ice boundary layers. J. Geophys. Res., Oceans 113.

Shaw, W., Stanton, T., McPhee, M., Morison, J., Martinson, D., 2009. Role of the upper ocean in the energy budget of Arctic sea ice during sheba. J. Geophys. Res., Oceans 114.

Sheen, K., Garabato, A.N., Brearley, J., Meredith, M., Polzin, K., Smeed, D., Forryan, A., King, B., Sallée, J.B., Laurent, L.S., et al., 2014. Eddy-induced variability in Southern Ocean abyssal mixing on climatic timescales. Nat. Geosci. 7, 577–582.

Shibley, N., Timmermans, M.L., 2019. The formation of double-diffusive layers in a weakly turbulent environment. J. Geophys. Res., Oceans 124, 1445–1458.

Shibley, N.C., Timmermans, M.L., Carpenter, J.R., Toole, J.M., 2017. Spatial variability of the Arctic Ocean's double-diffusive staircase. J. Geophys. Res., Oceans 122, 980–994. https://doi.org/10.1002/2016JC012419.

Sirevaag, A., Fer, I., 2009. Early spring oceanic heat fluxes and mixing observed from drift stations north of Svalbard. J. Phys. Oceanogr. 39, 3049–3069.

Sirevaag, A., Fer, I., 2012. Vertical heat transfer in the Arctic Ocean: the role of double-diffusive mixing. J. Geophys. Res., Oceans 117. https://doi.org/10.1029/2012jc007910.

Skagseth, Ø., Eldevik, T., Årthun, M., Asbjørnsen, H., Lien, V.S., Smedsrud, L.H., 2020. Reduced efficiency of the Barents sea cooling machine. Nat. Clim. Change, 1–6.

Smith, M., Stammerjohn, S., Persson, O., Rainville, L., Liu, G., Perrie, W., Robertson, R., Jackson, J., Thomson, J., 2018. Episodic reversal of autumn ice advance caused by release of ocean heat in the Beaufort sea. J. Geophys. Res., Oceans 123, 3164–3185.

Smith, M., Thomson, J., 2019. Ocean surface turbulence in newly formed marginal ice zones. J. Geophys. Res., Oceans 124, 1382–1398.

Spall, M.A., Pickart, R.S., Fratantoni, P.S., Plueddemann, A.J., 2008. Western Arctic shelfbreak eddies: formation and transport. J. Phys. Oceanogr. 38, 1644–1668.

Stanton, T.P., Shaw, W.J., Hutchings, J.K., 2012. Observational study of relationships between incoming radiation, open water fraction, and ocean-to-ice heat flux in the transpolar drift: 2002–2010. J. Geophys. Res., Oceans 117.

Stroeve, J., Holland, M.M., Meier, W., Scambos, R., Serreze, M., 2007. Arctic sea ice decline: faster than forecast. Geophys. Res. Lett. 34. https://doi.org/10.1029/2007GL029703.

Stroeve, J.C., Serreze, M.C., Holland, M.M., Kay, J.E., Malanik, J., Barrett, A.P., 2012. The Arctic's rapidly shrinking sea ice cover: a research synthesis. Clim. Change 110, 1005–1027.

Sundfjord, A., Fer, I., Kasajima, Y., Svendsen, H., 2007. Observations of turbulent mixing and hydrography in the marginal ice zone of the Barents sea. J. Geophys. Res., Oceans 112.

Testor, P., de Young, B., Rudnick, D.L., Glenn, S., Hayes, D., Lee, C.M., Pattiaratchi, C., Hill, K., Heslop, E., Turpin, V., Alenius, P., Barrera, C., Barth, J.A., Beaird, N., Bécu, G., Bosse, A., Bourrin, F., Brearley, J.A., Chao, Y., Chen, S., Chiggiato, J., Coppola, L., Crout, R., Cummings, J., Curry, B., Curry, R., Davis, R., Desai, K., DiMarco, S., Edwards, C., Fielding, S., Fer, I., Frajka-Williams, E., Gildor, H., Goni, G., Gutierrez, D., Haugan, P., Hebert, D., Heiderich, J., Henson, S., Heywood, K., Hogan, P., Houpert, L., Huh, S., Inall, E., Ishii, M., Ito, S.i., Itoh, S., Jan, S., Kaiser, J., Karstensen, J., Kirkpatrick, B., Klymak, J., Kohut, J., Krahmann, G., Krug, M., McClatchie, S., Marin, F., Mauri, E., Mehra, A., Meredith, P.M., Meunier, T., Miles, T., Morell, J.M., Mortier, L., Nicholson, S., O'Callaghan, J., O'Conchubhair, D., Oke, P., Pallàs-Sanz, E., Palmer, M., Park, J., Perivoliotis, L., Poulain, P.M., Perry, R., Queste, B., Rainville, L., Rehm, E., Roughan, M., Rome, N., Ross, T., Ruiz, S., Saba, G., Schaeffer, A., Schönau, M., Schroeder, K., Shimizu, Y., Sloyan, B.M., Smeed, D., Snowden, D., Song, Y., Swart, S., Tenreiro, M., Thompson, A., Tintore, J., Todd, R.E., Toro, C., Venables, H., Wagawa, T., Waterman, S., et al., 2019. OceanGliders: a component of the integrated GOOS. Front. Mar. Sci. 6. https://doi.org/10.3389/fmars.2019.00422.

Thomson, J., Rogers, W.E., 2014. Swell and sea in the emerging Arctic Ocean. Geophys. Res. Lett. 41, 3136–3140.

Timmermans, M.L., 2015. The impact of stored solar heat on Arctic sea ice growth. Geophys. Res. Lett. https://doi.org/10.1002/2015GL064541.1.

Timmermans, M.L., Garrett, C., Carmack, E., 2003. The thermohaline structure and evolution of deep waters in the Canada Basin, Arctic Ocean. Deep-Sea Res., Part 1 50, 1305–1321.

Timmermans, M.L., Jayne, S.R., 2016. The Arctic Ocean spices up. J. Phys. Oceanogr. 46, 1277–1284. https://doi.org/10.1175/JPO-D-16-0027.1.

Timmermans, M.L., Marshall, J., 2020. Understanding Arctic Ocean circulation: a review of ocean dynamics in a changing climate. J. Geophys. Res., Oceans 125.

Timmermans, M.L., Proshutinsky, A., Krishfield, R.A., Perovich, D.K., Richter-Menge, J.A., Stanton, T.P., Toole, J.M., 2011. Surface freshening in the Arctic Ocean's Eurasian basin: an apparent consequence of recent change in the wind-driven circulation. J. Geophys. Res., Oceans 116.

Timmermans, M.L., Toole, J., Krishfield, R., 2018. Warming of the interior Arctic Ocean linked to sea ice losses at the basin margins. Sci. Adv. 4, eaat6773.

Timmermans, M.L., Toole, J., Krishfield, R., Winsor, P., 2008a. Ice-tethered profiler observations of the double-diffusive staircase in the Canada basin thermocline. J. Geophys. Res., Oceans 113.

Timmermans, M.L., Toole, J., Proshutinsky, A., Krishfield, R., Plueddemann, A., 2008b. Eddies in the Canada basin, Arctic Ocean, observed from ice-tethered profilers. J. Phys. Oceanogr. 38, 133–145.

Toole, J., Krishfield, R., Timmermans, M.L., Proshutinsky, A., 2011. The ice-tethered profiler: Argo of the Arctic. Oceanography 24, 126–135. https://doi.org/10.5670/oceanog.2011.64. https://tos.org/oceanography/article/the-ice-tethered-profiler-argo-of-the-arctic.

Toole, J.M., Timmermans, M.L., Perovich, D.K., Krishfield, R.A., Proshutinsky, A., Richter-Menge, J.A., 2010. Influences of the ocean surface mixed layer and thermohaline stratification on Arctic sea ice in the central Canada Basin. J. Geophys. Res., Oceans 115.

Tsamados, M., Feltham, D.L., Schroeder, D., Flocco, D., Farrell, S.L., Kurtz, N., Laxon, S.W., Bacon, S., 2014. Impact of variable atmospheric and oceanic form drag on simulations of Arctic sea ice. J. Phys. Oceanogr. 44, 1329–1353.

Vlasenko, V., Stashchuk, N., Hutter, K., Sabinin, K., 2003. Nonlinear internal waves forced by tides near the critical latitude. Deep-Sea Res., Part 1 50, 317–338.

Walsh, D., Carmack, E., 2002. A note on the evanescent behaviour of Arctic thermohaline intrusions. J. Mar. Res. 60, 281–310.
Walsh, D., Carmack, E., 2003. The nested structure of Arctic thermohaline intrusions. Ocean Model. 5, 267–289.
Warner, J., Screen, J., Scaife, A., 2020. Linkages between Arctic sea ice and the extratropical atmospheric circulation explained by internal variability and tropical forcing. Geophys. Res. Lett. 47 (1), e2019GL085679.
Waterhouse, A.F., MacKinnon, J.A., Nash, J.D., Alford, M.H., Kunze, E., Simmons, H.L., Polzin, K.L., St. Laurent, L.C., Sun, O.M., Pinkel, R., Talley, L.D., Whalen, C.B., Huussen, T.N., Carter, G.S., Fer, I., Waterman, S., Naveira Garabato, A.C., Sanford, T.B., Lee, C.M., 2014. Global patterns of diapycnal mixing from measurements of the turbulent dissipation rate. J. Phys. Oceanogr. 44, 1854–1872. https://doi.org/10.1175/JPO-D-13-0104.1.
Wettlaufer, J., Worster, M.G., Huppert, H.E., 1997. The phase evolution of young sea ice. Geophys. Res. Lett. 24, 1251–1254.
Whalen, C.B., MacKinnon, J.A., Talley, L.D., Waterhouse, A.F., 2015. Estimating the mean diapycnal mixing using a finescale strain parameterization. J. Phys. Oceanogr. 45, 1174–1188. https://doi.org/10.1175/jpo-d-14-0167.1.
Whalen, C.B., Talley, L.D., MacKinnon, J.A., 2012. Spatial and temporal variability of global ocean mixing inferred from Argo profiles. Geophys. Res. Lett. 39, L18612. https://doi.org/10.1029/2012gl053196.
Woodgate, R.A., 2005. Pacific ventilation of the Arctic Ocean's lower halocline by upwelling and diapycnal mixing over the continental margin. Geophys. Res. Lett. 32. https://doi.org/10.1029/2005GL023999.
Woodgate, R.A., Aagaard, K., Muench, R., Gunn, J., Björk, G., Rudels, B., Roach, A., Schauer, U., 2001. The Arctic Ocean Boundary Current along the Eurasian slope and the adjacent Lomonosov Ridge: water mass properties, transport and transformations from moored instruments. Deep-Sea Res., Part 1 48, 1757–1792.
Woodgate, R.A., Aagaard, K., Swift, J.H., Smethie Jr, W.M., Falkner, K.K., 2007. Atlantic water circulation of Mendeleev Ridge and Chukchi Borderland from thermohaline intrusions and water mass properties. J. Geophys. Res. 112. https://doi.org/10.1029/2005JC003416.
Zhao, M., Timmermans, M.L., 2015. Vertical scales and dynamics of eddies in the Arctic Ocean's Canada basin. J. Geophys. Res., Oceans 120, 8195–8209.
Zhao, M., Timmermans, M.L., Cole, S., Krishfield, R., Proshutinsky, A., Toole, J., 2014. Characterizing the eddy field in the Arctic Ocean halocline. J. Geophys. Res., Oceans 119, 8800–8817. https://doi.org/10.1002/2014jc010488.
Zhao, M., Timmermans, M.L., Cole, S., Krishfield, R., Toole, J., 2016. Evolution of the eddy field in the Arctic Ocean's Canada Basin, 2005–2015. Geophys. Res. Lett. 43, 8106–8114. https://doi.org/10.1002/2016gl069671.
Zhao, M., Timmermans, M.L., Krishfield, R., Manucharyan, G., 2018. Partitioning of kinetic energy in the Arctic Ocean's Beaufort gyre. J. Geophys. Res., Oceans 123, 4806–4819.

Chapter 12

Mixing in the Southern Ocean

Sarah T. Gille[a], Katy L. Sheen[b], Sebastiaan Swart[c,d] and Andrew F. Thompson[e]

[a]Scripps Institution of Oceanography, University of California San Diego, La Jolla, CA, United States, [b]Penryn Campus, University of Exeter, Cornwall, England, United Kingdom, [c]Department of Marine Sciences, University of Gothenburg, Gothenburg, Sweden, [d]Department of Oceanography, University of Cape Town, Rondebosch, South Africa, [e]Environmental Science and Engineering, California Institute of Technology, Pasadena, CA, United States

12.1 Introduction

The Southern Ocean connects the components of the global ocean, and this makes it central to global ocean mixing. The eastward-flowing Antarctic Circumpolar Current (ACC) provides a horizontal link between the Atlantic, Indian, and Pacific Oceans. The Southern Ocean also connects low and high southern latitudes along the steeply tilted isopycnal surfaces of the ACC. Since water parcels preferentially mix along isopycnals (see Chapter 9), the Southern Ocean provides a key route between the ocean surface and the deep interior. In addition, the Southern Ocean connects ocean waters with the ice shelves that surround the Antarctic continent. Mixing processes on the Antarctic continental shelf help to determine the rate at which these ice shelves melt and how much freshwater enters the ocean. Any turbulent mixing that changes properties—whether resulting from buoyancy forcing at the ocean surface, processes associated with tides and bathymetry deep below the ocean surface, or interactions with the cryosphere—has the potential to have a disproportionate impact on global ocean properties, and therefore large-scale climate and ecosystems.

This chapter builds on earlier chapters to examine how a range of mixing processes shape the Southern Ocean circulation and climate. Our focus here is centred on the impact of physical processes on mixing rates. An overview of how Southern Ocean mixing influences the climate system is found in Chapter 2, and a review of the Southern Ocean's role in the large-scale ocean circulation is found in Chapter 3. Some of the mixing processes that are critical to Southern Ocean dynamics receive individual treatment in other chapters, and we have provided references to those in the text that follows. Here, Section 12.2, reviews the large-scale context framing Southern Ocean mixing. We then move through the water column, starting with turbulent mixing near the ocean surface in Section 12.3, regional and mesoscale interior mixing in Section 12.4, and small-scale turbulence in the ocean interior in Section 12.5. Each section presents the foundations underlying the science, and then highlights recent findings. A sidebar highlights how the Diapycnal and Isopycnal Mixing Experiment in the Southern Ocean (DIMES) brought together multiple observing capabilities and modelling strategies to advance understanding of Southern Ocean mixing. The chapter concludes with perspectives on emerging science challenges for the future.

12.2 Large-scale context: foundations

Large-scale circulation in the Southern Ocean is distinct from large-scale circulation in other parts of the global ocean because of the gap in continental boundaries in the latitude range of Drake Passage (56°–63°S). Eastward winds provide a continuous band of surface momentum forcing all the way around Antarctica. Because of the Coriolis effect, an eastward wind stress at the Southern Ocean surface results in northward Ekman transport of surface waters, carrying cold high-latitude surface water toward the subtropics (e.g. Toggweiler and Samuels, 1998). This Ekman advection establishes a north–south gradient in sea surface height, with elevated sea surface height to the north. The corresponding top-to-bottom meridional pressure gradient supports an eastward geostrophic flow, known as the Antarctic Circumpolar Current (ACC, black arrow in Fig. 12.1). The absence of topographic barriers above roughly 2000 m and the limited ability to sustain large-scale zonal pressure gradients above these depths allow the ACC to maintain mean meridional density gradients: density surfaces that are deep at the northern boundary of the ACC shoal to the south.

The surface water carried northward as Ekman transport is replaced with water that upwells within the ACC. The upwelling is attributed to Ekman pumping due to wind-stress divergence. This upwelled Circumpolar Deep Water (CDW) is transported and mixed through eddy motions along the Southern Ocean's steeply tilted isopycnals (light blue CDW

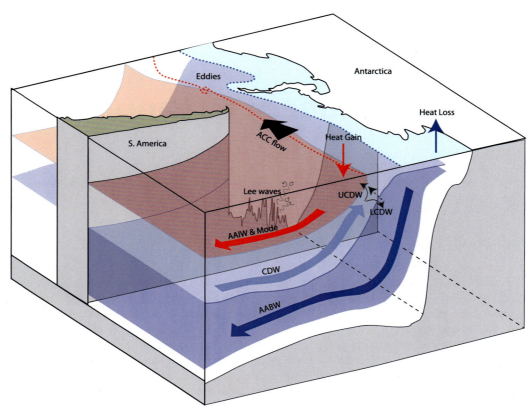

FIGURE 12.1 Schematic large-scale circulation in the Southern Ocean, showing the Antarctic Circumpolar Current (ACC) flowing eastward (black arrow), Circumpolar Deep Water (CDW) upwelling at about the same latitude as the ACC (light blue arrow), and the upper and lower cells of the meridional overturning circulation. Near the surface, Upper Circumpolar Deep Water (UCDW) advects northward and contributes to Antarctic Intermediate Water and mode water formation. Lower Circumpolar Deep Water (LCDW) loses heat near Antarctica and sinks to form Antarctic Bottom Water (AABW).

in Fig. 12.1) from mid-depth (around 2000–3000 m depth) in the subtropics to the surface within the Southern Ocean. Along-isopycnal pathways thus provide a direct connection between the surface and the mid-depth ocean interior. (See Chapter 9 for a more detailed exploration of isopycnal mixing.) The Southern Ocean is host to approximately 70% to 80% of upwelling of global deep waters to the sea surface (DeVries and Primeau, 2011; Talley, 2013). Along-isopycnal mixing plays a central role in the climate system: modelling studies have for many years highlighted the sensitivity of Southern Ocean circulation to the representation of this mixing (e.g. Redi, 1982; Danabasoglu et al., 1994; Danabasoglu and McWilliams, 1995; Griffies, 1998; Gent, 2016). Together interior upwelling and northward surface advection constitute the upper cell of the Southern Ocean's meridional overturning circulation (MOC) (Speer et al., 2000).

The lower cell of the Southern Ocean meridional overturning comprises denser CDW, known as Lower Circumpolar Deep Water (LCDW) that upwells across the ACC, but outcrops at the surface on the southern side of the ACC (Marshall and Speer, 2012). Water that upwells here loses heat to the atmosphere, while also increasing its salinity as a result of brine rejection during sea-ice formation to form Antarctic Bottom Water (AABW) that sinks to the sea floor (dark blue arrow in Fig. 12.1), providing a direct connection between the atmosphere and the deep ocean. Mixing processes in the lower cell occur in the subpolar gyres (e.g. Vernet et al., 2019), over the Antarctic continental slope (e.g. Stewart and Thompson, 2013; Heywood et al., 2014), and over the continental shelf, where they are modulated by tides (e.g. Padman et al., 2018; Stewart et al., 2018) as well as by complex topography and by interactions between sea ice, ice shelves, and the ocean (e.g. Gille et al., 2016).

The connection between the deep ocean and the surface is further supported by the Southern Ocean's low stratification. As temperatures drop, ocean density is progressively less dependent on temperature, and at temperatures colder than about 2 °C, salinity is the dominant controller of oceanic potential density (e.g. Nycander et al., 2015), such that surface-to-bottom temperature differences result in little difference in potential density. Cold temperatures of the Southern Ocean offer ideal conditions for low vertical stratification, particularly in regions such as the ACC latitudes, where freshwater inputs from precipitation or melting ice are minimal. As a result the Southern Ocean is generally weakly stratified and forms deep mixed layers (Dong et al., 2008; Holte et al., 2017; Speer et al., 2018) that allow small energy inputs to readily mix the ocean vertically.

The region also experiences significant horizontal stirring. The ACC consists of multiple narrow jets that roughly follow zonal pathways located between about 40° and 60°S. From north to south, the jets coincide with individual density gradients that are referred to as the Subantarctic Front, the Polar Front, and the southern ACC Front (Orsi et al., 1995). These frontal jets extend from the surface to the sea floor and are steered by topography. They can meander substantially, and at times they may split or merge so that there are more or less than three jets in some locations (Sokolov and Rintoul, 2009; Thompson and Sallée, 2012). The ACC jets are energetic enough to be baroclinically unstable, and interactions with topography are associated with flow instabilities that spawn transient eddies and heightened levels of eddy kinetic energy (EKE) (e.g. Meredith and Hogg, 2006; Hogg et al., 2008), influencing the Southern Ocean heat budget. In most of the global ocean, western boundary currents are responsible for the poleward transport of energy that is input by the sun in the tropics. However, in the Southern Ocean mean southward transport of heat is impeded by the absence of continental boundaries at the latitude of Drake Passage (55°–6°S), and instead eddies play a leading role in the poleward transport of heat south of 40°–50°S (e.g. deSzoeke and Levine, 1981; Keffer and Holloway, 1988; Johnson and Bryden, 1989; Ivchenko et al., 1996; Gille, 2003; Meijers et al., 2007; Volkov et al., 2010).

The ease with which the Southern Ocean mixes gives it a central role in the climate system. Recent observations indicate that the Southern Ocean is gaining heat more rapidly than other regions of the global ocean (e.g. Roemmich et al., 2015), although sea surface temperatures show evidence of cooling (Armour et al., 2016). The Southern Ocean has also been implicated as the major location for oceanic uptake of CO_2 (e.g. Caldeira and Duffy, 2000; Sabine et al., 2004; Gruber et al., 2019). Attribution studies suggest that the observed warming is due to heat uptake from the atmosphere within the Southern Ocean, rather than advection within the ocean (Swart et al., 2018a). This implies that upper-ocean turbulence mediates much of the long-term oceanic change, but that horizontal advection also matters. Climate simulations suggest that correctly representing the uptake of heat and carbon in the Southern Ocean is critical: multi-model assessments indicate that climate at the end of the century depends on the present structure of the ocean mixed layer, and thus the mixing within the upper ocean (e.g. Boé et al., 2009), and that ocean heat uptake is sensitive to the representation of air-sea fluxes (e.g. Huber and Zanna, 2017). Mixing processes not only close the global overturning circulation, but are also fundamentally important for oceanic exchange of carbon and heat (e.g. Watson and Naveira Garabato, 2006; Morrison et al., 2016), nutrient resupply for sustaining oceanic biological production (e.g. Pollard et al., 2006), and the melt rate of ice shelves (e.g. Silvano et al., 2018).

The Southern Ocean is unique in the ways that different mixing processes impact the region's circulation. For instance, nowhere else is such a broad range of density classes ventilated at the ocean surface. This supports a significant fraction of the water mass transformation associated with the global overturning circulation. The ACC also hosts strong deep-ocean currents (Chereskin et al., 2009; Fukamachi et al., 2010) and, through their interaction with large bathymetric features, generates distinct mixing processes. Finally, along-isopycnal mixing is the dominant process that delivers waters toward and away from key sites of water mass transformation at the ocean surface, over sloping topography, and on the continental shelf. In the remainder of this chapter, we offer an overview of these different components of Southern Ocean mixing and identify priorities for future research.

12.3 Upper cell: mixed-layer transformations

12.3.1 Foundations and setting

Surface mixing in the Southern Ocean is of particular importance to the climate system, because the cold, dense water that outcrops here has nearly the same potential density as deeper waters throughout the global ocean. The ventilation of these dense, near-surface waters means that surface mechanical and buoyancy forcing has the potential to set interior water properties.

Upper-ocean mixing processes maintain the surface mixed layer as the buffer layer, through which all exchanges of mass and energy pass between the ocean and atmosphere. The surface mixed layer is analogous to a 'vestibule' between the atmosphere and the ocean. Air-sea fluxes bring properties (e.g., heat, freshwater, and CO_2) into the mixed layer. Mechanisms at work in the mixed layer, such as adiabatic advection, stirring, turbulent diffusion, and convection, ultimately determine whether these properties are re-emitted to the atmosphere or enter the ocean interior—figuratively, whether properties pass through the vestibule into the ocean interior.

Characteristics of the Southern Ocean that are conducive to energetic upper-ocean mixing include, first, the presence of lateral gradients in temperature, salinity, or density, associated with hydrographic fronts and mesoscale eddies, and second,

304 Ocean Mixing

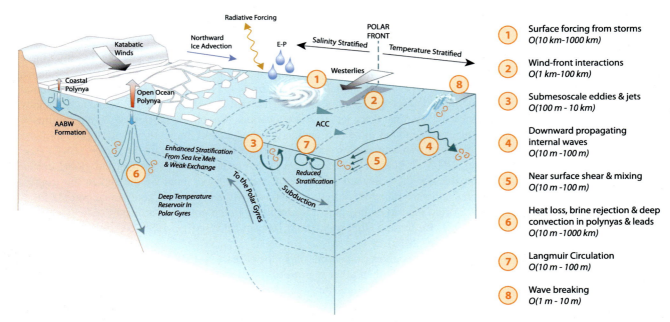

FIGURE 12.2 Conceptual view of the main processes driving mixing in the upper cell of the Southern Ocean. It includes mixing processes spanning the open-ocean ACC domains, dominated by surface winds and waves, to the sea ice and ice shelf-impacted regions of the Southern Ocean. An estimated scale for each mixing processes is indicated.

the weak stratification of the region that allows deep mixed layers to form seasonally (Dong et al., 2008; Holte et al., 2017), supporting vertical mixing between the upper ocean and the thermocline.

Overlying this energetic ocean environment is an atmosphere characterised by strong and variable winds that promote atmosphere–ocean interactions, e.g. via momentum, CO_2, and heat fluxes. The surface experiences a strong time-mean wind stress (momentum flux) associated with a band of maximum wind forcing (Lin et al., 2018) induced by transient storms, with near-surface wind speeds often exceeding 20 m s^{-1} (Yuan, 2004). Storms add a time-varying and rotational component to the background wind forcing, which can interact directly with the lateral density gradients and induce Ekman pumping and shear-driven mixing. In addition, seasonal air-sea and air-sea-ice fluxes of heat and freshwater change the buoyancy budget of the mixed layer, which as we will see later in this chapter, can enhance or arrest mixing processes within and below the thermocline, and are thus important for the ventilation of the lower thermocline. Fig. 12.2 encapsulates a broad range of processes that influence surface boundary-layer properties and their exchange with the interior.

12.3.2 Mixing in the surface boundary layer and connection to subsurface adiabatic stirring

12.3.2.1 Surface water mass transformation

A significant proportion of upper-ocean mixing and mixed-layer properties can be ascribed to relatively simple one-dimensional physical responses to surface momentum and buoyancy fluxes (Price et al., 1978; Large et al., 1994). Studies of surface boundary-layer turbulence have historically been weighted to Northern Hemisphere conditions. (For fundamentals, see Chapter 4 for discussion of surface layers and Chapter 5 for surface-generated internal waves.) More recent observations are allowing us to understand mixed-layer budgets in various regions of the Southern Ocean, including sea-ice impacted regions, albeit at relatively coarse resolution (Pellichero et al., 2017). Surface forcing mechanisms are exceptionally strong in the Southern Ocean and are characterised by distinct seasonal cycles of sea-ice freshwater and air-sea heat fluxes that are the primary drivers for surface salinity and temperature variations (Pellichero et al., 2017).

During austral winter, high levels of heat loss by the ocean can lead to deep convective mixing, resulting in deep mixed layers (Holte et al., 2012). This is particularly the case at the northern edge of the ACC (north of the Subantarctic Front), where mixed layers can exceed depths of 700 m in winter (Dong et al., 2008; Tamsitt et al., 2020). Heat loss events are intermittent, and most of the annual mixed-layer deepening appears to occur during a small number of events (Ogle et al., 2018). These strong air-sea exchanges lead to water-mass transformation at the surface and are a leading order driver of Sub-Antarctic Mode Water (SAMW) formation and its interannual variability (Naveira Garabato et al., 2009). In addition, the formation rate and characteristics of SAMW are also influenced by diapycnal mixing (Sloyan et al., 2010), eddy-induced transport and upwelling (Marshall, 1997; Sallée et al., 2010), and the large-scale Ekman transport of cool, low salinity water equatorward across the Subantarctic Front (Rintoul and England, 2002).

12.3.2.2 Submesoscale-induced mixing and stratification

Upper-ocean stratification and mixing rates are also impacted by complex, horizontal processes in response to fronts, eddies, and jets. These horizontal dynamics occur at small spatial scales that extend down to the submesoscale (0.1–10 km, hours–days; Fig. 12.2; see Chapter 8 for a more comprehensive description of submesoscale processes). Submesoscale mixing is enhanced by the factors that characterise the Southern Ocean's mixing favourable conditions: strong surface forcing, persistent lateral buoyancy gradients, weak vertical stratification, and deep mixed layers.

Initial understanding of the role of submesoscales in the Southern Ocean relied on submesoscale-resolving simulations for specific regions, including Drake Passage (Bachman et al., 2017) and Kerguelen Plateau (Rosso et al., 2014). These numerical studies showed that submesoscale motions directly impact upper-ocean stratification, for example, through mixed-layer shoaling with increased model resolution (Bachman et al., 2017), and through vertical exchanges between the surface layer and ocean interior (Rosso et al., 2014, 2015; Balwada et al., 2018; Su et al., 2018; Bachman and Klocker, 2020).

Observations have further highlighted the role of submesoscale instabilities. Using ship-based measurements along the edges of a large mesoscale eddy pinched off from the Polar Front, Adams et al. (2017) identified the presence of submesoscale instabilities that can lead to strong net subduction of water associated with finescale dense outcropping filaments. Submesoscale Rossby waves, found along the ACC, also drive intense vertical circulations (Taylor et al., 2018). These recent observations emphasise the importance of larger-scale adiabatic stirring at mesoscales (McWilliams, 2016) that catalyse submesoscale mixed-layer instabilities in the Southern Ocean, thereby providing a connection between the surface boundary layer and the ocean interior (Erickson et al., 2016), as biogeochemical tracers have demonstrated in the open ocean (Llort et al., 2018). Months-long glider-based surveys in the Southern Ocean have been used to estimate the presence and strength of submesoscale flows in the upper ocean. Mixed-layer eddies (MLE; Fig. 12.2) are expected to proliferate where mixed layers are deep in the Southern Ocean, particularly in the presence of sharp lateral buoyancy gradients. Slumping MLEs can rearrange horizontal buoyancy gradients to vertical stratification through an ageostrophic secondary circulation (see Chapter 8; Fox-Kemper et al., 2008). Observations show that MLEs can arrest mixing in the Southern Ocean and induce springtime mixed-layer stratification earlier than would be possible with surface buoyancy forcing alone (du Plessis et al., 2017). The impact of the submesoscale on upper-ocean stratification is also spatially heterogeneous due to its dependence on the background mesoscale eddy field. For instance, Viglione et al. (2018) found an abrupt transition in the strength of the submesoscale eddy field, delineated by the Shackleton Fracture Zone. Despite this topographic feature having a sill depth thousands of metres below the surface, it gives rise to different mesoscale stirring regimes that impact surface boundary layer dynamics, including the potential for more vigorous mixing processes, such as symmetric instability, occurring downstream of the fracture zone, but not upstream.

Submesoscale processes cannot be easily disentangled from the Southern Ocean's intense storm systems and associated wind stress at the surface (e.g., Klocker, 2018). Under the action of wind, submesoscale flows are influenced by finescale cross-frontal horizontal Ekman advection of water of different density (Ekman Buoyancy Flux, EBF; Fig. 12.2; Chapter 8). In the case of down-front wind conditions, cross-frontal Ekman flows that carry denser waters over the lighter side of the front can force convective instabilities and enhance mixing through small-scale turbulence. This can increase dissipation within and across the mixed layer by an order of magnitude compared to wind-driven shear mixing (Thomas, 2005). When the opposite occurs during up-front winds, light water is advected over the dense water on the other side of the front, mixing can be arrested, and the mixed layer can shoal. The Southern Ocean's prevailing westerly winds, oriented along the mean frontal flows associated with the ACC, make these wind-front processes particularly relevant (Johnson et al., 2016), with average down-front wind conditions occurring in the Southern Ocean and providing a significant destratifying and mixing flux to the surface ocean (du Plessis et al., 2019; Viglione et al., 2018; Giddy et al., 2021). This effect of EBF is large enough to delay the onset of springtime stratification by two months after the onset of seasonal surface heating and to eradicate half the stratification gained by surface buoyancy forcing (du Plessis et al., 2019). The major topographic features in the Southern Ocean regionally deflect the average frontal flow away from the prevailing wind direction, allowing for strong variability in mixing and rates of stratification across the open-ocean extent of the Southern Ocean. This heterogeneous characteristic of submesoscale processes and its impact on the timing of seasonal restratification is likely to have significant impacts on spring phytoplankton blooms and timing (Carranza and Gille, 2015; Swart et al., 2015; Pellichero et al., 2020).

Storms are known to induce energetic, "instantaneous", vertical mixing events in the Southern Ocean, with upper-ocean vertical eddy diffusivity, $K_z \sim O(10^{-1} \text{ m}^2 \text{ s}^{-1})$ (Cisewski et al., 2005; Forryan et al., 2015). This mixing contributes to energising and deepening the surface mixing layer. Wind-driven energy has also been shown to excite strong inertial motions within the upper ocean, which may last several days to weeks poststorm and result in enhanced shear-driven vertical mixing below the base of the mixing layer within a "mixing transition" layer (Forryan et al., 2015). In the presence of mesoscale and submesoscale variability, this wind-driven inertial energy is concentrated (e.g., Klein et al., 2004; Jing et al., 2011; Meyer et al., 2015) and may further impact the extent and magnitude of the transitional mixing layer (Alford et al., 2016). Ultimately, these small-scale features and their interactions with surface wind stress (storm-eddy interactions, Fig. 12.2) enhance the downward propagation of inertial energy into the subsurface ocean (Jing et al., 2011), and potentially induce the rapid breaking of near-inertial waves to produce intense vertical mixing in the Southern Ocean (Meyer et al., 2015) that can extend below the base of the mixed layer (Forryan et al., 2015). Further studies using both 1D and 3D model simulations have explored the interaction of mixing with mesoscale-to-submesoscale eddy stirring (Uchida et al., 2019, 2020) and the impact of the interior mixing due to inertial motion set by passing storms in the Southern Ocean (Jouanno et al., 2016; Nicholson et al., 2016, 2019; Whitt et al., 2019). These findings suggest that inertially driven subsurface mixing may persist for several days to weeks after a passing storm, and after the deepened active mixing layer has restabilised to the surface. Such intraseasonal mixing processes are proving important to biogeochemical cycles, with summer primary productivity in the Southern Ocean considerably enhanced and sustained, largely due to the supply of deep iron reservoirs below the mixed layer (Carranza and Gille, 2015; Swart et al., 2015; Whitt et al., 2017; Nicholson et al., 2019; Uchida et al., 2020).

Although these mechanisms are theorised and modelled to be important in the Southern Ocean, direct observations (Chapter 14) of these interactions are sparse. It thus remains uncertain how storm-driven mixing energy will alter the magnitude and shape of the mean upper-ocean vertical diffusion profiles, particularly in a dynamically complex ocean. Measurements of subsurface mixing are limited to microstructure data collected during summer periods (e.g., Forryan et al., 2015), emphasising the need to constrain these mixing rates in other seasons, especially in winter when mixing rates are expected to increase significantly.

12.3.2.3 Interactions between sea ice and Southern Ocean mixing

Sea-ice covers a vast portion of the Southern Ocean with a winter maximum extent exceeding 18 million km^2, and it is an important influence on surface boundary-layer processes. Observations of upper-ocean mixing and variability remain sparse in this region, contributing to both a poor mechanistic understanding of coupled ocean-ice-atmosphere processes and large uncertainty in model representations of sea-ice impact on larger-scale climate. Under sea ice or within regions of recent sea-ice melt, the surface sea-ice cap or surface freshwater lens are expected to suppress mixing and submesoscale energetics. However, sea-ice melt and associated freshwater fluxes (Abernathey et al., 2016; Haumann et al., 2016; Pellichero et al., 2018) can also strengthen lateral buoyancy gradients, stimulating submesoscale flows.

High-resolution observations using autonomous platforms at the edge of the Antarctic sea ice have revealed sub-kilometre sized submesoscale fronts, hypothesised to be driven by the stirring of the fresher surface layer by mesoscale eddies, which remain active throughout the spring-summer melt season (Swart et al., 2020; Giddy et al., 2021). These upper-ocean flows are shown to be intricately coupled to surface winds: strong winds (storms) increase shear stresses, which drive mixing and reduce lateral buoyancy gradients (Swart et al., 2020). Scientific tags attached to elephant seals have been used to identify the presence of submesoscale flows under sea ice in the middle of winter: Biddle and Swart (2020) postulated that during sea-ice formation, brine-rejection-induced convective mixing can erode the halocline and introduce lateral buoyancy gradients that catalyse submesoscale motions. Similarly, the presence of leads (gaps between sea-ice floes; e.g., Fons and Kurtz, 2019) allows for lateral heterogeneity in ocean-atmosphere interaction that adjusts the surface buoyancy and stresses, promoting both convective and submesoscale-induced mixing (Fig. 12.2). Whether these processes are intermittent or quasi-permanent, they are likely to generate vertical fluxes of heat and salt (Thomas et al., 2008) that will alter the buoyancy and heat budget of the upper ocean, further adjusting sea-ice melt and formation rates. This is especially relevant in the Antarctic marginal ice zone, and especially at the Antarctic shelf break, where warm and salty Circumpolar Deep Water (CDW) lies closely under the surface mixed layer, increasing the potential for water mass transformation. Furthermore, recent observations in the Ross Gyre reveal the presence of double-diffusive staircase structures underlying the surface mixed layer and the potential for these processes to regulate sea-ice thickness (Bebieva and Speer, 2019), suggesting the feedback between sea ice and mixing in these systems. (See Chapter 11 for further discussion on double diffusion in ice-covered regimes.)

Flow-topography interactions occurring under sea ice can enhance upward heat transport and generate eddies that exert divergent strain on the sea-ice cover (see Kurtakoti et al., 2018) causing the sea-ice thickness to decrease, and on sporadic occasions initiating open-water polynyas (Fig. 12.2), such as those documented in the Weddell Sea (Gordon, 1978; Holland, 2001). This, together with intense winter storm conditions and a frigid overlying atmosphere, results in a loss of surface buoyancy, generating vigorous convective mixing and overturning (Wilson et al., 2019), with heat and saline waters (CDW) lying just below the cold surface layer (Gordon, 1978). Most recently, the Maud Rise polynya in 2016 and 2017 was fortuitously sampled by Argo floats deployed as part of the Southern Ocean Carbon and Climate Observations and Modelling (SOCCOM) project. These observations showed periods of absent vertical stratification (near-zero buoyancy frequency), reflecting recent mixing with the winter surface waters, as well as intermediate layer water-mass intrusions, resembling submesoscale coherent vortices at significant depth (Campbell et al., 2019; Swart et al., 2018b). In addition, the effect of thermobaricity (Akitomo, 1999) may have modulated the depth of mixing by supplying additional kinetic energy to plumes descending across the pycnocline. In combination, the processes occurring in polynyas short-circuit conventional pathways to water-mass transformation and ventilation by connecting the surface boundary layer with the abyssal ocean through exceptionally high rates of deep (4000 m) winter convection (Martinson and Iannuzzi, 1998).

12.4 Interior mixing: regional and mesoscale processes

12.4.1 Foundations: Southern Ocean eddy pathways

The characteristics of lateral mixing in the Southern Ocean are apparent from any snapshot of two key physical properties: variability in the ocean's sea surface height (SSH) distribution and the tilt of Southern Ocean isopycnals as they plunge from the base of the surface boundary layer into the ocean interior at the ACC's northern boundary. These two properties are intricately linked and a signature of the importance of the eddy variability that dominates interior mixing in the Southern Ocean. At all longitudes, SSH increases broadly from south to north across the ACC with a slope of a couple millimetres per kilometre, sufficient to support the strongest current in the global circulation system. Yet this SSH slope, which provides information on the surface circulation through geostrophic balance, is far from uniform across the ACC. In certain locations, SSH undergoes a few abrupt jumps from south to north, separated by flat regions hundreds of kilometres wide. In other regions, the meridional profile of SSH is evocative of a washboard, rising northward in a series of bumps and wiggles. Measurements made from satellite altimeters have been the primary tool for studying mesoscale mixing in the Southern Ocean (e.g., Hughes and Ash, 2001), revealing how contours of SSH meander and pinch off in coherent vortices, periodically compress together and fan out to accommodate the complex bathymetry below.

The spatial distribution of SSH is mirrored in an inverse sense by the south-to-north deepening of density surfaces across the ACC. Although isopycnal surfaces are more difficult to observe than SSH, hydrographic transects show that they often appear as a staircase with regions of risers and steps providing a signature of the sinuous jets that comprise the ACC system. This complex topology of the Southern Ocean SSH and density field provides insight into the various mechanisms by which properties are stirred and advected laterally and vertically across the Southern Ocean and will be the primary focus of this section.

Throughout the ocean, mesoscale dynamics are critical for the adiabatic, or along-isopycnal, stirring of tracers (see Chapter 9). Whereas in most regions adiabatic stirring is synonymous with lateral transport, mesoscale processes take on an added importance in the Southern Ocean. The isopycnal tilt across the ACC provides a large source of available potential energy (APE) that sustains the relatively high levels of eddy kinetic energy (EKE) in the Southern Ocean. This density structure and the associated baroclinic transport of the ACC appears remarkably insensitive to climate forcing due to the interplay between the mechanical source of APE from the wind stress and baroclinic instability that persistently converts APE to kinetic energy in the form of mesoscale eddies (Hallberg and Gnanadesikan, 2006; Marshall et al., 2017). Critically for the climate system, the sustained tilting of isopycnals provides a pathway for adiabatic advection and stirring to transport tracers vertically through the water column, and thus ventilate the deep ocean interior in the absence of diabatic mixing of surface and subsurface waters (see also the more extended discussions in Chapters 2 and 3).

There are two mechanisms by which spatially and temporally varying eddies may transport a tracer C across the ACC within a density class with a thickness h. A meridional isopycnal flux can be decomposed as follows:

$$\frac{\overline{vhC}}{\overline{h}} = \overline{v'h'}\frac{\overline{C}}{\overline{h}} + \overline{v'C'} + \frac{\overline{v'h'C'}}{\overline{h}}, \tag{12.1}$$

where v is the meridional velocity, overbars indicate a zonal and/or temporal average over scales larger than the characteristic eddy scale, primes indicate a deviation from the average, and we have assumed that $\overline{v} = 0$. The triple correlation term on the right-hand side of (12.1) is typically assumed to be small. The first term on the right-hand side arises from the correlation of velocity fluctuations and buoyancy anomalies, or equivalently thickness anomalies, that gives rise to a mass or volume transport (Marshall and Radko, 2003; Marshall and Speer, 2012). When combined with the background tracer distribution, this gives rise to an eddy advective flux, typically represented in coarse resolution numerical simulations by the Gent and McWilliams (1990) parameterisation. Note that this eddy thickness flux supports both upwelling and downwelling branches of the overturning circulation, which in steady state, is determined by the combined distribution of the surface buoyancy and the surface buoyancy flux fields (Marshall, 1997; Marshall and Radko, 2003; Abernathey et al., 2016). The second term on the right of (12.1) indicates that even when the overturning circulation is weak (e.g., wind and eddy overturning circulations balance), eddy stirring along density surfaces will homogenise tracer distributions via down-gradient tracer fluxes (Abernathey and Ferreira, 2015), most commonly parameterised as an eddy diffusivity (Redi, 1982).

An important and active research question asks whether these density layers act as adiabatic pathways or whether they are 'leaky'. Diabatic mixing is discussed in Section 12.5, and is a key component of the Southern Ocean circulation, especially where strong frontal circulations interact with topographic features. Nevertheless, mixing processes in the upper part of the ACC are typically assumed to be largely adiabatic, although some studies have questioned this assumption (Sun et al., 2018). The case for weak modification in the upper ACC is provided by Nikurashin and Vallis (2011), who argued—from a scaling analysis—that the residence time of a fluid parcel as it traverses the ACC is short compared to the time scales associated with vertical displacement due to diapycnal mixing. In the following subsection, we explore the processes that enable tracers to spread along density surfaces. Though the discussion above emphasises a two-dimensional (depth-latitude) framework, the combination of a heterogeneous distribution of EKE with a strong zonal mean flow implies that mixing is an intermittent process, as water parcels trace out complicated three-dimensional trajectories spiralling both toward and away from Antarctica (Tamsitt et al., 2017).

12.4.2 Mixing and coherent structures: adiabatic recipes for Southern Ocean mixing

Progress on the dynamical properties of the ACC has taken advantage of its circumpolar nature, decomposing the flow into zonal averages and departures from this mean, as in (12.1) above. Strikingly, the distribution of EKE throughout the entire Southern Ocean is patchy and strongly influenced by the interaction of frontal currents with topographic features. Through the use of satellite altimetry data, key features of the EKE distribution have been studied (e.g., Gille and Kelly, 1996). The bathymetry of the Southern Ocean varies across a range of scales, but there are a discrete number of topographic features that enforce a significant deviation in the ACC's zonal flow. These include the Scotia Arc, the Southwest Indian Ridge, Kerguelen Plateau, Macquarie Ridge and Campbell Plateau, and the Pacific-Antarctic Ridge. At each of these locations, EKE is elevated in the lee of the topographic feature and suppressed in front of the feature (Witter and Chelton, 1998). The region of enhanced EKE typically extends 500 to 1000 km downstream of the topography, encompassing a region that is nearly an order of magnitude larger than a typical coherent mesoscale eddy (Lu and Speer, 2010). The mean flow in these elevated EKE regions is often associated with meridional displacements of the ACC streamlines (contours of SSH) that are largely steady, leading to their distinction as standing meanders that have been compared to atmospheric storm tracks (Abernathey and Cessi, 2014; Bischoff and Thompson, 2014; Chapman et al., 2015).

Explanations for the elevated EKE within standing meanders have focused on the role of baroclinic instability. Steering of the mean flow by bathymetry generates a standing meander that locally increases lateral buoyancy gradients in the vicinity of the ridge and provides an enhanced source of APE that feeds eddy growth (Fig. 12.3a; Abernathey and Cessi, 2014). Additionally, standing meanders can sustain a mean-flow configuration with a substantial meridional velocity component (Fig. 12.4). These non-zonal flows are susceptible to linear instabilities that grow faster than unstable modes of purely zonal flows (Smith, 2007). The distribution of EKE is also shaped by the propagation of the growing instability as it is advected downstream by the mean flow (Pierrehumbert, 1984; Bracco and Pedlosky, 2003). Enhancement of the lateral buoyancy gradients in and around standing meanders is a consequence of conservation of potential vorticity that squeezes the mean flow as it flows northward and over shallower bathymetry (Patmore et al., 2019). This structure not only increases the baroclinicity of the flow, but also results in strong horizontal velocity shear on the flanks of the frontal current that can lead to barotropically unstable flow configurations. Both Youngs et al. (2017) and Barthel et al. (2017) have argued that close to topographic features, barotropic or mixed barotropic-baroclinic instabilities can dominate over baroclinic instability. The distinction between EKE generated from these two different instabilities has implications for both the energy budget and the large-scale stratification of the ACC. For instance, the release of potential energy via baroclinic instability gives rise to eddy buoyancy fluxes that support the vertical transfer of momentum from the ocean surface to the seafloor via interfacial form stress (Treguier and McWilliams, 1990; Olbers et al., 2004) (Fig. 12.4), whereas barotropic instabilities are mediated by eddy momentum fluxes (Youngs et al., 2017). The localisation of eddy fluxes also impacts the closure of the Southern Ocean meridional overturning circulation. The large-scale balance between wind- and eddy-overturning cells that comprises the zonally averaged residual arises from localised regions where eddies dominate and isopycnal tilt

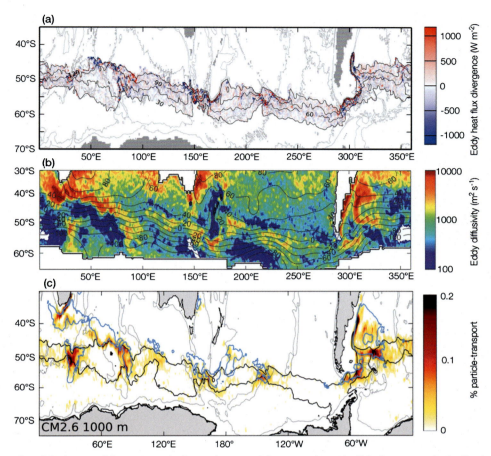

FIGURE 12.3 Examples of the impact of flow-topography interactions on mixing properties and adiabatic transport in the Southern Ocean. (a) The vertically integrated divergence of the transient eddy heat flux (W m^{-2}) in the ACC region, calculated from the Southern Ocean state estimate (SOSE). (b) Map of the eddy diffusivity [log10 (m^2 s^{-1})] for the full Southern Ocean. This estimate is derived from satellite altimetry-derived surface velocities and, through the use of both anomaly and time-mean sea surface heights measurements, includes an estimate of mean-flow suppression. This suppression is enhanced outside of standing meander regions. (c) Upwelling of passively advected particles across the 1000 m isobar from the CM2.6 numerical simulation. Upwelling, largely linked to lateral motion along tilted isopycnals, occurs preferentially in the lee of topographic features. Blue contours indicate regions of elevated eddy kinetic energy, and black contours bound the closed contours that pass through Drake Passage. Figure adapted from Abernathey and Cessi (2014); Ferrari and Nikurashin (2010); Tamsitt et al. (2017).

is relaxed, followed by regions where Ekman overturning dominates and isopycnals steepen again (Thompson and Naveira Garabato, 2014). Finally, Mazloff et al. (2013) argued that a third component, separate from the wind/eddy cells, arises from the outcropping of isopycnals at the sea surface and bathymetric features. This outcropping allows for the development of time-mean pressure gradients and a mean geostrophic gradients, which in the presence of strong boundary diabatic processes, can make a substantial contribution to the total overturning. Mazloff et al. (2013) suggested that this mean component is indeed dominant following a large cancellation of the other terms; it would be of interest to revisit this analysis using a fully eddy-resolving simulation.

The shaping of the EKE distribution by flow–topography interactions highlights the finding that lateral transport across the ACC is an intermittent process. An understanding of spatial variations in Southern Ocean mixing properties arose, in part, to reconcile observed patchy EKE distributions with traditional methods of defining Southern Ocean fronts as circumpolar features (Orsi et al., 1995; Sokolov and Rintoul, 2009). Marshall et al. (2006) used altimetry-derived surface velocities to advect a passive tracer and arrived at a near-surface eddy diffusivity estimate using an 'effective diffusivity' approach that quantifies tracer-contour complexity. This study found an eddy diffusivity that varied between 500 and 2000 m^2 s^{-1} with the magnitude of the eddy diffusivity inversely proportional to the strength of interior potential vorticity gradients. This approach provided a distribution of eddy diffusivities as a function of streamline, or SSH contour, but did not allow for along-stream variations in mixing properties. Ferrari and Nikurashin (2010) updated this approach by showing that strong mean flows can suppress the rate of eddy mixing and reduce the magnitude of the eddy diffusivity. Qualitatively, Ferrari

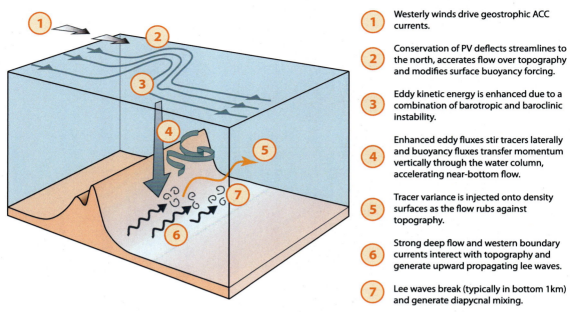

FIGURE 12.4 Schematic showing the pathway through which wind energy is transferred through the water column and converted into turbulent mixing through lee-wave generation and breaking. Arrow widths represent approximate relative power fluxes, with wind into geostrophic flow being roughly 1 TW globally. About half of all global lee-wave generation occurs in the Southern Ocean. Standing meanders are regions where isopycnal and diapycnal processes are closely linked as (adiabatic) eddy buoyancy fluxes, enhanced in the lee of topography, transfer momentum to the deep ocean.

and Nikurashin (2010) argued that a strong mean flow can effectively advect tracers downstream on timescales that are fast compared to the time needed to erode tracer gradients by the background eddy field. Quantitatively, this study provided an expression for the magnitude of the eddy diffusivity that accounted for the local magnitude of EKE and the background mean flow, both obtained from satellite altimetry observations, and was validated against the effective diffusivity approach. Their map of eddy diffusivities in the Southern Ocean revised earlier estimates (Keffer and Holloway, 1988; Stammer, 1998) to show that the suppression of the eddy diffusivity only occurs outside of standing meanders (Fig. 12.3b). Naveira Garabato et al. (2011) and Roach et al. (2016) observationally estimated the eddy diffusivity from repeat hydrographic section and Argo float trajectories, respectively, finding that in most regions, ACC jets suppress eddy diffusivities, but exceptions occur along transects near major bathymetric features: regions they labelled 'leaky jets'.

Progress on the quantification of an eddy diffusivity in the Southern Ocean coincided with a reevaluation of how to identify and classify Southern Ocean jets (Chapman, 2014). Though altimetric observations provided a comprehensive view of the multiple-jet structure of the ACC (Sokolov and Rintoul, 2009), these were still defined as circumpolar features, linked to contours of SSH. Thompson et al. (2010), using model output from a high-resolution ocean simulation, argued that near topographic features, coherent jets break down but re-emerge further downstream. The regions of jet rearrangement were collocated with the high EKE regions in the lee of topography. Both diagnosed eddy fluxes from idealised numerical simulations (MacCready and Rhines, 2001) as well as the advection of passive particles with satellite-derived surface velocities (Thompson and Sallée, 2012) offer evidence that standing meander regions are sites of enhanced meridional exchange. This finding has subsequently been supported by the analysis of cross-front transport of tracers and Lagrangian particles in high-resolution numerical models (Fig. 12.3c; Dufour et al., 2015; Tamsitt et al., 2017).

The previous discussion focused on mixing properties of the ACC, but mixing and stirring in the Southern Ocean's subpolar gyres and at the Antarctic margins are also critical as they regulate upwelling and downwelling branches of the overturning circulation's lower cell (Orsi et al., 1999; Gordon et al., 2001; Vernet et al., 2019). The two major Southern Hemisphere subpolar gyres, the Weddell and the Ross Gyres, are forced by a large-scale cyclonic wind-stress curl, and their interior dynamics follow Sverdrup balance, in which a poleward flow is closed by equatorward western boundary currents (Gordon et al., 1981). Yet, these gyres are also distinct from their low-latitude counterparts. First, the subpolar gyres are strongly constrained by topography along their northern and southern boundaries, not just to the east and west. Furthermore, these gyres participate in the global overturning circulation in ways that the subtropical and Northern Hemisphere subpolar gyres do not. In the gyres of the northern basins, most of the recirculation occurs above the thermocline and does not participate in the pole-to-pole overturning (Wolfe and Cessi, 2011). In contrast, the Southern Ocean subpolar gyres are the primary pathway by which volume and water mass properties are exchanged between the southern boundary of the ACC and Antarctica's continental shelf. The shelf is a site of strong water-mass transformation processes, including the formation of AABW, which fuels the ocean's deep overturning (see the following section). A new frontier in Southern Ocean research is the impact of mesoscale dynamics on both the long-term mean and variability of the subpolar gyres. Some of this work is influenced by Arctic studies: the Beaufort Gyre's mean density structure is set by eddy fluxes largely generated from boundary currents (Meneghello et al., 2017; Manucharyan and Spall, 2016). This process has also been highlighted in the Weddell Gyre as being key to setting gyre variability over seasonal to interannual timescales (Su et al., 2014). In Southern Ocean subpolar gyres, eddy generation is tightly linked to the dynamics of boundary currents over the continental shelf that also regulate heat transport toward Antarctica's floating ice shelves, although as the gyres' northern boundaries coincide with the ACC, lateral eddy fluxes are likely also critical for setting long-term properties (Roach and Speer, 2019).

The impact of gyre variability on the climate system relates to the ventilation of deep water masses, in particular in the Weddell Gyre, on interannual to seasonal time scales (Gordon et al., 2010). Variations in wind-stress curl have been hypothesised to vertically displace isopycnals in the gyre interior and modify the range of density classes that either incrop on topographic ridges to the north of the gyre or permit an adiabatic pathway into the southern ACC (Meredith et al., 2008). Su et al. (2014) introduced an idealised model of a subpolar gyre, borrowing ideas from the residual mean theory developed in zonal channels. Similar to zonally symmetric geometries, both mean (wind-driven) and eddy circulations contribute to the density structure of the gyre. In the gyre interior, isopycnals are displaced vertically by surface Ekman pumping, which leads to modifications in lateral density gradients and associated mesoscale eddy buoyancy fluxes near the gyre boundary. Su et al. (2014) tested various parameterisations of mesoscale eddy diffusivity, in particular accounting for mixing suppression by topographic slopes (Isachsen, 2011), and found that eddy variability can explain moored observations of phasing in temperature and salinity over the continental slope in response to a seasonally varying wind stress (Gordon et al., 2010; Dotto et al., 2018).

The transport of heat around the subpolar gyres also impacts the delivery of warm water toward the margins of Antarctica and ultimately to the continental shelf, where interactions with floating ice shelves and sea ice provide a closure of the overturning circulation. A key dynamical gateway for this poleward transport of heat is the continental shelf break, where topographic flow constraints over the shelf and slope combine with a strong frontal structure to establish the Antarctic Slope Front (ASF) and an associated along-slope current (Jacobs, 1991; Thompson et al., 2018). Traditionally, the exchange of heat and other tracers across the ASF was assumed to be dominated by Ekman transport in surface and bottom boundary layers (e.g., Gill, 1973), yet the advent of higher-resolution observations from ship-based, autonomous, and instrumented seal datasets revealed that (i) the ASF is a turbulent and variable circulation feature and (ii) eddy fluxes are critical for closing momentum, buoyancy, and tracer budgets. In perhaps the earliest study to argue for the importance of eddy mixing at the Antarctic margins, Nøst et al. (2011) used an insightful combination of observations and high-resolution numerical modelling to show that the inflow of warm CDW across the Eastern Weddell Sea continental shelf break is dominated by eddy transport. The interplay between wind and eddy overturning circulations was examined more explicitly in an idealised zonally symmetric modelling framework by Stewart and Thompson (2013), who highlighted the suppression of eddy mixing over the continental slope (Isachsen, 2011). This study, as well as a subsequent analysis of glider data in the northwestern Weddell Sea (Thompson et al., 2014), argued that the eddy-induced onshore transport of CDW was due to eddy thickness fluxes, $\overline{v'h'}$ (see (12.1)). The establishment of strong along-isopycnal temperature gradients due to the proximity of CDW and colder water masses on the shelf implies that eddy stirring can also contribute to the eddy heat flux. Indeed, Palóczy et al. (2018) found that stirring along-isopycnal surfaces dominated the heat transport across the continental slope in a 0.1° coupled ocean–sea-ice simulations. The importance of eddy transport is also supported by measurements from the West Antarctic Peninsula (WAP), which, due to its proximity to research stations and a LTER (Long-term ecological research) program, has been well observed. Here, mesoscale eddies have emerged as the most prominent mechanism of heat delivery to the WAP shelf (Dinniman et al., 2011; Martinson and McKee, 2012; Brearley et al., 2019), in particular at Marguerite Trough, where up to 50% of the lateral onshore heat transport of warm upper CDW to the WAP shelf was attributed to coherent eddies (Couto et al., 2017).

A consistent message from numerical simulations of the Antarctic margins is that resolution of scales smaller than the Rossby deformation radius on the shelf, typically less than 5 km, is required to fully represent eddy mixing and transport processes, (St-Laurent et al., 2013; Stewart and Thompson, 2015; Dinniman et al., 2016). Recent simulations, carried out over a short period of time, but with an extremely high 1/48° resolution, indicate that for almost all regions around Antarctica, the total transport of heat across the shelf break is well described by the eddy component, although this remarkably arises from a near-complete cancellation between large mean-flow and tidal contributions to the heat transport (Stewart et al., 2018). Tidal and eddy contributions were delimited by timescales shorter than and longer than 1 day, respectively. Though earlier studies have focused on wind forcing to explore modulation of mixing and eddy transport at the shelf break,

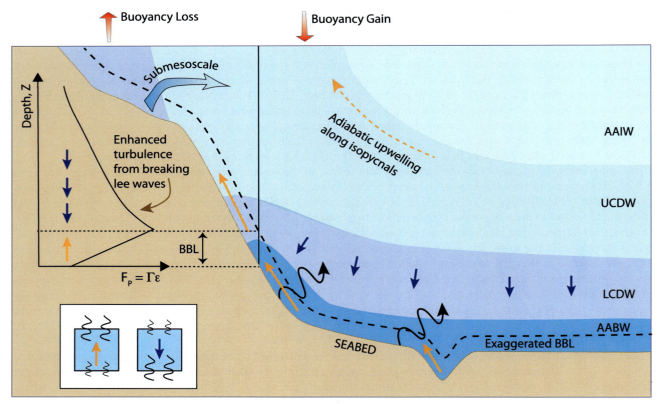

FIGURE 12.5 A depth-longitude section of the Southern Ocean, including the bottom-boundary layer (BBL, black dashed line). The BBL is shown with an exaggerated thickness to illustrate the effect of near-boundary mixing. Lee waves (black wiggly arrows) generated at rough topography form a region of enhanced turbulence and strong density flux, typically within 1 km of the seabed, as is apparent on the vertical turbulent density flux profile sketched on the left of the figure. The weakly stratified boundary layer is characterised by reduced stratification and density flux. Solid orange arrows indicate BBL diabatic along-boundary upwelling, and solid blue arrows show the diabatic sinking of waters in the ocean interior. Blue coloured bands represent water masses of Antarctic Intermediate Water (AAIW), Upper Circumpolar Deep Water (UCDW), Lower Circumpolar Deep Water (LCDW) and Antarctic Bottom Water (AABW), the boundaries of which delineate isopycnals (note they intersect the seabed at right-angles to satisfy the condition of zero buoyancy flux across the seafloor). Submesoscale flows can also catalyse near-boundary mixing, here represented by the blue shaded arrow transforming LCDW into lighter waters. Isopycnals of waters in the upper overturning cell (e.g., UCDW and AAIW) outcrop at the surface, allowing these lighter, adiabatically upwelled waters (orange dashed arrow) to undergo diabatic transformation by surface processes. The inset at bottom left illustrates how vertical variations in mixing modify a water parcel: stronger mixing on the upper face results in more buoyancy being gained from above as compared to buoyancy lost from below, a reduction of density, and upwelling of the parcel (and vice versa). Note that this schematic does not include AABW formation processes and the associated down-slope flow of bottom waters.

recent studies have highlighted how shelf processes can modify frontal properties at the shelf break and feed back on cross-slope transport (Hattermann, 2018; Daae et al., 2020).

12.5 Interior mixing: closing the budgets through turbulence at the smallest scales

12.5.1 Foundations

Isopycnals associated with the denser LCDW and AABW of the lower overturning branch (Fig. 12.1) do not outcrop in regions of positive surface buoyancy forcing. They instead rely on interior diabatic processes to converge buoyancy in the deep ocean, producing lighter waters that can upwell and return to the surface, as illustrated in Fig. 12.5 (Orsi et al., 1999, 2002; Naveira Garabato et al., 2007; Mashayek et al., 2015). Since AABW gradually flows northward, forming the deepest component of the global overturning circulation (e.g. Johnson, 2008), without cross-density exchanges, the global oceans would gradually fill up with cold dense waters, largely dissociated from the surface ocean and atmosphere. In this section, we outline the dominant processes responsible for the diabatic return of the bottom waters of the Southern Ocean, and we discuss the critical role that abyssal Southern Ocean mixing may play in controlling the overturning circulation, thereby influencing the global climate system (see also Chapter 2).

12.5.2 Recent findings: sub-surface diapycnal mixing pathways in the Southern Ocean

The dissipation of wind, tidal, and geothermal energy powers interior mixing and abyssal diapycnal upwelling in the ocean. In the Southern Ocean, the energy input from internal tides and geothermal heat fluxes is weak compared to that from standing internal waves, or lee waves (de Lavergne et al., 2016). Lee waves, generated by the interaction of bottom flow with seabed topography, transport energy from the large-scale, wind-driven circulation to turbulent scales when they break, giving rise to diapycnal mixing in the ocean interior (Fig. 12.4; see also Chapter 6; Bell, 1975; Thorpe, 2005; Cusack et al., 2017). Of the ~1 TW of wind power that is transferred into geostrophic ocean currents globally (Wunsch, 1998; von Storch et al., 2007), a significant fraction, estimated between 0.15 TW and 0.75 TW, is transmitted into lee waves (Nikurashin and Ferrari, 2011; Scott et al., 2011; Wright et al., 2014; Melet et al., 2014; Trossman et al., 2016; Yang et al., 2018). Lee-wave energy flux estimates are generated by applying linear lee-wave theory (Bell, 1975) to global distributions of topographic roughness, near-bottom stratification, and current speeds. Importantly, approximately half of the total global lee-wave energy flux occurs in the Southern Ocean, where the strong deep flow and intense eddy activity of the ACC interact with regions of rough topography (Gille and Kelly, 1996). Lee waves, characterised by short vertical scales, tend to break close to their generation sites and dissipate within 1 km of the topography, making this process an important source of abyssal mixing (see Chapter 6; Nikurashin and Ferrari, 2010b,a). More refined estimates of the proportion of lee-wave energy that dissipates locally is still an open question (Waterman et al., 2014; Kunze and Lien, 2019). The bottom-enhanced mixing signature associated with lee-wave breaking is evident in depth profiles of turbulent dissipation within energetic regions of the ACC (Fig. 12.6d; see also St. Laurent et al., 2013; Waterman et al., 2013; Sheen et al., 2013).

Southern Ocean abyssal mixing is by no means spatially or temporally homogeneous. Mixing hotspots are found in regions characterised by rough topography and intense mesoscale eddy flow, particularly at frontal locations: exactly the conditions where one might expect enhanced lee-wave generation. Further evidence of the role of lee-wave activity in generating abyssal mixing in the Southern Ocean is provided by the measurement of vigorous, upward-propagating, sub-inertial internal-wave energy at mixing hotspots (e.g., Waterman et al., 2013; Brearley et al., 2013; Sheen et al., 2013; Meyer et al., 2015, 2016). Such mixing hotspots include the Kerguelen Plateau, Drake Passage and the Scotia Sea (Naveira Garabato et al., 2004; Sloyan, 2005; St. Laurent et al., 2013; Waterman et al., 2013; Meyer et al., 2015; Sheen et al., 2013). Thus topographic features in the Southern Ocean shape surface-mixing properties (Section 12.3), cross-front transport in the interior (Section 12.4), as well as deep diabatic mixing.

In addition to lee waves having patchy spatial distributions, they also show temporally variable radiation and breaking, as indicated by mooring and other time-series data. This variability is due to changes in the strength and position of the background flow and to surface eddy activity over climatic timescales (Brearley et al., 2013; Sheen et al., 2014). Much of the observational evidence for the role of lee-wave-driven mixing in the Southern Ocean cited here is the output of two landmark field programs: the Southern Ocean FINEstructure (SOFINE) project and the Diapycnal and Isopycnal Mixing Experiment in the Southern Ocean (DIMES). See the segment of this chapter entitled DIMES case study box for an overview of the DIMES program.

The energy associated with breaking internal waves both produces heating through viscous dissipation ϵ, and changes the potential energy of the water column by mixing, represented by a turbulent buoyancy flux, $F_\rho = -(g/\rho_0)\overline{w'\rho'}$, where g is the acceleration due to gravity, ρ_0 is the mean density, and $\overline{w'\rho'}$ is the time-averaged correlation between the vertical velocity and density fluctuations. These two energy pathways are related by the mixing efficiency, $\Gamma = F_\rho/\epsilon$, which is generally taken to be a constant equal to 0.2 (Gregg et al., 2018). The diapycnal velocity induced by a buoyancy flux w^*, is related to both the vertical gradient of the turbulent buoyancy flux F_ρ, and the vertical stratification N^2, such that

$$w^* = -\frac{1}{N^2}\frac{\partial F_\rho}{\partial z} = -\frac{1}{N^2}\frac{\partial (\Gamma\epsilon)}{\partial z}. \tag{12.2}$$

Thus for a constant mixing efficiency, a dissipation ϵ profile that increases/decreases with depth acts to drive transport toward higher/lower densities (inset in Fig. 12.5; see also e.g. Polzin et al., 1997; St. Laurent et al., 2001; McDougall and Ferrari, 2017). On initial inspection, this relationship would suggest that the characteristic Southern Ocean bottom-intensified viscous dissipation associated with lee-wave generation and breaking would result in an overall cross-density downwelling, rather than lightening LCDWs and bottom waters, as required to close the overturning circulation. One approach to resolving this apparent inconsistency is to assume that the mixing efficiency Γ, decays to zero close to the ocean floor to satisfy a no density-flux boundary condition, $F_\rho = 0$ (Polzin et al., 1997; St. Laurent et al., 2001; Ferrari et al., 2016; de Lavergne et al., 2016). Although turbulent dissipation ϵ, may be enhanced near the seabed, due to the weak stratification, here there can be little density flux, i.e., you cannot mix water that is already well-mixed. The result is a narrow turbulent boundary layer, whereby $\Gamma\epsilon$ decreases with depth. The associated diapycnal flow w^* will act parallel to

314 Ocean Mixing

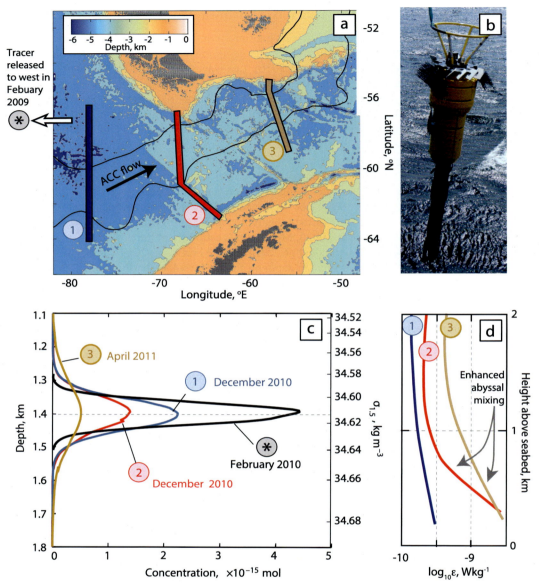

FIGURE 12.6 The DIMES field program. (a) The DIMES study region with location of three of the tracer and microstructure measurement sections marked: (1) Section at 78 W December 2010; (2) Western Drake Passage section December 2010; (3) Eastern Drake Passage section, April 2011. Background contours show bathymetry: note smooth seabed in the southeast Pacific and contrasting rough topography within Drake Passage. Black lines show rough positions of the Sub-Antarctic Front and the Polar Front. (b) Vertical microstructure profiler (VMP) being deployed in Drake Passage. Photograph from K. Sheen. (c) Mean depth profiles of tracer concentrations obtained from the three sections shows in panel a, plus a survey conducted in the southeast Pacific, 1 year after tracer release (black line). Figure adapted from Watson et al. (2014). (d) Representative profiles of section-averaged turbulent dissipation rates as measured by the VMP, based on data presented in Sheen et al. (2013). Note the enhanced mixing within the bottom 1 km. Panels c and d demonstrate the increased diapycnal mixing within the southwest Atlantic compared to the southeast Pacific.

the boundary (since isopycnals will intersect the ocean floor at right angles due to the no-flux boundary condition), and water within the bottom-boundary layer will cross to lighter density layers and upwell (Fig. 12.5). To assess the net abyssal diapycnal upwelling over basin scales, one must consider the shape of the ocean floor: the area of interaction between the seafloor and water of a particular density will determine the overall turbulent buoyancy flux experienced (de Lavergne et al., 2017; Holmes et al., 2018). Ferrari et al. (2016), for example, show that in numerical models the diapycnal upwelling of LCDW and AABW preferentially occurs where the associated density layers intersect with continental margins and abyssal ridges. The resultant boundary-layer upwelling is found to be sufficient to overcome the diapycnal downwelling within the stratified ocean interior, and hence close the abyssal overturning circulation. Parameterised mixing calculations by de Lavergne et al. (2016) support these results, reporting that lee waves in the Southern Ocean drive both upwelling of the bottom-most waters along topographic features and widespread downwelling in overlying water masses. Mashayek et al. (2017b) also challenged the assumption of a constant mixing efficiency above the bottom-boundary layer, demonstrating that potential variations in Γ result in a leading order difference to the abyssal branch of the overturning circulation. Despite recent progress in assessing the processes that drive upwelling in the Southern Ocean abyss, bottom-boundary layer observations are scarce, and theories are still very much actively developing. We refer the reader to Chapter 7 for a comprehensive discussion of the current thinking on mixing at the ocean's bottom boundary.

Lee-wave-enhanced, near-boundary dissipation within the Southern Ocean also influences tracer distributions. The overall mixing experienced by a tracer is dependent not only on the geographic distribution of mixing, but also the proportion of time spent in regions of varying mixing intensities. By considering lateral advection, stirring, and topographic steering of a tracer field, Mashayek et al. (2017a) showed that the tracer residency time spent in topographic mixing hotspots is enhanced in the Southern Ocean. This result explained the seemingly low values of observed 'snapshot' measurements of diapycnal mixing when compared to the time-averaged mixing experienced by a tracer. (See the DIMES case study box for further details.)

In addition to bottom-enhanced lee-wave breaking, several other mechanisms have been reported as important contributors to subsurface diabatic motion in the Southern Ocean. These mechanisms include tidal boundary mixing and entrainment into plumes (Silvester et al., 2014; Orsi et al., 2002; Daae et al., 2019). Mixing can also be catalysed by other near-boundary processes, such as topographically generated submesoscale flows, as found by Ruan et al. (2017) and later linked to enhanced turbulent dissipation by Naveira Garabato et al. (2019), along the Antarctic continental slope, resulting in the transformation of LCDW into lighter water masses (see also Chapter 8). Klymak (2018) and Klymak et al. (2021) also highlighted the potential importance of turbulence generated by near-bottom flow over topographic scales too large to emit propagating internal waves. The blocking, steering, and hydraulic effects associated with large topographic obstacles could represent a dominant pathway for the removal and dissipation of energy from the mean flow in the Southern Ocean. The inhomogeneous nature of interior mixing associated with large-amplitude topography (generally in the lee of large obstacles) presents a challenge to observational sampling.

In summary, a new picture is emerging of a complex pattern of abyssal mixing and tracer distribution within the Southern Ocean, set by the geography of both the stirring and steering of eddies and currents, and the topographically induced upwelling and interior downwelling in relation to water-mass distributions. In particular, the processes outlined above could indicate a more vigorous connection or 'short-circuiting' between the upper and lower branches of the overturning circulation than previously thought (Naveira Garabato et al., 2007; McDougall and Ferrari, 2017; Mashayek et al., 2017b,a).

12.6 Grand challenges

Targeted field campaigns and modelling studies of the past decade have substantially increased our understanding of turbulent processes in the Southern Ocean and of their impact on the Earth system as a whole. Despite recent advances, the processes that control the strength and pathways of the abyssal overturning, as well as the exchange between the upper and lower branches, remain unclear. Such knowledge gaps highlight the need for accurate mapping of both the spatial and the temporal variability of adiabatic mixing, and its driving processes, within the Southern Ocean. Field programs, such as DIMES, have been game-changers in advancing our understanding of Southern Ocean physics, both in terms of acquiring new measurements and in informing modelling studies (e.g., Tulloch et al., 2014; Chen et al., 2015; Ferrari et al., 2016). However, the remote, harsh conditions associated with the Southern Ocean, particularly during the winter months, make shipboard measurements challenging. Emerging technologies, such as autonomous instrumentation, gliders and floats, along with international collaborative programs will likely play a vital role in better constraining Southern Ocean mixing (Meredith et al., 2013).

Here we identify a few priorities for future research directions, targeting specific processes, i.e., upper-ocean turbulence and water-mass transformation, along-isopycnal adiabatic mixing, and interior mixing, and opportunities for probing links between these mixing processes to assess how the components of the system work together.

At the ocean surface, challenges centre around sampling small-scale and intermittent processes. This includes finding ways to sample regional variations throughout the year to probe how Southern Ocean basin asymmetries and latitudinal differences (Sallée et al., 2010) contribute to spatial variations in ocean mixing. Key processes that require further study include the impact of waves and Langmuir circulation on air-sea exchange as well as mechanisms that govern property exchange between the mixed layer and the thermocline. In particular, GCM simulations typically have mixed layers that are shallower, fresher, and warmer than observations, resulting in a stronger density contrast between the mixed layer and the thermocline below that is likely to impede vertical exchange (Belcher et al., 2012; Sallée et al., 2013). The marginal ice zone presents its own challenges, and there is a specific need to explore the impact of sea ice on the upper ocean and to obtain direct mixing estimates within and under sea ice. Observations are necessary to improve and validate the representation of these processes in climate models, which have traditionally had significant biases in polar regions (Heuzé et al., 2013).

Deep-water-mass conversion, especially along the bottom boundary, presents challenges that should receive as much attention as surface boundary layer turbulence. One challenge is that mixing associated with waves that break far away from their generation site, such as internal tides, makes tracking energy dissipation pathways difficult (Oka and Niwa, 2013; de Lavergne et al., 2016; Meyer et al., 2016). The recent work by Ferrari et al. (2016) suggests that a full picture of the abyssal circulation in the Southern Ocean will require detailed observations of turbulent boundary-layer fluxes in the ocean abyss. Autonomous vehicles have already proven useful in identifying active submesoscale flows (Ruan et al., 2017) and enhanced diapycnal mixing (Naveira Garabato et al., 2019) over sloping topography in the Southern Ocean. The challenge to better understand Southern Ocean abyssal mixing requires a multi-pronged approach, including more observations, improved theory of mixing parameterisations and further modelling studies, and tapping into the specific mechanisms and insights detailed in other chapters of this book.

Improved observations, theory, and models will feed into a more accurate mapping of the spatial patchiness and temporal evolution of interior and abyssal mixing in the Southern Ocean. Such developments will be key to determining the intricate water-mass transport pathways within our oceans and the strength of the abyssal overturning. In particular, setting the balance between topographically focused diapycnal upwelling and interior sinking will have implications for the diapycnal exchange of tracers and, in turn, ocean ventilation (Holmes et al., 2019), carbon storage, and marine biogeochemical cycling. Indeed, model outputs are highly sensitive to differing spatial distributions of Southern Ocean wave-breaking energy with radically different implications for ocean circulation (Simmons et al., 2004; Saenko et al., 2012; Melet et al., 2014; Mashayek et al., 2015, 2017a).

A final challenge is to develop a physical understanding of mixing that acknowledges the links and feedbacks between different mixing processes. Here, a few examples highlight contributions that build these connections.

- Naveira Garabato et al. (2016) used a small number of direct microstructure profiles to estimate the rates of both isopycnal and diapycnal (or isoneutral and dianeutral) mixing across the ACC as well as the magnitude of the overturning circulation. This study emphasised the sensitive link between small-scale mixing and the large-scale overturning that will have a significant impact on climate evolution in response to warming.
- Stewart and Hogg (2017) used an idealised numerical model with a topographic ridge to explore a similar link between smaller and larger scales. They showed that modification in AABW production on the continental shelf influences the outflow of this dense water across the ACC, which in turn modifies the density structure of the deep western boundary current on the downstream side of the topographic ridge, and influences eddy fluxes. This change has a signature in the SSH field, suggesting the potential for monitoring near-coast changes via altimetry in the future.
- Finally, Dufour et al. (2017) completed centennial-scale global simulations under preindustrial forcing at two resolutions. The high-resolution (0.1°) simulation produced intermittent polynyas in the Weddell Sea, whereas the coarser resolution simulation did not.

All of these studies offer insight into the sensitive balance between small-scale processes and large-scale circulations in the Southern Ocean, and they suggest approaches that should be further explored in future research. Understanding these interactions over decadal-to-centennial timescales is a critical challenge for the accurate prediction of future climate. Improvements in physical understanding and model representations of Southern Ocean mixing will directly impact the fidelity of projections for future ocean heat and carbon storage as well as the structure and strength of the MOC in different climate regimes.

DIMES case study box

Foundations

DIMES (The Diapycnal and Isopycnal Mixing Experiment in the Southern Ocean) was an extensive UK/US field program aimed at quantifying both the cross-density (diapycnal) and along density (isopycnal) mixing in the Southern Ocean. The project's motivation stemmed from a severe lack of *in situ* observations of Southern Ocean mixing and its driving processes, crucial to understanding the strength and structure of both the upper and lower limbs of the MOC. DIMES was located in the eastern Pacific and western Atlantic sectors of the ACC, providing a comparison of the mixing in two contrasting regions: the more quiescent, smooth-bottomed southeast Pacific; and the energetic Drake Passage and Scotia Sea, characterised by vigorous mesoscale eddy flow and rough topography (Fig. 12.6a).

The backbone of the DIMES program was a tracer release experiment. A parcel of inert, non-natural chemical CF_3SF_5 was released along a density surface at a depth of 1300 m in the southeast Pacific in 2009. Following the release, annual research cruises between 2009 and 2013 were scheduled to track the advection of the tracer as it progressed downstream through Drake Passage and the Scotia Sea as well as its subsequent dispersal both along and across density surfaces. The cross-density spread of the tracer was measured at multiple locations during each cruise, from which the mean diapycnal diffusivity experienced along the tracer path since its release could be computed (Ledwell et al., 2011; Watson et al., 2014). A total of 210 isopycnal-following "RAFOS" floats, tracked using acoustic sound sources, were deployed to evaluate along-isopycnal stirring and mixing processes (LaCasce et al., 2014; Balwada et al., 2016). Vertical profiles of 'on the spot' turbulent dissipation rates ϵ were also collected using microstructure profilers (Fig. 12.6b; Chapter 14). Microstructure profilers capture high-resolution (cm scale) vertical measurements of velocity shear, from which spectra can be computed and directly compared with turbulent theory via the Batchelor (1959) spectrum to extract dissipation rates. In DIMES, microstructure turbulence observations were complemented with collocated measurements of the kinetic and potential energy of internal waves. Internal-wave energy was extracted from velocity shear and strain spectra, computed from finescale vertical measurements, $O(1-100 m)$, of horizontal current flow, temperature, and salinity, as recorded by LADCP and CTD instruments. Moreover the direction of lee-wave propagation and characteristic frequency content can be deduced from such data (e.g., Waterman et al., 2014). Finescale measurements are too coarse to capture small-scale turbulent motions directly; however, ϵ may be derived from internal-wave energy spectra using finescale parameterisations (Polzin et al., 1995). Although approximate, the finescale parameterisation method is powerful as it enables cross-density mixing to be deduced from plentiful (and historical) hydrographic data as collected from CTDs (as opposed to scarce and difficult to obtain direct microstructure observations). In addition to tracer, floats, and microstructure, the DIMES experiment included a plethora of other observation platforms, including a two-year mooring array in Drake Passage (Brearley et al., 2013), EM-APEX floats (Kilbourne and Girton, 2015), and targeted deployment of eXpendable CTDs to assess turbulence proxies (Frants et al., 2013).

Key results

The DIMES data have fed into a range of both observational and modelling studies, many of which have been outlined in this chapter. Here we cover the key findings in relation to subsurface diapycnal and isopycnal mixing observational studies. Both tracer release and direct turbulence measurements from microstructure profiles showed an order of magnitude increase (estimates of turbulent dissipation ϵ ranged from 10^{-10} to 10^{-9} W kg^{-1}) in deep mixing between the southeast Pacific and southwest Atlantic (Fig. 12.6c,d; St. Laurent et al., 2013; Watson et al., 2014; Sheen et al., 2013). These results support the dominant role that topographically generated lee waves play in driving turbulent dissipation in the Southern Ocean. In addition, higher mixing rates were found to be associated with enhanced bottom currents, such as at frontal locations (St. Laurent et al., 2013; Brearley et al., 2013; Sheen et al., 2013). Mixing hotspots typically displayed enhanced turbulent dissipation within the bottom 1 km of the ocean, a characteristic signal of the topographic generation and breaking of internal waves (Fig. 12.6d). Mixing rates were also found to be modified by interior mesoscale features (Sheen et al., 2015).

Finescale measurements indicated that regions of enhanced mixing were associated with more energetic, relatively high-frequency internal waves that propagate away from the seafloor. In comparison, the upper-ocean internal-wave field was composed of downward propagating, near-inertial energy, characteristic of surface generated wind-driven waves (Sheen et al., 2013; Brearley et al., 2013). The DIMES data also supported other studies in that although lee-wave energy flux calculations and finescale estimations of mixing are generally consistent with direct turbulent measurements (see Chapter 6), they can over-predict the mixing measured directly with microstructure profilers under certain conditions, and hence parameterisation theory needs refinement (Frants et al., 2013; Waterman et al., 2013, 2014; Sheen et al., 2013; Takahashi and Hibiya, 2019).

DIMES was unprecedented in the number of microstructure profiles that were collected in the Southern Ocean (>75). However, measurements still fell short of capturing the full picture of abyssal mixing due to its spatial and temporal variability. Finescale parameterisations help to fill in the sampling gap. For example, the DIMES mooring array, which was deployed for a year in Drake Passage, and the repeat Drake Passage hydrographic section SR1B, which has been sampled annually since 1993, were used to show that Southern Ocean abyssal mixing is dependent on surface eddy energy, and hence potentially climatic perturbations in wind forcing (Sheen et al., 2014). Another approach to fill observational data gaps of abyssal mixing is through numerical model runs, informed by observations. Mashayek et al. (2017a) used a high-resolution ocean model to resolve an apparent inconsistency uncovered by DIMES: the cross-density mixing required for the observed vertical spreading of the DIMES tracer was an order of magnitude higher than that recorded by microstructure measurements of mixing at the mean tracer depth in regions of rough topography. The model showed that efficient tracer stirring and the long residency time spent by the tracer in stratified, near-boundary strong-mixing hotspots explained the large diapycnal diffusivities experienced by the tracer. It was found that geostrophic eddies and the mean flow tend to stir and trap the tracer near topography. The impact of uneven residency times in regions of high mixing on the tracer spread could not have been captured by spot microstructure measurements.

DIMES isopycnal mixing studies explored the hypothesis that the effective lateral diffusivity is small in the core of the ACC and enhanced at the 'steering level', where the westward propagation of Rossby waves is balanced by the eastward advection of the ACC (Smith and Marshall, 2009). Modelling studies, completed early in the project, highlighted considerable nuance in the spatio-temporal patterns of eddy mixing (e.g., Griesel et al., 2009, 2010; Abernathey et al., 2010; Klocker et al., 2012). To assess the effect of the mean flow on the vertical structure of lateral diffusivity, floats were targeted to drift at two depths, ∼700 m and ∼1800 m, with the tracer near 1300 m. Float observations implied a large range of effective eddy diffusivity values in the southeast Pacific that ranged from 2800 ± 600 m^2 s^{-1} at 700-m depth to 990 ± 200 m^2 s^{-1} at 1800-m depth. Even lower values were associated with cross-ACC diffusivity estimates, which were suppressed at shallower levels (690 ± 150 m^2 s^{-1} at 700-m depth as compared to 1000 ± 200 m^2 s^{-1} at 1800-m depth), consistent with theoretical predictions (LaCasce et al., 2014; Balwada et al., 2016). Notably, this signal of a mid-depth maximum in cross-ACC eddy diffusivity emerged at timescales greater than 6 months, which is significantly longer than the integral timescale of $O(10$ days$)$. An assessment of the tracer spread suggested a comparable diffusivity: 710 ± 260 m^2 s^{-1} (Tulloch et al., 2014). Finally, in the more energetic Scotia Sea, the cross-ACC diffusivity was larger, 1200 ± 500 m^2 s^{-1} (Balwada et al., 2016).

In addition to estimating the mesoscale lateral eddy diffusivity, studies also characterised eddy stirring at scales smaller than the dominant eddy scales (∼100 km). A much smaller eddy diffusivity, roughly 20 m^2 s^{-1}, was estimated at scales of ∼10 km (Boland et al., 2015; Balwada et al., 2021). Assessing the physical processes that give rise to these diffusivity values, Zika et al. (2020) and Balwada et al. (2021) argued that the eddy stirring at mid-depth in this region of the ACC is likely non-local, such that the largest eddies are primarily responsible for stirring and filamenting the tracers (Garrett, 1983).

Acknowledgements

STG acknowledges support from the U.S. National Science Foundation (NSF Grants PLR-1425989, OCE-1658001, and OPP-1936222). KS was supported by the UK National Environmental Research Council (NERC) for her work on the DIMES project. SS was supported by the Swedish Research Council (VR 2019-04400) and a Wallenberg Academy Fellowship (WAF 2015.0186). AFT acknowledges support from the U.S. National Science Foundation (NSF Grants OCE-1756956, OPP-1644172, OPP-2023259). We are grateful to Kevin Speer, Dhruv Balwada, and an anonymous reviewer for comments that improved this chapter.

Short bios of authors

- Sarah T. Gille is a professor of physical oceanography at Scripps Institution of Oceanography, University of California, San Diego. Previously, she completed her PhD in the MIT/Woods Hole oceanographic institution joint program, carried out postdocs at Scripps Institution of Oceanography and at the University of East Anglia, and served on the faculty of the University of California Irvine.
- Katy L. Sheen is a senior lecturer in physical geography at the University of Exeter. Before taking up her current position, she completed a PhD at the University of Cambridge and a postdoc at the University of Southampton, and worked as a senior scientist at the UK Met Office Hadley Centre.
- Sebastiaan Swart is an associate professor of physical oceanography at the University of Gothenburg, Sweden, where he is currently a Wallenberg Academy Fellow. He also holds the title of Honorary Research Associate at the U. Cape Town, S. Africa. He received his PhD jointly from the University of Cape Town and the Université de Bretagne and was a postdoc and scientist at the South African Council for Scientific & Industrial Research.
- Andrew F. Thompson is a professor of Environmental Science & Engineering at the California Institute of Technology. He completed a PhD at Scripps Institution of Oceanography, University of California San Diego, and was a postdoc at the University of East Anglia, the University of Cambridge, and the British Antarctic Survey.

References

Abernathey, R., Cessi, P., 2014. Topographic enhancement of eddy efficiency in baroclinic equilibration. J. Phys. Oceanogr. 44, 2107–2126. https://doi.org/10.1175/JPO-D-14-0014.1.

Abernathey, R., Ferreira, D., 2015. Southern Ocean isopycnal mixing and ventilation changes driven by winds. Geophys. Res. Lett. 42, 10357–10365. https://doi.org/10.1002/2015GL066238.

Abernathey, R., Marshall, J., Mazloff, M., Shuckburgh, E., 2010. Enhancement of mesoscale eddy stirring at steering levels in the Southern Ocean. J. Phys. Oceanogr. 40, 170–184. https://doi.org/10.1175/2009JPO4201.1.

Abernathey, R.P., Cerovecki, I., Holland, P.R., Newsom, E., Mazloff, M., Talley, L.D., 2016. Water-mass transformation by sea ice in the upper branch of the Southern Ocean overturning. Nat. Geosci. 9, 596–601. https://doi.org/10.1038/ngeo2749.

Adams, K.A., Hosegood, P., Taylor, J.R., Sallée, J.B., Bachman, S., Torres, R., Stamper, M., 2017. Frontal circulation and submesoscale variability during the formation of a Southern Ocean mesoscale eddy. J. Phys. Oceanogr. 47, 1737–1753. https://doi.org/10.1175/JPO-D-16-0266.1.

Akitomo, K., 1999. Open-ocean deep convection due to thermobaricity: 1. Scaling argument. J. Geophys. Res. 104, 5225–5234. https://doi.org/10.1029/1998JC900058.

Alford, M.H., MacKinnon, J.A., Simmons, H.L., Nash, J.D., 2016. Near-inertial internal gravity waves in the ocean. Annu. Rev. Mar. Sci. 8, 95–123. https://doi.org/10.1146/annurev-marine-010814-015746.

Armour, K.C., Marshall, J., Scott, J.R., Donohoe, A., Newsom, E.R., 2016. Southern Ocean warming delayed by circumpolar upwelling and equatorward transport. Nat. Geosci. 9, 549–554. https://doi.org/10.1038/ngeo2731.

Bachman, S.D., Klocker, A., 2020. Interaction of jets and submesoscale dynamics leads to rapid ocean ventilation. J. Phys. Oceanogr. 50, 2873–2883. https://doi.org/10.1175/JPO-D-20-0117.1.

Bachman, S.D., Taylor, J.R., Adams, K.A., Hosegood, P.J., 2017. Mesoscale and submesoscale effects on mixed layer depth in the Southern Ocean. J. Phys. Oceanogr. 47, 2173–2188. https://doi.org/10.1175/JPO-D-17-0034.1.

Balwada, D., LaCasce, J.H., Speer, K.G., Ferrari, R., 2021. Relative dispersion in the Antarctic Circumpolar Current. J. Phys. Oceanogr. 51, 553–574. https://doi.org/10.1175/JPO-D-19-0243.1.

Balwada, D., Smith, K.S., Abernathey, R., 2018. Submesoscale vertical velocities enhance tracer subduction in an idealized Antarctic Circumpolar Current. Geophys. Res. Lett. 45, 9790–9802. https://doi.org/10.1029/2018GL079244.

Balwada, D., Speer, K.G., LaCasce, J.H., Owens, W.B., Marshall, J., Ferrari, R., 2016. Circulation and stirring in the southeast Pacific Ocean and the Scotia Sea sectors of the Antarctic Circumpolar Current. J. Phys. Oceanogr. 46, 2005–2027. https://doi.org/10.1175/JPO-D-15-0207.1.

Barthel, A., Hogg, A.M., Waterman, S., Keating, S., 2017. Jet-topography interactions affect energy pathways to the deep Southern Ocean. J. Phys. Oceanogr. 47, 1799–1816. https://doi.org/10.1175/JPO-D-16-0220.1.

Batchelor, G.K., 1959. Small-scale variation of convected quantities like temperature in turbulent fluid Part 1. General discussion and the case of small conductivity. J. Fluid Mech. 5, 113–133. https://doi.org/10.1017/S002211205900009X.

Bebieva, Y., Speer, K., 2019. The regulation of sea ice thickness by double-diffusive processes in the Ross Gyre. J. Geophys. Res. 184, 7068–7081. https://doi.org/10.1029/2019JC015247.

Belcher, S.E., Grant, A.L.M., Hanley, K.E., Fox-Kemper, B., Van Roekel, L., Sullivan, P.P., Large, W.G., Brown, A., Hines, A., Calvert, D., Rutgersson, A., Pettersson, H., Bidlot, J.R., Janssen, P.A.E.M., Polton, J.A., 2012. A global perspective on Langmuir turbulence in the ocean surface boundary layer. Geophys. Res. Lett. 39. https://doi.org/10.1029/2012GL052932.

Bell Jr., T.H., 1975. Topographically generated internal waves in the open ocean. J. Geophys. Res. 80, 320–327. https://doi.org/10.1029/JC080i003p00320.

Biddle, L.C., Swart, S., 2020. The observed seasonal cycle of submesoscale processes in the Antarctic Marginal Ice Zone. J. Geophys. Res., Oceans 125, e2019JC015587. https://doi.org/10.1029/2019JC015587.

Bischoff, T., Thompson, A.F., 2014. Configuration of a Southern Ocean storm track. J. Phys. Oceanogr. 44, 3072–3078. https://doi.org/10.1175/JPO-D-14-0062.1.

Boé, J., Hall, A., Qu, X., 2009. Deep ocean heat uptake as a major source of spread in transient climate change simulations. Geophys. Res. Lett. 36, l22791. https://doi.org/10.1029/2009GL040845.

Boland, E.J.D., Shuckburgh, E., Haynes, P.H., Ledwell, J.R., Messias, M.J., Watson, A.J., 2015. Estimating a submesoscale diffusivity using a roughness measure applied to a tracer release experiment in the Southern Ocean. J. Phys. Oceanogr. 45, 247–257. https://doi.org/10.1175/JPO-D-14-0047.1.

Bracco, A., Pedlosky, J., 2003. Vortex generation by topography in locally unstable baroclinic flows. J. Phys. Oceanogr. 33, 207–219. https://doi.org/10.1175/1520-0485(2003)033<0207:VGBTIL>2.0.CO;2.

Brearley, J.A., Moffat, C., Venables, H.J., Meredith, M.P., Dinniman, M.S., 2019. The role of eddies and topography in the export of shelf waters from the West Antarctica Peninsula shelf. J. Geophys. Res. 124, 7718–7742. https://doi.org/10.1029/2018JC014679.

Brearley, J.A., Sheen, K.L., Naveira Garabato, A.C., Smeed, D.A., Waterman, S., 2013. Eddy-induced modulation of turbulent dissipation over rough topography in the Southern Ocean. J. Phys. Oceanogr. 43, 2288–2308. https://doi.org/10.1175/JPO-D-12-0222.1.

Caldeira, K., Duffy, P.B., 2000. The role of the Southern Ocean in uptake and storage of anthropogenic carbon dioxide. Science 287, 620–622. https://doi.org/10.1126/science.287.5453.620.

Campbell, E.C., Wilson, E.A., Moore, G.W.K., Riser, S.C., Brayton, C.E., Mazloff, M.R., Talley, L.D., 2019. Antarctic offshore polynyas linked to Southern Hemisphere climate anomalies. Nature 570, 319–325. https://doi.org/10.1038/s41586-019-1294-0.

Carranza, M.M., Gille, S.T., 2015. Southern Ocean wind-driven entrainment enhances satellite chlorophyll-a through the summer. J. Geophys. Res., Oceans 120, 304–323. https://doi.org/10.1002/2014JC010203.

Chapman, C.C., 2014. Southern Ocean jets and how to find them: improving and comparing common jet detection methods. J. Geophys. Res., Oceans 119, 4318–4339. https://doi.org/10.1002/2014JC009810.

Chapman, C.C., Hogg, A.M., Kiss, A.E., Rintoul, S.R., 2015. The dynamics of Southern Ocean storm tracks. J. Phys. Oceanogr. 45, 884–903. https://doi.org/10.1175/JPO-D-14-0075.1.

Chen, R., Gille, S.T., McClean, J.L., Flierl, G.R., Griesel, A., 2015. A multiwavenumber theory for eddy diffusivities and its application to the southeast Pacific (DIMES) region. J. Phys. Oceanogr. 45, 1877–1896. https://doi.org/10.1175/JPO-D-14-0229.1.

Chereskin, T.K., Donohue, K.A., Watts, D.R., Tracey, K.L., Firing, Y.L., Cutting, A.L., 2009. Strong bottom currents and cyclogenesis in Drake Passage. Geophys. Res. Lett. 36, L23602. https://doi.org/10.1029/2009GL040940.

Cisewski, B., Strass, V., Prandke, H., 2005. Upper-ocean vertical mixing in the Antarctic Polar Front Zone. Deep-Sea Res., Part 2, Top. Stud. Oceanogr. 52, 1087–1108. https://doi.org/10.1016/j.dsr2.2005.01.010.

Couto, N., Martinson, D.G., Kohut, J., Schofield, O., 2017. Distribution of Upper Circumpolar Deep Water on the warming continental shelf of the West Antarctic Peninsula. J. Geophys. Res., Oceans 122, 5306–5315. https://doi.org/10.1002/2017JC012840.

Cusack, J.M., Naveira Garabato, A.C., Smeed, D.A., Girton, J.B., 2017. Observation of a large lee wave in the Drake Passage. J. Phys. Oceanogr. 47, 793–810. https://doi.org/10.1175/JPO-D-16-0153.1.

Daae, K., Fer, I., Darelius, E., 2019. Variability and mixing of the Filchner Overflow Plume on the continental slope, Weddell Sea. J. Phys. Oceanogr. 49, 3–20. https://doi.org/10.1175/JPO-D-18-0093.1.

Daae, K., Hattermann, T., Darelius, E., Mueller, R., Naughten, K.A., Timmermann, R., Hellmer, H.H., 2020. Necessary conditions for warm inflow towards the Filchner Ice Shelf, Weddell Sea. Geophys. Res. Lett. 47, e2020GL089237. https://doi.org/10.1029/2020GL089237.

Danabasoglu, G., McWilliams, J.C., 1995. Sensitivity of the global ocean circulation to parameterizations of mesoscale tracer transports. J. Climate 8, 2967–2987. https://doi.org/10.1175/1520-0442(1995)008<2967:SOTGOC>2.0.CO;2.

Danabasoglu, G., McWilliams, J.C., Gent, P.R., 1994. The role of mesoscale tracer transports in the global ocean circulation. Science 264, 1123–1126. https://doi.org/10.1126/science.264.5162.1123.

deSzoeke, R.A., Levine, M.D., 1981. The advective flux of heat by mean geostrophic motions in the Southern Ocean. Deep-Sea Res. 28, 1057–1085. https://doi.org/10.1016/0198-0149(81)90048-0.

DeVries, T., Primeau, F., 2011. Dynamically and observationally constrained estimates of water-mass distributions and ages in the global ocean. J. Phys. Oceanogr. 41, 2381–2401. https://doi.org/10.1175/JPO-D-10-05011.1.

Dinniman, M.S., Asay-Davis, X.S., Galton-Fenzi, B.K., Holland, P.R., Jenkins, A., Timmermann, R., 2016. Modeling ice shelf/ocean interaction in Antarctica: a review. Oceanography 29, 144–153. https://doi.org/10.5670/oceanog.2016.106.

Dinniman, M.S., Klinck, J.M., Smith Jr., W.O., 2011. A model study of Circumpolar Deep Water on the West Antarctic Peninsula and Ross Sea continental shelves. Deep-Sea Res., Part 2 58, 1508–1523. https://doi.org/10.5670/oceanog.2016.106.

Dong, S., Sprintall, J., Gille, S.T., Talley, L., 2008. Southern Ocean mixed-layer depth from Argo float profiles. J. Geophys. Res. 113, C06013. https://doi.org/10.1029/2006JC004051.

Dotto, T.S., Naveira Garabato, A., Bacon, S., Tsamados, M., Holland, P.R., Hooley, J., Frajka-Williams, E., Ridout, A., Meredith, M.P., 2018. Variability of the Ross Gyre, Southern Ocean: drivers and responses revealed by satellite altimetry. Geophys. Res. Lett. 45, 6195–6204. https://doi.org/10.1029/2018GL078607.

Dufour, C.O., Griffies, S.M., de Souza, G.F., Frenger, I., Morrison, A.K., Palter, J.B., Sarmiento, J.L., Galbrait, E.D., Dunne, J.P., Anderson, W.G., Slater, R.D., 2015. Role of mesoscale eddies in cross-frontal transport of heat and biogeochemical tracers in the Southern Ocean. J. Phys. Oceanogr. 45, 3057–3081. https://doi.org/10.1175/JPO-D-14-0240.1.

Dufour, C.O., Morrison, A.K., Griffies, S.M., Frenger, I., Zanowski, H., Winton, M., 2017. Preconditioning of the Weddell Sea polynya by the ocean mesoscale and dense water overflows. J. Climate 30, 7719–7737. https://doi.org/10.1175/JCLI-D-16-0586.1.

Erickson, Z.K., Thompson, A.F., Cassar, N., Sprintall, J., Mazloff, M.R., 2016. An advective mechanism for deep chlorophyll maxima formation in southern Drake Passage. Geophys. Res. Lett. 43, 10846–10855. https://doi.org/10.1002/2016GL070565.

Ferrari, R., Mashayek, A., McDougall, T.J., Nikurashin, M., Campin, J.M., 2016. Turning ocean mixing upside down. J. Phys. Oceanogr. 46, 2239–2261. https://doi.org/10.1175/JPO-D-15-0244.1.

Ferrari, R., Nikurashin, M., 2010. Suppression of eddy mixing diffusivity across jets in the Southern Ocean. J. Phys. Oceanogr. 40, 1501–1519. https://doi.org/10.1175/2010JPO4278.1.

Fons, S.W., Kurtz, N.T., 2019. Retrieval of snow freeboard of Antarctic sea ice using waveform fitting of CryoSat-2 returns. Cryosphere 13, 861–878. https://doi.org/10.5194/tc-13-861-2019.

Forryan, A., Naveira Garabato, A.C., Polzin, K.L., Waterman, S., 2015. Rapid injection of near-inertial shear into the stratified upper ocean at an Antarctic Circumpolar Current front. Geophys. Res. Lett. 42, 3431–3441. https://doi.org/10.1002/2015GL063494.

Fox-Kemper, B., Ferrari, R., Hallberg, R., 2008. Parameterization of mixed layer eddies. Part I: theory and diagnosis. J. Phys. Oceanogr. 38, 1145–1165. https://doi.org/10.1175/2007JPO3792.1.

Frants, M., Damerell, G.M., Gille, S.T., Heywood, K.J., MacKinnon, J., Sprintall, J., 2013. An assessment of density-based finescale methods for estimating diapycnal diffusivity in the Southern Ocean. J. Atmos. Ocean. Technol. 30, 2647–2661. https://doi.org/10.1175/JTECH-D-12-00241.1.

Fukamachi, Y., Rintoul, S.R., Church, J.A., Aoki, S., Sokolov, S., Rosenberg, M.A., Wakatsuchi, M., 2010. Strong export of Antarctic Bottom Water east of the Kerguelen Plateau. Nat. Geosci. 3, 327–331. https://doi.org/10.1038/ngeo842.

Garrett, C., 1983. On the initial streakness of a dispersing tracer in two-and three-dimensional turbulence. Dyn. Atmos. Ocean. 7, 265–277. https://doi.org/10.1016/0377-0265(83)90008-8.

Gent, P.R., 2016. Effects of Southern Hemisphere wind changes on the meridional overturning circulation in ocean models. Annu. Rev. Mar. Sci. 8, 79–94. https://doi.org/10.1146/annurev-marine-122414-033929.

Gent, P.R., McWilliams, J.C., 1990. Isopycnal mixing in ocean circulation models. J. Phys. Oceanogr. 20, 150–155. https://doi.org/10.1175/1520-0485(1990)020<0150:IMIOCM>2.0.CO;2.

Giddy, I., Swart, S., du Plessis, M.D., Thompson, A.F., Nicholson, S.A., 2021. Stirring of sea-ice meltwater enhances submesoscale fronts in the Southern Ocean. J. Geophys. Res., Oceans. https://doi.org/10.1029/2020JC016814.

Gill, A.E., 1973. Circulation and bottom water production in the Weddell Sea. Deep-Sea Res. 20, 111–140. https://doi.org/10.1016/0011-7471(73)90048-X.

Gille, S.T., 2003. Float observations of the Southern Ocean: Part 2, eddy fluxes. J. Phys. Oceanogr. 33, 1182–1196. https://doi.org/10.1175/1520-0485(2003)033<1182:FOOTSO>2.0.CO;2.

Gille, S.T., Kelly, K.A., 1996. Scales of spatial and temporal variability in the Southern Ocean. J. Geophys. Res. 101, 8759–8773. https://doi.org/10.1029/96JC00203.

Gille, S.T., McKee, D.C., Martinson, D.G., 2016. Temporal changes in the Antarctic Circumpolar Current: implications for the Antarctic continental shelves. Oceanography 29. https://doi.org/10.5670/oceanog.2016.102.

Gordon, A.L., 1978. Deep Antarctic convection west of Maud Rise. J. Phys. Oceanogr. 8, 600–612. https://doi.org/10.1175/1520-0485(1978)008<0600:DACWOM>2.0.CO;2.

Gordon, A.L., Huber, B., McKee, D., Visbeck, M., 2010. A seasonal cycle in the export of bottom water from the Weddell Sea. Nat. Geosci. 3, 551–556. https://doi.org/10.1038/ngeo916.

Gordon, A.L., Martinson, D.G., Taylor, H.W., 1981. The wind-driven circulation in the Weddell-Enderby Basin. Deep-Sea Res. 28, 151–163. https://doi.org/10.1016/0198-0149(81)90087-X.

Gordon, A.L., Visbeck, M., Huber, B., 2001. Export of Weddell Sea deep and bottom water. J. Geophys. Res. 106, 9005–9017. https://doi.org/10.1029/2000JC000281.

Gregg, M., D'Asaro, E., Riley, J., Kunze, E., 2018. Mixing efficiency in the ocean. Annu. Rev. Mar. Sci. 10, 443–473. https://doi.org/10.1146/annurev-marine-121916-063643.

Griesel, A., Gille, S.T., Sprintall, J., McClean, J.L., LaCasce, J.H., Maltrud, M.E., 2010. Isopycnal diffusivities in the Antarctic Circumpolar Current inferred from Lagrangian floats in an eddying model. J. Geophys. Res. 115, C06006. https://doi.org/10.1029/2009JC005821.

Griesel, A., Gille, S.T., Sprintall, J., McClean, J.L., Maltrud, M.E., 2009. Assessing eddy heat flux and its parameterization: a wavenumber perspective from a 1/10° ocean simulation. Ocean Model. 29, 248–260. https://doi.org/10.1016/j.ocemod.2009.05.004.

Griffies, S.M., 1998. The Gent–McWilliams skew flux. J. Phys. Oceanogr. 28, 831–841. https://doi.org/10.1175/1520-0485(1998)028<0831:TGMSF>2.0.CO;2.

Gruber, N., Landschützer, P., Lovenduski, N.S., 2019. The variable Southern Ocean carbon sink. Annu. Rev. Mar. Sci. 11, 159–186. https://doi.org/10.1146/annurev-marine-121916-063407.

Hallberg, R., Gnanadesikan, A., 2006. The role of eddies in determining the structure and response of the wind-driven Southern Hemisphere overturning: results from the Modeling Eddies in the Southern Ocean (MESO) project. J. Phys. Oceanogr. 36, 2232–2252. https://doi.org/10.1175/JPO2980.1.

Hattermann, T., 2018. Antarctic thermocline dynamics along a narrow shelf with easterly winds. J. Phys. Oceanogr. 48, 2419–2443. https://doi.org/10.1175/JPO-D-18-0064.1.

Haumann, A., Gruber, N., Münnich, M., Frenger, I., Kern, S., 2016. Sea-ice transport driving Southern Ocean salinity and its recent trends. Nature 537, 89–92. https://doi.org/10.1038/nature1910.

Heuzé, C., Heywood, K.J., Stevens, D.P., Ridley, J.K., 2013. Southern Ocean bottom water characteristics in CMIP5 models. Geophys. Res. Lett. 40, 1409–1414. https://doi.org/10.1002/grl.50287.

Heywood, K.J., Schmidtko, S., Heuzé, C., Kaiser, J., Jickells, T.D., Queste, B.Y., Stevens, D.P., Wadley, M., Thompson, A.F., Fielding, S., Guihen, D., Creed, E., Ridley, J.K., Smith, W., 2014. Ocean processes at the Antarctic continental slope. Philos. Trans. R. Soc. A, Math. Phys. Eng. Sci. 372, 20130047. https://doi.org/10.1098/rsta.2013.0047.

Hogg, A.M.C., Meredith, M.P., Blundell, J.R., Wilson, C., 2008. Eddy heat flux in the Southern Ocean: response to variable wind forcing. J. Climate 21, 608–620. https://doi.org/10.1175/2007JCLI1925.1.

Holland, D.M., 2001. Explaining the Weddell Polynya - a large ocean eddy shed at Maud Rise. Science 292, 1697–1700. https://doi.org/10.1126/science.1059322.

Holmes, R.M., de Lavergne, C., McDougall, T.J., 2018. Ridges, seamounts, troughs, and bowls: topographic control of the dianeutral circulation in the abyssal ocean. J. Phys. Oceanogr. 48, 861–882. https://doi.org/10.1175/JPO-D-17-0141.1.

Holmes, R.M., Zika, J.D., Ferrari, R., Thompson, A.F., Newom, E.R., England, M.H., 2019. Atlantic Ocean heat transport enabled by Indo-Pacific heat uptake and mixing. Geophys. Res. Lett. 46, 13939–13949. https://doi.org/10.1029/2019GL085160.

Holte, J., Talley, L.D., Gilson, J., Roemmich, D., 2017. An Argo mixed layer climatology and database. Geophys. Res. Lett. 44, 5618–5626. https://doi.org/10.1002/2017GL073426.

Holte, J.W., Talley, L.D., Chereskin, T.K., Sloyan, B.M., 2012. The role of air-sea fluxes in Subantarctic Mode Water formation. J. Geophys. Res., Oceans 117. https://doi.org/10.1029/2011JC007798.

Huber, M.B., Zanna, L., 2017. Drivers of uncertainty in simulated ocean circulation and heat uptake. Geophys. Res. Lett. 44, 1402–1413. https://doi.org/10.1002/2016GL071587.

Hughes, C.W., Ash, E., 2001. Eddy forcing of the mean flow of the Southern Ocean. J. Geophys. Res. 106, 2713–2722. https://doi.org/10.1029/2000JC900332.

Isachsen, P.E., 2011. Baroclinic instability and eddy tracer transport across sloping bottom topography: how well does a modified eady model do in primitive equation simulations? Ocean Model. 39, 183–199. https://doi.org/10.1016/j.ocemod.2010.09.007.

Ivchenko, V.O., Richards, K.J., Stevens, D.P., 1996. The dynamics of the Antarctic Circumpolar Current. J. Phys. Oceanogr. 26, 753–774. https://doi.org/10.1175/1520-0485(1996)026<0753:TDOTAC>2.0.CO;2.

Jacobs, S.S., 1991. On the nature and significance of the Antarctic Slope Front. Mar. Chem. 35, 9–24. https://doi.org/10.1016/S0304-4203(09)90005-6.

Jing, Z., Wu, L., Li, L., Liu, C., Liang, X., Chen, Z., Hu, D., Liu, Q., 2011. Turbulent diapycnal mixing in the subtropical northwestern Pacific: spatial-seasonal variations and role of eddies. J. Geophys. Res., Oceans 116. https://doi.org/10.1029/2011JC007142.

Johnson, G.C., 2008. Quantifying Antarctic Bottom Water and North Atlantic Deep Water volumes. J. Geophys. Res., Oceans 113. https://doi.org/10.1029/2007JC004477.

Johnson, G.C., Bryden, H.L., 1989. On the size of the Antarctic Circumpolar Current. Deep-Sea Res. 36, 39–53. https://doi.org/10.1016/0198-0149(89)90017-4.

Johnson, L., Lee, C.M., D'Asaro, E.A., 2016. Global estimates of lateral springtime restratification. J. Phys. Oceanogr. 46, 1555–1573. https://doi.org/10.1175/JPO-D-15-0163.1.

Jouanno, J., Capet, X., Madec, G., Roullet, G., Klein, P., 2016. Dissipation of the energy imparted by mid-latitude storms in the Southern Ocean. Ocean Sci. 12, 743–769. https://doi.org/10.5194/os-12-743-2016.

Keffer, T., Holloway, G., 1988. Estimating Southern Ocean eddy flux of heat and salt from satellite altimetry. Nature 332, 624–626. https://doi.org/10.1038/332624a0.

Kilbourne, B.F., Girton, J.B., 2015. Quantifying high-frequency wind energy flux into near-inertial motions in the southeast Pacific. J. Phys. Oceanogr. 45, 369–386. https://doi.org/10.1175/JPO-D-14-0076.1.

Klein, P., Lapeyre, G., Large, W.G., 2004. Wind ringing of the ocean in presence of mesoscale eddies. Geophys. Res. Lett. 31. https://doi.org/10.1029/2004GL020274.

Klocker, A., 2018. Opening the window to the Southern Ocean: the role of jet dynamics. Sci. Adv. 4. https://doi.org/10.1126/sciadv.aao4719.

Klocker, A., Ferrari, R., LaCasce, J.H., Merrifield, S.T., 2012. Reconciling float-based and tracer-based estimates of lateral diffusivities. J. Mar. Res. 70, 569–602. https://doi.org/10.1357/002224012805262743.

Klymak, J.M., 2018. Nonpropagating form drag and turbulence due to stratified flow over large-scale abyssal hill topography. J. Phys. Oceanogr. 48, 2383–2395. https://doi.org/10.1175/JPO-D-17-0225.1.

Klymak, J.M., Balwada, D., Garabato, A.N., Abernathey, R., 2021. Parameterizing non-propagating form drag over rough bathymetry. J. Phys. Oceanogr. 51. https://doi.org/10.1175/JPO-D-20-0112.1.

Kunze, E., Lien, R.C., 2019. Energy sinks for lee waves in shear flow. J. Phys. Oceanogr. 49, 2851–2865. https://doi.org/10.1175/JPO-D-19-0052.1.

Kurtakoti, P., Veneziani, M., Stössel, A., Weijer, W., 2018. Preconditioning and formation of Maud Rise polynyas in a high-resolution Earth system model. J. Climate 31, 9659–9678. https://doi.org/10.1175/JCLI-D-18-0392.1.

LaCasce, J.H., Ferrari, R., Marshall, J., Tulloch, R., Balwada, D., Speer, K., 2014. Float-derived isopycnal diffusivities in the DIMES experiment. J. Phys. Oceanogr. 44, 764–780. https://doi.org/10.1175/JPO-D-13-0175.1.

Large, W.G., McWilliams, J.C., Doney, S., 1994. Oceanic vertical mixing: a review and a model with a nonlocal boundary layer parameterization. Rev. Geophys. 32, 363–403. https://doi.org/10.1029/94RG01872.

de Lavergne, C., Madec, G., Le Sommer, J., Nurser, A.J.G., Naveira Garabato, A.C., 2016. On the consumption of Antarctic Bottom Water in the abyssal ocean. J. Phys. Oceanogr. 46, 635–661. https://doi.org/10.1175/JPO-D-14-0201.1.

de Lavergne, C., Madec, G., Roquet, F., Holmes, R., McDougall, T.J., 2017. Abyssal ocean overturning shaped by seafloor distribution. Nature 551, 181–186. https://doi.org/10.1038/nature24472.

Ledwell, J.R., St. Laurent, L.C., Girton, J.B., Toole, J.M., 2011. Diapycnal mixing in the Antarctic Circumpolar Current. J. Phys. Oceanogr. 41, 241–246. https://doi.org/10.1175/2010JPO4557.1.

Lin, X., Zhai, X., Wang, Z., Munday, D.R., 2018. Mean, variability, and trend of Southern Ocean wind stress: role of wind fluctuations. J. Climate 31, 3557–3573. https://doi.org/10.1175/JCLI-D-17-0481.1.

Llort, J., Langlais, C., Matear, R., Moreau, S., Lenton, A., Strutton, P.G., 2018. Evaluating Southern Ocean carbon eddy-pump from biogeochemical-Argo floats. J. Geophys. Res., Oceans 123, 971–984. https://doi.org/10.1002/2017JC012861.

Lu, J., Speer, K., 2010. Topography, jets and eddy mixing in the Southern Ocean. J. Mar. Res. 68, 479–502. https://doi.org/10.1357/002224010794657227.

MacCready, P., Rhines, P.B., 2001. Meridional transport across a zonal channel: topographic localization. J. Phys. Oceanogr. 31, 1427–1439. https://doi.org/10.1175/1520-0485(2001)031<1427:MTAAZC>2.0.CO;2.

Manucharyan, G.E., Spall, M.A., 2016. Wind-driven freshwater buildup and release in the Beaufort Gyre constrained by mesoscale eddies. Geophys. Res. Lett. 43, 273–282. https://doi.org/10.1002/2015GL065957.

Marshall, D., 1997. Subduction of water masses in an eddying ocean. J. Mar. Res. 55, 201–222. https://doi.org/10.1357/0022240973224373.

Marshall, D.P., Ambaum, M.H.P., Maddison, J.R., Munday, D.R., Novak, L., 2017. Eddy saturation and friction control of the Antarctic Circumpolar Current. Geophys. Res. Lett. 44, 286–292. https://doi.org/10.1002/2016GL071702.

Marshall, J., Radko, T., 2003. Residual-mean solutions for the Antarctic Circumpolar Current and its associated overturning circulation. J. Phys. Oceanogr. 33, 2341–2354. https://doi.org/10.1175/1520-0485(2003)033<2341:RSFTAC>2.0.CO;2.

Marshall, J., Shuckburgh, E., Jones, H., Hill, C., 2006. Estimates and implications of surface eddy diffusivity in the Southern Ocean derived from tracer transport. J. Phys. Oceanogr. 36, 1806–1821. https://doi.org/10.1175/JPO2949.1.

Marshall, J., Speer, K., 2012. Closure of the meridional overturning circulation through Southern Ocean upwelling. Nat. Geosci. 5, 171–180. https://doi.org/10.1038/ngeo1391.

Martinson, D.G., Iannuzzi, R.A., 1998. Antarctic ocean-ice interaction: implications from ocean bulk property distributions in the Weddell Gyre. In: Jeffries, M.O. (Ed.), Antarctic Sea Ice: Physical Processes, Interactions and Variability, Vol. 74. Amer. Geophys. Union, pp. 243–271. https://dx.doi.org/10.1029/AR074p0243.

Martinson, D.G., McKee, D.C., 2012. Transport of warm Upper Circumpolar Deep Water onto the western Antarctic Peninsula continental shelf. Ocean Sci. 8, 433–442. https://doi.org/10.5194/os-8-433-2012.

Mashayek, A., Ferrari, R., Merrifield, S., Ledwell, J.R., Laurent, L.S., Garabato, A.N., 2017a. Topographic enhancement of vertical turbulent mixing in the Southern Ocean. Nat. Commun. 8. https://doi.org/10.1038/ncomms14197.

Mashayek, A., Ferrari, R., Nikurashin, M., Peltier, W.R., 2015. Influence of enhanced abyssal diapycnal mixing on stratification and the ocean overturning circulation. J. Phys. Oceanogr. 45, 2580–2597. https://doi.org/10.1175/JPO-D-15-0039.1.

Mashayek, A., Salehipour, H., Bouffard, D., Caulfield, C.P., Ferrari, R., Nikurashin, M., Peltier, W.R., Smyth, W.D., 2017b. Efficiency of turbulent mixing in the abyssal ocean circulation. Geophys. Res. Lett. 44, 6296–6306. https://doi.org/10.1002/2016GL072452.

Mazloff, M.R., Ferrari, R., Schneider, T., 2013. The force balance of the Southern Ocean overturning circulation. J. Phys. Oceanogr. 43, 1193–1208. https://doi.org/10.1175/JPO-D-12-069.1.

McDougall, T.J., Ferrari, R., 2017. Abyssal upwelling and downwelling driven by near-boundary mixing. J. Phys. Oceanogr. 47, 261–283. https://doi.org/10.1175/JPO-D-16-0082.1.

McWilliams, J.C., 2016. Submesoscale currents in the ocean. Proc. R. Soc. A, Math. Phys. Eng. Sci. 472, 20160117. https://doi.org/10.1098/rspa.2016.0117.

Meijers, A.J., Bindoff, N.L., Roberts, J.L., 2007. On the total, mean, and eddy heat and freshwater transports in the Southern Hemisphere of a 1/8° × 1/8° global ocean model. J. Phys. Oceanogr. 37, 277–295. https://doi.org/10.1175/JPO3012.1.

Melet, A., Hallberg, R., Legg, S., Nikurashin, M., 2014. Sensitivity of the ocean state to lee wave–driven mixing. J. Phys. Oceanogr. 44, 900–921. https://doi.org/10.1175/JPO-D-13-072.1.

Meneghello, G., Marshall, J., Cole, S.T., Timmermans, M.L., 2017. Observational inferences of lateral eddy diffusivity in the halocline of the Beaufort Gyre. Geophys. Res. Lett. 44, 12331–12338. https://doi.org/10.1002/2017GL075126.

Meredith, M.P., Hogg, A.M., 2006. Circumpolar response of Southern Ocean eddy activity to a change in the Southern Annular Mode. Geophys. Res. Lett. 33, L16608. https://doi.org/10.1029/2006GL026499.

Meredith, M.P., Naveira Garabato, A.C., Gordon, A.L., Johnson, G.C., 2008. Evolution of the deep and bottom waters of the Scotia Sea, Southern Ocean, during 1995–2005. J. Climate 21, 3327–3343. https://doi.org/10.1175/2007JCLI2238.1.

Meredith, M.P., Schofield, O., Newman, L., Urban, E., Sparrow, M., 2013. The vision for a Southern Ocean Observing System. Curr. Opin. Environ. Sustain. 5, 306–313. https://doi.org/10.1016/j.cosust.2013.03.002.

Meyer, A., Polzin, K.L., Sloyan, B.M., Phillips, H.E., 2016. Internal waves and mixing near the Kerguelen Plateau. J. Phys. Oceanogr. 46, 417–437. https://doi.org/10.1175/JPO-D-15-0055.1.

Meyer, A., Sloyan, B.M., Polzin, K.L., Phillips, H.E., Bindoff, N.L., 2015. Mixing variability in the Southern Ocean. J. Phys. Oceanogr. 45, 966–987. https://doi.org/10.1175/JPO-D-14-0110.1.

Morrison, A.K., Griffies, S.M., Winton, M., Anderson, W.G., Sarmiento, J.L., 2016. Mechanisms of Southern Ocean heat uptake and transport in a global eddying climate model. J. Climate 29, 2059–2075. https://doi.org/10.1175/JCLI-D-15-0579.1.

Naveira Garabato, A.C., Ferrari, R., Polzin, K.L., 2011. Eddy stirring in the Southern Ocean. J. Geophys. Res. 116, C09019. https://doi.org/10.1029/2010JC006818.

Naveira Garabato, A.C., Frajka-williams, E.E., Spingys, C.P., Legg, S., Polzin, K.L., Forryan, A., Abrahamsen, E.P., Buckingham, C.E., Griffies, S.M., Mcphail, S.D., Nicholls, K.W., Thomas, L., Meredith, M.P., 2019. Rapid mixing and exchange of deep-ocean waters in an abyssal boundary current. Proc. Natl. Acad. Sci. 116 (27), 13233–13238. https://doi.org/10.1073/pnas.1904087116.

Naveira Garabato, A.C., Jullion, L., Stevens, D.P., Heywood, K.J., King, B.A., 2009. Variability of Subantarctic mode water and Antarctic intermediate water in the Drake Passage during the late-twentieth and early-twenty-first centuries. J. Climate 22, 3661–3688. https://doi.org/10.1175/2009JCLI2621.1.

Naveira Garabato, A.C., Polzin, K.L., Ferrari, R., Zika, J.D., Forryan, A., 2016. A microscale view of mixing and overturning across the Antarctic Circumpolar Current. J. Phys. Oceanogr. 46, 233–254. https://doi.org/10.1175/JPO-D-15-0025.1.

Naveira Garabato, A.C., Polzin, K.L., King, B.A., Heywood, K.J., Visbeck, M., 2004. Widespread intense turbulent mixing in the Southern Ocean. Science 303, 210–213. https://doi.org/10.1126/science.1090929.

Naveira Garabato, A.C., Stevens, D.P., Watson, A.J., Roether, W., 2007. Short-circuiting of the overturning circulation in the Antarctic Circumpolar Current. Nature 447, 194–197. https://doi.org/10.1038/nature05832.

Nicholson, S.A., Lévy, M., Jouanno, J., Capet, X., Swart, S., Monteiro, P.M.S., 2019. Iron supply pathways between the surface and subsurface waters of the Southern Ocean: from winter entrainment to summer storms. Geophys. Res. Lett. 46, 14567–14575. https://doi.org/10.1029/2019GL084657.

Nicholson, S.A., Lévy, M., Llort, J., Swart, S., Monteiro, P.M.S., 2016. Investigation into the impact of storms on sustaining summer primary productivity in the sub-Antarctic Ocean. Geophys. Res. Lett. 43, 9192–9199. https://doi.org/10.1002/2016GL069973.

Nikurashin, M., Ferrari, R., 2010a. Radiation and dissipation of internal waves generated by geostrophic motions impinging on small-scale topography: application to the Southern Ocean. J. Phys. Oceanogr. 40, 2025–2042. https://doi.org/10.1175/2010JPO4315.1.

Nikurashin, M., Ferrari, R., 2010b. Radiation and dissipation of internal waves generated by geostrophic motions impinging on small-scale topography: theory. J. Phys. Oceanogr. 40, 1055–1074. https://doi.org/10.1175/2009JPO4199.1.

Nikurashin, M., Ferrari, R., 2011. Global energy conversion rate from geostrophic flows into internal lee waves in the deep ocean. Geophys. Res. Lett. 38. https://doi.org/10.1029/2011GL046576.

Nikurashin, M., Vallis, G., 2011. A theory of deep stratification and overturning circulation in the ocean. J. Phys. Oceanogr. 41, 485–502. https://doi.org/10.1175/2010JPO4529.1.

Nøst, O.A., Biuw, M., Tverberg, V., Lydersen, C., Hattermann, T., Zhou, Q., Smedsrud, L.H., Kovacs, K.M., 2011. Eddy overturning of the Antarctic Slope Front controls glacial melting in the Eastern Weddell Sea. J. Geophys. Res. 116, C11014. https://doi.org/10.1029/2011JC006965.

Nycander, J., Hieronymus, M., Roquet, F., 2015. The nonlinear equation of state of sea water and the global water mass distribution. Geophys. Res. Lett. 42, 7714–7721. https://doi.org/10.1002/2015GL065525.

Ogle, S.E., Tamsitt, V., Josey, S.A., Gille, S.T., Cerovečki, I., Talley, L.D., Weller, R.A., 2018. Episodic Southern Ocean heat loss and its mixed layer impacts revealed by the farthest South multiyear surface flux mooring. Geophys. Res. Lett. 45, 5002–5010. https://doi.org/10.1029/2017GL076909.

Oka, A., Niwa, Y., 2013. Pacific deep circulation and ventilation controlled by tidal mixing away from the sea bottom. Nat. Commun. 4, 2419. https://doi.org/10.1038/ncomms3419.

Olbers, D., Borowski, D., Volker, C., Wolff, J.O., 2004. The dynamical balance, transport, and circulation of the Antarctic Circumpolar Current. Antarct. Sci. 16, 439–470. https://doi.org/10.1017/S0954102004002251.

Orsi, A.H., Johnson, G.C., Bullister, J.L., 1999. Circulation, mixing, and production of Antarctic Bottom Water. Prog. Oceanogr. 43, 55–109. https://doi.org/10.1016/S0079-6611(99)00004-X.

Orsi, A.H., Smethie, W.M., Bullister, J.L., 2002. On the total input of Antarctic waters to the deep ocean: a preliminary estimate from chlorofluorocarbon measurements. J. Geophys. Res., Oceans 107. https://doi.org/10.1029/2001JC000976.

Orsi, A.H., Whitworth, T., Nowlin, W.D., 1995. On the meridional extent and fronts of the Antarctic Circumpolar Current. Deep-Sea Res., Part 1 42, 641–673. https://doi.org/10.1016/0967-0637(95)00021-W.

Padman, L., Siegfried, M.R., Fricker, H.A., 2018. Ocean tide influences on the Antarctic and Greenland ice sheets. Rev. Geophys. 56, 142–184. https://doi.org/10.1002/2016RG000546.

Palóczy, A., Gille, S.T., McClean, J.L., 2018. Oceanic heat delivery to the Antarctic continental shelf: large-scale, low-frequency variability. J. Geophys. Res. 123, 7678–7701. https://doi.org/10.1029/2018JC014345.

Patmore, R.D., Holland, P.R., Munday, D.R., Naveira Garabato, A.C., Stevens, D.P., Meredith, M.P., 2019. Topographic control of Southern Ocean gyres and the Antarctic Circumpolar Current: a barotropic perspective. J. Phys. Oceanogr. 49, 3221–3244. https://doi.org/10.1175/JPO-D-19-0083.1.

Pellichero, V., Boutin, J., Claustre, H., Merlivat, L., Sallée, J.B., Blain, S., 2020. Relaxation of wind stress drives the abrupt onset of biological carbon uptake in the Kerguelen bloom: a multisensor approach. Geophys. Res. Lett. 47, e2019GL085992. https://doi.org/10.1029/2019GL085992.

Pellichero, V., Sallée, J.B., Chapman, C.C., Downes, S.M., 2018. The Southern Ocean meridional overturning in the sea-ice sector is driven by freshwater fluxes. Nat. Commun. 9, 1789. https://doi.org/10.1038/s41467-018-04101-2.

Pellichero, V., Sallée, J.B., Schmidtko, S., Roquet, F., Charrassin, J.B., 2017. The ocean mixed layer under Southern Ocean sea-ice: seasonal cycle and forcing. J. Geophys. Res., Oceans 122, 1608–1633. https://doi.org/10.1002/2016JC011970.

Pierrehumbert, R.T., 1984. Local and global baroclinic instability of zonally varying flow. J. Atmos. Sci. 41, 2141–2162. https://doi.org/10.1175/1520-0469(1984)041<2141:LAGBIO>2.0.CO;2.

du Plessis, M., Swart, S., Ansorge, I.J., Mahadevan, A., 2017. Submesoscale processes promote seasonal restratification in the Subantarctic Ocean. J. Geophys. Res., Oceans 122, 2960–2975. https://doi.org/10.1002/2016JC012494.

du Plessis, M., Swart, S., Ansorge, I.J., Mahadevan, A., Thompson, A.F., 2019. Southern Ocean seasonal restratification delayed by submesoscale wind–front interactions. J. Phys. Oceanogr. 49, 1035–1053. https://doi.org/10.1175/JPO-D-18-0136.1.

Pollard, R., Tréguer, P., Read, J., 2006. Quantifying nutrient supply to the Southern Ocean. J. Geophys. Res., Oceans 111. https://doi.org/10.1029/2005JC003076.

Polzin, K.L., Toole, J.M., Ledwell, J.R., Schmitt, R.W., 1997. Spatial variability of turbulent mixing in the abyssal ocean. Science 276, 93–96. https://doi.org/10.1126/science.276.5309.93.

Polzin, K.L., Toole, J.M., Schmitt, R.W., 1995. Finescale parameterizations of turbulent dissipation. J. Phys. Oceanogr. 25, 306–328. https://doi.org/10.1175/1520-0485(1995)025<0306:FPOTD>2.0.CO;2.

Price, J.F., Mooers, C.N.K., Van Leer, J.C., 1978. Observation and simulation of storm-induced mixed-layer deepening. J. Phys. Oceanogr. 8, 582–599. https://doi.org/10.1175/1520-0485(1978)008<0582:OASOSI>2.0.CO;2.

Redi, M.H., 1982. Oceanic isopycnal mixing by coordinate rotation. J. Phys. Oceanogr. 12, 1154–1158. https://doi.org/10.1175/1520-0485(1982)012<1154:OIMBCR>2.0.CO;2.

Rintoul, S.R., England, M.H., 2002. Ekman transport dominates local air–sea fluxes in driving variability of Subantarctic mode water. J. Phys. Oceanogr. 32, 1308–1321. https://doi.org/10.1175/1520-0485.

Roach, C.J., Balwada, D., Speer, K., 2016. Horizontal mixing in the Southern Ocean from Argo float trajectories. J. Geophys. Res. 121, 5570–5586. https://doi.org/10.1002/2015JC011440.

Roach, C.J., Speer, K., 2019. Exchange of water between the Ross Gyre and ACC assessed by Lagrangian particle tracking. J. Geophys. Res., Oceans 124, 4631–4643. https://doi.org/10.1029/2018JC014845.

Roemmich, D., Church, J., Gilson, J., Monselesan, D., Sutton, P., Wijffels, S., 2015. Unabated planetary warming and its ocean structure since 2006. Nat. Clim. Change 5, 240–245. https://doi.org/10.1038/nclimate2513.

Rosso, I., Hogg, A.M., Kiss, A.E., Gayen, B., 2015. Topographic influence on submesoscale dynamics in the Southern Ocean. Geophys. Res. Lett. 42, 1139–1147. https://doi.org/10.1002/2014GL062720.

Rosso, I., Hogg, A.M., Strutton, P.G., Kiss, A.E., Matear, R., Klocker, A., van Sebille, E., 2014. Vertical transport in the ocean due to sub-mesoscale structures: impacts in the Kerguelen region. Ocean Model. 80, 10–23. https://doi.org/10.1016/j.ocemod.2014.05.001.

Ruan, X., Thompson, A.F., Flexas, M.M., Sprintall, J., 2017. Contribution of topographically generated submesoscale turbulence to Southern Ocean overturning. Nat. Geosci. 10, 840–845. https://doi.org/10.1038/ngeo3053.

Sabine, C.L., Feely, R.A., Gruber, N., Key, R.M., Lee, K., Bullister, J.L., Wanninkhof, R., Wong, C., Wallace, D.W.R., Tilbrook, B., et al., 2004. The oceanic sink for anthropogenic CO_2. Science 305, 367–371. https://doi.org/10.1126/science.1097403.

Saenko, O.A., Zhai, X., Merryfield, W.J., Lee, W.G., 2012. The combined effect of tidally and eddy-driven diapycnal mixing on the large-scale ocean circulation. J. Phys. Oceanogr. 42, 526–538. https://doi.org/10.1175/JPO-D-11-0122.1.

Sallée, J.B., Shuckburgh, E., Bruneau, N., Meijers, A.J.S., Bracegirdle, T.J., Wang, Z., Roy, T., 2013. Assessment of Southern Ocean water mass circulation and characteristics in CMIP5 models: historical bias and forcing response. J. Geophys. Res., Oceans 118, 1830–1844. https://doi.org/10.1002/jgrc.20135.

Sallée, J.b., Speer, K., Rintoul, S., 2010. Zonally asymmetric response of the Southern Ocean mixed-layer depth to the Southern Annular Mode. Nat. Geosci. 3, 273–279. https://doi.org/10.1038/ngeo812.

Scott, R.B., Goff, J.A., Naveira Garabato, A.C., Nurser, A.J.G., 2011. Global rate and spectral characteristics of internal gravity wave generation by geostrophic flow over topography. J. Geophys. Res., Oceans 116. https://doi.org/10.1029/2011JC007005.

Sheen, K.L., Brearley, J.A., Naveira Garabato, A.C., Smeed, D.A., Laurent, L.S., Meredith, M.P., Thurnherr, A.M., Waterman, S.N., 2015. Modification of turbulent dissipation rates by a deep Southern Ocean eddy. Geophys. Res. Lett. 42, 3450–3457. https://doi.org/10.1002/2015GL063216.

Sheen, K.L., Brearley, J.A., Naveira Garabato, A.C., Smeed, D.A., Waterman, S., Ledwell, J.R., Meredith, M.P., St. Laurent, L., Thurnherr, A.M., Toole, J.M., Watson, A.J., 2013. Rates and mechanisms of turbulent dissipation and mixing in the Southern Ocean: results from the diapycnal and isopycnal mixing experiment in the Southern Ocean (DIMES). J. Geophys. Res., Oceans 118, 2774–2792. https://doi.org/10.1002/jgrc.20217.

Sheen, K.L., Naveira Garabato, A.C., Brearley, J.A., Meredith, M.P., Polzin, K.L., Smeed, D.A., Forryan, A., King, B.A., Sallée, J.B., St. Laurent, L., Thurnherr, A.M., Toole, J.M., Waterman, S.N., Watson, A.J., 2014. Eddy-induced variability in Southern Ocean abyssal mixing on climatic timescales. Nat. Geosci. 7, 577–582. https://doi.org/10.1038/ngeo2200.

Silvano, A., Rintoul, S.R., Peña Molino, B., Hobbs, W.R., van Wijk, E., Aoki, S., Tamura, T., Williams, G.D., 2018. Freshening by glacial meltwater enhances melting of ice shelves and reduces formation of Antarctic Bottom Water. Sci. Adv. 4. https://doi.org/10.1126/sciadv.aap9467.

Silvester, J.M., Lenn, Y.D., Polton, J.A., Rippeth, T.P., Maqueda, M.M., 2014. Observations of a diapycnal shortcut to adiabatic upwelling of Antarctic Circumpolar Deep Water. Geophys. Res. Lett. 41, 7950–7956. https://doi.org/10.1002/2014GL061538.

Simmons, H.L., Jayne, S.R., Laurent, L.C.S., Weaver, A.J., 2004. Tidally driven mixing in a numerical model of the ocean general circulation. Ocean Model. 6, 245–263. https://doi.org/10.1016/S1463-5003(03)00011-8.

Sloyan, B., 2005. Spatial variability of mixing in the Southern Ocean. Geophys. Res. Lett. 32, L18603. https://doi.org/10.1029/2005GL023568.

Sloyan, B.M., Talley, L.D., Chereskin, T.K., Fine, R., Holte, J., 2010. Antarctic Intermediate Water and Subantarctic Mode Water formation in the southeast Pacific: the role of turbulent mixing. J. Phys. Oceanogr. 40, 1558–1574. https://doi.org/10.1175/2010JPO4114.1.

Smith, K.S., 2007. The geography of linear baroclinic instability in Earth's oceans. J. Mar. Res. 65, 655–683. https://doi.org/10.1357/002224007783649484.

Smith, K.S., Marshall, J., 2009. Evidence for enhanced eddy mixing at middepth in the Southern Ocean. J. Phys. Oceanogr. 39, 50–69. https://doi.org/10.1175/2008JPO3880.1.

Sokolov, S., Rintoul, S.R., 2009. Circumpolar structure and distribution of the Antarctic Circumpolar Current fronts: 1. Mean circumpolar paths. J. Geophys. Res. 114, C11018. https://doi.org/10.1029/2008JC005108.

Speer, K., Rintoul, S.R., Sloyan, B., 2000. The diabatic Deacon cell. J. Phys. Oceanogr. 30, 3212–3222. https://doi.org/10.1175/1520-0485(2000)030<3212:TDDC>2.0.CO;2.

Speer, K., Sallée, J.B., Pellichero, V., 2018. Antarctic mode water. J. Mar. Res. 76, 119–137.

St-Laurent, P., Klinck, J.M., Dinniman, M.S., 2013. On the role of coastal troughs in the circulation of warm Circumpolar Deep Water on Antarctic shelves. J. Phys. Oceanogr. 43, 51–64. https://doi.org/10.1175/JPO-D-11-0237.1.

St. Laurent, L., Naveira Garabato, A.C., Ledwell, J.R., Thurnherr, A.M., Toole, J.M., Watson, A.J., 2013. Turbulence and diapycnal mixing in Drake Passage. J. Phys. Oceanogr. 42, 2143–2152. https://doi.org/10.1175/JPO-D-12-027.1.

St. Laurent, L.C., Toole, J.M., Schmitt, R.W., 2001. Buoyancy forcing by turbulence above rough topography in the abyssal Brazil Basin. J. Phys. Oceanogr. 31, 3476–3495. https://doi.org/10.1175/1520-0485(2001)031<3476:BFBTAR>2.0.CO;2.

Stammer, D., 1998. On eddy characteristics, eddy transports, and mean flow properties. J. Phys. Oceanogr. 28, 727–739. https://doi.org/10.1175/1520-0485(1998)028<0727:OECETA>2.0.CO;2.

Stewart, A.L., Hogg, A.M., 2017. Reshaping the Antarctic Circumpolar Current via Antarctic Bottom Water export. J. Phys. Oceanogr. 47, 2577–2601. https://doi.org/10.1175/JPO-D-17-0007.1.

Stewart, A.L., Klocker, A., Menemenlis, D., 2018. Circum-Antarctic shoreward heat transport derived from an eddy- and tide-resolving simulation. Geophys. Res. Lett. 45, 834–845. https://doi.org/10.1002/2017GL075677.

Stewart, A.L., Thompson, A.F., 2013. Connecting Antarctic cross-slope exchange with Southern Ocean overturning. J. Phys. Oceanogr. 43, 1453–1471. https://doi.org/10.1175/JPO-D-12-0205.1.

Stewart, A.L., Thompson, A.F., 2015. Eddy-mediated transport of warm Circumpolar Deep Water across the Antarctic Shelf Break. Geophys. Res. Lett. 42, 432–440. https://doi.org/10.1002/2014GL062281.

von Storch, J.S., Sasaki, H., Marotzke, J., 2007. Wind-generated power input to the deep ocean: an estimate using a 1/10° general circulation model. J. Phys. Oceanogr. 37, 657–672. https://doi.org/10.1175/JPO3001.1.

Su, Z., Stewart, A.L., Thompson, A.F., 2014. An idealized model of Weddell Gyre export variability. J. Phys. Oceanogr. 44, 1671–1688. https://doi.org/10.1175/JPO-D-13-0263.1.

Su, Z., Wang, J., Klein, P., Thompson, A.F., Menemenlis, D., 2018. Ocean submesoscales as a key component of the global heat budget. Nat. Commun. 9, 775. https://doi.org/10.1038/s41467-018-02983-w.

Sun, S., Eisenman, I., Stewart, A.L., 2018. Does Southern Ocean surface forcing shape the global ocean overturning circulation? Geophys. Res. Lett. 45, 2413–2423. https://doi.org/10.1002/2017GL076437.

Swart, N.C., Gille, S.T., Fyfe, J.C., Gillett, N.P., 2018a. Recent Southern Ocean warming and freshening driven by greenhouse gas emissions and ozone depletion. Nat. Geosci. 11, 836–841. https://doi.org/10.1038/s41561-018-0226-1.

Swart, S., Campbell, E.C., Heuzé, C.H., Johnson, K., Lieser, J.L., Massom, R., Mazloff, M., Meredith, M., Reid, P., Sallée, J.B., Stammerjohn, S., 2018b. Return of the Maud Rise polynya: climate litmus or sea ice anomaly? [in "State of the Climate in 2017"]. Bull. Am. Meteorol. Soc. 99, S188–S198. https://doi.org/10.1175/2018BAMSStateoftheClimate.1.

Swart, S., du Plessis, M.D., Thompson, A.F., Biddle, L.C., Giddy, I., Linders, T., Mohrmann, M., Nicholson, S.A., 2020. Submesoscale fronts in the Antarctic marginal ice zone and their response to wind forcing. Geophys. Res. Lett. 47, e2019GL086649. https://doi.org/10.1029/2019GL086649.

Swart, S., Thomalla, S.J., Monteiro, P.M.S., 2015. The seasonal cycle of mixed layer dynamics and phytoplankton biomass in the Sub-Antarctic Zone: a high-resolution glider experiment. J. Mar. Syst. 147, 103–115. https://doi.org/10.1016/j.jmarsys.2014.06.002.

Takahashi, A., Hibiya, T., 2019. Assessment of finescale parameterizations of deep ocean mixing in the presence of geostrophic current shear: results of microstructure measurements in the Antarctic Circumpolar Current region. J. Geophys. Res., Oceans 124, 135–153. https://doi.org/10.1029/2018JC014030.

Talley, L.D., 2013. Closure of the global overturning circulation through the Indian, Pacific, and Southern Oceans: schematics and transports. Oceanography 26, 80–97. https://doi.org/10.5670/oceanog.2013.07.

Tamsitt, V., Cerovečki, I., Josey, S.A., Gille, S.T., Schulz, E., 2020. Mooring observations of air–sea heat fluxes in two Subantarctic Mode Water formation regions. J. Climate 33, 2757–2777. https://doi.org/10.1175/JCLI-D-19-0653.1.

Tamsitt, V., Drake, H., Morrison, A.K., Talley, L.D., Dufour, C.O., Gray, A.R., Griffies, S.M., Mazloff, M.R., Sarmiento, J.L., Wang, J., Weijer, W., 2017. Spiraling pathways of global deep waters to the surface of the Southern Ocean. Nat. Commun. 8, 172. https://doi.org/10.1038/s41467-017-00197-0.

Taylor, J.R., Bachman, S., Stamper, M., Hosegood, P., Adams, K., Sallee, J.B., Torres, R., 2018. Submesoscale Rossby waves on the Antarctic Circumpolar Current. Sci. Adv. 4. https://doi.org/10.1126/sciadv.aao2824.

Thomas, L.N., 2005. Destruction of potential vorticity by winds. J. Phys. Oceanogr. 35, 2457–2466. https://doi.org/10.1175/JPO2830.1.

Thomas, L.N., Tandon, A., Mahadevan, A., 2008. Submesoscale processes and dynamics. In: Hecht, M.W., Hasumi, H. (Eds.), Ocean Modeling in an Eddying Regime. American Geophysical Union (AGU), pp. 17–38. https://dx.doi.org/10.1029/177GM04.

Thompson, A.F., Haynes, P.H., Wilson, C., Richards, K.J., 2010. Rapid Southern Ocean front transitions in an eddy-resolving ocean GCM. Geophys. Res. Lett. 37, L23602. https://doi.org/10.1029/2010GL045386.

Thompson, A.F., Heywood, K.J., Schmidtko, S., Stewart, A.L., 2014. Eddy transport as a key component of the Antarctic overturning circulation. Nat. Geosci. 7, 879–884. https://doi.org/10.1038/ngeo2289.

Thompson, A.F., Naveira Garabato, A.C., 2014. Equilibration of the Antarctic Circumpolar Current by standing meanders. J. Phys. Oceanogr. 44, 1811–1828. https://doi.org/10.1175/JPO-D-13-0163.1.

Thompson, A.F., Sallée, J.B., 2012. Jets and topography: jet transitions and the impact on transport in the Antarctic Circumpolar Current. J. Phys. Oceanogr. 42, 956–972. https://doi.org/10.1175/JPO-D-11-0135.1.

Thompson, A.F., Stewart, A.L., Spence, P., Heywood, K.J., 2018. The Antarctic Slope Current in a changing climate. Rev. Geophys. 56, 741–770. https://doi.org/10.1029/2018RG000624.

Thorpe, S.A., 2005. The Turbulent Ocean. Cambridge University Press, Cambridge.

Toggweiler, J.R., Samuels, B., 1998. On the ocean's large-scale circulation near the limit of no vertical mixing. J. Phys. Oceanogr. 28, 1832–1852. https://doi.org/10.1175/1520-0485(1998)028<1832:OTOSLS>2.0.CO;2.

Treguier, A.M., McWilliams, J.C., 1990. Topographic influences on wind-driven, stratified flow in a β-plane channel: an idealized model for the Antarctic Circumpolar Current. J. Phys. Oceanogr. 20, 321–343. https://doi.org/10.1175/1520-0485(1990)020<0321:TIOWDS>2.0.CO;2.

Trossman, D.S., Arbic, B.K., Richman, J., Garner, S.T., Jayne, S., Wallcraft, A., 2016. Impact of topographic internal lee-wave drag on an eddying global ocean model. Ocean Model. 97, 109–128. https://doi.org/10.1016/j.ocemod.2015.10.013.

Tulloch, R., Ferrari, R., Jahn, O., Klocker, A., LaCasce, J., Ledwell, J.R., Marshall, J., Messias, M.J., Speer, K., Watson, A., 2014. Direct estimate of lateral eddy diffusivity upstream of Drake Passage. J. Phys. Oceanogr. 44, 2593–2616. https://doi.org/10.1175/JPO-D-13-0120.1.

Uchida, T., Balwada, D., Abernathey, R., McKinley, G., Smith, S., Lévy, M., 2019. The contribution of submesoscale over mesoscale eddy iron transport in the open Southern Ocean. J. Adv. Model. Earth Syst. 11, 3934–3958. https://doi.org/10.1029/2019MS001805.

Uchida, T., Balwada, D., Abernathey, R.P., McKinley, G.A., Smith, S.K., Lévy, M., 2020. Vertical eddy iron fluxes support primary production in the open Southern Ocean. Nat. Commun. 11, 1125. https://doi.org/10.1038/s41467-020-14955-0.

Vernet, M., Geibert, W., Hoppema, M., Brown, P.J., Haas, C., Helmer, H.H., Jokat, W., Jullion, L., Mazloff, M., Bakker, D.C.E., Brearley, J.A., Croot, P., Hattermann, T., Hauck, J., Hillenbrand, C.D., Hoppe, C.J.M., Huhn, O., Koch, B.P., Lechtenfeld, O.J., Meredith, M.P., Garabato, A.C.N., Nöthig, E.M., Peeken, I., van der Loeff, M.M.R., Schmidtko, S., Schröder, M., Strass, V.H., Torres-Valdés, S., Verdy, A., 2019. The Weddell Gyre, Southern Ocean: Present knowledge and future challenges. Annu. Rev. Fluid Mech. 57, 623–708. https://doi.org/10.1029/2018RG000604.

Viglione, G.A., Thompson, A.F., Flexas, M.M., Sprintall, J., Swart, S., 2018. Abrupt transitions in submesoscale structure in southern Drake Passage: glider observations and model results. J. Phys. Oceanogr. 48, 2011–2027. https://doi.org/10.1175/JPO-D-17-0192.1.

Volkov, D.L., Fu, L.-L., Lee, T., 2010. Mechanisms of the meridional heat transport in the Southern Ocean. Ocean Dyn. 60, 791–801. https://doi.org/10.1007/s10236-010-0288-0.

Waterman, S., Naveira Garabato, A.C., Polzin, K.L., 2013. Internal waves and turbulence in the Antarctic Circumpolar Current. J. Phys. Oceanogr. 43, 259–282. https://doi.org/10.1175/JPO-D-11-0194.1.

Waterman, S., Polzin, K.L., Naveira Garabato, A.C., Sheen, K.L., Forryan, A., 2014. Suppression of internal wave breaking in the Antarctic Circumpolar Current near topography. J. Phys. Oceanogr. 44, 1466–1492. https://doi.org/10.1175/JPO-D-12-0154.1.

Watson, A.J., Ledwell, J.R., Messias, M.J., King, B.A., Mackay, N., Meredith, M.P., Mills, B., Naveira Garabato, A.C., 2014. Rapid cross-density ocean mixing at mid-depths in the Drake Passage measured by tracer release. Nature 501, 408–411. https://doi.org/10.1038/nature12432.

Watson, A.J., Naveira Garabato, A.C., 2006. The role of Southern Ocean mixing and upwelling in glacial-interglacial atmospheric CO_2 change. Tellus, Ser. B Chem. Phys. Meteorol. 58, 73–87. https://doi.org/10.1111/j.1600-0889.2005.00167.x.

Whitt, D.B., Lévy, M., Taylor, J.R., 2017. Low-frequency and high-frequency oscillatory winds synergistically enhance nutrient entrainment and phytoplankton at fronts. J. Geophys. Res., Oceans 122, 1016–1041. https://doi.org/10.1002/2016JC012400.

Whitt, D.B., Nicholson, S.A., Carranza, M.M., 2019. Global impacts of subseasonal (<60 day) wind variability on ocean surface stress, buoyancy flux, and mixed layer depth. J. Geophys. Res., Oceans 124, 8798–8831. https://doi.org/10.1029/2019JC015166.

Wilson, E.A., Riser, S.C., Campbell, E.C., Wong, A.P.S., 2019. Winter upper-ocean stability and ice–ocean feedbacks in the sea ice–covered Southern Ocean. J. Phys. Oceanogr. 49, 1099–1117. https://doi.org/10.1175/JPO-D-18-0184.1.

Witter, D.L., Chelton, D.B., 1998. Eddy–mean flow interaction in zonal oceanic jet flow along zonal ridge topography. J. Phys. Oceanogr. 28, 2019–2039. https://doi.org/10.1175/1520-0485(1998)028<2019:EMFIIZ>2.0.CO;2.

Wolfe, C.L., Cessi, P., 2011. The adiabatic pole-to-pole overturning circulation. J. Phys. Oceanogr. 41, 1795–1810. https://doi.org/10.1175/2011JPO4570.1.

Wright, C.J., Scott, R.B., Ailliot, P., Furnival, D., 2014. Lee wave generation rates in the deep ocean. Geophys. Res. Lett. 41, 2434–2440. https://doi.org/10.1002/2013GL059087.

Wunsch, C., 1998. The work done by the wind on the oceanic general circulation. J. Phys. Oceanogr. 28, 2332–2340. https://doi.org/10.1175/1520-0485(1998)028<2332:TWDBTW>2.0.CO;2.

Yang, L., Nikurashin, M., Hogg, A.M., Sloyan, B.M., 2018. Energy loss from transient eddies due to lee wave generation in the Southern Ocean. J. Phys. Oceanogr. 48, 2867–2885. https://doi.org/10.1175/JPO-D-18-0077.1.

Youngs, M.K., Thompson, A.F., Lazar, A., Richards, K.J., 2017. ACC meanders, energy transfer, and mixed barotropic-baroclinic instability. J. Phys. Oceanogr. 47, 1291–1305. https://doi.org/10.1175/JPO-D-16-0160.1.

Yuan, X., 2004. High-wind-speed evaluation in the Southern Ocean. J. Geophys. Res., Atmos. 109. https://doi.org/10.1029/2003JD004179.

Zika, J.D., Sallée, J.B., Meijers, A.J.S., Naveira Garabato, A.C., Watson, A.J., Messias, M.J., King, B.A., 2020. Tracking the spread of a passive tracer through Southern Ocean water masses. Ocean Sci. 16, 323–336. https://doi.org/10.5194/os-16-323-2020.

Chapter 13

The crucial contribution of mixing to present and future ocean oxygen distribution

Marina Lévy[a], Laure Resplandy[b], Jaime B. Palter[c], Damien Couespel[a] and Zouhair Lachkar[d]

[a]Sorbonne Université, LOCEAN-IPSL (CNRS/IRD/MNHN), Paris, France, [b]Department of Geosciences and Princeton Environmental Institute, Princeton University, Princeton, NJ, United States, [c]Graduate School of Oceanography, University of Rhode Island, Narragansett, RI, United States, [d]Center for Prototype Climate Modelling, New-York University, Abu Dhabi, United Arab Emirates

13.1 Introduction

The choice of oxygen as an exemplary biogeochemical tracer to discuss how mixing processes shape marine biogeochemical cycling might seem odd at first sight. Indeed, unlike for other tracers, such as nutrients (Dufour et al., 2015; Letscher et al., 2016; Yamamoto et al., 2018), mixing is seldom a leading-order process in the oxygen balance (Couespel et al., 2019). In fact, the three main competing processes controlling the levels of dissolved oxygen in the ocean are air-sea exchange, advective transport and respiration. The transfer of atmospheric oxygen to the surface ocean is tied to temperature-dependent oxygen solubility, the capacity of the water to hold onto dissolved oxygen. Ocean ventilation is understood as a two-step process, the subduction of oxygen-rich waters from the mixed-layer to the ocean interior, followed by oxygen transport below the mixed layer. Finally, biological respiration is the consumption of dissolved oxygen, mostly by microorganisms that oxidise the organic matter produced at the sea surface as it sinks to the deep ocean. The layering of water masses with different degrees of oxygenation at the sea surface, transit times to the place where they are sampled, and biological respiration that occurs along their trajectory shape the vertical profile of oxygen in every locale (Ito and Deutsch, 2010). Mixing contributes to this balance as one of the processes contributing to total ventilation, in addition to purely advective transport.

Our intention in this chapter is to show that oxygen mixing is a primary player in controlling the volume and oxygen concentration of tropical Oxygen Minimum Zones (OMZs), and in the rate of global deoxygenation in response to anthropogenic warming, which are two crucial issues regarding present and future oxygen distributions in the ocean. OMZs have importance disproportionate to their size, because they are the main sites where biologically available nitrogen is removed from the ocean, and thereby made unavailable to fuel primary productivity. At the same time, denitrification emits nitrous dioxide, a potent greenhouse gas to the atmosphere (Lam et al., 2011; Babbin et al., 2015; Yang et al., 2020). Deoxygenation is emerging as a threat to marine life and could expand the volume of OMZs, and act as an amplifying feedback on global warming (IUCN, 2019). Here we review observational and model studies that have explored the role of ocean mixing in controlling oxygen budgets and their evolution in a warming climate, both globally, and, more specifically in OMZs.

We focus on diapycnal (mostly vertical) mixing in the upper ocean (Fox-Kemper et al., Chapter 4), and isopycnal mixing, which is dominantly controlled by mesoscale and sub-mesoscale eddies (Gula et al., Chapter 8). These eddies create sharp gradients that accelerate the irreversible mixing process (Abernathey et al., Chapter 9). Both of these turbulent processes are parameterised as diffusive transport in coarse resolution ocean models (Melet et al., Chapter 2). Fine resolution ocean models explicitly resolve the (sub-)mesoscale circulation and its associated mixing; in this case eddy mixing is often defined as advection by the eddying part of the flow, and referred to as eddy advection (Lévy et al., 2012). The degree to which eddy mixing is resolved depends on the grid resolution and diffusion operators used to parameterise subgrid scale processes.

In terms of models, we will use examples from Earth System Models (ESMs) with embedded biogeochemical components, which are global fully coupled ocean-atmosphere models, as well as regional ocean-biogeochemical models with restricted domain sizes and forced by prescribed atmospheric fluxes and lateral boundary conditions.

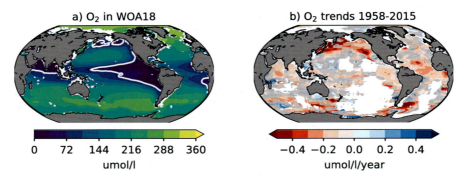

FIGURE 13.1 OMZs and deoxygenation. (a) Climatological mean dissolved oxygen concentrations between 200 m and 600 m (data from Word ocean atlas, 2018). The 60 μM contour is shown in white, and marks the boundary of major OMZs, which are located in the eastern North and South tropical Pacific ocean, the Northern Arabian sea and eastern tropical Atlantic. (b) Observed 200–600 m oxygen trend between 1958 and 2015 (data from Ito et al. (2017)). Red colours indicate deoxygenation. The trend is computed by linear regression; white regions are those where less than 10 years of data were available; regions where trends are not significantly different from zero at the 80% confidence level (p_values > 0.2) are shaded in grey.

13.2 Role of mixing in oxygen minimum zones

Oxygen minimum zones are O_2-deficient layers that occupy large volumes of the intermediate-depth tropical oceans (Stramma et al., 2008; Paulmier and Ruiz-Pino, 2009; Bianchi et al., 2012). The OMZs with the lowest oxygen concentrations are found in the eastern North and South tropical Pacific ocean and the Northern Arabian sea (Fig. 13.1). In the eastern tropical Atlantic, there are also well-studied OMZs, but these have relatively higher oxygen concentrations. Oxygen levels within the OMZ influence their biological and biogeochemical impacts: the transition to hypoxia (typically $O_2 < 60$ μM) leads to a major loss of biodiversity, and mortality events for many macro-organisms (Vaquer-Sunyer and Duarte, 2008), suboxia (typically $O_2 < 1$–5 μM) corresponds to the transition from O_2-based respiration to nitrate-based respiration and denitrification, and anoxia (typically $O_2 < 0.1$ μM) corresponds to the transition from nitrate-based respiration to sulfate reduction (Paulmier and Ruiz-Pino, 2009). Note that the thresholds for these transitions vary in the literature and are given here as an indication.

Hypoxic conditions in the water column are usually associated with low ventilation and high oxygen demand. OMZs are hence found in the poorly ventilated shadow zones of the subtropical oceans, where subsurface streamlines recirculate beneath a shallow mixed-layer with essentially no direct advective connection to the surface ocean (Luyten et al., 1983). In these domains, the residence time in the thermocline is at a maximum, and oxygen is slowly supplied mainly through mixing processes (Gnanadesikan et al., 2012). Near the surface, these regions are also biologically productive upwelling systems, producing organic matter that fuels intense bacterial respiration when it sinks through the thermocline. The strong and localised subsurface biological oxygen consumption generates strong spatial gradients in oxygen at the edges of the shadow zones, making mixing all the more important to the oxygen balance in major OMZs. Understanding the processes and quantifying the rates of along-isopycnal and diapycnal mixing is therefore crucial for assessing oxygen budgets in the OMZs. The dominant role of mixing as a source of oxygen in the OMZs is illustrated by the few examples below.

The OMZs in the eastern tropical Atlantic are unique among the OMZs in that they are rarely suboxic, with minimum O_2 concentrations typically above 40 μM (Fig. 13.1). The higher O_2 levels, relative to OMZs in the Pacific ocean, are likely the result of mixing with recently ventilated water masses. In the eastern tropical North Atlantic, just west of the coastal upwelling region offshore of Mauritania and Senegal, these water masses include the Labrador sea water and Mediterranean overflow water, which sit just beneath the oxygen minimum, and subtropical Central waters found to the west (Brandt et al., 2010, 2015), and Subantarctic mode water and Antarctic intermediate water to the south. The rates of mixing in the eastern tropical North Atlantic OMZ, have been quantified from a number of studies, using independent methods (Banyte et al., 2012, 2013; Hahn et al., 2014; Rudnickas et al., 2019). Mixing coefficients were recently evaluated from the dispersion of carefully ballasted isopycnal RAFOS floats, which drifted passively on an isopycnal surface, recording their position 4 times a day via acoustic tracking (Rudnickas et al., 2019). These floats were deployed in clusters at the core of the oxygen minimum (the 27.1 potential density surface, at about 500 m). Dispersion of the floats was used to estimate a coefficient of turbulent diffusivity, which grows with increasingly large spatial scales, plateauing at the mesoscale (nominally 100 km) at highly anisotropic values of 800 ± 310 m^2 s^{-1} in the meridional direction and 1410 ± 490 m^2 s^{-1} in the zonal direction.

The diffusivities diagnosed from the spreading of the isopycnal floats are consistent with those revealed from the spread of a patch of intentionally released chemical tracer (CF_3SF_5) at approximately 300 m over time (Banyte et al., 2012), which led to an estimate of isopycnal diffusivity of 500 m^2 s^{-1} in the meridional direction and 1200 m^2 s^{-1} in the zonal direction

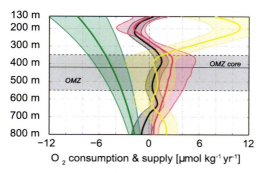

FIGURE 13.2 Tropical North Atlantic OMZ budget. Oxygen budget of the eastern tropical North Atlantic OMZ, derived from an extended observational program. Isopycnal meridional oxygen mixing (purple), vertical mixing (black), oxygen biological consumption (green), residual transport (the sum of zonal and meridional advection and zonal mixing, in yellow). The residual term is believed to be dominated by zonal advection, given evidence of narrow zonal jets providing oxygenated waters to the modelled region of the OMZ. All error estimates (coloured shadings) are the 95% confidence (except the isopycnal meridional eddy supply, where the error was estimated from both the error of the oxygen curvature (95% confidence) and the error of the eddy diffusivity, where a factor 2 was assumed). Meridional mixing is the strongest O_2 source in the core of the OMZ. The OMZ (between 350 and 550 m) is highlighted with grey shading, and its core (located at 420 m) is indicated by the black horizontal line. Adapted with permission from Brandt et al. (2015).

(Banyte et al., 2013). Furthermore, the vertical spreading of the chemical tracer allowed for the estimate of a diapycnal diffusivity coefficient of 10^{-5} m^2 s^{-1}. The anisotropy of the isopycnal diffusivity (also previously reported in other parts of the ocean, e.g., Spall et al. (1993)) supports the idea that zonally elongated mesoscale eddies (Rypina et al., 2012) play a major role in controlling mixing. It has also been suggested that the background potential vorticity (PV) distribution may shape the anisotropy by limiting mixing across PV contours (O'Dwyer et al., 2000). However, this hypothesis could not be confirmed in the eastern subtropical North Atlantic near the OMZ, where surface drifters cross mean PV contours, resulting in a poor correlation between the orientations of the mean PV contours and the direction of spreading (Rypina et al., 2012).

The quantification of the isopycnal and diapycnal diffusivities provide valuable constraints on oxygen budgets of the OMZs, which must take into account biological sources and sinks, advection, and mixing of oxygen. Given the climatological O_2 concentrations in the eastern tropical North Atlantic OMZ and the observationally constrained isopycnal and diapycnal diffusivity rates, it is possible to estimate the leading terms in the steady-state oxygen budget in the eastern tropical North Atlantic OMZ (Brandt et al., 2015; Rudnickas et al., 2019). Brandt et al. (2015) solved for these terms as the synthesis from a number of field programs and found that meridional mixing was the biggest supply term of oxygen at the core of the OMZ, balancing its removal via respiration (Fig. 13.2). Vertical mixing (black) and zonal advective and diffusive transport (solved as a residual), appear to be smaller supply terms, each providing about 15% of the supply by meridional mixing. The residual transport term is the dominant O_2 supply near the ocean's surface, but it is much smaller than the isopycnal mixing supply at the depth of the oxygen minimum. Meridional mixing is the biggest supply term, despite the fact that the meridional diffusivity coefficient is about half of the zonal coefficient (Banyte et al., 2013; Rudnickas et al., 2019). The reason is that meridional oxygen gradients are much stronger than zonal gradients (Fig. 13.1a). Zonal jets are mainly responsible for maintaining these sharp meridional oxygen gradients, as they advect high-oxygen waters eastward near the equator (Brandt et al., 2015).

The role that each term plays in the oxygen budgets of OMZs can be further assessed by comparing model simulations at different resolutions. Increasing model resolution generally allows for improved large-scale circulation, and also for explicit representation of eddy mixing. In the equatorial Pacific OMZ, for example, the large-scale equatorial circulation tied to the Equatorial Under Current (EUC) supplies oxygen to the upper part of the OMZ (top 300 m), and eddy transport achieves the final supply to the OMZ (Busecke et al., 2019). With a coarse resolution (1°) Earth system model, Busecke et al. (2019) noted an unrealistic deceleration of the EUC westward compared to observations (Fig. 13.3a–b, grey contour). As a result, the EUC in this model leaked large amounts of oxygen in the central Pacific, and the upper boundary of the OMZ was too deep (westward slanted upper boundary at 600 m in the central Pacific, Fig. 13.3a–b). In addition, the parameterised eddy transport used in this coarse resolution run remained confined to the edges of the OMZ (Fig. 13.3h). When the same model was run at finer resolution (1/10°), the EUC circulation and upper OMZ were more realistic (sustained eastward velocity across the basin and upper OMZ flat boundary at 300-m depth). Moreover, the presence of alternating equatorial zonal jets and intense eddy circulation in the fine resolution simulation enabled the presence of sharp meridional O_2 gradients and the transport by mixing of the oxygen carried by the EUC to the OMZ, reaching the OMZ core (defined here as $O_2 < 40$ μM, Fig. 13.3i).

332 Ocean Mixing

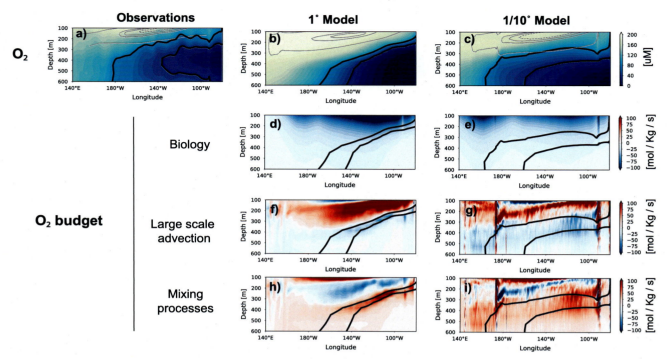

FIGURE 13.3 Pacific ocean OMZ along the equator. Oxygen concentration (colour) and Equatorial undercurrent (EUC) zonal velocities (20, 60, 70, and 80 cm s^{-1} in grey contours) from: (a) observations ((Bianchi et al., 2012) for O$_2$ and (Johnson et al., 2002) for velocities), (b) coarse 1° model (GFDL-CM1deg) and (c) eddying 1/10° model (GFDL-CM2.6). Oxygen budget associated with (d, e) biological activity, (f, g) large-scale advection, and (h, i) mixing (vertical diffusion, eddy mixing and eddy advection; resolved at 1/10° and parameterised at 1°). Thick black lines indicate O$_2$ contours of 80 μM and 40 μM. Fields are averaged between 1°S and 1°N. Adapted from Busecke et al. (2019).

The important role that eddy mixing plays in the oxygen balance was also established for the world's thickest OMZ in the Arabian sea. With a regional model of the North Indian ocean, whose horizontal grid size (1/12°) is fine enough to explicitly solve part of the mesoscale spectrum, Resplandy et al. (2012) have shown that mixing (which in their case included vertical diffusion and eddy advection) was the only oxygen source balancing the strong biological activity in the OMZ core (Fig. 13.4). In their model, 90% of the O$_2$ supply to the OMZ core was from eddy mixing: 46% from vertical eddy advection; 44% from horizontal eddy advection, and the other 10% from vertical diffusion. They also noted that, to a lesser extent, the mean large-scale advection actually removed O$_2$ from the OMZ core in their budget (as was also the case in the Pacific OMZ, Figs. 13.4c–d, 13.3g).

Lachkar et al. (2016) further examined the essential role of eddy mixing on the volume of the OMZ in a suite of regional model simulation of the Arabian sea with increasing horizontal resolution. They noted that, as resolution increased from 1/3° to 1/24° and eddy mixing was better resolved, the volume occupied by suboxic waters (defined in their case as O$_2$ < 4 μM) decreased by a factor close to two, in better agreement with observations (purple area in Fig. 13.5). Their analysis revealed that underestimating eddy mixing, as was occurring in the coarser resolution runs, caused a strong overestimation of the volume of suboxic waters. However, they also noted that eddy-driven ventilation (which provided an additional source of O$_2$ at subsurface) was partially offset by a biological feedback (which provided an additional sink of O$_2$ at subsurface). At fine resolution, the smaller extent of suboxic waters reduced denitrification (loss of nitrate due to microbial nitrate reduction in low O$_2$ environment), thereby increasing the supply of nitrate to the surface, enhancing biological production and amplifying respiration at subsurface. The consequence of this biological feedback was the increase of the volume of hypoxic waters (defined as 4 < O$_2$ < 60 μM, blue area in Fig. 13.5). Furthermore, they estimated the role of eddy mixing in the ventilation of the OMZ through the change in overturning circulation between their fine-resolution (1/24°) and coarse-resolution (1/3°) runs (Fig. 13.6). Ventilation by highly oxygenated waters (O$_2$ > 60 μM, in yellow) was quantified in terms of the volume of hypoxic water exported out of the Arabian sea (O$_2$ < 60 μM, in purple). The transport of these hypoxic waters out of the Arabian sea increased from 2 to 5 Sv when the resolution increased from 1/3° to 1/24° (Fig. 13.6a–c). This additional transport was essentially due to the eddy mixing term at 1/24°, which explains most of the differences in transport between the 1/3° and 1/24° runs (Fig. 13.6b–d).

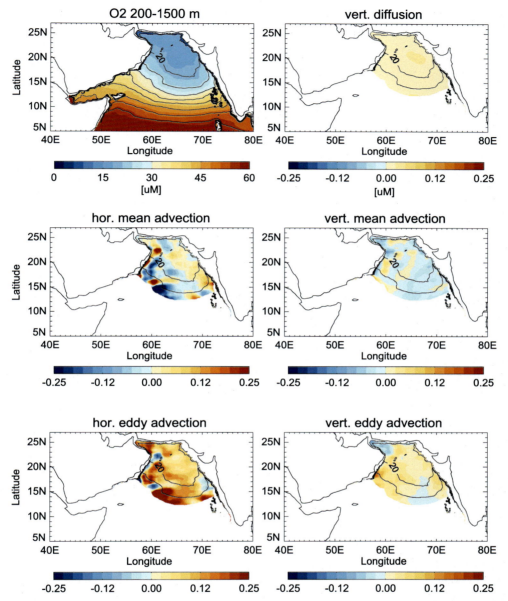

FIGURE 13.4 The Arabian sea OMZ budget. Oxygen budget between 200 and 1500 m in the Arabian sea OMZ, simulated with a forced eddy-resolving ocean regional model [μM/month]. (a) Equilibrated dissolved oxygen distribution, averaged between 200 and 1500 m. The 30 μM isoline delimits the OMZ. The oxygen tendency terms due to (b) vertical diffusion, (c) mean horizontal advection, (d) mean vertical advection, (e) horizontal eddy advection, and (f) vertical eddy advection. Only trends in the OMZ are shown. Mixing terms (vertical diffusion and eddy advection) are a source of dissolved oxygen to the OMZ (46% of the O$_2$ supply over the OMZ is from vertical eddy advection, 44% from horizontal eddy advection, 10% from vertical diffusion), while advective terms act on average as a sink (14% of O$_2$ removal are by mean vertical advection, 4% by mean horizontal advection, the remaining 82% by biological activity). Adapted from Resplandy et al. (2012).

One of the great uncertainties in ESMs is how they parameterise lateral eddy mixing (Melet et al., Chapter 2), and the degree to which these parameterisations, relative to the resolved circulation, hold the key to faithfully representing OMZs and accurately predicting their future change. Given that eddy mixing is the leading oxygen supply term to the OMZs (Gnanadesikan et al., 2013) and also that narrow currents, which are often unresolved in climate models, play a key role in setting the large-scale O$_2$ gradients (Brandt et al., 2015; Lachkar et al., 2016; Busecke et al., 2019; Rudnickas et al., 2019), it is not surprising that ESMs have difficulties in predicting the volume of OMZ (Bopp et al., 2013; Cabré et al., 2015; Bopp et al., 2017). Gnanadesikan et al. (2013) and Bahl et al. (2019) have explored the sensitivity of the global volume of hypoxic waters (which they defined as O$_2$ < 88 μM), simulated by an ESM to the choice of the isopycnal mixing coefficient (A_{REDI})

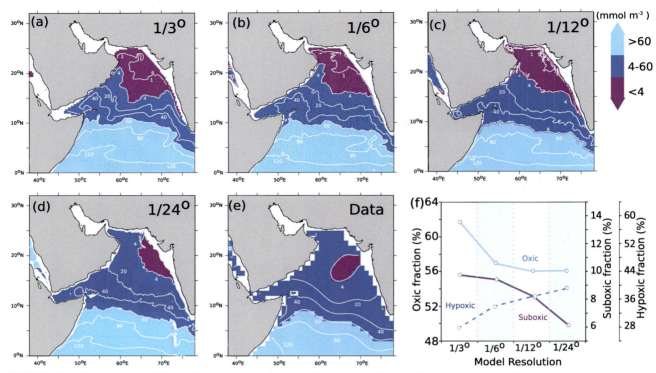

FIGURE 13.5 The Arabian sea OMZ sensitivity to model resolution. Horizontal distribution of dissolved oxygen at 250 m depth in winter (December–February) and volume fraction of the Arabian sea OMZ (defined as the volume of suboxic waters, in purple) in a series of regional model experiments run at increasing horizontal grid resolution at (a) 1/3°, (b) 1/6°, (c) 1/12°, (d) 1/24°, and (e) from World ocean atlas 2009 data set. (f) Oxic ($O_2 > 60$ μM), suboxic ($O_2 < 4$ μM), and hypoxic ($4 < O_2 < 60$ μM) volume fractions as simulated at different resolutions in the top 1000 m. Increased model resolution reduces the volume of suboxic waters, but due to a biological feedback through reduced denitrification, it expands the volume of hypoxic waters, and thus compresses habitats (i.e., compresses the volume of oxic waters). Reproduced from Lachkar et al. (2016).

within the range of values used by the current generation of ESMs. They found that low-mixing models simulate larger volumes of hypoxic waters than high-mixing models (Fig. 13.7B). This result is in line with the results of Lachkar et al. (2016) of expanding suboxic waters, but, because denitrification is absent in the Bahl et al. (2019) model, it occurs without the feedback of slowed denitrification increasing primary productivity and the associated respiration (Lachkar et al., 2016). The isopycnal mixing coefficient A_{REDI} is typically held constant in ocean models and is often isotropic. Gnanadesikan et al. (2013) and Bahl et al. (2019) have also attempted to use a more realistic, spatially-variable A_{REDI}, with very large values at the edges of boundary currents and very low values far from such currents, following Abernathey and Marshall (2013) (ABER2D, Fig. 13.7), or more simple zonally averaged version of the ABER2D parameterisation (ABERZONAL, Fig. 13.7). These two experiments yielded similar results than those with a large, and constant, A_{REDI} coefficient, and did not improve the fit to observations, suggesting that current parameterisations may represent eddy mixing too simplistically, or that the resolved circulation plays an important role that cannot be simply solved by improving the parameterisation of lateral mixing.

13.3 Role of mixing on global deoxygenation

One of the most dramatic transformations of the global ocean over the past fifty years is its loss of oxygen (Fig. 13.1b and Fig. 13.8). The last Intergovernmental Panel on Climate Change Special Report on the Ocean and Cryosphere in a changing climate (IPCC-SROC, 2019) assessment reports that over this period, the upper 1000 m of the ocean has lost about 2% of its oxygen reservoir (Ito et al., 2017; Schmidtko et al., 2017). Deoxygenation is projected to continue during the 21st century, unless greenhouse gas emissions are rapidly curtailed, as in the RCP2.6 scenario (Fig. 13.8) (Bopp et al., 2013; Cabré et al., 2015; Bopp et al., 2017; Hameau et al., 2020; Kwiatkowski, 2020; Li et al., 2020). For marine animals living in oceanic environments just at the edge of their metabolic oxygen needs the declining oxygen reservoir represents an immediate threat to survival (Vaquer-Sunyer and Duarte, 2008).

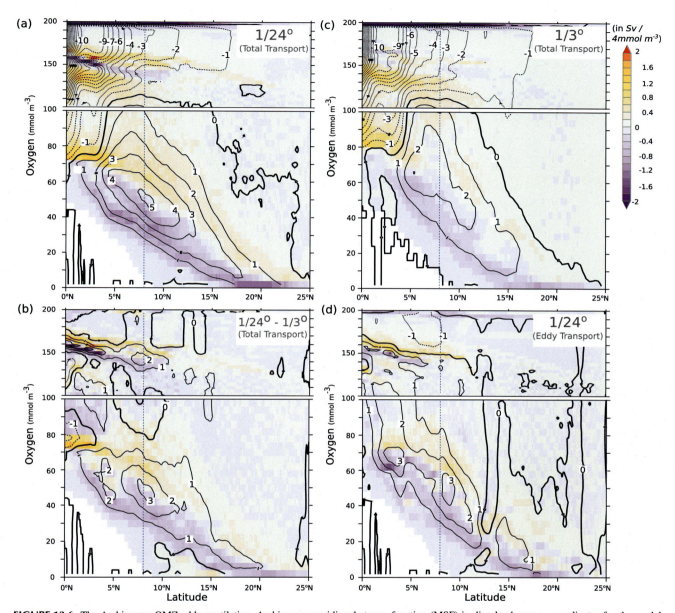

FIGURE 13.6 The Arabian sea OMZ eddy ventilation. Arabian sea meridional stream function (MSF) in dissolved oxygen coordinates for the model experiments shown in Fig. 13.5 (in Sv). (a) The 1/24° simulation, (b) the 1/3° simulation, (c) the difference between 1/24° and 1/3° simulations, and (d) the eddy component of the transport in the 1/24° simulation. Contour lines indicate the meridional stream function (positive indicates clockwise circulation). The colour shading shows the meridional component of the transport. The vertical blue dashed line indicates the latitude of the southern tip of India. Similarities of the MSF between panels b and c highlight the central role of eddy mixing in the compression of the OMZ as grid resolution is increased. Reproduced from Lachkar et al. (2016).

The drivers of deoxygenation remain poorly quantified (Oschlies et al., 2018). The easiest to quantify is the reduced solubility of oxygen as the ocean warms. In addition, warming can reduce exchanges between the surface and intermediate depths. Together, these changes slow the supply of oxygen to the ocean interior. On the other hand, declining exchanges between the surface and subsurface also reduces the supply of nutrients to the euphotic layer, limits primary production, and diminishes the organic matter export that fuels oxygen consumption by respiration in the subsurface layers. Deoxygenation results from an imbalance between the decreased oxygen consumption and decreased oxygen supply. The fraction of total oxygen decline due to solubility (ΔO_2^{sat}) can be evaluated from the changes in temperature and salinity of the seawater, and 40–50% of the contemporary oxygen loss has been attributed to the solubility effect (Helm et al., 2011; Bopp et al., 2013;

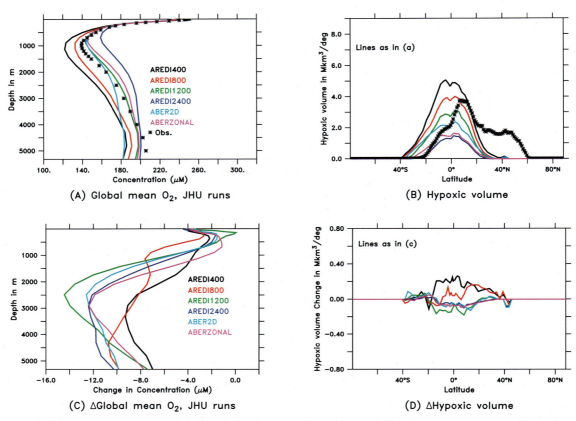

FIGURE 13.7 Global oxygen reservoir sensitivity to lateral diffusivity coefficient, A_{REDI}. (A) Global mean oxygen concentration profile during the preindustrial period in data (crosses) and a suite of the ESM2Mc ESM with different isopycnal mixing coefficients A_{REDI} (colour lines). In ABER2D and ABERZONAL, a spatially variable A_{REDI} was used, with relatively large values of A_{REDI}. (B) Volume of hypoxic waters (defined as $O_2 < 88$ μM). This panel highlights the critical role of the strength of eddy mixing in controlling the volume of OMZs. The volume is larger when lateral diffusivity is smaller. (C) Changes in total oxygen concentration under doubled CO_2. Deoxygenation is larger when lateral diffusivity is larger. (D) change in the hypoxic volume under doubled CO_2. The volume of hypoxic waters is increased for the smaller values of A_{REDI} (400 and 800 m^2 s^{-1}) and is decreased for all other cases. JHU = Johns Hopkins University. Reproduced with permission from Bahl et al. (2019).

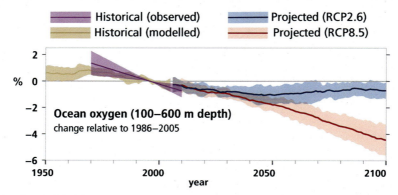

FIGURE 13.8 Historical and projected global deoxygenation. Observed (purple) and modelled (brown) historical changes in the ocean since 1950, and projected future changes under low (blue, RCP2.6) and high (red, RCP8.5) greenhouse gas emissions scenarios, of the global mean ocean oxygen content between 100 and 600 m. Assessed observational trends span 1970–2010 centred on 1996. The shading represents one standard deviation across models. Adapted from IPCC-SROC (2019).

Schmidtko et al., 2017). The remainder, apparent oxygen utilisation ($-\Delta AOU = \Delta O_2 - \Delta O_2^{sat}$) represents the change in oxygen concentration driven by changes in ocean dynamics and respiration.

Numerical models simulate a global average oxygen decline over the past several decades, but the multi-model average underestimates the rate of observed decline (Fig. 13.8). The models also misrepresent the spatial distribution of oxygen

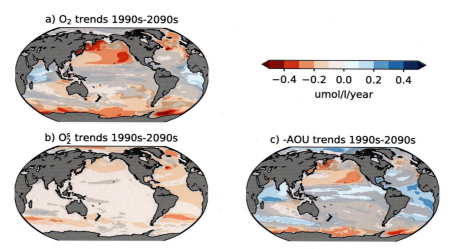

FIGURE 13.9 Projected deoxygenation trends. Deoxygenation trends between 200 and 600 m under RCP8.5 in the mean of nine simulations from different models of the CMIP5 framework in 2090–2099 relative to 1990–1999, for dissolved (a) O_2, (b) O_2^{sat} and (c) AOU. The trend is computed as the mean of the last decade minus the mean of the first decade. Grey shading marks low robustness between simulations (disagreement on sign of changes for more than one of the nine simulations).

decline, particularly in the tropics (Oschlies et al., 2017). ESMs project deoxygenation at the global scale of another 3.5–5.5% from the present to 2100 under the high-emissions scenario (RCP8.5), with stronger deoxygenation at mid- and high latitudes (Bopp et al., 2013; Cocco et al., 2013). These projections disagree with one another on the sign of the trend in the tropical oceans (i.e., in Fig. 13.9a the grey stippled regions show where at least one of the models do not agree on the sign of the future change; see Section 13.4 for discussion on OMZ response). At the global scale, ESMs project consistently reduced oxygen solubility tied to ocean warming at the global scale (ΔO_2^{sat}, Fig. 13.9b). This solubility effect is reinforced by a strong reduction in oxygen supply by ventilation at mid- and high latitudes ($-\Delta AOU$, Fig. 13.9c).

Assessing the drivers of deoxygenation quantitatively, and the uncertainties associated with each of them, is a necessary first step to understand the misfit between model and data, and the model spread. These drivers include changes in respiration, in ocean circulation, including the primarily wind-forced subtropical gyres, and/or changes in the overturning circulation that supplies oxygen-rich water to the deep ocean via convective mixing. The changing dynamics also encompasses smaller-scale processes, such as turbulent diapycnal and isopycnal mixing across oxygen gradients. In the following, we review two model studies that have attempted to tease apart the different factors involved in the change of AOU and have highlighted the leading role of mixing in the deoxygenation imbalance. Both studies arrive at similar conclusions about the importance of mixing in setting the pace of global deoxygenation in a warming world, using independent models and different methodologies.

One study (Palter and Trossman, 2018) manipulated a model framework to fully separate the roles of changing large-scale ocean circulation from changes in mixing. In the first set of simulations (run in triplicate to reduce the impact of internal variability on the results), the atmospheric CO_2 concentrations were increased at 1% per year, until they doubled at approximately year 70, and all components of the climate system evolved freely in response to this perturbation (V_{free}). The second set of simulations was identical, except that the ocean velocity field was held fixed at its pre-industrial values (V_{fixed}). Thus the role of parameterised mixing versus resolved circulation is revealed from the difference between V_{free} and V_{fixed} (Fig. 13.10). This model framework suggests that the resolved (large-scale) ocean circulation could not be blamed for the loss of oxygen from the ocean. In fact, circulation changes slightly slowed the global deoxygenation rate, both by reducing the pace of warming (Trossman et al., 2016) and stabilising mixed-layer depths, particularly in the Southern ocean (Fig. 13.10). In other words, in this model on multi-decadal timescales, changes in the parameterised mixing were responsible for essentially all of the deoxygenation not attributable to the effect of warming on oxygen solubility. Furthermore, the reduction in the oxygen consumption through respiration provided only a minor offset (less than 5%) to counterbalance the much larger loss of oxygen supply to the ocean interior.

Confirming the first-order role of mixing as a cause of global deoxygenation, Couespel et al. (2019) examined the oxygen budget in an ESM forced by a high-CO_2 emission scenario for the 21st century, using diagnostics of oxygen subduction. They partitioned physical dissolved oxygen exchanges between the surface and ocean interior into a component due to the kinematic flux linked to the subduction of water masses containing oxygen, and a diffusive flux due to oxygen mixing across the mixed-layer interface. For the sake of simplicity, these terms were named "kinematic subduction" of

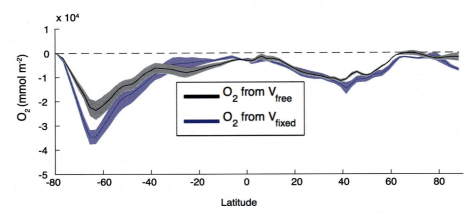

FIGURE 13.10 Zonally averaged deoxygenation: role of advection and mixing on O_2 content. Zonally averaged deoxygenation under a doubling of atmospheric CO_2 simulated with the Geophysical Fluid Dynamics Laboratory Climate Model version 2 (CM2.1) at coarse resolution (CM2Mc). Each curve represents a mean over an ensemble of 3 simulations. In the V_{free} set of experiments, changes in both advection and changes in mixing are accounted for. In V_{fixed}, only changes in mixing are accounted for. Comparison between the two sets reveals that the O_2 decline is due largely to changes in mixing, whereas changes in circulation slightly slow the deoxygenation, particularly in the Southern ocean, where the oxygen loss is the most pronounced. Reproduced from Palter and Trossman (2018).

oxygen and "diffusive subduction" of oxygen in Couespel et al. (2019), with the following definitions: the kinematic subduction of oxygen (S_{kin}, Eq. (13.2)) is the product of the kinematic subduction (S, Eq. (13.1)), in volume flux per unit area, as defined by Cushman-Roisin (1987), by the oxygen concentration (Eq. (13.2)); and the diffusive subduction of oxygen (S_{mix}, Eq. (13.3)) is defined as the transfer of oxygen by diffusion across the mixed-layer interface.

$$S = w_h + \vec{u_h} \cdot \vec{\nabla}_H h + \partial_t h \tag{13.1}$$

$$S_{kin} = S \times O_h \tag{13.2}$$

$$S_{mix} = (k_z \times \partial_z O)_h + k_l \times (\vec{\nabla_l} O)_h \cdot \vec{\nabla_l} h \tag{13.3}$$

Here, h is the mixed layer depth, u and w are the horizontal and vertical velocities at the base of the mixed layer, and k_z and k_l are the vertical and isopycnal mixing coefficients at that depth.

The kinematic flux (or "kinematic subduction", Eqs. (13.1) & (13.2)) includes advection through vertical and horizontal currents, entrainment/detrainment caused by temporal changes of the depth of the mixed layer, as well as advection by bolus velocities used to parameterise the flattening of isopycnals by eddies (Gent and McWilliams, 1990).

The diffusive flux (or "diffusive subduction", Eq. (13.3)) includes both vertical mixing and isopycnal mixing by eddies, which are proportional to oxygen gradients at the base of the mixed layer.

The changes between the beginning and the end of the 21st century reveal that, though kinematic subduction locally shows the largest amplitude of changes, these changes swing between large increases and decreases, and therefore sum to a small global reduction in the subduction of oxygen (Fig. 13.11). This balance is particularly apparent in the Southern ocean, where the S_{kin} changes are mainly a response to the southward shift of the westerly winds. In contrast, diffusive subduction is projected to decrease more uniformly over the globe. Therefore when globally integrated, negative changes in S_{mix} sum to over 75% of the reduction in the total subduction of O_2 (small inset in Fig. 13.11). Interestingly, the reduction in S_{mix} was linked to a reduction of the vertical oxygen gradient across the base of the surface mixed layer, rather than a reduction in mixing efficiency. The weakened vertical gradient can be explained by a reduction in oxygen undersaturation below the mixed layer, a direct consequence of the diminution in respiration following the reduction in nutrient supply by mixing and biological production at the surface. The globally integrated decline in respiration (green bar in Fig. 13.11 inset), which acts as a negative feedback on global deoxygenation, was much smaller than the decline in O_2 supply by subduction (the sum of the red and cyan bars in Fig. 13.11).

Regardless of the difference in methodologies, these two modelling studies come to similar conclusions about the drivers of global deoxygenation (Figs. 13.10 and 13.11). Under strong warming scenarios, both project on the order of 50 Tmol y^{-1} of oxygen loss from the global ocean, with the principal driver of this loss due to changes in mixing. Though these mixing changes are also accompanied by reduced global respiration in both models, this diminished sink term does not substantially offset the slowing O_2 supply to the ocean interior (inset in Fig. 13.11). The spatial patterns in Figs. 13.10 and 13.11 might be interpreted to suggest that these models—while having similar global rates and drivers of deoxygenation—are realising

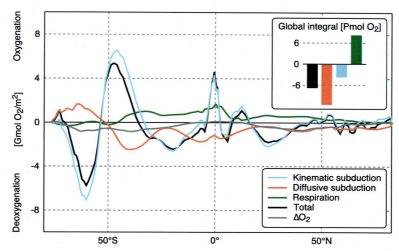

FIGURE 13.11 Zonally integrated deoxygenation: Role of advection, mixing and respiration on O$_2$ fluxes. Zonal integral of the projected change in kinematic oxygen subduction across the surface mixed layer (blue curve), of the projected change in oxygen mixing across the surface mixed layer (red curve, diffusive subduction), of the projected change in respiration below the surface mixed layer (green curve), and their sum (total) along the 21st century, in the IPSL RCP 8.5 CMIP5 simulations. Shown are cumulated fluxes over 110 years (1990 to 2099), and correspond to an average O$_2$ loss of 55 TM y^{-1}. Locally, projected changes in kinematic subduction prevail, particularly in the Southern ocean. However, when integrated globally (shown in small inset), global deoxygenation (total) is explained largely by a reduction in diffusive subduction (mixing), accompanied by a reduction in kinematic subduction, and partially offset by a reduction in respiration. Note that a zonal integral is shown here, and a zonal average is shown in Fig. 13.10. Adapted from Couespel et al. (2019).

the O$_2$ losses in different regions. However, Fig. 13.10 depicts the change in the vertically integrated O$_2$ reservoir, whereas Fig. 13.11 shows the change in the subduction flux of O$_2$ across the base of the mixed layer. The change in the O$_2$ reservoir suggests that the largest losses will be in the Southern ocean and northern hemisphere subtropical latitudes, which is in general agreement with the historical O$_2$ trends in Fig. 13.1 and ensemble-mean projected changes in Fig. 13.9.

13.4 Response of OMZ to global warming

A major concern is that OMZs may expand in the 21st century in response to ocean warning, threatening the survival of marine organisms that rely on dissolved oxygen for respiration, and affecting the biogeochemical cycling of carbon and nitrogen, potentially amplifying global warming (Stramma et al., 2008; Keeling et al., 2010; Fu et al., 2018; Lachkar et al., 2018; Resplandy, 2018; Lachkar et al., 2019). Observations indicate strong regional variations in the deoxygenation of the tropical oceans. OMZs have clearly expanded since the 1970s in the equatorial Pacific and Atlantic oceans (Stramma et al., 2008; Brandt et al., 2010), but with strong multi-decadal temporal variations (Deutsch et al., 2014). In the Arabian sea, the northern and western pars of the OMZ are expanding, whereas the southern part is shrinking (Banse et al., 2014; Piontkovski and Al-Oufi, 2014; Queste et al., 2018). ESMs project dramatically different changes in OMZ volume by 2100 (ranging from −2% to +16%) (Bopp et al., 2013; Cabré et al., 2015). Despite these differences, ESMs agree on many features that control deoxygenation in tropical OMZs. The solubility effect in tropical OMZs is partly offset by an increased oxygen supply by ventilation and a reduced biological oxygen consumption (Cabré et al., 2015; Bopp et al., 2017). It has also been suggested that tropical OMZ might shrink after 2100, following their initial expansion (Fu et al., 2018). The high uncertainties in OMZs evolution between ESMs largely arise from differences in the magnitude and timing of these ventilation and biological changes and how strongly they offset each other (Resplandy, 2018). For instance, the Pacific OMZ is more sensitive to changes in equatorial circulation in models that underestimate eddy-driven O$_2$ supply by mixing, which might explain the disparate future deoxygenation responses across models in this region (Shigemitsu et al., 2017; Busecke et al., 2019).

ESMs frequently project increased ventilation in tropical OMZs in the future, with diapycnal and isopycnal mixing accelerating the renewal of OMZ waters (Gnanadesikan et al., 2007; Bopp et al., 2017; Bahl et al., 2019). Here we report on two modelling studies that have shown that the fate of OMZs —whether they shrink or expand under projected warming —depends on the value of the mixing coefficient; one study focused on diapycnal mixing (Duteil and Oschlies, 2011), the other on isopycnal mixing (Bahl et al., 2019).

Bahl et al. (2019) varied the isopycnal mixing coefficient within the range used by current ESMs (constant coefficients from 400 to 2400 $m^2\ s^{-1}$, and two spatially variable coefficients with local values up to 10,000 $m^2\ s^{-1}$ in high-mixing areas). Under doubled atmospheric CO_2, the volume of hypoxic waters increased in models with low-mixing, but decreased in models with high-mixing and with spatially variable mixing coefficients (Fig. 13.7d). Surprisingly, these changes in hypoxic volume (largest increase in hypoxic volume in low-mixing models, Fig. 13.7d) were decoupled from the total amount of deoxygenation (largest decline in oxygen in high-mixing models, Fig. 13.7c).

Duteil and Oschlies (2011) varied the background value of vertical diffusivities within a range corresponding to observed values and to values currently used in ocean models (from 0.01 to 0.5 $cm^2\ s^{-1}$) in a coarse-resolution ESM, in which diapycnal mixing was parameterised as the sum of the regionally heterogeneous tidal and homogeneous background vertical mixing. Under projected 21st century CO_2 emissions, all of their model experiments predicted global deoxygenation, but the extent of suboxia (which they defined as $O_2 < 5\ \mu M$) expanded only for mixing rates higher than 0.2 $cm^2\ s^{-1}$, and declined for lower mixing rates (by up to $+/-$ 40% by 2100). In their experiments, both respiration and ventilation decreased with global warming. But when the vertical mixing coefficient exceeded 0.2 $cm^2\ s^{-1}$, the relative decline in respiration was smaller than the decline in physical supplies of oxygen, resulting in an expansion of the suboxic volume; in contrast at lower mixing intensities, the relative decline in respiration was larger than the decline in physical supplies, resulting in a reduction of the suboxic volume.

These two studies both report a non-monotonic response of the volume of suboxic and hypoxic waters to global warming against mixing coefficients, which reflects the fact that mixing affects the OMZ directly via ventilation and indirectly via biological feedbacks (respiration and denitrification).

In Fig. 13.12, we have attempted to conceptualise the possible responses of OMZs to global warming. The outcome for the OMZ depends on the balance between reduced oxygen supply (due to reduced mixing) and reduced oxygen consumption (due to reduced respiration). The respective changes in ventilation and respiration depend on multiple factors, such as the nutrient and oxygen gradients, the contribution of mixing to total transport for nutrient and oxygen, and the removal of nutrients by denitrification. Moreover, the changes due to mixing add to other responses to climate change, such as changes in large-scale advection of oxygen and of nutrients. Finally, the sign of the change in mixing intensity itself is highly uncertain and region-dependent (Gnanadesikan et al., 2007; Bopp et al., 2017; Bahl et al., 2019). This schematic shows how the OMZ can globally expand or shrink, but the outcome can also be different at the top and bottom of the OMZ, causing it to either deepen or get closer to the surface.

13.5 Conclusions and grand challenges

We have reviewed current knowledge on the role of mixing on the oxygen budget in the ocean, and its imbalance driven by climate change. The pace of global deoxygenation, as well as the volume and intensity of OMZs and their projected changes have been shown to be particularly sensitive to lateral and diapycnal mixing. Mixing directly influences ocean ventilation, but also indirectly influences oxygen via biological feedbacks (respiration and denitrification). Weaker mixing associated with global warming can reduce the supply of oxygen to the ocean interior, while simultaneously slowing the upward transport of nutrients that sustains biological production in the upper ocean and the consumption of that organic matter via microbial respiration at depth. The traditional view is that the effect of reduced ventilation is partly offset by this slowdown in biological respiration at depth. Yet, the net effect of the reduced ventilation and respiration, induced by mixing, strongly varies in space and probably in time, depending on oxygen levels and mixing rates. In ventilation regions at mid- and high latitudes, the decline in respiration is strongest near the surface, which can reduce the vertical oxygen gradient, thereby slowing the downward mixing of oxygen at the base of the mixed layer, and counter intuitively, reinforce the effect of deoxygenation through reduced ventilation (Couespel et al., 2019). In OMZs, where mixing rates are low, the supply and utilisation of oxygen are tightly coupled. As a result, the respiration changes can counterbalance and even exceed mixing-induced changes in ventilation (Duteil and Oschlies, 2011; Fu et al., 2018; Bahl et al., 2019). In addition, the respiration feedback in OMZs is reinforced by changes in microbial metabolic pathways between oxygen-based and nitrate-based (denitrification) respiration, stabilising oxygen levels and OMZs volume (Lachkar et al., 2016). Uncertainties in this subtle balance between mixing-induced ventilation and the associated biological feedbacks contribute to the strong bias in present simulated OMZ volume and to uncertainties in their future evolution (Resplandy, 2018).

Current ESMs often misrepresent observed oxygen patterns and, on average, underestimate historical deoxygenation. There are three main potential causes for these misrepresentations (Oschlies et al., 2017): unaccounted variations in respiratory oxygen demand, missing biogeochemical feedbacks, and/or unresolved transport processes. The reasons that a model misrepresents the spatial patterns of ocean oxygen, particularly the size and O_2 concentrations in OMZs, may be distinct from the causes of the mismatch in globally averaged rates of deoxygenation.

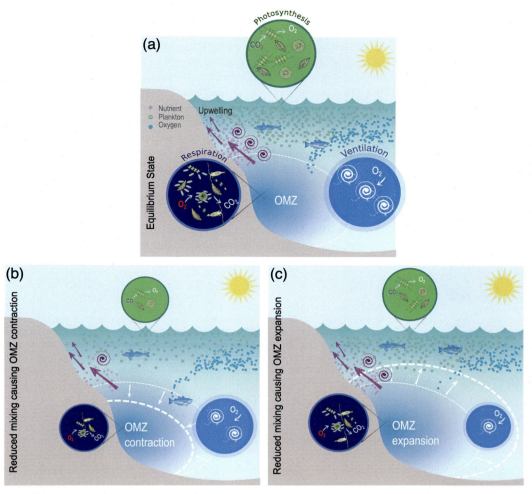

FIGURE 13.12 Possible responses of OMZs to reduced mixing associated with global warming. In this schematic representation of an OMZ, oxygen is replenished through physical processes (with intensity proportional to the light-blue circle, labelled "Ventilation"), a combination of advective (white arrows inside the light blue circle), and mixing (white swirls) processes. Nutrients are brought to the surface through upwelling (purple arrows) and vertical mixing (purple swirls). The intensity of photosynthesis depends on the intensity of the physical supply of nutrients through upwelling and mixing. For simplicity, we neglect denitrification here, so the magnitude of photosynthesis equals that of respiration in all panels (the photosynthesis and respiration circles have identical sizes). (a) The balanced state of an OMZ in which oxygen consumed by biological respiration is replenished through the ventilation of the ocean interior (the respiration and ventilation circles have identical sizes). In (b) and (c), mixing is assumed to be reduced in response to global warming, but with opposite consequences on the OMZ. Reduced mixing causes a reduction in both oxygen ventilation and nutrient supply, and hence a reduction in productivity and in biological respiration. This is reflected by the sizes of the photosynthesis, respiration and ventilation circles, which are smaller than in (a). In (b), the decline in respiration exceeds the decline in ventilation, resulting in the contraction of the OMZ. In (c) the reduction in ventilation exceeds the reduction in respiration, resulting in the expansion of the OMZ. The magnitude of these ventilation and respiration changes depends on changes in mixing, but also changes in the large-scale advection of oxygen and nutrients influencing the gradients upon which mixing is acting. Finally, the sign of change of mixing in response to global warming (decreased mixing versus increased mixing) is also uncertain and region-dependent.

The studies reviewed herein provide evidence that (sub-)mesoscale eddies strongly influence the volume of the lowest-oxygen water in the global OMZs. Therefore the parameterised representation of these eddies in coarse-resolution ESMs is critical to accurately simulate these regions and their evolution in the future. Current eddy mixing parameterisations rely on a diffusivity operator that acts on lateral oxygen gradients. Deoxygenation rates and the volume of OMZs are very sensitive to the value of the diffusivity coefficients of this operator, and neither varying this diffusivity across a wide range of constant values, nor representing its spatial variability (Abernathey and Marshall, 2013) was sufficient to bring the volume of hypoxic water simulated by a coarse resolution ESM in full agreement with observations (Gnanadesikan et al., 2013; Bahl et al., 2019). On the other hand, regional models show a significant improvement in their representation of OMZs when the grid resolution is increased (Lachkar et al., 2016; Busecke et al., 2019). Importantly, increasing grid resolution not only allows the model to explicitly resolve eddy-driven mixing, but also improves the large-scale circulation (Lévy et al., 2010; Busecke et al., 2019), and both aspects matter to the ventilation of the OMZ (Brandt et al., 2010; Resplandy et al., 2012; Rudnickas et al., 2019). Mixing depends not only on mixing strength, but also on oxygen gradients. Any factor that contributes to setting this gradient, such as the respiratory oxygen consumption or the large scale circulation, may lead to a misrepresentation, or improvement, in the mixing transport of oxygen. Finally, some important biogeochemical feedbacks, such as the switch from aerobic respiration to denitrification, depend on the strength of mixing (Lachkar et al., 2016). Therefore the three different causes for oxygen misrepresentation interact non-linearly, with the biogeochemical feedback being constrained by the amount of oxygen supplied by mixing, which itself is influenced by the biological oxygen demand.

In addition to playing a central role in the intensity and volume of OMZs, mixing also strongly influences the rate of global deoxygenation. Higher isopycnal diffusivity, or diffusivities that vary spatially, can change the total deoxygenation in response to warming by up to 50%. This sensitivity points to isopycnal mixing as an under-constrained parameterised process with high leverage to influence the deoxygenation (Bahl et al., 2019; Jones and Abernathey, 2019). Likewise, changes in diapycnal mixing can exert a strong influence on global oxygen levels, with the role of mixing across the base of the mixed layer, termed diffusive subduction (Couespel et al., 2019), contributing a large share to global average deoxygenation. Mixing appears to be a particularly strong driver when deep convection slows at high latitudes, denying the densest layers of the ocean interaction with the atmosphere (Palter and Trossman, 2018).

Given the threat posed to ocean ecosystems by deoxygenation, improving the parameterisation of mixing in climate models is critical to reliably project future changes in oceanic oxygen distribution.

Acknowledgements

We are grateful to Peter Brandt, Julius Busecke and Anand Gnanadesikan, who kindly shared with us or adapted figures from some of their previous work, and to Alessandro Tagliabue, who adapted a figure from the IPCC-SROC SPM report. We also thank Jan Zika and Ric Williams, who provided insightful reviews that helped to improve this chapter. This work was supported by ANR SOBUMS (ANR-16-CE01-0014). ZL and ML received support from New York University Abu Dhabi Institute's Center for Prototype Climate Modelling. LR gratefully acknowledges support of the High Meadows Environmental Institute "Climate and Energy Grand Challenge" and "Carbon Mitigation Initiative" and the Sloan Foundation Research Fellowship. JBP was supported by NSF-1736985.

References

Abernathey, R.P., Marshall, J., 2013. Global surface eddy diffusivities derived from satellite altimetry. J. Geophys. Res., Oceans 118 (2), 901–916. https://doi.org/10.1002/jgrc.20066.

Babbin, A.R., Bianchi, D., Jayakumar, A., Ward, B.B., 2015. Rapid nitrous oxide cycling in the suboxic ocean. Science 348 (6239), 1127–1129. https://doi.org/10.1126/science.aaa8380.

Bahl, A., Gnanadesikan, A., Pradal, M.A., 2019. Variations in ocean deoxygenation across Earth system models: isolating the role of parameterized lateral mixing. Glob. Biogeochem. Cycles 33 (6), 703–724. https://doi.org/10.1029/2018GB006121.

Banse, K., Naqvi, S.W.A., Narvekar, P.V., Postel, J.R., Jayakumar, D.A., 2014. Oxygen minimum zone of the open Arabian Sea: variability of oxygen and nitrite from daily to decadal timescales. Biogeosciences 11 (8), 2237–2261. https://doi.org/10.5194/bg-11-2237-2014-supplement.

Banyte, D., Tanhua, T., Visbeck, M., Wallace, D.W.R., Karstensen, J., Krahmann, G., Schneider, A., Stramma, L., Dengler, M., 2012. Diapycnal diffusivity at the upper boundary of the tropical North Atlantic oxygen minimum zone. J. Geophys. Res., Oceans 117 (C9). https://doi.org/10.1029/2011JC007762.

Banyte, D., Visbeck, M., Tanhua, T., Fischer, T., Krahmann, G., Karstensen, J., 2013. Lateral diffusivity from tracer release experiments in the tropical North Atlantic thermocline. J. Geophys. Res., Oceans 118 (5), 2719–2733. https://doi.org/10.1002/jgrc.20211.

Bianchi, D., Dunne, J.P., Sarmiento, J.L., Galbraith, E.D., 2012. Data-based estimates of suboxia, denitrification, and N2O production in the ocean and their sensitivities to dissolved O2. Glob. Biogeochem. Cycles 26 (2). https://doi.org/10.1029/2011GB004209.

Bopp, L., et al., 2013. Multiple stressors of ocean ecosystems in the 21st century: projections with CMIP5 models. Biogeosciences 10 (10), 6225–6245. https://doi.org/10.5194/bg-10-6225-2013.

Bopp, L., Resplandy, L., Untersee, A., Le Mezo, P., Kageyama, M., 2017. Ocean (de)oxygenation from the Last Glacial Maximum to the twenty-first century: insights from Earth System models. Philos. Trans. R. Soc. A, Math. Phys. Eng. Sci. 375 (2102), 20160323. https://doi.org/10.1098/rsta.2016.0323.

Brandt, P., et al., 2015. On the role of circulation and mixing in the ventilation of oxygen minimum zones with a focus on the eastern tropical North Atlantic. Biogeosciences 12 (2), 489–512. https://doi.org/10.5194/bg-12-489-2015.

Brandt, P., Hormann, V., Kortzinger, A., Visbeck, M., Krahmann, G., Stramma, L., Lumpkin, R., Schmid, C., 2010. Changes in the ventilation of the oxygen minimum zone of the tropical North Atlantic. J. Phys. Oceanogr. 40 (8), 1784–1801. https://doi.org/10.1175/2010JPO4301.1.

Busecke, J.J.M., Resplandy, L., Dunne, J.P., 2019. The equatorial undercurrent and the oxygen minimum zone in the Pacific. Geophys. Res. Lett. 46 (12), 6716–6725. https://doi.org/10.1029/2019GL082692.

Cabré, A., Marinov, I., Bernardello, R., Bianchi, D., 2015. Oxygen minimum zones in the tropical Pacific across CMIP5 models: mean state differences and climate change trends. Biogeosciences 12 (18), 5429–5454. https://doi.org/10.5194/bg-12-5429-2015.

Cocco, V., et al., 2013. Oxygen and indicators of stress for marine life in multi-model global warming projections. Biogeosciences 10 (3), 1849–1868. https://doi.org/10.5194/bg-10-1849-2013.

Couespel, D., Levy, M., Bopp, L., 2019. Major contribution of reduced upper ocean oxygen mixing to global ocean deoxygenation in an Earth system model. Geophys. Res. Lett. 46 (21), 12239–12249. https://doi.org/10.1029/2019GL084162.

Cushman-Roisin, B., 1987. Subduction. In: Muller, P., Henderson, D. (Eds.), Dynamics of the Oceanic Surface Mixed Layer. In: Hawaii Institute of Geophysics Special Publication, pp. 182–195.

Deutsch, C., et al., 2014. Centennial changes in North Pacific anoxia linked to tropical trade winds. Science 345 (6197), 665.

Dufour, C.O., et al., 2015. Role of mesoscale eddies in cross-frontal transport of heat and biogeochemical tracers in the Southern Ocean. J. Phys. Oceanogr. 45 (12), 3057–3081. https://doi.org/10.1175/JPO-D-14-0240.1.

Duteil, O., Oschlies, A., 2011. Sensitivity of simulated extent and future evolution of marine suboxia to mixing intensity. Geophys. Res. Lett. 38 (6), L06607. https://doi.org/10.1029/2011GL046877.

Fu, W., Primeau, F., Keith Moore, J., Lindsay, K., Randerson, J.T., 2018. Reversal of increasing tropical ocean hypoxia trends with sustained climate warming. Glob. Biogeochem. Cycles 32 (4), 551–564. https://doi.org/10.1002/2017GB005788.

Gent, P.R., McWilliams, J.C., 1990. Isopycnal mixing in ocean circulation models. J. Phys. Oceanogr. 20, 150–155.

Gnanadesikan, A., Bianchi, D., Pradal, M.-A., 2013. Critical role for mesoscale eddy diffusion in supplying oxygen to hypoxic ocean waters. Geophys. Res. Lett. 40 (19), 5194–5198. https://doi.org/10.1002/grl.50998.

Gnanadesikan, A., Dunne, J.P., John, J., 2012. Understanding why the volume of suboxic waters does not increase over centuries of global warming in an Earth System Model. Biogeosciences 9 (3), 1159–1172. https://doi.org/10.5194/bg-9-1159-2012.

Gnanadesikan, A., Russell, J.L., Zeng, F., 2007. How does ocean ventilation change under global warming? Ocean Sci. 3 (1), 43–53.

Hahn, J., Brandt, P., Greatbatch, R.J., Krahmann, G., Körtzinger, A., 2014. Oxygen variance and meridional oxygen supply in the Tropical North East Atlantic oxygen minimum zone. Clim. Dyn. 43 (11), 2999–3024. https://doi.org/10.1007/s00382-014-2065-0.

Hameau, A., Frölicher, T.L., Mignot, J., Joos, F., 2020. Is deoxygenation detectable before warming in the thermocline? Biogeosciences 17 (7), 1877–1895. https://doi.org/10.5194/bg-17-1877-2020.

Helm, K.P., Bindoff, N.L., Church, J.A., 2011. Observed decreases in oxygen content of the global ocean. Geophys. Res. Lett. 38 (23), L23602. https://doi.org/10.1029/2011GL049513.

IPCC-SROC, 2019. SROCC Summary for Policymakers. In: Portner, H.O., et al. (Eds.), IPCC Special Report on the Ocean and Cryosphere in a Changing Climate, pp. 1–36.

Ito, T., Deutsch, C., 2010. A conceptual model for the temporal spectrum of oceanic oxygen variability. Geophys. Res. Lett. 37 (3), L03601. https://doi.org/10.1029/2009GL041595.

Ito, T., Minobe, S., Long, M.C., Deutsch, C., 2017. Upper ocean O2 trends: 1958-2015. Geophys. Res. Lett. 44 (9), 4214–4223. https://doi.org/10.1002/2017GL073613.

IUCN, 2019. In: Laffoley, D., Baxter, J.M. (Eds.), Ocean deoxygenation: Everyone's problem.

Johnson, G.C., Sloyan, B.M., Kessler, W.S., McTaggart, K.E., 2002. Direct measurements of upper ocean currents and water properties across the tropical Pacific during the 1990s. Prog. Oceanogr. 52 (1), 31–61.

Jones, C.S., Abernathey, R.P., 2019. Isopycnal mixing controls deep ocean ventilation. Geophys. Res. Lett. 46 (22), 13144–13151. https://doi.org/10.1029/2019GL085208.

Keeling, R.F., Kortzinger, A., Gruber, N., 2010. Ocean deoxygenation in a warming world. Annu. Rev. Mar. Sci. 2 (1), 199–229. https://doi.org/10.1146/annurev.marine.010908.163855.

Kwiatkowski, L., 2020. Twenty-first century ocean warming, acidification, deoxygenation, and upper ocean nutrient decline from CMIP6 model projections. Biogeosci. Discuss., 1–43. https://doi.org/10.5194/bg-2020-16.

Lachkar, Z., Lévy, M., Smith, K.S., 2019. Strong intensification of the Arabian Sea oxygen minimum zone in response to Arabian Gulf warming. Geophys. Res. Lett. 46 (10), 5420–5429. https://doi.org/10.1029/2018GL081631.

Lachkar, Z., Levy, M., Smith, S., 2018. Intensification and deepening of the Arabian Sea oxygen minimum zone in response to increase in Indian monsoon wind intensity. Biogeosciences 15 (1), 159–186. https://doi.org/10.5194/bg-15-159-2018.

Lachkar, Z., Smith, S., Levy, M., Paulius, O., 2016. Eddies reduce denitrification and compress habitats in the Arabian Sea. Geophys. Res. Lett. 43 (17), 1–9. https://doi.org/10.1002/2016GL069876.

Lam, P., Jensen, M.M., Kock, A., Lettmann, K.A., Plancherel, Y., Lavik, G., Bange, H.W., Kuypers, M.M.M., 2011. Origin and fate of the secondary nitrite maximum in the Arabian Sea. Biogeosciences 8 (6), 1565–1577. https://doi.org/10.5194/bg-8-1565-2011.

Letscher, R.T., Primeau, F., Moore, J.K., 2016. Nutrient budgets in the subtropical ocean gyres dominated by lateral transport. Nat. Geosci. 9 (11), 815–819. https://doi.org/10.1038/ngeo2812.

Lévy, M., Iovino, D., Resplandy, L., Klein, P., Madec, G., Tréguier, A.M., Masson, S., Takahashi, K., 2012. Large-scale impacts of submesoscale dynamics on phytoplankton: local and remote effects. Ocean Model. 43–44 (C), 77–93. https://doi.org/10.1016/j.ocemod.2011.12.003.

Lévy, M., Klein, P., Tréguier, A.M., Iovino, D., Madec, G., Masson, S., Takahashi, K., 2010. Modifications of gyre circulation by sub-mesoscale physics. Ocean Model. 34 (1–2), 1–15. https://doi.org/10.1016/j.ocemod.2010.04.001.

Li, C., Huang, J., Ding, L., Liu, X., Yu, H., Huang, J., 2020. Increasing escape of oxygen from oceans under climate change. Geophys. Res. Lett., 1–13. https://doi.org/10.1029/2019GL086345.

Luyten, J., Pedlosky, J., Stommel, H., 1983. Climatic inferences from the ventilated thermocline. Clim. Change 5 (2), 183–191. https://doi.org/10.1007/BF00141269.

O'Dwyer, J., Williams, R.G., LaCasce, J., Speer, K.G., 2000. Does the potential vorticity distribution constrain the spreading of floats in the North Atlantic? J. Climate 30 (4), 721–732.

Oschlies, A., Brandt, P., Stramma, L., Schmidtko, S., 2018. Drivers and mechanisms of ocean deoxygenation. Nat. Geosci. 11 (7), 1–7. https://doi.org/10.1038/s41561-018-0152-2.

Oschlies, A., Duteil, O., Getzlaff, J., Koeve, W., Landolfi, A., Schmidtko, S., 2017. Patterns of deoxygenation: sensitivity to natural and anthropogenic drivers. Philos. Trans. R. Soc. A, Math. Phys. Eng. Sci. 375 (2102), 20160325. https://doi.org/10.1098/rsta.2016.0325.

Palter, J.B., Trossman, D.S., 2018. The sensitivity of future ocean oxygen to changes in ocean circulation. Glob. Biogeochem. Cycles 32 (5), 738–751. https://doi.org/10.1002/2017GB005777.

Paulmier, A., Ruiz-Pino, D., 2009. Oxygen minimum zones (OMZs) in the modern ocean. Prog. Oceanogr. 80 (3–4), 113–128. https://doi.org/10.1016/j.pocean.2008.08.001.

Piontkovski, S.A., Al-Oufi, H.S., 2014. The Omani shelf hypoxia and the warming Arabian Sea. Int. J. Environ. Stud. 72 (2), 256–264. https://doi.org/10.1080/00207233.2015.1012361.

Queste, B.Y., Vic, C., Heywood, K.J., Piontkovski, S.A., 2018. Physical controls on oxygen distribution and denitrification potential in the north west Arabian Sea. Geophys. Res. Lett., 1–11. https://doi.org/10.1029/2017GL076666.

Resplandy, L., 2018. Will ocean zones with low oxygen levels expand or shrink? Nature 557 (7705), 314–315. https://doi.org/10.1038/d41586-018-05034-y.

Resplandy, L., Lévy, M., Bopp, L., Echevin, V., Pous, S., Sarma, V.V.S.S., Kumar, D., 2012. Controlling factors of the oxygen balance in the Arabian Sea's OMZ. Biogeosciences 9 (12), 5095–5109. https://doi.org/10.5194/bg-9-5095-2012.

Rudnickas Jr., D., Palter, J., Hebert, D., Rossby, H.T., 2019. Isopycnal mixing in the North Atlantic oxygen minimum zone revealed by RAFOS floats. J. Geophys. Res., Oceans 124 (9), 6478–6497. https://doi.org/10.1029/2019JC015148.

Rypina, I.I., Kamenkovich, I., Berloff, P., Pratt, L.J., 2012. Eddy-induced particle dispersion in the near-surface North Atlantic. J. Phys. Oceanogr. 42 (12), 2206–2228. https://doi.org/10.1175/JPO-D-11-0191.1.

Schmidtko, S., Stramma, L., Visbeck, M., 2017. Decline in global oceanic oxygen content during the past five decades. Nature 542 (7641), 335–339. https://doi.org/10.1038/nature21399.

Shigemitsu, M., Yamamoto, A., Oka, A., Yamanaka, Y., 2017. One possible uncertainty in CMIP5 projections of low-oxygen water volume in the Eastern Tropical Pacific. Glob. Biogeochem. Cycles 31 (5), 804–820. https://doi.org/10.1002/2016GB005447.

Spall, M.A., Richardson, P.L., Price, J.F., 1993. Advection and eddy mixing in the Mediterranean salt tongue. J. Mar. Res. 51 (4), 797–818.

Stramma, L., Johnson, G.C., Sprintall, J., Mohrholz, V., 2008. Expanding oxygen-minimum zones in the tropical oceans. Science 320 (5876), 655–658. https://doi.org/10.1126/science.1153847.

Trossman, D.S., Palter, J.B., Merlis, T.M., Huang, Y., Xia, Y., 2016. Large-scale ocean circulation-cloud interactions reduce the pace of transient climate change. Geophys. Res. Lett. 43 (8), 3935–3943. https://doi.org/10.1002/2016GL067931.

Vaquer-Sunyer, R., Duarte, C.M., 2008. Thresholds of hypoxia for marine biodiversity. Proc. Natl. Acad. Sci. 105 (40), 15452–15457.

Yamamoto, A., Palter, J.B., Dufour, C.O., Griffies, S.M., Bianchi, D., Claret, M., Dunne, J.P., Frenger, I., Galbraith, E.D., 2018. Roles of the ocean mesoscale in the horizontal supply of mass, heat, carbon, and nutrients to the northern hemisphere subtropical gyres. J. Geophys. Res., Oceans 123 (10), 7016–7036. https://doi.org/10.1029/2018JC013969.

Yang, S., et al., 2020. Global reconstruction reduces the uncertainty of oceanic nitrous oxide emissions and reveals a vigorous seasonal cycle. Proc. Natl. Acad. Sci. USA 201921914.

Chapter 14

New technological frontiers in ocean mixing

Eleanor Frajka-Williams[a], J. Alexander Brearley[b], Jonathan D. Nash[c] and Caitlin B. Whalen[d]

[a]National Oceanography Centre, Southampton, United Kingdom, [b]British Antarctic Survey, Cambridge, United Kingdom, [c]College of Earth, Ocean and Atmospheric Science, Oregon State University, Corvallis, OR, United States, [d]Applied Physics Laboratory, University of Washington, Seattle, WA, United States

14.1 Introduction

The preceding chapters have outlined the advanced state of knowledge about ocean mixing and energetics, much of which is based on observations of mixing from traditional ship-based measurements. The onward influence of mixing on the large-scale ocean circulation and distribution of tracers in the ocean has been further developed through parameterisation of mixing in numerical simulations. Differences in mixing parameterisations lead to wide variations in the modelled large-scale mean ocean circulation (see Melet et al., Chapter 2). At its smallest scales, mixing results from molecular viscosity and diffusion working to homogenise (reduce gradients) in properties, such as temperature (scalar mixing) and velocity (dissipating energy). However, the intensity of ocean mixing is dominated by processes occurring at larger scales (than molecular) to generate the gradients in properties. One of the outstanding challenges in developing accurate parameterisations is the dearth of observations of these eddying processes—with a wide range in characteristic time- and space-scales, and dependence on proximity to interfaces—and their associated rates of turbulent kinetic energy dissipation.

Traditional measurements of ocean mixing are time-intensive, relying on profilers tended by ships and moving at relatively slow speeds. Surveys carried out with these profilers offer a localised snapshot of the turbulent conditions and mixing in the ocean. Instances where measurements have been repeated at the same location have shown that processes are intermittent in time and patchy in space, meaning that individual snapshots may misrepresent the overall magnitude of mixing. Additionally, ship-based measurements are limited in under-ice environments, close to the air-sea interface and near the seabed. Here we describe new frontiers in measuring turbulent dissipation and mixing in the ocean: technological advancements that offer the potential to greatly increase the availability of ocean-mixing measurements with opportunities in these critical environments. We focus in particular on deep-ocean regions, away from coastal areas and estuaries, where other techniques can be used.

We briefly summarise established methods for measuring oceanic mixing (Section 14.2), then give an overview of new methods and of the application of existing methods to new platforms (Section 14.3). In each case, we outline the methods and challenges in its application, and highlight circumstances under which the method would provide mixing measurements that are otherwise hard to achieve using traditional observations. Throughout, we focus primarily on the direct measurement or indirect estimation of the rate of turbulent dissipation ϵ or of dissipation of temperature variance χ, and on local or point measurements, rather than distributed estimates (e.g., tracers or inversions). We conclude with a summary and future outlook in Section 14.4.

14.2 Current and historical measurements of mixing

We summarise measurement techniques here to place the novel techniques (Section 14.3) in context. For more complete reviews and a history of turbulence measurements, we refer the reader to Gregg (1991); Lueck et al. (2002); Baumert et al. (2005).

Ocean mixing and turbulence can be estimated through measurements at microscale or finescale. Fluctuations at which entropy is generated by molecular processes are referred to as 'microstructure' and are $\mathcal{O}(0.1 - 1)$ cm. Finestructure is generally at larger scales [$\mathcal{O}(10 - 100)$ m] and may include velocity or scalar signatures of straining or mixing. Some of the earliest observations of ocean turbulence were made in the latter half of the 20th century using sensitive devices mounted on a submarine, a towed body behind a ship, or a free-fall profiler. These devices were used to measure small fluctuations in scalar quantities (temperature) or velocity associated with turbulent motions, requiring a very fast response time and a stable platform that allows the probe to sample undisturbed fluid. Turbulence can be estimated both directly through microscale measurements [$\mathcal{O}(1)$ cm] or estimated from finescale measurements [$\mathcal{O}(10-100)$ m]. Microscale measurements typically require specialised platforms free from vibration and fast-response sensors to measure tiny and high-frequency fluctuations in scalar (temperature or conductivity) or vector (velocity) quantities. Finescale measurements can be made using conventional hydrographic or velocity measurements, using platforms lowered from a research vessel.

The measurements of turbulence rely on several assumptions. Turbulence in the ocean can be both isotropic and anisotropic, but at the smallest scales is assumed to be isotropic (independent of direction). Theoretical spectra describe a cascade of energy from the larger scales and more energetic fluctuations to the smallest scales, where molecular diffusion and viscosity take over. This spectrum assumes steady state: that there is a continuous cascade of energy from the largest turbulent eddies (production scales) to dissipation scales. As described in Box 14.1, measuring mixing is focused on measurements of dissipation rates (ϵ or χ) and fluxes.

Direct measurement of turbulent dissipation at microscales

The current standard for direct measurements of turbulent dissipation ϵ is with an air-foil shear probe mounted on a free-fall vertical microstructure profiler or towed body as the platform (Lueck, 2005; Prandke, 2005). The air-foil shear probe measures fluctuations in the lateral water speed (normal to the platform's direction of travel) at a rate of about 512 Hz. With a known speed of the platform through the water (typically 0.5–2 m s^{-1}), the time series of velocity fluctuations can be converted to a velocity gradient spectrum in wavenumber space. Assuming isotropy—that the turbulence is independent of direction whether horizontal or vertical—the spectrum is fit to a theoretical spectrum Nasmyth (e.g., the Nasmyth spectrum for velocity shear, Nasmyth 1973) by varying ϵ for the Nasmyth spectrum until the best fit is reached. The spectrum is then integrated to estimate ϵ. It is necessary to fit to a theoretical spectrum as the observed spectrum does not extend to the highest wavenumbers of turbulence. Because the shear probes are very sensitive, vibrations of the platform can lead to contamination of the lateral water speed estimate, and the area forward of the sensor must be undisturbed. The typical noise floor for ϵ that can be measured is 10^{-10} W kg^{-1} using free-fall profilers or 10^{-9} W kg^{-1} for towed bodies.

Microscale measurements of temperature dissipation χ are made with a sensitive and fast response thermistor, and mounted on a similar platform. These measurements are again converted to a spectrum in wavenumber space and fitted with a theoretical spectrum for a passive scalar (the Batchelor spectrum, Batchelor, 1959). Microscale measurements of temperature and velocity both require specialised sensors as traditional temperature and velocity measurements in the ocean are not fast enough (the time-response is too long), nor do they resolve the small spatial scales required to measure microstructure (with temperature requiring smaller scales than velocity).

Free-fall profilers are typically not tethered to the ship, profiling downwards under negative buoyancy, then dropping a release weight at a pre-programmed depth or, if equipped with an altimeter, then at a prescribed height-above-bottom. They then return to the surface under positive buoyancy. Some free-fall profilers are tethered, profiling downwards with enough slack in the wire to not restrict the profiler's movement. They are then hauled back to the ship by the wire on a winch. Although profilers can measure through the full ocean depth (up to 6000 m), producing high-quality microstructure measurements of velocity shear and temperature gradients, they are labour-intensive, typically requiring a dedicated ship and limiting the space- and time-scales over which turbulence measurements can be made.

> **Box 14.1 Measuring mixing**
>
> In this chapter, we will primarily be discussing the turbulent dissipation of kinetic energy ϵ and of temperature variance χ. These estimates rely on a few fundamental equations, many of which can be found elsewhere in this book, but we summarise them here for convenience.
>
> **Mixing estimates from turbulent dissipation of kinetic energy**
>
> To estimate diffusivity κ from the turbulent dissipation rate of kinetic energy ϵ, we rely on the expectation that the flux of buoyancy in a stratified fluid can be estimated as a fraction of ϵ (Osborn, 1980),
>
> $$\kappa = \frac{\Gamma \epsilon}{N^2}, \tag{14.1}$$
>
> where Γ is the mixing efficiency and $N = -g/\rho_0(\partial \rho/\partial z)$, the buoyancy frequency. Note that though the value of 0.2 is often used for Γ, its actual value may vary. For further discussion, the reader is referred to Polzin and McDougall, Chapter 7, and Gregg et al. (2018).

The dissipation rate ϵ varies in magnitude from $\mathcal{O}(10^{-10})$ W/kg in the relatively quiescent open ocean, to $\mathcal{O}(10^{-6})$ W/kg in the shelf seas or energetic surface mixing layers. This large dynamic range—particularly the low turbulence values—requires measurement techniques that are low noise and high sensitivity.

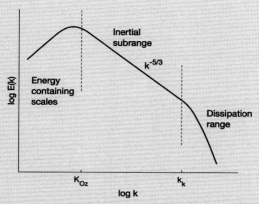

FIGURE 14.1 Kolmogorov turbulence spectrum showing the energy as a function of wavenumber. Microstructure measurements are in the inertial subrange, while the large-eddy method (Section 14.3.4) makes measurements of the energy-containing scales range.

The dissipation of turbulent kinetic energy (TKE) occurs at viscous scales, but the rate of dissipation can be illustrated in spectral space by the Kolmogorov spectrum. Within the inertial subrange (Fig. 14.1), TKE is cascaded to smaller scales until it is dissipated. The inertial subrange is bounded by the Kolmogorov and Ozmidov lengthscales. Between them, the Kolmogorov spectrum is given by

$$E(k) = c_k \epsilon^{2/3} k^{-5/3}, \tag{14.2}$$

where c_k is Kolmogorov's constant and k the wavenumber. The largest scales for turbulent motion are set by the Ozmidov length scale (L_{Oz}) for stratified turbulence,

$$L_{Oz} = \sqrt{\epsilon N^{-3}}. \tag{14.3}$$

The smallest scales (highest wavenumbers) are given by the Kolmogorov lengthscale (L_k), at which viscous at inertial forces are of the same order of magnitude,

$$L_k = \left(\frac{\nu^3}{\epsilon}\right)^{1/4}, \tag{14.4}$$

where ν is the kinematic viscosity and is $\mathcal{O}(10^{-6})$ m^2 s^{-1}. The Kolmogorov lengthscale gives the smallest scales over which turbulent motions occur as $L_k = 1\text{--}10$ cm, prescribing a desired spatial resolution for turbulence measurements. Direct measurements at this spatial resolution require sensors with a fast time-response relative to the speed with which the sensor or platform moves through the water.

Mixing estimates from turbulent dissipation of temperature variance

Similarly, κ can be estimated from the measurement of the dissipation of temperature variance χ as

$$\kappa = \frac{\chi}{2\langle dT/dz \rangle^2}, \tag{14.5}$$

where $\langle \cdot \rangle$ represents an average. Unlike Eq. (14.1), this formulation does not rely on any assumptions about the value of the mixing efficiency Γ. New approaches to estimating mixing from measurements of χ will be discussed in Section 14.3.4.

While much of this chapter will focus on the estimation of the dissipation of turbulent kinetic energy, the quantities we are really interested in are the consequences of mixing, i.e., the turbulent diapycnal fluxes of heat, salt, etc. These quantities are, however, very difficult to measure directly. They can be measured through tracer release experiments (Ledwell and Watson, 1988, 1991; Ledwell et al., 2011), where a tracer is released in a concentrated area with a specific density, then sampled at later periods to determine the diapycnal spread of the tracer away from its original density and location. We will not consider these methods further here.

14.3 Recent technological developments: novel methods

One of the primary advances that is enabled by these new platforms is an expansion of the number of measurements possible, rather than through a new sensor. In the examples below, we outline the methods of measuring shear microstructure on autonomous underwater vehicles (AUVs), temperature microstructure on autonomous platforms and a shipboard CTD (conductivity-temperature-depth) rosette, and applying finescale parameterisations and the large-eddy method to autonomous platforms. Additionally, bespoke platforms have been developed to measure turbulent dissipation near the air-sea interface, but will not be discussed further here (see Ward et al., 2014; Thomson, 2013). Throughout, these opportunities enable a larger number of measurements, or longer endurance measurements, which would not have been possible using traditional methods (e.g., vertical microstructure profilers).

In the past 20 years, the use of autonomous vehicles has moved from experimental to more operational, enabling less-expert users to deploy these vehicles in a range of areas. Underwater gliders have a range of several 1000 km and an endurance of 6–12 months when measuring a limited set of variables (temperature and salinity) and can span the top 1000 m of the ocean. The Argo profiling float array was begun in 2004 with the deployment of 1000 profiling floats; by 2007 the array was complete, with 3000 floats making profiles of temperature and salinity to 2000 m every 10 days. Both of these autonomous platforms offer new opportunities for oceanographic measurements, reducing the requirement for ship-based observations.

Even without the addition of specific turbulence sensors, both platform types have enabled estimates of turbulence using a finescale parameterisation approach (see Section 14.3.3) or a large-eddy approach (see Section 14.3.4). Early efforts to include turbulence-resolving sensors included using shear probes, Pitot tubes and an acoustic Doppler current profiler (ADCP) on autonomous underwater vehicles (AUVs) (Dhanak and Holappa, 1999). Many of the technologies now commercially available resemble this early solution. Microstructure sensor packages have now been developed for deployment on autonomous vehicles (e.g., the MicroRider by Rockland Scientific, Inc.), enabling these platforms to measure microstructure temperature and velocity, albeit with reduced endurance due to the power requirement of microstructure sensing (Section 14.3.1). New temperature/temperature-gradient sensors (χ pods) have been developed for moored applications, enabling microscale measurements on long duration (1-year) moorings, and reducing the contamination caused by mooring motion (Section 14.3.2). These new developments are outlined below.

14.3.1 Shear microstructure on AUVs

A key limitation of traditional measurements from vertical microstructure profilers is the limited number of samples that can be collected during an individual measurement campaign. Given the intense spatial and temporal patchiness of mixing, this is a severe limitation on constraining the mean value and variance of turbulent kinetic energy dissipation ϵ within a region. The possibility of using microstructure-equipped AUVs is exciting, as it potentially increases the number of measurements by an order of magnitude, and at lower cost than traditional ship-based operations. Novel AUVs, such as Autosub, also offer the possibility of measuring rates of TKE dissipation in highly challenging environments, for example, within deep canyons or under sea ice/ice shelves.

Shear microstructure on Slocum gliders

The most well-developed platform on which microstructure measurements have been made is the underwater glider. Compact sensor suites have been developed by Rockland Scientific for the Teledyne Slocum glider (Fig. 14.2a). Gliders, which dive as deep as 1000 m, are propelled by changes in vehicle buoyancy, which involves pumping oil internal to the pressure hull into an external bladder, at the same time as shifting the centre of mass of the vehicle so as to translate a component of the vertical buoyancy force into horizontal motion. Gliders typically fly between 20 and 30 degrees from horizontal, with the most efficient flight typically occurring around 26 degrees. In theory this means that gliders provide an excellent low-noise platform for obtaining measurements of shear and fast temperature, uncontaminated by, e.g., noise from a winch (as in tethered systems). The MicroRider unit has been adapted for the Slocum glider and is effectively a scaled-down variant of the vertical microstructure profiler (VMP), a more traditional freefall profiler. The MicroRider (Wolk et al., 2009), which is mounted on top of the nose of the Slocum, is a neutrally-buoyant, self-contained, relatively low-powered instrument (1–1.6 W) powered by the glider's battery, but with independent data storage (on a 16 or 64 Gb card). It is typically configured with two micro-temperature and two shear probes—a similar configuration to that used on free-fall profilers—protruding about 15 cm from the front of the vehicle. The use of both types of probes enables an independent check on the mixing efficiency (see Section 14.3.2).

Notable deployment successes have now been achieved using the Slocum MicroRider configuration. Initial testing of the instrument in Ashumet Pond (Wolk et al., 2009), albeit only in 12 m of water, demonstrated very low noise levels

FIGURE 14.2 Configuration of sensor suites for various autonomous vehicles (clockwise from top left): MicroRider suite for Slocum glider (an ocean microstructure glider, or OMG); Micropod suite for Seaglider; MicroRider suite for Autosub Long Range (ALR).

in both the shear and accelerometer data, with peaks at 28, 60 and 80 Hz (caused by the glider's tail fin assembly) not compromising the data quality. The use of a Goodman correction filter (Goodman et al., 2006) was proposed to remove coherent noise between the three accelerometer channels and the two shear channels, which relies on calculating a 3D covariance matrix between the accelerometer (representing body movement) and the shear channels, allowing the spectra to be 'cleaned'. During the testing, the probes revealed ϵ to be between 1×10^{-7} W kg^{-1} and 5×10^{-11} W kg^{-1} in the quiescent layer below the thermocline.

The first science papers using ocean microstructure glider (OMG) data appeared in 2014–2015. An OMG was deployed in the Faroe Bank channel in the North Atlantic for a one-week mission, collecting 152 profiles, many of which were collocated with 90 microstructure profiles from a conventional freefall profiler (Fer et al., 2014). This mission identified several recommendations to ensure good data quality, including deactivating the glider servo during flight to reduce vibrations, and the importance of an effective flight model for estimating the glider speed-through-water **V**. Speed-through-water is not measured directly by the glider, and thus has to be estimated using a flight model (Merckelbach et al., 2010, 2019). Since the shear scales with $\mathbf{V}^4$, inaccuracies result in error in the estimated turbulent dissipation. Fer et al. (2014) conclude that their glider estimates of dissipation are of comparable quality to the best VMP data, with a noise floor of less than 5×10^{-11} W kg^{-1}. In a follow-up study, Peterson and Fer (2014) demonstrate that temperature microstructure from the deployment can also be successfully used to calculate ϵ for values below 2×10^{-7} W kg^{-1}, though for higher dissipation values they find a consistent underestimate in using these techniques compared to shear methods.

Around the same time, a nine-day study with an OMG was carried out in the Celtic Sea, collecting 766 profiles (Palmer et al., 2015). They found different forcing mechanisms for the surface and bottom boundary layers, and that internal-wave energy was responsible for much of the interior mixing. In processing the data, Palmer et al. (2015) noted that use of the Goodman correction may cause an underestimate of the true dissipation rate, as the coherent noise removal that it entails may remove real signals that occur at the scale of the glider. However, other later studies have advocated continued use of the Goodman correction (Scheifele et al., 2018; Schultze et al., 2017; St. Laurent and Merrifield, 2017), largely to remove vibrational noise. It is notable that Palmer et al.'s study was in a relatively high-dissipation environment, with values of up to four orders of magnitude above background levels of 3–4 × 10^{-10} W kg^{-1}, which may explain why it was possible to omit the Goodman correction in this study, whereas others in lower dissipation environments have chosen to incorporate it.

We point to several other important studies incorporating OMG data. St. Laurent and Merrifield (2017) describe two projects (SPURS: Salinity Processes in the Upper-ocean Regional Study in the North Atlantic, and ASIRI: An Ocean Atmosphere Initiative for Bay of Bengal) which highlight the utility of glider microstructure in observing surface stable layers, a phenomenon not easily sampled with ship-based methods owing to disturbance by the vessel. Furthermore, Schultze et al. (2017) used two deployments of 12 and 17 days to identify a highly intermittent dissipation signal in the stratified thermocline region of the North Sea. More recently, Scheifele et al. (2018) used both shear and micro-temperature to estimate dissipation rates in the low-energy environment of the Beaufort Gyre in the Arctic, arguing that the noise floor of the shear measurements artificially skews the statistical distribution of dissipation below 10^{-10} W kg^{-1}. Instead they argue that micro-temperature measurements provide a better estimate of dissipation in low-turbulence environments, down to 2 × 10^{-12} W kg^{-1}. Lucas et al. (2019) also used a 9.5-day deployment of an OMG in the North Atlantic to quantify the response of the upper ocean to the passage of an autumn storm.

Shear microstructure on other autonomous platforms

There is a significantly less literature on microstructure-based methods from other autonomous platforms. A Micropod suite of sensors has been developed for the Seaglider autonomous underwater glider, a similar vehicle to the Slocum glider, but developed at the University of Washington and now produced by Huntington Ingalls (Fig. 14.2b). The Micropod, in contrast to the MicroRider, has two individual housings mounted either side of the CTD (Creed et al., 2015) and a dedicated pressure housing hosting the data logger inside the glider's flooded aft fairing. Each pod unit can contain either a shear probe or a thermistor, and the electronic board inside each unit is specific to either a thermistor or a shear sensor. Creed et al. (2015) describe the operation and initial results from the Micropod onboard the Seaglider, whereas Rainville et al. (2017) outline results from a six-month mission. Leadbitter et al. (2019) identified the potential to estimate ϵ using Thorpe scale methodology from the 512 Hz micro-temperature data from the Micropod, with resulting dissipation values comparing favourably to traditional CTD-based Thorpe scale estimates.

Microstructure sensors have also been fitted to other autonomous platforms on a more ad hoc basis. Autosub-based microstructure measurements have been used in under-ice environments, enabling estimates of turbulent fluxes under-ice shelves in Antarctica, which are key to understanding basal melt-rate processes (Kimura et al., 2016). More recently, a Rockland MicroRider was used onboard Autosub long range (ALR), a 3.6 m, 800 kg displacement AUV with a depth rating of 6000 m (Furlong et al., 2012; McPhail et al., 2019). The Rockland MicroRider unit in this case protrudes from the front, again to escape any flow distortion caused by the vehicle itself (Fig. 14.2c). Microstructure data were collected onboard ALR in the DynOPO (Dynamics of the Orkney Passage Outflow experiment) (Naveira Garabato et al., 2019) by using the ALR to complete a spatial survey along the western flank of Orkney Passage. One of the advantages of the ALR over a sawtooth-profiling vehicle, like Slocum, is that it uses downward-looking ADCPs to track its position when in range of the bottom, thus providing more accurate estimates of its position. In this way, ALR was able to complete a 3-day horizontal survey, covering 180 km without surfacing, giving a detailed spatial survey of near-bottom turbulent dissipation. Communications were found to introduce noise to the MicroRider measurements, and the authors suggest that dissipation rates of less than 1 × 10^{-9} W kg^{-1} might be difficult to detect, due to this and other vibrational issues with the vehicle.

A consistent challenge for the acquisition of AUV-based data is the assessment of spectral quality required to produce peer-review quality data. Authors have adopted a number of different methods for assessing spectral quality, most of which rely on analysing the difference between the observed spectra and the theoretical Nasmyth values, either via an algorithm (e.g. Schultze et al., 2017) or through laborious manual selection of spectra (e.g. Lucas et al., 2019). OMG data are increasingly being used in large field programs; however, dealing sensitively with the volume of data which is being produced poses a significant challenge. Greater automation in the processing of spectra is required to fulfil the potential of this technology. The increasing application of machine learning techniques in the environmental sciences offers a potential pathway forward (Hsieh, 2009), though the development of good training datasets will be required to apply such methods successfully.

Though autonomous platforms provide an avenue towards collecting longer endurance measurements of turbulent dissipation, one limitation is that microstructure data are not typically processed onboard the vehicle. This means that if the vehicle is lost or data are otherwise unable to be downloaded, that the dataset is lost. Additionally, because dissipation data are not processed mid-mission, it is not possible to use turbulence data to update the sampling plan for regions of interest. The raw data are very large for current transmission capability (roughly 1 Gb a day). However, new efforts are enabling processing or transmission of smaller data snippets to be transmitted back mid-mission, to allow evaluation of the sensor health and within mission adjustments to sampling (Cervelli and Haskins, 2018; Rainville et al., 2017).

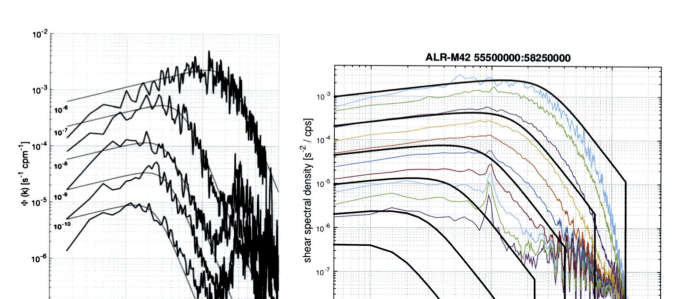

FIGURE 14.3 Example shear spectra for (a) MicroRider suite on Slocum glider; and (b) MicroRider for ALR from the Orkney passage. In (a), example spectra are given for different dissipation levels (in W kg^{-1}) collected using Slocum glider 390 (Beagle) in Loch Linnhe, West Scotland, equipped with Rockland Microrider 072. Data were collected on 21 and 22 November 2018 and accompanying theoretical Nasmyth spectra are plotted. High-frequency glider rudder noise can be seen above 10 cpm for lower dissipation levels, though this falls outside the band typically used for spectral integration. In (b), example spectra are for different levels (colours) from one of the ALR missions in the Orkney Passage. The theoretical Nasmyth spectra are plotted in black.

14.3.2 Temperature microstructure on new platforms

The widespread use of airfoil shear probes to characterise ocean turbulence has been driven by two factors: (1) the universal nature of the shear spectrum yielding robust estimates of ϵ from partially resolved spectra (see Fig. 14.3 and Moum et al., 1995) and (2) the turbulent kinetic energy (TKE) dissipation rate represents the irreversible sink of energy, and is an important constraint in tracing the flow of energy flow through the ocean from source to sink. However, ϵ is not a direct measure of ocean mixing, since it requires an understanding of the evolution of the turbulent event that is being sampled, the way that turbulent energy flows through either viscous dissipation or the increase of potential energy (through mixing), and assumptions about the instantaneous turbulent state sampled in relation to the complete evolution of a turbulent overturn (i.e., Smyth et al., 2001; Gregg et al., 2018).

A more direct measure of ocean mixing is obtained by measuring the dissipation rate of temperature variance χ, from which scalar diffusivities can be computed. Following Osborn and Cox (1972), we assume a steady-state balance between production of temperature variance ($P = \langle w'T' \rangle dT/dz$) and χ, where w is vertical velocity, T temperature, and primes denotes perturbations ($T' = T - \overline{T}$). By rewriting the turbulence flux in terms of a turbulent diffusivity (i.e., $\langle w'T' \rangle = K_T dT/dz$), we arrive at an explicit form for the diffusivity as

$$K_T = \frac{\chi}{2\langle dT/dz \rangle^2} \qquad (14.6)$$

without any assumptions about mixing efficiency. See Fig. 14.4.

Many of the earliest measurements of geophysical turbulence, i.e., the computation of ϵ, χ and K_T (Stewart and Grant, 1962), are derived from temperature microstructure including much of our early understanding of turbulence in lakes (Luketina and Imberger, 2000). However, to resolve the full spectrum of temperature variance, extremely slow profiling speeds are required, and these are not conducive to deep-ocean profiling. As a result, the measurement of χ from traditional turbulence profilers has largely fallen out of favour.

However, unlike airfoil shear probes, the sensing of microscale temperature is largely insensitive to platform vibration, making it highly suitable for deployment on moorings (Moum and Nash, 2009), or from standard shipboard CTD rosette (Goto et al., 2018), both of which have too much vibration for air-foil shear measurements of ϵ to be feasible. Moreover,

352 Ocean Mixing

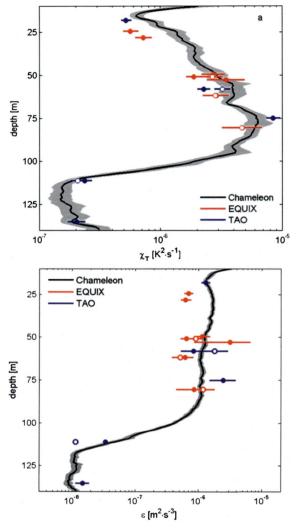

FIGURE 14.4 Computation of χ and ϵ as computed from moored χpods (symbols) as compared to that computed from the Chameleon free-falling microstructure profiler (lines and shading), demonstrating that moored measurements are unbiased. From Perlin and Moum (2012), ©American Meteorological Society. Used with permission.

the measurement of T requires very little energy, so multi-year time series can be obtained from a small package with minimal battery requirements. One such instrument is Oregon State University's χpods (Moum and Nash, 2009), moored mixing metres that consist of 1–2 FP07 thermistors, accelerometers, compass; more recently they have also included Pitot tubes that measure flow speed and TKE (Moum, 2015). Temperature microstructure has also been obtained on Argo-style profiling floats. Some of the first of these were obtained during the 2000/2002 Hawaiian Ocean Mixing Experiment. More recently, Lien et al. (2016) have developed profiling floats that measure both the finescale velocity gradients and microscale temperature gradients. One limitation of the microstructure temperature method to measure turbulent dissipation is that in cases where there is temperature variance not accompanied by density variance (i.e., compensated by salinity variance), then the temperature method may overestimate the dissipation rate since the compensated temperature variance is not dynamically active.

In addition to the moored deployment of sensors that measure χ, there is a recent effort to make the measurement of χ a standard part of the global repeat hydrography surveys using χ-pods installed on shipboard CTDs (Holmes et al., 2016; Goto et al., 2018). Such observations are used to complement the otherwise sparse global coverage of full-ocean depth mixing measurements (Waterhouse et al., 2014), and are intended to understand aspects of the global distribution of mixing and how it affects water-mass transformations and influences large-scale ocean circulation.

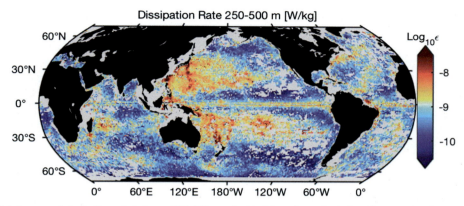

FIGURE 14.5 The global averaged dissipation rate between 250–500 m calculated from finescale strain estimates using Argo profiling floats. Adapted from Whalen et al. (2018).

χ pods have been deployed on the equatorial TAO (Tropical Atmosphere-Ocean) moorings almost continuously since 2006, and for extended duration elsewhere as well. They have been shown to provide estimates of ϵ and χ that agree with those from turbulence profilers to within a factor of two (Fig. 14.4 and Perlin and Moum, 2012). Multi-year deployments have revealed the first glimpses into how turbulence varies on annual to decadal timescales. For example, these long-time series of mixing have shown that annual cycles of sea surface cooling in the equatorial Pacific cold tongue are controlled by the turbulent heat flux (Moum et al., Chapter 10 and Moum et al. (2013)), and that mixing acts as a positive feedback mechanism to the development of El Niños and La Niñas (Warner and Moum, 2019).

14.3.3 Finescale parameterisations from autonomous platforms

Turbulent mixing varies dramatically across spatio-temporal scales and acts as a key driver for a range of dynamical processes from local energy budgets to the global overturning circulation. Microstructure measurements are useful for observing ocean turbulence on spatial scales of centimetres and temporal scales of hours. However, since microstructure observations require specialised, sensitive instrumentation, it is not practical to use them to observe basin or global scale variations in the diapycnal diffusivity and turbulent energy dissipation rate. Finescale parameterisations based on commonly made oceanographic measurements (Henyey et al., 1986; Gregg and Kunze, 1991; Polzin et al., 1995; Kunze et al., 2006; Polzin et al., 2014) are a practical way to estimate turbulence over the larger space and longer timescales needed to understand regional to global ocean dynamics and climate (Kunze, 2017b). These shear, strain, or shear/strain-based finescale parameterisations allow the estimation of dissipation using an indirect method applied to velocity and/or CTD data by assuming that the dissipation is the result of internal-wave breaking through internal wave-wave interactions (e.g. Gregg, 1989; Kunze et al., 2006).

Here we focus on the *strain*-based version of the finescale parameterisation applied to conductivity temperature and depth (CTD) measurements of strain between isopycnals collected using Argo profiling floats following the method described in Whalen et al. (2015). Since the Argo program consists of a global array of thousands of floats that typically profile every 10 days, the dataset is well suited to study variations in the dissipation rate on large spatial scales and long timescales. Applying the finescale parameterisation to millions of Argo float profiles reveals a complex geography of turbulent mixing throughout the globe (Fig. 14.5), similar to patterns observed by applying finescale parameterisation to repeat hydrography data (Kunze et al., 2006; Kunze, 2017a). Many of the concepts discussed here for the strain method are also relevant for *shear-strain* version of the finescale parameterisation, which additionally incorporates measurements of vertical shear from ADCPs.

Both strain and shear finescale methods rely on two major assumptions: (1) that internal waves dominate the observations of strain and/or shear on the observed vertical scales (often around 10–100 m); and (2) that the primary cause of turbulence in the observations is the down-scale transfer of energy from internal waves to turbulence, following the Garrett-Munk model. To satisfy the first assumption, parts of a strain or shear profile not dominated by internal waves are removed prior to using the parameterisation. For example, when applying the strain version of the parameterisation to Argo float profiles, the mixed layer, the transition layer, and the mode water portions of each profile are removed since they are associated with non-internal wave variance in strain (Whalen et al., 2015). To satisfy the second assumption, the parameterisation is only applied to regions where the internal-wave field is similar to the Garrett–Munk model, and places where mixing is not due to internal waves are excluded. Therefore the finescale parameterisation is not applicable in places such as submarine canyons, dense overflows, or the surface boundary layer (see also Thomas and Zhai, Chapter 5). In locations where these two assumptions are not met, χ pods can often be used to estimate the turbulent energy dissipation rate (Section 14.3.4). Finestructure estimates are preferable to χ pods when there is substantial centimetre scale variability in temperature and salinity, which has little or no density signature, causing the χ pod measurements to over-estimate the turbulent energy dissipation rate.

To estimate the dissipation rate using the strain or shear finescale parameterisation, profiles are selected that are dominated by internal waves, therefore satisfying the two assumptions listed above. When Argo float data is used each profile is divided into segments, typically 200 m in length. The strain $\xi_z = (N^2 - N_{fit}^2)/\overline{N^2}$ is calculated for each segment, where N^2 is the buoyancy frequency, N_{fit}^2 is a quadratic fit to the buoyancy frequency, and $\overline{N^2}$ is the segment-mean of the buoyancy frequency. Next, the strain variance $\langle \xi_z^2 \rangle$ is found by integrating the strain spectra of each segment starting at wavelengths of 100 m and continuing to the smallest wavelength between 40 and 10 m that satisfies $\langle \xi_z^2 \rangle < 0.2$ to avoid spectral saturation.

The strain variance is then used as one of the major components of the full finescale parameterisation to estimate the energy dissipation rate,

$$\epsilon = \epsilon_0 \frac{\overline{N^2}}{N_0^2} \frac{\langle \xi_z^2 \rangle^2}{\langle \xi_{zGM}^2 \rangle^2} h(R_\omega) L(f, N), \tag{14.7}$$

where $\langle \xi_{zGM}^2 \rangle$ is strain variance of the Garrett–Munk spectrum over the same integration range, and the constants are $N_0 = 5.24 \times 10^{-3}$ rad s^{-1}, and $\epsilon_0 = 6.73 \times 10^{-10}$ m^2 s^{-2}. The term $h(R_\omega)$ is a function of the shear-strain ratio R_ω; it takes a different form in the strain-only and shear-strain versions of the finescale parameterisation (Kunze et al., 2006). If shear information is available R_ω can be calculated directly, otherwise it is assumed constant. Since internal-wave dynamics vary with latitude, the correction $L(f, N)$ is included (Polzin et al., 1995; Gregg et al., 2003). For more details about the strain application to Argo profiles see Whalen et al. (2015), and for the general implementation, see Kunze et al. (2006) and Polzin et al. (2014).

Estimates of the dissipation rate using the finescale strain (Whalen et al., 2015) and shear-strain (Polzin et al., 1995) parameterisations agree with microstructure measurements within a factor of 2–3 in open-ocean conditions. Since microstructure measurements are on spatial scales of centimetres and temporal scales of seconds, it is imperative that they are sufficiently averaged, so that the spatio-temporal scales are equivalent to the finescale parameterisation estimates, i.e., spatial scales of a couple hundred metres and timescales of tens of hours to days (Whalen et al., 2015). Since ϵ is a log-normal variable (implying that the mean is larger than the majority of samples) spurious mismatches can be calculated if equivalent time- and lengthscales are not compared. Global observations of dissipation using the strain-based parameterisation have demonstrated that mesoscale eddies contribute to enhanced turbulent dissipation of wind-forced waves, particularly within anticyclonic eddies (Whalen et al., 2018). Such a conclusion would have been difficult to ascertain with certainty in the absence of Argo float-based dissipation estimates.

14.3.4 Large-eddy method using autonomous platforms and moorings

Ocean mixing can be estimated indirectly at finescales through the 'large-eddy method'. This method relies on the hypothesis that a steady cascade exists from turbulent production scales (the largest energy-containing eddies) through to viscous dissipation scales (Tennekes and Lumley, 1972). The method uses a simple scaling argument applied to the turbulent kinetic energy equation, where q' is the velocity scale of the largest eddies, and l is their lengthscale. Then the eddy timescale is given by $\tau \sim l/q'$, and the dissipation rate ϵ is estimated by the energy divided by time as

$$\epsilon \sim \frac{(q')^2}{\tau} \sim \frac{(q')^3}{l}. \tag{14.8}$$

In a highly turbulent coastal environment, Gargett (1999) estimated the turbulent velocity scale q' using vertical seawater velocity from an ADCP and the lengthscale l by considering the lengthscale over which q' reversed sign. For stratified flow, the eddy lengthscale can be estimated using the Ozmidov lengthscale (Eq. (14.3)). In contrast to direct measurements of turbulence, which aim to resolve the Kolmogorov scale (Eq. (14.4)), the large-eddy method uses measurements of the relatively large energy-containing scales, $\mathcal{O}(0.1 - 10)$ m, to infer dissipation (Gargett, 1999; Moum, 1996).

The method gives a scaling for the estimated turbulent dissipation e, but it requires a constant of proportionality c_ϵ to derive

$$e = c_\epsilon \frac{(q')^3}{l}. \tag{14.9}$$

This constant of proportionality has varied from 0.03 to 4 (Gargett, 1999), likely due to variations in how q' and l are defined. However, Gargett (1999) found that for a given method of determining q' and l that the constant of proportionality could be determined using a training dataset by comparison with microstructure profiles, then applied to later estimates of e in the same cruise and a later cruise. This method assumes that the turbulent velocity fluctuations q' can be separated from internal waves or other sources of velocity fluctuations.

Large-eddy method on gliders and floats

Autonomous platforms (gliders and floats) can sample for many months, allowing the possibility of measuring processes that vary on seasonal and longer timescales (Testor et al., 2019). These long endurance platforms offer the potential to estimate turbulence through an adaptation of the large-eddy method using vertical seawater velocity fluctuations to characterise turbulent eddies (Beaird et al., 2012; Cusack et al., 2017). This method can be applied using only the standard CTD sensors, enabling the long-endurance missions of autonomous gliders and floats to be achieved, while also providing estimates of turbulent dissipation. It can also be applied without assuming that the observed turbulent dissipation is coming from internal wave-wave interactions, as is assumed in the shear/strain-based finescale methods in Section 14.3.3, and so can be applied in nearly well-mixed regions.

To apply the large-eddy method to autonomous platforms, estimates of the turbulent velocity scale q' and lengthscale l are needed, as well as an estimate of the constant of proportionality c_ϵ. For the large-eddy method, the q' perturbations are intended to represent the velocity of isotropic turbulent eddies, and could in theory be estimated from horizontal or vertical seawater velocity. To determine l, the lengthscale of the largest energy-containing eddies, Beaird et al. (2012) used the Ozmidov lengthscale L_{Oz} (Eq. (14.3)), which assumes that in a stratified fluid, the large-eddy scales are related to the Ozmidov scale.

For gliders or a profiling float, the vertical velocity of the vehicle can be determined through differentiating their depth (measured from pressure) with time. This vertical motion results both from the vehicle's buoyancy anomaly relative to the surrounding water, and also from the vehicle being advected by the ambient seawater vertical velocity. Hydrodynamic models for Seagliders (Frajka-Williams et al., 2011) or floats (Cusack et al., 2017) can be used to estimate what the vertical position of the vehicle would be, assuming steady flight through still water. Differences between the vehicle's measured vertical velocity w_{meas} and the model-predicted vertical velocity w_{model} are treated as the vertical seawater velocity w,

$$w = w_{meas} - w_{model}. \tag{14.10}$$

The resultant estimate of w is for flows at or larger than the glider or float size [$\mathcal{O}(1)$ m], and is only valid when the flight satisfies the steady-flight assumption of the flight model. In particular, during parts of the dive, where the glider (or float) is accelerating, the steady-flight assumptions do not hold. In the case of the glider, a dive starts from rest at the surface and accelerates to steady flight before decelerating at apogee (the turning depth), and accelerating back to steady flight on the climb. On the climb, the glider's return to the surface does not involve a slow deceleration, but rather the glider tends to continue in steady flight until reaching the surface. Mid-profile (away from the surface and apogee), the glider is typically assumed to be in steady flight, though excessive glider maneuvers (e.g., roll maneuvers) may affect the vehicle's vertical velocity (Frajka-Williams et al., 2011). The parameters used in the vehicle flight model (e.g., lift and drag coefficients) can be tuned to improve estimates of seawater vertical water velocity (Frajka-Williams et al., 2011; Merckelbach et al., 2010; Cusack et al., 2017).

The flight model methods provide an estimate of seawater vertical velocity w, but a further assumption of the large-eddy method is that the q' scale be associated with turbulent eddies in the inertial subrange, and not internal waves. Several methods have been used to separate glider-derived vertical seawater velocity into that associated with turbulent eddies rather than internal waves, relying primarily on high-pass filters of the profile of vertical velocity (Beaird et al., 2012; Evans et al., 2018; Fernandez-Castro et al., 2020; Cusack et al., 2017). High-pass filtering the velocity profiles also removes biases in the flight model (Beaird et al., 2012), as the flight model tuning only affects the low-frequency fluctuations in w (Todd et al., 2017), i.e., at the scale of 1 km for a standard Seaglider diving to its maximum pressure rating.

The constant of proportionality c_ϵ is empirically determined through calibration against some other measure of turbulent dissipation. This constant of proportionality can take up differences between the eddy scale l and estimated scale L_{Oz}. Its

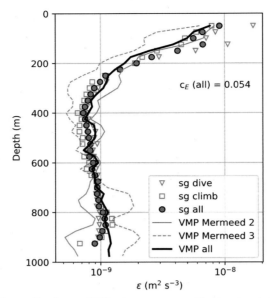

FIGURE 14.6 Comparison of log-averaged ϵ profiles from a VMP microstructure profiler (grey) and glider estimates using the large-eddy method (triangles for dives, squares for climbs and solid circles for both). The value for c_ϵ was obtained by least-squares minimisation of the difference between the log-averaged profiles. From Fernandez-Castro et al. (2020).

value may further vary depending on the choices made in high-pass filtering w, including fixed length filtering (Beaird et al., 2012; Fernandez-Castro et al., 2020; Cusack et al., 2017) or varying with N (Evans et al., 2018). Gargett (1999) showed that for a given method, c_ϵ could be appropriately chosen from a short training dataset, then meaningfully applied to the remainder of the data on the cruise and on a subsequent cruise with no change to the value of c_ϵ. The methods and Seagliders used in Beaird et al. (2012); Evans et al. (2018); Fernandez-Castro et al. (2020) yielded constants of $c_\epsilon = 0.37$, $c_\epsilon = 1.96$ and $c_\epsilon = 0.054$, whereas for the Argo profiling floats, the $c_\epsilon = 0.146$ and $c_\epsilon = 0.123$ (Cusack et al., 2017). It is not yet clear whether, once a piece of hardware is 'tuned' against a microstructure dataset, if it can be used with the same proportionality constant for a future dataset. These values of c_ϵ are within the range of previous applications of the large-eddy method (0.05–4, Gargett, 1999).

Applying the large-eddy method to autonomous platforms is subject to several limitations. In the case of the Argo float application, not all floats transmit the necessary engineering fields (Cusack et al., 2017). For both float and glider methods, the constant of proportionality c_ϵ means that historic glider datasets cannot be directly converted to estimates of turbulent dissipation. The sensitivity of the method to glider flight is another key limitation. Fernandez-Castro et al. (2020) was only able to use a subset of a glider dataset due to excessive roll maneuvers by the glider—before the flight model was tuned—contaminating estimates of glider w.

Despite these limitations, the large-eddy method applied to autonomous vehicles presents the opportunity to make long endurance, depth-resolved estimates of ϵ at relatively high horizontal resolution (3 km for gliders). In the northwest Atlantic, they were used to produce a full annual cycle of turbulent dissipation both within the surface mixed layer and the stratified water below, enabling a test of bulk parameterisations for mixed-layer dissipation driven by surface wind and buoyancy forcing (Evans et al., 2018). High-resolution sections within and outside of a mesoscale eddy were able to identify potential critical layers within the eddy and associated elevated dissipation rates below the eddy core (see Fig. 14.6 and Fernandez-Castro et al., 2020). The large-eddy method applied to two profiling floats measured the a lee wave in the Southern ocean at a ridge, along with an estimate of energy flux (from velocity measurements) and dissipation rates (Cusack et al., 2017). Accompanied by calibration profiles of turbulent dissipation (to determine c_ϵ), the method provides long endurance, high horizontal resolution and top 1000 m coverage of turbulent dissipation rate estimates, at no additional hardware or power cost. (In comparison, the power consumption of the MicroRider is 1.0–1.6 W when on, roughly equivalent to two or three times the power requirement of a glider's buoyancy engine.) It remains to be seen whether the constant of proportionality c_ϵ can be estimated some other way, or whether once determined for a particular platform, it can be used repeatedly on subsequent missions.

Large-eddy method on Lagrangian floats

The application of the large-eddy method on Lagrangian floats differs from the method above for gliders and profiling floats, in that vertical velocities are measured at high enough time-resolution to fit the Kolmogorov turbulence spectrum and estimate turbulence through integrating the spectrum. This method is not subject to the uncertainty associated with the constant of proportionality c_ϵ. Lagrangian floats are neutrally buoyant floats specially designed to follow a water parcel of roughly 1 m^3 volume (D'Asaro et al., 1996). They do so by measuring the ambient seawater density and then controlling their volume so that the float density matches ambient density, and using a square drogue to increase drag within the 1 m^3 volume. The advantage of a Lagrangian float is that it is not subject to flight relative to seawater, which is a current limitation of glider-based methods and the profiling-float methods mentioned above, thus enabling a more direct estimate of turbulence from onboard sensors than would otherwise be possible.

Turbulence measurements from these floats have been computed using vertical velocity spectra, where the vertical velocity is measured from both an onboard acoustic Doppler velocimeter (ADV) measuring at 25 Hz, thus resolving the inertial subrange (Fig. 14.1), and from the float measured vertical velocity using a pressure-sensor sampling at 1 Hz (Lien and D'Asaro, 2006). The ADV-based measurements agreed well with the anticipated $-5/3$ slope in the inertial subrange of the Kolmogorov turbulence spectrum (Fig. 14.1), though there were some minor discrepancies in spectra derived from the pressure sensor alone. However, these discrepancies only contributed an error of roughly 10%, and the derived time series of turbulence were well correlated between the ADV-based measurements and the pressure-sensor measurements. Further investigations have demonstrated that Lagrangian floats are also capable of measuring temperature variance using two vertically offset CTDs (χ, D'Asaro and Lien, 2007), and can be combined with high-frequency ADCPs (1000 kHz) to estimate turbulence during the profiling phase of the float (Shcherbina et al., 2018).

Large-eddy method on moored ADCPs: structure function

In contrast to the air-foil method of measuring velocity shear in the ocean, ADCPs measure water velocity from the Doppler frequency shift between a transmitted and received sound signal. In addition to the possibility of velocity measurements being used in finescale parameterisations to estimate turbulence, a recent adaptation of the 'structure function' method for measuring ϵ in the atmosphere has been modified for the ocean. This method uses high-frequency ADCPs and the same large-eddy scaling in Eq. (14.8) (Wiles et al., 2006; Guerra and Thomson, 2017). It is distinct from the original application of the large-eddy method to the ocean (Gargett, 1999), which used an eddy-resolving approach, but instead uses differences in along-beam velocities and the assumption of isotropic turbulence. The bin separation in the ADCP (the spatial resolution) needs to be within the inertial subrange of the turbulence spectrum. The structure function $D(z, r)$ is defined at a location z using velocity perturbations from the mean ($v' = v - \overline{v}$),

$$D(z, r) = \overline{(v'(z) - v'(z+r))^2}, \tag{14.11}$$

for two points separated by a distance r. The assumption is that the velocity difference is due to eddies with lengthscale r and velocity scale q', so that $D \sim (q')^2$. Using Eq. (14.8) allows a fit between the measured $D(z, r)$ and

$$D(z, r) = C_v^2 \epsilon^{2/3} r^{2/3}, \tag{14.12}$$

where C_v is constant. As for the glider method, this constant is empirically derived (ranging from 2.0 to 2.2). Finally, the structure function D is fit to an equation of the form

$$D(z, r) = M + A r^{2/3}, \tag{14.13}$$

where M and A are constants to find the value of A, which is used to estimate turbulent dissipation as

$$e = \frac{A^{3/2}}{C_v^3}. \tag{14.14}$$

Though this method still has a dependence on an empirically derived constant C_v, its range is smaller than that for the large-eddy method applied to glider vertical velocity data (0.005–1.96), and thus has lower uncertainties.

The resolution requirement for the velocity measurements (r to be between the Kolmogorov and Ozmidov scales) means that high-frequency ADCPs are needed (e.g., 600 kHz or 1.2 MHz), which tends to limit the vertical span over which turbulence is measured to a few metres. The scale requirement additionally restricts the applicability of the method in highly stratified environments, where the maximum value of r may be quite small (0.1–0.4 m in the seasonal thermocline).

Although the method was originally used on bottom-mounted ADCPs (Wiles et al., 2006), it has since been extended to mid-water ADCPs on moorings (Lucas et al., 2014) and to marginal ice zones (Smith and Thomson, 2019). The endurance is only limited by the power and memory of the ADCP, and so has been used in a year-long mooring deployment in the northeast Atlantic (Buckingham et al., 2019). These observations enabled a decomposition of the surface mixed layer turbulent dissipation (December to April) according to surface forcings, including buoyancy-forced convection, wind and wave forcing, Langmuir turbulence and submesoscale processes, demonstrating the importance of wind and wave forcing in the upper ocean (Buckingham et al., 2019). These methods are also being applied under sea ice in the Arctic, as described in Lenn et al. (Chapter 11) and (Smith and Thomson, 2019).

14.4 Future outlook

The current 'gold standard' of mixing measurements yield high-quality estimates of turbulent dissipation, but using labour-intensive, costly and relatively slow ship-based approaches, limiting the availability of microstructure measurements in the world's oceans. Recent developments and more widespread use of autonomous platforms have enabled broader coverage of the spatio-temporal variability of turbulence and mixing. Some of the methods use finescale parameterisations or large-eddy methods to make indirect estimates of turbulent dissipation, but based on standard CTD measurements from autonomous platforms. Others use new microstructure sensor packages developed for autonomous vehicles and providing high-quality shear or temperature-based dissipation estimates, but with some limitations in quality due to the vibrational characteristics of the vehicle or uncertainties in vehicle velocity, leading to a higher noise floor and potentially limiting their utility in quiescent environments.

An application of the strain-based finescale parameterisation to the global Argo profiling float array has enabled global estimates of turbulent dissipation below the surface mixed layer and in the top 2000 m (Section 14.3.3). These data have enabled the interrogation of mixing processes on global scales, e.g., associated with mesoscale eddies in every ocean basin or capturing seasonal variability of dissipation patterns. New sensor packages produce self-contained measurements of χ, enabling multi-year investigations into the dissipation of temperature variance associated with El Niños (3–5 years) on a mooring or spatial surveys on a standard CTD rosette (Section 14.3.2).

Autonomous underwater vehicles have opened windows both into wider space- and timescales, but also for targeted studies at higher resolution than Argo float data, including in such extreme and hard-to-reach environments as under-ice shelves and near-bottom along steep canyons. An application of the large-eddy method to glider CTD data has provided an annual cycle of turbulent dissipation in the top 1000 m in both the surface mixed layer and below, though subject to a constant of proportionality (Section 14.3.4). New microstructure sensor packages have been developed for underwater vehicles (Section 14.3.1). Their use is not yet routine, requiring close examination of spectra and correction for vehicle vibration and noise, but as methods develop further they will enable wider coverage of direct microstructure measurements in the inertial subrange.

The rapid and recent (2000–2020) development of autonomous platforms is now enabling orders of magnitude increases in the number of turbulence measurements. These new datasets underpin investigations into the long-timescale (>1 month) and broad spatial scale (mesoscale and larger) variability in turbulence and mixing processes, and may further enable comparisons between localised direct measurements and more integrative measures of mixing (e.g., tracer studies or large-scale inversions). As autonomous platforms gain even wider use, the issues noted above limiting the widespread application of new method (e.g., vehicle vibration noise, constants of proportionality) may be resolved, improving the quality of autonomous measurements of mixing.

14.5 Conclusions

New methods in measuring turbulent kinetic energy dissipation offer opportunities to bridge space-time gaps left by sparse and infrequent ship-based measurement methods, including the requirement for observations at high spatial resolution (Gula et al., Chapter 8), long timescale measurements (Moum et al., Chapter 10) and global scale measurements.

- High spatial resolution. AUVs instrumented with microstructure offer the potential to cover small spatial scales, enabling observationally based investigations of submesoscale interaction with turbulent dissipation recently identified as a gap in parameterisations.
- Long timescale observations. Moored estimates of temperature variance or turbulent dissipation enable long duration (a year and longer) deployments to capture seasonal and interannual variations of mixing previously missed by ship-based measurements.

- Global scale. Shear-based parameterisations applied to the global Argo dataset enable global scale estimates of dissipation by internal waves, albeit out of the mixed layer and away from topography.

Additionally, these new platforms can enable measurements closer to key geographic regions, including interfaces with topography (Polzin and McDougall, Chapter 7), the atmosphere (Fox-Kemper et al., Chapter 4), and ice (Lenn et al., Chapter 11 and Gille et al., Chapter 12).

- Bottom boundary layer. Though typical gliders only reach to 1000 m deep, they are able to profile to within about 5 m of the seabed and could potentially make microstructure measurements very near the bottom. Moored instruments, including bottom-mounted ADCPs, employing the structure function method have been used to make near-seabed measurements of ocean turbulence.
- Air-sea boundary layer. New autonomous profiling techniques enable microstructure profilers to measure in the top metres of the ocean, enabling investigations into the effect of turbulent dissipation on the surface mixed layer and surface temperatures, critical for the large-scale circulation.
- Sea-ice boundary layer. Moored observations of mixing enable under-ice measurements, not previously accessible by traditional means. AUV-mounted shear probes also offer this capability, though the observational campaigns to date are limited.

The measurement methods outlined above afford new opportunities to investigate turbulent kinetic energy dissipation, at space- and time-scales not previously accessible by ship-based measurements, and close to interfaces, where the importance of mixing to the large-scale circulation is anticipated to be outsized compared to the geographic size of these regions. Subject to the assumptions of each method, and the limitations of the platforms (e.g., reliability, onboard processing, etc., which are being actively worked on), the new technological frontiers offer the potential to address many of the observationally limited open questions in the field of ocean mixing.

References

Batchelor, G.K., 1959. Small-scale variation of convected quantities like temperature in turbulent fluid. Part 1. General discussion and the case of small conductivity. J. Fluid Mech. 5 (1), 113–133. https://doi.org/10.1017/S002211205900009X.

Baumert, H., Simpson, J., Sunderman, J., 2005. Marine Turbulence: Theories, Observations and Models. Results of the CARTUM Project. Cambridge University Press, Cambridge, UK, p. 644.

Beaird, N., Fer, I., Rhines, P., Eriksen, C., 2012. Dissipation of turbulent kinetic energy inferred from Seagliders: an application to the eastern Nordic Seas overflows. J. Phys. Oceanogr. 42, 2268–2282. https://doi.org/10.1175/JPO-D-12-094.1.

Buckingham, C.E., Lucas, N.S., Belcher, S.E., Rippeth, T.P., Grant, A.L.M., Le Sommer, J., Ajayi, A.O., Naveira Garabato, A.C., 2019. The contribution of surface and submesoscale processes to turbulence in the open ocean surface boundary layer. J. Adv. Model. Earth Syst. 11, 4066–4094. https://doi.org/10.1029/2019MS001801.

Cervelli, E., Haskins, T., 2018. MicroRider Proglet for Slocum glider. RSI Technical Note 44. Victoria, Canada.

Creed, E., Ross, W., Lueck, R., Stern, P., Douglas, W., Wolk, F., Hall, R., 2015. Integration of a RSI microstructure sensing package into a Seaglider. In: OCEANS 2015 - MTS/IEEE Washington, pp. 1–6.

Cusack, J.M., Naveira Garabato, A.C., Smeed, D.A., Girton, J.B., 2017. Observation of a large lee wave in the Drake Passage. J. Phys. Oceanogr. 47, 793–810. https://doi.org/10.1175/JPO-D-16-0153.1.

D'Asaro, E.A., Farmer, D.M., Osse, J.T., Dairiki, G.T., 1996. A Lagrangian float. J. Atmos. Ocean. Technol. 13 (6), 1230–1246. https://doi.org/10.1175/1520-0426(1996)013<1230:ALF>2.0.CO;2.

D'Asaro, E.A., Lien, R.-C., 2007. Measurement of scalar variance dissipation from Lagrangian floats. J. Atmos. Ocean. Technol. 24, 1066–1077. https://doi.org/10.1175/JTECH2031.1.

Dhanak, M.R., Holappa, K., 1999. An autonomous ocean turbulence measurement platform. J. Atmos. Ocean. Technol. 16 (11), 1506–1518. https://doi.org/10.1175/1520-0426(1999)016<1506:AAOTMP>2.0.CO;2.

Evans, D.G., Lucas, N.S., Hemsley, V., Frajka-Williams, E., Naveira Garabato, A.C., Martin, A., Painter, S.C., Inall, M.E., Palmer, M.R., 2018. Annual cycle of turbulent dissipation estimated from Seagliders. Geophys. Res. Lett. 45, 10,560–10,569. https://doi.org/10.1029/2018GL079966.

Fer, I., Peterson, A.K., Ullgren, J.E., 2014. Microstructure measurements from an underwater glider in the turbulent Faroe Bank Channel overflow. J. Atmos. Ocean. Technol. 31, 1128–1150. https://doi.org/10.1175/JTECH-D-13-00221.1.

Fernandez-Castro, B., Evans, D.G., Frajka-Williams, E., Vic, C., Naveira Garabato, A.C., 2020. Breaking of internal waves and turbulent dissipation in an anticyclonic mode water eddy. J. Phys. Oceanogr. https://doi.org/10.1175/JPO-D-19-0168.1.

Frajka-Williams, E., Eriksen, C.C., Rhines, P.B., Harcourt, R.R., 2011. Determining vertical water velocities from Seaglider. J. Atmos. Ocean. Technol. 28 (12), 1641–1656. https://doi.org/10.1175/2011JTECHO830.1.

Furlong, M.E., Paxton, D., Stevenson, P., Pebody, M., McPhail, S.D., Perrett, J., 2012. Autosub Long Range: a long range deep diving AUV for ocean monitoring. In: 2012 IEEE/OES Autonomous Underwater Vehicles (AUV), pp. 1–7.

Gargett, A.E., 1999. Velcro measurement of turbulent kinetic energy dissipation rate ϵ. J. Atmos. Ocean. Technol. 16, 1973–1993. https://doi.org/10.1175/1520-0426(1999)016<1973:VMOTKE>2.0.CO;2.

Goodman, L., Levine, E.R., Lueck, R.G., 2006. On measuring the terms of the turbulent kinetic energy budget from an AUV. J. Atmos. Ocean. Technol. 23, 977–990. https://doi.org/10.1175/JTECH1889.1.

Goto, Y., Yasuda, I., Nagasawa, M., 2018. Comparison of turbulence intensity from CTD-attached and free-fall microstructure profilers. J. Atmos. Ocean. Technol. 35.

Gregg, M.C., 1989. Scaling turbulent dissipation in the thermocline. J. Geophys. Res. 94 (C7), 9686–9698. https://doi.org/10.1029/JC094iC07p09686.

Gregg, M.C., 1991. The study of mixing in the ocean: a brief history. Oceanography 4, 39–45. https://doi.org/10.5670/oceanog.1991.21.

Gregg, M.C., D'Asaro, E.A., Riley, J.J., Kunze, E., 2018. Mixing efficiency in the ocean. Annu. Rev. Mar. Sci. 10 (1). https://doi.org/10.1146/annurev-marine-121916-063643.

Gregg, M., Kunze, E., 1991. Shear and strain in Santa Monica basin. J. Geophys. Res. 96 (C9), 16709–16719. https://doi.org/10.1029/91JC01385.

Gregg, M.C., Sanford, T.B., Winkel, D.P., 2003. Reduced mixing from the breaking of internal waves in equatorial waters. Nature 422, 513–515. https://doi.org/10.1038/nature01507.

Guerra, M., Thomson, J., 2017. Turbulence measurements from five-beam acoustic Doppler current profilers. J. Atmos. Ocean. Technol. 36 (6), 1267–1284. https://doi.org/10.1175/JTECH-D-16-0148.1.

Henyey, F.S., Wright, J., Flatte, S.M., 1986. Energy and action flow through the internal wave field - an eikonal approach. J. Geophys. Res. 91 (C7), 8487–8495. https://doi.org/10.1029/JC091iC07p08487.

Holmes, R.M., Moum, J.N., Thomas, L.N., 2016. Evidence for seafloor-intensified mixing by surface-generated equatorial waves. Geophys. Res. Lett. 43 (3), 1202–1210. https://doi.org/10.1002/2015GL066472.

Hsieh, W.W., 2009. Machine Learning Methods in the Environmental Sciences. Cambridge University Press, Cambridge, UK; New York.

Kimura, S., Jenkins, A., Dutrieux, P., Forryan, A., Garabato, A.C.N., Firing, Y., 2016. Ocean mixing beneath Pine Island Glacier ice shelf, West Antarctica. J. Geophys. Res. 121, 8496–8510. https://doi.org/10.1002/2016JC012149.

Kunze, E., 2017a. Internal-wave-driven mixing: global geography and budgets. J. Phys. Oceanogr. 47 (6), 1325–1345. https://doi.org/10.1175/JPO-D-16-0141.1.

Kunze, E., 2017b. The internal-wave-driven meridional overturning circulation. J. Phys. Oceanogr. 47 (11), 2673–2689. https://doi.org/10.1175/JPO-D-16-0142.1.

Kunze, E., Firing, E., Hummon, J.M., Chereskin, T.K., Thurnherr, A.M., 2006. Global abyssal mixing inferred from lowered ADCP shear and CTD strain profiles. J. Phys. Oceanogr. 36 (8), 1553–1576. https://doi.org/10.1175/JPO2926.1.

Leadbitter, P., Hall, R., Brearley, A., 2019. A methodology for Thorpe scaling 512 Hz fast thermistor data from buoyancy-driven gliders to estimate turbulent kinetic energy dissipation rate in the ocean. In: OCEANS 2019 MTS/IEEE Seattle, pp. 1–5.

Ledwell, J., Laurent, L.S., Girton, J., Toole, J., 2011. Diapycnal mixing in the Antarctic Circumpolar Current. J. Phys. Oceanogr. 41, 241–246. https://doi.org/10.1175/2010JPO4557.1.

Ledwell, J.R., Watson, A.J., 1988. The use of deliberately injected tracers for the study of diapycnal mixing in the ocean. In: Nihoul, J.C.J., Jamart, B.M. (Eds.), Small-Scale Turbulence and Mixing in the Ocean. In: Elsevier Oceanography Series, vol. 46. Elsevier, pp. 11–20.

Ledwell, J., Watson, A., 1991. The Santa Monica Basin tracer experiment: a study of diapycnal and isopycnal mixing. J. Geophys. Res. 96, 8695–8718. https://doi.org/10.1029/91JC00102.

Lien, R.-C., D'Asaro, E.A., 2006. Measurement of turbulent kinetic energy dissipation rate with a Lagrangian Float. J. Atmos. Ocean. Technol. 23 (7), 964–976. https://doi.org/10.1175/JTECH1890.1.

Lien, R.-C., Sanford, T.B., Carlson, J.A., Dunlap, J.H., 2016. Autonomous microstructure EM-APEX floats. Methods Oceanogr. 17, 282–295. https://doi.org/10.1016/j.mio.2016.09.003.

Lucas, N.S., Grant, A.L.M., Rippeth, T.P., Polton, J.A., Palmer, M.R., Brannigan, L., Belcher, S.E., 2019. Evolution of oceanic near-surface stratification in response to an autumn storm. J. Phys. Oceanogr. 49, 2961–2978. https://doi.org/10.1175/JPO-D-19-0007.1.

Lucas, N.S., Simpson, J., Rippeth, T., Old, C., 2014. Measuring turbulent dissipation using a tethered ADCP. J. Atmos. Ocean. Technol. 31, 1826–1837. https://doi.org/10.1175/JTECH-D-13-00198.1.

Lueck, R.G., 2005. Horizontal and vertical turbulence profilers. In: Baumert, H., Simpson, J., Sunderman, J. (Eds.), Marine Turbulence: Theories, Observations and Models. Results of the CARTUM Project. Cambridge University Press, Cambridge, UK, pp. 89–100.

Lueck, R.G., Wolk, F., Yamazaki, H., 2002. Oceanic velocity microstructure measurements in the 20th century. J. Oceanogr. 58, 153–174. https://doi.org/10.1023/A:1015837020019.

Luketina, D.A., Imberger, J., 2000. Determining turbulent kinetic energy dissipation from Batchelor curve fitting. J. Atmos. Ocean. Technol. 18 (1), 100–113.

McPhail, S., Templeton, R., Pebody, M., Roper, D., Morrison, R., 2019. Autosub Long Range AUV missions under the Filchner and Ronne ice shelves in the Weddell Sea, Antarctica - an engineering perspective. In: OCEANS 2019 - Marseille, pp. 1–8.

Merckelbach, L., Berger, A., Krahmann, G., Dengler, M., Carpenter, J.R., 2019. A dynamic flight model for Slocum gliders and implications for turbulence microstructure measurements. J. Atmos. Ocean. Technol. 36, 281–296. https://doi.org/10.1175/JTECH-D-18-0168.1.

Merckelbach, L., Smeed, D., Griffiths, G., 2010. Vertical water velocities from underwater gliders. J. Atmos. Ocean. Technol. 27, 547–563. https://doi.org/10.1175/2009JTECHO710.1.

Moum, J.N., 1996. Energy-containing scales of turbulence in the ocean thermocline. J. Geophys. Res. 101, 14095–14109. https://doi.org/10.1029/96JC00507.

Moum, J.N., 2015. Ocean speed and turbulence measurements using pitot-static tubes on moorings. J. Atmos. Ocean. Technol. 32 (7), 1400–1413.

Moum, J.N., Gregg, M.C., Lien, R.C., Carr, M., 1995. Comparison of turbulence kinetic energy dissipation rate estimates from two ocean microstructure profilers. J. Atmos. Ocean. Technol. 12 (2), 346–366.

Moum, J., Nash, J., 2009. Mixing measurements on an equatorial ocean mooring. J. Atmos. Ocean. Technol. 26 (2), 317–336.

Moum, J.N., Perlin, A., Nash, J.D., McPhaden, M.J., 2013. Seasonal sea surface cooling in the equatorial Pacific cold tongue controlled by ocean mixing. Nature 500 (7460), 64–67.

Nasmyth, P.W., 1973. Turbulence and microstructure in the upper ocean. Mem. Soc. R. Sci. Liege, Ser. 6 4, 47–56.

Naveira Garabato, A.C., Frajka-Williams, E., Spingys, C.P., Legg, S., Polzin, K.L., et al., 2019. Rapid mixing and exchange of deep-ocean waters in an abyssal boundary current. Proc. Natl. Acad. Sci. 116, 13233–13238. https://doi.org/10.1073/pnas.1904087116.

Osborn, T.R., 1980. Estimates of the local rate of vertical diffusion from dissipation measurements. J. Phys. Oceanogr. 10, 83–89. https://doi.org/10.1175/1520-0485(1980)010<0083:EOTLRO>2.0.CO;2.

Osborn, T.R., Cox, C.S., 1972. Oceanic fine structure. Geophys. Fluid Dyn. 3, 321–345.

Palmer, M.R., Stephenson, G.R., Inall, M.E., Balfour, C., Düsterhus, A., Green, J.A.M., 2015. Turbulence and mixing by internal waves in the Celtic Sea determined from ocean glider microstructure measurements. J. Mar. Syst. 144, 57–69. https://doi.org/10.1016/j.jmarsys.2014.11.005.

Perlin, A., Moum, J., 2012. Comparison of thermal variance dissipation rates from moored and profiling instruments at the equator. J. Atmos. Ocean. Technol. 29 (9), 1347–1362. https://doi.org/10.1175/JTECH-D-12-00019.1.

Peterson, A.K., Fer, I., 2014. Dissipation measurements using temperature microstructure from an underwater glider. In: Special Issue: Autonomous Marine Vehicles. Methods Oceanogr. 10, 44–69. https://doi.org/10.1016/j.mio.2014.05.002.

Polzin, K.L., Naveira Garabato, A.C., Huussen, T.N., Sloyan, B.M., Waterman, S.N., 2014. Finescale parameterizations of turbulent dissipation. J. Geophys. Res. 119 (2). https://doi.org/10.1002/2013JC008979.

Polzin, K.L., Toole, J.M., Schmitt, R.W., 1995. Finescale parameterizations of turbulent dissipation. J. Phys. Oceanogr. 25, 306–328. https://doi.org/10.1175/1520-0485(1995)025<0306:FPOTD>2.0.CO;2.

Prandke, H., 2005. Microstructure sensors. In: Baumert, H., Simpson, J., Sunderman, J. (Eds.), Marine Turbulence: Theories, Observations and Models. Results of the CARTUM Project. Cambridge University Press, Cambridge, UK, pp. 101–109.

Rainville, L., Gobat, J.I., Lee, C.M., Shilling, G.B., 2017. Multi-month dissipation estimates using microstructure from autonomous underwater gliders. Oceanography 30 (2), 49–50. https://doi.org/10.5670/oceanog.2017.219.

Scheifele, B., Waterman, S., Merckelbach, L., Carpenter, J.R., 2018. Measuring the dissipation rate of turbulent kinetic energy in strongly stratified, low-energy environments: a case study from the Arctic Ocean. J. Geophys. Res. 123, 5459–5480. https://doi.org/10.1029/2017JC013731.

Schultze, L.K.P., Merckelbach, L.M., Carpenter, J.R., 2017. Turbulence and mixing in a shallow shelf sea from underwater gliders. J. Geophys. Res. 122, 9092–9109. https://doi.org/10.1002/2017JC012872.

Shcherbina, A.Y., D'Asaro, E.A., Nylund, S., 2018. Observing finescale oceanic velocity structure with an autonomous Nortek acoustic Doppler current profiler. J. Atmos. Ocean. Technol. 35, 411–426. https://doi.org/10.1175/JTECH-D-17-0108.1.

Smith, M., Thomson, J., 2019. Ocean surface turbulence in newly formed marginal ice zones. J. Geophys. Res. 124, 1382–1398. https://doi.org/10.1029/2018JC014405.

Smyth, W.D., Moum, J.N., Caldwell, D.R., 2001. The efficiency of mixing in turbulent patches: inferences from direct simulations and microstructure observations. J. Phys. Oceanogr. 31, 1969–1992.

St. Laurent, L., Merrifield, S., 2017. Measurements of near-surface turbulence and mixing from autonomous ocean gliders. Oceanography 30, 116–125. https://doi.org/10.5670/oceanog.2017.231.

Stewart, R.W., Grant, H.L., 1962. Determination of the rate of dissipation of turbulent energy near the sea surface in the presence of waves. J. Geophys. Res. 67, 3177–3180.

Tennekes, H., Lumley, J.L., 1972. A First Course in Turbulence. MIT Press, Cambridge, MA.

Testor, P., de Young, B., Rudnick, D.L., Glenn, S., Hayes, D., et al., 2019. OceanGliders: a component of the integrated GOOS. Front. Mar. Sci. 6, 422. https://doi.org/10.3389/fmars.2019.00422.

Thomson, J., 2013. Wave breaking dissipation observed with "SWIFT" drifters. J. Atmos. Ocean. Technol. 29, 1866–1882. https://doi.org/10.1175/JTECH-D-12-00018.1.

Todd, R.E., Rudnick, D.L., Sherman, J.T., Owens, W.B., George, L., 2017. Absolute velocity estimates from autonomous underwater gliders equipped with Doppler current profilers. J. Atmos. Ocean. Technol. 34, 309–333. https://doi.org/10.1175/JTECH-D-16-0156.1.

Ward, B., Fristedt, T., Callaghan, A.H., Sutherland, G., Sanchez, X., Vialard, J., Doeschate, A., 2014. The air-sea interaction profiler (ASIP): an autonomous upwardly rising profiler for microstructure measurements in the upper ocean. J. Atmos. Ocean. Technol. 31 (10), 2246–2267. https://doi.org/10.1175/JTECH-D-14-00010.1.

Warner, S.J., Moum, J.N., 2019. Feedback of mixing to ENSO phase change. Geophys. Res. Lett. 46 (23), 13920–13927.

Waterhouse, A.F., MacKinnon, J.A., Nash, J.D., Alford, M.H., Kunze, E., Simmons, H.L., Polzin, K.L., St. Laurent, L.C., Sun, O.M., Pinkel, R., et al., 2014. Global patterns of diapycnal mixing from measurements of the turbulent dissipation rate. J. Phys. Oceanogr. 44 (7), 1854–1872.

Whalen, C.B., MacKinnon, J.A., Talley, L.D., 2018. Large-scale impacts of the mesoscale environment on mixing from wind-driven internal waves. Nat. Geosci. 11 (11), 842. https://doi.org/10.1038/s41561-018-0213-6.

Whalen, C.B., MacKinnon, J.A., Talley, L.D., Waterhouse, A.F., 2015. Estimating the mean diapycnal mixing using a finescale strain parameterization. J. Phys. Oceanogr. 45 (4), 1174–1188. https://doi.org/10.1175/JPO-D-14-0167.1.

Wiles, P., Rippeth, T., Simpson, J., Hendricks, P., 2006. A novel technique for measuring the rate of turbulent dissipation in the marine environment. Geophys. Res. Lett. 33, L21608. https://doi.org/10.1029/2006GL027050.

Wolk, F., Lueck, R.G., St. Laurent, L., 2009. Turbulence measurements from a glider. In: OCEANS 2009, pp. 1–6.

Index

A
Abyssal
  circulation, 51
  hills, 8, 122, 133, 134, 174
  mixing, 47, 50, 156, 268, 313
    Munk's, 146
  ocean, 12, 162
    circulation, 156
  overturning, 11, 49, 51, 56
    cell, 15, 49, 56
    circulation, 13, 50, 204, 315
  stratification, 12, 51, 95
Acoustic Doppler Current Profiler (ADCP), 348, 354, 357
Acoustic Doppler Velocimeter (ADV), 357
Adiabatic, 35
Advection
  chaotic, 223
  eastward, 317
  eddy, 329, 332
  Ekman, 301, 305
  northward surface, 302
  tracer variance, 242
  vertical, 99, 103, 104
Advective buoyancy flux, 191
Advective-diffusive process, 245
Ageostrophic secondary circulation (ASC), 185, 186, 201
Antarctic Bottom Water (AABW), 11, 12, 49, 171, 302
  lightening, 50
Antarctic Circumpolar Current (ACC), 12, 42, 55, 235, 301
Antarctic Intermediate Water (AAIW), 12, 15
Antarctic Slope Front (ASF), 311
Anthropogenic
  heat, 14
  tracers, 233, 234
    ocean, 234
Anticyclones, 104
Arctic, 275
  eddy, 285, 287, 291
    kinetic energy, 285
  freshwater, 291
  mixing, 276, 278–280
  sea ice, 275, 280, 292
  stratification, 278
  tides, 278
Arctic Observing Network (AON), 286

Area
  incrop, 161
  ocean, 55, 162
  surface, 76, 173, 282
Atlantic Meridional Overturning Circulation (AMOC), 51
Atlantic Water (AW), 275
Atmospheric winds, 97
Autonomous platforms, 353, 354
Autonomous underwater glider, 290
Autonomous Underwater Vehicle (AUV), 348
Autosub Long Range (ALR), 350
Available Potential Energy (APE), 307

B
Background
  diapycnal mixing, 13
  diffusivity, 160
  stratification, 284
  theoretical, 126, 130, 134
  turbulence, 81
  wind field, 96
  wind forcing, 304
Baroclinic
  eddies, 15
  instability, 42, 174, 175, 187, 194, 221
  tides, 96
  volume transport, 15
Barotropic tides, 21, 24, 96, 119
Basin-scale horizontal circulation, 53
Basin-scale overturning circulation, 49
Biogeochemical tracers, 204, 205, 234, 305
Bolus transport, 156, 162, 231, 232
Bottom
  injection, 195
  lightening, 47
  MLI wavelength, 195
Bottom Boundary Layer (BBL), 150, 156, 157, 163, 194
  baroclinic instability, 194
Bottom-intensified mixing, 155
Boundary conditions, 148
Boundary layer
  depth, 75, 82, 84
  downwelling, 169
  memory, 76
  ocean surface, 79, 80, 85
  process, 171
  turbulent, 68, 181, 313
Boundary mixing, 164

Bubbles, 76
Budget equations, 71
Bulk mixed layers, 83
Buoyancy
  effects, 76, 150
  forcing, 42, 169
  frequency, 78, 158, 181
  ocean, 7
  transport, 168, 220
Buoyant material, 205

C
Cabbeling, 10, 12, 48, 52, 56, 160, 243
Carbon uptake, 18
Chaotic advection, 223
Circulation
  abyssal, 51
    ocean, 156
    overturning, 13, 50, 204, 315
  basin-scale horizontal, 53
  basin-scale overturning, 49
  deep overturning, 36, 42
  diapycnal, 56, 57
  frontal vertical, 101, 102
  geostrophic, 201
  isopycnal, 43
  meridional, 40
  net, 50
  North Atlantic
    Deep Water, 51
    meridional overturning, 291
  ocean, 1, 40, 56, 84, 275, 337, 345
  physical, 243
  secondary, 12, 185, 199
  Stommel–Arons, 55, 163
  TTW, 199
Circumpolar Deep Water (CDW), 301, 306
Climate, 85
  models, 6, 19, 84, 267
Coefficient
  diffusivity, 242
  eddy diffusion, 245
  heat transfer, 280
  isopycnal
    diffusion, 220
    mixing, 333, 338, 340
Conical ocean, 159
Conservative tracers, 204
Consistent drag laws, 69
Continental margins, 133, 134

363

Convective process, 169, 175
Convoluted process, 153
Coordinate transformations, 149
Critical depth hypothesis, 66
Critical slope, 119
Critical turbulence hypothesis, 66
Cyclonic eddies, 285

D

Decreasing depth, 152
Deep
    cycle, 259
    nocturnal turbulence, 259
    ocean
        dissipation, 117
        floor, 120
        mixing, 98, 107
        Pacific, 240
    overturning circulations, 36, 42
    stratification, 123, 137, 138
    submesoscale currents, 205
    turbulence, 267
Densest ocean waters, 43, 49
Density
    gradients, 42
    intermediate, 151
    neutral, 44, 49, 220
    ocean, 7
    potential energy, 188
    stratification, 17, 67
    surface, 317
    transformations, 43
Depth
    boundary layer, 75, 82, 84
    ocean, 18, 119, 123, 186, 346
        infinite, 123
    range, 46, 51, 266, 289
Destratification, 202, 203
    downfront, 202
Diapycnal, 6, 316
    circulations, 56, 57
    diffusion, 226, 240
    diffusivity, 7, 8, 156, 162, 163, 225, 317, 331, 353
        background, 160
    direction, 218, 219, 226
    dissipation, 226
    downwelling, 47, 52, 56, 156, 314
    heat transport, 265
    mixing, 2, 6, 12, 24, 38, 39, 215
    transport, 156, 160, 161, 163
    turbulent, 337
    upwelling, 13, 43, 50, 52, 159
        peak, 50
        transport, 156
    velocity, 313
Diffusion
    diapycnal, 226, 240
    double, 287
    induced, 127, 128
    isopycnal, 228, 232, 233
    vertical, 202, 332

Diffusive transport, 329, 331
Diffusivity, 39, 51, 102, 146, 217, 237, 238, 240, 246
    coefficients, 242
    diapycnal, 7, 8, 156, 162, 163, 225, 317, 331, 353
        background, 160
    eddy, 217, 222, 228, 309, 310
        mesoscale, 221, 311
    effective, 309
    enhanced, 150
    isopycnal, 48, 57, 225
    isotropic, 40, 45, 51
    turbulence, 259
    turbulent, 146, 163
    vector, 38
    vertical, 10, 23, 111, 267, 340
DIMES, 316, 317
Dispersion, 205
    relative particle, 240
    single-particle, 238
Dissipation, 21, 80, 107, 200, 221
    deep ocean, 117
    diapycnal, 226
    enhanced, 108, 283
    interior, 108
    internal-tide
        far-field, 8
        near-field, 8
    near-bottom, 109
    near-surface, 107
    rates, 108, 129, 281, 347, 354
    ratio, 154
    scales, 154, 155, 346
    tidal, 21, 23
    turbulence, 67, 77, 80
    turbulent, 346, 347
    vertical, 65
        profile, 8, 9, 13
    viscous, 5, 11, 65, 107, 313, 351, 354
Double diffusion, 287
Downfront
    destratification, 202
    wind, 191, 195, 202
        stress, 191
Downward
    buoyancy flux, 46
    heat transport, 18
    propagating NIWs, 98
Downwelling, 157
    diapycnal, 47, 52, 56, 156, 314
    layer
        boundary, 169
        Ekman, 154, 169, 171, 175
Drifter methods, 238

E

Early Pliocene, 23
Earth System Model (ESM), 329
Eastward
    advection, 317
    wind, 301
        stress, 301

Eddy
    advection, 329, 332
    Arctic, 285, 287, 291
        kinetic energy, 285
    baroclinic, 15
    compensation, 15
    cyclonic, 285
    diffusion coefficient, 245
    diffusivity, 217, 222, 228, 309, 310
    Kinetic Energy (EKE), 188, 197, 203, 222, 229, 231, 303, 307
    mesoscale, 222
    mixing, 10, 15, 229, 244, 331–333
    saturation, 15
    stirring, 308, 311, 318
        flow field, 204
        transitions, 47
    tracer flux, 217, 228
    transport, 231, 242, 311, 331
    turbulent, 346, 355
Effect
    buoyancy, 76, 150
    mean-flow, 131, 132
    thermobaric, 48
Effective diffusivity, 309
Ekman
    advection, 301, 305
    pumping, 40, 42, 53
    transport, 15, 169, 301, 305
Ekman buoyancy flux (EBF), 191, 202, 305
Elastic scattering (ES), 127, 128
Energetic, 152, 153
    submesoscale currents, 183, 194
    turbulence, 52, 154, 279
Energy, 85
    dissipation, 8, 23
    exchange, 105
    wave, 132
Enhanced
    diapycnal mixing, 18, 19, 156, 205
    diffusivity, 150
    dissipation, 108, 283
    stratification, 264
    turbulence, 74, 84, 259
Entrainment, 65, 74, 78, 110
    layer, 68
Equatorial
    Kelvin waves, 264
    mixing, 265
    ocean, 257, 259, 260, 263, 265, 267
        Indian, 260
        Pacific, 268
    thermocline, 267
    trapping, 263
    Under Current (EUC), 331
    zone, 233
Equilibrium Climate Sensitivity (ECS), 18
Ertel potential vorticity, 189

F

Far-field internal-tide dissipation, 8
Fickian vertical mixing, 39
Finescale parameterisations, 267, 317, 348, 353
Finestructure methods, 129

Finite amplitude topography, 120, 124
Float-based methods, 238
Floats
    Lagrangian, 357
    subsurface, 238
Flow
    instabilities, 262
    ocean, 36, 280
    sub-inertial, 166, 168
Fluxes
    heat, 281, 288, 304
    nutrient, 20
    turbulent, 80
Foam, 76
Forced symmetric instability, 190
Forcing, 85
    buoyancy, 42, 169
    time-varying, 73
Foundations, 301, 303, 307, 312
Free convection layers, 71
Freshwater
    Arctic, 291
    fluxes, 42, 65, 68–70, 76
    riverine, 276
Friction, 168
Frontal
    arrest, 200
    restratification, 78
    submesoscale, 181, 230
    vertical circulations, 101, 102
    vertical motions, 101
Frontogenesis, 184, 227
    oceanic, 185
    process, 185
Fronts
    salinity, 70
    submesoscale, 184, 198, 203

G

General Circulation Model (GCM), 147, 171
Generalised Ocean Turbulence Model (GOTM), 187
Generation, 118
    internal wave reflection / internal tide, 164
Geostrophic
    circulation, 201
    transport, 151
    turbulence, 5, 38, 39, 41, 42, 47
Geothermal
    buoyancy flux, 158, 160
    heat, 288
    fluxes, 50, 313
    heating, 12, 35, 50
Glider, 290, 348, 355
    Slocum, 348, 350
    underwater, 348
    autonomous, 290
Global
    deoxygenation, 334
    distribution, 122, 125, 135
    Mean Thermosteric Sea Level (GMTSL), 19
    perspective, 85, 130, 133
    warming, 339
Gravitational Instability (GI), 190

H

Heat
    anthropogenic, 14
    balance, 11, 35, 57
    budget, 47, 265, 276, 288, 303, 306
    fluxes, 281, 288, 304
    geothermal, 288
    ocean, 18, 19, 290, 316
    oceanic, 290, 292
    surface, 40, 42, 69
    transfer, 35, 40, 267
        coefficient, 280
    transport, 6, 11, 53, 311
Heating, 11
    geothermal, 12, 35, 50
    net, 71
    surface, 35, 53, 78, 259
Homogeneous
    isopycnal layers, 40
    isotropic turbulence, 230
    turbulence, 245
Horizontal velocity, 69
Humidity, 70
Hypothesis
    critical depth, 66
    critical turbulence, 66

I

Ice-ocean interactions, 280
Impacts, 205
    mixing
        isopycnal, 216, 242
        ocean, 18
    submesoscale, 206
Incrop area, 161
Indian Deep Water (IDW), 12
Indian Ocean
    equatorial, 260
    tropical, 257
Induced Diffusion (ID), 127, 128
Inertia-gravity waves, 262
Inertial Instability (II), 190
Influence of vertical mixing, 198
Instability
    baroclinic, 42, 174, 175, 187, 194, 221
    bottom boundary layer baroclinic, 194
    flow, 262
    forced symmetric, 190
    marginal, 260
    mixed-layer baroclinic, 186
    of surface boundary layer fronts, 185
    parametric subharmonic, 128
    submesoscale, 195
    symmetric, 71, 183, 186, 189, 190
        forced, 190
Integral estimate of decay, 125
Interactions
    between sea ice and Southern Ocean mixing, 306
    ice-ocean, 280
    of near-inertial waves, 102
    wave-mean flow, 130
    wave-topography, 133
    wave-wave, 126, 129
    with frontal vertical circulations, 101
    with internal waves, 103
    with mean flows, 104
Interior
    dissipation, 108
    mixing, 307, 312
        ocean diapycnal, 7, 25
Intermediate density, 151
Intermittent process, 308, 309
Internal swash zone, 165
Internal tide, 46, 118, 128
    propagation, 125
Internal wave continuum, 283
Internal wave reflection / internal tide generation, 164
Internal wave strain, 108
International Polar Year (IPY), 278
Intra Thermocline Eddies (ITE), 206
Inverse methods, 240
Irreversible mixing, 38
Isopycnal, 38, 218
    circulation, 43
    coordinates, 150, 220, 232
    diffusion, 228, 232, 233
        coefficient, 220
    diffusivity, 48, 57, 225
    mixing, 38, 220, 227, 231, 233, 242, 244
        coefficient, 333, 338, 340
        eddy, 10
        impacts, 216, 242
        role, 17, 245
    stirring, 47, 48, 205, 221, 226
    submesoscale, 225
    tracer, 221, 228, 246
        variance, 226
Isotropic
    diffusivity, 40, 45, 51
    mixing, 39, 44
    turbulence, 66, 82, 357

K

K-profile parameterisations, 83
Kinetic Energy (KE), 103
    Eddy, 188, 197, 203, 222, 229, 231, 303, 307
Kolmogorov turbulence spectrum, 357

L

Laboratory experiments, 81
Lagrangian
    floats, 357
    observations, 246
    trajectory simulation, 241
Langmuir
    entrainment, 74
    turbulence, 74–76, 81–83, 111, 206, 358
Large Eddy Simulation (LES), 82, 169
Large-eddy method, 354, 357
Large-scale models, 265
Last Glacial Maximum (LGM), 23
Lateral
    mixing, 111, 221
    variation, 73

366 Index

Layer
    entrainment, 68
    free convection, 71
    mixed, 10, 66, 76, 98, 259
        bulk, 83
        slab, 83
    mixing, 66, 68
    surface, 68, 79, 82, 85
Leaky jets, 310
Lee waves, 9, 21, 117, 125
    quasi-steady, 123
Life-cycle, 184
Loss of balance, 227
Loss of shear, 80
Lower Circumpolar Deep Water (LCDW), 302

M

Marginal
    Ice Zone (MIZ), 280, 291, 292, 316
    instability, 260
    seas, 16
Marine ecosystems, 17
Mean flows, 105
Mean profiles, 79
Mean-flow effects, 131, 132
Meridional
    circulation, 40
    heat transport (MHT), 11, 12, 23
    Overturning Circulation (MOC), 6, 8, 11, 302
Mesoscale, 329
    baroclinic instability, 182
    coherent structures, 221
    currents, 45, 48, 131, 133
    eddy, 222
        diffusivity, 221, 311
            parameterisation, 231, 246
    mixing, 226, 245
        isopycnal, 222, 230, 242
    ocean, 98, 229
    processes, 307
    stirring, 38, 43, 47, 48, 226
    transport, 246, 247
    turbulence, 221, 222, 227, 230, 241
    variability, 98, 222
    velocities, 39, 221
    vorticity, 98
Methods
    drifter, 238
    finestructure, 129
    float-based, 238
    inverse, 240
    large-eddy, 354, 357
    novel, 348
    tracer-based, 234, 241
Metrics of mixing, 153
Microstructure
    measurements, 108, 353
    temperature, 351
Mixed Layer, 10, 66, 76, 98, 259
    baroclinic instability, 186
    Depth (MLD), 42
    Instability (MLI), 186
    transformations, 303

Mixing, 17, 35, 36, 39, 42, 43, 79, 152, 153, 205, 216, 330, 334, 345
    Arctic, 276, 278–280
    bottom boundary layer, 203
    bottom-intensified, 155
    boundary, 164
    deep ocean, 98, 107
    diapycnal, 2, 6, 12, 24, 38, 39, 215
    eddy, 10, 15, 229, 244, 331–333
    efficiency, 10, 262
    enhanced diapycnal, 18, 19, 156, 205
    equatorial, 265
    Fickian vertical, 39
    in the cold tongues, 259
    in the surface boundary layer, 304
    interior, 307, 312
    irreversible, 38
    isopycnal, 38, 220, 227, 231, 233, 242, 244
        eddy, 10
    isotropic, 39, 44
    lateral, 111, 221
    layers, 66, 68
    length, 235
    mesoscale, 226, 245
    metrics, 153
    non-breaking wave, 84
    observing, 79
    ocean, 1, 11, 18, 20, 21, 345, 354
    parameterisations, 129
    passive tracers, 43
    shear-driven, 9, 224
    Southern Ocean, 306, 308
    spurious numerical, 11
    submesoscale-induced, 305
    tidal, 14, 23, 50–52, 54, 168, 281, 292
    time-evolving, 11
    turbulent, 5, 8, 21, 36, 129, 137, 229, 283, 284, 301, 353
        ocean, 5
    vertical, 39, 84, 110, 111
Mixing-length theory, 228
Models
    1D boundary layer, 82
    climate, 6, 19, 84, 267
    large-scale, 265
    ocean, 84, 156
        general circulation, 95, 129
    one-dimensional, 149
Momentum, 75
    redistribution, 41
Monin–Obukhov scaling, 68
Moorings, 354
Munk regime, 51

N

Natural tracers, 234, 235, 245
Near-bottom dissipation, 109
Near-field internal-tide dissipation, 8
Near-inertial motions, 283
Near-Inertial Wave (NIW), 9, 95, 98, 102, 107
    trapping, 104
Near-surface dissipation, 107
Near-surface distinctions, 70

Net
    circulation, 50
    diapycnal
        transport, 160, 163
        upwelling, 46, 156, 159, 161, 163
    heating, 71
    surface
        fluxes, 71
        heat fluxes, 203
    transport, 15
    upwelling, 159, 161
        transport, 160, 161
    vertical flux, 205
Neutral
    density, 44, 49, 220
    stratification, 280
Non-breaking wave mixing, 84
Non-dissipative theories, 40
North Atlantic
    circulation
        gyre, 53
        meridional overturning, 291
    Deep Water (NADW), 11, 51
        circulation, 51
        upwelling, 52
    gyre, 53
    Oscillation (NAO), 97
    subtropical, 235
Northward
    Ekman transport, 15
    surface advection, 302
Nutrient fluxes, 20

O

Observational estimates, 134
Observations
    Lagrangian, 246
    turbulence, 81, 267, 278
Observing mixing, 79
Ocean
    abyssal, 12, 162
    anthropogenic heat, 18
    area, 55, 162
    biogeochemical cycles, 244
    buoyancy, 7
    circulation, 1, 40, 56, 84, 275, 337, 345
    climate models, 6, 7, 231
    conical, 159
    density, 7
    depth, 18, 119, 123, 186, 346
        infinite, 123
    dynamics, 21, 275, 336
    ecosystems, 244, 342
    equatorial, 257, 259, 260, 263, 265, 267
    floor, 8, 50, 119, 155, 161, 245, 313, 314
        sloping, 157
        topography, 161
    flows, 36, 280
    General Circulation Model (OGCM), 95, 129, 263
    heat, 18, 19, 290, 316
        balance, 48
        content, 18

exchanges, 292
uptake, 19, 20, 76, 257, 303
mesoscale, 98, 229
MHT, 13
Microstructure Glider (OMG), 349
mixing, 1, 11, 18, 20, 21, 345, 354
models, 84, 156
open, 16
polar, 275, 292
pycnocline, 225
Southern, 14
stratification, 9, 17, 21, 43, 84, 218
Surface Boundary Layer (OSBL), 79, 80, 85
tracers, 234
passive, 204
turbulence, 5, 152, 259, 351, 359
ventilation, 57, 224, 316, 329, 340
Oceanic
heat, 290, 292
fluxes, 290
paradigm, 167
pycnocline, 3
salinity, 70
turbulence, 38, 146
One-dimensional model, 149
One-dimensional solutions, 156
Open ocean, 16
Overturning
abyssal, 11, 49, 51, 56
cell, 185
abyssal, 15, 49, 56
circulation, 53, 56
meridional, 6, 8, 11, 302
Oxygen Minimum Zone (OMZ), 329, 330, 339

P

Pacific Deep Water (PDW), 12
Pacific Water (PW), 276
Parameterisations
finescale, 267, 317, 348, 353
K-profile, 83
mesoscale eddy, 231, 246
mixing, 129
Parametric Subharmonic Instability (PSI), 103, 127, 128
Particles, 204
Passive
tracers, 52, 148, 204, 244
conservative, 204
mixing, 43
spectrum, 204
Penetrating radiation, 71
Physical circulation, 243
Polar ocean, 275, 292
Potential Vorticity (PV), 71, 189, 331
Prediction, 148
Prevailing winds, 257, 260, 276
Process
advective-diffusive, 245
boundary layer, 171
convective, 169, 175
convoluted, 153
frontogenesis, 185
intermittent, 308, 309

mesoscale, 307
regional, 307
restratification, 67, 76
stirring, 220, 222, 228, 230
submesoscale, 193, 201
isopycnal mixing, 225
Propagation
internal tide, 125
wave, 131
Pycnocline, 108
ocean, 225
oceanic, 3

Q

Quasi-steady lee waves, 123

R

Reactive, 204
tracers, 205
Redistribution, 201, 204
momentum, 41
Refraction, 99
Regime
Munk, 51
topographic, 51
Regional processes, 307
Relative particle dispersion, 240
Restratification, 78, 201–203
frontal, 78
processes, 67, 76
submesoscale, 66, 68, 201
Reynolds-averaged tracer equations, 227
Richardson number, 10, 16, 154, 155, 187, 194, 257, 266, 267
Riverine freshwater, 276
Role
isopycnal mixing, 17, 245
submesoscale, 305
turbulence, 265
Rossby waves, 264
submesoscale, 305

S

Salinity, 70, 348, 354
flux, 70
fronts, 70
oceanic, 70
stratification, 66
subtropical, 242
variability, 83, 226
water masses, 57
Sea ice, 76, 306
Sea level rise, 19
Sea Surface Height (SSH), 183, 307
Sea Surface Temperature (SST), 42, 181
Seasonal timescales, 264
Second-moment closures, 83
Secondary circulation, 12, 185, 199
Self-organised criticality, 260
Sensible heating, 70, 73
Setting, 303
Shackleton Fracture Zone, 305
Shallow hemispheric cells, 53
Shallow water, 76

Shaping circulation, 43, 47
Shear microstructure on AUVs, 348
Shear-driven mixing, 9, 224
Shear-strain version, 353
Simulation
direct, 241
Lagrangian trajectory, 241
numerical, 129
Simulation-based estimates, 240
Single-particle dispersion, 238
Sinking transport, 160
Slab mixed layers, 83
Slocum glider, 348, 350
Slope
critical, 119
subcritical, 119
supercritical, 119
Sloping
bottom boundary, 156
bottom topography, 41, 162
isopycnals, 52, 156
topography, 10, 13, 150, 168, 171, 183, 282, 303, 316
Small amplitude topography, 120, 123
South Equatorial Current (SEC), 257
Southern Ocean, 14
eddy pathways, 307
mixing, 306, 308
upwelling, 52
Spectral-space view, 229
Spray, 76
Spurious numerical mixing, 11
Stable stratification, 69, 175, 192, 195, 259, 260
Statistical funnel, 245
Steeply sloping topography, 150
Steering level, 317
Stirring, 38, 181, 217, 225
eddy, 308, 311, 318
flow field, 204
transitions, 47
horizontal, 47
isopycnal, 47, 48, 205, 221, 226
mesoscale, 38, 43, 47, 48, 226
process, 220, 222, 228, 230
rates, 38
submesoscale, 234
subsurface adiabatic, 304
tracer, 39, 206, 307, 317
Stommel–Arons circulation, 55, 163
Strain-based version, 353
Straining, 100
Stratification, 305
abyssal, 12, 51, 95
anomalies, 78, 133
Arctic, 278
background, 284
deep, 123, 137, 138
density, 17, 67
effects, 181
enhanced, 264
neutral, 280
ocean, 9, 17, 21, 43, 84, 218
salinity, 66
stable, 66, 69, 175, 192, 195, 259, 260

temperature, 81
thermal, 20, 285
variable, 102, 159
vertical, 76, 156, 163, 182, 183, 192, 194, 201, 290, 302, 305, 307, 313
weak, 175, 283
Stratified
  Mixing Layer (SML), 150, 157, 163
  ocean currents, 118
  ocean interior, 6, 117, 315
  turbulence, 117, 262, 347
Strong zonal equatorial currents, 263
Sub-inertial flow, 166, 168
Sub-surface diapycnal mixing pathways, 313
Subcritical slope, 119
Subcritical topography, 122, 134
Subinertial
  internal tides, 118, 137
  waves, 104
Subinertial Mixed Layer (SML), 199
Submesoscale, 78, 181, 201
  baroclinic instability, 183, 189, 195
  Coherent Vortices (SCV), 183, 197, 205
  currents, 181, 182, 194, 205, 206
  dynamics, 78, 181–183, 205
  eddies, 10, 78, 187, 188, 198, 202, 204, 205
  frontal, 181, 230
    slumping, 183, 201
  fronts, 184, 198, 203
  isopycnal, 225
    mixing processes, 225
    stirring, 205, 225
  mixing, 215, 221, 305
  MLI, 186
  restratification, 66, 68, 201
  role, 305
  Rossby waves, 305
  signatures, 205
  stirring, 234
  turbulence, 2, 109, 111, 203
  vertical
    buoyancy flux, 202
    heat transport, 203
    velocities, 205
Submesoscale-induced mixing, 305
Subpolar gyres, 54
Subseasonal timescales, 264
Subsurface
  adiabatic stirring, 304
  floats, 238
  mixing, 264, 265, 306
  ventilation, 85
Subtropical
  Atlantic ocean, 206
  North Atlantic, 235
  salinity, 242
  thermocline, 42
Supercritical slope, 119
Surface
  area, 76, 173, 282
  boundary layer turbulence, 10
  buoyancy
    fluxes, 35, 42, 43, 70, 200, 202, 260, 308
    forcing, 48, 54, 305

cooling, 71, 73, 85, 191, 199, 200, 202
densities, 40, 42, 200, 317
Ekman pumping, 311
forcing, 71, 77, 191, 192
front, 193, 197
heat, 40, 42, 69
  fluxes, 35, 201
  loss, 35, 42
heating, 35, 53, 78, 259
submesoscale currents, 205
ventilation, 65
water, 183, 301
  mass transformation, 48, 304
waves, 73, 74, 84, 198, 206
wind, 42, 306
  stress, 40, 191, 200, 306
Symmetric Instability (SI), 71, 183, 186, 189, 190
  forced, 190

T

Tall topography, 124
Temperature
  dissipation, 346
  microstructure, 351
  stratification, 81
  variance, 80, 347
  virtual, 70
Theoretical background, 126, 130, 134
Theory
  mixing-length, 228
  non-dissipative, 40
Thermal stratification, 20, 285
Thermal wind, 187, 190, 191, 200
  balance, 101, 104, 190, 198, 199, 201
  shear, 107, 187, 189
  turbulent, 73, 198, 199
Thermobaric effect, 48
Thermobaricity, 10, 52, 56
Thermocline, 40, 53, 310, 316, 330
  equatorial, 267
  subtropical, 42
Thermohaline
  circulation, 5
  intrusions, 288
Thickness diffusion, 232
Tidal
  dissipation, 21, 23
  mixing, 14, 23, 50–52, 54, 168, 281, 292
Tides, 17, 21, 46
  Arctic, 278
  baroclinic, 96
  barotropic, 21, 24, 96, 119
  internal, 46, 118, 128
Time-evolving mixing, 11
Time-varying forcing, 73
Timescales
  interannual, 264
  seasonal, 264
  subseasonal, 264
Topographic
  regime, 51
  slope, 119
  wakes, 196

Topographically smooth, 267
Topography, 166
  amplitude
    finite, 120, 124
    small, 120, 123
  sloping, 10, 13, 150, 168, 171, 183, 282, 303, 316
  steeply, 150
  variable, 41, 159
Tracer
  anthropogenic, 233, 234
    ocean, 234
  biogeochemical, 204, 205, 234, 305
  conservative, 204
  distributions, 57, 315
  filaments, 222
  fluctuation, 217
  isopycnal, 221, 228, 246
  natural, 234, 235, 245
  passive, 52, 148, 204, 244
  reactive, 205
  release experiments, 234
  stirring, 39, 307, 317
  variance, 217, 226, 235
    advection, 242
Tracer-based methods, 234, 241
Trade winds, 260, 262
Transformation
  coordinate, 149
  density, 43
  mixed-layer, 303
  surface water mass, 304
  water-mass, 43
Transient Climate Response (TCR), 18
Transition zone, 68
Transport
  bolus, 156, 162, 231, 232
  buoyancy, 168, 220
  by coherent structures, 222
  diapycnal, 156, 160, 161, 163
    heat, 265
    upwelling, 156
  diffusive, 329, 331
  eddy, 231, 242, 311, 331
  Ekman, 15, 169, 301, 305
    northward, 15
  geostrophic, 151
  heat, 6, 11, 53, 311
  mesoscale, 246, 247
  net, 15
    diapycnal, 160, 163
    upwelling, 160, 161
  sinking, 160
  turbulent, 147, 148, 153
  upwelling, 158, 163
  vertical, 205
Tropical Instability Wave (TIW), 257, 269
Turbulence
  background, 81
  deep, 267
  diffusivity, 259
  dissipation, 67, 77, 80
    rate, 67, 81
  energetic, 52, 154, 279

enhanced, 74, 84, 259
geostrophic, 5, 38, 39, 41, 42, 47
homogeneous, 245
isotropic, 66, 82, 357
levels, 266, 279
measurements, 81, 163, 259, 345–347, 357, 358
mesoscale, 221, 222, 227, 230, 241
observations, 81, 267, 278
ocean, 5, 152, 259, 351, 359
oceanic, 38, 146
role, 265
stratified, 117, 262, 347
submesoscale, 2, 109, 111, 203
surface boundary layer, 10
wave-driven, 81
Turbulent
boundary layer, 68, 181, 313
buoyancy flux, 147, 150, 262, 313
diapycnal, 337
buoyancy flux, 158
diffusivity, 146, 163
vector, 38
dissipation, 347
rate, 125, 135, 317, 346, 356
eddies, 346, 355
energy dissipation rate, 353, 354
fluxes, 80
Kinetic Energy (TKE), 10, 11, 17, 107, 148, 259, 266, 347, 354
dissipation, 346
dissipation rates, 8, 285, 287
mixing, 5, 8, 21, 36, 129, 137, 229, 283, 284, 301, 353
ocean, 5
Thermal Wind (TTW), 73, 198, 199
circulation, 199
transport, 147, 148, 153
Turning latitudes, 263

U
Underwater gliders, 348
Upper-ocean gyres, 53

Upwelling, 171
diapycnal, 13, 43, 50, 52, 159
peak, 50
net, 159, 161
diapycnal, 46, 156, 161
Southern Ocean, 52
transport, 158, 163
waters, 15, 53

V
Variable
stratification, 102, 159
surface wind stresses, 9
topography, 41, 159
winds, 304
Ventilated thermocline theory, 40
Ventilation, 3, 40, 57, 155, 174, 303, 332
ocean, 57, 224, 316, 329, 340
processes, 175
subsurface, 85
surface, 65
Vertical
advection, 99, 103, 104
buoyancy flux, 174, 185, 188, 190, 201, 204
component, 36, 98, 119, 153, 158, 173, 174, 189
convection, 10
coordinate, 149
decay, 129
density
gradients, 201
stratification, 9
diffusion, 202, 332
diffusivity, 10, 23, 111, 267, 340
dissipation, 65
integration, 40, 55
Microstructure Profiler (VMP), 348
mixing, 39, 84, 110, 111
stratification, 76, 156, 163, 182, 183, 192, 194, 201, 290, 302, 305, 307, 313
stretching, 40, 162
transport, 205
Virtual temperature, 70
Viscous dissipation, 5, 11, 65, 107, 313, 351, 354

W
Water
shallow, 76
surface, 183, 301
upwelling, 15, 53
Water-mass transformations, 43
Wave-driven turbulence, 81
Wave-mean flow interactions, 130
Wave-topography interaction, 133
Wave-wave interactions, 126, 129
Waves, 73, 75
capture, 107
energy, 132
inertia-gravity, 262
lee, 9, 21, 117, 125
near-inertial, 9, 95, 98, 102, 107
propagation, 131
subinertial, 104
surface, 73, 74, 84, 198, 206
winds, 74, 81
Weak
stratification, 175, 283
topography approximation, 120, 122
Weather, 85
Wentzel Kramers Brillouin (WKB), 131
West Antarctic Peninsula (WAP), 311
Westerly Wind Burst (WWB), 257, 264
Western Equatorial Pacific (WEP), 262
Whitecaps, 76
Winds
atmospheric, 97
downfront, 191, 195, 202
eastward, 301
prevailing, 257, 260, 276
surface, 42, 306
thermal, 187, 190, 191, 200
trade, 260, 262
variable, 304
waves, 74, 81

Z
Zone
equatorial, 233
grey, 233
transition, 68

Printed in the United States
by Baker & Taylor Publisher Services